THE OXFORD ENGINEERING SCIENCE SERIES

10 P. PHAGEDORN: *Non-linear oscillations* (Second edition)
13 N.W. MURRAY: *Introduction to the theory of thin-walled structures*
14 R.I. TANNER: *Engineering rheology*
15 M.F. KANNINEN and C.H. POPELAR: *Advanced fracture mechanics*
18 K. HUSEYIN: *Multiple parameter stability theory and its applications*
19 R.N. BRACEWELL: *The Hartley transform*
22 C. SAMSON, M LE BORGNE, and B. ESPIAU: *Robot control: the task of function approach*
23 H.J. RAMM: *Fluid dynamics for the study of transonic flow*
24 R.R.A. SYMS: *Practical volume holography*
25 W.D. McCOMB: *The physics of fluid turbulence*
26 Z.P. BAZANT and L. CEDOLIN: *Stability of structures: elastic, inelastic, fracture and damage theories*
27 J.D. THORNTON: *Science and practice of liquid-liquid extraction* (Two volumes)
28 J. VAN BLADEL: *Singular electromagnetic fields and sources*
29 M.O. TOKHI and R.R. LEITCH: *Active noise control*
30 I.V. LINDELL: *Methods for electromagnetic field analysis*
31 J.A.C. KENTFIELD: *Nonsteady, one-dimensional, internal, compressible flows*
32 W.F. HOSFORD: *Mechanics of crystals and polycrystals*
33 G.S.H. LOCK: *The tubular thermosyphon: variations on a theme*
34 A. LINAN and F.A. WILLIAMS: *Fundamental aspects of combustion*
35 N. FACHE, D. DE ZUTTER, and F. OLYSLAGER: *Electromagnetic and circuit modelling of multiconductor transmission lines*
36 A.N. BERIS and B.J. EDWARDS: *Thermodynamics of flowing systems: with internal microstructure*
38 J.G. COLLIER and J.R. THOME: *Convective boiling and condensation* (Third edition)
39 I.I. GLASS and J.P. SISLIAN: *Nonstationary flows and shockwaves*
40 D.S. JONES: *Methods in electromagnetic wave propagation* (Second edition)
42 G.A. BIRD: *Molecular gas dynamics and the direct simulation of gas flows*
43 G.S.H. LOCK: *Latent heat transfer: an introduction to fundamentals*
44 C.E. BRENNEN: *Cavitation and bubble dynamics*
45 T.C.T. TING: *Anisotropic elasticity: theory and applications*
46 K. ISHIHARA: *Soil dynamics in earthquake engineering*
47 A. CHANDRA and S. MUKHERJEE: *Boundary element methods in manufacturing*

Anisotropic Elasticity

Theory and Applications

T. C. T. TING
Professor of Applied Mechanics
University of Illinois at Chicago

New York Oxford
OXFORD UNIVERSITY PRESS
1996

Oxford University Press

Oxford New York
Athens Auckland Bangkok Bombay
Calcutta Cape Town Dar es Salaam Delhi
Florence Hong Kong Istanbul Karachi
Kuala Lumpur Madras Madrid Melbourne
Mexico City Nairobi Paris Singapore
Taipei Tokyo Toronto

and associated companies in
Berlin Ibadan

Published by Oxford University Press, Inc.,
198 Madison Avenue, New York, New York 10016

Library of Congress Cataloging-in-Publication Data
Ting, T. C.-T. (Thomas Chi-Tsai), 1933–
Anisotropic elasticity: theory
and applications / T. C. T. Ting
p. cm. — (Oxford engineering science series)
Includes bibliographical references and index.
ISBN 0-19-507447-5
1. Composite materials—Mechanical properties.
2. Elasticity. 3. Anisotropy
I. Title II. Series.
TA418.9.C6T55 1995 620.1'1832—dc20 94-47372

1 3 5 7 9 8 6 4 2

Printed in the United States of America
on acid-free paper

To

Romana

PREFACE

In the early nineteen eighties when I was looking for a change to a new research topic, research on composite materials was in full blossom. For a fiber-reinforced laminated composite each layer can be modeled as an orthotropic material when referred to a certain coordinate system. Referred to a fixed coordinate system however each layer is mathematically a general anisotropic material. One needs a theory for general anisotropic elastic materials. In contrast to isotropic elastic materials that have only two elastic constants, anisotropic elastic materials may have as many as twenty-one elastic constants. When two-dimensional deformations are considered the analysis still requires fifteen elastic constants. With little background on isotropic elasticity, and after being intimidated by reading Muskhelishvili's book on two-dimensional problems in isotropic elasticity, I was reluctant and apprehensive in venturing into anisotropic elasticity. I believed that, as many of my fellow researchers in the mechanics community did, (i) one must know theory of isotropic elasticity before studying anisotropic elasticity and (ii) the analysis of anisotropic elasticity problems is much more complicated and difficult than that of isotropic elasticity problems. I soon discovered that, at least for two-dimensional deformations of anisotropic elastic solids to be presented in this book, nothing could be further from the truth if one employs the *Stroh formalism*. To this date I still know very little on isotropic elasticity. I ventured into anisotropic elasticity in the early nineteen eighties when I was nearly fifty. It shows that the subject is not difficult to learn, and that one is never too late to learn a new subject.

The *Stroh formalism* is mathematically elegant and technically powerful. The formalism is not new. The foundations of the formalism were laid down by Stroh in his two papers in 1958 and 1962, and it has been further developed ever since by many investigators. The formalism is well known in the physics, applied mathematics, and material sciences community. For some reason it has not been widely recognized in the mechanics community who did most work in composites. That the analysis of two-dimensional deformations for an anisotropic elastic material can be simpler than that for an isotropic elastic material is because isotropic elastic materials are more than a special group of anisotropic elastic materials. They are mathematically *degenerate* materials.

A mathematically degenerate problem usually requires *more* effort in obtaining the solution. As an illustration consider a linear third order ordinary differential equation for $y(x)$ with constant coefficients. If $e^{k_i x}$ (i=1,2,3) are the independent solutions with k_i being constants, the general solution is

$$y(x) = \sum_{i=1}^{3} c_i e^{k_i x} \tag{1}$$

where c_i are arbitrary constants. In the degenerate case when $k_2=k_1$ there are only two independent solutions. It requires *additional effort* to find the missing independent solution which is in this case $xe^{k_1 x}$. The general solution for the degenerate case $k_2=k_1\neq k_3$ is

$$y(x) = c_1 e^{k_1 x} + c_2 x e^{k_1 x} + c_3 e^{k_3 x}. \tag{2}$$

It should be pointed out that the solution (2) for the degenerate case cannot be written in a compact form as the solution (1) for the general case. This is precisely what happens to isotropic elastic materials. Isotropic elastic materials are a degenerate case of anisotropic elastic materials. The solution for an isotropic elastic material cannot be deduced from that for an anisotropic elastic material by specializing the material constants. *More effort* is required to deduce the solution for an isotropic elastic material. Moreover, the solution for an isotropic elastic material is in most cases not as compact and simple as that for an anisotropic elastic material.

One may ask: What happens to the fifteen elastic constants? Do they not appear in the solution, resulting in a lengthy and complicated expression? That is the beauty of Stroh's sextic formalism. The fifteen elastic constants are replaced by the complex eigenvalues p_α (α=1,2,3) and the Stroh eigenvectors $\boldsymbol{a}_\alpha$, $\boldsymbol{b}_\alpha$. The solution is written for the displacement $\mathbf{u}$, a 3-vector, and the stress function $\boldsymbol{\phi}$, also a 3-vector from which the stresses can be deduced by differentiation. There is no need to write explicitly the three components of displacement and the six components of stress. The mathematical expressions are simple and compact with the use of matrices. Since matrices are employed extensively in the book, Chapter 1 presents certain fundamentals of matrix algebra that are needed in the book.

I do not mean to give an impression that *every* elasticity problem is easier to solve when the material is anisotropic. For an infinite elastic medium with a hole subjected to a uniform loading at infinity, the solution can be obtained for a hole of arbitrary shape when the material is isotropic. For anisotropic materials the solution is available when the hole is an ellipse. It remains to this date a difficult problem for anisotropic elastic materials when the hole is not an ellipse (see Section 3.12 and the introduction to Chapter 10). Several attempts have been made and the results have been published, but the problem has not been solved correctly. What I am trying to convey to the reader is that, if an analytical solution is available, the solution is in general simpler and easier to derive for anisotropic elastic materials than for isotropic elastic materials. One may compare the solution for a crack in an infinite anisotropic elastic material subjected to a uniform loading given in (11.4-3), (11.4-4) against the corresponding solution for degenerate materials in (13.3-3), (13.3-4) and the solution for isotropic materials in (13.3-5), (13.3-6).

Chapter 2 presents linear stress-strain laws for general anisotropic elastic materials. Classifications of anisotropic materials according to the number of symmetry planes in the material are discussed. Also discussed are the restrictions on the prescription of elastic constants and the reduction in the number of elastic constants for each group of materials with certain material symmetry. The last section of the chapter provides a method for determining the symmetry planes of a material when such a plane exists.

The Stroh formalism is not needed for problems of antiplane deformations which are possible only for certain special anisotropic materials. This is studied in Chapter 3. The idea of stress function from which the stresses are obtained by differentiation is introduced here. Also introduced is the complex variable $z=x_1+px_2$ where p is any complex number. The Green's functions, the cracks, and elliptic inclusions are basic problems in applications. The solutions to these problems are presented. The reader is urged to read this chapter before they embark on Chapters 8, 10, and 11 on more general deformations of general anisotropic materials. An important subject is the mapping of a complex function $\zeta=f(z)$ where $z=x_1+px_2$. The mapping from the (x_1,x_2)-plane to the ζ-plane is in general *not conformal* unless $p=i$. This chapter should be referred to, and compared with, when one studies more general deformations in later chapters.

The Lekhnitskii formalism, a very popular formalism among the engineering community, is outlined in Chapter 4. The newer Stroh formalism is presented in Chapter 5. There is a similarity in the expression of the general solution between the two formalisms. However the similarity stops there. If Chapter 5 is the foundation of the Stroh formalism, Chapters 6 and 7 are the main structures of the formalism. Beside its mathematical elegance, one of the advantages of the Stroh formalism is the possibility of obtaining the solution in a real form through the identities presented in Chapters 6 and 7. As a result many physically interesting and unexpected phenomena, which have been shrouded in the complex form solutions, have been discovered (see Chapters 8 through 11). Most of the unexpected phenomena defy a physical explanation. Some of the results are unexpected even for isotropic materials.

Applications of the Stroh formalism to special subjects are presented in Chapters 8 through 12. Anisotropic materials are indeed interesting materials. A simple tensile stress applied to the material produces not only an extensional strain but also a shear strain. Likewise a pure shear stress applied to the material produces a shear strain and an extensional (or compression) strain. Consequently a loading that is symmetric (or skew-symmetric) with the x_2-axis, say, in general does not produce a deformation that is symmetric (or skew-symmetric) with the x_2-axis. When an isotropic material with a crack or an elliptic hole is re-oriented with respect to the crack or the elliptic hole, the solution remains unchanged. The same cannot be said of anisotropic materials. There are however surprises (Ting, 1992d). We will see in these chapters that there are cases in which anisotropic materials behave like isotropic materials.

The reader who likes to have a bird's-eye view of the Stroh formalism may go directly to the first six sections of Chapter 5. He may skip Chapters 6 and 7 to other chapters on applications to specific subjects. It may be necessary for him to return to Chapters 6 and 7 for derivation of various identities employed in applications.

It was stated earlier that the solution to degenerate materials, of which isotropic materials are a special case, cannot be deduced by specializing material constants in the solution for general anisotropic materials. In Chapter 13 we study modifications needed to produce the solutions for degenerate materials.

The Stroh formalism can be generalized in many ways. Extension of the formalism to steady state motion and surface waves presented in Chapter 12 is an example. In Chapter 14 we discuss generalization of the formalism to problems where the boundary conditions are prescribed not entirely in terms of displacement or surface traction. An example is the boundary conditions for a slippery surface. The boundary conditions are not in terms of the displacement vector $\mathbf{u}$ or the stress function $\boldsymbol{\phi}$, but a mixture of one component from $\mathbf{u}$ and two components from $\boldsymbol{\phi}$. This destroys the elegant sextic formalism of Stroh. A generalized formalism is proposed through which more general boundary conditions can be accommodated without destroying the elegant formalism. We also study generalization of the Stroh formalism to thermo-anisotropic elasticity and to piezoelectric materials.

The last chapter is an attempt to extend the Stroh formalism to three-dimensional problems. We study the Green's functions for a point force in the infinite space, in a half-space, and in two joined half-spaces.

Until the appearance of the workshop proceedings on *Modern Theory of Anisotropic Elasticity and Applications* edited by J. J.

Wu, T. C. T. Ting, and D. M. Barnett (1991) there were no books that presented the Stroh formalism. The treatment of Stroh's sextic formalism can be found in journals of various subjects, ranging from mechanics, material science, applied physics, to applied mathematics. The notations employed by authors are also quite varied. It is difficult for a beginner to study the subject without a book that presents the theory and applications all in one volume. It is hoped that this book will be useful for the beginners as well as the more advanced researchers who are interested in anisotropic elasticity.

My research on anisotropic elasticity in the early stage was supported by the then Army Materials and Mechanics Research Center, Watertown, MA. I am grateful to Mr. John F. Dignam and Dr. S. C. Chou of AMMRC for their encouragement and their foresight that the Stroh formalism should be further developed. The same foresight was envisioned by Drs. Jagdish Chandra and Julian J. Wu of the Mathematical Sciences Division of Army Research Office, Research Triangle Park, NC. The bulk of my research on anisotropic elasticity would have been impossible without their enthusiastic support. I am also grateful to Drs. Jim C. I. Chang and Spencer T. Wu of Air Force Office of Scientific Research for their effort in supporting my study on the extension of the Stroh formalism to thermo-anisotropic elasticity and piezoelectric materials. I learned Stroh's formalism not only from the literature, but also from researchers who have been in the field far ahead of me. In particular, I have learned a great deal from Professors David M. Barnett of Stanford University, Peter Chadwick of the University of East Anglia, and Jens Lothe of the University of Oslo. They have contributed extensively to the development of the Stroh formalism. On the subject of isotropic elasticity of which I know very little, I am indebted to Professor John Dundurs of Northwestern University for tirelessly answering my questions and showing me many relevant results in isotropic elasticity. This book would not have appeared without the inspiration of their work and many discussions I had with them.

Last, but not the least, I would like to thank my two sons, Albert (a computer engineer) and Bennett (a computer scientist), for their advice on computer hardware and software for writing the book. The figures in the book were drawn with the assistance of my former graduate student Dr. Gongpu Yan.

T. C. T. Ting

Chicago, Illinois
January, 1995

CONTENTS

7. TRANSFORMATION OF THE ELASTICITY MATRICES AND DUAL COORDINATE SYSTEMS

8. GREEN'S FUNCTIONS FOR INFINITE SPACE, HALF-SPACE, AND COMPOSITE SPACE

9. PARTICULAR SOLUTIONS, STRESS SINGULARITIES, AND STRESS DECAY

ANISOTROPIC ELASTICITY:
THEORY AND APPLICATIONS

Chapter 1

MATRIX ALGEBRA

We will present in this chapter some aspects of matrix algebra that are needed in this book. Most results presented here can be found in standard books on matrix algebra. Proofs are provided for those results that are either easily derived or not readily available elsewhere. For readers who have no knowledge of matrix algebra, this chapter is essential for the rest of the book. They may find the one-chapter treatment of matrix algebra in the book by Hildebrand (1954) helpful and informative. For readers who have some knowledge of matrix algebra this chapter can be skimmed or skipped altogether, depending on how familiar they are with the subject. If they want to find the proofs omitted in this chapter or want to devote more time on the subject, the books by Hohn (1965) and Pease (1965) are recommended. The notations employed in this chapter have no relations, in most cases, with the notations adopted in the rest of the book. This point should be kept in mind in referring back to this chapter.

1.1 Notations, Definitions, and Identities

A matrix is a rectangular array of numbers, variables, or functions. The *order* of a matrix is defined by the number of rows and columns in the matrix. If a matrix has n rows and m columns it is called an *n by m*, or $n \times m$ matrix. With few exceptions a matrix with more than one row and one column is denoted by a bold-face upper case letters **A**, **B**,... Thus the $n \times m$ matrix **A** is written as

$$\mathbf{A} = \begin{bmatrix} A_{11} & A_{12} & \cdots & A_{1m} \\ A_{21} & A_{22} & \cdots & A_{2m} \\ \vdots & \vdots & \ddots & \vdots \\ A_{n1} & A_{n2} & \cdots & A_{nm} \end{bmatrix} \tag{1.1-1}$$

in which A_{ij} denotes an element of the matrix. The first index i of A_{ij} specifies the row in which the element is located while the second index j specifies the column in which the element is located. When the rows and the columns of the matrix **A** are interchanged, we have the *transpose* of the matrix which is denoted by $\mathbf{A}^T$. Thus the transpose of (1.1-1) is

$$\mathbf{A}^T = \begin{bmatrix} A_{11} & A_{21} & \cdots & A_{n1} \\ A_{12} & A_{22} & \cdots & A_{n2} \\ \vdots & \vdots & \ddots & \vdots \\ A_{1m} & A_{2m} & \cdots & A_{nm} \end{bmatrix}, \tag{1.1-2}$$

In contrast to **A** that has the order $n \times m$, $\mathbf{A}^T$ has the order $m \times n$.

A vector with n components can be represented by an $n \times 1$ matrix. Since it contains only one column, it is called a *column matrix*. A vector or a column matrix will be denoted by a bold-face lower case letters **a**, **b**,... If

$$\mathbf{a} = \begin{bmatrix} a_1 \\ a_2 \\ \vdots \\ a_n \end{bmatrix},$$

its transpose

$$\mathbf{a}^T = [a_1, \ a_2, \ \ldots, \ a_n]$$

is an one-row matrix.

The addition of matrices **A** and **B** is written as

$$\mathbf{C} = \mathbf{A} + \mathbf{B}.$$

The elements C_{ij} are related to A_{ij} and B_{ij} by

$$C_{ij} = A_{ij} + B_{ij}. \tag{1.1-3}$$

Therefore two matrices must be of the same order before they can be added.

If **C** is the product of **A** and **B**, we write

$$\mathbf{C} = \mathbf{AB}.$$

The elements C_{ij} are related to A_{ij} and B_{ij} by

$$C_{ij} = A_{ik} B_{kj} \tag{1.1-4}$$

in which repeated indices imply summation. The product makes sense only when the number of columns in **A** is identical to the number of rows in **B**. If **A** is an $n \times m$ matrix and **B** is an $m \times n$ matrix, **AB** and **BA** have the order $n \times n$ and $m \times m$, respectively. Hence the matrix products **AB** and **BA** are different. When **A** and **B** are both of order $n \times n$, so are **AB** and **BA**. Nevertheless the product of **A** and **B** in general does not commute, i.e.,

$$\mathbf{AB} \neq \mathbf{BA}. \tag{1.1-5}$$

Therefore

$$
\begin{aligned}
(\mathbf{A}+\mathbf{B})^2 &= (\mathbf{A}+\mathbf{B})(\mathbf{A}+\mathbf{B}) \\
&= \mathbf{A}^2+\mathbf{AB}+\mathbf{BA}+\mathbf{A}^2 \\
&\neq \mathbf{A}^2+2\mathbf{AB}+\mathbf{B}^2,
\end{aligned}
$$

in general. We will study in Section 1.5 conditions under which the product of two $n\times n$ matrices commutes.

It can be shown that the transpose of the product of two matrices is equal to the product of the transposes of the two matrices in the reverse order, i.e.,

$$(\mathbf{AB})^T = \mathbf{B}^T\mathbf{A}^T. \tag{1.1-6}$$

The *scalar* product of two vectors **a** and **b** is denoted by

$$\mathbf{a}\cdot\mathbf{b} \quad \text{or} \quad \mathbf{a}^T\mathbf{b}\ .$$

The former is in vector notation and the latter in matrix notation. The *length* $|\mathbf{a}|$ of a vector **a** is

$$|\mathbf{a}| = (\mathbf{a}\cdot\mathbf{a})^{1/2} \quad \text{or} \quad |\mathbf{a}| = \left(\mathbf{a}^T\mathbf{a}\right)^{1/2}. \tag{1.1-7}$$

The length so obtained may not be real when **a** is complex. In fact the length may be zero when **a** is complex. An example is $\mathbf{a}=\mathbf{e}_1+i\mathbf{e}_2$ where $\mathbf{e}_1$ and $\mathbf{e}_2$ are real unit vectors orthogonal to each other. A complex vector **a** is called *isotropic* if $\mathbf{a}\cdot\mathbf{a}=0$ (Hayes, 1984). If **a** is not isotropic, and if **a** is an eigenvector which is unique up to a constant multiplier, the length of **a** can be made unity by choosing the constant multiplier properly. In some literature the length of a complex vector is defined as

$$(\bar{\mathbf{a}}\cdot\mathbf{a})^{1/2}$$

where the overbar denotes the complex conjugate. We will not employ this definition here.

An $n\times n$ matrix is called a *square* matrix of order n. Its transpose is also a square matrix of order n. The matrix is *symmetric* if

$$\mathbf{A} = \mathbf{A}^T. \tag{1.1-8}$$

A one-element matrix, i.e., an 1×1 matrix is obviously symmetric. A matrix is *skew-symmetric*, or *antisymmetric*, if

$$\mathbf{A} = -\mathbf{A}^T. \tag{1.1-9}$$

The elements $A_{11}, A_{22},\dots A_{66}$ of the matrix are called the *diagonal elements*. It is readily shown that the diagonal elements of a skew-symmetric matrix are zeros.

A *diagonal matrix* is a square matrix whose non-diagonal elements are zeros. It is a symmetric matrix. If

$$\Lambda = \begin{bmatrix} \lambda_1 & 0 & \dots & 0 \\ 0 & \lambda_2 & \dots & 0 \\ \vdots & \vdots & \ddots & \vdots \\ 0 & 0 & \dots & \lambda_n \end{bmatrix} \tag{1.1-10a}$$

we may write it as

$$\Lambda = \text{diag}\left[\lambda_1, \ \lambda_2, \ \dots \ \lambda_n\right] \tag{1.1-10b}$$

or

$$\Lambda = \langle \lambda_* \rangle. \tag{1.1-10c}$$

An *identity matrix* $\mathbf{I}$ is a diagonal matrix whose diagonal elements are ones. Thus

$$\mathbf{I} = \text{diag}\left[1, \ 1, \ \dots, \ 1\right].$$

It has the property that

$$\mathbf{AI} = \mathbf{A}, \qquad \mathbf{IA} = \mathbf{A}, \tag{1.1-11}$$

for any matrix $\mathbf{A}$. If $\mathbf{A}$ is an $n \times m$ matrix, the $\mathbf{I}$ in $(1.1\text{-}11)_1$ is a diagonal matrix of order m while the $\mathbf{I}$ in $(1.1\text{-}11)_2$ is a diagonal matrix of order n. Thus the order of the unit matrix $\mathbf{I}$ is self-determined from the equation it is in.

The *trace* of a square matrix $\mathbf{A}$, written as tr$\mathbf{A}$, is the sum of all its diagonal elements, i.e.,

$$\text{tr}\mathbf{A} = A_{11} + A_{22} + \dots + A_{nn}. \tag{1.1-12}$$

It is clear that

$$\text{tr}(\mathbf{A}^T) = \text{tr}\mathbf{A} \tag{1.1-13}$$

and, from the definition of the matrix product (1.1-4),

$$\text{tr}(\mathbf{AB}) = \text{tr}(\mathbf{BA}). \tag{1.1-14}$$

The matrices $\mathbf{A}$ and $\mathbf{B}$ in (1.1-14) need not be square matrices as long as $\mathbf{AB}$ and $\mathbf{BA}$ are.

The *determinant* $|\mathbf{A}|$ of a square matrix $\mathbf{A}$ is

$$|\mathbf{A}| = \delta_{ij \dots s} A_{i1} A_{j2} \dots A_{sn},$$

where again repeated indices imply summation. The *permutation symbol* $\delta_{ij \dots s}$ vanishes if any two subscripts are identical and assumes the value of +1 or −1 if the subscripts $i,j,...,s$ form an even or odd permutation, respectively. The two-index permutation symbol δ_{ij} is the *Kronecker delta.* It can be shown that

$$|\mathbf{A}^T| = |\mathbf{A}|, \tag{1.1-15}$$

and that

$$|\mathbf{AB}| = |\mathbf{A}| \cdot |\mathbf{B}|. \tag{1.1-16}$$

If $\mathbf{A}$ is a 3×3 matrix,

$$|\mathbf{A}| = \delta_{ijk} A_{i1} A_{j2} A_{k3} \tag{1.1-17a}$$

or

$$\delta_{pqr} |\mathbf{A}| = \delta_{ijk} A_{ip} A_{jq} A_{kr}.$$

Since

$$\delta_{pqr} \delta_{pqr} = 6,$$

we have an alternate expression

$$|\mathbf{A}| = \tfrac{1}{6} \delta_{ijk} \delta_{pqr} A_{ip} A_{jq} A_{kr}. \tag{1.1-17b}$$

The *cofactor* $\widehat{A}_{i1}$ is the coefficient of A_{i1} in (1.1-17a), i.e.,

$$\widehat{A}_{i1} = \delta_{ijk} A_{j2} A_{k3} = \delta_{ijk} \delta_{123} A_{j2} A_{k3} = \tfrac{1}{2} \delta_{ijk} \delta_{1qr} A_{jq} A_{kr}. \tag{1.1-18}$$

Hence

$$\widehat{A}_{ip} = \tfrac{1}{2} \delta_{ijk} \delta_{pqr} A_{jq} A_{kr}. \tag{1.1-19}$$

We also have the relation

$$\mathbf{A}\widehat{\mathbf{A}}^T = |\mathbf{A}| \mathbf{I} \tag{1.1-20}$$

where the matrix $\widehat{\mathbf{A}}^T$ is called the *adjoint matrix* of $\mathbf{A}$.

A useful identity is (Ting, 1970)

$$|\mathbf{A} + \alpha \mathbf{g} \mathbf{h}^T| = |\mathbf{A}| + \alpha \mathbf{g}^T \breve{\mathbf{A}} \mathbf{h} \tag{1.1-21}$$

where $\mathbf{A}$ is a square matrix of order n, $\mathbf{g}$ and $\mathbf{h}$ are n×1 column matrices, and α is a scalar.

A square matrix $\mathbf{A}$ is *singular* or *nonsingular* when $|\mathbf{A}|$ is zero or nonzero. If $\mathbf{A}$ is nonsingular, there exist matrices $\mathbf{C}$ and $\mathbf{D}$ such that

$$\mathbf{AC} = \mathbf{I}, \qquad \mathbf{DA} = \mathbf{I}.$$

If we post-multiply the second equation by $\mathbf{C}$ and make use of the first equation and (1.1-11), we have

$$\mathbf{D} = \mathbf{C}.$$

Therefore one has

$$\mathbf{AC} = \mathbf{I} = \mathbf{CA}. \tag{1.1-22}$$

This is a special case in which the matrix product commutes. The matrix **C** is the *inverse* of **A**, and is written as

$$\mathbf{C} = \mathbf{A}^{-1}.$$

Equations (1.1-22) are equivalent to

$$\mathbf{AA}^{-1} = \mathbf{I} = \mathbf{A}^{-1}\mathbf{A}. \tag{1.1-23}$$

It is easily shown that $\mathbf{A}^{-1}$ is symmetric if **A** is. When **A**, **B** are both nonsingular, so is **AB**. Let

$$\mathbf{U} = (\mathbf{AB})^{-1}.$$

From

$$(\mathbf{AB})\mathbf{U} = \mathbf{I}, \quad \mathbf{BU} = \mathbf{A}^{-1}, \quad \mathbf{U} = \mathbf{B}^{-1}\mathbf{A}^{-1},$$

we have

$$(\mathbf{AB})^{-1} = \mathbf{B}^{-1}\mathbf{A}^{-1}. \tag{1.1-24}$$

A square matrix **A** is *orthogonal*, or *rotational*, if

$$\mathbf{A}^{T} = \mathbf{A}^{-1}. \tag{1.1-25}$$

Therefore an orthogonal matrix satisfies the relations

$$\mathbf{AA}^{T} = \mathbf{I} = \mathbf{A}^{T}\mathbf{A}. \tag{1.1-26}$$

Taking the determinant of the matrices on both sides of the equalities we obtain

$$|\mathbf{A}|^2 = 1, \quad \text{or} \quad |\mathbf{A}| = \pm 1. \tag{1.1-27}$$

An orthogonal matrix is called *proper* orthogonal when its determinant is +1, and *improper* orthogonal when its determinant is –1.

When an orthogonal matrix **A** is real it has the following geometrical interpretation. Consider the equation

$$\mathbf{x}^* = \mathbf{Ax}$$

which transforms a real vector **x** into a new vector $\mathbf{x}^*$ through the matrix **A**. A different vector **y** will be transformed into $\mathbf{y}^*$ by

$$\mathbf{y}^* = \mathbf{Ay}.$$

When **A** is orthogonal, it is readily shown that

$$(\mathbf{x}^*)^T\mathbf{y}^* = (\mathbf{A}\mathbf{x})^T(\mathbf{A}\mathbf{y}) = \mathbf{x}^T\mathbf{A}^T\mathbf{A}\mathbf{y} = \mathbf{x}^T\mathbf{y}. \tag{1.1-28}$$

In the particular case when $\mathbf{y} = \mathbf{x}$, (1.1-28) tells us that the length of a vector is unchanged after the transformation. Using this result, (1.1-28) implies that the angle between any two vectors is unchanged after the transformation. Therefore a real orthogonal matrix transforms a space into another space without distorting the space. The transformation is a rigid body rotation if $\mathbf{A}$ is proper orthogonal. It is a rigid body rotation and a mirror reflection about a plane when $\mathbf{A}$ is improper orthogonal. If three vectors $\mathbf{a}$, $\mathbf{b}$, $\mathbf{c}$ form a right-hand triad before the transformation, the three vectors $\mathbf{a}^*$, $\mathbf{b}^*$, $\mathbf{c}^*$ after the transformation form a right-hand triad if $\mathbf{A}$ is proper orthogonal, and a left-hand triad if $\mathbf{A}$ is improper orthogonal.

1.2 Eigenvalues and Eigenvectors

The standard *eigenrelation* for a square matrix $\mathbf{A}$ which may be complex is

$$\mathbf{A}\mathbf{x} = \lambda\mathbf{x}. \tag{1.2-1}$$

λ is an eigenvalue and $\mathbf{x}$ is an eigenvector. Rewriting it as

$$(\mathbf{A} - \lambda\mathbf{I})\mathbf{x} = \mathbf{0} \tag{1.2-2}$$

we must have, for a nontrivial solution of $\mathbf{x}$,

$$|\mathbf{A} - \lambda\mathbf{I}| = 0. \tag{1.2-3}$$

Expansion of the determinant leads to the *eigenequation* or the *characteristic equation* for λ:

$$\lambda^n - c_1\lambda^{n-1} + \ldots + (-1)^n c_n = 0, \tag{1.2-4}$$

which provides n roots for the eigenvalues $\lambda = \lambda_1, \lambda_2, \ldots, \lambda_n$. The two coefficients c_1 and c_n have particularly simple relations with λ_i and $\mathbf{A}$. They are

$$\begin{aligned} c_1 &= \lambda_1 + \lambda_2 + \ldots + \lambda_n = \operatorname{tr}\mathbf{A}, \\ c_n &= \lambda_1\lambda_2\ldots\lambda_n = |\mathbf{A}|. \end{aligned} \tag{1.2-5}$$

For $n = 3$ the eigenequation is

$$\lambda^3 - c_1\lambda^2 + c_2\lambda - c_3 = 0, \tag{1.2-6a}$$

where

$$
\begin{aligned}
c_1 &= \lambda_1 + \lambda_2 + \lambda_3 = \text{ tr }\mathbf{A}, \\
c_2 &= \lambda_1\lambda_2 + \lambda_2\lambda_3 + \lambda_3\lambda_1 = \tfrac{1}{2}\{(\text{tr}\mathbf{A})^2 - \text{tr}(\mathbf{A}^2)\}, \\
c_3 &= \lambda_1\lambda_2\lambda_3 = |\mathbf{A}|.
\end{aligned} \tag{1.2-6b}
$$

The vector $\mathbf{x}$ in (1.2-1) is a *right eigenvector*. The *left eigenvector*, or the *reciprocal eigenvector*, $\mathbf{y}$ is defined by

$$\mathbf{y}^T\mathbf{A} = \lambda\mathbf{y}^T \quad \text{or} \quad \mathbf{A}^T\mathbf{y} = \lambda\mathbf{y}. \tag{1.2-7}$$

The eigenvalue λ of $(1.2\text{-}7)_2$ satisfies the equation

$$|\mathbf{A}^T - \lambda\mathbf{I}| = 0$$

which is identical to (1.2-3) in view of (1.1-15).

Let $\mathbf{x}_i$ and $\mathbf{y}_i$ $(i = 1,2,\ldots,n)$ be the right and left eigenvectors associated with the eigenvalue λ_i. From

$$\mathbf{A}\mathbf{x}_2 = \lambda_2\mathbf{x}_2, \quad \mathbf{y}_1^T\mathbf{A} = \lambda_1\mathbf{y}_1^T,$$

and pre-multiplying the first equation by $\mathbf{y}_1^T$ and post-multiplying the second equation by $\mathbf{x}_2$, we obtain

$$0 = (\lambda_1 - \lambda_2)\mathbf{y}_1^T\mathbf{x}_2.$$

Therefore

$$\mathbf{y}_1 \cdot \mathbf{x}_2 = 0, \quad \text{if } \lambda_1 \neq \lambda_2,$$

or, more generally,

$$\mathbf{y}_i \cdot \mathbf{x}_j = 0, \quad \text{if } \lambda_i \neq \lambda_j. \tag{1.2-8}$$

The left eigenvectors and the right eigenvectors associated with different eigenvalues are *orthogonal* to each other. If λ_i is a simple root, i.e., it is not a repeated root, $\mathbf{x}_i$ and $\mathbf{y}_i$ are unique up to a constant multiplier. We may *normalize* $\mathbf{x}_i$ and $\mathbf{y}_i$ such that

$$\mathbf{y}_i \cdot \mathbf{x}_i = 1, \quad i \text{ not summed}. \tag{1.2-9}$$

Since $\mathbf{x}_i$ and $\mathbf{y}_i$ each has a constant multiplier, (1.2-9) does not provide a unique solution for $\mathbf{x}_i$ and $\mathbf{y}_i$.

When $\mathbf{A}$ is symmetric the left eigenvector $\mathbf{y}$ and the right eigenvector $\mathbf{x}$ can be taken identical. If $\mathbf{A}$ is real and symmetric the eigenvalues λ's and the associated eigenvectors $\mathbf{x}$'s can be shown to be real. If λ and $\mathbf{x}$ were complex we have by taking the complex conjugate of (1.2-1),

$$\mathbf{A}\bar{\mathbf{x}} = \bar{\lambda}\bar{\mathbf{x}}.$$

Hence

$$\mathbf{x}^T\mathbf{A}\bar{\mathbf{x}} = \bar{\lambda}\mathbf{x}^T\bar{\mathbf{x}}$$

or, after transposing both sides and noticing that $\mathbf{A}^T = \mathbf{A}$,

$$\bar{\mathbf{x}}^T\mathbf{A}\mathbf{x} = \bar{\lambda}\bar{\mathbf{x}}^T\mathbf{x}. \tag{1.2-10}$$

If we pre-multiply (1.2-1) by $\bar{\mathbf{x}}^T$ and subtract (1.2-10) from the result we have

$$(\lambda - \bar{\lambda})\bar{\mathbf{x}}^T\mathbf{x} = 0.$$

Since $\bar{\mathbf{x}}^T\mathbf{x}$ is real and nonzero, $\lambda = \bar{\lambda}$, which means that λ is real.

When $\mathbf{A}$ is singular $\lambda = 0$ is an eigenvalue. The eigenvector associated with $\lambda = 0$ is called the *null vector*. If the matrix is not symmetric, the right and left null vectors are in general different.

Let $\mathbf{W}$ be a skew-symmetric matrix of order n, i.e.,

$$\mathbf{W} = -\mathbf{W}^T .$$

When n is an odd integer,

$$|\mathbf{W}| = (-1)^n|\mathbf{W}^T| = -|\mathbf{W}|$$

leads to

$$|\mathbf{W}| = 0.$$

Thus $\mathbf{W}$ is singular. The right null vector $\mathbf{w}$ satisfies the equation

$$\mathbf{W}\mathbf{w} = \mathbf{0}. \tag{1.2-11}$$

For $n = 3$, $\mathbf{W}$ can be written as

$$\mathbf{W} = \begin{bmatrix} 0 & w_3 & -w_2 \\ -w_3 & 0 & w_1 \\ w_2 & -w_1 & 0 \end{bmatrix} \tag{1.2-12}$$

and the right and left null vectors have the solution

$$\mathbf{w} = \begin{bmatrix} w_1 \\ w_2 \\ w_3 \end{bmatrix}. \tag{1.2-13}$$

The null vector $\mathbf{w}$ is the *axial vector* of the skew-symmetric matrix $\mathbf{W}$ (Chadwick, 1976a). $\mathbf{W}$ and $\mathbf{w}$ are related by

$$W_{ij} = \delta_{ijk}w_k, \quad w_k = \tfrac{1}{2}\delta_{kij}W_{ij}. \tag{1.2-14}$$

Unless $\mathbf{W} = \mathbf{0}$ the axial vector $\mathbf{w}$ given by (1.2-13) is unique except a constant multiplier.

> **Theorem 1.2-1** If two 3×3 skew-symmetric matrices $\mathbf{W}_1$ and $\mathbf{W}_2$ share the same axial vector, they differ by a constant multiplier k, i.e.,
>
> $$\mathbf{W}_1 = k\mathbf{W}_2.$$

1.3 Diagonalization of Simple and Semisimple Matrices

The right eigenvectors $\mathbf{x}$'s associated with distinct eigenvalues λ's are independent of each other. The same statement applies to the left eigenvectors $\mathbf{y}$'s. A matrix is *simple* if all its eigenvalues are distinct, i.e., all eigenvalues are simple roots of (1.2-3). The n right eigenvectors $\mathbf{x}_i$ are independent of each other and span an n-dimensional space, so do the n left eigenvectors $\mathbf{y}_i$. When normalized, (1.2-8) and (1.2-9) can be combined as

$$\mathbf{y}_i \cdot \mathbf{x}_j = \delta_{ij} \tag{1.3-1}$$

where δ_{ij} is the *Kronecker* delta. Introducing the $n \times n$ matrices

$$\begin{aligned} \mathbf{X} &= [\mathbf{x}_1,\ \mathbf{x}_2,\ \ldots,\ \mathbf{x}_n], \\ \mathbf{Y} &= [\mathbf{y}_1,\ \mathbf{y}_2,\ \ldots,\ \mathbf{y}_n], \end{aligned} \tag{1.3-2}$$

which are nonsingular, (1.3-1) is written as

$$\mathbf{Y}^T\mathbf{X} = \mathbf{I} = \mathbf{X}\mathbf{Y}^T. \tag{1.3-3}$$

Thus $\mathbf{Y}^T$ is the inverse, or the *reciprocal*, of $\mathbf{X}$ which provides a justification for calling $\mathbf{y}_i$ the *reciprocal* vector. The second equality in (1.3-3) follows from (1.1-23). Employing the notation of (1.1-10c), the eigenrelation (1.2-1) written for $\lambda = \lambda_1, \lambda_2, \ldots, \lambda_n$ can be combined into one equation as

$$\mathbf{A}\mathbf{X} = \mathbf{X}\langle\lambda_*\rangle. \tag{1.3-4}$$

Application of $(1.3\text{-}3)_{1,2}$ leads to the *diagonalization*

$$\mathbf{Y}^T\mathbf{A}\mathbf{X} = \langle\lambda_*\rangle \quad \text{and} \quad \mathbf{A} = \mathbf{X}\langle\lambda_*\rangle\mathbf{Y}^T. \tag{1.3-5}$$

When $\mathbf{A}$ is symmetric, $\mathbf{y}_i = \mathbf{x}_i$ and (1.3-1) becomes

$$\mathbf{x}_i \cdot \mathbf{x}_j = \delta_{ij}. \tag{1.3-6}$$

The eigenvectors $\mathbf{x}_1, \mathbf{x}_2, \ldots, \mathbf{x}_n$ are all unit vectors and form an *orthonormal* set. Equation (1.3-3) is replaced by

$$\mathbf{X}^T\mathbf{X} = \mathbf{I} = \mathbf{X}\mathbf{X}^T, \tag{1.3-7}$$

and (1.3-5) by

$$\mathbf{X}^T\mathbf{A}\mathbf{X} = \langle \lambda_* \rangle \quad \text{and} \quad \mathbf{A} = \mathbf{X}\langle \lambda_* \rangle \mathbf{X}^T. \tag{1.3-8}$$

The results obtained so far assume that the matrix **A** is simple, i.e., the eigenvalues λ's are distinct. When the eigenvalues λ's have a repeated root of multiplicity q, the eigenvectors associated with the repeated root may not have q independent eigenvectors. If they have q independent eigenvectors, and if the same situation happens to all other repeated roots, we have n independent eigenvectors for the matrix **A** of order n. A matrix of order n that has repeated eigenvalues and possesses n independent eigenvectors is called *semisimple*. If the matrix has less than n independent eigenvectors, it is called *nonsemisimple*.

For semisimple matrices the eigenvectors associated with a repeated eigenvalue are not unique. The nonuniqueness is more than a constant multiplier on the eigenvector. The right eigenvector $\mathbf{x}_i$ and the left eigenvector $\mathbf{y}_i$ associated with a repeated eigenvalue are not necessarily orthogonal to each other. However, due to the nonuniqueness, it is possible to choose a set of eigenvectors such that the orthogonality relations hold. We could also normalize these eigenvectors according to (1.2-9) (see also Section 12.11). Therefore (1.3-1)–(1.3-5) apply to semisimple matrices also. We will discuss the diagonalization of nonsemisimple matrices in the next section.

It can be shown that if **A** is real and symmetric, it is either simple or semisimple, i.e., there exit n independent eigenvectors regardless of whether **A** has a repeated eigenvalue or not. More generally, if **A** is complex and is *Hermitian* (Section 1.7), it is either simple or semisimple. However if **A** is complex and symmetric, it can be nonsemisimple. Scott (1993a,b) has shown that a complex symmetric matrix **A** has an isotropic eigenvector (Section 1.1) if and only if **A** is nonsemisimple.

1.4 Nonsemisimple Matrices

When a matrix **A** is nonsemisimple, it has a repeated eigenvalue λ of multiplicity q but does not have q independent eigenvectors associated with the repeated eigenvalue. In this case the missing eigenvectors are provided by the *generalized eigenvectors*.

Even though **A** does not have q independent eigenvectors, it must have at least one eigenvector, say $\mathbf{x}_1$, associated with a repeated eigenvalue λ. The generalized eigenvectors $\mathbf{x}_2, \mathbf{x}_3, \ldots,$ are determined by

$$\begin{aligned}
\mathbf{A}\mathbf{x}_1 &= \lambda\mathbf{x}_1, \\
\mathbf{A}\mathbf{x}_2 &= \lambda\mathbf{x}_2 + \mathbf{x}_1, \\
\mathbf{A}\mathbf{x}_3 &= \lambda\mathbf{x}_3 + \mathbf{x}_2, \\
&\vdots \\
\mathbf{A}\mathbf{x}_k &= \lambda\mathbf{x}_k + \mathbf{x}_{k-1}.
\end{aligned} \tag{1.4-1}$$

These equations can be rewritten as

$$\begin{aligned}
(\mathbf{A} - \lambda\mathbf{I})\mathbf{x}_1 &= \mathbf{0}, \\
(\mathbf{A} - \lambda\mathbf{I})^2\mathbf{x}_2 &= \mathbf{0}, \\
(\mathbf{A} - \lambda\mathbf{I})^3\mathbf{x}_3 &= \mathbf{0}, \\
&\vdots \\
(\mathbf{A} - \lambda\mathbf{I})^k\mathbf{x}_k &= \mathbf{0}.
\end{aligned} \tag{1.4-2}$$

Therefore $\mathbf{x}_2$ is a generalized eigenvector of *rank* 2 and $\mathbf{x}_k$ is a generalized eigenvector of *rank k*. Together with $\mathbf{x}_1$, they form a *chain of length* $k \le q$ and are independent of each other. They are not unique. If $\mathbf{x}_1, \mathbf{x}_2, \ldots, \mathbf{x}_k$ are solutions of (1.4-2), so are

$$\begin{aligned}
\hat{\mathbf{x}}_1 &= \mathbf{x}_1, \\
\hat{\mathbf{x}}_2 &= \mathbf{x}_2 + a_1\mathbf{x}_1, \\
\hat{\mathbf{x}}_3 &= \mathbf{x}_3 + a_1\mathbf{x}_2 + a_2\mathbf{x}_1 \\
&\vdots \\
\hat{\mathbf{x}}_k &= \mathbf{x}_k + a_1\mathbf{x}_{k-1} + a_2\mathbf{x}_{k-2} + \ldots + a_{k-1}\mathbf{x}_1,
\end{aligned}$$

where $a_1, a_2, \ldots, a_{k-1}$ are arbitrary constants. The subspace spanned by $\hat{\mathbf{x}}_1, \hat{\mathbf{x}}_2, \ldots, \hat{\mathbf{x}}_k$ are identical to the space spanned by $\mathbf{x}_1, \mathbf{x}_2, \ldots, \mathbf{x}_k$. We see that only the highest rank $\hat{\mathbf{x}}_k$ contains all the arbitrary constants. Once $\hat{\mathbf{x}}_k$ is chosen, the whole chain is determined. This suggests that we determine (1.4-2) in the reverse order from $\mathbf{x}_k$.

To find $\mathbf{x}_k$, we look for an integer k for which (Pease, 1965)

$$(\mathbf{A} - \lambda\mathbf{I})^k\mathbf{x}_k = \mathbf{0}, \quad (\mathbf{A} - \lambda\mathbf{I})^{k-1}\mathbf{x}_k \neq \mathbf{0}. \tag{1.4-3}$$

When an $\mathbf{x}_k$ exists, we determine the remaining vectors of the chain from (1.4-1) as

$$\begin{aligned}
\mathbf{x}_{k-1} &= (\mathbf{A} - \lambda\mathbf{I})\mathbf{x}_k, \\
\mathbf{x}_{k-2} &= (\mathbf{A} - \lambda\mathbf{I})\mathbf{x}_{k-1}, \\
&\vdots \\
\mathbf{x}_1 &= (\mathbf{A} - \lambda\mathbf{I})\mathbf{x}_2.
\end{aligned} \tag{1.4-4}$$

A repeated eigenvalue of multiplicity q has more than one chain when $k < q$. If (1.4-3) are satisfied by $k = k_1$ and k_2, we have two chains of length k_1 and k_2, respectively, associated with the repeated eigenvalue λ. A chain of length one is the one for which $\mathbf{x}_1$ of $(1.4\text{-}1)_1$ exists but $\mathbf{x}_2$ of $(1.4\text{-}1)_2$ does not exist. If all repeated eigenvalues have chains of length one, the matrix is semisimple.

The k equations (1.4-1) in the chain can be written in one equation as

$$\mathbf{A}[\mathbf{x}_1, \mathbf{x}_2, \ldots, \mathbf{x}_k] = [\mathbf{x}_1, \mathbf{x}_2, \ldots, \mathbf{x}_k]\mathbf{J}_k(\lambda), \tag{1.4-5}$$

where $\mathbf{J}_k(\lambda)$ is a $k \times k$ matrix with the repeated eigenvalue λ on the diagonal, 1 on the first super diagonal, and zeros elsewhere. Thus

$$\mathbf{J}_1(\lambda) = [\lambda],$$

$$\mathbf{J}_2(\lambda) = \begin{bmatrix} \lambda & 1 \\ 0 & \lambda \end{bmatrix},$$

$$\mathbf{J}_3(\lambda) = \begin{bmatrix} \lambda & 1 & 0 \\ 0 & \lambda & 1 \\ 0 & 0 & \lambda \end{bmatrix}, \tag{1.4-6}$$

$$\mathbf{J}_4(\lambda) = \begin{bmatrix} \lambda & 1 & 0 & 0 \\ 0 & \lambda & 1 & 0 \\ 0 & 0 & \lambda & 1 \\ 0 & 0 & 0 & \lambda \end{bmatrix},$$

$$\vdots$$

The matrix $\mathbf{J}_k(\lambda)$ is called the *upper Jordan block of order k*. When all chains associated with a repeated eigenvalue λ of multiplicity q are found, we should have q independent eigenvectors and generalized eigenvectors. With other repeated eigenvalues treated in the same manner, (1.4-5) for the full set can be written as

$$\mathbf{AX} = \mathbf{XJ} \tag{1.4-7}$$

where the $\mathbf{X}$ is defined in $(1.3\text{-}2)_1$ in which $\mathbf{x}_1, \mathbf{x}_2, \ldots, \mathbf{x}_n$ are now the eigenvectors and generalized eigenvectors and

$$\mathbf{J} = \begin{bmatrix} \mathbf{J}_{k_1}(\lambda_\alpha) & \mathbf{0} & \mathbf{0} & \cdots \\ \mathbf{0} & \mathbf{J}_{k_2}(\lambda_\beta) & \mathbf{0} & \cdots \\ \mathbf{0} & \mathbf{0} & \mathbf{J}_{k_3}(\lambda_\gamma) & \cdots \\ \vdots & \vdots & \vdots & \ddots \end{bmatrix}. \tag{1.4-8}$$

This is the *Jordan canonical form*. There is no loss in generality in arranging k_1, k_2, ... in the order of ascending rank, i.e.,

$$\begin{aligned} &1 \leq k_1 \leq k_2 \leq \ldots, \\ &k_1 + k_2 + \ldots = n. \end{aligned}$$

It should be pointed out that λ_α, λ_β, λ_γ,..., need not be distinct. One could have $\lambda_\alpha = \lambda_\gamma \neq \lambda_\beta$. It can be shown that $\mathbf{X}$ is nonsingular. Therefore its inverse, denoted by $\mathbf{Y}^T$, exists and we obtain from (1.4-7)

$$\mathbf{Y}^T\mathbf{A}\mathbf{X} = \mathbf{J} \quad \text{and} \quad \mathbf{A} = \mathbf{X}\mathbf{J}\mathbf{Y}^T. \tag{1.4-9}$$

These are the *diagonalization* of a nonsemisimple matrix $\mathbf{A}$.

The use of notation $\mathbf{Y}^T$ as the inverse of $\mathbf{X}$ is intentional. It can be shown that the columns of $\mathbf{Y}$ are the left eigenvectors and generalized left eigenvectors. The generalized left eigenvectors can be defined similar to the generalized right eigenvectors in (1.4-1) except that the order of the rank is reversed, i.e.,

$$\begin{aligned} \mathbf{A}^T\mathbf{y}_k &= \lambda\mathbf{y}_k, \\ \mathbf{A}^T\mathbf{y}_{k-1} &= \lambda\mathbf{y}_{k-1} + \mathbf{y}_k, \\ \mathbf{A}^T\mathbf{y}_{k-2} &= \lambda\mathbf{y}_{k-2} + \mathbf{y}_{k-1}, \\ &\vdots \\ \mathbf{A}^T\mathbf{y}_1 &= \lambda\mathbf{y}_1 + \mathbf{y}_2. \end{aligned} \tag{1.4-10}$$

Thus $\mathbf{y}_{k-1}$ is of rank 2 while $\mathbf{y}_1$ is of rank k.

It should be noted that the Jordan canonical form $\mathbf{J}$ is not a symmetric matrix. When $\mathbf{A}$ is complex, symmetric, and nonsemisimple, a symmetric canonical form can be obtained by a different approach (Scott, 1993a,b).

As an illustration consider the real matrix

$$\mathbf{A} = \begin{bmatrix} 1 & 1 & 0 & 0 & 0 \\ 0 & 1 & 0 & 0 & 0 \\ 0 & -2 & 2 & 1 & 0 \\ 0 & 2 & -1 & 1 & 1 \\ 0 & -2 & 1 & 0 & 0 \end{bmatrix}. \tag{1.4-11}$$

It is readily shown that

$$|\mathbf{A} - \lambda\mathbf{I}| = (1 - \lambda)^5$$

and hence $\lambda = 1$ is a repeated eigenvalue of multiplicity 5. With λ

= 1 we have

$$(\mathbf{A}-\lambda\mathbf{I})=\begin{bmatrix}0&1&0&0&0\\0&0&0&0&0\\0&-2&1&1&0\\0&2&-1&0&1\\0&-2&1&0&-1\end{bmatrix},$$

$$(\mathbf{A}-\lambda\mathbf{I})^2=\begin{bmatrix}0&0&0&0&0\\0&0&0&0&0\\0&0&0&1&1\\0&0&0&-1&-1\\0&0&0&1&1\end{bmatrix},$$

and

$$(\mathbf{A}-\lambda\mathbf{I})^3=\mathbf{0}.$$

We see that $(1.4\text{-}3)_1$ is satisfied for $k = 3$ with

$$\mathbf{x}_3^T=(a,\ b,\ c,\ d,\ e)$$

where a, b, c, d, and e are arbitrary constants. Equation $(1.4\text{-}3)_2$ is satisfied if $d + e \neq 0$. Let

$$\mathbf{x}_3^T=(0,\ 0,\ 0,\ 1,\ 0)$$

be the highest ranking member of the chain. The remaining members of the chain are obtained from (1.4-4) as

$$\mathbf{x}_2^T=(0,\ 0,\ 1,\ 0,\ 0),$$
$$\mathbf{x}_1^T=(0,\ 0,\ 1,\ -1,\ 1).$$

This completes the chain of length 3. Equations $(1.4\text{-}3)_{1,2}$ are also satisfied for $k = 2$ with

$$\mathbf{x}_2^{*T}=(0,\ 1,\ 0,\ 0,\ 0),$$

and (1.4-4) gives

$$\mathbf{x}_1^{*T}=(1,\ 0,\ -2,\ 2,\ -2).$$

This completes the second chain of length 2. In conformity with the ascending order of k in (1.4-8), we rename $\mathbf{x}_1^*$ and $\mathbf{x}_2^*$ of the second chain as $\mathbf{x}_1$ and $\mathbf{x}_2$, respectively, and $\mathbf{x}_1$, $\mathbf{x}_2$, $\mathbf{x}_3$ of the first chain as $\mathbf{x}_3$, $\mathbf{x}_4$, $\mathbf{x}_5$, respectively. Therefore,

$$\mathbf{X} = \begin{bmatrix} 1 & 0 & 0 & 0 & 0 \\ 0 & 1 & 0 & 0 & 0 \\ -2 & 0 & 1 & 1 & 0 \\ 2 & 0 & -1 & 0 & 1 \\ -2 & 0 & 1 & 0 & 0 \end{bmatrix}$$

and

$$\mathbf{J} = \begin{bmatrix} \mathbf{J}_2(1) & \mathbf{0} \\ \mathbf{0} & \mathbf{J}_3(1) \end{bmatrix}.$$

The inverse $\mathbf{Y}^T$ of $\mathbf{X}$ can be shown to be

$$\mathbf{Y}^T = \begin{bmatrix} 1 & 0 & 0 & 0 & 0 \\ 0 & 1 & 0 & 0 & 0 \\ 2 & 0 & 0 & 0 & 1 \\ 0 & 0 & 1 & 0 & -1 \\ 0 & 0 & 0 & 1 & 1 \end{bmatrix}.$$

It is readily verified that

$$\mathbf{y}_1^{*T} = (1,\ 0,\ 0,\ 0,\ 0),$$
$$\mathbf{y}_2^{*T} = (0,\ 1,\ 0,\ 0,\ 0),$$

and

$$\mathbf{y}_1^{T} = (2,\ 0,\ 0,\ 0,\ 1),$$
$$\mathbf{y}_2^{T} = (0,\ 0,\ 1,\ 0,\ -1),$$
$$\mathbf{y}_3^{T} = (0,\ 0,\ 0,\ 1,\ 1),$$

are, respectively, the chains of lengths 2 and 3 which satisfy (1.4-10) for $\lambda = 1$.

1.5 Commutative Matrices

Two square matrices are called *commutative* if

$$\mathbf{AB} = \mathbf{BA}.$$

It is obvious that all diagonal matrices are commutative. We also saw in (1.1-23) that $\mathbf{A}$ and $\mathbf{A}^{-1}$ are commutative.

Let $\mathbf{A}$ and $\mathbf{B}$ be simple or semisimple. Furthermore, let $\mathbf{A}$ and $\mathbf{B}$ have the same set of eigenvectors. By $(1.3\text{-}5)_2$ $\mathbf{A}$ and $\mathbf{B}$ can be diagonalized as

$$\mathbf{A} = \mathbf{X}\langle\lambda_*\rangle_A \mathbf{Y}^T, \qquad \mathbf{B} = \mathbf{X}\langle\lambda_*\rangle_B \mathbf{Y}^T,$$

where $\langle\lambda_*\rangle_A$ and $\langle\lambda_*\rangle_B$ are diagonal matrices whose diagonal elements are the eigenvalues of **A** and **B**, respectively. We then have

$$\begin{aligned} \mathbf{AB} &= \mathbf{X}\langle\lambda_*\rangle_A \mathbf{Y}^T \mathbf{X}\langle\lambda_*\rangle_B \mathbf{Y}^T = \mathbf{X}\langle\lambda_*\rangle_A \langle\lambda_*\rangle_B \mathbf{Y}^T, \\ \mathbf{BA} &= \mathbf{X}\langle\lambda_*\rangle_B \mathbf{Y}^T \mathbf{X}\langle\lambda_*\rangle_A \mathbf{Y}^T = \mathbf{X}\langle\lambda_*\rangle_B \langle\lambda_*\rangle_A \mathbf{Y}^T = \mathbf{X}\langle\lambda_*\rangle_A \langle\lambda_*\rangle_B \mathbf{Y}^T. \end{aligned} \tag{1.5-1}$$

Therefore **AB** = **BA**. It can be shown that the converse is also true, i.e., if **AB** = **BA**, **A** and **B** share the same set of eigenvectors.

> **Theorem 1.5–1** Let **A** and **B** be simple or semisimple. Then **AB**=**BA** if and only if they share the same set of eigenvectors.

If **A** and **B** are real and symmetric, they are necessarily simple or semisimple. Therefore a special case of Theorem 1.5-1 is:

> **Theorem 1.5–2** Let **A** and **B** be real and symmetric. Then **AB**=**BA** if and only if they share the same set of eigenvectors.

We now turn to the case when **A** and **B** are both nonsemisimple. First, from the definition of $\mathbf{J}_k(\lambda)$ in (1.4-6) we have

$$\mathbf{J}_1(a)\mathbf{J}_1(b) = [ab],$$

$$\mathbf{J}_2(a)\mathbf{J}_2(b) = \begin{bmatrix} ab & a{+}b \\ 0 & ab \end{bmatrix},$$

$$\mathbf{J}_3(a)\mathbf{J}_3(b) = \begin{bmatrix} ab & a{+}b & 1 \\ 0 & ab & a{+}b \\ 0 & 0 & ab \end{bmatrix},$$

$$\mathbf{J}_4(a)\mathbf{J}_4(b) = \begin{bmatrix} ab & a{+}b & 1 & 0 \\ 0 & ab & a{+}b & 1 \\ 0 & 0 & ab & a{+}b \\ 0 & 0 & 0 & ab \end{bmatrix},$$

$$\vdots$$

It is seen that $\mathbf{J}_k(a)\mathbf{J}_k(b)$ has ab on its diagonal, $a{+}b$ on its first super diagonal, 1 on its second super diagonal, and zeros elsewhere. The important point to observe is that $\mathbf{J}_k(a)$ and $\mathbf{J}_k(b)$ commute, as is evident from the expressions on the right sides of the equalities which are symmetric with a and b. This means that if $\mathbf{J}_A$ and $\mathbf{J}_B$ are the Jordan canonical forms for **A** and **B**, respectively, and if $\mathbf{J}_A$ and $\mathbf{J}_B$ have the same structure, i.e., the k_1, k_2,..., in (1.4-8) are the same for $\mathbf{J}_A$ and $\mathbf{J}_B$,

$$\mathbf{J}_A\mathbf{J}_B = \mathbf{J}_B\mathbf{J}_A. \tag{1.5-2}$$

Following the derivation of (1.5-1) we have:

> **Theorem 1.5–3** Let **A** and **B** be nonsemisimple. If the Jordan canonical forms for **A** and **B** have the same structure, (i.e., the k_1, k_2,..., in (1.4-8) are the same for **A** and **B**), and if **A** and **B** share the same set of eigenvectors and generalized eigenvectors, **A** and **B** commute.

1.6 Positive Definite Real Matrices

In this section we will consider real matrices only, even though some of the results obtained can be extended to complex matrices.

The *quadratic form* of a real square matrix **C** is

$$\mathbf{u}^T\mathbf{C}\mathbf{u} \tag{1.6-1}$$

where **u** is any nonzero vector or column matrix. The quadratic form is an 1×1 matrix. Since the transpose of an 1×1 matrix is identical to itself,

$$\mathbf{u}^T\mathbf{C}\mathbf{u} = (\mathbf{u}^T\mathbf{C}\mathbf{u})^T = \mathbf{u}^T\mathbf{C}^T\mathbf{u}.$$

From this it is clear that the quadratic form of a skew-symmetric matrix **W** vanishes, i.e.,

$$\mathbf{u}^T\mathbf{W}\mathbf{u} = 0, \quad \text{if } \mathbf{W} = -\mathbf{W}^T. \tag{1.6-2}$$

Therefore if the **C** in (1.6-1) is not symmetric, only the symmetric part of **C** contributes to the quadratic form.

A real symmetric matrix **C** is *positive definite* if

$$\mathbf{u}^T\mathbf{C}\mathbf{u} > 0 \tag{1.6-3}$$

for any nonzero real vector **u**. It is *positive semidefinite* if

$$\mathbf{u}^T\mathbf{C}\mathbf{u} \geq 0. \tag{1.6-4}$$

Since **C** is real and symmetric, its eigenvalues are real, and the right and left eigenvectors that are identical are also real (Section 1.2).

Let λ_i and $\mathbf{x}_i$ ($i = 1,2,\ldots,n$) be the eigenvalues and eigenvectors of $\mathbf{C}$, respectively. Employing the diagonalization $(1.3\text{-}8)_2$ we have

$$\mathbf{u}^T\mathbf{C}\mathbf{u} = \mathbf{u}^T\mathbf{X}\langle\lambda_*\rangle\mathbf{X}^T\mathbf{u} = (\mathbf{X}^T\mathbf{u})^T\langle\lambda_*\rangle(\mathbf{X}^T\mathbf{u}),$$

or

$$\mathbf{u}^T\mathbf{C}\mathbf{u} = \mathbf{z}^T\langle\lambda_*\rangle\mathbf{z} = \lambda_1 z_1^2 + \lambda_2 z_2^2 + \ldots + \lambda_n z_n^2, \qquad \mathbf{z} = \mathbf{X}^T\mathbf{u}. \tag{1.6-5}$$

That $\mathbf{X}^T$ is nonsingular means that if $\mathbf{u}$ is nonzero, so is $\mathbf{z}$. We therefore have the following theorem:

> **Theorem 1.6–1** A real symmetric matrix is positive definite if and only if all its eigenvalues are positive and nonzero. If some of the eigenvalues are zero, the matrix is positive semidefinite.

A consequence of the theorem is that, from $(1.2\text{-}5)_2$, the determinant of a real symmetric matrix must be positive and nonzero if the matrix is positive definite.

When we remove an equal number of rows and columns from a symmetric matrix $\mathbf{C}$ we obtain a square submatrix. The determinant of such a submatrix is called a *minor*. If the rows and columns removed have the same indices, the diagonal elements of the submatrix are the diagonal elements of the original matrix. The submatrix so obtained is called a *principal submatrix* and the determinant of the principal submatrix is a *principal minor*. The diagonal elements $C_{11}, C_{22}, \ldots, C_{nn}$ and the determinant $|\mathbf{C}|$ are principal minors. If the $r \times r$ principal submatrix is from the first r rows and the first r columns of $\mathbf{C}$, we have a *leading* principal submatrix. The determinant of a leading principal submatrix is a *leading principal minor*.

The condition (1.6-3) for a positive definite matrix $\mathbf{C}$ implies that all principal submatrices of $\mathbf{C}$ must be positive definite. This can be proved easily as follows. Consider the principal submatrix

$$\begin{bmatrix} C_{11} & C_{13} \\ C_{13} & C_{33} \end{bmatrix}. \tag{1.6-6}$$

By choosing

$$\mathbf{u}^T = [u_1,\ 0,\ u_3,\ \ldots,\ 0],$$

(1.6-3) can be written as

$$[u_1,\ u_3]\begin{bmatrix} C_{11} & C_{13} \\ C_{13} & C_{33} \end{bmatrix}\begin{bmatrix} u_1 \\ u_3 \end{bmatrix} > 0.$$

which shows that the 2×2 matrix is positive definite. A necessary condition for the 2×2 matrix to be positive definite is that its determinant is positive. Therefore (1.6-3) also implies that all principal minors of **C** must be positive..

There are 7 principal minors for a 3×3 symmetric matrix and 15 for a 4×4 symmetric matrix. The following theorem (Hohn, 1965) tells us that we do not have to inspect all principal minors to see if a real symmetric matrix is positive definite.

> **Theorem 1.6–2** A real symmetric matrix is positive definite if and only if the *leading* principal minors of the matrix are positive.

This means that a real symmetric matrix **C** of order n is positive definite if and only if the following n conditions are satisfied.

$$
\begin{gathered}
C_{11} > 0, \\
\begin{vmatrix} C_{11} & C_{12} \\ C_{12} & C_{22} \end{vmatrix} > 0, \\
\begin{vmatrix} C_{11} & C_{12} & C_{13} \\ C_{12} & C_{22} & C_{23} \\ C_{13} & C_{23} & C_{33} \end{vmatrix} > 0, \qquad (1.6\text{-}7) \\
\vdots \\
\begin{vmatrix} C_{11} & C_{12} & \cdots & C_{1n} \\ C_{12} & C_{22} & \cdots & C_{2n} \\ \vdots & \vdots & \ddots & \vdots \\ C_{1n} & C_{2n} & \cdots & C_{nn} \end{vmatrix} > 0.
\end{gathered}
$$

Theorem 1.6–2 provides an alternate to Theorem 1.6–1 for a set of necessary and sufficient conditions for a matrix to be positive definite. The determination of the eigenvalues of a matrix in general requires a numerical solution. Theorem 1.6–1 is therefore not very useful if the eigenvalues cannot be obtained explicitly in terms of the elements of the matrix.

The bounds on the quadratic form $\mathbf{u}^T\mathbf{C}\mathbf{u}$ of (1.6-5) for an arbitrarily given **u** can be set as follows. Let

$$\lambda_1 \geq \lambda_2 \geq \cdots \geq \lambda_n. \qquad (1.6\text{-}8)$$

Equation (1.6-5) can be written as

$$\mathbf{u}^T\mathbf{C}\mathbf{u} = \lambda_1|\mathbf{z}|^2 - (\lambda_1 - \lambda_2)z_2^2 - \ldots - (\lambda_1 - \lambda_n)z_n^2$$
$$= (\lambda_1 - \lambda_n)z_1^2 + \ldots + (\lambda_{n-1} - \lambda_n)z_{n-1}^2 + \lambda_n|\mathbf{z}|^2.$$

Since

$$|\mathbf{z}|^2 = (\mathbf{X}^T\mathbf{u})^T(\mathbf{X}^T\mathbf{u}) = \mathbf{u}^T\mathbf{u} = |\mathbf{u}|^2$$

we have

$$\lambda_n|\mathbf{u}|^2 \le \mathbf{u}^T\mathbf{C}\mathbf{u} \le \lambda_1|\mathbf{u}|^2. \tag{1.6-9}$$

Therefore, if **u** is normalized to a unit vector, (1.6-9) tells us that the quadratic form $\mathbf{u}^T\mathbf{C}\mathbf{u}$ for any unit vector **u** is bounded between the largest and the smallest eigenvalues of **C**. The matrix **C** need not be positive definite. Thus some of the λ's in (1.6-8) maybe negative.

When an $n\times n$ matrix **A** is singular its determinant as well as its minor may vanish. The *rank* of **A** is the size of the largest submatrix whose determinant is nonzero. The rank of a singular matrix is at most $n-1$. If the rank is $n-k$ there are k independent null vectors.

1.7 Hermitian Matrices

A complex matrix **M** is *Hermitian* if

$$\mathbf{M} = \overline{\mathbf{M}}^T \tag{1.7-1}$$

where the overbar denotes the complex conjugate. Writing **M** as

$$\mathbf{M} = \mathbf{D} + i\mathbf{W} \tag{1.7-2}$$

in which **D** and **W** are real and inserting into (1.7-1) yields

$$\mathbf{D} = \mathbf{D}^T, \quad \mathbf{W} = -\mathbf{W}^T. \tag{1.7-3}$$

Therefore the real part of a Hermitian matrix is symmetric while the imaginary part is skew-symmetric. A real symmetric matrix is a special case of Hermitian matrices. It is easily shown that if **M** is Hermitian, so are $-\mathbf{M}$, $\mathbf{M}^T$, $\overline{\mathbf{M}}$, and $\mathbf{M}^{-1}$, provided the inverse exists.

A complex matrix is *skew-Hermitian* if

$$\mathbf{M} = -\overline{\mathbf{M}}^T. \tag{1.7-4}$$

A real skew-symmetric matrix is a special case of skew-Hermitian matrices. Equation (1.7-4) can be rewritten as

$$(i\mathbf{M}) = (\overline{i\mathbf{M}})^T$$

which says that $i\mathbf{M}$ is Hermitian. Therefore a skew-Hermitian matrix is converted to a Hermitian matrix by a multiplication of i.

We define the quadratic form of a Hermitian as

$$\bar{\mathbf{u}}^T\mathbf{M}\mathbf{u} = (\bar{\mathbf{u}}^T\mathbf{M}\mathbf{u})^T = \mathbf{u}^T\mathbf{M}^T\bar{\mathbf{u}} = \mathbf{u}^T\overline{\mathbf{M}\mathbf{u}}. \tag{1.7-5}$$

The last expression is the complex conjugate of the first expression. Therefore the quadratic form $\bar{\mathbf{u}}^T\mathbf{M}\mathbf{u}$ is real for any Hermitian matrix $\mathbf{M}$.

The eigenrelation for a Hermitian matrix $\mathbf{M}$ is

$$\mathbf{M}\mathbf{x} = \lambda\mathbf{x}. \tag{1.7-6}$$

Premultiplied by $\bar{\mathbf{x}}^T$ we have

$$\bar{\mathbf{x}}^T\mathbf{M}\mathbf{x} = \lambda\bar{\mathbf{x}}^T\mathbf{x} \tag{1.7-7}$$

which indicates that the eigenvalue λ must be real because $\bar{\mathbf{x}}^T\mathbf{M}\mathbf{x}$ and $\bar{\mathbf{x}}^T\mathbf{x}$ are. Taking the complex conjugate of (1.7-6) and using (1.7-1) we have

$$\mathbf{M}^T\bar{\mathbf{x}} = \lambda\bar{\mathbf{x}} \quad \text{or} \quad \bar{\mathbf{x}}^T\mathbf{M} = \lambda\bar{\mathbf{x}}^T.$$

Therefore $\bar{\mathbf{x}}$ is the left eigenvector, which is the complex conjugate of the right eigenvector. From

$$\mathbf{M}\mathbf{x}_2 = \lambda_2\mathbf{x}_2, \qquad \bar{\mathbf{x}}_1^T\mathbf{M} = \lambda_1\bar{\mathbf{x}}_1^T, \tag{1.7-8}$$

we obtain, after premultiplying $(1.7\text{-}8)_1$ by $\bar{\mathbf{x}}_1^T$ and postmultiplying $(1.7\text{-}8)_2$ by $\mathbf{x}_2$,

$$0 = (\lambda_1 - \lambda_2)\bar{\mathbf{x}}_1^T\mathbf{x}_2$$

confirming the orthogonality relation (1.2-8).

A Hermitian matrix $\mathbf{M}$ is *positive definite* if

$$\bar{\mathbf{u}}^T\mathbf{M}\mathbf{u} > 0,$$

and positive semidefinite if

$$\bar{\mathbf{u}}^T\mathbf{M}\mathbf{u} \geq 0$$

for any nonzero complex vector $\mathbf{u}$. Writing $\mathbf{u}$ as

$$\mathbf{u} = \mathbf{a} + i\mathbf{b}$$

where $\mathbf{a}$ and $\mathbf{b}$ are real and imaginary parts of $\mathbf{u}$, it is readily shown that

$$\bar{\mathbf{u}}^T\mathbf{M}\mathbf{u} = \mathbf{a}^T\mathbf{D}\mathbf{a} + \mathbf{b}^T\mathbf{D}\mathbf{b} - 2\mathbf{a}^T\mathbf{W}\mathbf{b}. \tag{1.7-9}$$

Equation (1.7-9) tells us that a necessary condition for $\mathbf{M}$ to be positive definite is that the real part $\mathbf{D}$ be positive definite. This condition however is not sufficient.

We have shown in (1.7-7) that the eigenvalues of a Hermitian matrix are real. A necessary and sufficient condition for a Hermitian matrix to be positive definite is that its eigenvalues are positive and nonzero. If λ is an eigenvalue of a positive definite Hermitian matrix $\mathbf{M}$, it is also an eigenvalue of $\overline{\mathbf{M}}$ and $\mathbf{M}^T$, and λ^{-1} is an eigenvalue of $\mathbf{M}^{-1}$ and $\overline{\mathbf{M}}^{-1}$. Hence $\mathbf{M}^T$, $\overline{\mathbf{M}}$, $\mathbf{M}^{-1}$, and $\overline{\mathbf{M}}^{-1}$ are positive definite if $\mathbf{M}$ is.

In the remainder of this section we focus our attention on 3×3 Hermitian matrices. Let the skew-symmetric matrix $\mathbf{W}$ in (1.7-2) and its axial vector $\mathbf{w}$ be given by

$$\mathbf{W} = \begin{bmatrix} 0 & w_3 & -w_2 \\ -w_3 & 0 & w_1 \\ w_2 & -w_1 & 0 \end{bmatrix}, \qquad \mathbf{w} = \begin{bmatrix} w_1 \\ w_2 \\ w_3 \end{bmatrix}. \tag{1.7-10}$$

It was shown in (1.2-14) that $\mathbf{W}$ and $\mathbf{w}$ are related by

$$W_{ij} = \delta_{ijk} w_k, \quad w_k = \tfrac{1}{2}\delta_{kij} W_{ij}. \tag{1.7-11}$$

The last term in (1.7-9) can be written as

$$\mathbf{a}^T\mathbf{W}\mathbf{b} = a_i \delta_{ijk}\omega_k b_j = \begin{vmatrix} a_1 & b_1 & w_1 \\ a_2 & b_2 & w_2 \\ a_3 & b_3 & w_3 \end{vmatrix}.$$

This term vanishes if the three vectors $\mathbf{a}$, $\mathbf{b}$, and $\mathbf{w}$ are coplanar.

We next show that

$$\operatorname{tr}\mathbf{M} = \operatorname{tr}\mathbf{D}, \quad \operatorname{tr}(\mathbf{M}^2) = \operatorname{tr}(\mathbf{D}^2) + 2\mathbf{w}^T\mathbf{w}, \quad |\mathbf{M}| = |\mathbf{D}| - \mathbf{w}^T\mathbf{D}\mathbf{w}. \tag{1.7-12}$$

Equations $(1.7\text{-}12)_{1,2}$ follow from (1.7-2) and the identities

$$\operatorname{tr}\mathbf{W} = 0, \quad \operatorname{tr}(\mathbf{W}\mathbf{D}) = \mathbf{0} = \operatorname{tr}(\mathbf{D}\mathbf{W}), \quad \operatorname{tr}(\mathbf{W}^2) = -2\mathbf{w}^T\mathbf{w},$$

which can be established easily. As to $(1.7\text{-}12)_3$ we use (1.1-17a) to write

$$|\mathbf{M}| = \delta_{ijk}(D_{i1} + iW_{i1})(D_{j2} + iW_{j2})(D_{k3} + iW_{k3}).$$

Since $|\mathbf{M}|$ is real, keeping the real terms gives

$$|\mathbf{M}| = |\mathbf{D}| - \delta_{ijk} D_{i1} W_{j2} W_{k3} - \delta_{ijk} W_{i1} D_{j2} W_{k3} - \delta_{ijk} W_{i1} W_{j2} D_{k3}.$$

The second term can be written as

$$\delta_{ijk} D_{i1} W_{j2} W_{k3} = \begin{vmatrix} D_{11} & w_3 & -w_2 \\ D_{21} & 0 & w_1 \\ D_{31} & -w_1 & 0 \end{vmatrix} = D_{11} w_1^2 + D_{21} w_2 w_1 + D_{31} w_3 w_1 .$$

The remaining two terms can be determined similarly. Hence

$$|\mathbf{M}| = |\mathbf{D}| - D_{ik} w_i w_k \tag{1.7-13}$$

which is $(1.7\text{-}12)_3$.

Identities (1.7-12) provide relations between the eigenvalues of **M** and **D**. Let

$$d_1 \ge d_2 \ge d_3$$

be the eigenvalues of **D** and

$$\lambda_1 \ge \lambda_2 \ge \lambda_3$$

be the eigenvalues of **M**. It can be shown that

$$\lambda_1 + \lambda_2 + \lambda_3 = d_1 + d_2 + d_3 , \tag{1.7-14a}$$

$$\lambda_1^2 + \lambda_2^2 + \lambda_3^2 > d_1^2 + d_2^2 + d_3^2 , \tag{1.7-14b}$$

$$\lambda_1 \ge d_1 \ge \lambda_2 \ge d_3 \ge \lambda_3 . \tag{1.7-14c}$$

Equations (1.7-14a,b) follow directly from $(1.7\text{-}12)_{1,2}$ and (1.2-6b). To prove (1.7-14c) consider

$$f(\lambda) = |\mathbf{M} - \lambda\mathbf{I}| = |\mathbf{D} - \lambda\mathbf{I} + i\mathbf{W}| = |\mathbf{D} - \lambda\mathbf{I}| - \mathbf{w}^T(\mathbf{D} - \lambda\mathbf{I})\mathbf{w} \tag{1.7-15}$$

where use has been made of the identity $(1.7\text{-}12)_3$. By $(1.3\text{-}8)_2$ the diagonalization of **D** is

$$\mathbf{D} = \mathbf{X}\langle d_* \rangle \mathbf{X}^T$$

where the columns of **X** are the eigenvectors of **D**. Hence

$$f(\lambda) = |\mathbf{D} - \lambda\mathbf{I}| - \hat{\mathbf{w}}^T \langle d_* - \lambda \rangle \hat{\mathbf{w}} , \quad \hat{\mathbf{w}} = \mathbf{X}^T \mathbf{w} ,$$

or

$$f(\lambda) = (d_1 - \lambda)(d_2 - \lambda)(d_3 - \lambda) - [(d_1 - \lambda)\hat{w}_1^2 + (d_2 - \lambda)\hat{w}_2^2 + (d_3 - \lambda)\hat{w}_3^2] .$$

This leads to the inequalities in (1.7-14c) because

$$f(\infty) < 0, \quad f(d_1) \ge 0, \quad f(d_3) \le 0, \quad f(-\infty) > 0 .$$

From (1.7-14c) and $(1.7\text{-}12)_3$ we have the following theorem.

> **Theorem 1.7–1** A necessary and sufficient condition for a 3×3 Hermitian matrix $\mathbf{D}+i\mathbf{W}$ to be positive definite is that $\mathbf{D}$ be positive definite and
>
> $$|\mathbf{D}| > \mathbf{w}^T\mathbf{D}\mathbf{w}.$$

1.8 Eigenplane

When an eigenvector $\mathbf{x}$ is complex, it can be written as

$$\mathbf{x} = \mathbf{e}_1 + i\mathbf{e}_2$$

where $\mathbf{e}_1$ and $\mathbf{e}_2$ are, respectively, the real and imaginary parts of $\mathbf{x}$. A complex vector is also called a *bivector*. The real vectors $\mathbf{e}_1$ and $\mathbf{e}_2$ span a plane unless they are proportional. If they are, $\mathbf{x}$ is a real vector multiplied by a complex factor. Since an eigenvector is unique up to a constant multiplier, without loss in generality we may take $\mathbf{x}$ to be real whenever $\mathbf{e}_1$ and $\mathbf{e}_2$ are proportional. A complex vector is *genuinely complex* if its real and imaginary parts are not proportional. We assume in this section that $\mathbf{x}$ is genuinely complex so that $\mathbf{e}_1$ and $\mathbf{e}_2$ span a plane.

The vectors $\mathbf{e}_1$ and $\mathbf{e}_2$ are not unique. The non-uniqueness is more than a real constant multiplier. Let

$$\mathbf{x}^{\text{o}} = \mathbf{e}_1^{\text{o}} + i\mathbf{e}_2^{\text{o}}$$

be an eigenvector. Then $\mathbf{x}^{\text{o}}$ multiplied by a complex factor $\rho e^{-i\psi}$ where ρ and ψ are real and ρ is positive is also an eigenvector. Let $\rho = 1$ and

$$\mathbf{x} = e^{-i\psi}\mathbf{x}^{\text{o}} = \mathbf{e}_1 + i\mathbf{e}_2, \qquad (1.8\text{-}1)$$

where

$$\begin{aligned} \mathbf{e}_1 &= \mathbf{e}_1^{\text{o}}\cos\psi + \mathbf{e}_2^{\text{o}}\sin\psi, \\ \mathbf{e}_2 &= -\mathbf{e}_1^{\text{o}}\sin\psi + \mathbf{e}_2^{\text{o}}\cos\psi. \end{aligned} \qquad (1.8\text{-}2)$$

It is seen that $\mathbf{e}_1$ and $\mathbf{e}_2$ are not simply $\mathbf{e}_1^{\text{o}}$ and $\mathbf{e}_2^{\text{o}}$ multiplied by a real factor. Since $\mathbf{e}_1$ and $\mathbf{e}_2$ are both linear combinations of $\mathbf{e}_1^{\text{o}}$ and $\mathbf{e}_2^{\text{o}}$, $\mathbf{e}_1$ and $\mathbf{e}_2$ lie on the same plane spanned by $\mathbf{e}_1^{\text{o}}$ and $\mathbf{e}_2^{\text{o}}$. If $\rho \neq 1$, the real and imaginary parts of $\mathbf{x}$ still lie on the same plane. Therefore the plane is called the *eigenplane*. Regardless of the choice of the complex multiplier, the real and imaginary parts of a complex eigenvector lie on an eigenplane. It is clear that $\mathbf{x}$ and its complex conjugate $\overline{\mathbf{x}}$ span the same eigenplane.

As ψ varies, (1.8-2) show that $\mathbf{e}_1$ and $\mathbf{e}_2$ trace an ellipse, Fig. 1.1. The rotation of the vector $\mathbf{e}_1$ as ψ increases from zero is from $\mathbf{e}_1^o$ to $\mathbf{e}_2^o$. A pair of diameters in an ellipse are said to be *conjugates* if all chords parallel to one diameter are bisected by the other diameter. Therefore the tangent to the ellipse at the extremity of one diameter is parallel to the other diameter. It can be shown that $\mathbf{e}_1$ and $\mathbf{e}_2$ form a pair of *conjugate radii*. One could choose ψ such that $\mathbf{e}_1$ and $\mathbf{e}_2$ are orthogonal, and hence are the principal radii of the ellipse.

Let

$$\xi_o = |\mathbf{e}_1^o|, \quad \eta_o = |\mathbf{e}_2^o|, \quad \delta_o = \mathbf{e}_1^o \cdot \mathbf{e}_2^o, \tag{1.8-3}$$

where ξ_o and η_o are, respectively, the lengths of the vectors $\mathbf{e}_1^o$ and $\mathbf{e}_2^o$. We then have from (1.8-2)

$$\xi^2 = \xi_o^2 \cos^2 \psi + \eta_o^2 \sin^2 \psi + 2\delta_o \sin \psi \cos \psi,$$

$$\eta^2 = \xi_o^2 \sin^2 \psi + \eta_o^2 \cos^2 \psi - 2\delta_o \sin \psi \cos \psi.$$

Hence

$$\xi^2 + \eta^2 = \xi_o^2 + \eta_o^2 \tag{1.8-4}$$

which is *invariant* with ψ. Thus the sum of the square of the lengths of the conjugate radii is invariant with ψ.

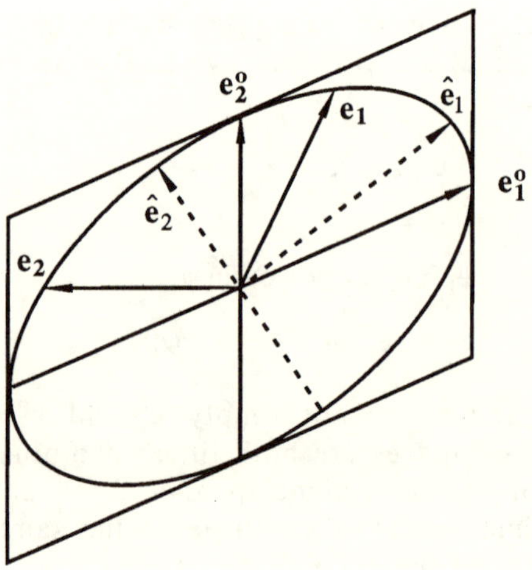

Fig. 1.1. The real and imaginary parts of a complex eigenvector $\mathbf{x}^o = \mathbf{e}_1^o + i\,\mathbf{e}_2^o$ span an eigenplane. The real and imaginary parts of $\mathbf{x} = e^{-i\psi}\mathbf{x}^o = \mathbf{e}_1 + i\,\mathbf{e}_2$, as ψ varies, describe an ellipse. They are a pair of conjugate radii of the ellipse.

Let $\hat{\mathbf{e}}_1$ and $\hat{\mathbf{e}}_2$ be the principal radii of the ellipse and

$$|\hat{\mathbf{e}}_1| \geq |\hat{\mathbf{e}}_2|. \tag{1.8-5}$$

From (1.8-2)

$$\mathbf{e}_1 \cdot \mathbf{e}_2 = -\tfrac{1}{2}(\xi_o^2 - \eta_o^2)\sin 2\psi + \delta_o \cos 2\psi.$$

If $\mathbf{e}_1 \cdot \mathbf{e}_2 = 0$ with $|\mathbf{e}_1| \geq |\mathbf{e}_2|$ at $\psi = \hat{\psi}$, $\mathbf{e}_1$ and $\mathbf{e}_2$ are the principal radii of the ellipse. The major axis is located at

$$\tan 2\hat{\psi} = \frac{2\delta_o}{\xi_o^2 - \eta_o^2}.$$

The condition (1.8-5) allows us to determine $\hat{\psi}$ uniquely if we limit

$$-\frac{\pi}{2} < \hat{\psi} \leq \frac{\pi}{2}.$$

It is not difficult to show that $\mathbf{e}_1^o$ and $\mathbf{e}_2^o$ are the principal axes of the ellipse if and only if $\mathbf{e}_1^o \cdot \mathbf{e}_2^o = 0$. We then have $\hat{\psi} = 0$, $\pi/2$, or arbitrary depending on whether $|\mathbf{e}_1^o|$ is larger, smaller, or equal to $|\mathbf{e}_2^o|$

Geometrical interpretations of the orthogonality of two bivectors and other subjects related to the algebra of bivectors can be found in Gibbs (1881), Synge (1966), Stone (1963), and Boulanger and Hayes (1991).

1.9 Square Roots of a Matrix

What are the *square roots* of a square matrix? In other words, what square matrices **B** of order n satisfy

$$\mathbf{B}^2 = \mathbf{A} \tag{1.9-1}$$

when a square matrix **A** is given? The matrix **B** is a square root of **A** and is denoted by $\sqrt{\mathbf{A}}$ or $\mathbf{A}^{1/2}$. Equation (1.9-1) is written as

$$\sqrt{\mathbf{A}}\sqrt{\mathbf{A}} = \mathbf{A} = \mathbf{A}^{1/2}\mathbf{A}^{1/2}. \tag{1.9-2}$$

When **A** is an identity matrix we have

$$\sqrt{\mathbf{A}}\sqrt{\mathbf{A}} = \mathbf{I}. \tag{1.9-3}$$

The obvious solutions to (1.9-3) are

$$\sqrt{\mathbf{A}} = \pm\mathbf{I} \tag{1.9-4}$$

but these two are not the only solutions if $n > 1$. For $n = 2$,

$$\sqrt{\mathbf{A}} = \frac{1}{\sqrt{2}}\begin{bmatrix} 1 & 1 \\ 1 & -1 \end{bmatrix}, \quad \begin{bmatrix} \sqrt{2} & -1 \\ 1 & -\sqrt{2} \end{bmatrix}, \quad \begin{bmatrix} i & 1 \\ 2 & -i \end{bmatrix}, \tag{1.9-5}$$

all satisfy (1.9-3). They are special solutions of

$$\sqrt{\mathbf{A}} = \pm\begin{bmatrix} \sqrt{1-ab} & a \\ b & -\sqrt{1-ab} \end{bmatrix} \tag{1.9-6}$$

in which a and b are arbitrary. Thus there are infinitely many solutions to (1.9-3). As shown in (1.9-5), square roots of a real symmetric matrix need not be real or symmetric. An example of a nontrivial real $\sqrt{\mathbf{A}}$ for (1.9-3) for $n = 6$ can be seen in (6.5-19) in Section 6.5.

How about the solutions to the following equation

$$\sqrt{\mathbf{A}}\sqrt{\mathbf{A}} = -\mathbf{I}\ ? \tag{1.9-7}$$

Again, the obvious solutions are

$$\sqrt{\mathbf{A}} = \pm i\mathbf{I}. \tag{1.9-8}$$

Less obvious solutions are

$$\sqrt{\mathbf{A}} = \pm\begin{bmatrix} \sqrt{ab-1} & a \\ -b & -\sqrt{ab-1} \end{bmatrix}. \tag{1.9-9}$$

In this case, not only $\sqrt{\mathbf{A}}$ can be real or complex, symmetric or non-symmetric, it can be skew-symmetric. Examples of a nontrivial real $\sqrt{\mathbf{A}}$ for (1.9-7) for n=6 can be found in (6.5-4), (8.8-41), and (8.8-43).

It can be shown from (1.9-2) that if μ is an eigenvalue of $\sqrt{\mathbf{A}}$, μ^2 is an eigenvalue of $\mathbf{A}$. Conversely, if λ is an eigenvalue of $\mathbf{A}$, the eigenvalue of $\sqrt{\mathbf{A}}$ can be $\sqrt{\lambda}$ or $-\sqrt{\lambda}$. Therefore there are two possible choices of eigenvalues for $\sqrt{\mathbf{A}}$ for each λ. When $\mathbf{A}$ is a 3×3 matrix with the eigenvalues λ_1, λ_2, and λ_3, the followings are eight possible sets of eigenvalues for $\sqrt{\mathbf{A}}$,

$$\left(\sqrt{\lambda_1},\ \sqrt{\lambda_2},\ \pm\sqrt{\lambda_3}\right), \tag{1.9-10a}$$

$$\left(\sqrt{\lambda_1},\ -\sqrt{\lambda_2},\ \pm\sqrt{\lambda_3}\right), \tag{1.9-10b}$$

$$\left(-\sqrt{\lambda_1},\ \sqrt{\lambda_2},\ \pm\sqrt{\lambda_3}\right), \tag{1.9-10c}$$

$$\left(-\sqrt{\lambda_1},\ -\sqrt{\lambda_2},\ \pm\sqrt{\lambda_3}\right). \tag{1.9-10d}$$

If $\mathbf{x}$ and $\mathbf{y}$ are, respectively, the right and left eigenvectors of $\sqrt{\mathbf{A}}$, they are also the right and left eigenvectors of $\mathbf{A}$. The converse however does not necessarily hold. In (1.9-4) any vector is an eigenvector of $\mathbf{A}=\mathbf{I}$ but not necessarily an eigenvector of $\sqrt{\mathbf{A}}$ shown in (1.9-5) or (1.9-6). When $\mathbf{A}$ is simple, i.e., when the eigenvalues $\lambda_1, \lambda_2, \ldots, \lambda_n$ are distinct, the converse holds. Employing the diagonalization (1.3-5),

$$\mathbf{A} = \mathbf{X}\langle\lambda_*\rangle\mathbf{Y}^T = \mathbf{X}\langle\lambda_*\rangle^{1/2}\langle\lambda_*\rangle^{1/2}\mathbf{Y}^T$$
$$= \mathbf{X}\langle\lambda_*\rangle^{1/2}\mathbf{Y}^T\mathbf{X}\langle\lambda_*\rangle^{1/2}\mathbf{Y}^T = \left\{\mathbf{X}\langle\lambda_*\rangle^{1/2}\mathbf{Y}^T\right\}^2$$

suggests that

$$\sqrt{\mathbf{A}} = \mathbf{X}\langle\lambda_*\rangle^{1/2}\mathbf{Y}^T \tag{1.9-11}$$

is a solution. The vectors $\mathbf{x}$ and $\mathbf{y}$ are unique except an arbitrary scalar multiplier. If $\mathbf{x}$ is multiplied by k, $\mathbf{y}$ is multiplied by k^{-1} so that $\mathbf{x}\cdot\mathbf{y}=1$. Since $\langle\lambda_*\rangle^{1/2}$ is a diagonal matrix, the multiplier k in a column of $\mathbf{X}$ is cancelled by the multiplier k^{-1} in the corresponding row of $\mathbf{Y}^T$. Therefore the expression (1.9-11) is independent of the multipliers on the eigenvectors and is unique when a particular set of $\langle\lambda_*\rangle^{1/2}$ is chosen. If $\mathbf{A}$ is a 3×3 matrix there are eight choices of $\langle\lambda_*\rangle^{1/2}$ whose diagonal elements are given in (1.9-10).

> **Theorem 1.9-1** The square root $\sqrt{\mathbf{A}}$ of a square matrix $\mathbf{A}$ is unique when $\mathbf{A}$ is simple and the eigenvalues of $\sqrt{\mathbf{A}}$ are specified.

What if $\mathbf{A}$ is nonsemisimple? We will discuss a particular case in which $\mathbf{A}$ is a 3×3 matrix with $\lambda_1=\lambda_2$ which form a chain of length two. We may consider $\mathbf{A}$ a limiting case of a simple matrix with $\lambda_1\to\lambda_2$, $\mathbf{x}_1\to\mathbf{x}_2$, and $\mathbf{y}_1\to\mathbf{y}_2$ (Ting and Hwu 1988). By Theorem 1.9-1 there are eight solutions for $\sqrt{\mathbf{A}}$ when $\mathbf{A}$ is simple. The eigenvalues of $\sqrt{\mathbf{A}}$ are given in (1.9-10). As $\lambda_1\to\lambda_2$, only (1.9-10a) and (1.9-10d) have repeated eigenvalues while (1.9-10b) and (1.9-10c) have distinct eigenvalues. In this book we will be interested in the eigenvalues given by (1.9-10a) and (1.9-10d) with $\lambda_1=\lambda_2$. For these four cases of interest the solutions for $\sqrt{\mathbf{A}}$ can be obtained as follows. From (1.4-9) $\mathbf{A}$ can be written as

$$\mathbf{A} = \mathbf{X}\mathbf{J}\mathbf{Y}^T = \mathbf{X}\mathbf{J}^{1/2}\mathbf{J}^{1/2}\mathbf{Y}^T$$
$$= \mathbf{X}\mathbf{J}^{1/2}\mathbf{Y}^T\mathbf{X}\mathbf{J}^{1/2}\mathbf{Y}^T = \left\{\mathbf{X}\mathbf{J}^{1/2}\mathbf{Y}^T\right\}^2$$

and hence

$$\sqrt{\mathbf{A}} = \mathbf{X}\mathbf{J}^{1/2}\mathbf{Y}^T. \tag{1.9-12}$$

In the above

$$\mathbf{J} = \begin{bmatrix} \lambda_1 & 1 & 0 \\ 0 & \lambda_1 & 0 \\ 0 & 0 & \lambda_3 \end{bmatrix},$$

and, setting $\eta = \frac{1}{2}\lambda_1^{-1/2}$,

$$\mathbf{J}^{1/2} = \begin{bmatrix} \sqrt{\lambda_1} & \eta & 0 \\ 0 & \sqrt{\lambda_1} & 0 \\ 0 & 0 & \pm\sqrt{\lambda_3} \end{bmatrix} \quad \text{or} \quad \begin{bmatrix} -\sqrt{\lambda_1} & -\eta & 0 \\ 0 & -\sqrt{\lambda_1} & 0 \\ 0 & 0 & \pm\sqrt{\lambda_3} \end{bmatrix}. \tag{1.9-13}$$

With (1.9-13), (1.9-12) provides the desired four solutions for $\sqrt{\mathbf{A}}$.

It should be pointed out that (1.9-12) is not a diagonalization of $\sqrt{\mathbf{A}}$ because $\mathbf{J}^{1/2}$ as given by (1.9-13) are not Jordan canonical forms. However, $\mathbf{J}^{1/2}$ of (1.9-13) can be written as

$$\mathbf{J}^{1/2} = \begin{bmatrix} 1 & 0 & 0 \\ 0 & \pm 2\sqrt{\lambda_1} & 0 \\ 0 & 0 & 1 \end{bmatrix} \begin{bmatrix} \pm\sqrt{\lambda_1} & 1 & 0 \\ 0 & \pm\sqrt{\lambda_1} & 0 \\ 0 & 0 & (\pm)\sqrt{\lambda_3} \end{bmatrix} \begin{bmatrix} 1 & 0 & 0 \\ 0 & \pm\eta & 0 \\ 0 & 0 & 1 \end{bmatrix}.$$

Let

$$\mathbf{X}^* = \mathbf{X}\begin{bmatrix} 1 & 0 & 0 \\ 0 & \pm 2\sqrt{\lambda_1} & 0 \\ 0 & 0 & 1 \end{bmatrix}, \quad \mathbf{Y}^* = \mathbf{Y}\begin{bmatrix} 1 & 0 & 0 \\ 0 & \pm\eta & 0 \\ 0 & 0 & 1 \end{bmatrix}. \tag{1.9-14}$$

It is easily shown that

$$\mathbf{X}^*(\mathbf{Y}^*)^T = \mathbf{X}\mathbf{Y}^T = \mathbf{I}, \quad (\mathbf{Y}^*)^T\mathbf{X}^* = \mathbf{Y}^T\mathbf{X} = \mathbf{I}.$$

Equation (1.9-12) can now be written in a diagonalized form

$$\sqrt{\mathbf{A}} = \mathbf{X}^*\begin{bmatrix} \pm\sqrt{\lambda_1} & 1 & 0 \\ 0 & \pm\sqrt{\lambda_1} & 0 \\ 0 & 0 & (\pm)\sqrt{\lambda_3} \end{bmatrix}(\mathbf{Y}^*)^T. \tag{1.9-15}$$

From $(1.9\text{-}14)_1$, the columns of $(\mathbf{x}_1, \mathbf{x}_2, \mathbf{x}_3)$ and $(\mathbf{x}_1, \pm 2\sqrt{\lambda_1}\mathbf{x}_2, \mathbf{x}_3)$ are, respectively, the right eigenvectors of $\mathbf{A}$ and $\sqrt{\mathbf{A}}$. The vectors $\mathbf{x}_2$ and $\pm 2\sqrt{\lambda_1}\mathbf{x}_2$ are the generalized right eigenvectors of $\mathbf{A}$ and $\sqrt{\mathbf{A}}$, respectively. Equation $(1.9\text{-}14)_2$ tells us that $(\mathbf{y}_1, \mathbf{y}_2, \mathbf{y}_3)$ and $(\mathbf{y}_1, \pm\eta\mathbf{y}_2, \mathbf{y}_3)$ are, respectively, the left eigenvectors of $\mathbf{A}$ and $\sqrt{\mathbf{A}}$. The vector $\mathbf{y}_1$ is the generalized left eigenvector for both $\mathbf{A}$ and $\sqrt{\mathbf{A}}$. The four solutions given by (1.9-12) and (1.9-13) are nonsemisimple.

Chapter 2

LINEAR ANISOTROPIC ELASTIC MATERIALS

The relations between stresses and strains in an anisotropic elastic material are presented in this chapter. A linear anisotropic elastic material can have as many as 21 elastic constants. This number is reduced when the material possesses a certain material symmetry. The number of elastic constants is also reduced, in most cases, when a two-dimensional deformation is considered. An important condition on elastic constants is that the strain energy must be positive. This condition implies that the 6×6 matrices of elastic constants presented herein must be positive definite.

2.1 Elastic Stiffnesses

Referring to a fixed rectangular coordinate system x_1, x_2, x_3, let σ_{ij} and ε_{ij} be the stress and strain, respectively, in an anisotropic elastic material. The stress-strain law can be written as

$$\sigma_{ij} = C_{ijks}\varepsilon_{ks} \tag{2.1-1}$$

in which C_{ijks} are the *elastic stiffnesses* which are components of a fourth rank tensor. They satisfy the full symmetry conditions

$$C_{ijks} = C_{jiks}, \quad C_{ijks} = C_{ijsk}, \quad C_{ijks} = C_{ksij}. \tag{2.1-2}$$

Before we present justifications for the three conditions in (2.1-2), we show that $(2.1\text{-}2)_1$ and $(2.1\text{-}2)_3$ imply $(2.1\text{-}2)_2$. Using $(2.1\text{-}2)_3$, $(2.1\text{-}2)_1$, and $(2.1\text{-}2)_3$ in that order we have

$$C_{ijks} = C_{ksij} = C_{skij} = C_{ijsk}$$

which proves $(2.1\text{-}2)_2$. Therefore the three conditions in (2.1-2) are written as

$$C_{ijks} = C_{jiks} = C_{ksij}. \tag{2.1-3}$$

One can also show that $(2.1\text{-}2)_2$ and $(2.1\text{-}2)_3$ imply $(2.1\text{-}2)_1$.

The first equation of (2.1-2) follows directly from the symmetry of the stress tensor $\sigma_{ij}=\sigma_{ji}$. The second equation of (2.1-2) does *not* follow directly from the symmetry of the strain tensor $\varepsilon_{ij}=\varepsilon_{ji}$. However, if the C_{ijks} in (2.1-1) do not satisfy $(2.1\text{-}2)_2$, we rewrite (2.1-1) as

$$\sigma_{ij} = \tfrac{1}{2} C_{ijks}\varepsilon_{ks} + \tfrac{1}{2} C_{ijks}\varepsilon_{ks} = \tfrac{1}{2} C_{ijks}\varepsilon_{ks} + \tfrac{1}{2} C_{ijsk}\varepsilon_{sk}$$

or, since $\varepsilon_{sk}=\varepsilon_{ks}$,

$$\sigma_{ij} = \tfrac{1}{2}(C_{ijks} + C_{ijsk})\varepsilon_{ks}. \tag{2.1-4}$$

The coefficients of ε_{ks} are symmetric with the subscripts ks. We can therefore redefine the coefficients of ε_{ks} in (2.1-4) as the new C_{ijks} which satisfy $(2.1\text{-}2)_2$.

The third equation follows from the consideration of strain energy. The strain energy W per unit volume of the material is

$$W = \int_0^{\varepsilon_{pq}} \sigma_{ij} d\varepsilon_{ij} = \int_0^{\varepsilon_{pq}} C_{ijks}\varepsilon_{ks} d\varepsilon_{ij}. \tag{2.1-5}$$

We demand that the integral be independent of the path ε_{ij} takes from 0 to ε_{pq}. If not, say path 1 yields a larger integral than path 2, one can consider loading the material from 0 to ε_{pq} through path 1, and unloading from ε_{pq} to 0 through the reverse of path 2. The energy gained is the difference between the W's for path 1 and path 2. If we repeat the process we can extract unlimited amount of energy from the material, which is physically impossible for a real material. Therefore the integral in (2.1-5) must be independent of the path taken by ε_{ij}, and W depends on the final strain ε_{pq} only. This implies that the integrand must be the total differential dW, i.e.,

$$C_{ijks}\varepsilon_{ks} d\varepsilon_{ij} = dW = \frac{\partial W}{\partial \varepsilon_{ij}} d\varepsilon_{ij}. \tag{2.1-6}$$

Since $d\varepsilon_{ij}$ is arbitrary we must have

$$\sigma_{ij} = C_{ijks}\varepsilon_{ks} = \frac{\partial W}{\partial \varepsilon_{ij}} \tag{2.1-7}$$

in which the first equality follows from (2.1-1). Differentiation of $(2.1\text{-}7)_2$ with ε_{ks} leads to

$$C_{ijks} = \frac{\partial^2 W}{\partial \varepsilon_{ks} \partial \varepsilon_{ij}}.$$

The double differentiations on the right are interchangeable. Therefore

$$C_{ijks} = C_{ksij}$$

is the condition for the integral in (2.1-5) to be independent of the loading path. This proves $(2.1\text{-}2)_3$.

With $(2.1\text{-}2)_3$, $(2.1\text{-}6)_1$ is written as

$$dW = C_{ijks}\varepsilon_{ks}d\varepsilon_{ij} = \tfrac{1}{2}d(C_{ijks}\varepsilon_{ij}\varepsilon_{ks}).$$

Hence

$$W = \tfrac{1}{2}C_{ijks}\varepsilon_{ij}\varepsilon_{ks} = \tfrac{1}{2}\sigma_{ij}\varepsilon_{ij}. \tag{2.1-8}$$

For the strain energy to be positive we must have

$$C_{ijks}\varepsilon_{ij}\varepsilon_{ks} > 0 \tag{2.1-9}$$

for any real, nonzero, symmetric tensor ε_{ij}.

The displacements u_i (i=1,2,3) are related to the strains by

$$\varepsilon_{ij} = \tfrac{1}{2}(u_{i,j} + u_{j,i}) \tag{2.1-10}$$

where the comma denotes differentiation. Let

$$\omega_{ij} = \tfrac{1}{2}(u_{i,j} - u_{j,i}) \tag{2.1-11}$$

which is skew-symmetric. We then have

$$u_{i,j} = \varepsilon_{ij} + \omega_{ij}. \tag{2.1-12}$$

It is readily shown that

$$C_{ijks}\omega_{ij} = 0, \quad C_{ijks}\omega_{ks} = 0. \tag{2.1-13}$$

The first equation follows from

$$C_{ijks}\omega_{ij} = C_{jiks}\omega_{ji} = -C_{jiks}\omega_{ij} = -C_{ijks}\omega_{ij}.$$

Similarly, one can prove the second equation. With (2.1-12) and (2.1-13) we have

$$C_{ijks}u_{i,j}u_{k,s} = C_{ijks}\varepsilon_{ij}\varepsilon_{ks}.$$

The condition (2.1-9) for the strain energy to be positive can therefore be replaced by the less restrictive condition

$$C_{ijks}\gamma_{ij}\gamma_{ks} > 0 \tag{2.1-14}$$

in which the γ_{ij} need not be symmetric.

2.2 Elastic Compliances

The inverse of (2.1-1) is written as

$$\varepsilon_{ij} = S_{ijks}\sigma_{ks} \tag{2.2-1}$$

where S_{ijks} are the *elastic compliances* which are components of a fourth rank tensor. They also possess the full symmetry

$$S_{ijks} = S_{jiks}, \quad S_{ijks} = S_{ijsk}, \quad S_{ijks} = S_{ksij}. \tag{2.2-2}$$

or, as in (2.1-3),

$$S_{ijks} = S_{jiks} = S_{ksij}. \tag{2.2-3}$$

The justifications of the first and second equations in (2.2-2) are similar to their counterparts in (2.1-2). The justification of $(2.2\text{-}2)_3$ also follows from the energy consideration. Integration by parts of $(2.1\text{-}5)_1$ leads to

$$\begin{aligned} W &= \sigma_{pq}\varepsilon_{pq} - \int_0^{\sigma_{pq}} \varepsilon_{ij} d\sigma_{ij} \\ &= \sigma_{pq}\varepsilon_{pq} - \int_0^{\sigma_{pq}} S_{ijks}\sigma_{ks} d\sigma_{ij}. \end{aligned}$$

If W depends on the final strain ε_{pq}, it depends on the final stress σ_{pq}. The last integral which represents the *complementary energy* must be independent of the path σ_{ij} takes from 0 to the final stress σ_{pq}. Following a similar argument for C_{ijks}, we deduce that $(2.2\text{-}2)_3$ must hold for the integral to be path independent.

For the strain energy to be positive, substitution of (2.2-1) into $(2.1\text{-}8)_2$ yields

$$S_{ijks}\sigma_{ij}\sigma_{ks} > 0. \tag{2.2-4}$$

The counterpart of (2.1-14) is

$$S_{ijks}\gamma_{ij}\gamma_{ks} > 0. \tag{2.2-5}$$

The σ_{ij} in (2.2-4) is symmetric while the γ_{ij} in (2.2-5) need not by symmetric.

2.3 Contracted Notations

Introducing the contracted notation (Voigt, 1910; Lekhnitskii, 1950; Jones, 1975; Christensen, 1979; see also Cowin et al., 1992)

$$\begin{aligned} &\sigma_{11} = \sigma_1, \quad \sigma_{22} = \sigma_2, \quad \sigma_{33} = \sigma_3, \\ &\sigma_{23} = \sigma_4, \quad \sigma_{31} = \sigma_5, \quad \sigma_{12} = \sigma_6, \end{aligned} \tag{2.3-1}$$

$$\begin{aligned} &\varepsilon_{11} = \varepsilon_1, \quad \varepsilon_{22} = \varepsilon_2, \quad \varepsilon_{33} = \varepsilon_3, \\ &2\varepsilon_{23} = \varepsilon_4, \quad 2\varepsilon_{31} = \varepsilon_5, \quad 2\varepsilon_{12} = \varepsilon_6, \end{aligned} \tag{2.3-2}$$

the stress-strain law (2.1-1) and (2.1-2) can be written as

$$\sigma_\alpha = C_{\alpha\beta}\varepsilon_\beta, \quad C_{\alpha\beta} = C_{\beta\alpha}, \tag{2.3-3a}$$

or, in matrix notation,

$$\boldsymbol{\sigma} = \mathbf{C}\boldsymbol{\varepsilon}, \quad \mathbf{C} = \mathbf{C}^T. \tag{2.3-3b}$$

In the above $\boldsymbol{\sigma}$ and $\boldsymbol{\varepsilon}$ are 6×1 column matrices and **C** is the 6×6 symmetric matrix given by

$$\mathbf{C} = \begin{bmatrix} C_{11} & C_{12} & C_{13} & C_{14} & C_{15} & C_{16} \\ & C_{22} & C_{23} & C_{24} & C_{25} & C_{26} \\ & & C_{33} & C_{34} & C_{35} & C_{36} \\ & & & C_{44} & C_{45} & C_{46} \\ & & & & C_{55} & C_{56} \\ & & & & & C_{66} \end{bmatrix}. \tag{2.3-4}$$

Only the upper triangle of the matrix is shown above since **C** is symmetric. The transformation between C_{ijks} and $C_{\alpha\beta}$ is accomplished by replacing the subscripts ij (or ks) by α (or β) using the following rules:

$$\begin{array}{ccc} ij \text{ (or } ks) & \leftrightarrow & \alpha \text{ (or } \beta) \\ 11 & \leftrightarrow & 1 \\ 22 & \leftrightarrow & 2 \\ 33 & \leftrightarrow & 3 \\ 23 \text{ or } 32 & \leftrightarrow & 4 \\ 31 \text{ or } 13 & \leftrightarrow & 5 \\ 12 \text{ or } 21 & \leftrightarrow & 6 \end{array} \tag{2.3-5a}$$

We may write the transformation in (2.3-5a) as

$$\alpha = \begin{cases} i, & \text{if } i = j, \\ 9 - i - j, & \text{if } i \neq j, \end{cases} \qquad \beta = \begin{cases} k, & \text{if } k = s, \\ 9 - k - s, & \text{if } k \neq s. \end{cases} \tag{2.3-5b}$$

The presence of the factor 2 in $(2.3\text{-}2)_{4,5,6}$ but not in $(2.3\text{-}1)_{4,5,6}$ is necessary for the **C** to be symmetric. Had we omitted the factor 2 in $(2.3\text{-}2)_{4,5,6}$ the 6×6 matrix **C** would not be symmetric.

With (2.3-1) and (2.3-2), the stress-strain law (2.2-1) in contracted notation is

$$\varepsilon_\alpha = s_{\alpha\beta}\sigma_\beta, \quad s_{\alpha\beta} = s_{\beta\alpha}, \tag{2.3-6a}$$

or

$$\boldsymbol{\varepsilon} = \mathbf{s}\boldsymbol{\sigma}, \quad \mathbf{s} = \mathbf{s}^T, \tag{2.3-6b}$$

where

$$\mathbf{s} = \begin{bmatrix} s_{11} & s_{12} & s_{13} & s_{14} & s_{15} & s_{16} \\ & s_{22} & s_{23} & s_{24} & s_{25} & s_{26} \\ & & s_{33} & s_{34} & s_{35} & s_{36} \\ & & & s_{44} & s_{45} & s_{46} \\ & & & & s_{55} & s_{56} \\ & & & & & s_{66} \end{bmatrix}. \tag{2.3-7}$$

Again, only the upper triangle is shown for the matrix **s**. We employ the lower case **s** for the 6×6 matrix because the upper case **S** is reserved for something else (Section 5.4). The transformation between S_{ijks} and $s_{\alpha\beta}$ is similar to that between C_{ijks} and $C_{\alpha\beta}$ except the following:

$$\begin{aligned} S_{ijks} &= s_{\alpha\beta}, \quad \text{if both } \alpha,\beta \le 3, \\ 2S_{ijks} &= s_{\alpha\beta}, \quad \text{if either } \alpha \text{ or } \beta \le 3, \\ 4S_{ijks} &= s_{\alpha\beta}, \quad \text{if both } \alpha,\beta > 3. \end{aligned} \tag{2.3-8}$$

The strain energy W is, from $(2.1\text{-}8)_2$, $(2.3\text{-}3b)_1$, and $(2.3\text{-}6b)_1$,

$$W = \tfrac{1}{2}\boldsymbol{\sigma}^T\boldsymbol{\varepsilon} = \tfrac{1}{2}\boldsymbol{\varepsilon}^T\mathbf{C}\boldsymbol{\varepsilon} = \tfrac{1}{2}\boldsymbol{\sigma}^T\mathbf{s}\boldsymbol{\sigma}. \tag{2.3-9}$$

For the W to be positive we must have

$$\boldsymbol{\varepsilon}^T\mathbf{C}\boldsymbol{\varepsilon} > 0 \tag{2.3-10}$$

or

$$\boldsymbol{\sigma}^T\mathbf{s}\boldsymbol{\sigma} > 0. \tag{2.3-11}$$

Equations (2.3-10) and (2.3-11) indicate that the matrices **C** and **s** are both positive definite. Substitution of $(2.3\text{-}6b)_1$ into $(2.3\text{-}3b)_1$ yields

$$\mathbf{Cs} = \mathbf{I} = \mathbf{sC}. \tag{2.3-12}$$

The second equality follows from the first equality which says that the **C** and **s** are the inverses of each other, and hence their products commute.

2.4 Reduced Elastic Compliances

For the most part of this book we will be concerned with two-dimensional deformations of anisotropic elastic bodies for which $\varepsilon_3=0$. When $\varepsilon_3=0$, the stress-strain law $(2.3\text{-}3a)_1$ becomes

$$\sigma_\alpha = \sum_{\beta\neq 3} C_{\alpha\beta}\varepsilon_\beta, \qquad \alpha = 1,2,\ldots,6. \tag{2.4-1}$$

Ignoring the equation for σ_3 we may write (2.4-1) as

$$\boldsymbol{\sigma}^{\mathrm{o}} = \mathbf{C}^{\mathrm{o}}\boldsymbol{\varepsilon}^{\mathrm{o}}, \qquad \mathbf{C}^{\mathrm{o}} = (\mathbf{C}^{\mathrm{o}})^T, \tag{2.4-2}$$

where

$$\begin{aligned}(\boldsymbol{\sigma}^{\mathrm{o}})^T &= [\sigma_1, \sigma_2, \sigma_4, \sigma_5, \sigma_6],\\ (\boldsymbol{\varepsilon}^{\mathrm{o}})^T &= [\varepsilon_1, \varepsilon_2, \varepsilon_4, \varepsilon_5, \varepsilon_6],\end{aligned} \tag{2.4-3}$$

and

$$\mathbf{C}^{\mathrm{o}} = \begin{bmatrix} C_{11} & C_{12} & C_{14} & C_{15} & C_{16} \\ & C_{22} & C_{24} & C_{25} & C_{26} \\ & & C_{44} & C_{45} & C_{46} \\ & & & C_{55} & C_{56} \\ & & & & C_{66} \end{bmatrix} \tag{2.4-4}$$

The matrix $\mathbf{C}^{\mathrm{o}}$ is obtained from $\mathbf{C}$ by deleting the third row and third column of $\mathbf{C}$. Since $\mathbf{C}^{\mathrm{o}}$ is a principal submatrix of $\mathbf{C}$, it is positive definite. It contains 15 independent elastic constants.

The stress-strain law $(2.3\text{-}6\text{a})_1$ for $\varepsilon_3=0$ is

$$\varepsilon_3 = 0 = s_{3\beta}\sigma_\beta.$$

Solving for σ_3,

$$\sigma_3 = -\frac{1}{s_{33}}\sum_{\beta\neq 3} s_{3\beta}\sigma_\beta, \tag{2.4-5}$$

and substituting σ_3 into $(2.3\text{-}6\text{a})_1$ leads to

$$\varepsilon_\alpha = \sum_{\beta\neq 3} s'_{\alpha\beta}\sigma_\beta, \tag{2.4-6}$$

where

$$s'_{\alpha\beta} = s_{\alpha\beta} - \frac{s_{\alpha 3}s_{3\beta}}{s_{33}} = s'_{\beta\alpha}. \tag{2.4-7}$$

The $s'_{\alpha\beta}$ are the *reduced elastic compliances*. It is clear that

$$s'_{\alpha 3} = 0, \quad s'_{3\beta} = 0, \quad \alpha,\beta = 1,2,\ldots,6. \tag{2.4-8}$$

Therefore there is no need to exclude $\beta=3$ in (2.4-6). Using the notation of (2.4-3), (2.4-6) can be written as

$$\boldsymbol{\varepsilon}^{\mathrm{o}} = \mathbf{s}'\boldsymbol{\sigma}^{\mathrm{o}}, \qquad \mathbf{s}' = (\mathbf{s}')^T, \tag{2.4-9}$$

where **s'** is the 5×5 symmetric matrix given by

$$\mathbf{s}' = \begin{bmatrix} s'_{11} & s'_{12} & s'_{14} & s'_{15} & s'_{16} \\ & s'_{22} & s'_{24} & s'_{25} & s'_{26} \\ & & s'_{44} & s'_{45} & s'_{46} \\ & & & s'_{55} & s'_{56} \\ & & & & s'_{66} \end{bmatrix}. \tag{2.4-10}$$

Like $\mathbf{C}^o$, **s'** contains 15 independent constants. Substitution of (2.4-9) into (2.4-2) yields

$$\sigma^o = \mathbf{C}^o \mathbf{s}' \sigma^o$$

which tells us that, since σ^o is arbitrary,

$$\mathbf{C}^o \mathbf{s}' = \mathbf{I} = \mathbf{s}' \mathbf{C}^o. \tag{2.4-11}$$

Therefore $\mathbf{C}^o$ and **s'** are the inverses of each other. The positive definite of $\mathbf{C}^o$ indicates that **s'** is also positive definite.

An alternate proof that $\mathbf{C}^o$ and **s'** are positive definite is to write the strain energy of (2.3-9) as, noting that $\varepsilon_3=0$,

$$W = \tfrac{1}{2}\sigma^T \varepsilon = \tfrac{1}{2}(\sigma^o)^T \varepsilon^o. \tag{2.4-12}$$

Use of (2.4-2) and (2.4-9) gives

$$W = \tfrac{1}{2}(\varepsilon^o)^T \mathbf{C}^o \varepsilon^o = \tfrac{1}{2}(\sigma^o)^T \mathbf{s}' \sigma^o. \tag{2.4-13}$$

If W is positive for any nonzero σ^o or ε^o, the $\mathbf{C}^o$ and **s'** must be positive definite.

The relation (2.4-11) is a property of elastic constants. Its validity should be independent of whether $\varepsilon_3=0$ or not. To prove (2.4-11) without assuming $\varepsilon_3=0$, we write $C^o_{\alpha\beta}$ in terms of $C_{\alpha\beta}$ as

$$C^o_{\alpha\beta} = C_{\alpha\beta} - \delta_{\alpha 3} C_{3\beta} - C_{\alpha 3}\delta_{3\beta} + \delta_{\alpha 3} C_{33} \delta_{3\beta}$$

where $\delta_{\alpha\beta}$ is the Kronecker delta. From (2.4-7) and (2.3-12) it is easily verified that

$$C^o_{\alpha\beta} s'_{\beta\gamma} = \delta_{\alpha\gamma} - \delta_{\alpha 3}\delta_{3\gamma}. \tag{2.4-14}$$

Written as 6×6 matrices, $C^o_{\alpha\beta}$, $s'_{\beta\gamma}$, and $(\delta_{\alpha\gamma} - \delta_{\alpha 3}\delta_{3\gamma})$ have only zero elements in the third rows and third columns. Equation (2.4-14) is equivalent to $(2.4\text{-}11)_1$ when all matrices are reduced to 5×5 matrices by deleting the third rows and third columns.

2.5 Material Symmetry

The 6×6 matrices **C** and **s** contain 21 independent elastic constants. The number of independent constants is reduced when the material possesses a certain material symmetry.

Written as C_{ijks}, the elastic stiffnesses are components of a fourth rank tensor. Under an orthogonal transformation

$$x_i^* = \Omega_{ij}x_j \quad \text{or} \quad \mathbf{x}^* = \Omega\mathbf{x} \tag{2.5-1}$$

in which Ω is an orthogonal matrix (Section 1.1) which satisfies the relation

$$\Omega_{ij}\Omega_{kj} = \delta_{ik} = \Omega_{ji}\Omega_{jk} \tag{2.5-2a}$$

or

$$\Omega\Omega^T = \mathbf{I} = \Omega^T\Omega, \tag{2.5-2b}$$

the elastic compliances C^*_{ijks} referred to the x_i^* coordinate system are

$$C^*_{ijks} = \Omega_{ip}\Omega_{jq}\Omega_{kr}\Omega_{st}C_{pqrt}. \tag{2.5-3}$$

An identical equation can be written for S_{ijks}. When $C^*_{ijks} = C_{ijks}$, i.e.,

$$C_{ijks} = \Omega_{ip}\Omega_{jq}\Omega_{kr}\Omega_{st}C_{pqrt}, \tag{2.5-4}$$

the material is said to possess a *symmetry* with respect to Ω.

An anisotropic material possesses the symmetry of *central inversion* if (2.5-4) is satisfied for

$$\Omega = \begin{bmatrix} -1 & 0 & 0 \\ 0 & -1 & 0 \\ 0 & 0 & -1 \end{bmatrix} = -\mathbf{I}. \tag{2.5-5}$$

It is obvious that (2.5-4) is satisfied by the Ω given in (2.5-5) for any C_{ijks}. Therefore all anisotropic elastic materials have the symmetry of central inversion.

If Ω is a proper orthogonal, the transformation (2.5-1) represents a rigid body rotation about an axis. When (2.5-4) is satisfied by a proper orthogonal, the material possesses a *rotational symmetry*. For instance,

$$\Omega^{(r)}(\theta) = \begin{bmatrix} \cos\theta & \sin\theta & 0 \\ -\sin\theta & \cos\theta & 0 \\ 0 & 0 & 1 \end{bmatrix} \tag{2.5-6}$$

represents a rotation about the x_3-axis an angle θ.

An orthogonal transformation Ω is a *reflection* if

$$\boldsymbol{\Omega} = \mathbf{I} - 2\mathbf{n}\mathbf{n}^T \tag{2.5-7a}$$

where **n** is a unit vector normal to the reflection plane. If **m** is any vector on the plane,

$$\boldsymbol{\Omega}\mathbf{n} = -\mathbf{n}, \qquad \boldsymbol{\Omega}\mathbf{m} = \mathbf{m}. \tag{2.5-7b}$$

Thus a vector normal to the reflection plane reverses its direction after the transformation while a vector on the reflection plane remains unchanged. When (2.5-4) is satisfied by the $\boldsymbol{\Omega}$ of (2.5-7a), the material is said to possess a *symmetry plane*. For example, let

$$\mathbf{n}^T = [\cos\theta, \ \sin\theta, \ 0].$$

The symmetry plane contains the x_3-axis. The $\boldsymbol{\Omega}$ of (2.5-7a) has the expression

$$\boldsymbol{\Omega}(\theta) = \begin{bmatrix} -\cos 2\theta & -\sin 2\theta & 0 \\ -\sin 2\theta & \cos 2\theta & 0 \\ 0 & 0 & 1 \end{bmatrix}, \quad -\tfrac{\pi}{2} < \theta \le \tfrac{\pi}{2}, \tag{2.5-8}$$

which is an improper orthogonal matrix. Since θ and $\theta+\pi$ represent the same plane, θ is limited to the range shown in $(2.5\text{-}8)_2$. At $\theta=0$,

$$\boldsymbol{\Omega}(0) = \begin{bmatrix} -1 & 0 & 0 \\ 0 & 1 & 0 \\ 0 & 0 & 1 \end{bmatrix}. \tag{2.5-9}$$

This represents a reflection about the plane $x_1=0$. When (2.5-4) is satisfied by (2.5-9), the material has a symmetry plane at $x_1=0$. If (2.5-4) is satisfied by (2.5-8) for any θ, the material is *transversely isotropic*. The x_3-axis is the *axis of symmetry*.

Two extreme cases of anisotropic elastic materials are *triclinic materials* and *isotropic materials*. A triclinic material possesses no rotational symmetry or a plane of reflection symmetry. An isotropic material possesses infinitely many rotational symmetries and planes of reflection symmetry. For isotropic materials it can be shown that

$$C_{ijks} = \lambda\delta_{ij}\delta_{ks} + \mu(\delta_{ik}\delta_{js} + \delta_{is}\delta_{jk}), \tag{2.5-10}$$

where λ and μ are the Lamé constants, satisfies (2.5-4) for any orthogonal $\boldsymbol{\Omega}$.

If we multiply both sides of (2.5-4) by

$$\Omega_{ia}\Omega_{jb}\Omega_{kc}\Omega_{sd}$$

and make use of (2.5-2a), we have

$$C_{abcd} = \Omega_{ia}\Omega_{jb}\Omega_{kc}\Omega_{sd}C_{ijks} \tag{2.5-11}$$

This implies that the material is symmetric with respect to the orthogonal matrix $\Omega^T = \Omega^{-1}$.

> **Theorem 2.5–1** If an anisotropic elastic material possesses a material symmetry with the orthogonal matrix Ω, it possesses the material symmetry with $\Omega^T = \Omega^{-1}$.

This means that, if the material has the symmetry with rotation about the x_3-axis an angle θ, it also has the symmetry with rotation about the x_3-axis an angle $-\theta$.

Let the material possess a symmetry with Ω' and Ω'', i.e.,

$$C_{ijks} = \Omega'_{ia}\Omega'_{jb}\Omega'_{kc}\Omega'_{sd}C_{abcd}, \tag{2.5-12}$$

and

$$C_{abcd} = \Omega''_{ap}\Omega''_{bq}\Omega''_{cr}\Omega''_{dt}C_{pqrt}. \tag{2.5-13}$$

Insertion of (2.5-13) into (2.5-12) leads to

$$C_{ijks} = (\Omega'_{ia}\Omega''_{ap})(\Omega'_{jb}\Omega''_{bq})(\Omega'_{kc}\Omega''_{cr})(\Omega'_{sd}\Omega''_{dt})C_{pqrt}.$$

This is in the form of (2.5-4).

> **Theorem 2.5–2** If an anisotropic elastic material possesses a symmetry with Ω' and Ω'', it possesses a symmetry with $\Omega = \Omega'\Omega''$.

The orthogonal matrix $\Omega(\theta)$ of (2.5-8) represents a reflection symmetry with a plane whose normal lies on the (x_1, x_2) plane making an angle of θ with the x_1-axis. It can be shown that

$$\begin{aligned} \Omega(-\theta) &= \Omega(0)\Omega(\theta)\Omega(0), \\ \Omega(2\theta) &= \Omega(\theta)\Omega(0)\Omega(\theta). \end{aligned} \tag{2.5-14}$$

By Theorem 2.5–2, if a material has symmetry planes at $\theta=0$ and θ_0, say, it also has symmetry planes at $\theta=-\theta_0$ and $2\theta_0$. Repeated applications of (2.5-14) show that the material has symmetry planes at $\theta=k\theta_0$ where k is any integer, positive or negative. If θ_0 is an irrational number times π, the material has symmetry plane at any θ. This suggests that, to test if a material is transversely isotropic in experiment, all one has to do is to see if the material has the reflection symmetry with the planes at $\theta=0$ and θ_0 where θ_0 is any irrational number times π. We will show in the next section that the material is transversely isotropic if it has reflection symmetry at $\theta=0$ and θ_0, where θ_0 is any angle not equal to $\pi/4$, $\pi/3$, or $\pi/2$.

Theorems 2.5–1 and 2.5–2, which remain valid for nonlinear materials (Truesdell and Noll, 1965), are useful in determining the structure of C_{ijks} when the material possesses symmetry planes.

2.6 Matrix C for Materials with Symmetry Planes

Depending on the number of rotations and/or reflection symmetry a crystal possesses, Voigt (1910) has classified crystals into 32 classes. (See also Love, 1927; Musgrave, 1970; Gurtin, 1972; Cowin and Mehrabadi, 1987; and Mehrabadi and Cowin, 1990). In terms of the 6×6 matrix **C** however there are only 8 basic groups. For a non-crystalline material the structure of **C** can also be represented by one of the 8 basic groups. We list below the 8 basic groups for **C** according to the number of symmetry planes the group has. The derivation of each group will be given later. Consideration of rotational symmetry does not change the structure of **C** in each group. Without loss in generality we choose the symmetry plane (or planes) to coincide with the coordinate planes whenever possible. If the matrix **C** referred to a different coordinate system is desired, we use (2.5-3) to obtain the new matrix **C**. (In some literature the structure of **C** referred to a rotated coordinate system is considered as a separate group.) We will therefore employ the orthogonal matrix Ω (2.5-8) which represents a reflection with respect to a plane whose normal is on the (x_1, x_2) plane making an angle θ with the x_1-axis. We will also employ the orthogonal matrix

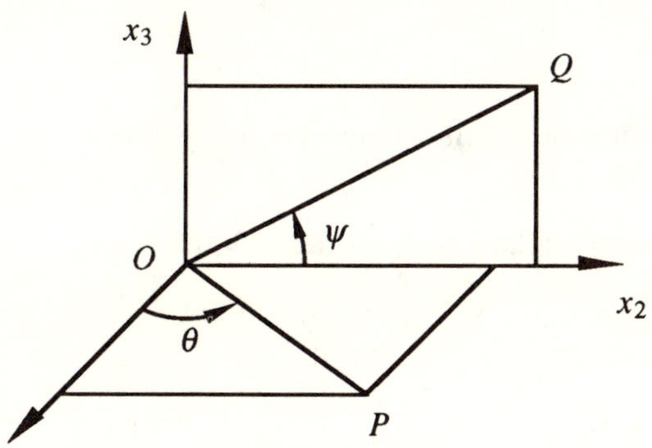

Fig. 2.1 The vector OP and OQ are, respectively, the normals to planes of reflection symmetry defined in (2.5-8) and (2.6-1).

$$\hat{\Omega}(\psi) = \begin{bmatrix} 1 & 0 & 0 \\ 0 & -\cos 2\psi & -\sin 2\psi \\ 0 & -\sin 2\psi & \cos 2\psi \end{bmatrix}, \qquad -\tfrac{\pi}{2} < \psi \le \tfrac{\pi}{2} \tag{2.6-1}$$

which represents a reflection with respect to a plane whose normal is on the (x_2,x_3) plane making an angle ψ with the x_2-axis, Fig. 2.1. The plane $x_2=0$ can be represented by either $\theta=\pi/2$ or $\psi=0$. Only the upper triangle of the matrix **C** is shown. The symbols employed follow those of Musgrave (1970).

$*$	indicates a possibly nonzero element
$*$—$*$	indicates the two elements are equal
$*$—$\bar{*}$	indicates the two elements are equal but opposite in signs
$\otimes$	indicates $C_{66} = \frac{1}{2}(C_{11} - C_{12})$
m	indicates the number of independent elastic constants

(I) **Triclinic Materials**. No symmetry planes exist.

$$\mathbf{C} = \begin{bmatrix} * & * & * & * & * & * \\ & * & * & * & * & * \\ & & * & * & * & * \\ & & & * & * & * \\ & & & & * & * \\ & & & & & * \end{bmatrix}, \quad m=21. \tag{2.6-2}$$

(II) **Monoclinic Materials**. One symmetry plane.

(a) Symmetry plane at $x_1=0$, i.e., $\theta=0$.

$$\mathbf{C} = \begin{bmatrix} * & * & * & * & 0 & 0 \\ & * & * & * & 0 & 0 \\ & & * & * & 0 & 0 \\ & & & * & 0 & 0 \\ & & & & * & * \\ & & & & & * \end{bmatrix}, \quad m=13. \tag{2.6-3a}$$

(b) Symmetry plane at $x_2=0$, i.e., $\theta=\pi/2$ or $\psi=0$.

$$\mathbf{C}=\begin{bmatrix} * & * & * & 0 & * & 0 \\ & * & * & 0 & * & 0 \\ & & * & 0 & * & 0 \\ & & & * & 0 & * \\ & & & & * & 0 \\ & & & & & * \end{bmatrix}, \quad \text{m}=13. \tag{2.6-3b}$$

(c) Symmetry plane at $x_3=0$, i.e., $\psi=\pi/2$.

$$\mathbf{C}=\begin{bmatrix} * & * & * & 0 & 0 & * \\ & * & * & 0 & 0 & * \\ & & * & 0 & 0 & * \\ & & & * & * & 0 \\ & & & & * & 0 \\ & & & & & * \end{bmatrix}, \quad \text{m}=13. \tag{2.6-3c}$$

(III) **Orthotropic (or Rhombic) Materials**. The three coordinate planes $\theta=0$, $\pi/2$, and $\psi=\pi/2$ are the symmetry planes.

$$\mathbf{C}=\begin{bmatrix} * & * & * & 0 & 0 & 0 \\ & * & * & 0 & 0 & 0 \\ & & * & 0 & 0 & 0 \\ & & & * & 0 & 0 \\ & & & & * & 0 \\ & & & & & * \end{bmatrix}, \quad \text{m}=9. \tag{2.6-4}$$

(IV) **Trigonal Materials**. Three symmetry planes at $\theta=0$ and $\pm\pi/3$.

$$\mathbf{C}=\begin{bmatrix} * & * & * & * & 0 & 0 \\ & * & * & * & 0 & 0 \\ & & * & 0 & 0 & 0 \\ & & & * & 0 & 0 \\ & & & & * & * \\ & & & & & \otimes \end{bmatrix}, \quad \text{m}=6. \tag{2.6-5}$$

(V) **Tetragonal Materials.** Five symmetry planes at $\theta=0$, $\pm\pi/4$, $\pi/2$, and $\psi=\pi/2$.

$$\mathbf{C}=\begin{bmatrix} \ast & \ast & \ast & 0 & 0 & 0 \\ & \ast & \ast & 0 & 0 & 0 \\ & & \ast & 0 & 0 & 0 \\ & & & \ast & 0 & 0 \\ & & & & \ast & 0 \\ & & & & & \ast \end{bmatrix}, \quad m=6. \tag{2.6-6}$$

(VI) **Transversely Isotropic (or Hexagonal) Materials.** The symmetry planes are the $x_3=0$ plane and any plane that contains the x_3-axis. The x_3-axis is the axis of symmetry.

$$\mathbf{C}=\begin{bmatrix} \ast & \ast & \ast & 0 & 0 & 0 \\ & \ast & \ast & 0 & 0 & 0 \\ & & \ast & 0 & 0 & 0 \\ & & & \ast & 0 & 0 \\ & & & & \ast & 0 \\ & & & & & \otimes \end{bmatrix}, \quad m=5. \tag{2.6-7}$$

(VII) **Cubic Materials.** Nine planes of symmetry whose normals are on the three coordinate axes and on the coordinate planes making an angle $\pi/4$ with the coordinate axes.

$$\mathbf{C}=\begin{bmatrix} \ast & \ast & \ast & 0 & 0 & 0 \\ & \ast & \ast & 0 & 0 & 0 \\ & & \ast & 0 & 0 & 0 \\ & & & \ast & 0 & 0 \\ & & & & \ast & 0 \\ & & & & & \ast \end{bmatrix}, \quad m=3. \tag{2.6-8}$$

(VIII) **Isotropic Materials.** Any plane is a symmetry plane.

$$\mathbf{C}=\begin{bmatrix} \ast & \ast & \ast & 0 & 0 & 0 \\ & \ast & \ast & 0 & 0 & 0 \\ & & \ast & 0 & 0 & 0 \\ & & & \otimes & 0 & 0 \\ & & & & \otimes & 0 \\ & & & & & \otimes \end{bmatrix}, \quad m=2. \tag{2.6-9a}$$

If λ and μ are the *Lamé constants*, (2.6-9a) has the expression

$$\mathbf{C} = \begin{bmatrix} \lambda+2\mu & \lambda & \lambda & 0 & 0 & 0 \\ & \lambda+2\mu & \lambda & 0 & 0 & 0 \\ & & \lambda+2\mu & 0 & 0 & 0 \\ & & & \mu & 0 & 0 \\ & & & & \mu & 0 \\ & & & & & \mu \end{bmatrix}, \quad m=2. \tag{2.6-9b}$$

μ is also called the *shear modulus*. While the number of nonzero elements in $\mathbf{C}$ may increase when different coordinate systems are employed, the number of independent elastic constants m does not depend on the choice of the coordinate systems.

We will now show how the structure of $\mathbf{C}$ in each group is determined. It should be noted that, by Theorem 2.5–2, the number of symmetry planes required in determining the structure of $\mathbf{C}$ is usually less than the number of the symmetry planes that exist in the material. For example, for orthotropic materials (Group III) the symmetry planes at $\theta=0$ and $\pi/2$ implies that the group also has a symmetry plane at $\psi=\pi/2$. This is so because

$$\hat{\Omega}(\tfrac{\pi}{2}) = -\Omega(0)\Omega(\tfrac{\pi}{2}).$$

The two Ω's on the right represent reflection with planes at $\theta=0$ and $\pi/2$ while the minus sign represents, by (2.5-5), the symmetry of central inversion which is possessed by all materials. In fact we will see that all materials including isotropic materials require no more than three symmetry planes to determine the structure of $\mathbf{C}$.

We begin with monoclinic materials with the symmetry plane at $x_1=0$. This means that (2.5-4) is satisfied by the Ω given in (2.5-9). If we write (2.5-9) as

$$\Omega_{ij} = \begin{cases} \delta_{ij}, & \text{if } i \neq 1 \\ -\delta_{ij}, & \text{if } i = 1 \end{cases}$$

where δ_{ij} is the Kronecker delta, (2.5-4) reduces to

$$C_{ijks} = (-1)^N C_{ijks} \tag{2.6-10}$$

in which N is the number of ones appearing in the subscripts $ijks$. We see that (2.6-10) is a trivial identity for N=0, 2, and 4, and

$$C_{ijks} = 0 \tag{2.6-11}$$

for N=1 and 3. Thus C_{ijks} vanishes when the number 1 appears once or three times in $ijks$. Due to the symmetry (2.1-2) for C_{ijks}, the following eight $ijks$ are the only possible choices for which N=1 or 3.

$$ijks = 3313,\ 3312,\ 2313,\ 2312,\ 1113,\ 2213,\ 1112,\ 2212. \qquad (2.6\text{-}12)$$

Using contracted notations, (2.6-11) and (2.6-12) lead to

$$C_{35} = C_{36} = C_{45} = C_{46} = C_{15} = C_{25} = C_{16} = C_{26} = 0 \qquad (2.6\text{-}13)$$

which verify the zero elements of **C** in (2.6-3a). The structure of **C** for (2.6-3b) and (2.6-3c) can be determined similarly.

We next consider the existence of a second plane of symmetry whose Ω is given by (2.5-8) with $\theta \neq 0$. By writing (2.5-8) as

$$\Omega = \Omega''\Omega',$$

$$\Omega'' = \begin{bmatrix} -\cos\theta & -\sin\theta & 0 \\ -\sin\theta & \cos\theta & 0 \\ 0 & 0 & 1 \end{bmatrix},$$

$$\Omega' = \begin{bmatrix} \cos\theta & \sin\theta & 0 \\ -\sin\theta & \cos\theta & 0 \\ 0 & 0 & 1 \end{bmatrix},$$

it is seen that Ω' and Ω'' are both orthogonal matrices and that Ω'' is symmetric. Moreover, from the relation

$$\Omega''_{ij} = \begin{cases} \Omega'_{ij}, & \text{if } i \neq 1 \\ -\Omega'_{ij}, & \text{if } i = 1 \end{cases}$$

(2.5-4) can be written as

$$\Omega''_{ip}\Omega''_{jq}\Omega''_{kr}\Omega''_{st}C_{pqrt} = \Omega'_{ip}\Omega'_{jq}\Omega'_{kr}\Omega'_{st}C_{pqrt}$$

or

$$\Omega'_{ip}\Omega'_{jq}\Omega'_{kr}\Omega'_{st}C_{pqrt}C_{pqrt} = (-1)^N\,\Omega'_{ip}\Omega'_{jq}\Omega'_{kr}\Omega'_{st}C_{pqrt}$$

where N is the number of ones appearing in $ijks$. This is similar to (2.6-10). Following the same arguments in deriving (2.6-11) and (2.6-12) we obtain

$$\Omega'_{ip}\Omega'_{jq}\Omega'_{kr}\Omega'_{st}C_{pqrt} = 0 \qquad (2.6\text{-}14)$$

for the eight $ijks$ given in (2.6-12).

We consider a symmetry plane at $\sin\theta \neq 0$ in addition to the symmetry plane at θ=0. The latter has given us (2.6-13). For the first $ijks$ in (2.6-12), (2.6-14) reduces to

$$\Omega'_{1r}C_{33r3} = 0$$

or

$$C_{35}\cos\theta + C_{34}\sin\theta = 0.$$

Since $C_{35}=0$ by (2.6-13) and $\sin\theta \neq 0$,

$$C_{34} = 0. \tag{2.6-15}$$

For the next two sets of *ijks* in (2.6-12), (2.6-14) gives

$$(C_{13} - C_{23})\cos\theta = 0,$$
$$(C_{44} - C_{55})\cos\theta = 0.$$

These mean that

$$C_{13} = C_{23} \text{ and } C_{44} = C_{55}, \text{ if } \cos\theta \neq 0. \tag{2.6-16}$$

The next three sets of *ijks* in (2.6-12) lead to a system of three equations for C_{56}, C_{24}, and C_{14},

$$C_{56} = (2C_{56} - C_{24} + C_{14})\cos^2\theta,$$

$$-C_{24} = (2C_{56} - C_{24} + C_{14})\cos^2\theta,$$

$$C_{14} = (2C_{56} - C_{24} + C_{14})\cos^2\theta.$$

The right sides of the equations are identical, implying that

$$C_{56} = -C_{24} = C_{14} = \gamma, \text{ say.}$$

All three equations can then be written as

$$\gamma = 4\gamma\cos^2\theta.$$

We therefore have the results that

$$\begin{aligned} C_{56} &= -C_{24} = C_{14}, && \text{if } \cos^2\theta = \tfrac{1}{4}, \\ C_{56} &= -C_{24} = C_{14} = 0, && \text{if } \cos^2\theta \neq \tfrac{1}{4}. \end{aligned} \tag{2.6-17}$$

The last two sets of *ijks* in (2.6-12) provide two equations which, by adding and subtracting the two equations, can be written as

$$(C_{11} - C_{22})\cos\theta = 0,$$
$$\{C_{12} + 2C_{66} - \tfrac{1}{2}(C_{11} + C_{22})\}\cos\theta\cos 2\theta = 0.$$

These two equations mean that

$$\begin{aligned} &C_{11} = C_{22} \text{ if } \cos 2\theta = 0, \\ &C_{11} = C_{22} \text{ and } C_{66} = \tfrac{1}{2}(C_{11} - C_{12}) \text{ if } \cos\theta \neq 0 \neq \cos 2\theta. \end{aligned} \tag{2.6-18}$$

The structure of **C** for Groups III–VI as shown in (2.6-4)–(2.6-7) can now be determined from (2.6-15)–(2.6-18) as follows:

Group III: $\theta=0$ and $\pi/2$.

Group IV: $\theta=0$ and $\pi/3$.

Group V: $\theta=0$ and $\pi/4$.

Group VI: (i) $\theta=0$ and θ_o, $\theta_o\neq\pi/2$, $\pi/3$, $\pi/4$, or
(ii) $\theta=0$, $\pi/3$, and $\pi/4$, or
(iii) $\theta=0$, $\pi/3$, and $\pi/2$, or
(iv) $\theta=0$, $\pi/3$, and $\psi=\pi/2$.

It is seen that two planes of symmetry are sufficient to determine the structure of **C** for Groups III–VI. Group VI has four choices. The first choice (i) requires only two planes of symmetry. The second choice (ii) can be verified easily by combining the structure of **C** shown in (2.6-5) and (2.6-6). Likewise, the third choice (iii) is the result of combining (2.6-5) and (2.6-4), and the fourth choice (iv) is obtained by combining (2.6-5) and (2.6-3c).

To determine the structure of **C** for the remaining two groups we consider the transformation (2.6-1). It can be shown that

$$\hat{\Omega} = \hat{\Omega}''\hat{\Omega}'$$

where

$$\hat{\Omega}'' = \begin{bmatrix} 1 & 0 & 0 \\ 0 & -\cos\psi & -\sin\psi \\ 0 & -\sin\psi & \cos\psi \end{bmatrix},$$

$$\hat{\Omega}' = \begin{bmatrix} 1 & 0 & 0 \\ 0 & \cos\psi & \sin\psi \\ 0 & -\sin\psi & \cos\psi \end{bmatrix}.$$

Both $\hat{\Omega}'$ and $\hat{\Omega}''$ are orthogonal matrices and $\hat{\Omega}''$ is symmetric. Observing that

$$\hat{\Omega}''_{ij} = \begin{cases} \hat{\Omega}'_{ij}, & \text{if } i \neq 2 \\ -\hat{\Omega}'_{ij}, & \text{if } i = 2 \end{cases}$$

and following the arguments leading to (2.6-14) we have

$$\hat{\Omega}'_{ip}\hat{\Omega}'_{jq}\hat{\Omega}'_{kr}\hat{\Omega}'_{st}C_{pqrt} = 0 \qquad (2.6\text{-}19)$$

whenever the number of twos appearing in $ijks$ is one or three. Like (2.6-12) there are only eight possible choices which are

$$ijks = 1121, 1123, 1213, 1222, 1233, 2223, 2333, 1323. \tag{2.6-20}$$

We will assume that the material has a symmetry plane at θ=0, and see what we can get when it also has a symmetry plane at a $\psi \neq 0$. The derivation is similar to the one that leads to (2.6-15)–(2.6-18). It can be shown that (2.6-19) is automatically satisfied by four of the $ijks$ listed in (2.6-20). The remaining four $ijks$ give us

$$\begin{aligned} 2C_{14}\cos 2\psi &= (C_{12} - C_{13})\sin 2\psi, \\ 2C_{56}\cos 2\psi &= (C_{66} - C_{55})\sin 2\psi, \end{aligned} \tag{2.6-21}$$

$$\begin{aligned} 2(C_{24} + C_{34})\cos 2\psi &= (C_{22} - C_{33})\sin 2\psi, \\ 4(C_{24} - C_{34})\cos 4\psi &= (C_{12} + C_{33} - 2C_{23} - 4C_{44})\sin 4\psi. \end{aligned} \tag{2.6-22}$$

Equations $(2.6\text{-}21)_{1,2}$ are, respectively, from $ijks$=1123 and 1213. Equations $(2.6\text{-}22)_{1,2}$ are the results of adding and subtracting two equations obtained from applying $ijks$=2223 and 2333. Ignoring the case ψ=0 which is equivalent to $\theta=\pi/2$ studied earlier, we obtain from (2.6-21) and (2.6-22)

$$\begin{aligned} &C_{14} = C_{24} = C_{34} = C_{56} = 0, \text{ if } \psi = \tfrac{1}{2}\pi, \\ &C_{12} = C_{13},\ C_{55} = C_{66},\ C_{22} = C_{33},\ C_{24} = C_{34}, \text{ if } \psi = \tfrac{1}{4}\pi. \end{aligned} \tag{2.6-23}$$

Application of the conditions for $\psi = \pi/2$ in (2.6-23) to (2.6-5) leads to (2.6-7). This was the fourth choice of symmetry planes for Group VI discussed earlier. Use of the conditions for $\psi=\pi/4$ in (2.6-23) on (2.6-6) and (2.6-5) determines the structure of **C** for Groups VII and VIII, respectively.

Group VII: θ=0, $\pi/4$, and $\psi=\pi/4$.

Group VIII: θ=0, $\pi/3$, and $\psi=\pi/4$.

It is remarkable that, for isotropic materials, it takes only three planes of symmetry to reduce the number of elastic constants from 21 to 2.

2.7 Matrices s and s' for Materials with Symmetry Planes

Like C_{ijks}, the elastic compliances S_{ijks} are components of a fourth rank tensor. Under the orthogonal transformation (2.5-1) S^*_{ijks} in the new coordinate system are related to S_{ijks} by

$$S^*_{ijks} = \Omega_{ip}\Omega_{jq}\Omega_{kr}\Omega_{st}S_{pqrt}. \tag{2.7-1}$$

This is identical to (2.5-3). Therefore the analysis presented in the previous section for C_{ijks} apply to S_{ijks}. The structure of the matrix **C** appearing in (2.6-2)–(2.6-9a) remains valid for the matrix **s** with the following modifications required by (2.3-8). The relation

$$C_{56} = -C_{24} = C_{14}$$

in (2.6-5) is replaced by

$$\tfrac{1}{2}s_{56} = -s_{24} = s_{14} \tag{2.7-2}$$

and the C_{66} element in (2.6-5), (2.6-7), and (2.6-9a) is replaced by

$$s_{66} = 2(s_{11} - s_{12}). \tag{2.7-3}$$

In engineering applications the matrix **s** for isotropic materials is written as

$$\mathbf{s} = \frac{1}{E}\begin{bmatrix} 1 & -\nu & -\nu & 0 & 0 & 0 \\ & 1 & -\nu & 0 & 0 & 0 \\ & & 1 & 0 & 0 & 0 \\ & & & 2(1+\nu) & 0 & 0 \\ & & & & 2(1+\nu) & 0 \\ & & & & & 2(1+\nu) \end{bmatrix}, \tag{2.7-4}$$

where

$$E = \frac{\mu(3\lambda + 2\mu)}{\lambda + \mu}, \qquad \nu = \frac{\lambda}{2(\lambda + \mu)} \tag{2.7-5a}$$

are, respectively, the Young's modulus and Poisson ratio. It can be shown that

$$\lambda = \frac{E\nu}{(1+\nu)(1-2\nu)}, \qquad \mu = \frac{E}{2(1+\nu)}. \tag{2.7-5b}$$

The reduced elastic compliances $s'_{\alpha\beta}$ is defined in (2.4-7) in terms of $s_{\alpha\beta}$. As shown in (2.4-10) $s'_{\alpha\beta}$ can be written as a 5×5 symmetric matrix **s'**. It is not difficult to show that in Groups IV–VIII where the relations

$$s_{11} = s_{22}, \quad s_{44} = s_{55} (= s_{66})$$

hold for **s**, the corresponding relations

$$s'_{11} = s'_{22}, \quad s'_{44} = s'_{55} (= s'_{66}) \tag{2.7-6}$$

hold for **s'**. Likewise, the relations (2.7-2) and (2.7-3), which apply to the matrix **s** for Group IV and Groups IV, VI, VIII, respectively, are replaced by the same relations

$$\tfrac{1}{2}s'_{56} = -s'_{24} = s'_{14}, \tag{2.7-7}$$

$$s'_{66} = 2(s'_{11} - s'_{12}), \tag{2.7-8}$$

for the matrix **s'**. The net result is that the structure of the 5×5 matrix **s'** for each group can be obtained from the matrix for **s** in the group by deleting the third row and third column. In addition it can be shown that $s'_{\alpha\beta} = s_{\alpha\beta}$ for certain elements of $s'_{\alpha\beta}$. For Group II, monoclinic materials,

$$s'_{\alpha\beta} = s_{\alpha\beta} \text{ for } \alpha,\beta = \begin{cases} 5 \text{ or } 6, \text{ if } x_1 = 0 \text{ is the symmetry plane,} \\ 4 \text{ or } 6, \text{ if } x_2 = 0 \text{ is the symmetry plane,} \\ 4 \text{ or } 5, \text{ if } x_3 = 0 \text{ is the symmetry plane.} \end{cases} \tag{2.7-9}$$

For Groups III–VIII,

$$s'_{\alpha\beta} = s_{\alpha\beta} \text{ for } \alpha,\ \beta = 4,\ 5, \text{ or } 6. \tag{2.7-10}$$

The structure of **s'** for isotropic materials

$$\mathbf{s}' = \frac{1}{2\mu}\begin{bmatrix} 1-\nu & -\nu & 0 & 0 & 0 \\ & 1-\nu & 0 & 0 & 0 \\ & & 2 & 0 & 0 \\ & & & 2 & 0 \\ & & & & 2 \end{bmatrix}. \tag{2.7-11}$$

2.8 Transformation of C and s

We had stated earlier that the structure of **C** (or **s**) for each symmetry group in (2.6-2)–(2.6-8) is for the coordinate system specified in the group. For a different coordinate system we apply (2.5-3) to obtain the structure of **C** referred to the new coordinate system. The zero elements in the original matrix **C** may no longer be zero in the new matrix **C**. In applications the choice of the coordinate system is very often dictated by the boundary conditions of the problem and hence may not coincide with the symmetry planes of the material. Therefore transformation of the matrix **C** to a different coordinate system becomes necessary. The transformation law (2.5-3) is for C_{ijks}. It is not convenient for the 6×6 matrix $C_{\alpha\beta}$. We will drive the transformation law for $C_{\alpha\beta}$.

The stresses σ_{ij}, being the components of a second rank tensor, transform according to

$$\sigma^*_{ij} = \Omega_{ik}\Omega_{js}\sigma_{ks}. \tag{2.8-1}$$

With the contracted notation (2.3-1) this can be written in matrix notation as (Bond, 1943; Auld, 1973)

$$\boldsymbol{\sigma}^* = \mathbf{K}\boldsymbol{\sigma}, \tag{2.8-2}$$

where

$$\mathbf{K} = \begin{bmatrix} \mathbf{K}_1 & 2\mathbf{K}_2 \\ \mathbf{K}_3 & \mathbf{K}_4 \end{bmatrix}, \tag{2.8-3}$$

$$\mathbf{K}_1 = \begin{bmatrix} \Omega_{11}^2 & \Omega_{12}^2 & \Omega_{13}^2 \\ \Omega_{21}^2 & \Omega_{22}^2 & \Omega_{23}^2 \\ \Omega_{31}^2 & \Omega_{32}^2 & \Omega_{33}^2 \end{bmatrix}, \tag{2.8-4a}$$

$$\mathbf{K}_2 = \begin{bmatrix} \Omega_{12}\Omega_{13} & \Omega_{13}\Omega_{11} & \Omega_{11}\Omega_{12} \\ \Omega_{22}\Omega_{23} & \Omega_{23}\Omega_{21} & \Omega_{21}\Omega_{22} \\ \Omega_{32}\Omega_{33} & \Omega_{33}\Omega_{31} & \Omega_{31}\Omega_{32} \end{bmatrix}, \tag{2.8-4b}$$

$$\mathbf{K}_3 = \begin{bmatrix} \Omega_{21}\Omega_{31} & \Omega_{22}\Omega_{32} & \Omega_{23}\Omega_{33} \\ \Omega_{31}\Omega_{11} & \Omega_{32}\Omega_{12} & \Omega_{33}\Omega_{13} \\ \Omega_{11}\Omega_{21} & \Omega_{12}\Omega_{22} & \Omega_{13}\Omega_{23} \end{bmatrix}, \tag{2.8-4c}$$

$$\mathbf{K}_4 = \begin{bmatrix} \Omega_{22}\Omega_{33}+\Omega_{23}\Omega_{32} & \Omega_{23}\Omega_{31}+\Omega_{21}\Omega_{33} & \Omega_{21}\Omega_{32}+\Omega_{22}\Omega_{31} \\ \Omega_{32}\Omega_{13}+\Omega_{33}\Omega_{12} & \Omega_{33}\Omega_{11}+\Omega_{31}\Omega_{13} & \Omega_{31}\Omega_{12}+\Omega_{32}\Omega_{11} \\ \Omega_{12}\Omega_{23}+\Omega_{13}\Omega_{22} & \Omega_{13}\Omega_{21}+\Omega_{11}\Omega_{23} & \Omega_{11}\Omega_{22}+\Omega_{12}\Omega_{21} \end{bmatrix}. \tag{2.8-4d}$$

The inverse of (2.8-2) is

$$\boldsymbol{\sigma} = \mathbf{K}^{-1}\boldsymbol{\sigma}^*. \tag{2.8-5}$$

To find $\mathbf{K}^{-1}$ it is best writing (2.8-1) as

$$\sigma_{ij} = \Omega_{ki}\Omega_{sj}\sigma_{ks}^* \tag{2.8-6}$$

and cast (2.8-6) in the form of (2.8-5). It can be shown (Ting, 1987) that

$$(\mathbf{K}^{-1})^T = \begin{bmatrix} \mathbf{K}_1 & \mathbf{K}_2 \\ 2\mathbf{K}_3 & \mathbf{K}_4 \end{bmatrix}. \tag{2.8-7}$$

The transformation of the strains ε_{ij} is identical to (2.8-1) for the stresses σ_{ij}, i.e.,

$$\varepsilon_{ij}^* = \Omega_{ik}\Omega_{js}\varepsilon_{ks}. \tag{2.8-8}$$

However, the transformation of ε_{ij} in contracted notation is not

(2.8-2) due to the different laws for ε_{ij} shown in (2.3-2). It can be shown from (2.8-8) and (2.3-2) that (Lekhnitskii, 1950)

$$\boldsymbol{\varepsilon}^* = \begin{bmatrix} \mathbf{K}_1 & \mathbf{K}_2 \\ 2\mathbf{K}_3 & \mathbf{K}_4 \end{bmatrix} \boldsymbol{\varepsilon},$$

or, by (2.8-7),

$$\boldsymbol{\varepsilon}^* = (\mathbf{K}^{-1})^T \boldsymbol{\varepsilon}. \tag{2.8-9}$$

From

$$\boldsymbol{\sigma} = \mathbf{C}\,\boldsymbol{\varepsilon}, \quad \boldsymbol{\varepsilon} = \mathbf{s}\,\boldsymbol{\sigma},$$

$$\boldsymbol{\sigma}^* = \mathbf{C}^* \boldsymbol{\varepsilon}^*, \quad \boldsymbol{\varepsilon}^* = \mathbf{s}^* \boldsymbol{\sigma}^*,$$

and the transformation laws (2.8-5) and (2.8-9), it is easily shown that

$$\mathbf{C}^* = \mathbf{K}\mathbf{C}\mathbf{K}^T, \quad \mathbf{s}^* = (\mathbf{K}^{-1})^T \mathbf{s}\,\mathbf{K}^{-1}. \tag{2.8-10}$$

These are the transformation laws for **C** and **s**.

The special case in which the transformation is a rotation about the x_3-axis an angle θ,

$$\boldsymbol{\Omega} = \begin{bmatrix} \cos\theta & \sin\theta & 0 \\ -\sin\theta & \cos\theta & 0 \\ 0 & 0 & 1 \end{bmatrix}.$$

It can be shown that

$$\mathbf{K} = \begin{bmatrix} m^2 & n^2 & 0 & 0 & 0 & 2mn \\ n^2 & m^2 & 0 & 0 & 0 & -2mn \\ 0 & 0 & 1 & 0 & 0 & 0 \\ 0 & 0 & 0 & m & -n & 0 \\ 0 & 0 & 0 & n & m & 0 \\ -mn & mn & 0 & 0 & 0 & m^2-n^2 \end{bmatrix},$$

$$\mathbf{K}^{-1} = \begin{bmatrix} m^2 & n^2 & 0 & 0 & 0 & -2mn \\ n^2 & m^2 & 0 & 0 & 0 & 2mn \\ 0 & 0 & 1 & 0 & 0 & 0 \\ 0 & 0 & 0 & m & n & 0 \\ 0 & 0 & 0 & -n & m & 0 \\ mn & -mn & 0 & 0 & 0 & m^2-n^2 \end{bmatrix},$$

where

$$m = \cos\theta, \qquad n = \sin\theta.$$

For second rank tensors such as σ_{ij} or ε_{ij}, it is known that

$$\sigma_{ii}, \quad \sigma_{ij}\sigma_{ij}, \text{ and } \quad |\sigma_{ij}|$$

are invariants under orthogonal transformations. For the fourth rank tensors C_{ijks} it is readily shown from (2.5-3) and (2.5-2a) that

$$C^*_{iikk} = C_{iikk}, \qquad C^*_{ijij} = C_{ijij}.$$

Hence,

$$C_{iikk} = C_{11} + C_{22} + C_{33} + 2(C_{12} + C_{23} + C_{31}),$$

$$C_{ijij} = C_{11} + C_{22} + C_{33} + 2(C_{44} + C_{55} + C_{66}),$$

are invariants under orthogonal transformations (Hirth and Lothe, 1968; Juretschke, 1974). These are linear in $C_{\alpha\beta}$. Ting (1987) has presented several nonlinear invariants under any orthogonal transformations. When the transformation is a rotation about the x_3-axis, additional linear invariants (Lekhnitskii, 1950; Hearmon, 1961; Tsai and Pagano, 1980; Tsai and Hahn, 1980) and second order invariants (Tsai and Hahn, 1980; Ting, 1987) are obtained.

2.9 Restrictions on Elastic Constants

The strong ellipticity condition which ensures that the governing differential equations for elastostatics problems be completely elliptic is

$$C_{ijks}a_i b_j a_k b_s > 0 \tag{2.9-1}$$

for any real vectors a_i and b_i. For isotropic materials, inserting

$$C_{ijks} = \lambda\delta_{ij}\delta_{ks} + \mu(\delta_{ik}\delta_{js} + \delta_{is}\delta_{jk}) \tag{2.9-2}$$

into (2.9-1) yields

$$(\lambda + 2\mu)(\mathbf{a}\cdot\mathbf{b})^2 + \mu[(\mathbf{a}\cdot\mathbf{a})(\mathbf{b}\cdot\mathbf{b}) - (\mathbf{a}\cdot\mathbf{b})^2] > 0.$$

The coefficients of $(\lambda+2\mu)$ and μ are greater than or equal to zero. Therefore the strong elliptic conditions for isotropic materials are

$$\lambda + 2\mu > 0, \qquad \mu > 0. \tag{2.9-3}$$

The strong convexity condition which is equivalent to the positive definiteness of the strain energy is

$$C_{ijks}\varepsilon_{ij}\varepsilon_{ks} > 0. \tag{2.9-4}$$

Substitution of (2.9-2) into the above leads to

$$\lambda(\varepsilon_{ii})^2 + 2\mu(\varepsilon_{ij}\varepsilon_{ij}) > 0$$

or

$$(\lambda + \tfrac{2}{3}\mu)(\varepsilon_{ii})^2 + \tfrac{2}{3}\mu[3(\varepsilon_{ij}\varepsilon_{ij}) - (\varepsilon_{ii})^2] > 0.$$

Without loss in generality, we take the principal axes of the strains to be the coordinate axes so that ε_{ij} is a diagonal matrix with the diagonal elements ε_1, ε_2, and ε_3. We then have

$$(\lambda + \tfrac{2}{3}\mu)(\varepsilon_1 + \varepsilon_2 + \varepsilon_3)^2 + \tfrac{2}{3}\mu[(\varepsilon_1 - \varepsilon_2)^2 + (\varepsilon_2 - \varepsilon_3)^2 + (\varepsilon_3 - \varepsilon_1)^2] > 0.$$

The coefficients of $(\lambda + \frac{2}{3}\mu)$ and μ are greater than or equal to zero. Hence,

$$\infty > \lambda + \tfrac{2}{3}\mu > 0, \quad \infty > \mu > 0, \tag{2.9-5}$$

are the necessary and sufficient conditions for the strain energy to be positive for isotropic materials. The left inequalities for $(\lambda + \frac{2}{3}\mu)$ and μ are obtained from (2.7-5b) and (2.9-8) to be derived later.

We see that (2.9-5) implies (2.9-3) but the reverse does not hold. In other words, the positive definiteness of strain energy function imposes a more stringent limitation on elastic constants than the strong elliptic condition. This is true also for general anisotropic materials. To prove this statement for general anisotropic materials we first notice that (2.9-4) remains valid if ε_{ij} is not symmetric. This is so because the skew-symmetric part of ε_{ij} contributes nothing to the inequality (Section 2.1). Now, when (2.9-1) is satisfied by a choice of a_i and b_i, we take $\varepsilon_{ij}=a_i b_j$ so that (2.9-4) is satisfied. On the other hand, if (2.9-4) is satisfied by an ε_{ij}, it is not always possible to rewrite ε_{ij} as $a_i b_j$. If we regard $a_i b_j$ as a matrix, it is a singular matrix because

$$(a_i b_j)x_j = 0$$

for any vector x_i that is perpendicular to b_i. In contrast, ε_{ij} need not be singular. Therefore (2.9-4) imposes a more stringent limitation on elastic constants than (2.9-1).

In the contracted notation (2.9-4) is equivalent to (2.3-10) which implies that the 6×6 matrix **C** is positive definite. If **C** is positive definite, all its principal minors are positive (Section 1.6), i.e.,

$$C_{ii} > 0 \quad (i \text{ not summed}),$$

$$\begin{vmatrix} C_{ii} & C_{ij} \\ C_{ij} & C_{jj} \end{vmatrix} > 0 \quad (i, j, \text{ not summed}),$$

$$\begin{vmatrix} C_{ii} & C_{ij} & C_{ik} \\ C_{ij} & C_{jj} & C_{jk} \\ C_{ik} & C_{jk} & C_{kk} \end{vmatrix} > 0 \quad (i, j, k, \text{ not summed}), \tag{2.9-6}$$

$$\vdots$$

where i, j, k are distinct integers which can have any value from 1 to 6. According to Theorem 1.6–2, the *necessary and sufficient conditions* for the $\mathbf{C}$ to be positive definite are the positivity of its *leading principal minors*, i.e.,

$$C_{11} > 0,$$

$$\begin{vmatrix} C_{11} & C_{12} \\ C_{12} & C_{22} \end{vmatrix} > 0,$$

$$\begin{vmatrix} C_{11} & C_{12} & C_{13} \\ C_{12} & C_{22} & C_{23} \\ C_{13} & C_{23} & C_{33} \end{vmatrix} > 0, \tag{2.9-7}$$

$$\vdots$$

$$\begin{vmatrix} C_{11} & C_{12} & C_{13} & C_{14} & C_{15} & C_{16} \\ C_{12} & C_{22} & C_{23} & C_{24} & C_{25} & C_{26} \\ C_{13} & C_{23} & C_{33} & C_{34} & C_{35} & C_{36} \\ C_{14} & C_{24} & C_{34} & C_{44} & C_{45} & C_{46} \\ C_{15} & C_{25} & C_{35} & C_{45} & C_{55} & C_{56} \\ C_{16} & C_{26} & C_{36} & C_{46} & C_{56} & C_{66} \end{vmatrix} > 0.$$

From (2.3-11) and the discussions in Section 2.4, the 6×6 matrix $\mathbf{s}$ and the 5×5 matrices $\mathbf{C}^o$ and $\mathbf{s}'$ are positive definite. The inequalities (2.9-6) and (2.9-7) apply to these matrices. Applications of (2.9-7) to the matrices $\mathbf{s}$ of (2.7-4) and $\mathbf{s}'$ of (2.7-11) show that these two matrices are positive definite if

$$\infty > E > 0, \qquad -1 < \nu < \tfrac{1}{2}. \tag{2.9-8}$$

It should be noted that E must be bounded. Otherwise a nonzero stress can induce zero strains and hence zero strain energy, contradicting the assumption of positive strain energy.

It should also be noted that inequalities (2.9-1) or (2.9-4) for two-dimensional deformations may yield a different set of inequalities from those for three-dimensional deformations. As an illustration take the positive strain energy condition (2.9-4) and consider a cubic material. This means that the matrix **C** of (2.6-8) is positive definite. Application of (2.9-7) leads to

$$\begin{gathered} C_{11} > 0, \quad C_{11}^2 - C_{12}^2 > 0, \\ (C_{11} - C_{12})^2 (C_{11} + 2C_{12}) > 0, \quad C_{44} > 0. \end{gathered} \tag{2.9-9}$$

They are equivalent to

$$C_{11} + 2C_{12} > 0, \quad C_{11} - C_{12} > 0, \quad C_{44} > 0, \tag{2.9-10}$$

because the first three inequalities in (2.9-9) can be replaced by

$$C_{11} + C_{12} > 0, \quad C_{11} - C_{12} > 0, \quad C_{11} + 2C_{12} > 0,$$

which in turn can be replaced by the first two inequalities in (2.9-10). For two-dimensional deformations for which $\varepsilon_3=0$, the third row and the third column of (2.6-9) are not needed. The reduced 5×5 matrix $\mathbf{C}^o$ is positive definite. Application of (2.9-7) yields

$$C_{11} > 0, \quad C_{11}^2 - C_{12}^2 > 0, \quad C_{44} > 0$$

which are equivalent to

$$C_{11} + C_{12} > 0, \quad C_{11} - C_{12} > 0, \quad C_{44} > 0. \tag{2.9-11}$$

The first inequalities in (2.9-10) and (2.9-11) are different. The quantities $C_{11} + 2C_{12}$ and $C_{11} + C_{12}$ can be identified, respectively, as the *bulk moduli* for three-dimensional and two-dimensional deformations. Inequalities (2.9-11) are less restrictive than those in (2.9-10).

2.10 Determination of Symmetry Planes

The classification of the materials according to the number of symmetry planes presented in Section 2.6 is based on the assumption that we have the knowledge on the number and the locations of the planes of symmetry a given material has. This is not the case when we are presented with an unknown material. We are then forced to determine elastic constants of the material referred to an arbitrarily chosen coordinate system. The result is that, if there exists a symmetry plane, it may not be one of the coordinate planes. Consequently all elements of the matrix **C** can be nonzero. The problem is to locate the symmetry planes if they exist when the matrix **C** is given.

When a plane of symmetry exists, (2.5-4) is satisfied by the $\mathbf{\Omega}$ given in (2.5-7a), i.e.,

$$\Omega_{ij} = \delta_{ij} - 2n_i n_j \tag{2.10-1}$$

where **n** is a unit vector normal to the plane of symmetry. The matrix Ω_{ij} has the properties given in (2.5-7b) where **m** is any vector perpendicular to **n**. It also has the properties

$$n_i \Omega_{ij} = -n_j, \quad m_i \Omega_{ij} = m_j. \tag{2.10-2}$$

Cowin and Mehrabadi (1987) have shown that a set of necessary and sufficient conditions for **n** to be a unit normal vector to a plane of symmetry is

$$C_{ijkk} n_j = (C_{pqtt} n_p n_q) n_i, \tag{2.10-3}$$

$$C_{isks} n_k = (C_{pqrq} n_p n_r) n_i, \tag{2.10-4}$$

$$C_{ijks} n_j n_s n_k = (C_{pqrt} n_p n_q n_r n_t) n_i, \tag{2.10-5}$$

$$C_{ijks} m_j m_s n_k = (C_{pqrt} n_p m_q n_r m_t) n_i. \tag{2.10-6}$$

We need (2.10-1) and (2.10-2) in deriving the four equations (2.10-3)–(2.10-6). Equation (2.10-3) is obtained by multiplying both sides of (2.5-4) by n_j and setting $s=k$. If we multiply (2.5-4) by n_k and set $j=s$, we have (2.10-4). Equation (2.10-5) and (2.10-6) are derived by multiplying (2.5-4) by $n_j n_k n_s$ and $m_j n_k, m_s$, respectively. All repeated indices imply summation in these derivations.

Cowin (1989) (see also Norris, 1989) later showed that (2.10-5) and (2.10-6) are necessary and sufficient conditions. In fact (2.10-3) and (2.10-6), or (2.10-4) and (2.10-6), are also necessary and sufficient conditions. To see this let the plane $x_1=0$ be the plane of symmetry so that

$$n_i = \delta_{i1}, \quad m_i = \delta_{i2} \cos\theta + \delta_{i3} \sin\theta, \tag{2.10-7}$$

where θ is an arbitrary constant. Substitution of $(2.10\text{-}7)_1$ into (2.10-3)–(2.10-5) leads to

$$C_{i1kk} = C_{11tt} \delta_{i1}, \quad C_{is1s} = C_{1q1q} \delta_{i1}, \quad C_{i111} = C_{1111} \delta_{i1}.$$

For i=2, 3 they give

$$C_{21kk} = C_{31kk} = 0, \quad C_{2s1s} = C_{3s1s} = 0, \quad C_{2111} = C_{3111} = 0,$$

or, using the contracted notation,

$$C_{16}+C_{26}+C_{36}=C_{15}+C_{25}+C_{35}=0, \tag{2.10-8}$$

$$C_{16}+C_{26}+C_{45}=C_{15}+C_{46}+C_{35}=0, \tag{2.10-9}$$

$$C_{16}=C_{15}=0. \tag{2.10-10}$$

Substitution of $(2.10\text{-}7)_2$ into (2.10-6) with (i) $\theta=0$, (ii) $\theta=\pi/2$, and (iii) arbitrary θ yields, respectively,

$$C_{i212}=C_{1212}\delta_{i1}, \quad C_{i313}=C_{1313}\delta_{i1},$$

$$C_{i213}+C_{i312}=(C_{1213}+C_{1312})\delta_{i1}.$$

For i=2 ,3, and using the contracted notation we obtain

$$C_{26}=C_{46}=C_{45}=C_{35}=C_{25}=C_{36}=0. \tag{2.10-11}$$

Equations (2.10-8)–(2.10-11) are, respectively, a specialization of (2.10-3)–(2.10-6) when the unit normal **n** is along the x_1-axis. If x_1=0 is a plane of symmetry, we must have by (2.6-3a)

$$C_{15}=C_{16}=C_{25}=C_{26}=C_{35}=C_{36}=C_{45}=C_{46}=0. \tag{2.10-12}$$

It is clear that (2.10-10) and (2.10-11) are equivalent to (2.10-12), so are (2.10-8) and (2.10-11), and (2.10-9) and (2.10-11). However, (2.10-8)–(2.10-10) do not imply (2.10-12).

Equations (2.10-3)–(2.10-6) tell us that **n** is an eigenvector of the 3×3 symmetric matrices **U**, **V**, **Q**(**n**), and **Q**(**m**) whose elements are

$$U_{ij}=C_{ijkk}, \quad V_{ik}=C_{isks}, \quad Q_{ik}(\mathbf{n})=C_{ijks}n_j n_s. \tag{2.10-13}$$

We may therefore state a modified Cowin–Mehrabadi theorem as follows.

> **Theorem 2.10–1** An anisotropic elastic material with given elastic compliances C_{ijks} has a plane of symmetry if and only if **n** is an eigenvector of (i) **Q**(**n**) and **Q**(**m**), (ii) **U** and **Q**(**m**), or (iii) **V** and **Q**(**m**). The vector **n** is normal to the plane of symmetry while **m** is any vector on the plane of symmetry.

The conditions in (i) are due to Cowin (1989). The tensor **Q**(**n**) is the acoustic tensor for waves propagating in the direction of **n**. When **n** is an eigenvector of **Q**(**n**) the wave is *longitudinal*, with **n** being the *specific direction*. Kolodner (1966) has shown that there exist at least three distinct specific directions in an anisotropic elastic material. If **n** is an eigenvector of **Q**(**m**) the wave is *transversal*, with **m** being the propagation direction and **n** the *specific axis*. The conditions in (i) mean that necessary and sufficient conditions for **n** to be normal to a plane of symmetry are that it be a specific direction and a specific axis. Physical interpretations of the plane of symmetry can also be made in terms of static loading (Hayes and Norris, 1992).

Theorem 2.10–1 is not suitable for determining **n** because the matrix **Q**(**m**) depends on **m** which, in turns, depends on **n**. The following theorem is more useful for computing **n**.

> **Theorem 2.10–2** An anisotropic elastic material has a plane of symmetry if and only if the normal **n** to the plane of symmetry is a common eigenvector of **U** and **V**, and satisfies
>
> $$C_{ijks} m_i n_j n_k n_s = 0, \tag{2.10-14}$$
>
> $$C_{ijks} m_i m_j m_k n_s = 0, \tag{2.10-15}$$
>
> for any two independent vectors $\mathbf{m}^{(\alpha)}$ (α=1,2) on the plane of symmetry that do not form an angle a multiple of $\pi/3$.

In the above, (2.10-14) and (2.10-15) need to be satisfied by only two independent vectors on the plane of symmetry that do not form an angle a multiple of $\pi/3$. To prove the theorem let **n** be along the x_1-axis so that $n_i = \delta_{i1}$ and let the two **m**'s be given by

$$\begin{aligned} m_i^{(1)} &= \delta_{i2} \cos\theta + \delta_{i3} \sin\theta, \\ m_i^{(2)} &= \delta_{i2} \cos\psi + \delta_{i3} \sin\psi, \end{aligned} \tag{2.10-16}$$

where θ and ψ are arbitrary constants subjected to the condition

$$\theta - \psi \neq \tfrac{1}{3}k\pi, \quad k \text{ integer}. \tag{2.10-17}$$

Substitution of (2.10-16) into (2.10-14) yields

$$\begin{aligned} C_{16}\cos\theta + C_{15}\sin\theta = 0, \\ C_{16}\cos\psi + C_{15}\sin\psi = 0. \end{aligned}$$

Since the determinant of the coefficient matrix for $[C_{15}, C_{16}]$ is $\sin(\theta - \psi)$ which is nonzero in view of (2.10-17),

$$C_{15} = C_{16} = 0. \tag{2.10-18}$$

When $n_i = \delta_{i1}$ is a common eigenvector of **U** and **V**, (2.10-8) and (2.10-9) apply. Together with (2.10-18) we have

$$C_{36} = C_{45} = -C_{26}, \quad C_{25} = C_{46} = -C_{35}. \tag{2.10-19}$$

Finally, application of (2.10-16) to (2.10-15) and making use of (2.10-19) lead to

$$\cos\theta(2\cos 2\theta - 1)C_{26} - \sin\theta(2\cos 2\theta + 1)C_{35} = 0,$$
$$\cos\psi(2\cos 2\psi - 1)C_{26} - \sin\psi(2\cos 2\psi + 1)C_{35} = 0.$$

The determinant of the coefficient matrix of $[C_{26}, C_{35}]$ can be shown to be

$$\sin(\theta - \psi)[4\cos^2(\theta - \psi) - 1]$$

which is nonzero in view of (2.10-17). Hence

$$C_{26} = C_{35} = 0. \tag{2.10-20}$$

Equations (2.10-18)–(2.10-20) are equivalent to (2.10-12). This completes the proof.

In the contracted notation **U** and **V** have the expressions

$$\mathbf{U} = \begin{bmatrix} C_{11}+C_{12}+C_{13} & C_{16}+C_{26}+C_{36} & C_{15}+C_{25}+C_{35} \\ & C_{12}+C_{22}+C_{23} & C_{14}+C_{24}+C_{34} \\ & & C_{13}+C_{23}+C_{33} \end{bmatrix}$$
$$\mathbf{V} = \begin{bmatrix} C_{11}+C_{55}+C_{66} & C_{16}+C_{26}+C_{45} & C_{15}+C_{35}+C_{46} \\ & C_{22}+C_{44}+C_{66} & C_{24}+C_{34}+C_{56} \\ & & C_{33}+C_{44}+C_{55} \end{bmatrix} \tag{2.10-21}$$

Only the upper triangles are shown for **U** and **V** because they are symmetric. We will show that there is no need to find the eigenvectors of **U** and **V** in order to see if **U** and **V** have an eigenvector in common. If **n** is an eigenvector of **U** and **V**, we write

$$\mathbf{Un} = u\mathbf{n}, \quad \mathbf{Vn} = v\mathbf{n}, \tag{2.10-22}$$

where u and v are the corresponding eigenvalues. This means that

$$\mathbf{VUn} = uv\mathbf{n}, \quad \mathbf{UVn} = uv\mathbf{n}. \tag{2.10-23}$$

Subtracting the two equations leads to

$$\mathbf{Wn} = \mathbf{0}, \tag{2.10-24}$$

where, noticing that **U** and **V** are symmetric,

$$\mathbf{W} = \mathbf{UV} - \mathbf{VU} = \mathbf{UV} - (\mathbf{UV})^T. \tag{2.10-25}$$

Thus **n** is a null vector of the skew-symmetric matrix **W**. When **UV** is symmetric, we have $\mathbf{W}=\mathbf{0}$ or $\mathbf{UV}=\mathbf{VU}$, implying that the product of $\mathbf{U}$ and $\mathbf{V}$ commutes. By Theorem 1.5–2, $\mathbf{U}$ and $\mathbf{V}$ share the same set of eigenvectors. If $\mathbf{UV}$ is not symmetric, let

$$\mathbf{W}=\begin{bmatrix} 0 & w_3 & -w_2 \\ -w_3 & 0 & w_1 \\ w_2 & -w_1 & 0 \end{bmatrix}.$$

The unnormalized null vector $\mathbf{n}$ can be taken as

$$\mathbf{n}^T=[w_1,\ w_2,\ w_3]. \tag{2.10-26}$$

If the $\mathbf{n}$ given by (2.10-26) is an eigenvector of $\mathbf{U}$, it is an eigenvector of $\mathbf{V}$, and vice versa. To see this, suppose that the $\mathbf{n}$ of (2.10-26) satisfies $(2.10\text{-}22)_1$. Premultiplied by $\mathbf{V}$ leads to

$$\mathbf{VUn}=u\mathbf{Vn}, \quad \text{or} \quad (\mathbf{UV}-\mathbf{W})\mathbf{n}=u\mathbf{Vn},$$

by (2.10-25). In view of (2.10-24) we have

$$\mathbf{U}(\mathbf{Vn})=u(\mathbf{Vn}).$$

Comparison with $(2.10\text{-}22)_1$ tells us that $\mathbf{Vn}$ is proportional to $\mathbf{n}$, which is exactly $(2.10\text{-}22)_2$.

> **Theorem 2.10–3** If $\mathbf{UV}$ is symmetric, $\mathbf{U}$ and $\mathbf{V}$ share the same set of eigenvectors. If $\mathbf{UV}$ is not symmetric, $\mathbf{U}$ and $\mathbf{V}$ have one common eigenvector when the $\mathbf{n}$ of (2.10-26) is an eigenvector of $\mathbf{U}$ or $\mathbf{V}$. Otherwise there is no common eigenvector.

In conjunction with Theorem 2.10–2 it is clear that $\mathbf{UV}$ must be symmetric if a material possesses more than one symmetry plane. If $\mathbf{UV}$ is not symmetric, the material can have at most one symmetry plane. It should be noted that $\mathbf{UV}$ can be symmetric for materials with one or no symmetry plane.

In closing this chapter we note that the transformation of coordinate systems employed here is restricted to orthogonal coordinate systems. If one allows the transformation for the full linear group, Olver (1988, 1992) showed that all anisotropic materials are equivalent to an orthotropic material.

Chapter 3

ANTIPLANE DEFORMATIONS

As a starter for anisotropic elastostatics we study special two-dimensional deformations of anisotropic elastic bodies, namely, *antiplane deformations*. Not all anisotropic elastic materials are capable of an antiplane deformation. When they are, the *inplane displacement* and the *antiplane displacement* are uncoupled. The deformations due to inplane displacement are *plane strain deformations*. Associated with plane strain deformations are *plane stress deformations*. After defining these special deformations in Sections 3.1 and 3.2 we present some basic solutions of antiplane deformations. They provide useful references for more general deformations we will study in Chapters 8, 10, and 11. The derivation and motivation in solving more general deformations in those chapters become more transparent if the reader reads this chapter first. The solutions obtained in those chapters reduce to the solutions presented here when the materials are restricted to special materials and the deformations are limited to antiplane deformations.

3.1 Uncoupling of Inplane and Antiplane Displacements

In a fixed rectangular coordinate system x_i (i=1, 2, 3), let u_i, σ_{ij}, and ε_{ij} be the displacement, stress, and strain, respectively. The strain-displacement relations and the equations of equilibrium are

$$\varepsilon_{ij} = \tfrac{1}{2}(u_{i,j} + u_{j,i}), \tag{3.1-1}$$

$$\sigma_{ij,j} = 0, \tag{3.1-2}$$

in which repeated indices imply summation and a comma stands for differentiation. The stress-strain laws for an anisotropic elastic material can be written as

$$\sigma_{ij} = C_{ijks}\varepsilon_{ks} \quad \text{or} \quad \varepsilon_{ij} = S_{ijks}\sigma_{ks}, \tag{3.1-3}$$

where C_{ijks} and S_{ijks} are, respectively, the elastic stiffnesses and compliances. They possess the full symmetry and satisfy the requirement that the strain energy is positive (Sections 2.1 and 2.2). Insertion of (3.1-1) into $(3.1\text{-}3)_1$, and then into (3.1-2) yields

$$\sigma_{ij} = C_{ijks}u_{k,s}, \tag{3.1-4}$$

$$C_{ijks}u_{k,sj} = 0. \tag{3.1-5}$$

For two-dimensional deformations for which the displacement u_i depends on x_1 and x_2 only, (3.1-5) is written explicitly as

$$C_{i1k1}u_{k,11} + C_{i2k2}u_{k,22} + (C_{i1k2} + C_{i2k1})u_{k,12} = 0 \tag{3.1-6}$$

which provides three differential equations for the three displacement components u_1, u_2, and u_3. If the *antiplane displacement* u_3 can be uncoupled from the *inplane displacements* u_1 and u_2, we may consider antiplane deformations and inplane deformations separately. The inplane deformations are equivalent to *plane strain deformations* in isotropic elasticity.

Consider the *plane strain deformations*

$$u_i = u_1(x_1,x_2), \quad u_2 = u_2(x_1,x_2), \quad u_3 = 0. \tag{3.1-7}$$

Employing the contracted notation (Section 2.3), (3.1-6) for i=1,2,3 have the expressions

$$\begin{aligned} &C_{11}u_{1,11} + C_{66}u_{1,22} + 2C_{16}u_{1,12} \\ &\quad + C_{16}u_{2,11} + C_{26}u_{2,22} + (C_{12} + C_{66})u_{2,12} = 0, \end{aligned} \tag{3.1-8a}$$

$$\begin{aligned} &C_{16}u_{1,11} + C_{26}u_{1,22} + (C_{66} + C_{12})u_{1,12} \\ &\quad + C_{66}u_{2,11} + C_{22}u_{2,22} + 2C_{26}u_{2,12} = 0, \end{aligned} \tag{3.1-8b}$$

$$\begin{aligned} &C_{15}u_{1,11} + C_{46}u_{1,22} + (C_{56} + C_{14})u_{1,12} \\ &\quad + C_{56}u_{2,11} + C_{24}u_{2,22} + (C_{25} + C_{46})u_{2,12} = 0. \end{aligned} \tag{3.1-8c}$$

There are three equations for the two displacements u_1 and u_2. The third equation vanishes identically if we let

$$C_{14} = C_{15} = C_{24} = C_{25} = C_{46} = C_{56} = 0. \tag{3.1-9}$$

We cannot choose the first or the second equation because it contains the constants C_{11}, C_{22}, and C_{66} which are positive and nonzero. Equations (3.1-9) are satisfied by monoclinic materials with the symmetry plane at x_3 =0 as shown in (2.6-3c). In fact (3.1-9) represents a material more general than monoclinic materials with the symmetry plane at x_3=0 because the latter require C_{34}=C_{35}=0 also.

Next consider the *antiplane deformations* defined by

$$u_1 = u_2 = 0, \quad u_3 = u(x_1,x_2), \text{ say.} \tag{3.1-10}$$

Equation (3.1-6) for i=1,2,3 are

$$C_{15}u_{,11}+C_{46}u_{,22}+(C_{14}+C_{56})u_{,12}=0,$$
$$C_{56}u_{,11}+C_{24}u_{,22}+(C_{46}+C_{25})u_{,12}=0,$$
$$C_{55}u_{,11}+C_{44}u_{,22}+2C_{45}u_{,12}=0.$$

There are three differential equations for the single displacement u. If we impose the conditions (3.1-9) the first two equations vanish identically. Thus an anisotropic elastic material that can sustain a plane strain deformation can also sustain an antiplane deformation.

For a general anisotropic elastic material an antiplane deformation or a plane strain deformation in general does not exist. All three displacement components are coupled, and have to be considered simultaneously even though they depend on x_1 and x_2 only. A deformation in which all three displacement components are coupled and depend on x_1 and x_2 only is called a *generalized plane strain deformation*. The antiplane displacement u_3 is not zero, nor the shear strains ε_{13} and ε_{23}. However, ε_{33}=0 for generalized plane strain deformations because $u_{3,3}$=0.

The results presented here apply also to inhomogeneous anisotropic materials (Horgan and Miller, 1994) and to elastodynamics. It can be extended to small deformations superposed on finite deformations (Woo and Shield, 1962).

3.2 Plane Stress Deformations

Consider a thin plate of thickness 2δ occupying the region $-\delta \le x_3 \le \delta$ that is subjected to a loading parallel to the x_3=0 plane. The surfaces $x_3=\pm\delta$ of the plate are traction-free, i.e.,

$$\sigma_{31}=\sigma_{32}=\sigma_{33}=0, \quad \text{at } x_3=\pm\delta. \qquad (3.2\text{-}1)$$

If the plate is a general anisotropic elastic material the *inplane stresses* σ_{11}, σ_{22}, σ_{12} would generate the shear strains ε_{31} and ε_{32} so that the u_3 displacement at x_3=0 would not be zero. This means that the mid-plane of the plate would not remain plane after the inplane loading. We will therefore consider monoclinic materials with the symmetry plane at x_3=0 for which

$$C_{14}=C_{15}=C_{24}=C_{25}=C_{34}=C_{35}=C_{46}=C_{56}=0. \qquad (3.2\text{-}2)$$

The displacements in the plate are symmetric with the mid-planex_3=0 if the stresses are. This means that if we define the average displacements $\tilde{u}_i$ by

$$\tilde{u}_i = \frac{1}{2\delta}\int_{-\delta}^{\delta} u_i dx_3$$

and define

$$[u_i] = \frac{u_i(x_1, x_2, \delta) - u_i(x_1, x_2, -\delta)}{2\delta},$$

we have

$$[u_1] = [u_2] = 0, \quad \tilde{u}_3 = 0. \tag{3.2-3}$$

With (3.2-3) it can be shown that

$$\frac{1}{2\delta}\int_{-\delta}^{\delta} u_{k,s} dx_3 = \begin{cases} \tilde{u}_{k,s}, & \text{for } k, s \neq 3, \\ [u_3] = \tilde{\varepsilon}_3, & \text{for } k = s = 3, \\ 0, & \text{otherwise.} \end{cases}$$

Hence integration of (3.1-4) with x_3 yields

$$\tilde{\sigma}_{ij} = C_{ijk1}\tilde{u}_{k,1} + C_{ijk2}\tilde{u}_{k,2} + C_{ij33}\tilde{\varepsilon}_3, \quad k \neq 3. \tag{3.2-4}$$

The summation with k is for k=1 and 2, not 3.

The equation of equilibrium (3.1-2) for i=3 is

$$\sigma_{31,1} + \sigma_{32,2} + \sigma_{33,3} = 0. \tag{3.2-5}$$

If we apply this at $x_3 = \pm\delta$ we have by $(3.2\text{-}1)_{1,2}$

$$\sigma_{33,3} = 0, \quad \text{at } x_3 = \pm\delta.$$

This and the fact that σ_{33}=0 at $x_3 = \pm\delta$ tell us that σ_{33} at $x_3 \neq \pm\delta$ is small if it is not zero. Therefore the average stress $\tilde{\sigma}_{33}$ can be approximated as zero. Equation (3.2-4) for $\tilde{\sigma}_{33}$ is

$$0 = C_{33k1}\tilde{u}_{k,1} + C_{33k2}\tilde{u}_{k,2} + C_{3333}\tilde{\varepsilon}_3, \quad k \neq 3. \tag{3.2-6}$$

Solving for $\tilde{\varepsilon}_3$ and inserting it into (3.2-4) leads to

$$\tilde{\sigma}_{ij} = C'_{ijks}\tilde{u}_{k,s} \qquad (k, s \neq 3) \tag{3.2-7}$$

where

$$C'_{ijks} = C_{ijks} - \frac{C_{ij33}C_{33ks}}{C_{3333}} \tag{3.2-8a}$$

are the *reduced elastic stiffnesses*. In the contracted notation we have

$$C'_{\alpha\beta} = C_{\alpha\beta} - \frac{C_{\alpha 3}C_{3\beta}}{C_{33}}. \tag{3.2-8b}$$

Integrating (3.1-2) with x_3 and using (3.2-1) yields

$$\tilde{\sigma}_{ij,j} = 0 \quad (j \neq 3)$$

or, by (3.2-7),

$$C'_{ijks}\tilde{u}_{k,sj} = 0 \quad (j,k,s \neq 3). \tag{3.2-9}$$

Equations (3.2-7) and (3.2-9) are identical to (3.1-4) and (3.1-5) when the latter are specialized for plane strain deformations for which u_3=0 and u_1 and u_2 depend on x_1 and x_2 only. Therefore (3.1-8) apply here if $C_{\alpha\beta}$ and u_k in (3.1-8) are replaced, respectively, by $C'_{\alpha\beta}$ and $\tilde{u}_k$. With (3.2-2) and (3.2-8b), it is easily shown that (3.1-9) remain valid for $C'_{\alpha\beta}$. Hence (3.1-8c) vanishes identically and (3.1-8a,b) give us two differential equations for the average displacements $\tilde{u}_1$ and $\tilde{u}_2$.

We thus have the result that the problems of plane stress deformations are mathematically identical to those of plane strain deformations if $C_{\alpha\beta}$, u_k, and σ_{ij} are replaced by $C'_{\alpha\beta}$, $\tilde{u}_k$, and $\tilde{\sigma}_{ij}$, respectively. There are however several differences. (i) The theory of plane strain is an exact theory whereas that of plane stress is an approximate theory unless $\tilde{\sigma}_{33}$ is exactly zero. (ii) The plane strain theory requires that those $C_{\alpha\beta}$ listed in (3.1-9) vanish. The plane stress theory requires in addition C_{34}=C_{35}=0 as shown in (3.2-2). (iii) The stress component σ_{33} is nonzero in general for plane strain deformations whereas in plane stress deformations $\tilde{\sigma}_{33}$ is assumed zero. Finally, (iv) the strain ε_3 vanishes for plane strain deformations. For plane stress deformations, $\tilde{\varepsilon}_3$ obtained from (3.2-6) is in general nonzero.

A different plane strain deformation was found in one-component surface waves for special anisotropic elastic materials for which u_1 and u_3 (not u_2) depend on x_1 and x_2 while u_2 vanishes everywhere. The displacements, which do not depend on x_3, are polarized on planes parallel to x_2=0. The only nonzero stresses are σ_{11}, σ_{33}, and σ_{13}. Thus the stresses are also polarized on planes parallel to x_2=0 (Section 12.3). Therefore one-component surface waves have the unusual feature that the deformations are both plane stress and plane strain. Of course, the plane strain deformation in one-component surface waves is different from the one defined in the previous section.

3.3 General solutions for Antiplane Deformations

In the following sections we consider antiplane deformations of special anisotropic materials whose elastic stiffnesses satisfy the conditions

$$C_{14} = C_{15} = C_{24} = C_{25} = C_{46} = C_{56} = 0. \qquad (3.3\text{-}1)$$

Monoclinic materials with the plane of symmetry at $x_3=0$ more than satisfy these conditions because they also possess the properties $C_{34}=C_{35}=0$ that are not required in (3.3-1).

For antiplane deformations,

$$u_1 = u_2 = 0, \quad u_3 = u(x_1, x_2), \qquad (3.3\text{-}2)$$

and the stress-strain law (3.1-4) reduces to

$$\sigma_{ij} = C_{ij31}u_{,1} + C_{ij32}u_{,2}.$$

Using the contracted notation for C_{ijks} the nonzero stresses are

$$\begin{aligned} \sigma_{31} &= C_{55}u_{,1} + C_{45}u_{,2}, \\ \sigma_{32} &= C_{45}u_{,1} + C_{44}u_{,2}, \\ \sigma_{33} &= C_{35}u_{,1} + C_{34}u_{,2}. \end{aligned} \qquad (3.3\text{-}3)$$

The equations of equilibrium (3.1-2) are automatically satisfied for i=1,2 while for i=3,

$$\sigma_{31,1} + \sigma_{32,2} = 0. \qquad (3.3\text{-}4)$$

Insertion of (3.3-3) into (3.3-4) leads to

$$C_{55}u_{,11} + 2C_{45}u_{,12} + C_{44}u_{,22} = 0. \qquad (3.3\text{-}5)$$

This is a homogeneous second order differential equation for u. The general solution for u depends on a linear combination of x_1 and x_2. Without loss in generality we let the coefficient of x_1 in the linear combination to be unity and let

$$u = \frac{-i}{2\mu} f(z), \qquad (3.3\text{-}6)$$

$$z = x_1 + px_2. \qquad (3.3\text{-}7)$$

In the above f is an arbitrary function of its argument z, $i=\sqrt{-1}$, p is a constant to be determined and

$$\mu = \sqrt{C_{44}C_{55} - C_{45}^2} > 0 \qquad (3.3\text{-}8)$$

which is real and nonzero because μ^2 is a principal minor of the 6×6 matrix $C_{\alpha\beta}$ (Section 2.9). Substitution of (3.3-6) into (3.3-5) yields

$$C_{55} + 2pC_{45} + p^2 C_{44} = 0 \qquad (3.3\text{-}9)$$

which provides two roots for p. One of them is

$$p = \frac{-C_{45} + i\mu}{C_{44}}, \tag{3.3-10}$$

and the other is its complex conjugate $\bar{p}$. For isotropic materials $C_{45}=0$ and $C_{44}=C_{55}=\mu$, giving $p=i$ and μ the *shear modulus*. The general solution is obtained by a linear superposition of two solutions of the form (3.3-6) associated with p and $\bar{p}$. For the displacement u to be real, we let

$$u = \frac{1}{2\mu}\left\{-if(z) + i\bar{f}(\bar{z})\right\}$$

or

$$u = \frac{1}{\mu}\operatorname{Im} f(z) \tag{3.3-11}$$

where Im stands for the imaginary part.

From (3.3-9) and (3.3-10) we have

$$C_{55} + pC_{45} = -p(C_{45} + pC_{44}) = -ip\mu .$$

The stresses σ_{31} and σ_{32} obtained from (3.3-3) and (3.3-11) are

$$\sigma_{31} = \frac{1}{\mu}\operatorname{Im}\left\{(C_{55} + pC_{45})\frac{df(z)}{dz}\right\} = -\operatorname{Re}\left\{p\frac{df(z)}{dz}\right\},$$

$$\sigma_{32} = \frac{1}{\mu}\operatorname{Im}\left\{(C_{45} + pC_{44})\frac{df(z)}{dz}\right\} = \operatorname{Re}\left\{\frac{df(z)}{dz}\right\}.$$

Hence σ_{31} and σ_{32} can be rewritten as (Wu, 1991c)

$$\sigma_{31} = -\phi_{,2}, \quad \sigma_{32} = \phi_{,1}, \tag{3.3-12a}$$

or

$$\sigma_{3j} = \delta_{3ij}\phi_{,i} \tag{3.3-12b}$$

where

$$\phi = \operatorname{Re} f(z)$$

is the *stress function*. In the above δ_{kij} is the permutation symbol and Re stands for the real part.

In summary, the general solutions for the displacement u and the stress function ϕ for an antiplane deformation are (Ting, 1995c)

$$u = \frac{1}{\mu}\operatorname{Im} f(z), \quad \phi = \operatorname{Re} f(z), \tag{3.3-13}$$

or

$$\phi + i\mu u = f(z). \tag{3.3-14}$$

The stresses are obtained from (3.3-12) and $(3.3\text{-}3)_3$ by differen tiation. An alternative to $(3.3\text{-}3)_3$ for the stress component σ_{33} is to employ (2.4-5). Noting that $\sigma_{11}=\sigma_{22}=\sigma_{12}=0$, we have

$$\sigma_{33}=\frac{-1}{s_{33}}(s_{34}\sigma_{32}+s_{35}\sigma_{31}). \tag{3.3-15}$$

In applications all we have to do is to determine the complex function $f(z)$ such that the prescribed boundary conditions are satisfied. Since the differential equation (3.3-5) is linear, we may superimpose more than one solution of the form (3.3-13) or (3.3-14) to satisfy the boundary conditions. This is illustrated in the remainder of this chapter.

When the boundary conditions are prescribed in terms of surface tractions, they are related to ϕ in the following way. Let Γ be the boundary and s be the arclength measured along Γ such that, when one faces the direction of increasing s, the material lies on the right side. If **t** is the traction vector,

$$t_i=\sigma_{ij}n_j$$

where **n** is the unit outward normal to Γ. For antiplane deformations the only nonzero component of **t** is

$$t_3=\sigma_{31}n_1+\sigma_{32}n_2=-\phi_{,2}n_1+\phi_{,1}n_2$$

where the second equality follows from (3.3-12a). It is readily shown that

$$n_1=-\frac{dx_2}{ds},\qquad n_2=\frac{dx_1}{ds},$$

so that

$$t_3=\frac{d\phi}{ds}. \tag{3.3-16}$$

This is a generalization of (3.3-12a). If the boundary Γ is a traction-free surface, ϕ=constant. In most cases we may set $\phi=0$ without loss in generality. If the boundary is an interface between two dissimilar materials, the continuity of surface traction means that $d\phi/ds$ is continuous at Γ. This means that the *difference* in ϕ across Γ is constant. Again, we may set the difference to zero so that ϕ is continuous across Γ.

3.4 The Complex Functions $\ln z$ and $\sqrt{z^2-a^2}$

The two most frequently encountered complex functions in the solutions to elasticity problems are

$$\ln z \quad \text{and} \quad \sqrt{z^2-a^2}\,.$$

Before we consider these functions we study the mapping of the complex variable z.

Let p' and p'' be, respectively, the real and imaginary parts of p,

$$p = p' + ip''. \tag{3.4-1}$$

The complex variable

$$z = x_1 + px_2 = (x_1 + p'x_2) + ip''x_2 \tag{3.4-2}$$

is a mapping from the (x_1,x_2)-plane to the z-plane. The mapping can be regarded as a deformation. The (x_1,x_2)-plane is sheared in the x_1-direction with the shear strain p' and stretched (or contracted) in the x_2-direction with the stretch ratio p''. Hence the x_1-axis does not move. Straight lines deform into straight lines and parallel lines deform into parallel lines. The point (0,1) is moved to (p',p''), Fig. 3.1. Any point other than (0,1) is moved in the direction parallel to the movement of (0,1) to (p',p''). The magnitude of the movement is proportional to its distance from the x_1-axis. Thus a straight line is rotated (and stretched or contracted) about its intersection with the x_1-axis. Clearly, a square is mapped into a parallelogram. A unit circle with its center at the origin is mapped to an ellipse with the vectors $\mathbf{e}_1^{\circ}$, $\mathbf{e}_2^{\circ}$ shown in Fig. 3.1 being the conjugate radii (Section 1.2).

In the polar coordinate system

$$x_1 = r\cos\theta, \quad x_2 = r\sin\theta, \tag{3.4-3}$$

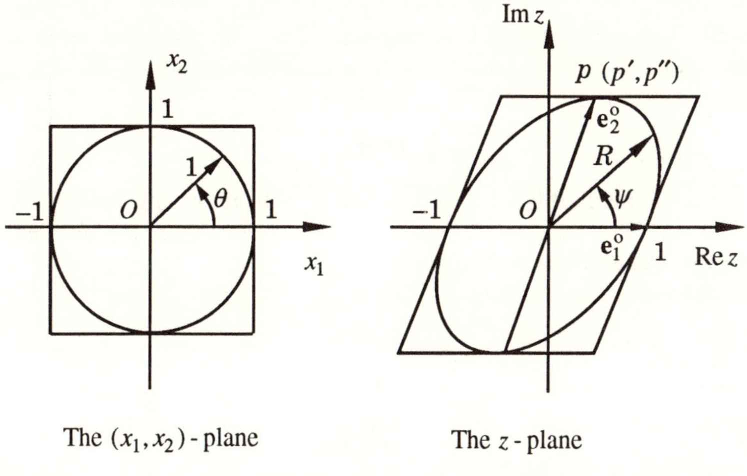

Fig. 3.1 $z=x_1+px_2$ where $p=p'+ip''$ transforms a unit circle to an ellipse.

(3.4-2) is written as

$$z = r\{(\cos\theta + p'\sin\theta) + i(p''\sin\theta)\} = rRe^{i\psi} \tag{3.4-4}$$

where the new polar coordinate (R,ψ) in the z-plane is

$$R\cos\psi = \cos\theta + p'\sin\theta, \quad R\sin\psi = p''\sin\theta. \tag{3.4-5}$$

Equations (3.4-5) show that a unit circle in the (x_1,x_2)-plane is mapped into an ellipse in the z-plane. The ellipse reduces to a circle when $p=i$. It is readily verified that

$$\left.\begin{aligned} &R=1, \quad \psi=0, \quad at\ \theta=0,\\ &R=\sqrt{p'^2+p''^2}, \quad \psi=\pm\frac{\pi}{2}-\tan^{-1}\left(\frac{p'}{p''}\right), \quad at\ \theta=\pm\frac{\pi}{2},\\ &R=1, \quad \psi=\pm\pi, \quad at\ \theta=\pm\pi. \end{aligned}\right\} \tag{3.4-6}$$

From (3.4-4) we obtain

$$\ln z = \ln(rR) + i\psi. \tag{3.4-7}$$

This means that, by (3.4-6),

$$\ln z = \begin{cases} \ln r, & \text{at } \theta = 0,\\ \ln r \pm i\pi, & \text{at } \theta = \pm\pi, \end{cases} \tag{3.4-8}$$

and hence

$$(\ln z)_{\theta=\pi} - (\ln z)_{\theta=-\pi} = 2\pi i. \tag{3.4-9}$$

It should be noted that (3.4-8) and (3.4-9) are independent of p. The function $\ln z$ is a multi-valued function which increases its value by $2\pi i$ each time θ increases by 2π. In order to maintain a unique value for $\ln z$, we introduce a *branch cut* along the negative x_1-axis so that the range of θ is limited to

$$-\pi \le \theta \le \pi. \tag{3.4-10}$$

We could instead of (3.4-10) introduce a branch cut in any direction. With (3.4-10) $\ln z$ is continuous in the (x_1,x_2)-plane except across the negative x_1-axis.

From (3.4-4) the square root of z is

$$\sqrt{z} = \sqrt{rR}e^{i\psi/2}. \tag{3.4-11}$$

Therefore by (3.4-6)

$$\sqrt{z} = \begin{cases} \sqrt{r} & \text{at } \theta = 0,\\ \pm i\sqrt{r} & \text{at } \theta = \pm\pi. \end{cases} \tag{3.4-12}$$

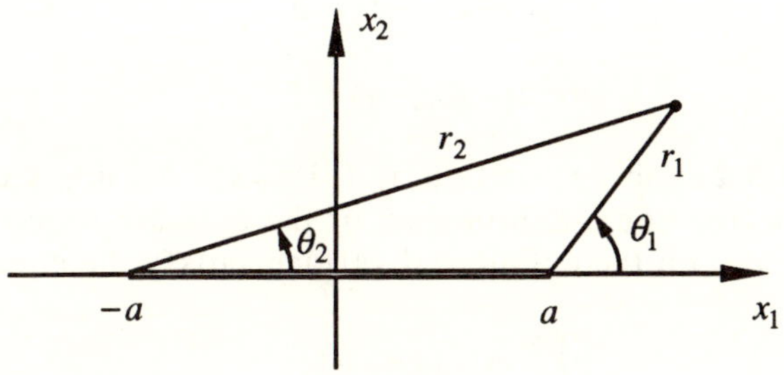

Fig. 3.2 The complex function $\sqrt{z^2-a^2}$ requires a branch cut $-a<x_1<a$ on the x_1-axis.

If we introduce a branch cut along the negative x_1-axis, $\sqrt{z}$ is continuous everywhere except across the branch cut.

Consider now the function

$$\sqrt{z^2-a^2}=\sqrt{z-a}\sqrt{z+a}. \tag{3.4-13}$$

Employing the notations r_1, θ_1, r_2, θ_2, depicted in Fig. 3.2 and using (3.4-11), (3.4-13) can be written as

$$\sqrt{z^2-a^2}=\sqrt{r_1R_1r_2R_2}\,e^{\frac{i}{2}(\psi_1+\psi_2)}$$

where, by (3.4-5),

$$R_k\cos\psi_k=\cos\theta_k+p'\sin\theta_k,\quad R_k\sin\psi_k=p''\sin\theta_k,\quad (k=1,2).$$

With θ_1, θ_2 in the range defined in (3.4-10) and by virtue of (3.4-6)

$$\sqrt{z^2-a^2}=\begin{cases}\pm\sqrt{x_1^2-a^2}\,, & \text{at } x_2=0,\ \ \pm x_1>a,\\ \pm i\sqrt{a^2-x_1^2}\,, & \text{at } x_2=\pm 0,\ \ |x_1|<a.\end{cases} \tag{3.4-14}$$

The function $\sqrt{z^2-a^2}$ is continuous everywhere except across the branch cut $x_2=0$, $|x_1|<a$. Again, it is important to note that (3.4-12) and (3.4-14) are independent of p.

3.5 Green's Functions for Infinite Space and Half-Space

Consider an infinite anisotropic elastic material subjected to a line of constant force f per unit length acting on, and in the direction of, the x_3-axis. In addition there is a screw dislocation of magnitude b along the x_3-axis. Taking the negative x_1-axis as the slip plane of the screw dislocation, the boundary conditions for the problem at hand are

$$u(r,\pi)-u(r,-\pi)=\mathrm{b}, \tag{3.5-1}$$

$$\phi(r,\pi)-\phi(r,-\pi)=\mathrm{f}. \tag{3.5-2}$$

Equation (3.5-2) can be verified as follows. Taking a cylinder of radius r as a free body, equilibrium of the cylinder requires that the shear stress σ_{3r} on the cylindrical surface satisfy the equation

$$\int_{-\pi}^{\pi}\sigma_{3r}r d\theta=-\mathrm{f}. \tag{3.5-3}$$

From (3.3-16) we have

$$\sigma_{3r}=-\frac{d\phi}{rd\theta}$$

so that (3.5-3) leads to (3.5-2).

For the solution (3.3-14) to satisfy (3.5-1) and (3.5-2), the function $f(z)$ must increase its value when θ increases by 2π. The complex function ln$z$ discussed in the previous section has the desired property. Therefore the solution is

$$\phi+i\mu u=\frac{1}{2\pi i}q^{\infty}\ln z \tag{3.5-4}$$

where q^{∞} is a constant which, upon application of (3.5-1) and (3.5-2) and use of (3.4-9), has the expression

$$q^{\infty}=\mathrm{f}+i\mu\mathrm{b}. \tag{3.5-5}$$

Equating the real and imaginary parts on both sides of (3.5-4) we have

$$u=\frac{-1}{2\pi\mu}\mathrm{Re}\{q^{\infty}\ln z\},\quad \phi=\frac{1}{2\pi}\mathrm{Im}\{q^{\infty}\ln z\} \tag{3.5-6}$$

or, by (3.4-7) and (3.5-5),

$$u=\frac{1}{2\pi}\left\{\mathrm{b}\psi-\frac{\mathrm{f}}{\mu}\ln(rR)\right\},\quad \phi=\frac{1}{2\pi}\{\mathrm{f}\psi+\mu\mathrm{b}\ \ln(rR)\}. \tag{3.5-7}$$

With (3.5-5), (3.5-4) or (3.5-6) is the Green's function for an infinite space subjected to an antiplane force f and a screw dislocation b at the origin r=0. Equations (3.5-7) provide a real form expression for the Green's function. Had we chosen a slip plane not on the negative x_1-axis, u and ϕ differ by constants in some regions of the (x_1,x_2)-plane.

Two solutions of the form (3.5-4) can be employed to construct Green's functions for the half-space $x_2\geq 0$ subjected to an antiplane force f and a screw dislocation b located at

$$x_1 = 0, \quad x_2 = d > 0. \tag{3.5-8}$$

The surface x_2=0 can be traction-free or rigidly clamped. Consider the solution

$$\phi + i\mu u = \frac{1}{2\pi i}\left\{q^\infty \ln(z - pd) + q \ln(z - \bar{p}d)\right\} \tag{3.5-9}$$

where q^∞ is given in (3.5-5) and q is a constant to be determined. The first term on the right is the Green's function for the infinite space with the antiplane force f and the screw dislocation b applied at the location (3.5-8). The second term is also the Green's function for the infinite space whose antiplane force and screw dislocation are located outside the half-space as we will show below. Therefore the antiplane force and screw dislocation of the second term are the *image singularities*.

If x_2=0 is a traction-free surface, ϕ=0 at x_2=0. The real part of (3.5-9) for x_2=0 is

$$0 = \text{Im}\left\{q^\infty \ln(x_1 - pd)\right\} + \text{Im}\left\{q \ln(x_1 - \bar{p}d)\right\}. \tag{3.5-10}$$

The following relations

$$\text{Re}\{f(z)\} = \text{Re}\{\bar{f}(\bar{z})\}, \quad \text{Im}\{f(z)\} = -\text{Im}\{\bar{f}(\bar{z})\}, \tag{3.5-11}$$

which are trivial, will be useful in the sequel. Application of (3.5-11)$_2$ to the first term on the right of (3.5-10) yields

$$0 = -\text{Im}\left\{\bar{q}^\infty \ln(x_1 - \bar{p}d)\right\} + \text{Im}\left\{q \ln(x_1 - \bar{p}d)\right\}.$$

Therefore (Wu, 1991c)

$$q = \bar{q}^\infty = \text{f} - i\mu\text{b}. \tag{3.5-12}$$

If x_2=0 is a rigid (or clamped) surface, u=0 at x_2=0. A similar derivation with the use of (3.5-11)$_1$ gives

$$q = -\bar{q}^\infty = -\text{f} + i\mu\text{b}. \tag{3.5-13}$$

From (3.5-12) and (3.5-13) we conclude that the antiplane force and the screw dislocation for the image singularities are (f,–b) when x_2=0 is a free surface and (–f,b) when x_2=0 is a rigid surface. The location $(\hat{x}_1, \hat{x}_2)$ of the image singularities is determined by

$$\hat{z} - \bar{p}d = 0, \quad \text{or} \quad \hat{x}_1 + p\hat{x}_2 - \bar{p}d = 0, \tag{3.5-14}$$

from which we have

$$\hat{x}_1 = 2p'd, \quad \hat{x}_2 = -d. \tag{3.5-15}$$

Since $\hat{x}_2<0$ the image singularities are located outside the half-space. It is the point A in Fig. 3.3. Its location is not the mirror image of D where the real singularities are located unless $p'=0$. Notice that p'', which is the imaginary part of p, plays no roles in (3.5-15).

According to Fig. 3.1 and the discussion following (3.4-2), the point D is mapped to pd while the point A is mapped to $\bar{p}d$. Hence A and D are *mirror images* of each other in the z-plane. From the properties of the mapping discussed earlier, the line connecting A to $\bar{p}d$ is parallel to the line connecting D to pd, and the lines connecting D to A and pd to $\bar{p}d$ intersect at the x_1-axis. These suggest a graphical solution for locating the image singularity A. We will see the graphical solution in more complicated problems later in Section 3.6 and in Chapter 8. It should be noted that the location of image singularities depends only on elastic constants. It does not depend on the nature of the singularities nor on the boundary conditions at $x_2=0$ (Ting, 1992b).

The complex form solution (3.5-9) can be converted into a real form. We will consider the conversion in the next section when we study a more general case of the Green's function for bimaterials.

Consider the half-space that is subjected to the screw dislocation b only. The *image force* or the Peach–Koehler (1950) force F

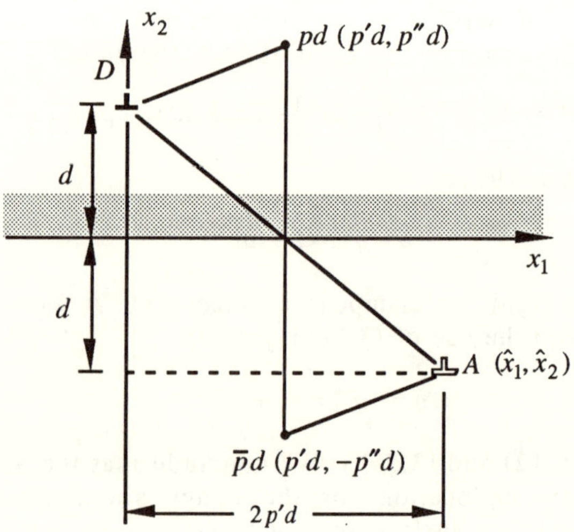

Fig. 3.3 The half-space is subjected to an antiplane force f and a screw dislocation b at the point D. The image singularities are at the point A. With the complex variable $z=x_1+px_2$, D is mapped to pd while A is mapped to $\bar{p}d$. pd and $\bar{p}d$ are mirror images of each other.

exerted on the dislocation is

$$F = -\mathrm{b}\sigma_{31}^{d} = \mathrm{b}\phi_{,2}^{\mathrm{d}} \tag{3.5-16}$$

where σ_{31}^{d} is the stress derived from the second term in $(3.5\text{-}9)_2$ evaluated at $(x_1,x_2)=(0,d)$. F is in the direction of the positive x_2-axis if it is positive. We have

$$F = \frac{\mathrm{b}}{2\pi d}\mathrm{Im}\frac{pq}{p-\overline{p}} = \frac{\mathrm{b}}{4\pi d}\mathrm{Im}\frac{(p'+ip'')q}{ip''}. \tag{3.5-17}$$

If the surface $x_2=0$ is traction-free, insertion of (3.5-12) with f=0 yields

$$F = -\frac{\mu \mathrm{b}^2}{4\pi d} \tag{3.5-18}$$

which is negative. The image force attracts the dislocation to the free surface. If $x_2=0$ is a rigid surface, use of (3.5-13) with f=0 gives us

$$F = \frac{\mu \mathrm{b}^2}{4\pi d} \tag{3.5-19}$$

which is positive. The image force is repulsive for a fixed boundary. These agree with the special cases of isotropic materials (Head, 1953; Dundurs, 1969; Mura, 1987).

The *surface Green's function* for the half-space refers to the solution for the half-space with the surface $x_2=0$ being traction-free and subjected to an antiplane force f at the surface $x_1=x_2=0$. The solution can be obtained by letting $d=0$ in (3.5-9) and b=0 in (3.5-12). We have

$$\phi + i\mu u = \frac{\mathrm{f}}{\pi i}\ln z, \tag{3.5-20a}$$

or

$$u = \frac{-\mathrm{f}}{\mu\pi}\mathrm{Re}(\ln z) = \frac{-\mathrm{f}}{\mu\pi}\ln(rR), \quad \phi = \frac{\mathrm{f}}{\pi}\mathrm{Im}(\ln z) = \frac{\mathrm{f}}{\pi}\psi. \tag{3.5-20b}$$

In closing this section we point out that (3.5-9) with the q given by $(3.5\text{-}12)_1$ or $(3.5\text{-}13)_1$ can be rewritten in a compact form with the use of (3.5-11). If $x_2=0$ is a free surface,

$$\begin{aligned} u &= \frac{-1}{2\pi\mu}\mathrm{Re}\{q^{\infty}\ln[(z-pd)(\overline{z}-pd)]\}, \\ \phi &= \frac{1}{2\pi}\mathrm{Im}\{q^{\infty}\ln[(z-pd)(\overline{z}-pd)^{-1}]\}. \end{aligned} \tag{3.5-21}$$

If $x_2=0$ is a rigid surface,

$$u = \frac{-1}{2\pi\mu}\mathrm{Re}\{q^{\infty}\ln[(z-pd)(\bar{z}-pd)^{-1}]\},$$
$$\phi = \frac{1}{2\pi}\mathrm{Im}\{q^{\infty}\ln[(z-pd)(\bar{z}-pd)]\}. \tag{3.5-22}$$

3.6 Green's Function for Bimaterials

Let two dissimilar anisotropic elastic half-spaces be bonded together along $x_2=0$. The materials in $x_2>0$ ($\pi>\theta>0$) and in $x_2<0$ ($-\pi<\theta<0$) are designated as material 1 and material 2, respectively. An antiplane force f and a screw dislocation of magnitude b are applied at $(x_1,x_2)=(0,d)$, $d>0$, which is located in material 1. Following (3.5-9) the solution for material 1 can be written as

$$\phi^{(1)}+i\mu_1 u^{(1)} = \frac{1}{2\pi i}\{q_1^{\infty}\ln(z^{(1)}-p^{(1)}d)+q_1\ln(z^{(1)}-\bar{p}^{(1)}d)\} \tag{3.6-1}$$

and that for material 2 as

$$\phi^{(2)}+i\mu_2 u^{(2)} = \frac{1}{2\pi i}\{q_2\ln(z^{(2)}-p^{(1)}d)\}. \tag{3.6-2}$$

The superscripts (1), (2) and the subscripts 1, 2 refer to materials 1 and 2, respectively,

$$z^{(k)} = x_1 + p^{(k)}x_2, \quad k=1,2, \tag{3.6-3}$$

q_1, q_2 are constants to be determined, and

$$q_1^{\infty} = \mathrm{f} + i\mu_1\mathrm{b}.$$

The continuity of displacements and surface tractions at $x_2=0$ mean that

$$u^{(1)} = u^{(2)}, \quad \phi^{(1)} = \phi^{(2)}, \quad \text{at } x_2=0. \tag{3.6-4}$$

Substitution of (3.6-1) and (3.6-2) into (3.6-4), and use of (3.5-11) yields

$$(\mu_2/\mu_1)(q_1^{\infty}+\bar{q}_1) = q_2, \quad q_1^{\infty}-\bar{q}_1 = q_2. \tag{3.6-5}$$

Hence

$$q_1 = -K\bar{q}_1^{\infty}, \quad q_2 = (1+K)q_1^{\infty}, \tag{3.6-6}$$

where

$$K = \frac{\mu_2-\mu_1}{\mu_2+\mu_1}, \quad -1<K<1. \tag{3.6-7}$$

With (3.6-6), (3.6-1) and (3.6-2) are the Green's functions for the bimaterial with an antiplane force f and a screw dislocation b applied at $(x_1,x_2)=(0,d)$ (Wu, 1991c).

The first term of (3.6-1) is the Green's function for the infinite space occupied by material 1. The singularities of the first term are located at $(x_1,x_2)=(0,d)$. They are the antiplane force f and the screw dislocation b. The singularities of the second term are at the location given by (3.5-15). They are located at $x_2<0$ (the point A in Fig. 3.4) and represent *image singularities* for material 1. The magnitudes of the antiplane force $f^{(1)}$ and the screw dislocation $b^{(1)}$ are, after setting $q_1^\infty = f + i\mu_1 b$ in $(3.6\text{-}6)_1$,

$$f^{(1)} = -Kf, \qquad b^{(1)} = Kb. \tag{3.6-8}$$

The solution (3.6-2) for material 2 is the Green's function for an infinite space occupied by material 2. The location $(\hat{x}_1,\hat{x}_2)$ of the singularities are determined by

$$\hat{z}^{(2)} - p^{(1)}d = 0, \quad \text{or} \quad \hat{x}_1 + p^{(2)}\hat{x}_2 - p^{(1)}d = 0, \tag{3.6-9a}$$

from which we obtain

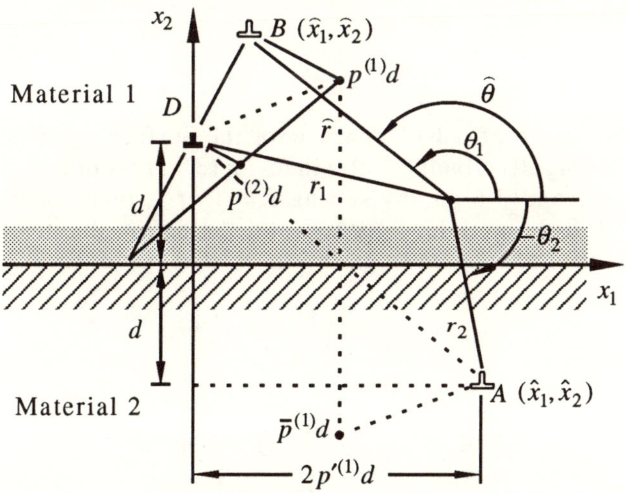

Fig. 3.4 The bimaterial is subjected to an antiplane force f and a screw dislocation b at the point D. The image singularities for material 1 are located at A while the image singularities for material 2 are located at B. The points D, A are mapped to $p^{(1)}d$, $\bar{p}^{(1)}d$, respectively, by the complex variable $z=x_1+p^{(1)}x_2$ while B is mapped to $p^{(1)}d$ by the complex variable $z=x_1+p^{(2)}x_2$. $p^{(1)}d$ and $\bar{p}^{(1)}d$ are mirror images of each other.

$$\frac{\hat{x}_1}{d} = p'^{(1)} - \frac{p''^{(1)}}{p''^{(2)}} p'^{(2)}, \quad \frac{\hat{x}_2}{d} = \frac{p''^{(1)}}{p''^{(2)}}. \tag{3.6-9b}$$

Since $p''^{(1)}$ and $p''^{(2)}$ are positive and nonzero, $\hat{x}_2>0$. This assures us that the image singularities for material 2 are located outside material 2. They are at the point B in Fig. 3.4. The magnitudes of the antiplane force $\mathrm{f}^{(2)}$ and the screw dislocation $\mathrm{b}^{(2)}$ are, by (3.6-6)$_2$,

$$\mathrm{f}^{(2)} = (1+K)\mathrm{f}, \quad \mathrm{b}^{(2)} = (1-K)\mathrm{b}. \tag{3.6-10}$$

According to the discussion of Fig. 3.3 $p^{(1)}d$ and $\bar{p}^{(1)}d$ are, respectively, the mapping of D and A from the (x_1,x_2)-plane to the z-plane through the complex variable $x_1+p^{(1)}x_2$. By (3.6-9a) the point $p^{(1)}d$ is the mapping of B from the (x_1,x_2)-plane to the z-plane through the complex variable $x_1+p^{(2)}x_2$. Therefore the real and image singularities for materials 1 and 2 are *mirror images* of each other in the z-plane. Following the graphical solution of point A depicted in Fig. 3.3 it can be shown that the lines connecting B to $p^{(1)}d$ and D to $p^{(2)}d$ are parallel to each other, and the lines connecting B to D and $p^{(1)}d$ to $p^{(2)}d$ intersect at the x_1-axis.

The image force F that exerts on the dislocation is given by (3.5-16) in which σ_{31}^d is the stress determined from the second term in (3.6-1) evaluated at $(x_1,x_2)=(0,d)$. It can be shown that F is given by (3.5-17) if p and q are replaced by $p^{(1)}$ and q_1, respectively. Inserting q_1 from (3.6-6)$_1$ with f=0 in (3.5-17) yields

$$F = \frac{\mu_1 K}{4\pi d}\mathrm{b}^2. \tag{3.6-11}$$

The special cases of a half-space with the surface $x_2=0$ being traction-free or rigidly clamped obtained in (3.5-18) and (3.5-19) can be deduced from (3.6-11) by setting $K=-1$ or 1, respectively.

When $d=0$ the antiplane force f and the screw dislocation b are on the interface. By letting $d=0$ in (3.6-1) and (3.6-2), it can shown that

$$\phi^{(k)} + i\mu_k u^{(k)} = \frac{1}{2\pi i} q^{(k)} \ln z^{(k)} \tag{3.6-12}$$

for k=1,2 where

$$\begin{aligned} q^{(1)} &= q_1^\infty + q_1 = (1-K)(\mathrm{f} + i\mu_2 \mathrm{b}), \\ q^{(2)} &= q_2 = (1+K)(\mathrm{f} + i\mu_1 \mathrm{b}). \end{aligned} \tag{3.6-13}$$

The complex form solution (3.6-1) and (3.6-2) can be converted into a real form. Making use of (3.4-7) we have

$$\ln(z^{(1)} - p^{(1)}d) = \ln(r_1 R_1) + i\psi_1,$$

$$\ln(z^{(1)} - \bar{p}^{(1)}d) = \ln(r_2 R_2) + i\psi_2,$$

$$\ln(z^{(2)} - p^{(1)}d) = \ln(\hat{r}\hat{R}) + i\hat{\psi},$$

where, by (3.4-5),

$$R_k \cos\psi_k = \cos\theta_k + p'^{(1)}\sin\theta_k, \quad R_k \sin\psi_k = p''^{(1)}\sin\theta_k,$$

$$\hat{R}\cos\hat{\psi} = \cos\hat{\theta} + p'^{(2)}\sin\hat{\theta}, \quad \hat{R}\sin\hat{\psi} = p''^{(2)}\sin\hat{\theta},$$

and r_1, r_2, $\hat{r}$, θ_1, θ_2, $\hat{\theta}$ are defined in Fig. 3.4. With q_1, q_2 given by (3.6-6) the real and imaginary parts of (3.6-1) have the expressions

$$\begin{aligned} \phi^{(1)} &= \frac{\mathrm{f}}{2\pi}(\psi_1 - K\psi_2) + \frac{\mu_1 \mathrm{b}}{2\pi}[\ln(r_1 R_1) + K\ln(r_2 R_2)], \\ u^{(1)} &= \frac{-\mathrm{f}}{2\pi\mu_1}[\ln(r_1 R_1) - K\ln(r_2 R_2)] + \frac{\mathrm{b}}{2\pi}(\psi_1 + K\psi_2), \end{aligned} \tag{3.6-14}$$

while the real and imaginary parts of (3.6-2) are

$$\begin{aligned} \phi^{(2)} &= \frac{\mathrm{f}}{2\pi}(1-K)\hat{\psi} + \frac{\mu_2 \mathrm{b}}{2\pi}(1-K)\ln(\hat{r}\hat{R}), \\ u^{(2)} &= \frac{-\mathrm{f}}{2\pi\mu_2}(1+K)\ln(\hat{r}\hat{R}) + \frac{\mathrm{b}}{2\pi}(1-K)\hat{\psi}. \end{aligned} \tag{3.6-15}$$

The Green's function for the half-space given in (3.5-9) can be converted into a real form from (3.6-14) by setting $K=-1$ if $x_2=0$ is a free boundary, and $K=1$ if $x_2=0$ is a fixed boundary. When the bimaterial is subjected to the screw dislocation b only, the displacements are given by

$$u^{(1)} = \frac{\mathrm{b}}{2\pi}(\psi_1 + K\psi_2), \quad u^{(2)} = \frac{\mathrm{b}}{2\pi}(1-K)\hat{\psi}.$$

In the special case of isotropic materials $\psi_1 = \hat{\psi} = \hat{\theta} = \theta_1$ and $\psi_2 = \theta_2$ so that the solution recovers the one obtained by Dundurs (1969). It should be noted that θ_1, θ_2 defined here differ from θ_1, θ_2 defined by Dundurs (1969) by $\pi/2$.

3.7 Mapping of an Ellipse to a Circle for $z=x_1+px_2$

When the boundary of an anisotropic elastic material has the shape of an ellipse it is more convenient to transform the ellipse to a circle before solving the problem. The mapping of an ellipse to a circle is well-known for the complex variable $z=x_1+px_2$ when $p=i$, but little is known when $p\neq i$ (see Lekhnitskii, 1950; Hwu and Yen, 1993).

Let the geometry of the ellipse Γ be given by

$$\Gamma: \quad x_1 = a\cos\psi, \quad x_2 = b\sin\psi, \tag{3.7-1}$$

where ψ is a real parameter and $a\geq b>0$ are the principal radii of the ellipse. The italic b here should not be confused with the screw dislocation b. When $a=b$ the ellipse is a circle. We will see that no simplification is gained by considering $a=b$ when $p\neq i$.

Consider the mapping

$$z = c\zeta + d\zeta^{-1} \tag{3.7-2}$$

where c and d are constants to be chosen such that the ellipse is mapped into a circle of unit radius. Thus when x_1 and x_2 are given by (3.7-1), $\zeta=e^{i\psi}$. Hence

$$a\cos\psi + pb\sin\psi = c(\cos\psi + i\sin\psi) + d(\cos\psi - i\sin\psi)$$

from which we obtain

$$c = \tfrac{1}{2}(a-ipb), \quad d = \tfrac{1}{2}(a+ipb). \tag{3.7-3}$$

Notice that when $a=b$ we still need the mapping (3.7-2) unless $p=i$. The mapping is unnecessary when $p=ia/b$ so that $c=a$ and $d=0$. One of the solutions for ζ in (3.7-2) is

$$\zeta = \frac{z+\sqrt{z^2-(a^2+p^2b^2)}}{a-ipb}. \tag{3.7-4}$$

This solution has the property that $|\zeta|\to\infty$ as $|z|\to\infty$. The other solution denoted by τ is

$$\tau = \frac{z-\sqrt{z^2-(a^2+p^2b^2)}}{a-ipb}. \tag{3.7-5}$$

This solution gives $|\tau|\to 0$ as $|z|\to\infty$. Since the inverse ζ^{-1} of (3.7-4) is

$$\zeta^{-1} = \frac{z-\sqrt{z^2-(a^2+p^2b^2)}}{a+ipb}, \tag{3.7-6}$$

(3.7-5) can be written as

$$\tau = t\zeta^{-1} \tag{3.7-7}$$

where

$$t = \frac{d}{c} = \frac{a+ipb}{a-ipb}, \quad |t|<1. \tag{3.7-8}$$

Letting $p=p'+ip''$ where p', p'' are real, the inequality follows from

$$\frac{|d|^2}{|c|^2} = \frac{|a+ipb|^2}{|a-ipb|^2} = \frac{(a-p''b)^2+(p'b)^2}{(a+p''b)^2+(p'b)^2} < 1 \tag{3.7-9}$$

because a, b, and p'' are all positive and nonzero. Therefore every point $Z=(x_1,x_2)$ in the (x_1,x_2)-plane is mapped to two points ζ and τ given by (3.7-4) and (3.7-7) in the ζ-plane. In particular, the point $(x_1,x_2)=(a,0)$ in the (x_1,x_2)-plane is mapped to $\zeta=1$ and t in the ζ-plane, Fig. 3.5.

The mapping (3.7-2) is one-to-one for points outside the ellipse in the (x_1,x_2)-plane to points outside the unit circle in the ζ-plane if the roots of

$$\frac{dz}{d\zeta} = c - d\zeta^{-2} = 0$$

are located inside the unit circle $|\zeta|<1$. The roots are $\pm\tilde{\zeta}$ where

$$\tilde{\zeta} = \sqrt{d/c} = \sqrt{t}, \quad |t|<|\tilde{\zeta}|<1. \tag{3.7-10}$$

The inequalities follow from (3.7-9) and $(3.7\text{-}10)_{1,2}$. $|t|=|\tilde{\zeta}|$ only when they both vanish. That $\pm\tilde{\zeta}$ are inside the unit circle assures us that the mapping is one-to-one for points outside the ellipse. However the mapping is *not conformal*. The angle between any two line elements in the (x_1,x_2)-plane is in general different from the angle between the corresponding elements in the ζ-plane.

The square root term in (3.7-4) requires a branch cut in the (x_1,x_2)-plane for the function to be single-valued. The branch points are at $\pm\tilde{z}$ where

$$\tilde{x}_1 + p\tilde{x}_2 = \tilde{z} = \sqrt{a^2+p^2b^2} = 2\sqrt{cd}. \tag{3.7-11}$$

The first equality allows us to compute the physical branch points $\tilde{Z}=(\tilde{x}_1,\tilde{x}_2)$ and $-\tilde{Z}$. The last equality follows from (3.7-3). Insertion of $(3.7\text{-}10)_1$ into (3.7-2) also leads to (3.7-11). Thus the branch points $\pm\tilde{Z}$ are mapped to $\pm\tilde{\zeta}$ in the ζ-plane. Since $\pm\tilde{\zeta}$ are inside the unit circle, $\pm\tilde{Z}$ are inside the ellipse.

The relations between t, $\tilde{\zeta}$, and $\tilde{Z}$ can be given more explicitly as follows. Let

$$2d = a+ipb = (a-p''b)+ip'b = \rho_1^2 e^{2i\psi_1},$$

$$2c = a-ipb = (a+p''b)-ip'b = \rho_2^2 e^{-2i\psi_2},$$

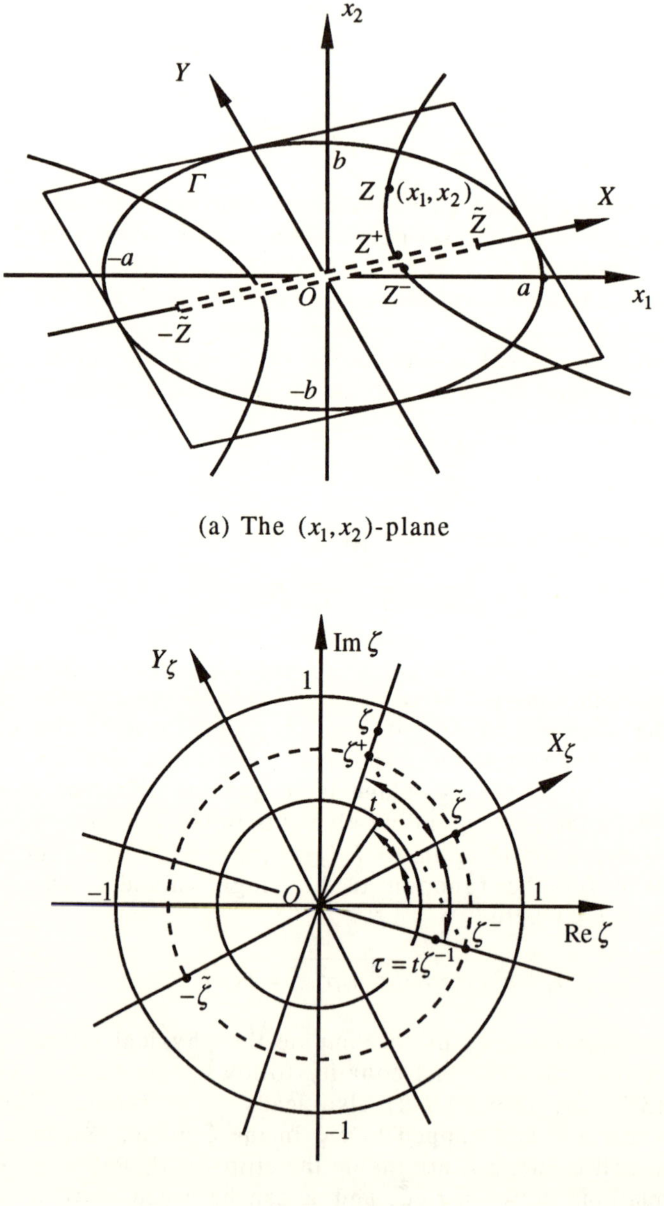

(a) The (x_1,x_2)-plane

(b) The ζ-plane

Fig. 3.5 The mapping $z=c\zeta+d\zeta^{-1}$ maps the entire (x_1,x_2)-plane to the region outside or inside the $\tilde{\zeta}$-circle in the ζ-plane. Concentric circles and radial lines in the ζ-plane are confocal ellipses and hyperbolas in the (x_1,x_2)-plane when referred to the oblique coordinates (X,Y).

where ρ_1, ρ_2, ψ_1, ψ_2 are real and

$$0 \le 2\psi_1 \le \pi, \quad 0 \le 2\psi_2 \le \pi/2, \quad \text{if } p' > 0,$$

$$0 \ge 2\psi_1 \ge -\pi, \quad 0 \ge 2\psi_2 \ge -\pi/2, \quad \text{if } p' < 0,$$

$$|\psi_1| \ge |\psi_2|.$$

We then have

$$t = d/c = (\rho_1/\rho_2)^2 e^{2i(\psi_1+\psi_2)},$$

$$\tilde{\zeta} = \sqrt{d/c} = (\rho_1/\rho_2) e^{i(\psi_1+\psi_2)},$$

$$\tilde{z} = 2\sqrt{cd} = \rho_1\rho_2 e^{i(\psi_1-\psi_2)}.$$

The point $\tilde{Z}$=($\tilde{x}_1$, $\tilde{x}_2$) associated with $\tilde{z}$ is determined from

$$\tilde{x}_1 + p\tilde{x}_2 = \rho_1\rho_2 e^{i(\psi_1-\psi_2)}$$

or

$$\tilde{x}_1 + p'\tilde{x}_2 = \rho_1\rho_2 \cos(\psi_1 - \psi_2), \quad p''\tilde{x}_2 = \rho_1\rho_2 \sin(\psi_1 - \psi_2).$$

The locations of t, $\pm\tilde{\zeta}$, and $\pm\tilde{Z}$ are shown in Fig. 3.5 for p'>0. The polar angle of t is twice the polar angle of $\tilde{\zeta}$. The radial line $O\tilde{\zeta}$ not only bisects the polar angle of Ot, by virtue of (3.7-7) it also bisects the angle between the radial lines $O\zeta$ and $O\tau$.

After dividing by $2\sqrt{cd}$, (3.7-2) can be cast in a canonical form

$$\frac{z}{\tilde{z}} = \frac{1}{2}\left(\frac{\zeta}{\tilde{\zeta}} + \frac{\tilde{\zeta}}{\zeta}\right) \tag{3.7-12}$$

and (3.7-4) is written as

$$\frac{\zeta}{\tilde{\zeta}} = \frac{z}{\tilde{z}} + \sqrt{\left(\frac{z}{\tilde{z}}\right)^2 - 1}. \tag{3.7-13}$$

When ζ is replaced by $\tau = t\zeta^{-1}$, the right side of (3.7-12) remains the same because t=$\tilde{\zeta}^2$. This confirms the statement made earlier that every point Z is mapped to two points, ζ and τ. In the ζ-plane the circle that passes through $\tilde{\zeta}$ with its center at the origin is called the *$\tilde{\zeta}$-circle*. The points on two sides of the branch cut connecting $-\tilde{Z}$ and $\tilde{Z}$ in the (x_1,x_2)-plane are mapped to the $\tilde{\zeta}$-circle. Indeed, if we let $z = \tilde{z}\cos v$, $0 \le v \le 2\pi$, (3.7-13) reduces to

$$\zeta/\tilde{\zeta} = \cos v + i \sin v$$

which is the $\tilde{\zeta}$-circle in the ζ-plane. The points Z^+, Z^- on the opposite sides of the branch cut are mapped to ζ^+, ζ^- on the circle with the line connecting ζ^+, ζ^- perpendicular to the diameter connecting $-\tilde{\zeta}$ to $\tilde{\zeta}$.

The circle that passes through the point t with its center at the origin in the ζ-plane is called the *t-circle*. The transformation (3.7-4) maps a point outside or inside the ellipse in the (x_1,x_2)-plane, respectively, to a point outside the unit circle or inside the annular region between the unit circle and the $\tilde{\zeta}$-circle in the ζ-plane. The transformation (3.7-5) maps a point outside or inside the ellipse in the (x_1,x_2)-plane, respectively, to a point inside the t-circle or inside the annular region between the t-circle and the $\tilde{\zeta}$-circle in the ζ-plane. Therefore the entire (x_1,x_2)-plane with the branch cut is mapped to the region outside or inside the $\tilde{\zeta}$-circle in the ζ-plane. The (x_1,x_2)-plane can be regarded as two infinite sheets, one on top of the other, with the same branch cut. The positive side of the branch cut on the top sheet is connected to the negative side of the branch cut on the bottom sheet and vice versa. The two sheets which cross each other at the branch cut form a *Riemann surface* (Churchill, 1948). One can move continuously from the top sheet to the bottom sheet and back to the top sheet through the connection at the branch cut. If the top sheet is mapped into the region outside the $\tilde{\zeta}$-circle, the bottom sheet is mapped into the region inside the $\tilde{\zeta}$-circle.

We will show that concentric circles with the center at the origin and radial lines in the ζ-plane are ellipses and hyperbolas in the (x_1,x_2)-plane. Let

$$\zeta = \eta e^{i\psi}, \quad \tilde{\zeta} = \tilde{\eta} e^{i\tilde{\psi}}, \tag{3.7-14}$$

$$\begin{aligned} z &= x_1 + p x_2 = (x_1 + p' x_2) + i p'' x_2 = R e^{i\sigma}, \\ \tilde{z} &= \tilde{x}_1 + p \tilde{x}_2 = (\tilde{x}_1 + p' \tilde{x}_2) + i p'' \tilde{x}_2 = \tilde{R} e^{i\tilde{\sigma}}, \end{aligned} \tag{3.7-15}$$

where η, $\tilde{\eta}$, ψ, $\tilde{\psi}$, R, $\tilde{R}$, σ, and $\tilde{\sigma}$ are real. The real and imaginary parts of the canonical form (3.7-12) are

$$\begin{aligned} \frac{X}{s} &= \frac{R}{\tilde{R}}\cos(\sigma - \tilde{\sigma}) = \frac{1}{2}\left(\frac{\eta}{\tilde{\eta}} + \frac{\tilde{\eta}}{\eta}\right)\cos(\psi - \tilde{\psi}), \\ \frac{Y}{s} &= \frac{R}{\tilde{R}}\sin(\sigma - \tilde{\sigma}) = \frac{1}{2}\left(\frac{\eta}{\tilde{\eta}} - \frac{\tilde{\eta}}{\eta}\right)\sin(\psi - \tilde{\psi}), \end{aligned} \tag{3.7-16}$$

where

$$s^2 = \tilde{x}_1^2 + \tilde{x}_2^2.$$

Thus $2s$ is the length of the branch cut. X and Y can be expressed in terms of x_1 and x_2 by writing the real and imaginary parts of z in

(3.7-15) as

$$R\cos\sigma = x_1 + p'x_2, \quad R\sin\sigma = p''x_2.$$

It follows from (3.7-16) that

$$X = (s/\tilde{R})[(x_1 + p'x_2)\cos\tilde{\sigma} + p''x_2\sin\tilde{\sigma}],$$

$$Y = (s/\tilde{R})[p''x_2\cos\tilde{\sigma} - (x_1 + p'x_2)\sin\tilde{\sigma}].$$

Elimination of ψ or η in (3.7-16) leads to

$$\frac{X^2}{A^2} + \frac{Y^2}{B^2} = 1, \tag{3.7-17}$$

$$\frac{X^2}{\{s\cos(\psi - \tilde{\psi})\}^2} - \frac{Y^2}{\{s\sin(\psi - \tilde{\psi})\}^2} = 1, \tag{3.7-18}$$

where

$$A = \frac{s}{2}\left(\frac{\eta}{\tilde{\eta}} + \frac{\tilde{\eta}}{\eta}\right), \quad B = \frac{s}{2}\left(\frac{\eta}{\tilde{\eta}} - \frac{\tilde{\eta}}{\eta}\right).$$

These equations show that the circles η=constants and the radial lines ψ=constants in the ζ-plane are, respectively, ellipses and hyperbolas in the (x_1,x_2)-plane. Referred to the oblique coordinate system (X,Y), the ellipses and hyperbolas are confocal with the focal points at $-\tilde{Z}$ and $\tilde{Z}$. The X-axis is along the branch cut while the Y-axis is in general not perpendicular to the X-axis. The portions of the X- and Y-axes inside the ellipse are a pair of conjugate diameters of the ellipse (Section 1.8). From (3.7-16) $Y=0$ when $\psi=\tilde{\psi}$ and $X=0$ when $\psi=\tilde{\psi}+\pi/2$. Therefore the mapping of the X- and Y-axes in the ζ-plane are the X_ζ- and Y_ζ-axes which are perpendicular to each other.

One could visualize how ellipses and hyperbolas in the (x_1,x_2)-plane are mapped to circles and radial lines in the ζ-plane from the following observation. The mapping (3.7-2) can be accomplished in two steps. In the first step we map the (x_1,x_2)-plane to the z-plane. According to Fig. 3.1 and the discussion following (3.4-2), this mapping can be regarded as applying a shear deformation in the x_1-direction with the shear strain p' and an extension or contraction in the x_2-direction with the stretch ratio p''. The ellipses and hyperbolas deform into new ellipses and hyperbolas. The ellipse Γ with the principal radii a, b in the (x_1,x_2)-plane is mapped into a new ellipse in the z-plane with $\mathbf{e}_1^{\circ}=(a,0)$, $\mathbf{e}_2^{\circ}=(bp',bp'')$ being the conjugate radii. The principal axes $\hat{\mathbf{e}}_1$ and $\hat{\mathbf{e}}_2$ of this new ellipse can be determined (Section 1.8). They are the mapping of the X-

and Y-axes from the (x_1,x_2)-plane. The second step is the mapping of the new ellipse in the z-plane to the unit circle in the ζ-plane. This step is the well-known *conformal mapping*. The ellipses and hyperbolas in the z-plane are mapped to circles and radial lines in the ζ-plane.

When one follows a radial line in the ζ-plane from infinity towards the origin, the corresponding path in the (x_1,x_2)-plane follows a hyperbola from infinity to the branch cut on the top sheet of the Riemann surface, and continues to the bottom sheet of the Riemann surface on the same hyperbola to infinity. The radial lines $O\zeta$ and $O\tau$ in Fig. 3.5(b) correspond to the same hyperbola in the (x_1,x_2)-plane, but on both sheets of the Riemann surface. As we stated earlier the point $(x_1,x_2)=(a,0)$ in the (x_1,x_2)-plane is mapped to $\zeta=1$ and t in the ζ-plane. Therefore the real ζ-axis and the radial line Ot in the ζ-plane correspond to the same hyperbola that passes through $(x_1,x_2)=(a,0)$ but on both sheets of the Riemann surface in the (x_1,x_2)-plane.

3.8 Infinite Space with an Elliptic Hole or an Elliptic Rigid Inclusion

We consider in this section the problems of an infinite anisotropic material with an elliptic hole or an elliptic rigid inclusion. The case when the ellipse is filled with an anisotropic elastic material will be discussed in the next two sections.

(i) **The Green's Functions.** Let the infinite space with an elliptic hole or an elliptic rigid inclusion be subjected to an antiplane force f and a screw dislocation b applied at

$$(x_1,x_2)=(x_1^{\text{o}},x_2^{\text{o}}) \tag{3.8-1}$$

as shown in Fig. 3.6. The notation b for screw dislocation should not be confused with the italic b for the minor principal axis of the ellipse. Following the solutions (3.5-9) for the half-space let (Hwu and Yen, 1991, 1992)

$$\phi+i\mu u=\frac{1}{2\pi i}\left\{q^{\infty}\ln(\zeta-\zeta_{\text{o}})+q\ln(\zeta^{-1}-\bar{\zeta}_{\text{o}})\right\} \tag{3.8-2}$$

where q^{∞} is given in (3.5-5), q is a constant to be determined, and

$$\zeta_{\text{o}}=\frac{z_{\text{o}}+\sqrt{z_{\text{o}}^2-(a^2+p^2b^2)}}{a-ipb},\quad z_{\text{o}}=x_1^{\text{o}}+px_2^{\text{o}}. \tag{3.8-3}$$

Since $(x_1^{\text{o}},x_2^{\text{o}})$ is outside the ellipse, $|\zeta_{\text{o}}|>1$. The first term on the right of (3.8-2) takes care of the prescribed singularities at (x_1^o, x_2^o). The second term is included to satisfy the boundary conditions at the elliptic boundary.

At the elliptic boundary Γ,

$$\zeta = e^{i\psi}, \quad \zeta^{-1} = e^{-i\psi}. \tag{3.8-4}$$

If the ellipse is a void ϕ=0 at Γ. The real part of (3.8-2) is

$$0 = \mathrm{Im}\{q^\infty \ln(e^{i\psi} - \zeta_o)\} + \mathrm{Im}\{q \ln(e^{-i\psi} - \bar{\zeta}_o)\}.$$

Using $(3.5\text{-}11)_2$ the first term can be replaced by the negative of its complex conjugate, and hence

$$q = \bar{q}^\infty = \mathrm{f} - i\mu\mathrm{b}. \tag{3.8-5}$$

If the ellipse is a rigid body u=0 at Γ. A similar analysis leads to

$$q = -\bar{q}^\infty = -\mathrm{f} + i\mu\mathrm{b}. \tag{3.8-6}$$

These are identical to (3.5-12) and (3.5-13), respectively.

Analogous to (3.5-21) and (3.5-22) we may rewrite (3.8-2) as

$$\left.\begin{aligned} u &= \frac{-1}{2\pi\mu}\mathrm{Re}\left\{q^\infty \ln\left((\zeta - \zeta_o)(\bar{\zeta}^{-1} - \zeta_o)\right)\right\}, \\ \phi &= \frac{1}{2\pi}\mathrm{Im}\left\{q^\infty \ln\left((\zeta - \zeta_o)(\bar{\zeta}^{-1} - \zeta_o)^{-1}\right)\right\}. \end{aligned}\right\} \tag{3.8-7}$$

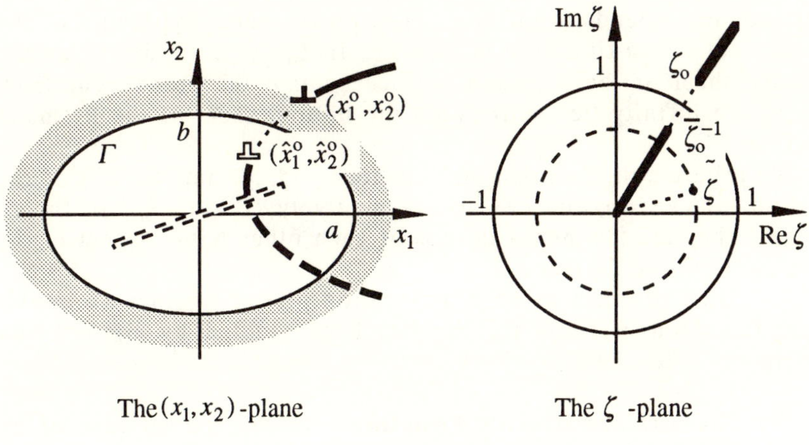

Fig. 3.6 The infinite medium with an elliptic hole is subjected to an antiplane force f and a screw dislocation b at (x_1^o, x_2^o). The image singularities are at $(\hat{x}_1^o, \hat{x}_2^o)$. Their mappings in the ζ-plane are ζ_o and $\bar{\zeta}_o^{-1}$, respectively, which are mirror images of each other with respect to the unit circle.

if the ellipse is a void and

$$u = \frac{-1}{2\pi\mu}\mathrm{Re}\left\{q^{\infty}\ln\left((\zeta-\zeta_o)(\bar{\zeta}^{-1}-\zeta_o)^{-1}\right)\right\},$$
$$\phi = \frac{1}{2\pi}\mathrm{Im}\left\{q^{\infty}\ln\left((\zeta-\zeta_o)(\bar{\zeta}^{-1}-\zeta_o)\right)\right\}. \tag{3.8-8}$$

if the ellipse is a rigid body.

The logarithmic terms in (3.8-2) require two branch cuts for the functions to be single-valued. If the branch cut for the first term extends from $\zeta=\zeta_o$ to infinity along the radial line in the ζ-plane, the corresponding branch cut in the (x_1,x_2)-plane is a hyperbola extending from (x_1^o,x_2^o) to infinity, Fig. 3.6. The second term is

$$\ln(\zeta^{-1}-\bar{\zeta}_o) = \ln(-\bar{\zeta}_o) + \left\{\ln(\zeta-\bar{\zeta}_o^{-1}) - \ln\zeta\right\}. \tag{3.8-9}$$

The points ζ_o and $\bar{\zeta}_o^{-1}$ lie on the same radial line with their radial distances the inverses of each other. Hence they are the mirror images of each other with respect to the unit circle. The singularities of (3.8-9) are at $\zeta=0$ and $\bar{\zeta}_o^{-1}$. Therefore the branch cut in the ζ-plane is a finite cut connecting $\zeta=0$ and $\bar{\zeta}_o^{-1}$. In the (x_1,x_2)-plane the *image singularities* are at infinity and at $(\hat{x}_1^o,\hat{x}_2^o)$ which is determined from (3.7-2) by setting $\zeta=\bar{\zeta}_o^{-1}$, i.e.,

$$\hat{x}_1^o + p\hat{x}_2^o = c\bar{\zeta}_o^{-1} + d\bar{\zeta}_o. \tag{3.8-10}$$

If $|\bar{\zeta}_o^{-1}| > |\tilde{\zeta}|$ as shown in Fig. 3.6, the branch cut for the image singularity begins at $(\hat{x}_1^o,\hat{x}_2^o)$ inside the ellipse, passes through the branch cut that connecting $-\tilde{Z}$ to $\tilde{Z}$ to the bottom sheet of the Riemann surface, and extends to infinity. It is on the same hyperbola of the branch cut of the first term in (3.8-2). If $|\bar{\zeta}_o^{-1}| < |\tilde{\zeta}|$, $(\hat{x}_1^o,\hat{x}_2^o)$ is on the bottom sheet of the Riemann surface so that the branch cut from $(\hat{x}_1^o,\hat{x}_2^o)$ to infinity lies entirely on the bottom sheet of the Riemann surface.

We reiterate the fact that the locations (x_1^o,x_2^o) and $(\hat{x}_1^o,\hat{x}_2^o)$ of the real and image singularities are, respectively, ζ_o and $\bar{\zeta}_o^{-1}$ in the ζ-plane. They are the mirror images of each other with respect to the unit circle in the ζ-plane.

A related problem of edge dislocation pile-up against a circular rigid inclusion of finite rigidity has been studied by Barnett and Tetelman (1966, 1967).

(ii) **The Surface Green's Function.** In the special case of an elliptic hole subjected to an antiplane force f at the hole surface (Fig. 3.7) we let

$$x_1^o = a\cos\psi_o, \quad x_2^o = b\sin\psi_o, \tag{3.8-11}$$

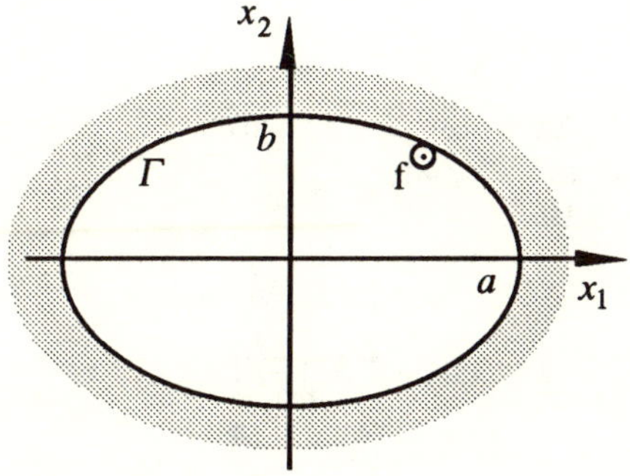

Fig. 3.7 The infinite space with an elliptic hole is subjected to an antiplane force f at the surface of the hole.

where ψ_o is a constant. Hence

$$\zeta_o = e^{i\psi_o}. \tag{3.8-12}$$

Inserting (3.8-12) into (3.8-7) with q^∞=f we have

$$\begin{aligned} u &= \frac{-\mathrm{f}}{2\pi\mu}\mathrm{Re}\left\{\ln\left((\zeta - e^{i\psi_o})(\bar{\zeta}^{-1} - e^{i\psi_o})\right)\right\} \\ &= \frac{-\mathrm{f}}{2\pi\mu}\mathrm{Re}\left\{\ln\left((\zeta - e^{i\psi_o})(\zeta^{-1} - e^{-i\psi_o})\right)\right\}, \\ \phi &= \frac{\mathrm{f}}{2\pi}\,\mathrm{Im}\left\{\ln\left((\zeta - e^{i\psi_o})(\bar{\zeta}^{-1} - e^{i\psi_o})^{-1}\right)\right\} \\ &= \frac{\mathrm{f}}{2\pi}\,\mathrm{Im}\left\{\ln\left((\zeta - e^{i\psi_o})(\zeta^{-1} - e^{-i\psi_o})\right)\right\}. \end{aligned} \tag{3.8-13}$$

The second equalities follow from application of (3.5-11). This is the surface Green's function for the elliptic hole subjected to an antiplane force f at the surface of the elliptic hole. It should be noted that the applied singularities at (x_1^o, x_2^o) and the image singularities at $(\hat{x}_1^o, \hat{x}_2^o)$ coalesce to the same point in this case.

(iii) **Uniform Loading at Infinity.** When the infinite anisotropic elastic solid with an elliptic hole is subjected to a uniform loading σ_{31}^∞ and σ_{32}^∞ at infinity, we may consider the solution in the form

$$\phi + i\mu u = q\zeta - \bar{q}\zeta^{-1} \tag{3.8-14}$$

in which q is a constant to be determined. The traction-free condition ϕ=0 at the elliptic boundary is satisfied. If the elliptic hole is a rigid body, let

$$\phi + i\mu u = q\zeta + \bar{q}\zeta^{-1} \tag{3.8-15}$$

so that u=0 at the elliptic boundary is satisfied. It remains to find q so that the stresses at infinity are σ_{31}^{∞} and σ_{32}^{∞}.

From (3.7-4) and (3.7-6),

$$\zeta = z/c, \quad \zeta^{-1} = 0, \quad \text{as } |z| \to \infty.$$

Therefore the ϕ in (3.8-14) and (3.8-15) reduce to

$$\phi = \mathrm{Re}(qz/c),$$

as $|z| \to \infty$. From (3.3-12a), the stresses at infinity are

$$\phi_{,1} = \sigma_{32}^{\infty}, \quad \phi_{,2} = -\sigma_{31}^{\infty},$$

or

$$\mathrm{Re}(q/c) = \sigma_{32}^{\infty}, \quad \mathrm{Re}(pq/c) = -\sigma_{31}^{\infty}. \tag{3.8-16}$$

With p given by (3.3-10) it can be shown that

$$q/c = \sigma_{32}^{\infty} + i\mu u_{,1}^{\infty} \tag{3.8-17}$$

where, by (3.3-3),

$$u_{,1}^{\infty} = \mu^{-2}(C_{44}\sigma_{31}^{\infty} - C_{45}\sigma_{32}^{\infty}) \tag{3.8-18}$$

is twice the strain ε_{31} at infinity.

(iv) **Arbitrary Loading on the Elliptic Hole.** Let the traction t_3 be prescribed on the surface Γ of the elliptic hole. From (3.3-16),

$$t_3 = \frac{d\phi}{\rho d\psi} \tag{3.8-19}$$

where

$$(\rho d\psi)^2 = (dx_1)^2 + (dx_2)^2,$$

or, by (3.7-1),

$$\rho = \sqrt{(a \sin\psi)^2 + (b\cos\psi)^2}. \tag{3.8-20}$$

The stresses at infinity vanish. The solution for the problem can be constructed from

$$\phi + i\mu u = q_o \ln\zeta + \sum_{k=1}^{\infty} q_k \zeta^{-k} \tag{3.8-21}$$

where q_k (k=0,1,...) are constants to be determined.

It is clear that $\phi_{,1}$ and $\phi_{,2}$ vanish at infinity so that the stresses indeed vanish at infinity. At the elliptic boundary Γ, (3.8-21) gives

$$\begin{aligned}\phi(\psi) &= \mathrm{Re}\left(iq_o\psi + \sum_{k=1}^{\infty} q_k e^{-i\psi} \right) \\ &= -q_o''\psi + \sum_{k=1}^{\infty}\left(q_k' \cos k\psi + q_k'' \sin k\psi\right)\end{aligned} \tag{3.8-22}$$

where q_k' and q_k'' (k=0,1,2,...) are, respectively, the real and imaginary parts of q_k. Substitution of (3.8-22) into (3.8-19) yields

$$\rho t_3 = -q_o'' + \sum_{k=1}^{\infty} k\left(-q_k' \sin k\psi + q_k'' \cos k\psi\right)$$

which is in the form of a Fourier series. Therefore (Sneddon, 1951)

$$q_o'' = \frac{-1}{2\pi}\int_0^{2\pi} \rho t_3 d\psi, \tag{3.8-23}$$

$$\begin{aligned} q_k' &= \frac{-1}{k\pi}\int_0^{2\pi} \rho t_3 \sin k\psi d\psi, \\ q_k'' &= \frac{1}{k\pi}\int_0^{2\pi} \rho t_3 \cos k\psi d\psi, \end{aligned} \tag{3.8-24}$$

(k=1,2,...). Since $\ln\zeta = i\psi$ at the elliptic boundary, the displacement u of (3.8-21) is single-valued if

$$q_o' = 0.$$

Together with (3.8-23) and combining the two equations in (3.8-24) into one, we have finally,

$$\begin{aligned} q_o &= \frac{-i}{2\pi}\int_0^{2\pi} \rho t_3 d\psi, \\ q_k &= \frac{i}{k\pi}\int_0^{2\pi} \rho t_3 e^{ik\psi} d\psi, \quad (k = 1,2,\ldots). \end{aligned} \tag{3.8-25}$$

The solution (3.8-21) is now complete.

3.9 Anisotropic Elliptic Body

Let an anisotropic elastic material occupy an elliptic region. The transformation (3.7-4), namely,

$$\zeta = \frac{z + \sqrt{z^2 - (a^2 + p^2 b^2)}}{a - ipb} \tag{3.9-1}$$

maps the ellipse into an annual region in the ζ-plane between the unit circle and the $\tilde{\zeta}$-circle, Fig. 3.5. Due to the presence of the square root term in (3.9-1) there is a branch cut inside the ellipse. The mapping is one-to-one except for points on the branch cut. The branch cut can be eliminated by adding to (3.9-1)

$$t\zeta^{-1} = \frac{z - \sqrt{z^2 - (a^2 + p^2 b^2)}}{a - ipb} \tag{3.9-2}$$

deduced from (3.7-7) and (3.7-5) where, by (3.7-8),

$$t = \frac{d}{c} = \frac{a + ipb}{a - ipb}, \quad |t| < 1. \tag{3.9-3}$$

Thus the function

$$\zeta + t\zeta^{-1} = z/c \tag{3.9-4}$$

has no branch cut in the (x_1, x_2)-plane. More generally, it is easily shown that the function $f_k(z)$ defined by (Lekhnitskii, 1950)

$$\zeta^k + (t\zeta^{-1})^k = f_k(z), \tag{3.9-5}$$

k being a positive integer, is a polynomial in z of degree k when ζ and ζ^{-1} of (3.7-4) and (3.7-6) are employed. Hence $f_k(z)$ has no branch cut inside the ellipse. In this section we employ (3.9-5) to solve the problem of an elliptic body subjected to antiplane forces.

(i) **The Green's Function**. Let the boundary Γ of the ellipse be rigidly clamped so that

$$\mathrm{u} = 0 \quad \text{on } \Gamma \tag{3.9-6}$$

and an antiplane force f be applied at

$$(x_1, x_2) = (x_1^o, x_2^o)$$

inside the elliptic body, Fig. 3.8. Consider the solution

$$\phi + i\mu(u - u_o) = \frac{\mathrm{f}}{2\pi i}\left\{ \ln\left(\frac{z - z_o}{z - \hat{z}_o}\right) + \sum_{k=1}^{\infty} \frac{1}{k} h_k \left(\zeta^k + (t\zeta^{-1})^k\right) \right\} \tag{3.9-7}$$

in which u_o and h_k are constants and, using (3.7-2),

$$x_1^o + p x_2^o = z_o = c\zeta_o + d\zeta_o^{-1}, \tag{3.9-8}$$

$$\hat{x}_1^o + p\hat{x}_2^o = \hat{z}_o = c\bar{\zeta}_o^{-1} + d\bar{\zeta}_o. \tag{3.9-9}$$

Equations (3.9-8), (3.9-9), in that order, provide ζ_o, $\hat{x}_1^o$, and $\hat{x}_2^o$. We could also determine ζ_o from (3.8-3). The logarithmic term on the right of (3.9-7) represents two Green's functions for the infinite space due to the antiplane force f applied at (x_1^o, x_2^o) (Section 3.5) and an antiplane force –f applied at $(\hat{x}_1^o, \hat{x}_2^o)$. The point $(\hat{x}_1^o, \hat{x}_2^o)$ is the image singularity. The corresponding point $\bar{\zeta}_o^{-1}$ in the ζ-plane is located outside the unit circle, and lies on the same radial line as the point ζ_o does, Fig. 3.8. Its radial distance is the inverse of that of ζ_o. Thus ζ_o and $\bar{\zeta}_o^{-1}$ are the mirror images of each other with respect to the unit circle. The infinite series in (3.9-7) are needed because the image singularity alone is not sufficient to satisfy the boundary condition (3.9-6). Making use of the transformation (3.7-2) we have

$$z - z_o = c(\zeta - \zeta_o) + d(\zeta^{-1} - \zeta_o^{-1})$$

$$= c(\zeta - \zeta_o)(\zeta - \tau_o)\zeta^{-1}, \quad \tau_o = t\zeta_o^{-1}, \tag{3.9-10}$$

$$z - \hat{z}_o = c(\zeta - \bar{\zeta}_o^{-1}) + d(\zeta^{-1} - \bar{\zeta}_o)$$

$$= c(\zeta - \bar{\zeta}_o^{-1})(\zeta - \hat{\tau}_o)\zeta^{-1}, \quad \hat{\tau}_o = t\bar{\zeta}_o. \tag{3.9-11}$$

Hence

$$\ln(z - z_o) - \ln(z - \hat{z}_o) = \left\{\ln(\zeta - \zeta_o) - \ln(\zeta - \bar{\zeta}_o^{-1})\right\}$$

$$+\left\{\ln(\zeta - \tau_o) - \ln(\zeta - \hat{\tau}_o)\right\}. \tag{3.9-12}$$

The right side of the equation tells us that the two branch cuts needed for the logarithmic functions to be single-valued are

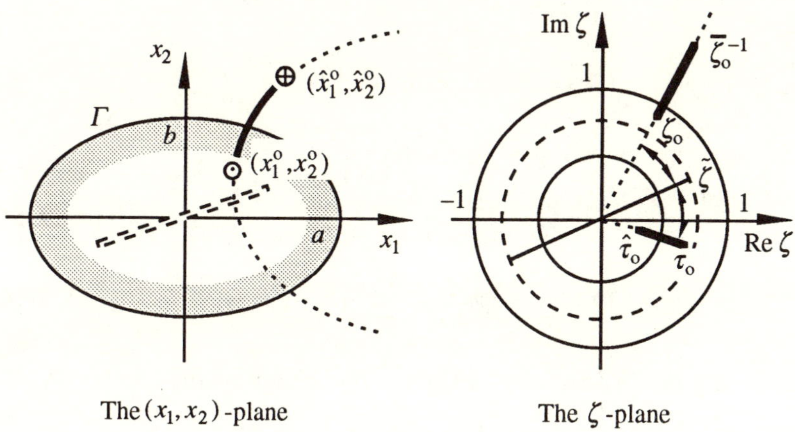

Fig. 3.8 The elliptic body is subjected to an antiplane force f at (x_1^o, x_2^o) with its boundary Γ being fixed. The image singularity is at $(\hat{x}_1^o, \hat{x}_2^o)$.

between the points ζ_o and $\overline{\zeta}_o^{-1}$, and between τ_o and $\hat{\tau}_o$. As discussed in Section 3.7 the point (x_1^o, x_2^o) is mapped to two points, ζ_o and τ_o, while the point $(\hat{x}_1^o, \hat{x}_2^o)$ is mapped into $\overline{\zeta}_o^{-1}$ and $\hat{\tau}_o$. Hence the two corresponding branch cuts in the (x_1, x_2)-plane are the same hyperbola connecting (x_1^o, x_2^o) to $(\hat{x}_1^o, \hat{x}_2^o)$, one on the top sheet while the other on the bottom sheet of the Riemann surface. The singularities at τ_o and $\hat{\tau}_o$ in the ζ-plane are at (x_1^o, x_2^o) and $(\hat{x}_1^o, \hat{x}_2^o)$ in the (x_1, x_2)-plane but on the bottom sheet of the Riemann surface. These singularities are *ghost singularities*. Unlike image singularities, ghost singularities never coalesce with the applied singularities in the limit the applied singularities approach the elliptic boundary because they remain on the bottom sheet of the Riemann surface.

To determine the unknown constants u_o and h_k in (3.9-7) we write the right side of (3.9-12) as

$$\left\{\ln(-\overline{\zeta}_o\zeta) + \ln(1-\zeta_o\zeta^{-1}) - \ln(1-\overline{\zeta}_o\zeta)\right\}$$
$$+\left\{\ln(1-\tau_o\zeta^{-1}) - \ln(1-\hat{\tau}_o\zeta^{-1})\right\}.$$

At the elliptic boundary Γ, $\zeta = e^{i\psi}$ so that

$$\ln(z-z_o) - \ln(z-\hat{z}_o) = i\psi + \ln(-\overline{\zeta}_o) + 2i\,\mathrm{Im}\left\{\ln(1-\zeta_o e^{-i\psi})\right\}$$
$$+\ln(1-\tau_o e^{-i\psi}) - \ln(1-\hat{\tau}_o e^{-i\psi}).$$

Making use of (3.5-11) the imaginary parts of (3.9-7) evaluated at Γ can be written as

$$\mu(u-u_o) = \frac{-\mathrm{f}}{2\pi}\mathrm{Re}\left\{\ln(-\zeta_o) + \ln(1-\tau_o e^{-i\psi}) - \ln(1-\hat{\tau}_o e^{-i\psi})\right\}$$
$$-\frac{\mathrm{f}}{2\pi}\mathrm{Re}\sum_{k=1}^{\infty}\frac{1}{k}(\overline{h}_k + t^k h_k)e^{-ik\psi}.$$

The boundary condition $u=0$ is satisfied if

$$u_o = \frac{\mathrm{f}}{2\pi\mu}\mathrm{Re}\left\{\ln(-\zeta_o)\right\}, \qquad (3.9\text{-}13)$$

$$\ln(1-\tau_o e^{-i\psi}) - \ln(1-\hat{\tau}_o e^{-i\psi}) + \sum_{k=1}^{\infty}\frac{1}{k}(\overline{h}_k + t^k h_k)e^{-ik\psi} = 0. \qquad (3.9\text{-}14)$$

Employing the infinite series for the logarithmic function

$$\ln(1-x) = -\sum_{k=1}^{\infty}\frac{1}{k}x^k \qquad (3.9\text{-}15)$$

(3.9-14) is reduced to

$$\bar{h}_k + t^k h_k = \eta_k, \tag{3.9-16}$$

$$\eta_k = \tau_o^k - \hat{\tau}_o^k = t^k(\zeta_o^{-k} - \bar{\zeta}_o^k). \tag{3.9-17}$$

Elimination of $\bar{h}_k$ from (3.9-16) and its complex conjugate

$$h_k + \bar{t}^k \bar{h}_k = \bar{\eta}_k$$

yields

$$h_k = \frac{\bar{\eta}_k - \bar{t}^k \eta_k}{1-(t\bar{t})^k}. \tag{3.9-18}$$

With h_k provided by (3.9-18), the infinite series in (3.9-7) can be shown to be bounded (Section 3.10). It should be noted that when t=0 (which occurs if $p=ia/b$), η_k and h_k vanish in view of (3.9-17) and (3.9-18). The infinite series in (3.9-7) vanishes, and we have a closed form solution.

(ii) **Two Equal and Opposite Forces on the Elliptic Body.** Let the elliptic body be subjected to an antiplane force f at (x_1^o, x_2^o) and another antiplane force –f be applied at $(-x_1^o, -x_2^o)$. The elliptic boundary is traction-free, Fig. 3.9(a). It is not difficult to show that the solution is

$$\phi + i\mu u = \frac{\mathrm{f}}{2\pi i}\left\{\ln\left(\frac{(z-z_o)(z-\hat{z}_o)}{(z+z_o)(z+\hat{z}_o)}\right) - \sum_{k=1}^{\infty}\frac{2}{k}h_k\left(\zeta^k + (t\zeta^{-1})^k\right)\right\} \tag{3.9-19}$$

in which

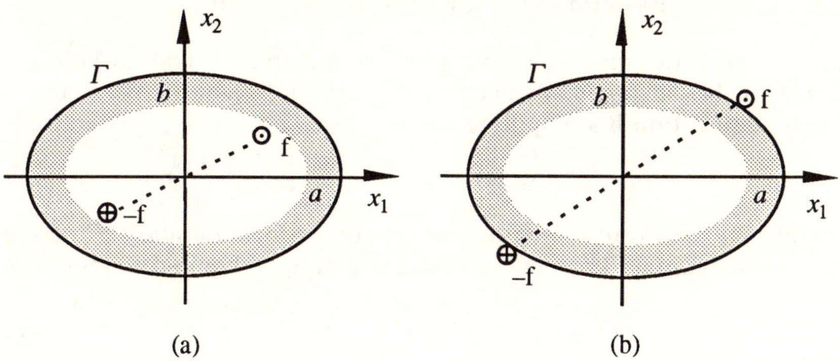

Fig. 3.9 The elliptic body is subjected to two equal and opposite antiplane forces with the elliptic boundary Γ being traction-free.

$$h_k = \frac{\overline{\eta}_k + \bar{t}^k \eta_k}{1-(t\bar{t})^k}, \tag{3.9-20}$$

$$\eta_k = \tau_o^k + \hat{\tau}_o^k = t^k(\zeta_o^{-k} + \overline{\zeta}_o^k).$$

The constant u_o has been ignored since it represents a rigid body motion.

In the special case (x_1^o, x_2^o) is on the elliptic boundary (Fig. 3.9(b)), we let

$$x_1^o = a\cos\psi_o, \quad x_2^o = b\sin\psi_o.$$

One then obtains

$$\zeta_o = e^{i\psi_o} = \overline{\zeta}_o^{-1}, \quad z_o = \hat{z}_o,$$

and (3.9-19) is reduced to

$$\phi + i\mu u = \frac{\mathrm{f}}{\pi i}\left\{\ln\left(\frac{z-z_o}{z+z_o}\right) - \sum_{k=1}^{\infty}\frac{1}{k}h_k\left(\zeta^k + (t\zeta^{-1})^k\right)\right\} \tag{3.9-21}$$

where

$$h_k = 2\frac{(\bar{t}e^{i\psi_o})^k + (t\bar{t}e^{-i\psi_o})^k}{1-(t\bar{t})^k}. \tag{3.9-22}$$

As before, the infinite series in (3.9-19) and (3.9-21) vanish when $p=ia/b$.

The associated problem of a circular elastic body subjected to two diametrically loaded forces has been studied by Lekhnitskii (1950) for orthotropic materials.

3.10 Green's Functions for an Elliptic Inclusion

Let the infinite space be occupied by material 1 and an elliptic inclusion be occupied by material 2. An antiplane force f and a screw dislocation b are applied at

$$(x_1, x_2) = (x_1^o, x_2^o)$$

which may be located outside, inside, or at the boundary Γ of the elliptic inclusion. The two materials are rigidly bonded together so that

$$u^{(1)} = u^{(2)}, \quad \phi^{(1)} = \phi^{(2)} \quad \text{on } \Gamma \tag{3.10-1}$$

where the superscripts (1) and (2) (and later the subscripts 1 and 2) refer to materials 1 and 2, respectively. Thus the relation

$$q_k^\infty = \mathrm{f} + i\mu_k \mathrm{b}, \quad (k=1,2) \tag{3.10-2}$$

refers to material 1 or 2 when k=1 or 2. We also have

$$c^{(k)} = \tfrac{1}{2}(a - ip^{(k)}b), \quad d^{(k)} = \tfrac{1}{2}(a + ip^{(k)}b), \quad \mathrm{t}^{(k)} = d^{(k)}/c^{(k)},$$

$$\begin{aligned} x_1 + p^{(k)}x_2 &= z^{(k)} = c^{(k)}\zeta^{(k)} + d^{(k)}\zeta^{(k)^{-1}}, \\ x_1^{\mathrm{o}} + p^{(k)}x_2^{\mathrm{o}} &= z_{\mathrm{o}}^{(k)} = c^{(k)}\zeta_{\mathrm{o}}^{(k)} + d^{(k)}\zeta_{\mathrm{o}}^{(k)^{-1}}, \end{aligned} \tag{3.10-3}$$

and

$$\begin{aligned} \zeta^{(k)} &= \frac{z^{(k)} + \sqrt{z^{(k)^2} - (a^2 + p^{(k)^2}b^2)}}{a - ip^{(k)}b}, \\ \zeta_{\mathrm{o}}^{(k)} &= \frac{z_{\mathrm{o}}^{(k)} + \sqrt{z_{\mathrm{o}}^{(k)^2} - (a^2 + p^{(k)^2}b^2)}}{a - ip^{(k)}b}. \end{aligned} \tag{3.10-4}$$

Combining the knowledge of the Green's functions for an elliptic hole and for an elliptic body presented in the previous sections it is fairly straight forward to construct the solutions for the present problem (Ting, 1995b).

(i) $(x_1^{\mathrm{o}}, x_2^{\mathrm{o}})$ **outside the inclusion.**

With reference to (3.8-2) and (3.9-7) the solution can be written as

$$\phi^{(1)} + i\mu_1 u^{(1)} = \frac{1}{2\pi i}\left\{ q_1^\infty \ln(\zeta^{(1)} - \zeta_{\mathrm{o}}^{(1)}) + q_1 \ln(\zeta^{(1)^{-1}} - \bar{\zeta}_{\mathrm{o}}^{(1)}) + \sum_{k=1}^{\infty} \frac{1}{k} g_k (\zeta^{(1)})^{-k} \right\} \tag{3.10-5}$$

for material 1 outside the ellipse and

$$\begin{aligned} (\phi^{(2)} - \phi_{\mathrm{o}}^{(2)}) + i\mu_2(u^{(2)} - u_{\mathrm{o}}^{(2)}) &= \frac{1}{2\pi i} q_2 \ln(z^{(2)} - \hat{z}_0^{(2)}) \\ &+ \frac{1}{2\pi i}\sum_{k=1}^{\infty} \frac{1}{k} h_k \left\{ (\zeta^{(2)})^k + (t^{(2)}\zeta^{(2)^{-1}})^k \right\} \end{aligned} \tag{3.10-6}$$

for material 2 inside the ellipse. $u_{\mathrm{o}}^{(2)}$, $\phi_{\mathrm{o}}^{(2)}$, q_1, q_2, g_k, h_k are constants to be determined, and

$$\hat{x}_1^{\mathrm{o}} + p^{(2)}\hat{x}_2^{\mathrm{o}} = \hat{z}_{\mathrm{o}}^{(2)} = c^{(2)}\zeta_{\mathrm{o}}^{(1)} + d^{(2)}\zeta_{\mathrm{o}}^{(1)^{-1}}. \tag{3.10-7}$$

Equation $(3.10\text{-}7)_2$ determines $\hat{z}_{\mathrm{o}}^{(2)}$ while $(3.10\text{-}7)_1$ defines $(\hat{x}_1^{\mathrm{o}}, \hat{x}_2^{\mathrm{o}})$. The first term on the right of (3.10-5) takes care of the antiplane force f and the screw dislocation b applied at (x_1^o, x_2^o) in material 1. The second term has singularities at $\zeta=0$ and $(\bar{\zeta}^{(1)})^{-1}$. They are the image singularities for material 1. $\zeta=0$ corresponds to (x_1, x_2) at infinity, while $(\hat{x}_1^o, \hat{x}_2^o)$ associated with $(\bar{\zeta}^{(1)})^{-1}$ is determined by

$$\hat{x}_1^o + p^{(1)}\hat{x}_2^o = \hat{z}_o^{(1)} = c^{(1)}\bar{\zeta}_o^{(1)^{-1}} + d^{(1)}\bar{\zeta}_o^{(1)}. \tag{3.10-8}$$

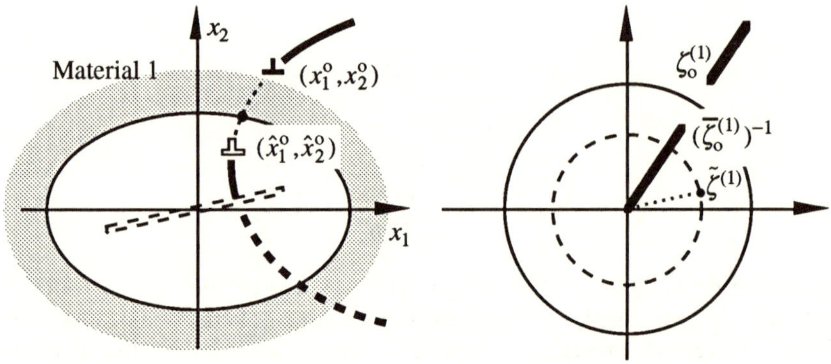

The (x_1, x_2)-plane The ζ-plane

(a) The branch cuts for material 1 outside the ellipse.

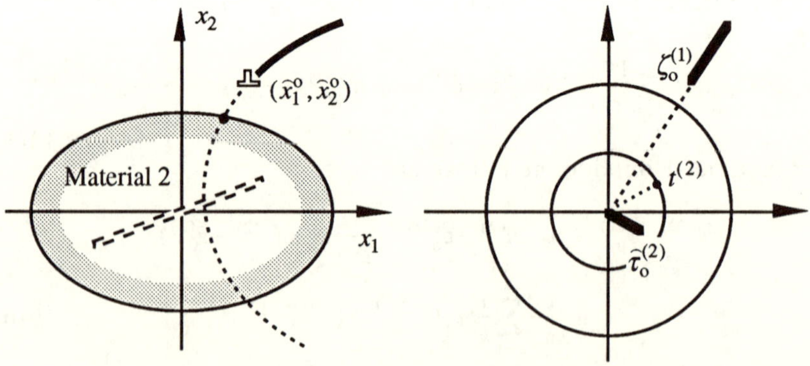

The (x_1, x_2)-plane The ζ-plane

(b) The branch cuts for material 2 inside the ellipse

Fig. 3.10 An anisotropic elliptic inclusion is subjected to an antiplane force f and a screw dislocation b at (x_1^o, x_2^o) in material 1 outside the ellipse. The image singularities for materials 1 and 2 are at $(\hat{x}_1^o, \hat{x}_2^o)$ and $(\tilde{x}_1^o, \tilde{x}_2^o)$, respectively. The hyperbolic branch cuts for materials 1 and 2 are different, but they intersect the ellipse at the same point.

It is inside the ellipse. The first term on the right of (3.10-6) is the Green's function for the infinite space occupied by material 2. The singularity of this term is at $(\tilde{x}_1^o, \tilde{x}_2^o)$ defined in (3.10-7). It is outside the ellipse, and is the image singularity for material 2. The branch cuts for materials 1 and 2 required for the logarithmic functions to be single-valued are shown in Fig. 3.10. The reader is referred to Figs. 3.6 and 3.8 for the construction of these branch cuts. The geometrical relations between the locations of real and image singularities for materials 1 and 2 become more transparent if one looks at their locations in the ζ-plane. First, the point (x_1^o, x_2^o) is mapped to $\zeta_o^{(1)}$ with $p = p^{(1)}$ in the complex variable $z = x_1 + px_2$. The mirror image of $\zeta_o^{(1)}$ with respect to the unit circle is $(\bar{\zeta}_o^{(1)})^{-1}$. Its inverse mapping to the (x_1, x_2)-plane is $(\hat{x}_1^o, \hat{x}_2^o)$, again with $p = p^{(1)}$. The point $\zeta_o^{(1)}$ is also the image singularity for material 2. Its inverse mapping to the (x_1, x_2)-plane is $(\tilde{x}_1^o, \tilde{x}_2^o)$ with $p = p^{(2)}$ in the complex variable $z = x_1 + px_2$. Because of different p, (x_1^o, x_2^o) and $(\tilde{x}_1^o, \tilde{x}_2^o)$ are not the same point even though they both correspond to the same point $\zeta_o^{(1)}$ in the ζ-plane. For the same reason the hyperbola that passes through (x_1^o, x_2^o) and $(\hat{x}_1^o, \hat{x}_2^o)$ for material 1 is not the same hyperbola that passes through $(\tilde{x}_1^o, \tilde{x}_2^o)$ for material 2. However both hyperbolas intersect the ellipse at the same point. It should be noted that while $(\tilde{x}_1^o, \tilde{x}_2^o)$ depends on elastic constants of both materials, $(\hat{x}_1^o, \hat{x}_2^o)$ depends only on elastic constants of material 1 in which the antiplane force f and the screw dislocation b are applied.

Making use of (3.10-3) and (3.10-7) to write

$$z^{(2)} - \tilde{z}_o^{(2)} = c^{(2)}(\zeta^{(2)} - \zeta_o^{(1)})(1 - \tilde{\tau}_o^{(2)}\zeta^{(2)^{-1}}), \quad \tilde{\tau}_o^{(2)} = t^{(2)}\zeta_o^{(1)^{-1}}, \quad (3.10\text{-}9)$$

(3.10-5) and (3.10-6) evaluated at the elliptic boundary are

$$\phi^{(1)} + i\mu_1 u^{(1)} = \frac{1}{2\pi i}\left\{ q_1^\infty \ln(e^{i\psi} - \zeta_o^{(1)}) + q_1 \ln(e^{-i\psi} - \bar{\zeta}_o^{(1)}) + \sum_{k=1}^{\infty} \frac{1}{k} g_k e^{-ik\psi} \right\},$$

$$(\phi^{(2)} - \phi_o^{(2)}) + i\mu_2(u^{(2)} - u_o^{(2)}) = \frac{1}{2\pi i} q_2 \left\{ \ln(c^{(2)}) + \ln(e^{i\psi} - \zeta_o^{(1)}) \right.$$

$$\left. + \ln(1 - \tilde{\tau}_o^{(2)} e^{-i\psi}) \right\} + \frac{1}{2\pi i} \sum_{k=1}^{\infty} \frac{1}{k} h_k \left\{ e^{ik\psi} + (t^{(2)})^k e^{-ik\psi} \right\}.$$

Imposition of the continuity conditions (3.10-1) leads to

$$\phi_o^{(2)} + i\mu_2 u_o^{(2)} = \frac{-1}{2\pi i} q_2 \ln(c^{(2)}), \quad (3.10\text{-}10)$$

$$(\mu_2/\mu_1)(q_1^\infty + \bar{q}_1) = q_2, \quad q_1^\infty - \bar{q}_1 = q_2, \quad (3.10\text{-}11)$$

and, using (3.9-15),

$$
\begin{aligned}
(\mu_2/\mu_1)g_k &= -\eta_k + \bar{h}_k + (t^{(2)})^k h_k, \\
g_k &= -\eta_k - \bar{h}_k + (t^{(2)})^k h_k,
\end{aligned}
\tag{3.10-12}
$$

where

$$\eta_k = q_2(\hat{\tau}_o^{(2)})^k. \tag{3.10-13}$$

Equations (3.10-11) are identical to (3.6-5) whose solutions are given in (3.6-6), i.e.,

$$q_1 = -K\bar{q}_1^\infty, \quad q_2 = (1+K)q_1^\infty, \tag{3.10-14}$$

where

$$K = \frac{\mu_2 - \mu_1}{\mu_2 + \mu_1}, \quad -1<K<1.$$

Elimination of η_k between $(3.10\text{-}12)_{1,2}$ leads to

$$h_k = \frac{K}{1-K}\bar{g}_k. \tag{3.10-15}$$

With (3.10-15), $(3.10\text{-}12)_2$ is reduced to

$$g_k = -(1-K)\eta_k + K(t^{(2)})^k \bar{g}_k.$$

Solving for g_k from the above equation and its complex conjugate

$$\bar{g}_k = -(1-K)\bar{\eta}_k + K(\bar{t}^{(2)})^k g_k$$

one obtains

$$g_k = -(1-K)\frac{\eta_k + K(t^{(2)})^k \bar{\eta}_k}{1 - K^2(t^{(2)}\bar{t}^{(2)})^k}. \tag{3.10-16}$$

The unknown constants in (3.10-5) and (3.10-6) have all been determined.

We will show that the infinite series in (3.10-5) and (3.10-6) are bounded. With (3.10-13) and (3.10-16), the infinite series in (3.10-5) can be written as

$$\sum_{k=1}^{\infty}\frac{1}{k}g_k(\zeta^{(1)})^{-k} = -(1-K)\left\{q_2F\left(\hat{\tau}_o^{(2)}(\zeta^{(1)})^{-1}\right) + K\bar{q}_2F\left(\bar{\hat{\tau}}_o^{(2)}\hat{\tau}^{(2)}\right)\right\}$$

where $\hat{\tau}^{(2)} = t^{(2)}(\zeta^{(1)})^{-1}$ and

$$F(x) = \sum_{k=1}^{\infty}\frac{x^k}{k\left\{1 - K^2(t^{(2)}\bar{t}^{(2)})^k\right\}}.$$

Since $|t^{(2)}|<1$,

$$|F(x)| \le \sum_{k=1}^{\infty} \frac{|x|^k}{k\{1-K^2(t^{(2)}\bar{t}^{(2)})^k\}} < \sum_{k=1}^{\infty} \frac{|x|^k}{k\{1-K^2(t^{(2)}\bar{t}^{(2)})\}} = \frac{-\ln(1-|x|)}{1-K^2(t^{(2)}\bar{t}^{(2)})}$$

which is bounded for $x=\bar{\tau}_o^{(2)}(\zeta^{(1)})^{-1}$ and $\bar{\bar{\tau}}_o^{(2)}\bar{\tau}^{(2)}$. For these x, $|x|<|t^{(2)}|$ for $|\zeta^{(1)}|>1$ so that

$$|F(x)| < \frac{-\ln(1-|t^{(2)}|)}{1-K^2(t^{(2)}\bar{t}^{(2)})}.$$

Similarly one can prove that the infinite series in (3.10-6) is bounded.

Proper account of image singularities is important because the image singularities become real singularities in the limit the antiplane force f and the screw dislocation b approach the elliptic boundary. If the image singularities are not properly accounted for, the infinite series in (3.10-5) and (3.10-6) may diverge, or converge slowly, for points near the image singularities when f and b approach the elliptic boundary. This is indeed the case as can be seen in the numerical computation by Yen, et al. (1995).

(ii) (x_1^o, x_2^o) **inside the inclusion.**

In constructing the solution for the antiplane force f and the screw dislocation b applied inside the inclusion, care must be exercised so that the solutions for the present case and the case considered in (i) reduce to the same solution in the limit f and b are applied at the elliptic boundary. Therefore consider the solution

$$\phi^{(1)}+i\mu_1 u^{(1)} = \frac{1}{2\pi i}\left\{q_1^\infty \ln(\zeta^{(1)}-\zeta_o^{(2)}) + q_1 \ln(\zeta^{(1)^{-1}}-\zeta_o^{(2)^{-1}}) + \sum_{k=1}^{\infty}\frac{1}{k}g_k(\zeta^{(1)})^{-k}\right\} \tag{3.10-17}$$

for material 1 outside the ellipse and

$$(\phi^{(2)}-\phi_o^{(2)})+i\mu_2(u^{(2)}-u_o^{(2)}) = \frac{1}{2\pi i}\left\{q_2^\infty \ln(z^{(2)}-z_o^{(2)}) + (q_2-q_2^\infty)\ln(z^{(2)}-\hat{z}_o^{(2)})\right\}$$
$$+\frac{1}{2\pi i}\sum_{k=1}^{\infty}\frac{1}{k}h_k\left\{(\zeta^{(2)})^k + (t^{(2)}\zeta^{(2)^{-1}})^k\right\} \tag{3.10-18}$$

for material 2 inside the ellipse. Again, $u_o^{(2)}$, $\phi_o^{(2)}$, q_1, q_2, g_k, h_k are constants to be determined, and

$$\hat{x}_1^o + p^{(2)}\hat{x}_2^o = \hat{z}_o^{(2)} = c^{(2)}\bar{\zeta}_o^{(2)^{-1}} + d^{(2)}\bar{\zeta}_o^{(2)} \tag{3.10-19}$$

from which $(\hat{x}_1^o, \hat{x}_2^o)$ can be determined. $(\hat{x}_1^o, \hat{x}_2^o)$ is located outside the ellipse. The first term on the right of (3.10-18) takes care of the applied antiplane force f and the screw dislocation b at (x_1^o, x_2^o). The second term is the image singularity for material 2 with the singularity located at $(\hat{x}_1^o, \hat{x}_2^o)$. The image singularities for material 1 are, from the first two terms of (3.10-17), at ζ=0 and $\zeta_o^{(2)}$. In the (x_1, x_2)-plane they are at infinity and at $(\tilde{x}_1^o, \tilde{x}_2^o)$ determined by

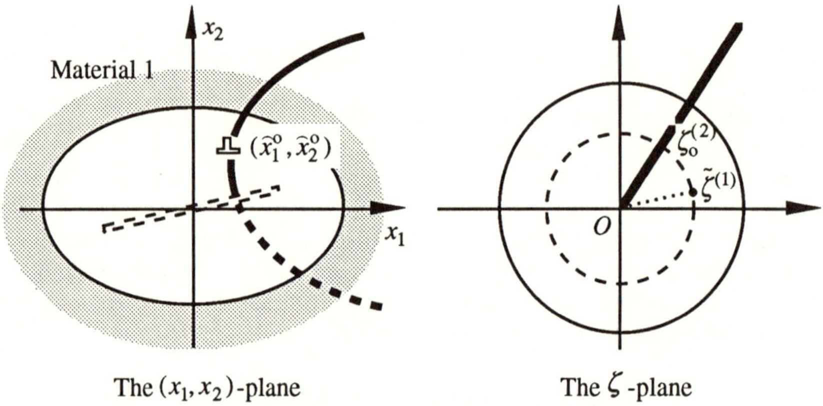

(a) The branch cuts for material 1 outside the ellipse.

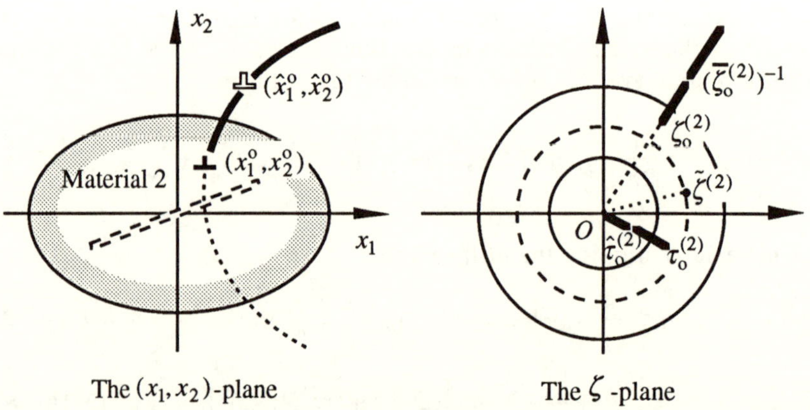

(b) The branch cuts for material 2 inside the ellipse

Fig. 3.11 An anisotropic elliptic inclusion is subjected to an antiplane force f and a screw dislocation b at (x_1^o, x_2^o) in material 2 inside the ellipse. The image singularities for materials 2 and 1 are at $(\hat{x}_1^o, \hat{x}_2^o)$ and $(\tilde{x}_1^o, \tilde{x}_2^o)$, respectively. The hyperbolic branch cuts for materials 1 and 2 are different, but they intersect the ellipse at the same point.

$$\widehat{x}_1^{\mathrm{o}} + p^{(1)}\widehat{x}_2^{\mathrm{o}} = \widehat{z}_{\mathrm{o}}^{(1)} = c^{(1)}\zeta_{\mathrm{o}}^{(2)} + d^{(1)}\zeta_{\mathrm{o}}^{(2)^{-1}}. \tag{3.10-20}$$

It is located inside the ellipse. The branch cuts for materials 1 and 2 are shown in Fig. 3.11. As explained in Case (i) the geometrical relations between the locations of real and image singularities for materials 1 and 2 become transparent if one looks at their mapping in the ζ-plane. In the ζ-plane they are mirror images of each other with respect to the unit circle. The points $(x_1^{\mathrm{o}}, x_2^{\mathrm{o}})$ and $(\widehat{x}_1^{\mathrm{o}}, \widehat{x}_2^{\mathrm{o}})$ both correspond to the same point $\zeta_{\mathrm{o}}^{(2)}$ in the ζ-plane, but they are not the same point in the (x_1, x_2)-plane. This is so because of different p in the complex variables $z = x_1 + px_2$. For the same reason the hyperbola that passes through $(x_1^{\mathrm{o}}, x_2^{\mathrm{o}})$ and $(\hat{x}_1^{\mathrm{o}}, \hat{x}_2^{\mathrm{o}})$ for material 2 is not the same hyperbola that passes through $(\widehat{x}_1^{\mathrm{o}}, \widehat{x}_2^{\mathrm{o}})$ for material 1. Both hyperbolas intersect the ellipse at the same point. It should be noted that while $(\widehat{x}_1^{\mathrm{o}}, \widehat{x}_2^{\mathrm{o}})$ depends on elastic constants of both materials, $(\hat{x}_1^{\mathrm{o}}, \hat{x}_2^{\mathrm{o}})$ depends only on elastic constants of material 2 in which the antiplane force f and the screw dislocation b are applied.

Making use of (3.10-3) and (3.10-19) to write

$$\zeta^{(1)} - \zeta_{\mathrm{o}}^{(2)} = \zeta^{(1)}\left(1 - \zeta_{\mathrm{o}}^{(2)}\zeta^{(1)^{-1}}\right),$$

$$\zeta^{(1)^{-1}} - \zeta_{\mathrm{o}}^{(2)^{-1}} = -\zeta_{\mathrm{o}}^{(2)^{-1}}\left(1 - \zeta_{\mathrm{o}}^{(2)}\zeta^{(1)^{-1}}\right),$$

$$z^{(2)} - z_{\mathrm{o}}^{(2)} = c^{(2)}\zeta^{(2)}\left(1 - \zeta_{\mathrm{o}}^{(2)}\zeta^{(2)^{-1}}\right)\left(1 - \tau_{\mathrm{o}}^{(2)}\zeta^{(2)^{-1}}\right), \quad \tau_{\mathrm{o}}^{(2)} = t^{(2)}\zeta_{\mathrm{o}}^{(2)^{-1}}, \tag{3.10-21}$$

$$z^{(2)} - \hat{z}_{\mathrm{o}}^{(2)} = -c^{(2)}\overline{\zeta}_{\mathrm{o}}^{(2)^{-1}}\left(1 - \overline{\zeta}_{\mathrm{o}}^{(2)}\zeta^{(2)}\right)\left(1 - \hat{\tau}_{\mathrm{o}}^{(2)}(\zeta^{(2)})^{-1}\right), \quad \hat{\tau}_{\mathrm{o}}^{(2)} = t^{(2)}\overline{\zeta}_{\mathrm{o}}^{(2)}, \tag{3.10-22}$$

(3.10-17) and (3.10-18) evaluated at the elliptic boundary are

$$\phi^{(1)} + i\mu_1 u^{(1)} = \frac{1}{2\pi i}\Big\{q_1^\infty\left(i\psi + \ln(1 - \zeta_0^{(2)}e^{-i\psi})\right) + q_1\left(-\ln(-\zeta_0^{(2)})\right.$$
$$\left.\left. + \ln(1 - \zeta_0^{(2)}e^{-i\psi})\right)\right\} + \frac{1}{2\pi i}\sum_{k=1}^{\infty}\frac{1}{k}g_k e^{-ik\psi},$$

$$(\phi^{(2)} - \phi_{\mathrm{o}}^{(2)}) + i\mu_2(u^{(2)} - u_{\mathrm{o}}^{(2)}) = \frac{1}{2\pi i}\Big\{q_2^\infty\left(i\psi + \ln(c^{(2)}) + \ln(1 - \zeta_{\mathrm{o}}^{(2)}e^{-i\psi})\right.$$
$$\left. + \ln(1 - \tau_{\mathrm{o}}^{(2)}e^{-i\psi})\right) + (q_2 - q_2^\infty)\left(\ln(c^{(2)}) - \ln(-\overline{\zeta}_{\mathrm{o}}^{(2)}) + \ln(1 - \overline{\zeta}_{\mathrm{o}}^{(2)}e^{i\psi})\right.$$
$$\left.\left. + \ln(1 - \hat{\tau}_{\mathrm{o}}^{(2)}e^{-i\psi})\right)\right\} + \frac{1}{2\pi i}\sum_{k=1}^{\infty}\frac{1}{k}h_k\left\{e^{ik\psi} + (t^{(2)})^k e^{-ik\psi}\right\}.$$

Imposition of the continuity conditions (3.10-1) yields

$$\mu_2 u_o^{(2)} = \frac{1}{2\pi}\mathrm{Re}\left\{q_2 \ln(c^{(2)}) + [(\mu_2/\mu_1)q_1 - (\bar{q}_2 - \bar{q}_2^\infty)]\ln(-\zeta_o^{(2)})\right\},$$
$$\phi_o^{(2)} = \frac{-1}{2\pi}\mathrm{Im}\left\{q_2 \ln(c^{(2)}) + [q_1 + \bar{q}_2 - \bar{q}_2^\infty]\ln(-\zeta_o^{(2)})\right\}, \tag{3.10-23}$$

$$(\mu_2/\mu_1)(q_1^\infty + q_1) = q_2^\infty + \bar{q}_2 - \bar{q}_2^\infty,$$
$$q_1^\infty + q_1 = q_2^\infty - \bar{q}_2 + \bar{q}_2^\infty, \tag{3.10-24}$$

and

$$(\mu_2/\mu_1)g_k = -\eta_k + \bar{h}_k + (t^{(2)})^k h_k,$$
$$g_k = -\eta_k - \bar{h}_k + (t^{(2)})^k h_k, \tag{3.10-25}$$

where

$$\eta_k = (q_2 - q_2^\infty)(\hat{\tau}_o^{(2)})^k + q_2^\infty(\tau_o^{(2)})^k. \tag{3.10-26}$$

From $(3.10\text{-}24)_{1,2}$,

$$(\mu_2/\mu_1)q_1 - (\bar{q}_2 - \bar{q}_2^\infty) = q_2^\infty - (\mu_2/\mu_1)q_1^\infty = -[(\mu_2/\mu_1) - 1]\mathrm{f},$$
$$q_1 + \bar{q}_2 - \bar{q}_2^\infty = q_2^\infty - q_1^\infty = i(\mu_2 - \mu_1)\mathrm{b},$$

so that $(3.10\text{-}23)_{1,2}$ can be combined into one equation as

$$\phi_o^{(2)} + i\mu_2 u_o^{(2)} = \frac{-1}{2\pi i}\left\{q_2 \ln(c^{(2)}) - [(\mu_2/\mu_1) - 1]\bar{q}_1^\infty \,\mathrm{Re}[\ln(-\zeta_o^{(2)})]\right\}. \tag{3.10-27}$$

When (x_1^o, x_2^o) is at the elliptic boundary we will see in Case (iii) below that the last term vanishes because $\ln(-\zeta_o^{(2)})$ is purely imaginary. As to $(3.10\text{-}24)_{1,2}$, noting that

$$q_2^\infty - \bar{q}_2^\infty = 2i\mu_2\mathrm{b}, \quad q_2^\infty + \bar{q}_2^\infty = 2\mathrm{f},$$

they can be reduced to (3.10-11). Hence q_1 and q_2 are given by (3.10-14). Finally, $(3.10\text{-}25)_{1,2}$ are identical to $(3.10\text{-}12)_{1,2}$ so that the solutions for g_k and h_k are provided in (3.10-15) and (3.10-16) in which η_k is defined in (3.10-26). Although η_k in (3.10-26) and (3.10-13) are different, we will see below that they become identical when (x_1^o, x_2^o) is on the elliptic boundary.

The special case of an edge dislocation inside a circular isotropic inclusion was studied by Dundurs and Sendeckyj (1965b).

(iii) (x_1^o, x_2^o) on the elliptic boundary.

When (x_1^o, x_2^o) is on the elliptic boundary,

$$x_1^{\mathrm{o}} = a\cos\psi_{\mathrm{o}}, \quad x_2^{\mathrm{o}} = b\sin\psi_{\mathrm{o}},$$

where ψ_{o} is a real constant. We then have

$$\zeta_{\mathrm{o}}^{(1)} = \zeta_{\mathrm{o}}^{(2)} = e^{i\psi_{\mathrm{o}}}, \quad \overline{\zeta}_{\mathrm{o}}^{(1)} = (\zeta_{\mathrm{o}}^{(2)})^{-1} = e^{-i\psi_{\mathrm{o}}},$$

$$z_{\mathrm{o}}^{(2)} = \hat{z}_{\mathrm{o}}^{(2)} = \widetilde{z}_{\mathrm{o}}^{(2)}, \quad \tau_{\mathrm{o}}^{(2)} = \hat{\tau}_{\mathrm{o}}^{(2)} = \widetilde{\tau}_{\mathrm{o}}^{(2)} = t^{(2)}e^{-i\psi_{\mathrm{o}}}.$$

It is easily seen that, in the limit $(x_1^{\mathrm{o}}, x_2^{\mathrm{o}})$ is on the elliptic boundary, $u_{\mathrm{o}}^{(2)}$, $\phi_{\mathrm{o}}^{(2)}$, q_1, q_2, g_k, and h_k in cases (i) and (ii) are identical. Therefore (3.10-5) and (3.10-6), or (3.10-17) and (3.10-18) apply to the present case.

When $t^{(2)}=0$ (which arises when $p^{(2)}=ia/b$) η_k vanishes so do g_k and h_k. The infinite series that appeared in the solutions vanish. Therefore all solutions presented have a closed form. This is the case as long as $p^{(2)}=ia/b$ for material 2 inside the inclusion. Material 1 outside the inclusion can be arbitrary. In the special case of a circular inclusion with isotropic materials inside and outside the inclusion, the results presented here recover that obtained by Dundurs (1967a).

Closed form solutions are also possible when

$$\mu_1 = \mu_2, \quad \text{or} \quad K=0.$$

This does not mean that materials 1 and 2 are identical because, by (3.3-10), the eigenvalue p can be different. If $(x_1^{\mathrm{o}}, x_2^{\mathrm{o}})$ is outside the elliptic inclusion it can be shown that

$$q_1 = 0, \quad q_2 = q_1^{\infty}, \quad h_k = 0, \quad g_k = -q_1^{\infty}(\widetilde{\tau}_{\mathrm{o}}^{(2)})^k, \tag{3.10-28}$$

so that the infinite series in (3.10-6) vanishes while the infinite series in (3.10-5) can be written with the use of (3.9-15) as

$$\sum_{k=1}^{\infty}\frac{1}{k}g_k(\zeta^{(1)})^{-k} = -q_1^{\infty}\sum_{k=1}^{\infty}\frac{1}{k}\left(\widetilde{\tau}_{\mathrm{o}}^{(2)}(\zeta^{(1)})^{-1}\right)^k = q_1^{\infty}\ln\left(1-\widetilde{\tau}_{\mathrm{o}}^{(2)}(\zeta^{(1)})^{-1}\right). \tag{3.10-29}$$

If $(x_1^{\mathrm{o}}, x_2^{\mathrm{o}})$ is inside the elliptic inclusion, observing that $q_2^{\infty} = q_1^{\infty}$ it can be shown that (3.10-28) and (3.10-29) remain valid if $\widetilde{\tau}_{\mathrm{o}}^{(2)}$ is replaced by $\tau_{\mathrm{o}}^{(2)}$. Therefore (3.10-17) and (3.10-18) also have a closed form solution when $\mu_1 = \mu_2$ (Ting, 1995b).

3.11 A Crack in an Infinite Space and Bimaterials

A crack is a special case of an ellipse with the minor axis b=0. Therefore the solutions to crack problems can be deduced from the corresponding solutions to the problems for an elliptic hole presented in the previous section. The t-circle and the $\tilde{\zeta}$-circle in Fig. 3.5(b) coincide with the unit circle. All image singularities are located on the second sheet of the Riemann surface (Section 3.7).

Let a crack of length $2a$ be located at

$$-a < x_1 < a, \quad x_2 = 0.$$

The function ζ and ζ^{-1} in (3.7-4) and (3.7-6) are replaced by

$$\zeta = \left\{z + \sqrt{z^2 - a^2}\right\} a^{-1}, \quad \zeta^{-1} = \left\{z - \sqrt{z^2 - a^2}\right\} a^{-1} \tag{3.11-1}$$

When the solutions contain $\ln\zeta$ or $\ln(\zeta^{-1})$, the factor a^{-1} on the right of (3.11-1) can be ignored because they only add constant terms to the solutions. A constant term for the displacement means a rigid body motion while a constant term for the stress function provides no stresses. We now discuss solutions to various problems below.

(i) **The Green's Functions.** For an antiplane force f and a screw dislocation b applied at (x_1^o, x_2^o), Fig. 3.12(a), the solution is, by (3.8-2),

$$\phi + i\mu u = \frac{1}{2\pi i}\left\{q^\infty \ln(\zeta - \zeta_o) + q\ln(\zeta^{-1} - \bar{\zeta}_o)\right\} \tag{3.11-2}$$

where

$$q^\infty = \mathrm{f} + i\mu\,\mathrm{b},$$

$$\zeta_o = \left\{z_o + \sqrt{z_o^2 - a^2}\right\} a^{-1}, \quad z_o = x_1^o + p x_2^o.$$

In (3.11-2) ζ and ζ^{-1} are given in (3.11-1), $q = \bar{q}^\infty$ if the crack

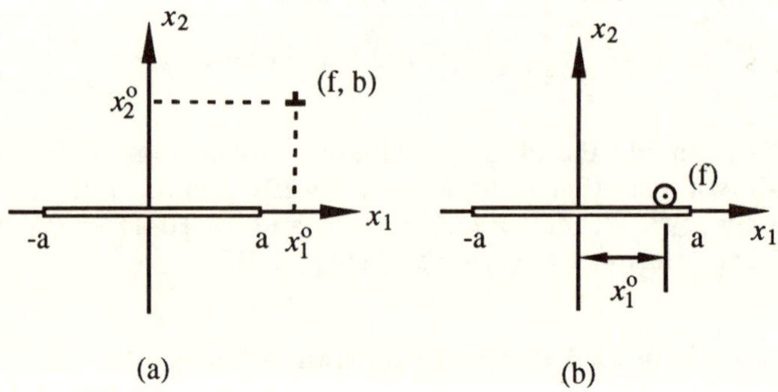

Fig. 3.12 The infinite space with a crack is subjected to (a) an antiplane force f and a screw dislocation b at (x_1^o, x_2^o) or (b) an antiplane force f at the crack surface.

surface is traction-free, and $q=-\overline{q}^{\infty}$ if the crack surface is rigidly clamped, i.e., if the crack is a rigid line inclusion.

(ii) **The Surface Green's Function**. Let an antiplane force f be applied at a point on the crack surface (Fig. 3.12(b)). We have

$$x_1^o = a\cos\psi_o, \quad x_2^o = 0, \quad \zeta_o = e^{i\psi_o}, \quad q^{\infty} = \overline{q}^{\infty} = \mathrm{f}.$$

If f is on the upper (or lower) crack surface, $0 \le \psi_o \le \pi$ (or $-\pi < \psi_o < 0$). The solution (3.11-2) reduces to

$$\phi + i\mu u = \frac{\mathrm{f}}{2\pi i}\ln\left[(\zeta - e^{i\psi_o})(\zeta^{-1} - e^{-i\psi_o})\right]. \tag{3.11-3}$$

The real and imaginary parts of (3.11-3) can be written in the form shown in (3.8-13).

(iii) **Uniform Loading at Infinity**. When the material is subjected to a uniform loading σ_{31}^{∞} and σ_{32}^{∞} at infinity, the solutions are given by (3.8-14) or (3.8-15) depending on whether the crack surface is traction-free or fixed with no displacement. The expressions

$$q\zeta - \overline{q}\zeta^{-1} \quad \text{and} \quad q\zeta + \overline{q}\zeta^{-1}$$

appearing in these two equations can be simplified and written explicitly for the problem at hand. However it is perhaps more transparent if we derive the solution for the present problem directly instead of considering the crack as a degenerate case of an ellipse. The solution can be written as

$$\phi + i\mu u = q^{\infty}z + q\left\{\sqrt{z^2 - a^2} - z\right\} \tag{3.11-4}$$

where q^{∞} and q are constants to be determined. The term associated with q^{∞} is the solution for a uniform stress when the crack is absent, and the term associated with q is the *disturbed* solution due to the presence of the crack. As $|z| \to \infty$ the q terms vanish, assuring us that u and ϕ vanish at infinity for the disturbed solution. Thus the q^{∞} terms must satisfy the conditions that the uniform stresses are σ_{31}^{∞} and σ_{32}^{∞}, i.e.,

$$\mathrm{Re}(q^{\infty}) = \sigma_{32}^{\infty}, \quad \mathrm{Re}(pq^{\infty}) = -\sigma_{31}^{\infty}.$$

This is identical to (3.8-16). By (3.8-17) we have

$$q^{\infty} = \sigma_{32}^{\infty} + i\mu u_{,1}^{\infty} \tag{3.11-5}$$

where, by (3.8-18),

$$u_{,1}^{\infty} = 2\varepsilon_{31}^{\infty} = \mu^{-2}(C_{44}\sigma_{31}^{\infty} - C_{45}\sigma_{32}^{\infty}).$$

If the crack is a rigid line inclusion u=0 at the crack. It is readily shown from (3.11-4) that

$$q = i\mu u_{,1}^{\infty}. \tag{3.11-6}$$

The displacement at points on the x_1-axis outside the rigid line is

$$u = \pm u_{,1}^{\infty}\sqrt{x_1^2 - a^2} \quad \text{for } \pm x_1 > a \tag{3.11-7}$$

which depends only on $u_{,1} = 2\varepsilon_{31}$ at infinity. The surface traction σ_{32} on the surface x_2 =0 is

$$\sigma_{32} = \phi_{,1} = \begin{cases} \sigma_{32}^{\infty}, & \text{for } |x_1| > a, \\ \sigma_{32}^{\infty} \pm \mu u_{,1}^{\infty}\dfrac{x_1}{\sqrt{a^2 - x_1^2}}, & \text{for } x_2 = \pm 0,\ |x_1| < a. \end{cases} \tag{3.11-8}$$

The shear stress σ_{31} at x_2=0 can be shown to be

$$\sigma_{31} = -\phi_{,2} = \begin{cases} \sigma_{31}^{\infty} + \dfrac{\mu^2 u_{,1}^{\infty}}{C_{44}}\left\{\dfrac{|x_1|}{\sqrt{x_1^2 - a^2}} - 1\right\}, & \text{for } |x_1| > a, \\ \dfrac{\mu^2 u_{,1}^{\infty}}{C_{44}}\left\{\dfrac{\pm C_{45}x_1}{\mu\sqrt{a^2 - x_1^2}} - 1\right\}, & \text{for } x_2 = \pm 0,\ |x_1| < a. \end{cases} \tag{3.11-9}$$

Thus there is a square root stress singularity at the crack tip unless $u_{,1}^{\infty}$ vanishes. Notice that σ_{31} at the tip of the rigid line inclusion is not singular when C_{45}=0.

If the crack surface is traction-free, setting ϕ=0 in (3.11-4) leads to

$$q = \sigma_{32}^{\infty}. \tag{3.11-10}$$

The displacement at points on the x_1-axis is

$$u = \begin{cases} u_{,1}^{\infty}x_1, & \text{for } \pm x_1 > a, \\ u_{,1}^{\infty}x_1 \pm \dfrac{1}{\mu}\sigma_{32}^{\infty}\sqrt{a^2 - x_1^2}, & \text{for } x_2 = \pm 0,\ |x_1| < a, \end{cases} \tag{3.11-11}$$

Therefore the *crack opening* Δu is

$$\Delta u = u^+ - u^- = \frac{2}{\mu}\sigma_{32}^{\infty}\sqrt{a^2 - x_1^2}. \tag{3.11-12}$$

The surface traction σ_{32} on the surface $x_2=0$ outside the crack is

$$\sigma_{32} = \phi_{,1} = \sigma_{32}^{\infty} \frac{|x_1|}{\sqrt{x_1^2 - a^2}}, \quad \text{for } |x_1| > a \tag{3.11-13}$$

which is independent of the material constants. There is a square root stress singularity at the crack tip unless σ_{32}^{∞} vanishes. The stress σ_{31} along the crack is

$$\sigma_{31} = \begin{cases} \sigma_{31}^{\infty} + \dfrac{C_{45}\sigma_{32}^{\infty}}{C_{44}} \left\{ \dfrac{|x_1|}{\sqrt{x_1^2 - a^2}} - 1 \right\}, & \text{for } |x_1| > a, \\[2ex] \sigma_{31}^{\infty} - \dfrac{\sigma_{32}^{\infty}}{C_{44}} \left\{ \dfrac{\pm \mu x_1}{\sqrt{a^2 - x_1^2}} - C_{45} \right\}, & \text{for } x_2 = \pm 0,\ |x_1| < a. \end{cases} \tag{3.11-14}$$

(iv) **An Interfacial Crack.** Let two dissimilar anisotropic elastic half-spaces be bonded together along $x_2=0$. The materials in $x_2>0$ and $x_2<0$ are designated as materials 1 and 2, respectively. An interfacial crack of length $2a$ is located at $x_2=0$, $|x_1|<a$, and is subjected to a uniform surface traction $-\sigma_{32}^{\infty}$ at the crack surface. Let the subscripts 1, 2 or the superscripts (1), (2) denote quantities in materials 1 and 2, respectively. Consider the solutions

$$\phi^{(1)} + i\mu_1 u^{(1)} = \sigma_{32}^{\infty} \left\{ \sqrt{z^{(1)2} - a^2} - z^{(1)} \right\} \tag{3.11-15}$$

for material 1 and

$$\phi^{(2)} + i\mu_2 u^{(2)} = \sigma_{32}^{\infty} \left\{ \sqrt{z^{(2)2} - a^2} - z^{(2)} \right\} \tag{3.11-16}$$

for material 2. They are the second term in (3.11-4) with $q = \sigma_{32}^{\infty}$. It is easily shown that

$$u^{(1)} = u^{(2)} = 0 \quad \text{at } x_2 = 0,\ |x_1| > a,$$

$$\sigma_{32}^{(1)} = \sigma_{32}^{(2)} = \begin{cases} -\sigma_{32}^{\infty} & \text{at } x_2 = 0,\ |x_1| < a, \\[2ex] \sigma_{32}^{\infty} \left(\dfrac{|x_1|}{\sqrt{x_1^2 - a^2}} - 1 \right) & \text{at } x_2 = 0,\ |x_1| > a. \end{cases}$$

Hence all boundary conditions of the problem are satisfied. The crack opening Δu can be shown to be

$$\Delta u = \left(\mu_1^{-1} + \mu_2^{-1} \right) \sigma_{32}^{\infty} \sqrt{a^2 - x_1^2}\,.$$

3.12 Infinite Space With a Hole of Arbitrary Shape

Consider a simply connected region whose boundary Γ is of arbitrary shape. The region outside Γ is mapped into the region outside a unit circle in the ζ plane through the transformation

$$z = x_1 + px_2 = w(\zeta). \tag{3.12-1}$$

For the transformation to be one-to-one, the roots of

$$\frac{d}{d\zeta} w(\zeta) = 0 \tag{3.12-2}$$

must lie inside the circle $|\zeta| < 1$. For sufficiently large $|\zeta|$, $w(\zeta)$ may be approximated by (Muskhelishvili, 1953)

$$z = c_o\zeta + c_1\zeta^{-1} + c_2\zeta^{-2} + \dots \tag{3.12-3}$$

where c_o, c_1, c_2,... are complex constants. If ζ is replaced by ζ^{-1}, the region outside Γ is mapped into the region inside the unit circle. To map the region inside Γ into the region inside the unit circle, we remove the $c_o\zeta$ term and replace ζ^{-1}, ζ^{-2},... by ζ, ζ^2,...

Let p' and p'' be the real and imaginary parts of p, respectively, and write

$$z = X_1 + iX_2 = w(\zeta) \tag{3.12-4}$$

where

$$X_1 = \operatorname{Re} z = x_1 + p'x_2, \quad X_2 = \operatorname{Im} z = p''x_2. \tag{3.12-5}$$

It is known that any given curve in the z-plane can be mapped into a circle of unit radius, although it is not a simple task to determine $w(\zeta)$ for an arbitrarily given curve. For anisotropic materials for which $p \neq i$, we first map the curve Γ in the (x_1,x_2)-plane to a curve in the z-plane through (3.12-5), and then map the curve in the z-plane to a unit circle in the ζ-plane. As an illustration, let the hole in the (x_1,x_2)-plane be a rectangle of sides $2a$ and $2b$, Fig. 3.13. It is mapped into a parallelogram in the z-plane. To map the parallelogram into a unit circle in the ζ-plane we employ the *Schwartz–Christoffel* transformation (Churchill, 1948) which gives

$$z = A\int^{\zeta} (t - e^{i\psi_1})^{k_1} (t - e^{i\psi_2})^{k_2} (t - e^{i\psi_3})^{k_3} (t - e^{i\psi_4})^{k_4} t^{-2} dt \tag{3.12-6}$$

where A is a complex constant,

$$k_1 = k_3 = \frac{1}{2} + \frac{1}{\pi}\tan^{-1}(p'/p''),$$

$$k_2 = k_4 = \frac{1}{2} - \frac{1}{\pi}\tan^{-1}(p'/p''),$$

$$\psi_3 = \pi + \psi_1, \quad \psi_4 = \pi + \psi_2.$$

Without loss in generality we let $\psi_2 = \pi - \psi_1$ and write (3.12-6) as

$$z = A\int^{\zeta}(t^2 - e^{2i\psi_1})^{k_1}(t^2 - e^{-2i\psi_1})^{k_2}t^{-2}dt. \tag{3.12-7}$$

The constants A and ψ_1 can be chosen such that points 1 and 2 in the (x_1,x_2)-plane are mapped to points 1 and 2 on the unit circle.

When the function $w(\zeta)$ of (3.12-1) is found, and assuming that the inverse

$$\zeta = w^{-1}(z) \tag{3.12-8}$$

can be obtained, the solutions presented in Section 3.8 for an elliptic hole apply to a hole of arbitrary shape if we replace the ζ appeared there by (3.12-8). Thus the Green's function (3.8-2), the surface Green's function (3.8-13), the solutions (3.8-14) and (3.8-15) for uniform loading at infinity, and the solution (3.8-21) for

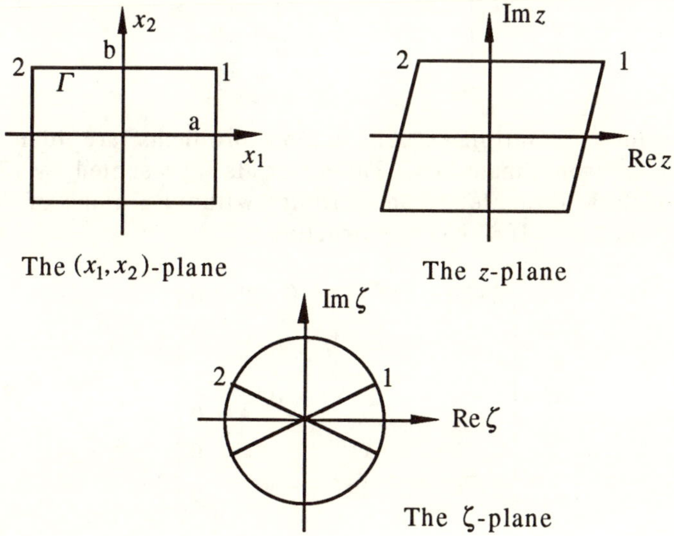

Fig. 3.13 With the complex variable $z = x_1 + p\,x_2$ a rectangle in the (x_1, x_2)-plane is first mapped into a parallelogram in the z-plane. It is then mapped into a circle in the ζ-plane.

arbitrary loading at the surface of the hole remain valid for a hole of arbitrary shape.

3.13 Remarks

From (3.1-9), a plane strain or an antiplane deformation is possible for an anisotropic material if its elastic stiffness matrix **C** has the structure

$$\mathbf{C}=\begin{bmatrix} * & * & * & 0 & 0 & * \\ & * & * & 0 & 0 & * \\ & & * & * & * & * \\ & & & * & * & 0 \\ & & & & * & 0 \\ & & & & & * \end{bmatrix} \tag{3.13-1}$$

where only the upper triangle is shown. Monoclinic materials with the symmetry plane at $x_3=0$ are a special case of (3.13-1). If **C** has the structure

$$\mathbf{C}=\begin{bmatrix} * & * & * & 0 & 0 & * \\ & * & * & 0 & 0 & * \\ & & * & * & * & * \\ & & & \mu & 0 & 0 \\ & & & & \mu & 0 \\ & & & & & * \end{bmatrix} \tag{3.13-2}$$

the solutions to antiplane deformation problems are identical to those for isotropic materials. The materials represented by (3.13-2) can be triclinic (completely anisotropic) without a plane of symmetry (Section 2.6). If **C** has the structure

$$\mathbf{C}=\begin{bmatrix} \lambda+2\mu & \lambda & * & 0 & 0 & 0 \\ & \lambda+2\mu & * & 0 & 0 & 0 \\ & & * & * & * & * \\ & & & * & * & 0 \\ & & & & * & 0 \\ & & & & & \mu \end{bmatrix} \tag{3.13-3}$$

where λ and μ are the Lamé constants, the solutions to plane strain problems are identical to those for isotropic materials. Again, the materials represented by (3.13-3) can be triclinic.

In summary, the solutions available in the literature for isotropic materials are valid for more general materials given by (3.13-2) for antiplane deformations and by (3.13-3) for plane strain deformations. The corresponding elastic compliance matrix **s**, which is the inverse of **C** given in (3.13-2) or (3.13-3) in general has no zero elements (Ting, 1994; see also Section 4.7).

3.14 A Theorem

In studying two-dimensional deformations in which the displacements depend on x_1 and x_2 only, a differential equation of the form

$$U_{,1}+V_{,2}=0 \tag{3.14-1}$$

where U and V are functions of x_1 and x_2 appears. An example is seen in (3.3-4). The following theorem is useful in dealing with (3.14-1).

> **Theorem 3.14–1** Let $U(x_1,x_2)$ and $V(x_1,x_2)$ satisfy (3.14-1). Then there exists a potential function Φ such that U and V are given by
>
> $$U=-\Phi_{,2}, \quad V=\Phi_{,1}. \tag{3.14-2}$$

It is clear that (3.14-2) satisfies (3.14-1). To show the completeness of the representation (3.14-2), suppose that U and V satisfy (3.14-1) and consider (Chadwick and Smith, 1974)

$$\Phi(x_1,x_2)=\int_{a_1}^{x_1} V(\xi,a_2)d\xi-\int_{a_2}^{x_2} U(x_1,\eta)d\eta \tag{3.14-3}$$

where a_1 and a_2 are arbitrary constants. By a direct differentiation and employing (3.14-1), it is readily shown that (3.14-3) leads to (3.14-2)

Equation (3.3-12a) could have been obtained by an application of the theorem to (3.3-4). Theorem 3.14–1 will be useful in Chapters 4 and 5.

Chapter 4

The Lekhnitskii Formalism

A two-dimensional deformation means that the displacements u_i (i= 1,2,3) or the stresses σ_{ij} depend on x_1 and x_2 only. Among several formalisms for two-dimensional deformations of anisotropic elastic materials the Lekhnitskii (1950, 1957) formalism is the oldest, and has been extensively employed by the engineering community. The Lekhnitskii formalism essentially generalizes the Muskhelishvili (1953) approach for solving two-dimensional deformations of isotropic elastic materials. The formalism begins with the stresses and assumes that they depend on x_1 and x_2 only. The Stroh formalism, to be introduced in the next chapter, starts with the displacements and assumes that they depend on x_1 and x_2 only. Therefore the Lekhnitskii formalism is in terms of the reduced elastic compliances while the Stroh formalism is in terms of the elastic stiffnesses. It should be noted that Green and Zerna (1960) also proposed a formalism for two-dimensional deformations of anisotropic elastic materials. Their approach however is limited to materials that possess a symmetry plane at x_3=0. The derivations presented below do not follow exactly those of Lekhnitskii.

4.1 The Airy Function χ and the Stress Function ψ

When the stresses σ_{ij} are independent of x_3 the equations of equilibrium (3.1-2) can be written as

$$\sigma_{i1,1} + \sigma_{i2,2} = 0, \tag{4.1-1}$$

i=1,2,3. This is in the form of (3.14-1). By theorem 3.14–1 there exist potential functions φ_i (i=1,2,3) such that

$$\sigma_{i1} = -\varphi_{i,2}, \quad \sigma_{i2} = \varphi_{i,1}. \tag{4.1-2}$$

The fact that $\sigma_{21} = \sigma_{12}$ means that

$$\varphi_{1,1} + \varphi_{2,2} = 0. \tag{4.1-3}$$

This is again in the form of (3.14-1), and thus there exists a potential function χ such that

$$\varphi_1 = -\chi_{,2}, \quad \varphi_2 = \chi_{,1}.$$

In conjunction with (4.1-2) and setting $\varphi_3 = -\psi$ we obtain

$$\sigma_{11} = \chi_{,22}, \quad \sigma_{22} = \chi_{,11}, \quad \sigma_{12} = -\chi_{,12}, \qquad \sigma_{32} = -\psi_{,1}, \quad \sigma_{31} = \psi_{,2}. \tag{4.1-4}$$

The function χ is the Airy function while ψ is the stress function. When the stresses are represented by (4.1-4) in terms of χ and ψ, the equations of equilibrium (4.1-1) are automatically satisfied. It remains to determine the functions χ and ψ. It should be noted that the stress component σ_{33} is missing in (4.1-1) and (4.1-4).

4.2 Displacements for Two-Dimensional Stresses

The Lekhnitskii formalism assumes that the stresses σ_{ij} are two-dimensional. This is different from the assumption that the displacements u_i are two-dimensional (Chapter 5). When u_i (i=1,2,3) depend on x_1 and x_2 only, so do σ_{ij}. The converse however does not hold. When σ_{ij} depend on x_1 and x_2 only the displacements u_i may depend on x_3. The additional terms in u_i that depend on x_3 are determined as follows.

Instead of u_1, u_2, u_3, let u, v w be the displacement components and write the strain-displacement relations (3.1-1) as

$$\varepsilon_{11} = u_{,1}, \quad \varepsilon_{22} = v_{,2}, \quad \varepsilon_{33} = w_{,3}, \qquad 2\varepsilon_{23} = v_{,3} + w_{,2}, \quad 2\varepsilon_{13} = u_{,3} + w_{,1} \quad 2\varepsilon_{12} = u_{,2} + v_{,1}. \tag{4.2-1}$$

Since σ_{ij} are independent of x_3 so are ε_{ij}. Hence the right sides of (4.2-1) are independent of x_3. From $(4.2\text{-}1)_{1,2,3}$ we obtain

$$\begin{aligned} u(x_1,x_2,x_3) &= u_1(x_1,x_2) + U(x_2,x_3), \\ v(x_1,x_2,x_3) &= u_2(x_1,x_2) + V(x_1,x_3), \\ w(x_1,x_2,x_3) &= u_3(x_1,x_2) + x_3 W(x_1,x_2), \end{aligned} \tag{4.2-2}$$

where u_i (i=1,2,3) and W depend on x_1, x_2 only, U on x_2, x_3, and V on x_1, x_3. Use of (4.2-2) in $(4.2\text{-}1)_{4,5}$ and remember that ε_{ij} do not depend on x_3,

$$V_{,3} + x_3 W_{,2}, \quad U_{,3} + x_3 W_{,1}, \quad U_{,2} + V_{,1} \tag{4.2-3}$$

must be independent of x_3. The first two imply that $U_{,3}$ and $V_{,3}$ are linear in x_3, $W_{,2}$ depends on x_1, and $W_{,1}$ depends on x_2. Hence

$$U = -\tfrac{1}{2} x_3^2 (A + D x_2) + x_3 f(x_2),$$

$$V = -\tfrac{1}{2} x_3^2 (B + D x_1) + x_3 g(x_1),$$

$$W = Ax_1 + Bx_2 + C + Dx_1x_2,$$

where A, B, C, D are arbitrary constants and f and g are functions of x_2 and x_1, respectively. The last one in (4.2-3) tells us that

$$-Dx_3^2 + x_3 f'(x_2) + x_3 g'(x_1)$$

in which the prime denotes differentiation is independent of x_3. Therefore

$$D = 0, \quad f = -(\omega x_2 + \gamma_2), \quad g = \omega x_1 + \gamma_1,$$

where γ_1, γ_2, and ω are constants. In summary, the displacements for two-dimensional stresses have the representation

$$\begin{aligned} u(x_1,x_2,x_3) &= u_1(x_1,x_2) - \tfrac{1}{2}Ax_3^2 - \omega x_2 x_3, \\ v(x_1,x_2,x_3) &= u_2(x_1,x_2) - \tfrac{1}{2}Bx_3^2 + \omega x_1 x_3, \\ w(x_1,x_2,x_3) &= u_3(x_1,x_2) + x_3(Ax_1 + Bx_2 + C). \end{aligned} \tag{4.2-4}$$

The $-\gamma_2 x_3$ term in the first equation and the $\gamma_1 x_3$ term in the second equation have been deleted by adding the following rigid body rotation

$$u = \gamma_2 x_3 - \gamma_3 x_2, \quad v = \gamma_3 x_1 - \gamma_1 x_3, \quad w = \gamma_1 x_2 - \gamma_2 x_1,$$

and by allowing the terms containing x_1 and x_2 to be absorbed in $u_i(x_1,x_2)$ $(i=1,2,3)$. We may still superimpose on the right sides of (4.2-4) a rigid body translation and rotation without altering the stresses. The x_3 terms in (4.2-4) are the differences between the assumptions that the stresses or the displacements are two-dimensional.

The strains obtained by inserting (4.2-4) into (4.2-1) are

$$\begin{aligned} &\varepsilon_{11} = u_{1,1}, \quad \varepsilon_{22} = u_{2,2}, \quad \varepsilon_{33} = Ax_1 + Bx_2 + C, \\ &2\varepsilon_{23} = u_{3,2} + \omega x_1, \quad 2\varepsilon_{13} = u_{3,1} - \omega x_2, \quad 2\varepsilon_{12} = u_{1,2} + u_{2,1}. \end{aligned} \tag{4.2-5}$$

The strain ε_{33} is linear in x_1 and x_2, implying that the deformation is a bending about the line $Ax_1+Bx_2+C=0$. The deformation associated with ω is a torsion about the x_3-axis. The compatibility conditions (Sokolnikoff, 1956) for (4.2-5) are

$$\varepsilon_{13,2} - \varepsilon_{23,1} = -\omega, \quad \varepsilon_{11,22} + \varepsilon_{22,11} - 2\varepsilon_{12,12} = 0. \tag{4.2-6}$$

The first equation is obtained by eliminating u_3 between $(4.2\text{-}5)_4$ and $(4.2\text{-}5)_5$. The second equation is derived from $(4.2\text{-}5)_{1,2,6}$ by eliminating u_1 and u_2.

4.3 Differential Equations for χ and ψ

In the contracted notation the strain-stress relations are

$$\varepsilon_\alpha = s_{\alpha\beta}\sigma_\beta \tag{4.3-1}$$

where $s_{\alpha\beta}$ are the elastic compliances. For $\alpha = 3$,

$$\varepsilon_3 = s_{3\beta}\sigma_\beta \tag{4.3-2}$$

from which we obtain

$$\sigma_3 = \frac{\varepsilon_3}{s_{33}} - \frac{1}{s_{33}}\sum_{\beta\neq 3} s_{3\beta}\sigma_\beta . \tag{4.3-3}$$

Elimination of σ_3 in (4.3-1) leads to

$$\varepsilon_\alpha = s'_{\alpha\beta}\sigma_\beta + \frac{s_{\alpha 3}}{s_{33}}\varepsilon_3 \tag{4.3-4}$$

where

$$s'_{\alpha\beta} = s_{\alpha\beta} - \frac{s_{\alpha 3}s_{3\beta}}{s_{33}} \tag{4.3-5}$$

are the *reduced elastic compliances*. It is clear that $s'_{\alpha\beta}$ is symmetric and that

$$s'_{\alpha 3} = s'_{3\alpha} = 0 \quad (\alpha = 1,2,\ldots,6). \tag{4.3-6}$$

Therefore there is no need to exclude β=3 in the repeated indices in (4.3-4).

The strains in terms of χ and ψ are obtained from (4.3-4) and (4.1-4) as

$$\varepsilon_\alpha = s'_{\alpha 1}\chi_{,22} + s'_{\alpha 2}\chi_{,11} - s'_{\alpha 4}\psi_{,1} + s'_{\alpha 5}\psi_{,2} - s'_{\alpha 6}\chi_{,12} + \frac{s_{\alpha 3}}{s_{33}}\varepsilon_3 \tag{4.3-7}$$

where ε_3 is given by $(4.2\text{-}5)_3$. Using the contracted notation the compatibility conditions (4.2-6) are

$$\varepsilon_{5,2} - \varepsilon_{4,1} = -2\omega, \quad \varepsilon_{1,22} + \varepsilon_{2,11} - \varepsilon_{6,12} = 0.$$

Insertion of (4.3-7) into the above leads to

$$L_3\chi + L_2\psi = \Delta, \qquad L_4\chi + L_3\psi = 0, \tag{4.3-8}$$

where

$$\Delta = -2\omega + \frac{1}{s_{33}}(s_{43}A - s_{53}B) \tag{4.3-9}$$

and the operators L_2, L_3, and L_4 are

$$L_2 = s'_{55}\frac{\partial^2}{\partial x_2^2} - 2s'_{45}\frac{\partial^2}{\partial x_2 \partial x_1} + s'_{44}\frac{\partial^2}{\partial x_1^2},$$

$$L_3 = s'_{15}\frac{\partial^3}{\partial x_2^3} - (s'_{14}+s'_{56})\frac{\partial^3}{\partial x_2^2 \partial x_1} + (s'_{25}+s'_{46})\frac{\partial^3}{\partial x_2 \partial x_1^2} - s'_{24}\frac{\partial^3}{\partial x_1^3}, \quad (4.3\text{-}10)$$

$$L_4 = s'_{11}\frac{\partial^4}{\partial x_2^4} - 2s'_{16}\frac{\partial^4}{\partial x_2^3 \partial x_1} + (2s'_{12}+s'_{66})\frac{\partial^4}{\partial x_2^2 \partial x_1^2} - 2s'_{26}\frac{\partial^4}{\partial x_2 \partial x_1^3} + s'_{22}\frac{\partial^4}{\partial x_1^4}.$$

The Δ in (4.3-8) is the inhomogeneous term which is a constant according to (4.3-9).

Equations (4.3-8) are the differential equations for χ and ψ. By setting

$$\hat{\psi} = \psi - \frac{\Delta}{2s'_{44}}x_1^2, \quad (4.3\text{-}11)$$

they are reduced to a system of homogeneous differential equations

$$L_3\chi + L_2\hat{\psi} = 0, \qquad L_4\chi + L_3\hat{\psi} = 0. \quad (4.3\text{-}12)$$

To find a general solution of (4.3-12) we eliminate $\hat{\psi}$ between the two equations to obtain

$$(L_2L_4 - L_3L_3)\chi = 0. \quad (4.3\text{-}13)$$

The same equation applies to $\hat{\psi}$. This is a homogeneous differential equation of sixth order. A general solution depends on a composite variable which is a linear combination of x_1 and x_2. Without loss in generality we set the coefficient of x_1 in the linear combination to be unity. Hence let

$$\chi(x_1, x_2) = F(z), \quad (4.3\text{-}14)$$

$$z = x_1 + px_2, \quad (4.3\text{-}15)$$

where p is a constant to be determined. Substitution of (4.3-14) into (4.3-13) yields

$$l_2(p)l_4(p) - l_3(p)l_3(p) = 0 \quad (4.3\text{-}16)$$

in which

$$l_2(p) = s'_{55}p^2 - 2s'_{45}p + s'_{44},$$

$$l_3(p) = s'_{15}p^3 - (s'_{14}+s'_{56})p^2 + (s'_{25}+s'_{46})p - s'_{24}, \quad (4.3\text{-}17)$$

$$l_4(p) = s'_{11}p^4 - 2s'_{16}p^3 + (2s'_{12}+s'_{66})p^2 - 2s'_{26}p + s'_{22}.$$

The $\hat{\psi}$ is obtained from (4.3-12) as

$$\hat{\psi}(x_1, x_2) = \lambda F'(z), \tag{4.3-18}$$

$$\lambda(p) = -\frac{l_3(p)}{l_2(p)} = -\frac{l_4(p)}{l_3(p)}. \tag{4.3-19}$$

In (4.3-18) the prime denotes differentiation with z. The second equality in (4.3-19) is equivalent to (4.3-16).

Equation (4.3-16) is a sextic equation in p. Therefore there are six roots for p which will be shown to be complex in the next section. They are the eigenvalues of elastic constants.

4.4 Eigenvalues p

The six roots p of (4.3-16) are the eigenvalues of elastic constants. $l_2(p)$, $l_4(p)$, and $l_2(p)l_4(p) - l_3(p)l_3(p)$ are polynomials in p of degree two, four, and six, respectively. We will show that they cannot have a real root if the strain energy is positive.

First we rearrange the rows and columns of the 5×5 matrix $\mathbf{s}'$ and divide it into four submatrices as

$$\mathbf{s}' = \left[\begin{array}{ccc|cc} s'_{11} & s'_{12} & s'_{16} & s'_{14} & s'_{15} \\ & s'_{22} & s'_{26} & s'_{24} & s'_{25} \\ & & s'_{66} & s'_{64} & s'_{65} \\ \hline & & & s'_{44} & s'_{45} \\ & & & & s'_{55} \end{array}\right] = \left[\begin{array}{c|c} \mathbf{s}'_4 & \mathbf{s}'_3 \\ \hline \mathbf{s}'^T_3 & \mathbf{s}'_2 \end{array}\right]. \tag{4.4-1}$$

In this rearranged form $\mathbf{s}'$ remains positive definite. Therefore the submatrices $\mathbf{s}'_4$ and $\mathbf{s}'_2$ are positive definite. It is easily verified that the polynomials l_2, l_3, and l_4 can be written as

$$l_2(p) = \mathbf{h}_2^T \mathbf{s}'_2 \mathbf{h}_2, \quad l_3(p) = \mathbf{h}_4^T \mathbf{s}'_3 \mathbf{h}_2, \quad l_4(p) = \mathbf{h}_4^T \mathbf{s}'_4 \mathbf{h}_4, \tag{4.4-2}$$

where

$$\mathbf{h}_2^T = (-1, \ p), \quad \mathbf{h}_4^T = (p^2, \ 1, \ -p).$$

Since $\mathbf{h}_2$ and $\mathbf{h}_4$ are nonzero, l_2 and l_4 are nonzero for any real p. Therefore l_2 and l_4 cannot have a real root p.

With l_2 known to be nonzero for any real p we write

$$l_2 l_4 - l_3 l_3 = l_2 (l_4 + \lambda l_3) \tag{4.4-3}$$

where $\lambda = -l_3 / l_2$ is bounded. Choosing

$$\mathbf{h}^T = (p^2, \ 1, \ -p, \ -\lambda, \ \lambda p),$$

it can be shown that

$$\mathbf{h}^T\mathbf{s}'\mathbf{h} = l_4 + 2\lambda l_3 + \lambda^2 l_2 = l_4 + \lambda l_3$$

which tells us that $l_4 + \lambda l_3$ and hence $l_2 l_4 - l_3 l_3$ are positive and nonzero for a real p. Therefore $l_2 l_4 - l_3 l_3$ cannot have a real root.

The sextic equation (4.3-16) has real coefficients. Since the roots cannot be real they consist of three pairs of complex conjugates. If p_α ($\alpha = 1,2,...,6$) are the roots, we let

$$p_{\alpha+3} = \bar{p}_\alpha, \quad \operatorname{Im} p_\alpha > 0, \quad (\alpha = 1,2,3), \tag{4.4-4}$$

where Im denotes the imaginary part. For materials that possess certain symmetry planes the sextic equation has a simple expression. We consider some special cases below.

(1) **Monoclinic Materials with the Symmetry Plane at x_1=0.** For this case the structure of $\mathbf{s}'$ is identical to that of $\mathbf{C}$ shown in (2.6-3a). Hence

$$\begin{aligned} l_2(p) &= s'_{55}p^2 + s'_{44}, \\ l_3(p) &= -(s'_{14} + s'_{56})p^2 - s'_{24}, \\ l_4(p) &= s'_{11}p^4 + (2s'_{12} + s'_{66})p^2 + s'_{22}, \end{aligned} \tag{4.4-5}$$

and the sextic equation is a polynomial of degree three in p^2. If the three roots of p^2 are all real, they must be negative since p cannot be real. We then have

$$p_1 = i\beta_1, \quad p_2 = i\beta_2, \quad p_3 = i\beta_3, \tag{4.4-6}$$

where β_1, β_2, and β_3 are real and positive. If only one root of p^2, say p_3^2, is real the other two roots p_1^2 and p_2^2 must be complex conjugates. We then have

$$p_1 = \alpha + i\beta, \quad p_2 = -\alpha + i\beta, \quad p_3 = i\beta_3, \tag{4.4-7}$$

in which α is real, and β and β_3 are real and positive.

(2) **Monoclinic Materials with the Symmetry Plane at x_2=0.** From (2.6-3b) the $l_2(p)$ and $l_4(p)$ are identical to those in (4.4-5) while

$$l_3(p) = s'_{15}p^3 + (s'_{25} + s'_{46})p. \tag{4.4-8}$$

The sextic equation is again a polynomial of degree three in p^2. Hence the eigenvalues have the representations (4.4-6) or (4.4-7).

(3) **Monoclinic Materials with the Symmetry Plane at x_3=0.** For these materials the $\mathbf{s}'_3$ in (4.4-1) vanishes identically. Therefore

the coefficients of the polynomial $l_3(p)$ all vanish and the sextic equation reduces to

$$l_2(p) = 0, \quad l_4(p) = 0. \tag{4.4-9}$$

If p_3 denotes the root of $l_2(p)=0$ with positive imaginary part, we have

$$p_3 = \frac{1}{s'_{55}}\left\{s'_{45} + i\sqrt{s'_{44}s'_{55} - s'_{45}s'_{45}}\right\}. \tag{4.4-10}$$

p_1 and p_2 are the roots of

$$s'_{11}p^4 - 2s'_{16}p^3 + (2s'_{12} + s'_{66})p^2 - 2s'_{26}p + s'_{22} = 0 \tag{4.4-11}$$

with positive imaginary parts. They have the representations

$$p_1 = \alpha_1 + i\beta_1, \quad p_2 = \alpha_2 + i\beta_2, \tag{4.4-12}$$

where α_1 and α_2 are real, and β_1 and β_2 are real and positive.

(4) **Transversely Isotropic Materials with the Axis of Symmetry at the x_3-axis.** The structure of **s**′ for these materials is identical to that shown in (2.6-7) with $s'_{66} = 2(s'_{11} - s'_{12})$. It is readily shown that the sextic equation reduces to

$$(p^2 + 1)^3 = 0. \tag{4.4-13}$$

Therefore

$$p_1 = p_2 = p_3 = i. \tag{4.4-14}$$

(5) **Isotropic Materials.** The sextic equation reduces to (4.4-13) and the eigenvalues are given by (4.4-14).

It should be pointed out that the triple eigenvalues (4.4-14) are not limited to isotropic materials and transversely isotropic materials with the axis of symmetry at the x_3-axis. As we will see in the next section there are many anisotropic materials whose eigenvalues are given by (4.4-14). In fact there are infinitely many anisotropic materials whose three eigenvalues can be equal and arbitrary.

4.5 Anisotropic Materials with $p_1 = p_2 = p_3$

Isotropic materials and transversely isotropic materials with the axis of symmetry at the x_3-axis are not the only materials for which

$$p_1 = p_2 = p_3 = i. \tag{4.5-1}$$

Since the eigenvalues p depend on the reduced elastic compliances $s'_{\alpha\beta}$, any anisotropic materials whose $s'_{\alpha\beta}$ are identical to $s'_{\alpha\beta}$ of isotropic materials (Section 4.7) have three identical eigenvalues given by (4.5-1). We will show that there are anisotropic materials whose $s'_{\alpha\beta}$ are different from those of isotropic materials but whose eigenvalues are given by (4.5-1).

If the eigenvalues are given by (4.5-1) the sextic equation (4.3-16) must have the form

$$(p^2+1)^3 = p^6+3p^4+3p^2+1=0. \tag{4.5-2}$$

This means that the coefficients of the p^5, p^3, and p terms in (4.3-16) must vanish, the coefficients of the p^6 term and the constant term are identical, and the coefficients of the p^4 and p^2 terms are three times the coefficients of the p^6 term. These relations impose six constraints on $s'_{\alpha\beta}$ if the eigenvalues p are given by (4.5-1). An important question is whether it is possible to find a positive definite 5×5 matrix $s'_{\alpha\beta}$ that satisfies all these six constraints.

For monoclinic materials with the symmetry plane at $x_1=0$ the first three constraints are automatically satisfied. The remaining three constraints can be written as, after assuming $s'_{44}=s'_{55}$,

$$(s'_{24})^2 = (s'_{22}-s'_{11})s'_{55},$$
$$s'_{14}+s'_{56}=s'_{24}, \qquad 2s'_{12}+s'_{66}=s'_{22}+s'_{11}.$$

They are satisfied by the following 5×5 matrix

$$\mathbf{s}' = s'_{55}\begin{bmatrix} \gamma & \nu & \psi-\varepsilon & 0 & 0 \\ & \alpha\gamma & \psi & 0 & 0 \\ & & 1 & 0 & 0 \\ & & & 1 & \varepsilon \\ & & & & \eta \end{bmatrix}, \tag{4.5-3}$$

$$\psi = \pm\sqrt{\gamma(\alpha-1)}, \qquad \eta=\gamma(\alpha+1)-2\nu,$$

in which α, γ, ν, and ε are arbitrary real constants. It can be shown (Ting, 1994a) that the matrix $\mathbf{s}'$ is positive definite if

$$\left.\begin{aligned} &s'_{55}>0, \quad \gamma>0, \quad \alpha\ge 1, \\ &\gamma(\alpha\gamma^2-\nu^2) > [\psi\nu+\alpha\gamma(\varepsilon-\psi)]^2. \end{aligned}\right\} \tag{4.5-4}$$

The last inequality can be written as

$$\left.\begin{aligned} &[\hat\psi\hat\nu+\sqrt{\alpha}(\hat\varepsilon-\hat\psi)]^2 < 1-\hat\nu, \\ &\hat\varepsilon=\frac{\varepsilon}{\sqrt{\gamma}}, \quad \hat\nu=\frac{\nu}{\gamma\sqrt{\alpha}}, \quad \hat\psi=\pm\sqrt{\alpha-1}. \end{aligned}\right\} \tag{4.5-5}$$

For a given α, the inequality in (4.5-5) represents the region inside an ellipse in the $(\hat{\varepsilon},\hat{v})$-plane. Thus it is possible to have anisotropic materials with a positive definite $\mathbf{s}'$ such that the eigenvalues are given by (4.5-1). In the special case when $v=\varepsilon=0$, the last two inequalities in (4.5-4) are simplified to

$$\frac{\sqrt{5}+1}{2} > \alpha \geq 1.$$

We next show that, if p is an eigenvalue of an anisotropic material with a positive definite $\mathbf{s}'$, it is always possible to find a new positive definite $\mathbf{s}'$ such that the new material possesses the eigenvalue βp where β is an arbitrary positive real constant. To this end, we observe that when p is replaced by βp, the polynomial $l_2(p)$ in (4.3-17) becomes $\beta^2 l_2(p)$ if s'_{45} and s'_{44} are replaced, respectively, by $\beta s'_{45}$ and $\beta^2 s'_{44}$. A similar operation can be applied to $l_3(p)$ and $l_4(p)$ in (4.3-17) so that they become $\beta^3 l_3(p)$ and $\beta^4 l_4(p)$, respectively. The sextic equation (4.3-16) remains the same with these operations. Thus if p is an eigenvalue of a material with positive definite $\mathbf{s}'$, βp is an eigenvalue of the new material whose $\mathbf{s}'$ is given by

$$\begin{bmatrix} s'_{11} & \beta^2 s'_{12} & \beta s'_{14} & s'_{15} & \beta s'_{16} \\ & \beta^4 s'_{22} & \beta^3 s'_{24} & \beta^2 s'_{25} & \beta^3 s'_{26} \\ & & \beta^2 s'_{44} & \beta s'_{45} & \beta^2 s'_{46} \\ & & & s'_{55} & \beta s'_{56} \\ & & & & \beta^2 s'_{66} \end{bmatrix}. \tag{4.5-6}$$

This matrix is equivalent to

$$\Lambda \mathbf{s}' \Lambda, \quad \Lambda = \mathrm{diag}[1,\ \beta^2,\ \beta,\ 1,\ \beta],$$

indicating that it is positive definite if $\mathbf{s}'$ is. Thus if an anisotropic materials with a positive definite matrix $\mathbf{s}'$ is known to possess three identical eigenvalues given by (4.5-1), the anisotropic material with positive definite $\mathbf{s}'$ given by (4.5-6) possesses the following three identical eigenvalues

$$p_1 = p_2 = p_3 = i\beta. \tag{4.5-7}$$

Finally, we show that by rotating the material whose $\mathbf{s}'$ is given by (4.5-6) an angle about the x_3-axis, the three identical eigenvalues p can assume any complex number, not necessarily purely imaginary. Rotating the materials about the x_3-axis an angle $-\theta$ with the coordinate axes being held stationary is equivalent to rotating the coordinate axes about the x_3-axis an angle θ with the material being held stationary. It will be shown in Section 7.1 that the eigenvalue $p(\theta)$ referred to the rotated coordinate system is related to the original p by

$$p(\theta)=\frac{p\cos\theta-\sin\theta}{p\sin\theta+\cos\theta}.$$

If $p=i\beta$ is an eigenvalue in the original coordinate system,

$$p(\theta)=\frac{i\beta\cos\theta-\sin\theta}{i\beta\sin\theta+\cos\theta}, \tag{4.5-8}$$

is an eigenvalue in the rotated coordinate system. Equation (4.5-8) allows us to determine β and θ for any prescribed complex number $p(\theta)$. Thus we can obtain an anisotropic material with prescribed three identical eigenvalues of any complex number by rotating the material whose $\mathbf{s}'$ is given by (4.5-6) an angle θ about the x_3-axis.

4.6 Representation of General Solutions

In this section we assume that p_1, p_2, and p_3 are distinct. We will ignore the terms that depend on x_3. They can be treated by a particular solution. Hence we assume that

$$A=B=C=\omega=0. \tag{4.6-1}$$

This means that $\varepsilon_{33}=0$ by $(4.2\text{-}5)_3$ and the displacements of (4.2-4) are given by u_1, u_2, and u_3 which represent a generalized plane strain deformation (Section 3.1).

When (4.6-1) hold, the Δ of (4.3-9) vanishes and $\hat{\psi}=\psi$ by (4.3-11). With (4.3-14) and (4.3-18) the stresses and strains given by (4.1-4) and (4.3-7) have the expressions

$$\begin{aligned}\sigma_{11}=p^2 f'(z),\quad \sigma_{22}=f'(z),\quad \sigma_{12}=-pf'(z),\\ \sigma_{32}=-\lambda f'(z),\quad \sigma_{31}=\lambda pf'(z),\end{aligned} \tag{4.6-2}$$

and

$$\varepsilon_\alpha=\xi_\alpha(p)f'(z) \tag{4.6-3}$$

where

$$\xi_\alpha(p)=p^2 s'_{\alpha1}-ps'_{\alpha6}+s'_{\alpha2}+\lambda(p)(ps'_{\alpha5}-s'_{\alpha4}), \tag{4.6-4}$$

$$f(z)=F'(z). \tag{4.6-5}$$

With the left sides of (4.2-5) given by (4.6-3), $(4.2\text{-}5)_{1,2,4}$ can be integrated to obtain

$$u_1=\xi_1(p)f(z),$$

$$u_2 = p^{-1}\xi_2(p)f(z), \tag{4.6-6}$$

$$u_3 = p^{-1}\xi_4(p)f(z).$$

The additional terms that arose from integration can be ignored without loss in generality (Lekhnitskii, 1950; see also Chapter 5). Substitution of (4.6-6) into $(4.2\text{-}5)_{5,6}$ leads to (4.3-16) and hence $(4.2\text{-}6)_{5,6}$ are automatically satisfied.

The general solutions for the stresses and displacements are obtained by superimposing six solutions of the form (4.6-2) and (4.6-6) associated with six eigenvalues p. For monoclinic materials with the symmetry plane at $x_3=0$, the coefficients of the polynomial $l_3(p)$ all vanish and the sextic equation reduces to two equations shown in (4.4-9). Let p_1 and p_2 be the roots of $l_4(p)=0$ and p_3 be the roots of $l_2(p)=0$. Furthermore, let the three roots be distinct. We then have from $(4.3\text{-}19)_1$ $\lambda(p)=0$ for p_1 and p_2, and from $(4.3\text{-}19)_2$ $\lambda(p)=\infty$ for p_3.

To avoid the situation at which $\lambda(p)$ becomes infinity at p_3, we replace $f(z)$ by $\lambda^{-1}f(z)$ for $p=p_3$ and denote $\lambda^{-1}(p_3)$ by λ_3 in superposing the six solutions. We therefore obtain from (4.6-2) the general solution for stresses as

$$\begin{aligned}
\sigma_{11} &= p_1^2 f_1'(z_1) + p_2^2 f_2'(z_2) + p_3^2\lambda_3 f_3'(z_3) \\
&\quad + \bar{p}_1^2 f_4'(\bar{z}_1) + \bar{p}_2^2 f_5'(\bar{z}_2) + \bar{p}_3^2\bar{\lambda}_3 f_6'(\bar{z}_3), \\
\sigma_{22} &= f_1'(z_1) + f_2'(z_2) + \lambda_3 f_3'(z_3) \\
&\quad + f_4'(\bar{z}_1) + f_5'(\bar{z}_2) + \bar{\lambda}_3 f_6'(\bar{z}_3), \\
\sigma_{12} &= -p_1 f_1'(z_1) - p_2 f_2'(z_2) - p_3\lambda_3 f_3'(z_3) \\
&\quad - \bar{p}_1 f_4'(\bar{z}_1) - \bar{p}_2 f_5'(\bar{z}_2) - \bar{p}_3\bar{\lambda}_3 f_6'(\bar{z}_3), \\
\sigma_{32} &= -\lambda_1 f_1'(z_1) - \lambda_2 f_2'(z_2) - f_3'(z_3) \\
&\quad - \bar{\lambda}_1 f_4'(\bar{z}_1) - \bar{\lambda}_2 f_5'(\bar{z}_2) - f_6'(\bar{z}_3), \\
\sigma_{31} &= p_1\lambda_1 f_1'(z_1) + p_2\lambda_2 f_2'(z_2) + p_3 f_3'(z_3) \\
&\quad + \bar{p}_1\bar{\lambda}_1 f_4'(\bar{z}_1) + \bar{p}_2\bar{\lambda}_2 f_5'(\bar{z}_2) + \bar{p}_3 f_6'(\bar{z}_3),
\end{aligned} \tag{4.6-7}$$

where $f_1, f_2, \ldots, f_6$ are arbitrary functions of their arguments,

$$z_\alpha = x_1 + p_\alpha x_2, \quad (\alpha = 1,2,3), \tag{4.6-8}$$

and

$$\lambda_\alpha = -\frac{l_3(p_\alpha)}{l_2(p_\alpha)} = -\frac{l_4(p_\alpha)}{l_3(p_\alpha)}, \quad \text{for } \alpha = 1,2,$$
$$\lambda_3 = -\frac{l_2(p_3)}{l_3(p_3)} = -\frac{l_3(p_3)}{l_4(p_3)}. \tag{4.6-9}$$

The stress component σ_{33} is, by setting $\varepsilon_3 = 0$ in (4.3-3),

$$\sigma_{33} = -\frac{1}{s_{33}}\{s_{31}\sigma_{11} + s_{32}\sigma_{22} + s_{34}\sigma_{23} + s_{35}\sigma_{13} + s_{36}\sigma_{12}\}. \tag{4.6-10}$$

As to the displacements we have from (4.6-6)

$$\begin{aligned}
u_1 &= \xi_1(p_1)f_1(z_1) + \xi_1(p_2)f_2(z_2) + \eta_1(p_3)f_3(z_3) \\
&\quad + \xi_1(\bar{p}_1)f_4(\bar{z}_1) + \xi_1(\bar{p}_2)f_5(\bar{z}_2) + \bar{\eta}_1(\bar{p}_3)f_6(\bar{z}_6), \\
u_2 &= p_1^{-1}\xi_2(p_1)f_1(z_1) + p_2^{-1}\xi_2(p_2)f_2(z_2) + p_3^{-1}\eta_2(p_3)f_3(z_3) \\
&\quad + \bar{p}_1^{-1}\xi_2(\bar{p}_1)f_4(\bar{z}_1) + \bar{p}_2^{-1}\xi_2(\bar{p}_2)f_5(\bar{z}_2) + \bar{p}_3^{-1}\bar{\eta}_2(\bar{p}_3)f_6(\bar{z}_6), \\
u_3 &= p_1^{-1}\xi_4(p_1)f_1(z_1) + p_2^{-1}\xi_4(p_2)f_2(z_2) + p_3^{-1}\eta_4(p_3)f_3(z_3) \\
&\quad + \bar{p}_1^{-1}\xi_4(\bar{p}_1)f_4(\bar{z}_1) + \bar{p}_2^{-1}\xi_4(\bar{p}_2)f_5(\bar{z}_2) + \bar{p}_3^{-1}\bar{\eta}_4(\bar{p}_3)f_6(\bar{z}_6),
\end{aligned} \tag{4.6-11}$$

where, for α=1,2,4,

$$\begin{aligned}
\xi_\alpha(p_\beta) &= p_\beta^2 s'_{\alpha 1} - p_\beta s'_{\alpha 6} + s'_{\alpha 2} + \lambda_\beta(p_\beta s'_{\alpha 5} - s'_{\alpha 4}), \\
\eta_\alpha(p_3) &= \lambda_3(p_3^2 s'_{\alpha 1} - p_3 s'_{\alpha 6} + s'_{\alpha 2}) + (p_3 s'_{\alpha 5} - s'_{\alpha 4}).
\end{aligned} \tag{4.6-12}$$

Instead of the stress components presented in (4.6-7) we may introduce the stress function φ_i so that σ_{ij} in (4.6-7) are recovered from φ_i by using (4.1-2). It can be shown that

$$\begin{aligned}
\varphi_1 &= -p_1 f_1(z_1) - p_2 f_2(z_2) - p_3\lambda_3 f_3(z_3) \\
&\quad - \bar{p}_1 f_4(\bar{z}_1) - \bar{p}_2 f_5(\bar{z}_2) - \bar{p}_3\bar{\lambda}_3 f_6(\bar{z}_3), \\
\varphi_2 &= f_1(z_1) + f_2(z_2) + \lambda_3 f_3(z_3) \\
&\quad + f_4(\bar{z}_1) + f_5(\bar{z}_2) + \bar{\lambda}_3 f_6(\bar{z}_3), \\
\varphi_3 &= -\lambda_1 f_1(z_1) - \lambda_2 f_2(z_2) - f_3(z_3) \\
&\quad - \bar{\lambda}_1 f_4(\bar{z}_1) - \bar{\lambda}_2 f_5(\bar{z}_2) - f_6(\bar{z}_3).
\end{aligned} \tag{4.6-13}$$

The general solutions (4.6-7), (4.6-11), and (4.6-13) apply when the eigenvalues p are distinct. The degenerate case in which p has a repeated root is discussed in Chapter 13. In applications we may choose the functions f_4, f_5, f_6 the complex conjugates of f_1, f_2, f_3, respectively, so that the solutions given by (4.6-7), (4.6-11), and (4.6-13) are real.

4.7 Dependence of Solutions on Elastic Constants

The solutions for generalized plane strain deformations presented in (4.6-7) and (4.6-11) depend on the 15 reduced elastic compliances $s'_{\alpha\beta}$. From (4.3-5) the elastic compliances

$$s_{\alpha\beta} = s'_{\alpha\beta} + \frac{s_{\alpha 3}s_{3\beta}}{s_{33}} \tag{4.7-1}$$

can be obtained by assuming $s_{\alpha 3}$ (α = 1,2,...,6) arbitrarily. Therefore every solution of generalized plane strain deformations applies to a group of anisotropic elastic materials given by (4.7-1) in which $s_{\alpha 3}$ are arbitrary. If the 5×5 matrix $\mathbf{s}'$ is positive definite, the 6×6 matrix $s_{\alpha\beta}$ is positive definite provided

$$s_{33} > 0. \tag{4.7-2}$$

This is so because, from (4.7-1),

$$\sum_{\alpha=1}^{6}\sum_{\beta=1}^{6} s_{\alpha\beta}\eta_\alpha\eta_\beta = \sum_{\alpha=1}^{6}\sum_{\beta=1}^{6} s'_{\alpha\beta}\eta_\alpha\eta_\beta + \frac{1}{s_{33}}\left\{\sum_{\alpha=1}^{6} s_{\alpha 3}\eta_\alpha\right\}^2.$$

With (4.7-2) the right side is always positive. It vanishes only when all η_α vanish, confirming that $s_{\alpha\beta}$ is positive definite.

It should be noted that if the 6×6 matrix $s_{\alpha\beta}$ is positive definite, the 5×5 matrix $\mathbf{s}'$ is necessarily positive definite. When $s_{\alpha\beta}$ is positive definite, so is its inverse $\mathbf{C}$. The 5×5 matrix $\mathbf{C}^{\circ}$ of (2.4-4) which is a principle submatrix of $\mathbf{C}$ is certainly positive definite. By (2.4-11) $\mathbf{s}'$ is the inverse of $\mathbf{C}^{\circ}$ and hence it is positive definite.

A particularly interesting special case is when $\mathbf{s}'$ is identical to that of an isotropic material. By letting

$$s_{\alpha 3} = \frac{\gamma_\alpha \gamma_3}{2\mu}$$

where γ_α are arbitrary real constants with $\gamma_3 \neq 0$ and μ is the shear modulus we have

$$s_{\alpha\beta}=\frac{1}{2\mu}\begin{bmatrix} 1-\nu+\gamma_1^2 & -\nu+\gamma_1\gamma_2 & \gamma_1\gamma_3 & \gamma_1\gamma_4 & \gamma_1\gamma_5 & \gamma_1\gamma_6 \\ & 1-\nu+\gamma_2^2 & \gamma_2\gamma_3 & \gamma_2\gamma_4 & \gamma_2\gamma_5 & \gamma_2\gamma_6 \\ & & \gamma_3^2 & \gamma_3\gamma_4 & \gamma_3\gamma_5 & \gamma_3\gamma_6 \\ & & & 2+\gamma_4^2 & \gamma_4\gamma_5 & \gamma_4\gamma_6 \\ & & & & 2+\gamma_5^2 & \gamma_5\gamma_6 \\ & & & & & 2+\gamma_6^2 \end{bmatrix} \tag{4.7-3}$$

in which ν is the Poisson's ratio. The $s_{\alpha\beta}$ of (4.7-3) represents isotropic materials when $\gamma_1=\gamma_2=-\nu\gamma_3$, $\gamma_3=(1+\nu)^{-1/2}$, and $\gamma_4=\gamma_5=\gamma_6=0$. If γ_α are chosen arbitrarily, the materials represented by (4.7-3) can be completely anisotropic without a plane of symmetry (Section 2.10). Yet *the solutions to two-dimensional elastostatics problems and steady-state motions such as surface waves for the materials given by* (4.7-3) *are identical to the solutions for isotropic materials* (Ting, 1994; see also Section 3.13). The inverse **C** of the matrix $s_{\alpha\beta}$ in (4.7-3) will be given in (5.10-4).

In closing this section we point out the following invariant properties on elastic constants under traction loading at the boundary. The solution (4.6-7) for the stress depends on elastic constants through p_α and λ_α (α=1,2,3). According to (4.3-16) and (4.6-9) they are governed by the polynomials $l_2(p)$, $l_3(p)$, and $l_4(p)$. Consider now two anisotropic materials subjected to the same loading at the boundary. Let $s'^{(1)}_{\alpha\beta}$ and $s'^{(2)}_{\alpha\beta}$ be the reduced elastic compliances of the two materials Inspection of (4.3-17) for $l_2(p)$, $l_3(p)$, $l_4(p)$ tells us that the stress in the two materials are identical if

$$s'^{(2)}_{\alpha\beta}=m\,s'^{(1)}_{\alpha\beta}$$

for all α, β except

$$\begin{aligned} 2s'^{(2)}_{12}&=2ms'^{(1)}_{12}+a, & s'^{(2)}_{66}&=ms'^{(1)}_{66}-a, \\ s'^{(2)}_{14}&=ms'^{(1)}_{14}+d, & s'^{(2)}_{56}&=ms'^{(1)}_{56}-d, \\ s'^{(2)}_{25}&=ms'^{(1)}_{25}+e, & s'^{(2)}_{46}&=ms'^{(1)}_{46}-e, \end{aligned} \tag{4.7-4}$$

where m, a, d, and e are arbitrary constants. This is a generalization of the invariant property for isotropic materials first discovered by Cherkaev et al. (1992). Extensions of the result for multiply connected domains and to isotropic composites have been given by Thorpe and Jasiuk (1992). Dundurs and Markenscoff (1993) have generalized the result to inhomogeneous isotropic materials. For general inhomogeneous anisotropic materials for which $s'^{(1)}_{\alpha\beta}$ and

$s'^{(2)}_{\alpha\beta}$ depend on x_1 and x_2, it can be shown (Dundurs and Ting, 1994) that (4.7-4) remain valid except that the first two equations have the following more general expressions

$$\begin{aligned} 2s'^{(2)}_{12} &= 2ms'^{(1)}_{12} + a + bx_1 + cx_2, \\ s'^{(2)}_{66} &= 2ms'^{(1)}_{66} - a - bx_1 - cx_2. \end{aligned} \tag{4.7-5}$$

In the above m, a, b, and c are arbitrary constants.

More results relating the dependence of the stress solutions on elastic constants will be given in Section 5.10.

Chapter 5

THE STROH FORMALISM

In this chapter we study Stroh's sextic formalism for two-dimensional deformations of an anisotropic elastic body. The Stroh formalism can be traced to the work of Eshelby, Read, and Shockley (1953). We therefore present the latter first. Not all results presented in this chapter are due to Stroh (1958, 1962). Nevertheless we name the sextic formalism after Stroh because he laid the foundations for researchers who followed him. The derivation of Stroh's formalism is rather simple and straightforward. The general solution resembles that obtained by the Lekhnitskii formalism. However, the resemblance between the two formalisms stops there. As we will see in the rest of the book, the Stroh formalism is indeed mathematically elegant and technically powerful in solving two-dimensional anisotropic elasticity problems. The possibility of extending the formalism to three-dimensional deformations is explored in Chapter 15.

5.1 The Eshelby–Reid–Shockley Formalism

In a fixed rectangular coordinate system x_i (i=1,2,3) let u_i and σ_{ij} be, respectively, the displacement and stress in an anisotropic elastic material. The stress-strain laws and the equations of equilibrium are

$$\sigma_{ij} = C_{ijks} u_{k,s}, \tag{5.1-1}$$

$$C_{ijks} u_{k,sj} = 0, \tag{5.1-2}$$

in which a comma denotes differentiation, repeated indices imply summation and C_{ijks} are the elastic stiffnesses which are assumed to possess the full symmetry

$$C_{ijks} = C_{jiks} = C_{ksij} = C_{ijsk}. \tag{5.1-3}$$

It is shown in Section 2.1 that $(5.1\text{-}3)_1$ and $(5.1\text{-}3)_2$ imply $(5.1\text{-}3)_3$. For two-dimensional deformations in which u_i (i=1,2,3) depends on x_1 and x_2 only, (5.1-2) is a homogeneous second-order differential equation in two independent variables x_1 and x_2. A general solution for u_i depends on one composite variable that is a linear combination of x_1 and x_2. Without loss in generality we choose the coefficient of x_1 in the linear combination to be unity and let

$$u_i = a_i f(z), \quad \text{or} \quad \mathbf{u} = \boldsymbol{a} f(z) \tag{5.1-4}$$

where

$$z = x_1 + p x_2. \tag{5.1-5}$$

In the above f is an arbitrary function of z, and p and a_i are unknown constants to be determined. Differentiation of (5.1-4) with x_s gives

$$u_{k,s} = (\delta_{s1} + p\delta_{s2}) a_k f'(z) \tag{5.1-6}$$

in which the prime denotes differentiation with the argument z and δ_{si} is the Kronecker delta. Differentiating once more with x_j we find that (5.1-2) is satisfied if

$$C_{ijks}(\delta_{j1} + p\delta_{j2})(\delta_{s1} + p\delta_{s2}) a_k = 0 \tag{5.1-7}$$

or

$$\left\{C_{i1k1} + p(C_{i1k2} + C_{i2k1}) + p^2 C_{i2k2}\right\} a_k = 0.$$

This can be written in matrix notation as

$$\left\{\mathbf{Q} + p(\mathbf{R} + \mathbf{R}^T) + p^2 \mathbf{T}\right\} \boldsymbol{a} = \mathbf{0} \tag{5.1-8}$$

where **Q**, **R**, and **T** are 3×3 matrices whose components are

$$Q_{ik} = C_{i1k1}, \quad R_{ik} = C_{i1k2}, \quad T_{ik} = C_{i2k2}. \tag{5.1-9}$$

It is seen that **Q** and **T** are symmetric and positive definite if the strain energy is positive (Section 2.9). For a nontrivial solution of $\boldsymbol{a}$ we must have

$$\left|\mathbf{Q} + p(\mathbf{R} + \mathbf{R}^T) + p^2 \mathbf{T}\right| = 0 \tag{5.1-10}$$

which gives six roots for the eigenvalue p. The associated eigenvector $\boldsymbol{a}$ is determined from (5.1-8). The stresses σ_{ij} are obtained by inserting (5.1-6) into (5.1-1). Making use of (5.1-9) we obtain

$$\begin{aligned} \sigma_{i1} &= (Q_{ik} + pR_{ik}) a_k f'(z), \\ \sigma_{i2} &= (R_{ki} + pT_{ik}) a_k f'(z). \end{aligned} \tag{5.1-11}$$

The p of (5.1-10) cannot be real. If p is real, (5.1-7) multiplied by a_i and summed with i leads to

$$C_{ijks}(\delta_{j1} + p\delta_{j2}) a_i (\delta_{s1} + p\delta_{s2}) a_k = 0.$$

By choosing

$$u_{i,j} = (\delta_{j1} + p\delta_{j2}) a_i, \quad u_{k,s} = (\delta_{s1} + p\delta_{s2}) a_k,$$

we obtain

$$C_{ijks}u_{i,j}u_{k,s} = 0$$

which violates the condition that the strain energy is positive definite (Section 2.1).

Since the coefficients of the sextic equation for p arising from (5.1-10) are real there are three pairs of complex conjugates for p. If p_α, $\boldsymbol{a}_\alpha$ (α=1,2,...,6) are the eigenvalues and the associated eigenvectors, we let

$$\operatorname{Im} p_\alpha > 0, \quad p_{\alpha+3} = \bar{p}_\alpha, \quad \boldsymbol{a}_{\alpha+3} = \bar{\boldsymbol{a}}_\alpha \quad (\alpha = 1,2,3), \tag{5.1-12}$$

where Im stands for the imaginary part and the overbar denotes the complex conjugate. Assuming that p are distinct, the general solution obtained by superposing six solutions of the form (5.1-4) is

$$\mathbf{u} = \sum_{\alpha=1}^{3} \left\{ \boldsymbol{a}_\alpha f_\alpha(z_\alpha) + \bar{\boldsymbol{a}}_\alpha f_{\alpha+3}(\bar{z}_\alpha) \right\}, \tag{5.1-13}$$

where f_α (α=1,2,...,6) are arbitrary functions of their arguments and

$$z_\alpha = x_1 + p_\alpha x_2. \tag{5.1-14}$$

Likewise, the general solution for the stresses obtained from (5.1-11) can be written as

$$\begin{aligned} \mathbf{t}_1 &= \sum_{\alpha=1}^{3} \left\{ (\mathbf{Q} + p_\alpha \mathbf{R}) \boldsymbol{a}_\alpha f'_\alpha(z_\alpha) + (\mathbf{Q} + \bar{p}_\alpha \mathbf{R}) \bar{\boldsymbol{a}}_\alpha f'_{\alpha+3}(\bar{z}_\alpha) \right\} \\ \mathbf{t}_2 &= \sum_{\alpha=1}^{3} \left\{ (\mathbf{R}^T + p_\alpha \mathbf{T}) \boldsymbol{a}_\alpha f'_\alpha(z_\alpha) + (\mathbf{R}^T + \bar{p}_\alpha \mathbf{T}) \bar{\boldsymbol{a}}_\alpha f'_{\alpha+3}(\bar{z}_\alpha) \right\} \end{aligned} \tag{5.1-15}$$

in which the components of the vectors $\mathbf{t}_1$, $\mathbf{t}_2$ are

$$(\mathbf{t}_1)_i = \sigma_{i1}, \quad (\mathbf{t}_2)_i = \sigma_{i2}. \tag{5.1-16}$$

The stress component σ_{33} is determined from (2.4-5), which is the condition for ε_{33}=0.

5.2 Eigenvalues p

Employing the contracted notation for C_{ijks} introduced in Section 2.3 the three matrices **Q**, **R**, and **T** defined in (5.1-9) have the expressions

$$\mathbf{Q}=\begin{bmatrix} C_{11} & C_{16} & C_{15} \\ C_{16} & C_{66} & C_{56} \\ C_{15} & C_{56} & C_{55} \end{bmatrix},\quad \mathbf{R}=\begin{bmatrix} C_{16} & C_{12} & C_{14} \\ C_{66} & C_{26} & C_{46} \\ C_{56} & C_{25} & C_{45} \end{bmatrix},\quad \mathbf{T}=\begin{bmatrix} C_{66} & C_{26} & C_{46} \\ C_{26} & C_{22} & C_{24} \\ C_{46} & C_{24} & C_{44} \end{bmatrix}. \tag{5.2-1}$$

Hence (5.1-8) can be written as

$$\begin{bmatrix} C_{11}+2pC_{16}+p^2C_{66} & C_{16}+p(C_{12}+C_{66})+p^2C_{26} & C_{15}+p(C_{14}+C_{56})+p^2C_{46} \\ C_{16}+p(C_{12}+C_{66})+p^2C_{26} & C_{66}+2pC_{26}+p^2C_{22} & C_{56}+p(C_{46}+C_{25})+p^2C_{24} \\ C_{15}+p(C_{14}+C_{56})+p^2C_{46} & C_{56}+p(C_{46}+C_{25})+p^2C_{24} & C_{55}+2pC_{45}+p^2C_{44} \end{bmatrix}\mathbf{a}=\mathbf{0}. \tag{5.2-2}$$

A sextic equation for the eigenvalues p is obtained by setting the determinant of the 3×3 matrix on the left to zero.

The sextic equation for p is simplified for special materials such as monoclinic materials with the symmetry plane at one of the coordinate planes. We consider these and other special materials below.

(1) **Monoclinic Materials with the Symmetry Plane at x_1=0.** For this case (5.2-2) reduces to

$$\begin{bmatrix} C_{11}+p^2C_{66} & p(C_{12}+C_{66}) & p(C_{14}+C_{56}) \\ p(C_{12}+C_{66}) & C_{66}+p^2C_{22} & C_{56}+p^2C_{24} \\ p(C_{14}+C_{56}) & C_{56}+p^2C_{24} & C_{55}+p^2C_{44} \end{bmatrix}\mathbf{a}=\mathbf{0}. \tag{5.2-3}$$

The sextic equation is a polynomial of degree three in p^2. If the three roots of p^2 are all real, they must be negative since p cannot be real. We then have

$$p_1=i\beta_1,\quad p_2=i\beta_2,\quad p_3=i\beta_3, \tag{5.2-4}$$

where β_1, β_2, and β_3 are real and positive. If only one root of p^2, say p_3^2, is real the other two roots p_1^2 and p_2^2 must be complex conjugates. We then have

$$p_1=\alpha+i\beta,\quad p_2=-\alpha+i\beta,\quad p_3=i\beta_3, \tag{5.2-5}$$

in which α is real, and β and β_3 are real and positive.

(2) **Monoclinic Materials with the Symmetry Plane at x_2=0.** In this case (5.2-2) reduces to

$$\begin{bmatrix} C_{11}+p^2C_{66} & p(C_{12}+C_{66}) & C_{15}+p^2C_{46} \\ p(C_{12}+C_{66}) & C_{66}+p^2C_{22} & p(C_{46}+C_{25}) \\ C_{15}+p^2C_{46} & p(C_{46}+C_{25}) & C_{55}+p^2C_{44} \end{bmatrix} \boldsymbol{a} = \mathbf{0}. \tag{5.2-6}$$

The sextic equation is again a polynomial of degree three in p^2. Hence the eigenvalues have the representations (5.2-4) or (5.2-5).

(3) **Monoclinic Materials with the Symmetry Plane at x_3=0.** Equation (5.2-2) now becomes

$$\begin{bmatrix} C_{11}+2pC_{16}+p^2C_{66} & C_{16}+p(C_{12}+C_{66})+p^2C_{26} & 0 \\ C_{16}+p(C_{12}+C_{66})+p^2C_{26} & C_{66}+2pC_{26}+p^2C_{22} & 0 \\ 0 & 0 & C_{55}+2pC_{45}+p^2C_{44} \end{bmatrix} \boldsymbol{a} = \mathbf{0}. \tag{5.2-7}$$

The sextic equation for p decomposes into a quadratic equation

$$C_{55} + 2pC_{45} + p^2C_{44} = 0 \tag{5.2-8}$$

and a quartic equation

$$\left(C_{11}+2pC_{16}+p^2C_{66}\right)\left(C_{11}+2pC_{16}+p^2C_{66}\right)-\left[C_{16}+p(C_{12}+C_{66})+p^2C_{26}\right]^2 = 0. \tag{5.2-9}$$

If p_3 denotes the root of (5.2-8) with positive imaginary part,

$$p_3 = \frac{1}{C_{44}}\left\{-C_{45} + i\sqrt{C_{44}C_{55} - C_{45}^2}\right\}. \tag{5.2-10}$$

The roots p_1, p_2 of (5.2-9) with positive imaginary parts have the representations

$$p_1 = \alpha_1 + i\beta_1, \quad p_2 = \alpha_2 + i\beta_2, \tag{5.2-11}$$

where α_1 and α_2 are real, and β_1 and β_2 are real and positive.

(4) **Transversely Isotropic Materials with the Axis of Symmetry at the x_3-axis.** For this case we have

$$\begin{bmatrix} \lambda+2\mu+\mu p^2 & (\lambda+\mu)p & 0 \\ (\lambda+\mu)p & \mu+(\lambda+2\mu)p^2 & 0 \\ 0 & 0 & C_{44}(1+p^2) \end{bmatrix} \boldsymbol{a} = \mathbf{0} \tag{5.2-12}$$

where λ and μ are constants. It is easily verified that

$$p_1 = p_2 = p_3 = i \tag{5.2-13}$$

and that, at $p=i$, (5.2-12) becomes

$$\begin{bmatrix} (\lambda+\mu) & (\lambda+\mu)i & 0 \\ (\lambda+\mu)i & -(\lambda+\mu) & 0 \\ 0 & 0 & 0 \end{bmatrix} \boldsymbol{a} = \mathbf{0}. \tag{5.2-14}$$

The three columns of the 3×3 matrix are proportional to each other, yielding only two independent eigenvectors $\boldsymbol{a}$ associated with the triple eigenvalues (5.2-13). The system is mathematically degenerate.

(5) **Isotropic Materials.** Equation (5.2-12) applies to isotropic materials when λ and μ are identified as the Lamé constants and $C_{44}=\mu$. Equations (5.2-13) and (5.2-14) remain valid for isotropic materials. The eigenvectors $\boldsymbol{a}$ for isotropic materials are identical to those for transversely isotropic materials with the axis of symmetry at the x_3-axis. In fact the solutions to two-dimensional elasticity problems for isotropic materials and transversely isotropic materials with the axis of symmetry at the x_3-axis are identical except that the constant C_{44} is μ for the former and arbitrary for the latter.

It should be pointed out that the triple eigenvalues (5.2-13) are not limited to isotropic materials and transversely isotropic materials with the axis of symmetry at the x_3-axis. There are infinitely many anisotropic materials whose eigenvalues are given by (5.2-13) (Ting, 1994; see also Section 4.5).

For the special materials considered above explicit solutions of p in radicals can be obtained. For general anisotropic materials or for materials with the symmetry planes that are not on the coordinate planes, the sextic equation for p is not solvable in radical, not even for cubic materials which have only three elastic constants (Head, 1979a,b).

5.3 The Sextic Formalism of Stroh

The stresses shown in (5.1-11) can be rewritten as

$$\sigma_{i1} = -pb_i f'(z), \qquad \sigma_{i2} = b_i f'(z), \tag{5.3-1}$$

where, in matrix notation,

$$\boldsymbol{b} = (\mathbf{R}^T + p\mathbf{T})\boldsymbol{a} = -\frac{1}{p}(\mathbf{Q} + p\mathbf{R})\boldsymbol{a}. \tag{5.3-2}$$

The second equality in (5.3-2) follows from (5.1-8). Introducing the stress function

$$\varphi_i = b_i f(z) \quad \text{or} \quad \boldsymbol{\phi} = \boldsymbol{b} f(z), \tag{5.3-3}$$

(5.3-1) are equivalent to

$$\sigma_{i1} = -\varphi_{i,2}, \quad \sigma_{i2} = \varphi_{i,1}. \tag{5.3-4}$$

It is sufficient therefore to consider the stress function $\boldsymbol{\phi}$ because the stresses can be obtained from (5.3-4) by differentiation. The two equations in (5.3-4) can be combined into one equation as

$$\sigma_{ij} = \delta_{3kj}\varphi_{i,k}.$$

The fact that $\sigma_{21}=\sigma_{12}$ means that, by (5.3-4),

$$\varphi_{1,1} + \varphi_{2,2} = 0 \tag{5.3-5}$$

and, by (5.3-3),

$$b_1 + pb_2 = 0. \tag{5.3-6}$$

Indeed, (5.3-6) can be verified directly by using (5.2-1) when b_1 is determined from $(5.3\text{-}2)_1$ and b_2 from $(5.3\text{-}2)_2$.

The general solution for the stress function $\boldsymbol{\phi}$ is obtained by superposing six solutions of the form (5.3-3) associated with six eigenvalues p_α. Together with (5.1-13) we have

$$\begin{aligned} \mathbf{u} &= \sum_{\alpha=1}^{3}\left\{\boldsymbol{a}_\alpha f_\alpha(z_\alpha) + \bar{\boldsymbol{a}}_\alpha f_{\alpha+3}(\bar{z}_\alpha)\right\}, \\ \boldsymbol{\phi} &= \sum_{\alpha=1}^{3}\left\{\boldsymbol{b}_\alpha f_\alpha(z_\alpha) + \bar{\boldsymbol{b}}_\alpha f_{\alpha+3}(\bar{z}_\alpha)\right\}. \end{aligned} \tag{5.3-7}$$

In the above $\boldsymbol{b}_\alpha$ is related to $\boldsymbol{a}_\alpha$ through (5.3-2) and

$$\boldsymbol{b}_{\alpha+3} = \bar{\boldsymbol{b}}_\alpha \quad (\alpha = 1,2,3). \tag{5.3-8}$$

Equations (5.3-7) are the *sextic formalism* due to Stroh (1958, 1962). The vectors $\boldsymbol{a}_\alpha$, $\boldsymbol{b}_\alpha$ are the Stroh eigenvectors. The displacements are given by $(5.3\text{-}7)_1$ while the stresses are obtained from $(5.3\text{-}7)_2$ and (5.3-4) by differentiation. The only stress component missing is σ_{33}. It is determined in terms of other stress components using (2.4-5), which is the condition for $\varepsilon_{33}=0$.

In most applications the arbitrary functions f_α appearing in (5.3-7) have the same function form. We may therefore let

$$f_\alpha(z_\alpha)=f(z_\alpha)q_\alpha, \quad f_{\alpha+3}(\bar{z}_\alpha)=\bar{f}(\bar{z}_\alpha)\bar{q}_\alpha, \tag{5.3-9}$$

(α=1,2,3) where q_α are arbitrary complex constants. The second equation is useful for obtaining real form solutions for **u** and $\boldsymbol{\phi}$. Equations (5.3-7) can then be written as

$$\mathbf{u}=2\,\mathrm{Re}\{\mathbf{A}\langle f(z_*)\rangle\mathbf{q}\}, \quad \boldsymbol{\phi}=2\,\mathrm{Re}\{\mathbf{B}\langle f(z_*)\rangle\mathbf{q}\}. \tag{5.3-10}$$

In the above Re stands for the real part, **A** and **B** are 3×3 complex matrices defined by

$$\mathbf{A}=[\boldsymbol{a}_1,\boldsymbol{a}_2,\boldsymbol{a}_3], \quad \mathbf{B}=[\boldsymbol{b}_1,\boldsymbol{b}_2,\boldsymbol{b}_3] \tag{5.3-11}$$

and $\langle f(z_*)\rangle$ is the diagonal matrix

$$\langle f(z_*)\rangle=\mathrm{diag}[f(z_1),f(z_2),f(z_3)]. \tag{5.3-12}$$

When **q** is replaced by $-i\mathbf{q}$, (5.3-10) leads to the alternate form

$$\mathbf{u}=2\,\mathrm{Im}\{\mathbf{A}\langle f(z_*)\rangle\mathbf{q}\}, \quad \boldsymbol{\phi}=2\,\mathrm{Im}\{\mathbf{B}\langle f(z_*)\rangle\mathbf{q}\}. \tag{5.3-13}$$

For a given problem all one has to do is to determine the unknown function $f(z_*)$ and the complex vector **q**.

Other representations of the general solutions are obtained as follows. Let the 3-vectors $[f(z)]$, $[\hat{f}(\bar{z})]$ be defined by

$$[f(z)]=\begin{bmatrix} f_1(z_1) \\ f_2(z_2) \\ f_3(z_3) \end{bmatrix}, \quad [\hat{f}(\bar{z})]=\begin{bmatrix} f_4(\bar{z}_1) \\ f_5(\bar{z}_2) \\ f_6(\bar{z}_3) \end{bmatrix}. \tag{5.3-14}$$

Equation (5.3-7) can be written as

$$\begin{bmatrix} \mathbf{u} \\ \boldsymbol{\phi} \end{bmatrix}=\begin{bmatrix} \mathbf{A} & \overline{\mathbf{A}} \\ \mathbf{B} & \overline{\mathbf{B}} \end{bmatrix}\begin{bmatrix} [f(z)] \\ [\hat{f}(\bar{z})] \end{bmatrix}. \tag{5.3-15}$$

When the columns of **A** and **B** are properly normalized we will show in Section 5.5 that

$$\begin{bmatrix} \mathbf{B}^T & \mathbf{A}^T \\ \overline{\mathbf{B}}^T & \overline{\mathbf{A}}^T \end{bmatrix}\begin{bmatrix} \mathbf{A} & \overline{\mathbf{A}} \\ \mathbf{B} & \overline{\mathbf{B}} \end{bmatrix}=\begin{bmatrix} \mathbf{I} & \mathbf{0} \\ \mathbf{0} & \mathbf{I} \end{bmatrix}.$$

Hence (5.3-15) is equivalent to

$$\begin{bmatrix} \mathbf{B}^T & \mathbf{A}^T \\ \overline{\mathbf{B}}^T & \overline{\mathbf{A}}^T \end{bmatrix}\begin{bmatrix} \mathbf{u} \\ \boldsymbol{\phi} \end{bmatrix}=\begin{bmatrix} [f(z)] \\ [\hat{f}(\bar{z})] \end{bmatrix} \tag{5.3-16}$$

from which we have the representation (Yeh et al., 1993a,b)

$$\mathbf{A}^T\boldsymbol{\phi} + \mathbf{B}^T\mathbf{u} = [f(z)]. \tag{5.3-17}$$

We also have

$$\boldsymbol{\phi} + i\mathbf{M}^T\mathbf{u} = (\mathbf{A}^T)^{-1}[f(z)] \tag{5.3-18}$$

where

$$\mathbf{M} = -i\mathbf{B}\mathbf{A}^{-1}.$$

When the explicit expressions for **A**, **B**, and **M** to be derived in (6.3-7), (6.3-8), (6.4-4) are employed for special anisotropic materials under antiplane deformations, (5.3-17) and (5.3-18) both reduce to the general solution (3.3-14) for antiplane deformations.

5.4 The Stress Function ϕ and the Airy Function χ

Let **t** be the traction vector at a point on the boundary Γ of an anisotropic elastic material and **n** be the unit outward normal to Γ so that (Fig. 5.1)

$$t_i = \sigma_{ij}n_j.$$

If s is the arc length measured along the boundary Γ as shown,

$$n_1 = -\frac{dx_2}{ds}, \quad n_2 = \frac{dx_1}{ds}, \quad n_3 = 0. \tag{5.4-1}$$

Using (5.3-4) and (5.4-1) differentiation of $\varphi_i(x_1,x_2)$ along the boundary Γ is

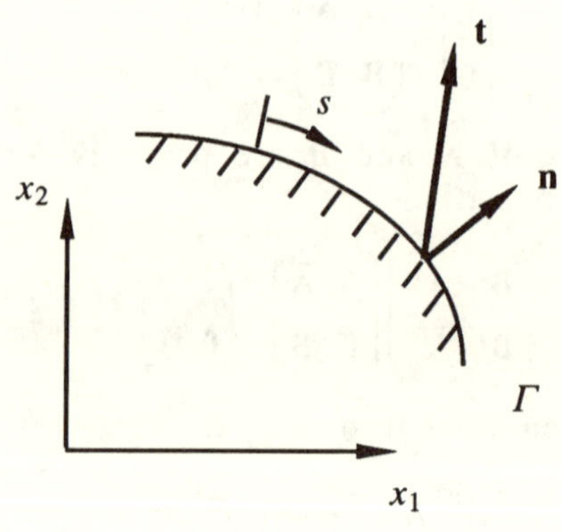

Fig. 5.1 The surface traction **t** on a curved boundary Γ with a unit outward normal vector **n**. It is related to the stress function ϕ by (5.4-2).

$$\frac{d}{ds}\varphi_i = \varphi_{i,1}\frac{dx_1}{ds} + \varphi_{i,2}\frac{dx_2}{ds}$$
$$= \sigma_{i1}n_1 + \sigma_{i2}n_2 = \sigma_{ij}n_j.$$

Hence

$$t_i = \frac{d}{ds}\varphi_i \quad \text{or} \quad \mathbf{t} = \frac{d}{ds}\boldsymbol{\phi}. \tag{5.4-2}$$

Equation (5.4-2) reduces to $(5.3\text{-}4)_{1,2}$ when Γ is a plane parallel to the x_1=0 plane or the x_2=0 plane. In applying (5.4-2), s should be chosen such that when one faces the direction of increasing s the material lies on the right side.

Let $s_2 > s_1$ be two points on Γ. The difference $\Delta\boldsymbol{\phi}$ between $\boldsymbol{\phi}(s_2)$ and $\boldsymbol{\phi}(s_1)$ is

$$\Delta\boldsymbol{\phi} = \boldsymbol{\phi}(s_2) - \boldsymbol{\phi}(s_1) = \int_{s_1}^{s_2}\mathbf{t}(s)ds, \tag{5.4-3}$$

which represents the resultant force of surface tractions acting on Γ between s_1 and s_2. If Γ encloses a region and there is a concentrated force $\mathbf{f}$ inside the region, the equilibrium of the body demands that

$$\Delta\boldsymbol{\phi} = \int_\Gamma \mathbf{t}(s)ds = -\mathbf{f}. \tag{5.4-4}$$

This means that if we traverse Γ counter-clockwise one complete circle, $\boldsymbol{\phi}$ increases by the amount $\mathbf{f}$. Therefore $\boldsymbol{\phi}$ must be a multiple-valued function. A Riemann surface with a proper cut has to be introduced to maintain a unique solution for $\boldsymbol{\phi}$. If no concentrated forces are inside the region of Γ, $\boldsymbol{\phi}$ is a single-valued function.

Equation (5.3-5) is in the form of (3.14-1). According to Theorem 3.14–1 there exists a function, say χ, such that

$$\varphi_1 = -\chi_{,2}, \quad \varphi_2 = \chi_{,1}. \tag{5.4-5}$$

Hence

$$\chi(z) = \int^z \varphi_2(\lambda)d\lambda = b_2\int^z f(\lambda)d\lambda. \tag{5.4-6}$$

It is readily shown from (5.4-5) and (5.3-4) that χ is the *Airy function* (Section 4.1).

The moment about the x_3-axis due to the traction $\mathbf{t}$ on Γ is, using (5.4-2) and (5.4-5),

$$\int_\Gamma (x_1t_2 - x_2t_1)ds = \int_\Gamma x_1d\varphi_2 - x_2d\varphi_1$$
$$= \int_\Gamma d(x_1\varphi_2 - x_2\varphi_1) - \int_\Gamma(\varphi_2dx_1 - \varphi_1dx_2)$$

$$= \int_\Gamma d(x_1\varphi_2 - x_2\varphi_1 - \chi)$$

Thus the moment due to surface tractions acting between z_1 and z_2 is

$$\Delta(x_1\varphi_2 - x_2\varphi_1 - \chi) = \Delta(x_1\varphi_2 - x_2\varphi_1) - \int_{z_1}^{z_2} \varphi_2(\lambda)d\lambda. \qquad (5.4\text{-}7)$$

The general solution for the displacement u_i given by (5.1-4) tells us that it is codirectional with the vector a_i. The general solution φ_i given by $(5.3\text{-}3)_1$ together with $(5.4\text{-}2)_1$ implies that the surface traction t_i is codirectional with the vector b_i. Hence the Stroh eigenvectors a_i, b_i represent the directions of the displacement u_i and traction t_i, respectively. Although a_i and b_i are complex they do have a real physical interpretation (Section 5.8).

5.5 Orthogonality and Closure Relations

What distinguishes the Stroh formalism from others is that the vectors $\boldsymbol{a}_\alpha$ and $\boldsymbol{b}_\alpha$ for different α are related. The complex matrices **A** and **B** defined in (5.3-11) possess certain properties that are important in applications.

The two equations in (5.3-2) can be cast in the form

$$\begin{bmatrix} -\mathbf{Q} & \mathbf{0} \\ -\mathbf{R}^T & \mathbf{I} \end{bmatrix} \begin{bmatrix} \boldsymbol{a} \\ \boldsymbol{b} \end{bmatrix} = p \begin{bmatrix} \mathbf{R} & \mathbf{I} \\ \mathbf{T} & \mathbf{0} \end{bmatrix} \begin{bmatrix} \boldsymbol{a} \\ \boldsymbol{b} \end{bmatrix}, \qquad (5.5\text{-}1)$$

where **I** is the 3×3 identity matrix. Noting that **T** is positive definite and $\mathbf{T}^{-1}$ exists, it is easily shown that

$$\begin{bmatrix} \mathbf{0} & \mathbf{T}^{-1} \\ \mathbf{I} & -\mathbf{R}\mathbf{T}^{-1} \end{bmatrix} \begin{bmatrix} \mathbf{R} & \mathbf{I} \\ \mathbf{T} & \mathbf{0} \end{bmatrix} = \begin{bmatrix} \mathbf{I} & \mathbf{0} \\ \mathbf{0} & \mathbf{I} \end{bmatrix}. \qquad (5.5\text{-}2)$$

Therefore pre-multiplication on both sides of (5.5-1) by the first matrix on the left of (5.5-2) reduces (5.5-1) to the following standard eigenrelation

$$\mathbf{N}\boldsymbol{\xi} = p\boldsymbol{\xi}, \qquad (5.5\text{-}3)$$

$$\mathbf{N} = \begin{bmatrix} \mathbf{N}_1 & \mathbf{N}_2 \\ \mathbf{N}_3 & \mathbf{N}_1^T \end{bmatrix}, \qquad \boldsymbol{\xi} = \begin{bmatrix} \boldsymbol{a} \\ \boldsymbol{b} \end{bmatrix}, \qquad (5.5\text{-}4)$$

$$\mathbf{N}_1 = -\mathbf{T}^{-1}\mathbf{R}^T, \quad \mathbf{N}_2 = \mathbf{T}^{-1}, \quad \mathbf{N}_3 = \mathbf{R}\mathbf{T}^{-1}\mathbf{R}^T - \mathbf{Q}. \qquad (5.5\text{-}5)$$

It is seen that $\mathbf{N}_2$ and $\mathbf{N}_3$ are symmetric and that $\mathbf{N}_2$ is positive definite. We will show in Section 5.3 that $-\mathbf{N}_3$ is positive semidefinite. The 6×6 matrix $\mathbf{N}$ is the *fundamental elasticity matrix* first introduced by Ingebrigtsen and Tonning (1969). It should be noted that $\mathbf{N}_3$ has the dimensions of stress, $\mathbf{N}_2$ has the dimension of compliance and $\mathbf{N}_1$ is dimensionless.

Since $\mathbf{N}$ is not symmetric the ξ in (5.5-3) is a right eigenvector. The left eigenvector denoted by η satisfies the eigenrelation

$$\mathbf{N}^T\eta = p\eta. \tag{5.5-6}$$

Introducing the 6×6 constant matrix

$$\hat{\mathbf{I}} = \begin{bmatrix} \mathbf{0} & \mathbf{I} \\ \mathbf{I} & \mathbf{0} \end{bmatrix}, \qquad \hat{\mathbf{I}} = \hat{\mathbf{I}}^T = \hat{\mathbf{I}}^{-1}, \tag{5.5-7}$$

it is easily shown that $\hat{\mathbf{I}}\mathbf{N}$ is symmetric, i.e.,

$$\hat{\mathbf{I}}\mathbf{N} = (\hat{\mathbf{I}}\mathbf{N})^T = \mathbf{N}^T\hat{\mathbf{I}}. \tag{5.5-8}$$

From (5.5-3)

$$\hat{\mathbf{I}}\mathbf{N}\xi = p\hat{\mathbf{I}}\xi$$

and, by (5.5-8),

$$\mathbf{N}^T(\hat{\mathbf{I}}\xi) = p(\hat{\mathbf{I}}\xi).$$

The left eigenvector η can therefore assume the form

$$\eta = \hat{\mathbf{I}}\xi = \begin{bmatrix} \boldsymbol{b} \\ \boldsymbol{a} \end{bmatrix}. \tag{5.5-9}$$

Thus the Stroh eigenvectors $\boldsymbol{a}$ and $\boldsymbol{b}$ which form the right eigenvector ξ and the left eigenvector η of the fundamental elasticity matrix $\mathbf{N}$ are in the reverse order.

It is known that the left and right eigenvectors associated with different eigenvalues are orthogonal to each other, i.e.,

$$\eta_\alpha \cdot \xi_\beta = 0, \qquad \text{if } p_\alpha \neq p_\beta. \tag{5.5-10}$$

The vector ξ given by (5.5-3), and hence the vector η by (5.5-9), is unique up to an arbitrary multiplier. Assuming that p_α are distinct, we normalize ξ_α such that

$$\eta_\alpha \cdot \xi_\beta = \delta_{\alpha\beta} \tag{5.5-11}$$

or, by $(5.5\text{-}4)_2$ and (5.5-9),

$$\boldsymbol{b}_\alpha^T\boldsymbol{a}_\beta + \boldsymbol{a}_\alpha^T\boldsymbol{b}_\beta = \delta_{\alpha\beta} \tag{5.5-12}$$

where $\delta_{\alpha\beta}$ is the Kronecker delta. In view of $(5.1\text{-}12)_3$, (5.3-8), and the matrices **A**, **B** defined in (5.3-11), (5.5-12) is written as

$$\begin{bmatrix} \mathbf{B}^T & \mathbf{A}^T \\ \overline{\mathbf{B}}^T & \overline{\mathbf{A}}^T \end{bmatrix} \begin{bmatrix} \mathbf{A} & \overline{\mathbf{A}} \\ \mathbf{B} & \overline{\mathbf{B}} \end{bmatrix} = \begin{bmatrix} \mathbf{I} & \mathbf{0} \\ \mathbf{0} & \mathbf{I} \end{bmatrix} \tag{5.5-13}$$

or

$$\begin{aligned} \mathbf{B}^T\mathbf{A} + \mathbf{A}^T\mathbf{B} = \mathbf{I} = \overline{\mathbf{B}}^T\overline{\mathbf{A}} + \overline{\mathbf{A}}^T\overline{\mathbf{B}}, \\ \mathbf{B}^T\overline{\mathbf{A}} + \mathbf{A}^T\overline{\mathbf{B}} = \mathbf{0} = \overline{\mathbf{B}}^T\mathbf{A} + \overline{\mathbf{A}}^T\mathbf{B}. \end{aligned} \tag{5.5-14}$$

These are the *orthogonality* relations. The two 6×6 matrices on the left of (5.5-13) are the inverses of each other, and hence their products commute, i.e.,

$$\begin{bmatrix} \mathbf{A} & \overline{\mathbf{A}} \\ \mathbf{B} & \overline{\mathbf{B}} \end{bmatrix} \begin{bmatrix} \mathbf{B}^T & \mathbf{A}^T \\ \overline{\mathbf{B}}^T & \overline{\mathbf{A}}^T \end{bmatrix} = \begin{bmatrix} \mathbf{I} & \mathbf{0} \\ \mathbf{0} & \mathbf{I} \end{bmatrix} \tag{5.5-15}$$

or

$$\begin{aligned} \mathbf{A}\mathbf{B}^T + \overline{\mathbf{A}\mathbf{B}}^T = \mathbf{I} = \mathbf{B}\mathbf{A}^T + \overline{\mathbf{B}\mathbf{A}}^T, \\ \mathbf{A}\mathbf{A}^T + \overline{\mathbf{A}\mathbf{A}}^T = \mathbf{0} = \mathbf{B}\mathbf{B}^T + \overline{\mathbf{B}\mathbf{B}}^T. \end{aligned} \tag{5.5-16}$$

These are the *closure* relations. Equations $(5.5\text{-}16)_{1,2}$ tell us that the real part of $\mathbf{AB}^T$ is $\frac{1}{2}\mathbf{I}$, while $(5.5\text{-}16)_{3,4}$ imply that $\mathbf{AA}^T$ and $\mathbf{BB}^T$ are purely imaginary. Hence the three matrices **S**, **H**, and **L** defined by

$$\mathbf{S} = i(2\mathbf{AB}^T - \mathbf{I}), \quad \mathbf{H} = 2i\mathbf{AA}^T, \quad \mathbf{L} = -2i\mathbf{BB}^T \tag{5.5-17}$$

are real. We will show in Section 7.2 that they are tensors of rank two when the transformation is a rotation about the x_3-axis. It is clear that **H** and **L** are symmetric. It will be shown (Section 7.6) that they are positive definite. Therefore the matrices **A** and **B** are nonsingular (see also Section 12.4).

The matrix products $\mathbf{AB}^{\mathrm{T}}$, $\mathbf{AA}^T$, and $\mathbf{BB}^T$ appear often in the final solutions to two-dimensional anisotropic elasticity problems. They can be replaced in terms of the real tensors **S**, **H**, and **L** through (5.5-17). Therefore it is possible to obtain the solutions in a real form for many problems. As will be shown in Sections 6.4 and 7.6 there are means to compute **S**, **H**, and **L** explicitly.

The eigenrelation (5.5-3) for $p=p_1$, p_2, p_3 can be combined into one equation as

$$\mathbf{N}\begin{bmatrix} \mathbf{A} & \overline{\mathbf{A}} \\ \mathbf{B} & \overline{\mathbf{B}} \end{bmatrix} = \begin{bmatrix} \mathbf{A} & \overline{\mathbf{A}} \\ \mathbf{B} & \overline{\mathbf{B}} \end{bmatrix} \begin{bmatrix} \mathbf{P} & \mathbf{0} \\ \mathbf{0} & \overline{\mathbf{P}} \end{bmatrix} \tag{5.5-18}$$

where

$$\mathbf{P} = \langle p_* \rangle = \mathrm{diag}[p_1, p_2, p_3]. \tag{5.5-19}$$

By (5.5-15) we have

$$\mathbf{N} = \begin{bmatrix} \mathbf{A} & \overline{\mathbf{A}} \\ \mathbf{B} & \overline{\mathbf{B}} \end{bmatrix} \begin{bmatrix} \mathbf{P} & \mathbf{0} \\ \mathbf{0} & \overline{\mathbf{P}} \end{bmatrix} \begin{bmatrix} \mathbf{B}^T & \mathbf{A}^T \\ \overline{\mathbf{B}}^T & \overline{\mathbf{A}}^T \end{bmatrix} \tag{5.5-20}$$

which is the diagonalization of **N**.

The derivations presented so far assume that the eigenvalues p_α are distinct or, if they are not, there exist six independent eigenvectors ξ_α. In other words the derivations are valid if the 6×6 matrix **N** in (5.5-3) is simple or semisimple (Section 1.3). A semisimple matrix **N** can be found for certain transversely isotropic materials whose axis of symmetry is not on the x_3-axis (Tanuma, 1996). If **N** is nonsemisimple, the general solutions (5.3-7), (5.3-10), and (5.3-13) need to be modified. The orthogonality relations (5.5-13) and the closure relations (5.5-15) are not valid. Anisotropic elastic materials for which **N** is non-semisimple are called *degenerate materials*. They are degenerate in the mathematical sense, not necessarily in the physical sense. Isotopic materials are a special group of degenerate materials for which $p=i$ is an eigenvalue of multiplicity three but there are only two independent eigenvectors for ξ (Section 5.2). Although **S**, **H**, and **L** defined in (5.5-17) are not valid for degenerate materials, Barnett and Lothe (1973) devised an integral formalism to compute these matrices directly from the elastic stiffnesses C_{ijks} (Section 7.6). Thus the need for calculating the eigenvalues p_α and the eigenvectors ξ_α is circumvented, and the problems associated with degenerate materials disappear. **S**, **H**, and **L** are therefore called the *Barnett–Lothe tensors*. The tensorial nature of **S**, **H**, and **L** will be studied in Section 7.2.

It should be pointed out that the problems associated with degenerate materials are not unique with the Stroh formalism. Other formalisms have the same problems. Modifications required for the general solutions and for the Barnett–Lothe tensors **S**, **H**, **L** will be discussed in Sections 13.2 and 5.9, respectively.

5.6 Positive Definite Hermitian Matrices

We have pointed out that the products

$$2i\mathbf{A}\mathbf{A}^T \quad \text{and} \quad -2i\mathbf{B}\mathbf{B}^T$$

appearing in (5.5-17) are real, symmetric, and positive definite. They are special cases of positive definite Hermitian matrices in that they are real, not complex. From $(5.5\text{-}14)_4$ Stroh (1958) observed that $\overline{\mathbf{B}}^T\mathbf{A}$ is skew-Hermitian (Section 1.7). This means that

$$i\overline{\mathbf{B}}^T\mathbf{A} = -i\overline{\mathbf{A}}^T\mathbf{B}$$

is Hermitian. We will show that it is also positive definite, and that it is but one of several positive definite Hermitian matrices that involve products of matrices associated with **A** and **B**.

The *impedance tensor* **M** and its inverse $\mathbf{M}^{-1}$ defined by (Ingebrigtsen and Tonning, 1969; Lothe and Barnett, 1976a)

$$\mathbf{M} = -i\mathbf{B}\mathbf{A}^{-1}, \quad \mathbf{M}^{-1} = i\mathbf{A}\mathbf{B}^{-1}, \tag{5.6-1}$$

are Hermitian because (5.6-1) are equivalent to

$$\mathbf{M} = \mathbf{B}(i\overline{\mathbf{B}}^T\mathbf{A})^{-1}\overline{\mathbf{B}}^T, \quad \mathbf{M}^{-1} = \mathbf{A}(-i\overline{\mathbf{A}}^T\mathbf{B})^{-1}\overline{\mathbf{A}}^T. \tag{5.6-2}$$

Since $i\overline{\mathbf{B}}^T\mathbf{A} = -i\overline{\mathbf{A}}^T\mathbf{B}$ are Hermitian, so are their inverses. Hence **M** and its inverse $\mathbf{M}^{-1}$ are Hermitian. We will show in Section 6.6 that **M** and $\mathbf{M}^{-1}$ are positive definite. By (5.6-2), $(i\overline{\mathbf{B}}^T\mathbf{A})^{-1}$ is also positive definite because **B** is nonsingular. Thus $i\overline{\mathbf{B}}^T\mathbf{A}$ is positive definite Hermitian as we stated earlier.

Equation $(5.6\text{-}1)_1$ means

$$\mathbf{B} = i\mathbf{M}\mathbf{A}$$

or

$$\boldsymbol{b}_\alpha = i\mathbf{M}\boldsymbol{a}_\alpha \quad (\alpha = 1,2,3). \tag{5.6-3}$$

The relation between $\boldsymbol{b}_\alpha$ and $\boldsymbol{a}_\alpha$ is independent of α. It offers an alternative to (5.3-2) in which the relation between $\boldsymbol{b}_\alpha$ and $\boldsymbol{a}_\alpha$ depends on α. As indicated in (5.6-3) **M** effectively transforms the displacement related vector $\boldsymbol{a}_\alpha$ to the traction related vector $\boldsymbol{b}_\alpha$, justifying the term impedance tensor for **M**.

The matrices

$$i\mathbf{B}^{-1}\overline{\mathbf{B}} \quad \text{and} \quad -i\mathbf{A}^{-1}\overline{\mathbf{A}} \tag{5.6-4}$$

are positive definite Hermitian. In fact they are orthogonal matrices (Ting, 1993). From

$$i\mathbf{B}^{-1}\overline{\mathbf{B}} = i\mathbf{B}^T(\mathbf{B}\mathbf{B}^T)^{-1}\overline{\mathbf{B}} = 2\mathbf{B}^T\mathbf{L}^{-1}\overline{\mathbf{B}}$$

and, employing $(5.5\text{-}16)_4$,

$$(i\mathbf{B}^{-1}\overline{\mathbf{B}})(i\mathbf{B}^{-1}\overline{\mathbf{B}})^T = -\mathbf{B}^{-1}\overline{\mathbf{B}}\overline{\mathbf{B}}^T(\mathbf{B}^{-1})^T = \mathbf{B}^{-1}\mathbf{B}\mathbf{B}^T(\mathbf{B}^{-1})^T = \mathbf{I},$$

the positive definite and orthogonal properties of $i\mathbf{B}^{-1}\overline{\mathbf{B}}$ are verified. Following the same procedures, $-i\mathbf{A}^{-1}\overline{\mathbf{A}}$ can be shown to be positive definite orthogonal Hermitian.

Making use of (5.5-17) and the identities

$$\mathbf{B}\mathbf{A}^{-1} = (\mathbf{A}\mathbf{B}^T)^T(\mathbf{A}\mathbf{A}^T)^{-1}, \quad \mathbf{A}\mathbf{B}^{-1} = (\mathbf{A}\mathbf{B}^T)(\mathbf{B}\mathbf{B}^T)^{-1}, \tag{5.6-5}$$

which can be verified easily, (5.6-1) is written as

$$\mathbf{M} = -i\mathbf{B}\mathbf{A}^{-1} = \mathbf{H}^{-1} + i\mathbf{H}^{-1}\mathbf{S},$$
$$\mathbf{M}^{-1} = i\mathbf{A}\mathbf{B}^{-1} = \mathbf{L}^{-1} - i\mathbf{S}\mathbf{L}^{-1}. \tag{5.6-6}$$

Since the imaginary part of a Hermitian is skew-symmetric we have

$$\mathbf{H}^{-1}\mathbf{S} + \mathbf{S}^T\mathbf{H}^{-1} = \mathbf{0} = \mathbf{S}\mathbf{L}^{-1} + \mathbf{L}^{-1}\mathbf{S}^T. \tag{5.6-7}$$

In fact we had used $(5.6\text{-}7)_1$ in writing $(5.6\text{-}6)_2$. By inserting (5.6-6) into the relation $\mathbf{M}\mathbf{M}^{-1} = \mathbf{I}$ we obtain

$$(\mathbf{H}^{-1} + i\mathbf{H}^{-1}\mathbf{S})(\mathbf{L}^{-1} - i\mathbf{S}\mathbf{L}^{-1}) = \mathbf{I}.$$

The imaginary parts cancel each other while the real parts give

$$\mathbf{H}^{-1}\mathbf{L}^{-1} + \mathbf{H}^{-1}\mathbf{S}\mathbf{S}\mathbf{L}^{-1} = \mathbf{I}$$

or

$$\mathbf{H}\mathbf{L} - \mathbf{S}\mathbf{S} = \mathbf{I}. \tag{5.6-8}$$

Thus the three real tensors **S**, **H**, and **L** are related. More identities connecting **S**, **H**, and **L** will be presented in Section 6.5. It should be noted that the impedance tensor **M** alone is sufficient to determine **S**, **H**, and **L**. When **M** is given **H** and **S** are computed from the real and imaginary parts of **M** using $(5.6\text{-}6)_2$. **L** is then deduced from (5.6-8).

The orthogonality relations (5.5-13) and the closure relations (5.5-15) are useful in applications. So are the positive definite Hermitian matrices and the identities (5.6-6)–(5.6-8) presented here. These features do not exist in other formalisms.

5.7 The Matrix Differential Equation

For two-dimensional deformations studied here the equations of equilibrium are

$$\sigma_{i1,1} + \sigma_{i2,2} = 0. \tag{5.7-1}$$

From Theorem 3.14–1 there exists a stress function φ_i such that

$$\sigma_{i1} = -\varphi_{i,2}, \quad \sigma_{i2} = \varphi_{i,1}, \tag{5.7-2}$$

which were obtained in (5.3-4) by a different approach. Since

$$\sigma_{i1} = C_{i1k1}u_{k,1} + C_{i1k2}u_{k,2} = Q_{ik}u_{k,1} + R_{ik}u_{k,2},$$
$$\sigma_{i2} = C_{i2k1}u_{k,1} + C_{i2k2}u_{k,2} = R_{ki}u_{k,1} + T_{ik}u_{k,2},$$

(5.7-2) can be written as

$$\begin{aligned} \mathbf{Q}\,\mathbf{u}_{,1} + \mathbf{R}\mathbf{u}_{,2} &= -\boldsymbol{\phi}_{,2}, \\ \mathbf{R}^T\mathbf{u}_{,1} + \mathbf{T}\mathbf{u}_{,2} &= \boldsymbol{\phi}_{,1}, \end{aligned} \tag{5.7-3a}$$

or

$$\begin{bmatrix} -\mathbf{Q} & \mathbf{0} \\ -\mathbf{R}^T & \mathbf{I} \end{bmatrix} \begin{bmatrix} \mathbf{u}_{,1} \\ \boldsymbol{\phi}_{,1} \end{bmatrix} = \begin{bmatrix} \mathbf{R} & \mathbf{I} \\ \mathbf{T} & \mathbf{0} \end{bmatrix} \begin{bmatrix} \mathbf{u}_{,2} \\ \boldsymbol{\phi}_{,2} \end{bmatrix}. \tag{5.7-3b}$$

Employing the identity (5.5-2) we obtain (Malen, 1971; Barnett and Lothe, 1973; Chadwick and Smith, 1977)

$$\mathbf{N}\boldsymbol{\varpi}_{,1} = \boldsymbol{\varpi}_{,2}, \quad \boldsymbol{\varpi} = \begin{bmatrix} \mathbf{u} \\ \boldsymbol{\phi} \end{bmatrix}, \tag{5.7-4}$$

or

$$\begin{aligned} \mathbf{u}_{,2} &= \mathbf{N}_1\mathbf{u}_{,1} + \mathbf{N}_2\boldsymbol{\phi}_{,1}, \\ \boldsymbol{\phi}_{,2} &= \mathbf{N}_3\mathbf{u}_{,1} + \mathbf{N}_1^T\boldsymbol{\phi}_{,1}. \end{aligned} \tag{5.7-5}$$

This is the *matrix differential equation* for two-dimensional deformations of an anisotropic elastic solid. A general solution for $\boldsymbol{\varpi}$ is

$$\boldsymbol{\varpi} = \boldsymbol{\xi} f(z), \quad z = x_1 + p x_2. \tag{5.7-6}$$

On substituting (5.7-6) into $(5.7\text{-}4)_1$ we recover the eigenrelation (5.5-3).

The matrix differential equation (5.7-4) is obtained without the differentiation of elastic constants C_{ijks}. Therefore (5.7-4) remains valid for inhomogeneous materials for which C_{ijks} can depend on x_1 and x_2. Of course the general solution (5.7-6) does not apply to inhomogeneous materials.

When x_2=0 is an interface between two materials $(5.7\text{-}4)_1$ or (5.7-5) allows us to determine the *anti-plane* differentiation $\boldsymbol{\varpi}_{,2}$ in terms of the *in-plane* differentiation $\boldsymbol{\varpi}_{,1}$ (Hill, 1983). If $\mathbf{u}$ and $\boldsymbol{\phi}$ are continuous across the interface so are $\mathbf{u}_{,1}$ and $\boldsymbol{\phi}_{,1}$. Equations (5.7-5) then provide us $\mathbf{u}_{,2}$ and $\boldsymbol{\phi}_{,2}$ in one material when $\mathbf{u}_{,1}$ and $\boldsymbol{\phi}_{,1}$ in the other material are known. From (5.3-4) and (5.1-16) we rewrite (5.7-5) as

$$\begin{aligned} \mathbf{u}_{,2} &= \mathbf{N}_1\mathbf{u}_{,1} + \mathbf{N}_2\mathbf{t}_2, \\ -\mathbf{t}_1 &= \mathbf{N}_3\mathbf{u}_{,1} + \mathbf{N}_1^T\mathbf{t}_2. \end{aligned} \tag{5.7-7}$$

The vector $\mathbf{t}_2$ is the traction vector on the surface x_2=0 while $\mathbf{t}_1$ is the *hoop stress vector*, which is the traction vector acting on the surface perpendicular to x_2=0. For an interface for which $\mathbf{u}$ and $\mathbf{t}_2$ are continuous (5.7-7) leads to

$$
\begin{aligned}
[\mathbf{u}_{,2}] &= [\mathbf{N}_1]\mathbf{u}_{,1} + [\mathbf{N}_2]\mathbf{t}_2, \\
-[\mathbf{t}_1] &= [\mathbf{N}_3]\mathbf{u}_{,1} + [\mathbf{N}_1^T]\mathbf{t}_2,
\end{aligned}
\tag{5.7-8}
$$

in which the square brackets denote the discontinuity in the quantity across the interface $x_2=0$.

If there is only one material occupying the space $x_2 > 0$ for which $x_2=0$ is a *free surface* or a *rigid (clamped) surface*, we have

$$\mathbf{t}_1 = -\mathbf{N}_3\mathbf{u}_{,1}, \quad \text{for a free surface,} \tag{5.7-9}$$

$$\mathbf{t}_1 = -\mathbf{N}_1^T\mathbf{t}_2, \quad \text{for a rigid surface.} \tag{5.7-10}$$

Equation (5.7-9) follows from $(5.7\text{-}7)_2$ by setting $\mathbf{t}_2=\mathbf{0}$ while (5.7-10) is obtained from $(5.7\text{-}7)_2$ by applying (5.7-14) derived below. Equations (5.7-9) and (5.7-10) are useful in obtaining the hoop stress vector $\mathbf{t}_1$ when $\mathbf{u}$ along the free surface, or $\mathbf{t}_2$ on the rigid surface, is known. When $x_2=0$ is a traction free surface, the stress component σ_{12} and σ_{21} must vanish. This means that the second component of the hoop stress vector $\mathbf{t}_1$ in (5.7-9) should vanish for any $\mathbf{u}_{,1}$. This is assured because the matrix $\mathbf{N}_3$ has the structure (Section 6.1)

$$\mathbf{N}_3 = \begin{bmatrix} * & 0 & * \\ 0 & 0 & 0 \\ * & 0 & * \end{bmatrix}, \tag{5.7-11}$$

where the asterisk $*$ denotes a possibly nonzero element.

The fact that $x_2=0$ is a rigid surface does not necessarily mean that $\mathbf{u}=\mathbf{0}$ at $x_2=0$. The displacement at $x_2=0$ can have a rigid body translation and rotation. Assuming that $x_2 < 0$ consists of a rigid body, the displacement in $x_2 < 0$ is

$$\mathbf{u} = \mathbf{u}_o + \omega \begin{bmatrix} -x_2 \\ x_1 \\ 0 \end{bmatrix} \tag{5.7-12}$$

where ω and $\mathbf{u}_o$ are constants. Therefore

$$\mathbf{u}_{,1} = \omega \begin{bmatrix} 0 \\ 1 \\ 0 \end{bmatrix} \tag{5.7-13}$$

which applies to $x_2 \le 0$. With (5.7-11) and (5.7-13) it is clear that

$$\mathbf{N}_3\mathbf{u}_{,1} = \mathbf{0}, \tag{5.7-14}$$

proving that $(5.7\text{-}7)_2$ leads to (5.7-10).

5.8 Physical Meanings of *p*, *a*, and *b*

The general solutions for the displacement **u** and the stress function ϕ as given in (5.3-7) can be made real by letting $f_{\alpha+3}$ the complex conjugate of f_α (α=1,2,3). We then have

$$\mathbf{u} = \sum_{\alpha=1}^{3}\left\{\boldsymbol{a}_\alpha f_\alpha(z_\alpha) + \bar{\boldsymbol{a}}_\alpha \bar{f}_\alpha(\bar{z}_\alpha)\right\}, \quad \boldsymbol{\phi} = \sum_{\alpha=1}^{3}\left\{\boldsymbol{b}_\alpha f_\alpha(z_\alpha) + \bar{\boldsymbol{b}}_\alpha \bar{f}_\alpha(\bar{z}_\alpha)\right\}, \quad (5.8\text{-}1)$$

where

$$z_\alpha = x_1 + p_\alpha x_2.$$

The eigenvalues p_α and the eigenvectors $\boldsymbol{a}_\alpha$ and $\boldsymbol{b}_\alpha$ depend on the elastic stiffnesses C_{ijks} only. Once they are determined C_{ijks} are no longer needed. Therefore p_α, $\boldsymbol{a}_\alpha$, and $\boldsymbol{b}_\alpha$ can be regarded as material constants even though they are complex-valued.

The general solutions (5.8-1) consist of three *independent* real solutions of the form

$$\mathbf{u} = \boldsymbol{a} f(z) + \bar{\boldsymbol{a}} \bar{f}(\bar{z}), \quad \boldsymbol{\phi} = \boldsymbol{b} f(z) + \bar{\boldsymbol{b}} \bar{f}(\bar{z}). \qquad (5.8\text{-}2)$$

Let

$$\boldsymbol{a} = \boldsymbol{a}' + i\boldsymbol{a}'', \quad \boldsymbol{b} = \boldsymbol{b}' + i\boldsymbol{b}'',$$

where $\boldsymbol{a}'$, $\boldsymbol{a}''$, $\boldsymbol{b}'$, and $\boldsymbol{b}''$ are real. The real vectors $\boldsymbol{a}'$ and $\boldsymbol{a}''$ span a plane if $\boldsymbol{a}$ is *genuinely complex*, i.e., if $\boldsymbol{a}'$ and $\boldsymbol{a}''$ are not proportional to each other. The plane spanned by $\boldsymbol{a}'$ and $\boldsymbol{a}''$ is called an *eigenplane* (Section 1.8) or *the plane* $\boldsymbol{a}$. It is clear that the plane $\boldsymbol{a}$ and the plane $\bar{\boldsymbol{a}}$ are the same plane. Likewise, the plane $\boldsymbol{b}$ and the plane $\bar{\boldsymbol{b}}$ are the same plane. Equations (5.8-2) tell us that the displacement **u** at any point (x_1,x_2) is polarized on the plane $\boldsymbol{a}$ while the stress function $\boldsymbol{\phi}$ is polarized on the plane $\boldsymbol{b}$. The latter means that, from (5.4-2), the surface traction vector **t** is polarized on the plane $\boldsymbol{b}$.

For general anisotropic elastic materials there are three polarization planes (the planes $\boldsymbol{a}_1$, $\boldsymbol{a}_2$, $\boldsymbol{a}_3$) for the displacement **u** and three polarization planes (the planes $\boldsymbol{b}_1$, $\boldsymbol{b}_2$, $\boldsymbol{b}_3$) for the surface traction **t**. If one of complex vectors, say $\boldsymbol{a}_3$, is not genuinely complex, $\boldsymbol{a}_3$ can be taken as a real vector. The displacement **u** associated with $\boldsymbol{a}_3$ is then polarized along this real vector. This happens to monoclinic materials with the symmetry plane at x_3=0. The vectors $\boldsymbol{a}_3$ and $\boldsymbol{b}_3$ are both real and are both in the direction parallel to the x_3-axis. The displacement associated with $\boldsymbol{a}_3$ is the antiplane displacement and the stresses associated with $\boldsymbol{b}_3$ are the antiplane stresses. As to the planes $\boldsymbol{a}_1$ and $\boldsymbol{a}_2$ for monoclinic materials with the symmetry plane at $x_3=0$, they coincide with the plane $x_3=0$, so do the planes $\boldsymbol{b}_1$ and $\boldsymbol{b}_2$. Therefore the displacements associated with $\boldsymbol{a}_1$, $\boldsymbol{a}_2$ are inplane displacements and the stresses associated with $\boldsymbol{b}_1$, $\boldsymbol{b}_2$ are inplane stresses.

As to the complex eigenvalues p_α, no physical meanings have been found that can be attributed directly to p_α. However it has been shown in Sections 3.5, 3.6, 3.8–3.10, and will be shown in Chapters 8 and 10 that the complex eigenvalues p_α are solely responsible for the locations of the image singularities for Green's functions for half-spaces, bimaterials, and elliptic inclusions. The locations of the image singularities do not depend on the nature of the singularity prescribed, nor on the boundary conditions.

5.9 Nonsemisimple N

When $\mathbf{N}$ is nonsemisimple, say $p_1=p_2$, there is only one eigenvector ξ_1 associated with $p_1=p_2$. The eigenrelations for $p_1=p_2$ are

$$\begin{aligned} \mathbf{N}\xi_1 &= p_1\xi_1, \\ \mathbf{N}\xi_2 &= p_1\xi_2 + \xi_1, \end{aligned} \tag{5.9-1}$$

where ξ_2 is the generalized eigenvector (Section 1.4). The left eigenvectors associated with $p_1=p_2$ satisfy the eigenrelations

$$\begin{aligned} \mathbf{N}^T\eta_1 &= p_1\eta_1 + \eta_2, \\ \mathbf{N}^T\eta_2 &= p_1\eta_2, \end{aligned} \tag{5.9-2}$$

in which η_1, not η_2, is the generalized eigenvector. The ξ_1, ξ_2, η_1, and η_2 are not unique. We will show how ξ_1, ξ_2, η_1, and η_2 can be chosen such that the orthogonality relations

$$\eta_2^T\xi_1 = 0, \quad \eta_1^T\xi_2 = 0, \quad \eta_1^T\xi_1 = 1, \quad \eta_2^T\xi_2 = 1, \tag{5.9-3}$$

are satisfied. When $(5.9\text{-}1)_{1,2}$ are multiplied by η_1^T and use is made of (5.9-2) we obtain $(5.9\text{-}3)_1$ and

$$\eta_2^T\xi_2 = \eta_1^T\xi_1.$$

Therefore we only have to satisfy $(5.9\text{-}3)_2$ and $(5.9\text{-}3)_3$ or $(5.9\text{-}3)_4$. Multiplication of (5.9-1) by $\hat{\mathbf{I}}$ defined in (5.5-7) and with the aid of the relation (5.5-8) we have

$$\begin{aligned} \mathbf{N}^T(\hat{\mathbf{I}}\xi_1) &= p_1(\hat{\mathbf{I}}\xi_1), \\ \mathbf{N}^T(\hat{\mathbf{I}}\xi_2) &= p_1(\hat{\mathbf{I}}\xi_2) + (\hat{\mathbf{I}}\xi_1). \end{aligned}$$

Comparison with (5.9-2) suggests that we may set

$$\eta_1 = \hat{\mathbf{I}}\xi_2, \quad \eta_2 = \hat{\mathbf{I}}\xi_1, \tag{5.9-4}$$

so that (5.9-3) are written as

$$\xi_1^T\hat{\mathbf{I}}\xi_1 = 0, \quad \xi_2^T\hat{\mathbf{I}}\xi_2 = 0, \quad \xi_2^T\hat{\mathbf{I}}\xi_1 = 1, \quad \xi_1^T\hat{\mathbf{I}}\xi_2 = 1. \tag{5.9-5}$$

It is seen that $(5.9\text{-}5)_3$ and $(5.9\text{-}5)_4$ are equivalent. We need to satisfy $(5.9\text{-}5)_2$ and $(5.9\text{-}5)_4$. Noting that if ξ_1^{o} and ξ_2^{o} are solutions of (5.9-1), so are

$$\xi_1 = k_1\xi_1^{\text{o}}, \quad \xi_2 = k_1\xi_2^{\text{o}} + k_2\xi_1^{\text{o}}, \tag{5.9-6}$$

where k_1 and k_2 are constants. Upon substitution into $(5.9\text{-}5)_4$ and by virtue of $(5.9\text{-}5)_1$ which applies to ξ_1^{o} and ξ_2^{o} one obtains

$$k_1^2 = \left((\xi_1^{\text{o}})^T\hat{\mathbf{I}}\xi_2^{\text{o}}\right)^{-1}. \tag{5.9-7}$$

Insertion of $(5.9\text{-}6)_2$ into $(5.9\text{-}5)_2$ and with the aid of $(5.9\text{-}5)_1$ and (5.9-7),

$$k_2 = -\tfrac{1}{2}k_1^3\left((\xi_2^{\text{o}})^T\hat{\mathbf{I}}\xi_2^{\text{o}}\right). \tag{5.9-8}$$

This completes the normalization of ξ_1 and ξ_2. The η_1 and η_2 are determined by (5.9-4).

The eigenrelations for $p_1=p_2$ and p_3 can be written as

$$\mathbf{N}[\xi_1, \xi_2, \xi_3] = [\xi_1, \xi_2, \xi_3]\mathbf{J}, \tag{5.9-9}$$

$$\mathbf{N}^T[\eta_1, \eta_2, \eta_3] = [\eta_1, \eta_2, \eta_3]\mathbf{J}^T, \tag{5.9-10}$$

where ξ_2 and η_1 are the generalized eigenvectors and

$$\mathbf{J} = \begin{bmatrix} p_1 & 1 & 0 \\ 0 & p_1 & 0 \\ 0 & 0 & p_3 \end{bmatrix} \tag{5.9-11}$$

is a Jordan canonical form (Section 1.4). By (5.9-4),

$$[\eta_1, \eta_2, \eta_3] = \hat{\mathbf{I}}[\xi_2, \xi_1, \xi_3] = \hat{\mathbf{I}}[\xi_1, \xi_2, \xi_3]\Gamma, \tag{5.9-12}$$

where

$$\Gamma = \begin{bmatrix} 0 & 1 & 0 \\ 1 & 0 & 0 \\ 0 & 0 & 1 \end{bmatrix} = \Gamma^T. \tag{5.9-13}$$

If we express the 6×1 vector ξ by the two 3×1 vectors $\boldsymbol{a}$ and $\boldsymbol{b}$ and employ the notation (5.3-11) we have

$$[\boldsymbol{\xi}_1, \boldsymbol{\xi}_2, \boldsymbol{\xi}_3] = \begin{bmatrix} \mathbf{A} \\ \mathbf{B} \end{bmatrix}, \quad [\boldsymbol{\eta}_1, \boldsymbol{\eta}_2, \boldsymbol{\eta}_3] = \begin{bmatrix} \mathbf{B}\Gamma \\ \mathbf{A}\Gamma \end{bmatrix}. \tag{5.9-14}$$

Similar expressions can be written for the eigenvectors associated with $p_4 = p_5$ and p_6 which are complex conjugates of $p_1 = p_2$ and p_3. The orthogonality relations between $\boldsymbol{\xi}_\alpha$ and $\boldsymbol{\eta}_\alpha$ now take the form

$$\begin{bmatrix} \Gamma\mathbf{B}^T & \Gamma\mathbf{A}^T \\ \Gamma\overline{\mathbf{B}}^T & \Gamma\overline{\mathbf{A}}^T \end{bmatrix} \begin{bmatrix} \mathbf{A} & \overline{\mathbf{A}} \\ \mathbf{B} & \overline{\mathbf{B}} \end{bmatrix} = \mathbf{I}. \tag{5.9-15}$$

The two 6×6 matrices on the left are the inverses of each other, and hence

$$\begin{bmatrix} \mathbf{A} & \overline{\mathbf{A}} \\ \mathbf{B} & \overline{\mathbf{B}} \end{bmatrix} \begin{bmatrix} \Gamma\mathbf{B}^T & \Gamma\mathbf{A}^T \\ \Gamma\overline{\mathbf{B}}^T & \Gamma\overline{\mathbf{A}}^T \end{bmatrix} = \mathbf{I} \tag{5.9-16}$$

or

$$\begin{aligned} \mathbf{A}\Gamma\mathbf{B}^T + \overline{\mathbf{A}}\Gamma\overline{\mathbf{B}}^T &= \mathbf{I} = \mathbf{B}\Gamma\mathbf{A}^T + \overline{\mathbf{B}}\Gamma\overline{\mathbf{A}}^T, \\ \mathbf{A}\Gamma\mathbf{A}^T + \overline{\mathbf{A}}\Gamma\overline{\mathbf{A}}^T &= \mathbf{I} = \mathbf{B}\Gamma\mathbf{B}^T + \overline{\mathbf{B}}\Gamma\overline{\mathbf{B}}^T. \end{aligned} \tag{5.9-17}$$

These are the closure relations. Using the arguments following (5.5-16) the three matrices **S**, **H**, and **L** defined by

$$\mathbf{S} = i(2\mathbf{A}\Gamma\mathbf{B}^T - \mathbf{I}), \quad \mathbf{H} = 2i\mathbf{A}\Gamma\mathbf{A}^T, \quad \mathbf{L} = -2i\mathbf{B}\Gamma\mathbf{B}^T, \tag{5.9-18}$$

are real. It is clear that **H** and **L** are symmetric. We will show in Section 7.6 that they are positive definite.

From (5.9-9) and (5.9-14)$_1$

$$\mathbf{N}\begin{bmatrix} \mathbf{A} & \overline{\mathbf{A}} \\ \mathbf{B} & \overline{\mathbf{B}} \end{bmatrix} = \begin{bmatrix} \mathbf{A} & \overline{\mathbf{A}} \\ \mathbf{B} & \overline{\mathbf{B}} \end{bmatrix} \begin{bmatrix} \mathbf{J} & \mathbf{0} \\ \mathbf{0} & \overline{\mathbf{J}} \end{bmatrix} \tag{5.9-19}$$

or, by (5.9-16),

$$\mathbf{N} = \begin{bmatrix} \mathbf{A} & \overline{\mathbf{A}} \\ \mathbf{B} & \overline{\mathbf{B}} \end{bmatrix} \begin{bmatrix} \mathbf{J} & \mathbf{0} \\ \mathbf{0} & \overline{\mathbf{J}} \end{bmatrix} \begin{bmatrix} \Gamma\mathbf{B}^T & \Gamma\mathbf{A}^T \\ \Gamma\overline{\mathbf{B}}^T & \Gamma\overline{\mathbf{A}}^T \end{bmatrix}. \tag{5.9-20}$$

This is the diagonalization of **N** when it is nonsemisimple.

The above derivations are taken from Ting and Hwu (1988) (see also Sections 13.1 and 13.2) where a more general situation was investigated. The matrix **N** was assumed to be almost nonsemisimple, i.e., p_1 and p_2 are almost identical, so are $\boldsymbol{\xi}_1$ and $\boldsymbol{\xi}_2$. This causes a problem in that the magnitudes of the normalized eigenvectors $\boldsymbol{\xi}_1$ and $\boldsymbol{\xi}_2$ are very large, and become infinity at $p_1 = p_2$. When

$p_1 \neq p_2$, the eigenrelations for p_1 and p_2 are

$$\mathbf{N}\xi_1 = p_1\xi_1, \quad \mathbf{N}\xi_2 = p_2\xi_2, \tag{5.9-21}$$

from which we have

$$\mathbf{N}(\xi_2 - \xi_1) = p_2\xi_2 - p_1\xi_1 = p_2(\xi_2 - \xi_1) + (p_2 - p_1)\xi_1$$

or

$$\mathbf{N}\xi' = p_2\xi' + \xi_1, \tag{5.9-22}$$

$$\xi' = \frac{\xi_2 - \xi_1}{p_2 - p_1}. \tag{5.9-23}$$

The vector ξ' is bounded. As $p_2 \to p_1$ and $\xi_2 \to \xi_1$, (5.9-22) becomes

$$\mathbf{N}\xi' = p_1\xi' + \xi_1 \tag{5.9-24}$$

which is identical to $(5.9\text{-}1)_2$ if we identify ξ' as the generalized eigenvector ξ_2. Notice that (5.9-24) is the differentiation of $(5.9\text{-}21)_1$ with p_1, assuming that ξ_1 depends on p_1. A similar operation on (5.1-8) and (5.3-2) gives

$$\begin{gathered} \left\{\mathbf{Q} + p_2(\mathbf{R} + \mathbf{R}^T) + p_2^2\mathbf{T}\right\}\boldsymbol{a}' = -\left\{(\mathbf{R} + \mathbf{R}^T) + (p_1 + p_2)\mathbf{T}\right\}\boldsymbol{a}_1, \\ \boldsymbol{b}' = (\mathbf{R}^T + p_2\mathbf{T})\,\boldsymbol{a}' + \mathbf{T}\,\boldsymbol{a}_1 = -\frac{1}{p_2}(\mathbf{Q} + p_1\mathbf{R})\,\boldsymbol{a}' + \frac{1}{p_1 p_2}\mathbf{Q}\,\boldsymbol{a}_1. \end{gathered} \tag{5.9-25}$$

At $p_2 = p_1$, (5.9-25) provide an alternate to $(5.9\text{-}1)_2$ for determining the generalized eigenvector $\xi_2 = (\mathbf{a}', \mathbf{b}')$. The presentation by Ting and Hwu allows the matrix $\mathbf{N}$ to be simple or nonsemisimple so that there is a smooth transition from a simple $\mathbf{N}$ to a nonsemisimple $\mathbf{N}$.

We will not discuss the degenerate case in which $p_1 = p_2 = p_3$ with only one eigenvector ξ_1. The question is open whether there exists an anisotropic material that possesses such a degeneracy. Even for isotropic materials for which $p_1 = p_2 = p_3 = i$, there exists two independent eigenvectors ξ_1 and ξ_3 associated with the triple eigenvalues.

5.10 Dependence of Solutions on Elastic Constants

The first part of this section complement the first part of Section 4.7. The second part gives additional information on the dependence of stress solutions on elastic constants.

The general solutions for generalized plane strain deformations presented in (5.1-13) and (5.1-15) depend on the 15 elastic stiffnesses shown in (5.2-1). They can be represented by the 5×5 sym metric matrix $\mathbf{C}^o$ given in (2.4-4). They do not depend on $C_{3\alpha}$ (α=1,2,...,6). Therefore any solution of generalized plane strain deformations applies to a group of anisotropic elastic materials whose $\mathbf{C}^o$ is identical but whose $C_{3\alpha}$ are different. If $\mathbf{C}^o$ is positive definite, we may choose $C_{3\alpha}$ such that $\mathbf{C}$ is positive definite. To do this, first observe that the determinant of $\mathbf{C}$ remains the same if we move the third column to sixth column and the third row to sixth row. The re-arranged matrix $\mathbf{C}$ can be written as

$$\mathbf{C} = \begin{bmatrix} \mathbf{C}^o & \mathbf{h} \\ \mathbf{h}^T & C_{33} \end{bmatrix} = \begin{bmatrix} \mathbf{C}^o & \mathbf{0} \\ \mathbf{h}^T & 1 \end{bmatrix} \begin{bmatrix} \mathbf{I} & (\mathbf{C}^o)^{-1}\mathbf{h} \\ \mathbf{0} & C_{33} - \mathbf{h}^T(\mathbf{C}^o)^{-1}\mathbf{h} \end{bmatrix} \tag{5.10-1}$$

where

$$\mathbf{h}^T = (C_{13}, C_{23}, C_{34}, C_{35}, C_{36}).$$

Taking the determinant on both sides of (5.10-1) leads to

$$|\mathbf{C}| = |\mathbf{C}^o|\{C_{33} - \mathbf{h}^T(\mathbf{C}^o)^{-1}\mathbf{h}\}. \tag{5.10-2}$$

If $\mathbf{C}^o$ is positive definite, so is $\mathbf{C}$ provided $|\mathbf{C}|>0$ (Theorem 1.6–2). Therefore if we choose

$$C_{33} > \mathbf{h}^T(\mathbf{C}^o)^{-1}\mathbf{h} = \mathbf{h}^T\mathbf{s}'\mathbf{h} \tag{5.10-3}$$

where $\mathbf{s}'$ is the 5×5 matrix given in (2.4-10) and (2.4-11), we have a positive definite $\mathbf{C}$.

A particularly interesting special case is when $\mathbf{C}^o$ is identical to that of isotropic materials. We then have

$$\mathbf{C} = \begin{bmatrix} \lambda+2\mu & \lambda & C_{13} & 0 & 0 & 0 \\ & \lambda+2\mu & C_{23} & 0 & 0 & 0 \\ & & C_{33} & C_{34} & C_{35} & C_{36} \\ & & & \mu & 0 & 0 \\ & & & & \mu & 0 \\ & & & & & \mu \end{bmatrix} \tag{5.10-4}$$

where λ and μ are Lamé constants and $C_{3\alpha}$ (α=1,2,...,6) are arbitrary real constants (see also Section 4.7). It can be shown that (5.10-3) for the $\mathbf{C}$ given by (5.10-4) becomes

$$C_{33} > \frac{1}{4\mu(\lambda+\mu)}\{\lambda(C_{13}-C_{23})^2+2\mu(C_{13}^2+C_{23}^2)\} + \frac{1}{\mu}\{C_{34}^2+C_{35}^2+C_{36}^2\}. \tag{5.10-5}$$

Thus if λ and μ satisfy the conditions (2.9-5) and $C_{3\alpha}$ (α=1,2,...,6) are chosen such that (5.10-5) is satisfied, the $\mathbf{C}$ in (5.10-4) is posi tive definite. The materials represented by (5.10-4) can be completely anisotropic without a plane of symmetry. Nevertheless the solutions to two-dimensional elastostatics problems and steady-state motions such as surface waves for the materials given by (5.10-4) are identical to the solutions for isotropic materials.

From the fact that $\mathbf{C}$ in (5.10-4) and $s_{\alpha\beta}$ in (4.7-3) are the inverses of each other, $C_{3\alpha}$ can be determined in terms of γ_α as (Ting, 1994a)

$$C_{13} = -\frac{1}{\gamma_3}\{\lambda(\gamma_1+\gamma_2)+2\mu\gamma_1\},$$

$$C_{23} = -\frac{1}{\gamma_3}\{\lambda(\gamma_1+\gamma_2)+2\mu\gamma_2\},$$

$$C_{33} = \frac{1}{\gamma_3^2}\{2\mu(1+\gamma_1^2+\gamma_2^2)+\mu(\gamma_4^2+\gamma_5^2+\gamma_6^2)+\lambda(\gamma_1+\gamma_2)^2\}, \tag{5.10-6}$$

$$C_{33} = \frac{1}{\gamma_3^2}\{2\mu(1+\gamma_1^2+\gamma_2^2)+\mu(\gamma_4^2+\gamma_5^2+\gamma_6^2)+\lambda(\gamma_1+\gamma_2)^2\},$$

$$C_{3\alpha} = -\mu\frac{\gamma_\alpha}{\gamma_3}, \qquad \alpha = 4,5,6.$$

Since $s_{\alpha\beta}$ in (4.7-3) is positive definite for arbitrary γ_α with $\gamma_3 \neq 0$, the same can be said about the matrix $\mathbf{C}$ of (5.10-4) with $C_{3\alpha}$ given by (5.10-6).

The solution for the stress does not depend on elastic constants as much as the solution for the displacement does. It is known for isotropic materials that the solution for the stress does not depend on elastic constants if

(i) the deformation is plane strain, i.e., u_1, u_2 depend on x_1, x_2 only while $u_3=0$,

(ii) the boundary conditions are prescribed in terms of tractions, and

(iii) the Michell (1899) condition is satisfied, i.e., the integration of the surface tractions on any closed curve vanishes.

Condition (i) is possible for monoclinic materials with the symmetry plane at $x_3=0$. Only six out of thirteen elastic constants are needed for these materials under a plane strain deformation. The plane strain solution is in terms of two complex variables $z_1=x_1+p_1x_2$ and $z_2=x_1+p_2x_2$. The real and imaginary parts of p_1, p_2 account for four elastic constants. It can be shown (Ting, 1995a) that the solution for the stress does not depend on elastic constants other than p_1 and p_2 when conditions (ii) and (iii) are satisfied.

Consider now a bimaterial that consists of two dissimilar homogeneous isotropic media bonded together along their interface. The geometry of the interface can be rather arbitrary. There are two elastic constants each in the two materials, resulting in a total of four elastic constants. The solution for the stress depends on these four elastic constants, but they can be reduced to three by a dimensional analysis. Dundurs (1967a,b, 1970) has proved that the solution for the stress depends on two composite elastic constants (known as Dundurs constants) provided conditions (i), (ii), (iii) are satisfied. For bimaterials that consist of two dissimilar monoclinic materials with symmetry plane at $x_3=0$, Ting (1995a) has proved that the solution for the stress depends on two composite elastic constants α and β in addition to p_1 and p_2 in the two materials. They reduce to Dundurs constants in the isotropic limit.

5.11 A. N. Stroh (1926–1962)

Not all results presented in this chapter are due to Stroh. Nevertheless his 1958 and 1962 papers laid the foundations of a new formalism for two-dimensional deformations of anisotropic elastic materials. His influence reaches almost every chapter we will present in the rest of the book. A fact seldom mentioned in the literature is that Stroh, in his 1958 paper on dislocations and cracks, also presented an alternate formalism in terms of the reduced elastic compliances $s'_{\alpha\beta}$. Unknown to him at that time the alternate formalism recovered some of Lekhnitskii's results. In particular, he reproduced Lekhnitskii's sextic equation in his equation (74). He was the first person to obtain explicit expressions in terms of $s'_{\alpha\beta}$ of the matrices $\mathbf{N}_3$ of (5.5-5) and $\mathbf{N}_3^{(-1)}$ to be introduced in (6.2-4) for general anisotropic elastic materials, and the inverse of the tensor $\mathbf{L}$ of $(5.5\text{-}17)_3$ for monoclinic materials with the symmetry plane at $x_3=0$. These will be studied in Chapter 6. His 1962 paper on steady state problems to be discussed in Chapter 12 has provided great impetus to the study of surface waves in anisotropic elastic materials. In but a few paragraphs he showed that the condition for determining the speed of a subsonic surface wave is far less restrictive than originally speculated by Synge (1956). It is amazing that he accomplished so much in those two papers at the time when the formalism was still in its infancy.

The reader perhaps wants to know more about the master who has inspired so many researchers in the field. The following biography and the list of publications of Stroh are based on information provided by Professor P. Chadwick of the University of East Anglia and by the M.I.T. Museum Archives through Professor Ali S. Argon. Professor Argon was also instrumental in securing a photo of Stroh from the M.I.T. Museum with the assistance of Ms Kara Schneiderman.

Biography of A. N. Stroh

Professor A. N. Stroh was born in Queenstown, South Africa on April 4, 1926. He attended Queens College (1934-42) and Rhodes University College (1943-47) where he was a lecturer in physics from 1947 to 1950. He received B.Sc. (1945) in mathematics and physics with distinction and M.Sc. (1947) in physics with distinction from the University of South Africa. Stroh moved to England in 1951 and did research at H. H. Wills Physical Laboratory at the University of Bristol. He was initially supervised by J. D. Eshelby and later, when Eshelby left Bristol in 1952, by N. F. (later Sir Nevill) Mott. Stroh received his Ph. D. from Bristol in 1953. In 1954-55 he was on a D.S.I.R. Senior Research Award at the Cavendish Laboratory in Cambridge. He then moved to University of Sheffield as a lecturer in Physics. During this period he became a Fellow of the Institute of Physics. Stroh came to Massachusetts Institute of Technology in 1958 as a Sloan Foreign Post-doctoral Research Fellow, and became an Associate Professor of Mechanical Engineering in 1959. In 1961-62 he was on leave from his department to do research in the Insulation Laboratory. On a rainy afternoon of Friday, September 21, 1962, on his way to a new post at the Boeing Scientific Research Laboratories in Seattle, Stroh was killed when his car skidded and went over an embankment on a highway near Steamboat Springs, Colorado. He was survived by his mother, Mrs. Iris M. Stroh of Queenstown, South Africa.

Stroh was never married. He was active in mountaineering, and was a member of the Appalachian Mountain Club. He had a brush with death in the summer of 1961 while a member of a mountain climbing team in British Columbia.

Publications of A. N. Stroh

1. J. D. Eshelby and A. N. Stroh, "Dislocations in thin plates," *Phil. Mag.* **42**, 1401-1405, 1951.
2. F. C. Frank and A. N. Stroh, "On the theory of kinking," *Proc. Phys. Soc.* B**65**, 811-821 1952.
3. A. N. Stroh, "The mean shear stress in an array of dislocations and latent hardening," *Proc. Phys. Soc.* B**66**, 2-6, 1953.
4. A. N. Stroh, "A theoretical calculation of the stored energy in a work-hardening material," *Proc. Roy. Soc. Lond.* A**218**, 391-400, 1953.
5. A. N. Stroh, "Constrictions and jogs in extended dislocations," *Proc. Phys. Soc.* B**67**, 427-436, 1954.
6. A. N. Stroh, "The formation of cracks as a result of plastic flow," *Proc. Roy. Soc. Lond.* A**223**, 404-414, 1954.

Alan Neil Stroh (1926–1962)

7. A. N. Stroh, "The formation of cracks in plastic flow. II," *Proc. Roy. Soc. Lond.* A**232**, 548-560, 1955.
8. A. N. Stroh, "Brittle fracture and yielding," *Phil. Mag.* **46**, 968-972, 1955.
9. A. N. Stroh, "The strength of Lomer–Cottrell sessile dislocations," *Phil. Mag.* **1**, 489-502, 1956.
10. A. N. Stroh, "The existence of microcracks after cold-work," *Phil. Mag.* **2**, 1-4, 1957.
11. B. A. Bilby, L. R. T. Gardner, and A. N. Stroh, "Continuous distributions of dislocations and the theory of plasticity," *Proc. 9th Int. Congress Appl. Mech.*, VIII, 35-44, 1957.
12. A. N. Stroh, "A theory of the fracture of metals," *Adv. Physics* **6**, 418-465, 1957.
13. A. N. Stroh, "The cleavage of metal single crystals," *Phil. Mag.* **3**, 597-606, 1958.
14. A. N. Stroh, "Dislocations and cracks in anisotropic elasticity," *Phil. Mag.* **3**, 625-646, 1958.
15. A. N. Stroh, "Crack nucleation in body-centered cubic metals," in *Fracture*, ed. by B. L. Averbach, D. K. Felbeck, G. T. Hahn, and D. A. Thomas, The Technology Press of MIT and Wiley, New York, 117-122, 1959.
16. A. N. Stroh, "A simple model of a propagating crack," *J. Mech. Phys. Solids* **8**, 119-122, 1960.
17. A. N. Stroh, "The deformation of a tilt boundary under applied forces," *Acta Metallurgica* **9**, 315-319, 1961.
18. A. N. Stroh, "Steady state problems in anisotropic elasticity," *J. Math. Phys.* **41**, 77-103, 1962.
19. A. N. Stroh, "Force on a moving dislocation," *Phys. Rev.* **128**, 55-61, 1962.

Chapter 6

THE STRUCTURES AND IDENTITIES OF THE ELASTICITY MATRICES

The matrices **Q**, **R**, **T**, **A**, **B**, $\mathbf{N}_1$, $\mathbf{N}_2$, $\mathbf{N}_3$, **S**, **H**, **L**, and **M** introduced in the previous chapter are the elasticity matrices. They depend on elastic constants only, and appear frequently in the solutions to two-dimensional problems. The matrices **A**, **B**, and **M** are complex while the others are real. We present their structures and identities relating them in this chapter. In Chapter 7 we will show that **A** and **B** are tensors of rank one and **S**, **H**, **L**, and **M** are tensors of rank two when the transformation is a rotation about the x_3-axis. Readers who are not interested in how the structures of these matrices and the identities relating them are derived may skip this chapter. They may return to this chapter when they read later chapters on applications where the results presented here are employed.

6.1 The Structure of $\mathbf{N}_i$

The three 3×3 real matrices $\mathbf{N}_i$ (i= 1,2,3) defined in (5.5-5) are

$$\mathbf{N}_1 = -\mathbf{T}^{-1}\mathbf{R}^T, \quad \mathbf{N}_2 = \mathbf{T}^{-1}, \quad \mathbf{N}_3 = \mathbf{R}\mathbf{T}^{-1}\mathbf{R}^T - \mathbf{Q}, \tag{6.1-1}$$

where

$$Q_{ik} = C_{i1k1}, \quad R_{ik} = C_{i1k2}, \quad T_{ik} = C_{i2k2}. \tag{6.1-2}$$

The matrices **Q**, **R**, **T** are shown in (5.2-1) in the contracted notation $C_{\alpha\beta}$. Since **Q** and **T** are symmetric and positive definite, $\mathbf{N}_2$ and $\mathbf{N}_3$ are symmetric and $\mathbf{N}_2$ is positive definite. We will show later that $-\mathbf{N}_3$ is positive semidefinite. The elements of $\mathbf{N}_1$ and $\mathbf{N}_3$ have complicated expressions in terms of $C_{\alpha\beta}$. In terms of the reduced elastic compliances $s'_{\alpha\beta}$ however they have a simpler expression.

We first show that certain elements of $\mathbf{N}_1$ and $\mathbf{N}_3$ are zero. From (6.1-2) and the symmetry of C_{ijks},

$$Q_{i2} = R_{i1}, \quad R_{2k} = T_{1k}. \tag{6.1-3}$$

Therefore the second column (also the second row) of **Q** is identical to the first column of **R**, and the second row of **R** is identical to the first row (and the first column) of $\mathbf{T}$. These are also evident from (5.2-1). Observing that $\mathbf{T}$ is symmetric the second column of $\mathbf{N}_1$ is

$$(\mathbf{N}_1)_{i2} = -(\mathbf{T}^{-1})_{ik} R_{2k} = -(\mathbf{T}^{-1})_{ik} T_{k1} = -\delta_{i1} \tag{6.1-4}$$

where δ_{ij} is the Kronecker delta. Therefore $\mathbf{N}_1$ has the structure

$$-\mathbf{N}_1 = \begin{bmatrix} * & 1 & * \\ * & 0 & * \\ * & 0 & * \end{bmatrix} \tag{6.1-5}$$

in which the asterisk $*$ denotes a possibly nonzero element. As to $\mathbf{N}_3$, we have from $(6.1\text{-}1)_3$ and $(6.1\text{-}1)_1$

$$\mathbf{N}_3 = -\mathbf{R}\mathbf{N}_1 - \mathbf{Q}.$$

The second column of $\mathbf{N}_3$ is, from (6.1-4) and $(6.1\text{-}3)_1$,

$$(\mathbf{N}_3)_{i2} = -R_{ik}(\mathbf{N}_1)_{k2} - Q_{i2} = R_{ik}\delta_{k1} - R_{i1} = 0.$$

In view of the symmetry of $\mathbf{N}_3$ it has the structure

$$\mathbf{N}_3 = \begin{bmatrix} * & 0 & * \\ 0 & 0 & 0 \\ * & 0 & * \end{bmatrix}. \tag{6.1-6}$$

We will now present $\mathbf{N}_1$, $\mathbf{N}_2$, and $\mathbf{N}_3$ in terms of the reduced elastic compliances. The starting point is the identity $(2.4\text{-}11)_1$ which, after rearranging the columns and rows, is written as

$$\left[\begin{array}{cc|ccc} C_{11} & C_{15} & C_{16} & C_{12} & C_{14} \\ C_{15} & C_{55} & C_{56} & C_{25} & C_{45} \\ \hline C_{16} & C_{56} & C_{66} & C_{26} & C_{46} \\ C_{12} & C_{25} & C_{26} & C_{22} & C_{24} \\ C_{14} & C_{45} & C_{46} & C_{24} & C_{44} \end{array}\right] \left[\begin{array}{cc|ccc} s'_{11} & s'_{15} & s'_{16} & s'_{12} & s'_{14} \\ s'_{15} & s'_{55} & s'_{56} & s'_{25} & s'_{45} \\ \hline s'_{16} & s'_{56} & s'_{66} & s'_{26} & s'_{46} \\ s'_{12} & s'_{25} & s'_{26} & s'_{22} & s'_{24} \\ s'_{14} & s'_{45} & s'_{46} & s'_{24} & s'_{44} \end{array}\right] = \mathbf{I}. \tag{6.1-7}$$

or

$$\left[\begin{array}{ccc|ccc} C_{11} & C_{16} & C_{15} & C_{16} & C_{12} & C_{14} \\ C_{16} & C_{66} & C_{56} & C_{66} & C_{26} & C_{46} \\ C_{15} & C_{56} & C_{55} & C_{56} & C_{25} & C_{45} \\ \hline C_{16} & C_{66} & C_{56} & C_{66} & C_{26} & C_{46} \\ C_{12} & C_{26} & C_{25} & C_{26} & C_{22} & C_{24} \\ C_{14} & C_{46} & C_{45} & C_{46} & C_{24} & C_{44} \end{array}\right] \left[\begin{array}{ccc|ccc} s'_{11} & 0 & s'_{15} & s'_{16} & s'_{12} & s'_{14} \\ 0 & 0 & 0 & 0 & 0 & 0 \\ s'_{15} & 0 & s'_{55} & s'_{56} & s'_{25} & s'_{45} \\ \hline s'_{16} & 0 & s'_{56} & s'_{66} & s'_{26} & s'_{46} \\ s'_{12} & 0 & s'_{25} & s'_{26} & s'_{22} & s'_{24} \\ s'_{14} & 0 & s'_{45} & s'_{46} & s'_{24} & s'_{44} \end{array}\right] = \left[\begin{array}{ccc|ccc} 1 & 0 & 0 & 0 & 0 & 0 \\ 0 & 0 & 0 & 1 & 0 & 0 \\ 0 & 0 & 1 & 0 & 0 & 0 \\ \hline 0 & 0 & 0 & 1 & 0 & 0 \\ 0 & 0 & 0 & 0 & 1 & 0 \\ 0 & 0 & 0 & 0 & 0 & 1 \end{array}\right].$$

in which the 5×5 matrices in (6.1-7) are expanded to 6×6 matrices by inserting new second rows and second columns in $\mathbf{C}^o$ and $\mathbf{s}'$. The modified relation can be written as

$$\begin{bmatrix} \mathbf{Q} & \mathbf{R} \\ \mathbf{R}^T & \mathbf{T} \end{bmatrix} \begin{bmatrix} \mathbf{q}_2 & \mathbf{r}_2 \\ \mathbf{r}_2^T & \mathbf{t} \end{bmatrix} = \begin{bmatrix} \mathbf{I}-\mathbf{I}_2 & \mathbf{I}_{12}^T \\ \mathbf{0} & \mathbf{I} \end{bmatrix}, \tag{6.1-8}$$

where

$$\mathbf{q}_2 = \begin{bmatrix} s'_{11} & 0 & s'_{15} \\ 0 & 0 & 0 \\ s'_{15} & 0 & s'_{55} \end{bmatrix}, \quad \mathbf{r}_2 = \begin{bmatrix} s'_{16} & s'_{12} & s'_{14} \\ 0 & 0 & 0 \\ s'_{56} & s'_{25} & s'_{45} \end{bmatrix}, \quad \mathbf{t} = \begin{bmatrix} s'_{66} & s'_{26} & s'_{46} \\ s'_{26} & s'_{22} & s'_{24} \\ s'_{46} & s'_{24} & s'_{44} \end{bmatrix}, \tag{6.1-9}$$

$$\mathbf{I}_{12} = \begin{bmatrix} 0 & 1 & 0 \\ 0 & 0 & 0 \\ 0 & 0 & 0 \end{bmatrix}, \tag{6.1-10}$$

and

$$\mathbf{I}_2 = \mathrm{diag}[0,\ 1,\ 0]. \tag{6.1-11}$$

The matrix $\mathbf{t}$ is obtained from $\mathbf{T}$ by replacing $C_{\alpha\beta}$ by $s'_{\alpha\beta}$. The matrix $\mathbf{r}_2$ is obtained from $\mathbf{R}$ by replacing $C_{\alpha\beta}$ by $s'_{\alpha\beta}$ and setting the second row to zero. The same operation is applied to $\mathbf{Q}$ to obtain $\mathbf{q}_2$ except that the second column as well as the second row of $\mathbf{q}_2$ is set to zero. Trivial identities are

$$\mathbf{I}_{12}\mathbf{r}_2 = \mathbf{0}, \quad \mathbf{I}_{12}\mathbf{q}_2 = \mathbf{0}. \tag{6.1-12}$$

Employing the relation

$$\begin{bmatrix} \mathbf{I} & \mathbf{N}_1^T \\ \mathbf{0} & \mathbf{N}_2 \end{bmatrix} \begin{bmatrix} \mathbf{Q} & \mathbf{R} \\ \mathbf{R}^T & \mathbf{T} \end{bmatrix} = \begin{bmatrix} -\mathbf{N}_3 & \mathbf{0} \\ -\mathbf{N}_1 & \mathbf{I} \end{bmatrix} \tag{6.1-13}$$

which is easily verified, (6.1-8) becomes

$$\begin{bmatrix} -\mathbf{N}_3 & \mathbf{0} \\ -\mathbf{N}_1 & \mathbf{I} \end{bmatrix} \begin{bmatrix} \mathbf{q}_2 & \mathbf{r}_2 \\ \mathbf{r}_2^T & \mathbf{t} \end{bmatrix} = \begin{bmatrix} \mathbf{I} & \mathbf{N}_1^T \\ \mathbf{0} & \mathbf{N}_2 \end{bmatrix} \begin{bmatrix} \mathbf{I}-\mathbf{I}_2 & \mathbf{I}_{12} \\ \mathbf{0} & \mathbf{I} \end{bmatrix}$$

or

$$\begin{aligned} -\mathbf{N}_3\mathbf{q}_2 &= \mathbf{I}-\mathbf{I}_2, & -\mathbf{N}_3\mathbf{r}_2 &= \mathbf{I}_{12}^T + \mathbf{N}_1^T, \\ -\mathbf{N}_1\mathbf{q}_2 + \mathbf{r}_2^T &= \mathbf{0}, & -\mathbf{N}_1\mathbf{r}_2 + \mathbf{t} &= \mathbf{N}_2. \end{aligned} \tag{6.1-14}$$

With the structure of $\mathbf{N}_1$ and $\mathbf{N}_3$ given in (6.1-5) and (6.1-6) and making use of $(6.1\text{-}12)_1$, explicit expressions of $\mathbf{N}_3$, $\mathbf{N}_1$, and $\mathbf{N}_2$ in that order can be determined from $(6.1\text{-}14)_{1,2,4}$ as

$$\mathbf{N}_3 = -\mathbf{q}_2^{-1} = \frac{-1}{\Delta}\begin{bmatrix} s'_{55} & 0 & -s'_{15} \\ 0 & 0 & 0 \\ -s'_{15} & 0 & s'_{11} \end{bmatrix}, \tag{6.1-15}$$

$$-\mathbf{N}_1 = \mathbf{I}_{12} - \mathbf{r}_2^T\mathbf{q}_2^{-1} = \begin{bmatrix} r_6 & 1 & s_6 \\ r_2 & 0 & s_2 \\ r_4 & 0 & s_4 \end{bmatrix}, \tag{6.1-16}$$

$$\mathbf{N}_2 = \mathbf{t} - \mathbf{r}_2^T\mathbf{q}_2^{-1}\mathbf{r}_2. \tag{6.1-17}$$

In the above $\mathbf{q}_2^{-1}$ is a *pseudo inverse* of $\mathbf{q}_2$ and

$$\begin{aligned} \Delta &= s'_{11}s'_{55} - s'_{15}s'_{15}, \\ r_\alpha &= \frac{1}{\Delta}(s'_{15}s'_{5\alpha} - s'_{55}s'_{1\alpha}), \\ s_\alpha &= \frac{1}{\Delta}(s'_{15}s'_{1\alpha} - s'_{11}s'_{5\alpha}), \end{aligned} \tag{6.1-18}$$

(α = 6,2,4). It is easily shown that $(6.1\text{-}14)_3$ is automatically satisfied when $(6.1\text{-}16)_1$ and $(6.1\text{-}12)_2$ are employed. The expression of $\mathbf{N}_3$ has been obtained by Stroh (1958) and that of $\mathbf{N}_1$ and $\mathbf{N}_3$ were given by Ting (1988a), both using different derivations.

The matrix

$$\begin{bmatrix} s'_{55} & s'_{15} \\ s'_{15} & s'_{11} \end{bmatrix},$$

being a principal submatrix of $s'_{\alpha\beta}$, is positive definite (Section 1.6) and hence its eigenvalues are positive and nonzero. Therefore the matrix

$$\begin{bmatrix} s'_{55} & -s'_{15} \\ -s'_{15} & s'_{11} \end{bmatrix},$$

which has the same eigenvalues as has the one above, is positive definite. Thus $-\mathbf{N}_3$ given by (6.1-15) is positive semidefinite.

The $\mathbf{N}_1$, $\mathbf{N}_2$, and $\mathbf{N}_3$ presented in (6.1-15)–(6.1-17) form an interesting contrast with the ones given in (6.1-1). If we solve for $\mathbf{T}$, $\mathbf{R}$, and $\mathbf{Q}$ in that order from $(6.1\text{-}1)_{2,1,3}$ we have

$$\mathbf{T} = \mathbf{N}_2^{-1}, \quad \mathbf{R} = -\mathbf{N}_1^T\mathbf{N}_2^{-1}, \quad \mathbf{Q} = \mathbf{N}_1^T\mathbf{N}_2^{-1}\mathbf{N}_1 - \mathbf{N}_3. \tag{6.1-19}$$

6.2 The Structure of $\mathbf{N}_i^{(-1)}$

Following the derivation of the eigenrelation (5.5-3) for $\mathbf{N}$ we may derive the eigenrelation for $\mathbf{N}^{-1}$, and obtain an explicit expression of $\mathbf{N}^{-1}$ in terms of $\mathbf{Q}$, $\mathbf{R}$, and $\mathbf{T}$. We begin with (5.5-1). In place of the identity (5.5-2) we employ

$$\begin{bmatrix} -\mathbf{Q}^{-1} & \mathbf{0} \\ -\mathbf{R}^T\mathbf{Q}^{-1} & \mathbf{I} \end{bmatrix} \begin{bmatrix} -\mathbf{Q} & \mathbf{0} \\ -\mathbf{R}^T & \mathbf{I} \end{bmatrix} = \begin{bmatrix} \mathbf{I} & \mathbf{0} \\ \mathbf{0} & \mathbf{I} \end{bmatrix} \tag{6.2-1}$$

which can be verified easily. Multiplication of both sides of (5.5-1) by the first 6×6 matrix on the left of (6.2-1) we obtain

$$\mathbf{N}^{-1}\xi = \frac{1}{p}\xi, \tag{6.2-2}$$

$$\mathbf{N}^{-1} = \begin{bmatrix} \mathbf{N}_1^{(-1)} & \mathbf{N}_2^{(-1)} \\ \mathbf{N}_3^{(-1)} & \mathbf{N}_1^{(-1)T} \end{bmatrix}, \quad \xi = \begin{bmatrix} \mathbf{a} \\ \mathbf{b} \end{bmatrix}, \tag{6.2-3}$$

$$\mathbf{N}_1^{(-1)} = -\mathbf{Q}^{-1}\mathbf{R}, \quad \mathbf{N}_2^{(-1)} = -\mathbf{Q}^{-1}, \quad \mathbf{N}_3^{(-1)} = \mathbf{T} - \mathbf{R}^T\mathbf{Q}^{-1}\mathbf{R}. \tag{6.2-4}$$

Thus $-\mathbf{N}_2^{(-1)}$ is positive definite. We will show later that $\mathbf{N}_3^{(-1)}$ is positive semidefinite.

Using (6.1-3) and the derivation of (6.1-5) and (6.1-6), it can be shown that $-\mathbf{N}_1^{(-1)}$ and $\mathbf{N}_3^{(-1)}$ have the structure

$$-\mathbf{N}_1^{(-1)} = \begin{bmatrix} 0 & * & * \\ 1 & * & * \\ 0 & * & * \end{bmatrix}, \tag{6.2-5}$$

$$\mathbf{N}_3^{(-1)} = \begin{bmatrix} 0 & 0 & 0 \\ 0 & * & * \\ 0 & * & * \end{bmatrix}. \tag{6.2-6}$$

The elements denoted by an asterisk $*$ also have simple expressions in terms of the reduced elastic compliances. The derivation is similar to that of $\mathbf{N}_1$ and $\mathbf{N}_3$. We employ instead of (6.1-8)

$$\begin{bmatrix} \mathbf{Q} & \mathbf{R} \\ \mathbf{R}^T & \mathbf{T} \end{bmatrix} \begin{bmatrix} \mathbf{q} & \mathbf{r}_1 \\ \mathbf{r}_1^T & \mathbf{t}_1 \end{bmatrix} = \begin{bmatrix} \mathbf{I} & \mathbf{0} \\ \mathbf{I}_{12} & \mathbf{I}-\mathbf{I}_1 \end{bmatrix}, \tag{6.2-7}$$

where

$$\mathbf{q} = \begin{bmatrix} s'_{11} & s'_{16} & s'_{15} \\ s'_{16} & s'_{66} & s'_{56} \\ s'_{15} & s'_{56} & s'_{55} \end{bmatrix}, \quad \mathbf{r}_1 = \begin{bmatrix} 0 & s'_{12} & s'_{14} \\ 0 & s'_{26} & s'_{46} \\ 0 & s'_{25} & s'_{45} \end{bmatrix}, \quad \mathbf{t}_1 = \begin{bmatrix} 0 & 0 & 0 \\ 0 & s'_{22} & s'_{24} \\ 0 & s'_{24} & s'_{44} \end{bmatrix}. \tag{6.2-8}$$

$\mathbf{I}_{12}$ is defined in (6.1-10), and

$$\mathbf{I}_1 = \text{diag}[1, \ 0 \ \ 0]. \tag{6.2-9}$$

The matrices $\mathbf{r}_1$ and $\mathbf{t}_1$ have the properties

$$\mathbf{r}_1\mathbf{I}_{12} = \mathbf{0}, \qquad \mathbf{t}_1\mathbf{I}_{12} = \mathbf{0}. \tag{6.2-10}$$

Also, instead of (6.1-13) we employ the identity

$$\begin{bmatrix} \mathbf{N}_1^{(-1)T} & \mathbf{I} \\ \mathbf{N}_2^{(-1)} & \mathbf{0} \end{bmatrix} \begin{bmatrix} \mathbf{Q} & \mathbf{R} \\ \mathbf{R}^T & \mathbf{T} \end{bmatrix} = \begin{bmatrix} \mathbf{0} & \mathbf{N}_3^{(-1)} \\ -\mathbf{I} & \mathbf{N}_1^{(-1)} \end{bmatrix}$$

so that (6.2-7) can be written as

$$\begin{bmatrix} \mathbf{0} & \mathbf{N}_3^{(-1)} \\ -\mathbf{I} & \mathbf{N}_1^{(-1)} \end{bmatrix} \begin{bmatrix} \mathbf{q} & \mathbf{r}_1 \\ \mathbf{r}_1^T & \mathbf{t}_1 \end{bmatrix} = \begin{bmatrix} \mathbf{N}_1^{(-1)T} & \mathbf{I} \\ \mathbf{N}_2^{(-1)} & \mathbf{0} \end{bmatrix} \begin{bmatrix} \mathbf{I} & \mathbf{0} \\ \mathbf{I}_{12} & \mathbf{I}-\mathbf{I}_1 \end{bmatrix}$$

or

$$\begin{aligned} &\mathbf{N}_3^{(-1)}\mathbf{t}_1 = \mathbf{I} - \mathbf{I}_1, \qquad \mathbf{N}_3^{(-1)}\mathbf{r}_1^T = \mathbf{N}_1^{(-1)T} + \mathbf{I}_{12}, \\ &\mathbf{N}_1^{(-1)}\mathbf{t}_1 - \mathbf{r}_1 = \mathbf{0}, \qquad \mathbf{N}_1^{(-1)}\mathbf{r}_1^T - \mathbf{q} = \mathbf{N}_2^{(-1)}. \end{aligned} \tag{6.2-11}$$

From (6.2-5), (6.2-6), and $(6.2\text{-}11)_{1,2,4}$ we have

$$\mathbf{N}_3^{(-1)} = \mathbf{t}_1^{-1} = \frac{1}{\Delta'} \begin{bmatrix} 0 & 0 & 0 \\ 0 & s'_{44} & -s'_{24} \\ 0 & -s'_{24} & s'_{22} \end{bmatrix}, \tag{6.2-12}$$

$$-\mathbf{N}_1^{(-1)} = \mathbf{I}_{12}^T - \mathbf{r}_1\mathbf{t}_1^{-1} = \begin{bmatrix} 0 & r'_1 & s'_1 \\ 1 & r'_6 & s'_6 \\ 0 & r'_5 & s'_5 \end{bmatrix}, \tag{6.2-13}$$

$$\mathbf{N}_2^{(-1)} = -\mathbf{q} + \mathbf{r}_1\mathbf{t}_1^{-1}\mathbf{r}_1^T. \tag{6.2-14}$$

In the above $\mathbf{t}_1^{-1}$ is a *pseudo inverse* of $\mathbf{t}_1$ and

$$\Delta' = s'_{22}s'_{44} - s'_{24}s'_{24},$$

$$r'_\alpha = \frac{1}{\Delta'}(s'_{24}s'_{4\alpha} - s'_{44}s'_{2\alpha}), \tag{6.2-15}$$

$$s'_\alpha = \frac{1}{\Delta'}(s'_{24}s'_{2\alpha} - s'_{22}s'_{4\alpha}),$$

(α = 1,6,5). It is easily shown from $(6.2\text{-}10)_2$ and (6.2-13) that $(6.2\text{-}11)_3$ is automatically satisfied. The matrix $\mathbf{N}_3^{(-1)}$ has a similar property as has $-\mathbf{N}_3$ in that it is positive semidefinite.

The representations of $\mathbf{N}_3^{(-1)}$ in (6.2-12) and $-\mathbf{N}_1^{(-1)}$ in (6.2-13) have been obtained, respectively, by Stroh (1958) and Ting (1988a) using different derivations.

When (6.2-4) are solved for **Q**, **R**, and **T** we have

$$\mathbf{Q} = -(\mathbf{N}_2^{(-1)})^{-1}, \qquad \mathbf{R} = (\mathbf{N}_2^{(-1)})^{-1}\mathbf{N}_1^{(-1)},$$
$$\mathbf{T} = \mathbf{N}_3^{(-1)} - (\mathbf{N}_1^{(-1)})^T(\mathbf{N}_2^{(-1)})^{-1}\mathbf{N}_1^{(-1)}, \tag{6.2-16}$$

which provide an alternate to (6.1-19). By equating **Q**, **R**, **T**, between the two equations we obtain

$$\mathbf{N}_2^{(-1)} = \left\{\mathbf{N}_3 - \mathbf{N}_1^T\mathbf{N}_2^{-1}\mathbf{N}_1\right\}^{-1}, \quad \mathbf{N}_1^{(-1)} = -\left\{\mathbf{N}_3 - \mathbf{N}_1^T\mathbf{N}_2^{-1}\mathbf{N}_1\right\}^{-1}\mathbf{N}_1^T\mathbf{N}_2^{-1},$$

$$\mathbf{N}_3^{(-1)} = \mathbf{N}_2^{-1} + \mathbf{N}_2^{-1}\mathbf{N}_1\left\{\mathbf{N}_3 - \mathbf{N}_1^T\mathbf{N}_2^{-1}\mathbf{N}_1\right\}^{-1}\mathbf{N}_1^T\mathbf{N}_2^{-1}.$$

6.3 Explicit Expressions of A and B

The complex matrices **A** and **B** are defined in (5.3-11). The columns of **A** are eigenvectors of the 3×3 matrix shown in (5.1-8) and the columns of **B** are related to those of **A** through (5.3-2). Alternately, the columns of **A** and **B** together are the eigenvectors of the fundamental elasticity matrix **N** shown in (5.5-4). Explicit expressions of these eigenvectors in terms of the elastic stiffnesses $C_{\alpha\beta}$ are cumbersome to derive. In terms of the reduced elastic compliances $s'_{\alpha\beta}$, however, they are easily obtained by simply comparing Stroh's general solutions with Lekhnitskii's general solutions.

Stroh's general solutions for the displacement **u** and the stress function ϕ are presented in (5.3-7) while the corresponding solutions by Lekhnitskii are given in (4.6-11) and (4.6-13). By comparing these two solutions we have

$$\mathbf{A} = \begin{bmatrix} k_1\xi_1(p_1) & k_2\xi_1(p_2) & k_3\eta_1(p_3) \\ k_1p_1^{-1}\xi_2(p_1) & k_2p_2^{-1}\xi_2(p_2) & k_3p_3^{-1}\eta_2(p_3) \\ k_1p_1^{-1}\xi_4(p_1) & k_2p_2^{-1}\xi_4(p_2) & k_3p_3^{-1}\eta_4(p_3) \end{bmatrix}, \tag{6.3-1}$$

$$\mathbf{B} = \begin{bmatrix} -k_1 p_1 & -k_2 p_2 & -k_3 p_3 \lambda_3 \\ k_1 & k_2 & k_3 \lambda_3 \\ -k_1 \lambda_1 & -k_2 \lambda_2 & -k_3 \end{bmatrix}, \tag{6.3-2}$$

where λ_α, $\xi_\alpha(p_\beta)$, $\eta_\alpha(p_3)$ are defined in (4.6-9) and (4.6-12), and k_i (i=1,2,3) are the normalization factors. From (5.5-12) the normalization factors k_i satisfy the relation

$$2k_i^2\{p_i^{-1}\xi_2(p_i) - p_i\xi_1(p_i) - \lambda_i p_i^{-1}\xi_4(p_i)\} = 1 \tag{6.3-3a}$$

for i=1,2 and

$$2k_3^2\{\lambda_3[p_3^{-1}\eta_2(p_3) - p_3\eta_1(p_3)] - p_3^{-1}\xi_4(p_3)\} = 1. \tag{6.3-3b}$$

We will see that in many occasions the normalization factors k_i are not needed or redundant. The inverse of **B** can be shown to be

$$\mathbf{B}^{-1} = \frac{1}{\Delta}\begin{bmatrix} -k_1^{-1}(1-\lambda_2\lambda_3) & -k_1^{-1}(p_2 - \lambda_2\lambda_3 p_3) & -k_1^{-1}\lambda_3(p_2 - p_3) \\ k_2^{-1}(1-\lambda_1\lambda_3) & k_2^{-1}(p_1 - \lambda_1\lambda_3 p_3) & k_2^{-1}\lambda_3(p_1 - p_3) \\ k_3^{-1}(\lambda_1 - \lambda_2) & k_3^{-1}(\lambda_1 p_2 - \lambda_2 p_1) & -k_3^{-1}(p_1 - p_2) \end{bmatrix} \tag{6.3-4}$$

where

$$\Delta = (p_1 - p_2) + \lambda_3\{(\lambda_1 p_2 - \lambda_2 p_1) - p_3(\lambda_1 - \lambda_2)\}. \tag{6.3-5}$$

The expressions of **A** and **B** are simplified substantially for monoclinic materials with the symmetry plane at x_3 =0. For these materials λ_1=λ_2=λ_3=0 (Section 4.6) and $\xi_4(p_1)$=$\xi_4(p_2)$=$\eta_1(p_3)$=$\eta_2(p_3)$=0. Moreover, from (4.6-12) and (4.4-10),

$$p_3^{-1}\eta_4(p_3) = s'_{45} - p_3^{-1}s'_{44} = i\mu^{-1}$$

where

$$\mu = \{s'_{44}s'_{55} - s'_{45}s'_{45}\}^{-1/2} = \{C_{44}C_{55} - C_{45}C_{45}\}^{1/2}. \tag{6.3-6}$$

The second equality is proved by Ting (1992a). μ is the shear modulus when the material is isotropic. Therefore

$$\mathbf{A} = \begin{bmatrix} k_1\xi_1(p_1) & k_2\xi_1(p_2) & 0 \\ k_1 p_1^{-1}\xi_2(p_1) & k_2 p_2^{-1}\xi_2(p_2) & 0 \\ 0 & 0 & ik_3\mu^{-1} \end{bmatrix}, \tag{6.3-7}$$

$$\mathbf{B} = \begin{bmatrix} -k_1 p_1 & -k_2 p_2 & 0 \\ k_1 & k_2 & 0 \\ 0 & 0 & -k_3 \end{bmatrix}, \tag{6.3-8}$$

$$\mathbf{B}^{-1} = \frac{1}{p_1 - p_2} \begin{bmatrix} -k_1^{-1} & -p_2 k_1^{-1} & 0 \\ k_2^{-1} & p_1 k_2^{-1} & 0 \\ 0 & 0 & (p_2 - p_1) k_3^{-1} \end{bmatrix}, \tag{6.3-9}$$

where

$$\xi_\alpha(p) = s'_{\alpha 1} p^2 - s'_{\alpha 6} p + s'_{\alpha 2} \tag{6.3-10}$$

and p_1, p_2 are the roots of (4.4-11), namely,

$$s'_{11} p^4 - 2 s'_{16} p^3 + (2 s'_{12} + s'_{66}) p^2 - 2 s'_{26} p + s'_{22} = 0, \tag{6.3-11}$$

with positive imaginary parts. We may also use (5.2-9) to compute p_1 and p_2.

Equations (6.3-7)–(6.3-9) were first obtained by Suo (1990) and Ting (1992a).

6.4 Explicit Expressions of S, H, and L

With explicit expressions of **A** and **B** presented in (6.3-1) and (6.3-2) the three real matrices **S**, **H**, and **L** defined in (5.5-17) can be determined. However, a direct substitution of **A** and **B** into (5.5-17) would lead to a very unwieldy algebraic calculation because of the presence of the normalization factors k_i (Dongye and Ting, 1989). An alternate approach is to employ $(5.6\text{-}6)_4$ which is written as

$$-\mathbf{A}\mathbf{B}^{-1} = \mathbf{S}\mathbf{L}^{-1} + i\mathbf{L}^{-1}. \tag{6.4-1}$$

When **A** of (6.3-1) is multiplied by $\mathbf{B}^{-1}$ of (6.3-4), it is seen that the normalization factors canceled each other. In other words, there is no need to introduce the normalization factors in computing $-\mathbf{A}\mathbf{B}^{-1}$. Equation (6.4-1) tells us that the real and imaginary parts of $-\mathbf{A}\mathbf{B}^{-1}$ are $\mathbf{S}\mathbf{L}^{-1}$ and $\mathbf{L}^{-1}$, respectively. Once $\mathbf{S}\mathbf{L}^{-1}$ and $\mathbf{L}^{-1}$ are determined, we compute **L** from $\mathbf{L}^{-1}$ and **S** from

$$\mathbf{S} = (\mathbf{S}\mathbf{L}^{-1})\mathbf{L}. \tag{6.4-2}$$

The matrix **H** is determined by using (5.6-8), i.e.,

$$\mathbf{H} = \mathbf{L}^{-1} + \mathbf{S}(\mathbf{S}\mathbf{L}^{-1}). \tag{6.4-3}$$

With this approach the algebra is much simplified. There is no need to calculate the Stroh eigenvectors $\mathbf{a}_\alpha$ and $\mathbf{b}_\alpha$ although we still need to compute the eigenvalues p_α. Explicit expressions of **S**, **H**, **L** for general anisotropic materials and, in particular, for monoclinic materials with the symmetry plane at $x_1=0$ or $x_2=0$ have been presented by Wei and Ting (1994).

The procedure outlined above is particularly simple for monoclinic materials with the symmetry plane at $x_3=0$ (Suo, 1990; Ting, 1992a). For these materials **A** and $\mathbf{B}^{-1}$ are given by (6.3-7) and (6.3-9) and we have

$$-\mathbf{AB}^{-1} = \begin{bmatrix} s'_{11}(p_1+p_2)-s'_{16} & s'_{11}p_1p_2-s'_{12} & 0 \\ s'_{12}-s'_{22}(p_1p_2)^{-1} & s'_{26}-s'_{22}(p_1^{-1}+p_2^{-1}) & 0 \\ 0 & 0 & i\mu^{-1} \end{bmatrix}. \tag{6.4-4}$$

Noting that the roots p_1, p_2, $\bar{p}_1$, and $\bar{p}_2$ are related to the coefficients of the quartic equation in (6.3-11), one has

$$\begin{aligned} \mathrm{Re}(p_1+p_2) &= s'_{16}/s'_{11}, \\ (p_1+p_2)(\bar{p}_1+\bar{p}_2)+(p_1p_2+\bar{p}_1\bar{p}_2) &= (2s'_{12}+s'_{66})/s'_{11}, \\ \mathrm{Re}(p_1^{-1}+p_2^{-1}) &= s'_{26}/s'_{22}, \\ p_1p_2\bar{p}_1\bar{p}_2 &= s'_{22}/s'_{11}, \end{aligned} \tag{6.4-5}$$

where Re stands for the real part. Hence

$$\begin{aligned} s'_{11}(p_1+p_2)-s'_{16} &= is'_{11}\,\mathrm{Im}(p_1+p_2), \\ s'_{26}-s'_{22}(p_1^{-1}+p_2^{-1}) &= -is'_{22}\,\mathrm{Im}(p_1^{-1}+p_2^{-1}), \\ s'_{12}-s'_{22}(p_1p_2)^{-1} &= s'_{12}-s'_{11}\bar{p}_1\bar{p}_2, \end{aligned} \tag{6.4-6}$$

in which Im denotes the imaginary part. Using $(6.4\text{-}5)_4$,

$$s'_{22}(p_1^{-1}+p_2^{-1}) = s'_{11}p_1p_2\bar{p}_1\bar{p}_2(p_1^{-1}+p_2^{-1}) = s'_{11}\bar{p}_1\bar{p}_2(p_1+p_2)$$

and $(6.4\text{-}6)_2$ is replaced by

$$s'_{26}-s'_{22}(p_1^{-1}+p_2^{-1}) = is'_{11}\,\mathrm{Im}\{p_1p_2(\bar{p}_1+\bar{p}_2)\}.$$

Therefore (6.4-4) is equivalent to

$$-\mathbf{AB}^{-1} = \begin{bmatrix} is'_{11}\,\mathrm{Im}(p_1+p_2) & s'_{11}p_1p_2-s'_{12} & 0 \\ s'_{12}-s'_{11}\bar{p}_1\bar{p}_2 & is'_{11}\,\mathrm{Im}\{p_1p_2(\bar{p}_1+\bar{p}_2)\} & 0 \\ 0 & 0 & i\mu^{-1} \end{bmatrix}. \tag{6.4-7}$$

Let

$$p_1+p_2=a+ib, \quad p_1p_2=c+id,$$

$$b>0, \quad e=ad-bc>0, \tag{6.4-8}$$

$$g=(s'_{12}/s'_{11})-c>0, \quad 1>s=g(be-d^2)^{-1/2}>0.$$

The justification of the inequalities will be given shortly. The real and imaginary parts of (6.4-7) are

$$\mathbf{SL}^{-1}=gs'_{11}\begin{bmatrix}0 & -1 & 0\\ 1 & 0 & 0\\ 0 & 0 & 0\end{bmatrix}, \tag{6.4-9}$$

$$\mathbf{L}^{-1}=s'_{11}\begin{bmatrix}b & d & 0\\ d & e & 0\\ 0 & 0 & (\mu s'_{11})^{-1}\end{bmatrix}. \tag{6.4-10}$$

The explicit expression of $\mathbf{L}^{-1}$ was first obtained by Stroh (1958) using a different derivation. Hence

$$\mathbf{L}=\frac{s^2}{s'_{11}g^2}\begin{bmatrix}e & -d & 0\\ -d & b & 0\\ 0 & 0 & \mu s'_{11}g^2s^{-2}\end{bmatrix} \tag{6.4-11a}$$

and, using (6.4-2) and (6.4-3),

$$\mathbf{S}=\frac{s^2}{g}\begin{bmatrix}d & -b & 0\\ e & -d & 0\\ 0 & 0 & 0\end{bmatrix}, \quad \mathbf{H}=s'_{11}(1-s^2)\begin{bmatrix}b & d & 0\\ d & e & 0\\ 0 & 0 & \chi\end{bmatrix} \tag{6.4-11b,c}$$

where

$$\chi=[\mu s'_{11}(1-s^2)]^{-1}.$$

Since **L** and **H** are positive definite (6.4-11a,c) tell us that e, b, $be-d^2$, and $1-s^2$ are positive and nonzero. The inequalities in (6.4-8) are justified if $g>0$ or, by (6.4-11b), $S_{21}>0$. It has been shown (Ting, 1992a) that $S_{21}>0$ for monoclinic materials with the symmetry plane at $x_3=0$. For general anisotropic materials we will show in (8.12-41) that $S_{21}-S_{12}>0$. It should be pointed out that a, b, c, d, e, g, and h are all dimensionless real constants. From (6.4-5) they are related to $s'_{\alpha\beta}$ by

$$a = s'_{16}/s'_{11}, \qquad a^2 + b^2 + 2c = (2s'_{12} + s'_{66})/s'_{11},$$
$$ac + bd = s'_{26}/s'_{11}, \qquad c^2 + d^2 = s'_{22}/s'_{11}. \tag{6.4-12}$$

For isotropic materials $a=d=0$, $c=-1$, $b=e=2$, $h=1/4$, and

$$s'_{11} = \frac{1-\nu}{2\mu}, \qquad s'_{12} = \frac{-\nu}{2\mu},$$

where μ and ν are the shear modulus and Poisson's ratio, respectively. Hence

$$\mathbf{S} = s\begin{bmatrix} 0 & -1 & 0 \\ 1 & 0 & 0 \\ 0 & 0 & 0 \end{bmatrix}, \quad \mathbf{H} = \frac{1}{\mu\kappa}\begin{bmatrix} 1-s^2 & 0 & 0 \\ 0 & 1-s^2 & 0 \\ 0 & 0 & \kappa \end{bmatrix}, \quad \mathbf{L} = \mu\begin{bmatrix} \kappa & 0 & 0 \\ 0 & \kappa & 0 \\ 0 & 0 & 1 \end{bmatrix}, \tag{6.4-13}$$

in which

$$\kappa = \frac{1}{1-\nu}, \qquad s = \frac{1-2\nu}{2(1-\nu)}. \tag{6.4-14}$$

Since $-1<\nu<\frac{1}{2}$ by (2.9-8),

$$2 > \kappa > \tfrac{1}{2}, \qquad \tfrac{3}{4} > s > 0. \tag{6.4-15}$$

The readers are reminded that **S**, **H**, and **L** defined in (5.5-17) are not valid for degenerate materials for which isotropic materials are a special case. Nevertheless **S**, **H**, and **L** shown above for isotropic materials are valid. If the correct expressions (5.9-18) were employed, **S**, **H**, and **L** computed for isotropic materials agree with the ones given above (Ting and Hwu, 1988).

6.5 Identities Relating S, H, and L

We have shown in (5.5-17) that the three matrices **S**, **H**, and **L** defined by

$$\mathbf{S} = i(2\mathbf{A}\mathbf{B}^T - \mathbf{I}), \quad \mathbf{H} = 2i\mathbf{A}\mathbf{A}^T, \quad \mathbf{L} = -2i\mathbf{B}\mathbf{B}^T, \tag{6.5-1}$$

are real. This means that the right sides of the equations can be replaced by their complex conjugates. If we replace one-halves of the right sides by their complex conjugates we obtain alternate expressions

$$\mathbf{S} = i(\mathbf{A}\mathbf{B}^T - \overline{\mathbf{A}\mathbf{B}}^T), \quad \mathbf{H} = i(\mathbf{A}\mathbf{A}^T - \overline{\mathbf{A}\mathbf{A}}^T), \quad \mathbf{L} = -i(\mathbf{B}\mathbf{B}^T - \overline{\mathbf{B}\mathbf{B}}^T). \tag{6.5-2}$$

Equation (6.5-2) can be cast in the form

$$\begin{bmatrix} \mathbf{S} & \mathbf{H} \\ -\mathbf{L} & \mathbf{S}^T \end{bmatrix} = \begin{bmatrix} \mathbf{A} & \overline{\mathbf{A}} \\ \mathbf{B} & \overline{\mathbf{B}} \end{bmatrix} \begin{bmatrix} i\mathbf{I} & \mathbf{0} \\ \mathbf{0} & -i\mathbf{I} \end{bmatrix} \begin{bmatrix} \mathbf{B}^T & \mathbf{A}^T \\ \overline{\mathbf{B}}^T & \overline{\mathbf{A}}^T \end{bmatrix}. \tag{6.5-3}$$

Making use of the orthogonality relation (5.5-13) and the closure relations (5.5-15) it can be shown that

$$\begin{bmatrix} \mathbf{S} & \mathbf{H} \\ -\mathbf{L} & \mathbf{S}^T \end{bmatrix} \begin{bmatrix} \mathbf{S} & \mathbf{H} \\ -\mathbf{L} & \mathbf{S}^T \end{bmatrix} = -\begin{bmatrix} \mathbf{I} & \mathbf{0} \\ \mathbf{0} & \mathbf{I} \end{bmatrix}. \tag{6.5-4}$$

If **N** is nonsemisimple (6.5-1) is replaced by (5.9-18), i.e.,

$$\mathbf{S} = i(2\mathbf{A}\Gamma\mathbf{B}^T - \mathbf{I}), \quad \mathbf{H} = 2i\mathbf{A}\Gamma\mathbf{A}^T, \quad \mathbf{L} = -2i\mathbf{B}\Gamma\mathbf{B}^T, \tag{6.5-5}$$

When one-halves of the right sides are replaced by their complex conjugates we have the alternate expressions

$$\begin{aligned} \mathbf{S} &= i(\mathbf{A}\Gamma\mathbf{B}^T - \overline{\mathbf{A}}\Gamma\overline{\mathbf{B}}^T), \\ \mathbf{H} &= i(\mathbf{A}\Gamma\mathbf{A}^T - \overline{\mathbf{A}}\Gamma\overline{\mathbf{A}}^T), \; . \\ \mathbf{L} &= -i(\mathbf{B}\Gamma\mathbf{B}^T - \overline{\mathbf{B}}\Gamma\overline{\mathbf{B}}^T). \end{aligned} \tag{6.5-6}$$

It should be stressed that **A** and **B** in (6.5-6) are different from their counterparts in (6.5-2). By writing (6.5-6) as

$$\begin{bmatrix} \mathbf{S} & \mathbf{H} \\ -\mathbf{L} & \mathbf{S}^T \end{bmatrix} = \begin{bmatrix} \mathbf{A} & \overline{\mathbf{A}} \\ \mathbf{B} & \overline{\mathbf{B}} \end{bmatrix} \begin{bmatrix} i\mathbf{I} & \mathbf{0} \\ \mathbf{0} & -i\mathbf{I} \end{bmatrix} \begin{bmatrix} \Gamma\mathbf{B}^T & \Gamma\mathbf{A}^T \\ \Gamma\overline{\mathbf{B}}^T & \Gamma\overline{\mathbf{A}}^T \end{bmatrix}, \tag{6.5-7}$$

and by virtue of (5.9-15) and (5.9-16), (6.5-4) remains valid for nonsemisimple **N**. It is interesting to point out that (6.5-3) and (6.5-7) are the diagonalization of the 6×6 matrix

$$\tilde{\mathbf{N}} = \begin{bmatrix} \mathbf{S} & \mathbf{H} \\ -\mathbf{L} & \mathbf{S}^T \end{bmatrix}$$

whose eigenvalues are $\pm i$, each of multiplicity three. Regardless of whether **N** is simple, semisimple, or nonsemisimple, $\tilde{\mathbf{N}}$ is always semisimple (see Section 6.10).

Carrying out the matrix multiplication in (6.5-4) we have

$$\mathbf{HL} - \mathbf{SS} = \mathbf{I} = \mathbf{LH} - \mathbf{S}^T\mathbf{S}^T, \tag{6.5-8}$$

$$\mathbf{SH} + \mathbf{HS}^T = \mathbf{0} = \mathbf{LS} + \mathbf{S}^T\mathbf{L}. \tag{6.5-9}$$

Since **H** and **L** are symmetric (6.5-9) tell us that **SH** and **LS** are skew-symmetric. When $(6.5\text{-}9)_1$ is premultiplied by $\mathbf{H}^{-1}$ and postmultiplied by $\mathbf{H}^{-1}$ we obtain

$$\mathbf{H}^{-1}\mathbf{S}+\mathbf{S}^{T}\mathbf{H}^{-1}=\mathbf{0}. \qquad (6.5\text{-}10)$$

A similar operation on $(6.5\text{-}9)_2$ by $\mathbf{L}^{-1}$ leads to

$$\mathbf{S}\mathbf{L}^{-1}+\mathbf{L}^{-1}\mathbf{S}^{T}=\mathbf{0}. \qquad (6.5\text{-}11)$$

Hence $\mathbf{H}^{-1}\mathbf{S}$ and $\mathbf{S}\mathbf{L}^{-1}$ are also skew-symmetric. Equations $(6.5\text{-}8)_1$, (6.5-10), and (6.5-11) are identical to (5.6-8) and $(5.6\text{-}7)_{1,2}$, respectively.

By the Cayley–Hamilton theory a matrix satisfies its own characteristic equation. We will show in Section 6.6 that $\mathbf{S}$ satisfies

$$\mathbf{S}^{3}+s^{2}\mathbf{S}=\mathbf{0} \qquad (6.5\text{-}12)$$

where

$$1>s=\left\{-\tfrac{1}{2}\mathrm{tr}(\mathbf{S}^{2})\right\}^{1/2}>0. \qquad (6.5\text{-}13)$$

If $\mathbf{w}$ is the right null vector of $\mathbf{S}$, it is the right null vector of $\mathbf{LS}$ and $\mathbf{H}^{-1}\mathbf{S}$. Since $\mathbf{LS}$ and $\mathbf{H}^{-1}\mathbf{S}$ are 3×3 skew-symmetric matrices it is also the left null vector of $\mathbf{LS}$ and $\mathbf{H}^{-1}\mathbf{S}$. By Theorem 1.2–1 $\mathbf{LS}$ and $\mathbf{H}^{-1}\mathbf{S}$ are proportional to each other. By the same argument if $\mathbf{h}$ is the left null vector of $\mathbf{S}$, it is the left (also the right) null vector of $\mathbf{SH}$ and $\mathbf{SL}^{-1}$. Therefore $\mathbf{SH}$ and $\mathbf{SL}^{-1}$ are also proportional to each other. The proportionality relations can be written explicitly as

$$\mathbf{LS}=(1-s^{2})\mathbf{H}^{-1}\mathbf{S}, \quad \mathbf{SH}=(1-s^{2})\mathbf{SL}^{-1}. \qquad (6.5\text{-}14)$$

To prove (6.5-14) we employ (6.5-12) and replace the $\mathbf{S}^2$ portion of $\mathbf{S}^3$ by $\mathbf{HL}-\mathbf{I}$ obtained from $(6.5\text{-}8)_1$. We have

$$(\mathbf{HL}-\mathbf{I})\mathbf{S}+s^{2}\mathbf{S}=\mathbf{0}$$

and

$$\mathbf{S}(\mathbf{HL}-\mathbf{I})+s^{2}\mathbf{S}=\mathbf{0}.$$

These equations can be rewritten in the form given in (6.5-14). Identities (6.5-14) provide the relations

$$\mathbf{HLS}=\mathbf{SHL}=(1-s^{2})\mathbf{S}. \qquad (6.5\text{-}15)$$

Noting that $\mathbf{H}^{-1}\mathbf{S}$ and $\mathbf{SL}^{-1}$ are skew-symmetric $(6.5\text{-}8)_1$ is equivalent to

$$\mathbf{H}^{-1}-\mathbf{L}=\mathbf{S}^{T}\mathbf{H}^{-1}\mathbf{S}, \quad \mathbf{L}^{-1}-\mathbf{H}=\mathbf{SL}^{-1}\mathbf{S}^{T}. \qquad (6.5\text{-}16)$$

Since $\mathbf{H}^{-1}$ and $\mathbf{L}^{-1}$ are positive definite while $\mathbf{S}$ is singular (Section 6.6), $\mathbf{H}^{-1}-\mathbf{L}$ and $\mathbf{L}^{-1}-\mathbf{H}$ are positive semidefinite. By virtue of (6.5-9) we write $(6.5\text{-}8)_2$ as

$$\mathbf{L}^{-1}\mathbf{H}^{-1} + \mathbf{S}\mathbf{L}^{-1}\mathbf{H}^{-1}\mathbf{S} = \mathbf{I}$$

or, from $(6.5\text{-}14)_1$,

$$\mathbf{L}^{-1}\mathbf{H}^{-1} + \tau^2\mathbf{S}\mathbf{S} = \mathbf{I} \tag{6.5-17}$$

where

$$\tau = \frac{1}{\sqrt{1-s^2}}. \tag{6.5-18}$$

Together with the fact that $\mathbf{H}^{-1}\mathbf{S}$ and $\mathbf{S}\mathbf{L}^{-1}$ are skew-symmetric we have

$$\begin{bmatrix} \tau\mathbf{S}^T & \mathbf{H}^{-1} \\ \mathbf{L}^{-1} & \tau\mathbf{S} \end{bmatrix}\begin{bmatrix} \tau\mathbf{S}^T & \mathbf{H}^{-1} \\ \mathbf{L}^{-1} & \tau\mathbf{S} \end{bmatrix} = \begin{bmatrix} \mathbf{I} & \mathbf{0} \\ \mathbf{0} & \mathbf{I} \end{bmatrix}, \tag{6.5-19}$$

which forms an interesting contrast with (6.5-4).

Equation (6.5-12) allows us to replace $\mathbf{S}^3$ by $-s^2\mathbf{S}$ and $\mathbf{S}^4$ by $-s^2\mathbf{S}^2$. It is readily shown that

$$(\mathbf{I} - \eta^2\mathbf{S}^2)\left(\mathbf{I} + \frac{\eta^2}{1+\eta^2 s^2}\mathbf{S}^2\right) = \mathbf{I}$$

where η is an arbitrary constant, not necessarily real. Hence

$$(\mathbf{I} - \eta^2\mathbf{S}^2)^{-1} = \mathbf{I} + \frac{\eta^2}{1+\eta^2 s^2}\mathbf{S}^2. \tag{6.5-20}$$

In particular, for $\eta = i$ and 1,

$$(\mathbf{I} \pm \mathbf{S}^2)^{-1} = \mathbf{I} \mp \frac{1}{1 \mp s^2}\mathbf{S}^2. \tag{6.5-21}$$

Observing that

$$(\mathbf{I} - \eta^2\mathbf{S}^2) = (\mathbf{I} + \eta\mathbf{S})(\mathbf{I} - \eta\mathbf{S})$$

we have from (6.5-20)

$$(\mathbf{I} \pm \eta\mathbf{S})^{-1} = (\mathbf{I} \mp \eta\mathbf{S})\left(\mathbf{I} + \frac{\eta^2}{1+\eta^2 s^2}\mathbf{S}^2\right)$$

or, after replacing $\mathbf{S}^3$ by $-s^2\mathbf{S}$,

$$(\mathbf{I} \pm \eta\mathbf{S})^{-1} = \mathbf{I} \mp \frac{\eta}{1+\eta^2 s^2}\mathbf{S}(\mathbf{I} \mp \eta\mathbf{S}). \tag{6.5-22}$$

For $\eta = i$, (6.5-22) recovers the identities obtained by Chadwick and Ting (1987).

6.6 The Structure of S, H, and L

Writing the 3×3 real matrix **S** as

$$\mathbf{S} = \mathbf{W}\mathbf{L}, \quad \mathbf{W} = \mathbf{S}\mathbf{L}^{-1}, \tag{6.6-1}$$

S is the product of a skew-symmetric matrix **W** by a positive definite matrix **L**. We could also write

$$\mathbf{S} = \mathbf{L}^{-1}(\mathbf{L}\mathbf{S})$$

so that **S** is again the product of a skew-symmetric matrix **LS** by a positive definite matrix $\mathbf{L}^{-1}$, but in the reverse order. In particular, **S** can be a skew-symmetric matrix if **L** is a multiple of the identity matrix **I**.

Since **W** is skew-symmetric and hence is singular, taking the determinant on both sides of $(6.6\text{-}1)_1$ yields

$$|\mathbf{S}| = 0. \tag{6.6-2}$$

Also, from (1.1-13) and (1.1-14),

$$\begin{aligned} \operatorname{tr}\mathbf{S} &= \operatorname{tr}(\mathbf{W}\mathbf{L})^T = \operatorname{tr}(\mathbf{L}\mathbf{W}^T) \\ &= -\operatorname{tr}(\mathbf{L}\mathbf{W}) = -\operatorname{tr}(\mathbf{W}\mathbf{L}) = -\operatorname{tr}\mathbf{S}. \end{aligned}$$

Hence

$$\operatorname{tr}\mathbf{S} = 0. \tag{6.6-3}$$

With (6.6-2) and (6.6-3) the characteristic equation (1.2-6a) for **S** is

$$\lambda^3 + s^2\lambda = 0 \tag{6.6-4}$$

where

$$1 > s = \left\{-\tfrac{1}{2}\operatorname{tr}(\mathbf{S}^2)\right\}^{1/2} > 0. \tag{6.6-5}$$

By the Cayley–Hamilton theory,

$$\mathbf{S}^3 + s^2\mathbf{S} = \mathbf{0}. \tag{6.6-6}$$

The inequality $s<1$ will be shown in (6.6-24). To show that s is real and positive, let L_1, L_2, and L_3 be the eigenvalues of **L** which are positive and nonzero. Employing (1.1-14) and the diagonalization $(1.3\text{-}8)_2$ for **L**,

$$\begin{aligned} -\tfrac{1}{2}\operatorname{tr}(\mathbf{S}^2) &= -\tfrac{1}{2}\operatorname{tr}\left\{\mathbf{W}\mathbf{X}\langle L_*\rangle\mathbf{X}^T\mathbf{W}\mathbf{X}\langle L_*\rangle\mathbf{X}^T\right\} \\ &= -\tfrac{1}{2}\operatorname{tr}\left\{\mathbf{X}^T\mathbf{W}\mathbf{X}\langle L_*\rangle\mathbf{X}^T\mathbf{W}\mathbf{X}\langle L_*\rangle\right\} = -\tfrac{1}{2}\operatorname{tr}\left\{\left(\hat{\mathbf{W}}\langle L_*\rangle\right)^2\right\} \end{aligned}$$

where

$$\hat{\mathbf{W}} = \mathbf{X}^T \mathbf{W} \mathbf{X}.$$

Since **W** is skew-symmetric so is $\hat{\mathbf{W}}$. By replacing $\hat{\mathbf{W}}$ in the form of (1.2-12) and carrying out the matrix multiplication, it can be shown that

$$-\tfrac{1}{2}\mathrm{tr}\left\{\left(\hat{\mathbf{W}}\langle L_*\rangle\right)^2\right\} = L_2 L_3 \hat{w}_1^2 + L_3 L_1 \hat{w}_2^2 + L_1 L_2 \hat{w}_3^2$$

which is positive and nonzero unless $\mathbf{W}=\mathbf{0}$. $\mathbf{W}=\mathbf{0}$ if and only if $\mathbf{S}=\mathbf{0}$. There is no direct proof available that shows $\mathbf{S}\neq\mathbf{0}$. An indirect proof that $S_{21}-S_{12}>0$, and hence $\mathbf{S}\neq\mathbf{0}$ will be given in (8.12-41).

An alternate proof without invoking the diagonalization of **L** (see also Chadwick and Ting, 1987) is to write **W** in the form of (1.2-14) so that the **S** of $(6.6\text{-}1)_1$ is

$$S_{ik} = \delta_{ipt} w_t L_{pk}.$$

Hence

$$(\mathbf{S}^2)_{ij} = S_{ik} S_{kj} = \delta_{ipt} w_t L_{pk} \delta_{kqr} w_r L_{qj}.$$

From $(1.1\text{-}18)_1$ and noting that **L** is symmetric,

$$\delta_{kqr} L_{pk} L_{qj} = \delta_{spj} \widehat{L}_{rs}$$

where

$$\widehat{L}_{rs} = (\mathbf{L}^{-1})_{rs} |\mathbf{L}|$$

is the adjoint of L_{rs}. Making use of the identity (Chadwick, 1976a; Spencer, 1980)

$$\delta_{ipt}\delta_{spj} = \delta_{is}\delta_{jt} - \delta_{ij}\delta_{st}$$

we have

$$(\mathbf{S}^2)_{ij} = \left\{(\mathbf{L}^{-1})_{ir} w_j w_r - \delta_{ij} (\mathbf{L}^{-1})_{rt} w_r w_t\right\} |\mathbf{L}| \tag{6.6-7}$$

and

$$s^2 = -\tfrac{1}{2}\mathrm{tr}(\mathbf{S}^2) = (\mathbf{w}^T \mathbf{L}^{-1} \mathbf{w}) |\mathbf{L}|.$$

Since $\mathbf{L}^{-1}$ is positive definite, the right side is always positive and nonzero.

The characteristic equation (6.6-4) tells us that the eigenvalues are $\pm is$ and 0. Let the associated eigenvectors be $\mathbf{e}_1 \mp i\mathbf{e}_2$ and $\mathbf{e}_3$, i.e.,

$$\mathbf{S}(\mathbf{e}_1 \mp i\mathbf{e}_2) = \pm is(\mathbf{e}_1 \mp i\mathbf{e}_2), \qquad \mathbf{S}\mathbf{e}_3 = \mathbf{0}. \tag{6.6-8a}$$

Equating the real and imaginary parts of $(6.6\text{-}8a)_1$, (6.6-8a) can be written as

$$\mathbf{S}\mathbf{e}_1 = s\mathbf{e}_2, \quad \mathbf{S}\mathbf{e}_2 = -s\mathbf{e}_1, \quad \mathbf{S}\mathbf{e}_3 = \mathbf{0}. \tag{6.6-8b}$$

A consequence of $(6.6\text{-}8b)_{1,2}$ is

$$\mathbf{S}^2\mathbf{e}_i = -s^2\mathbf{e}_i, \quad i = 1, 2. \tag{6.6-9}$$

The matrix **S** has three distinct eigenvalues. Therefore the three eigenvectors are independent of each other. The independence of $(\mathbf{e}_1 + i\mathbf{e}_2)$ and $(\mathbf{e}_1 - i\mathbf{e}_2)$ imply that $\mathbf{e}_1$ and $\mathbf{e}_2$ are independent of each other. Equation (6.6-9) is satisfied by $\mathbf{e}_1$ and $\mathbf{e}_2$. It provides two independent solutions which span an eigenplane for $(\mathbf{e}_1 \mp i\mathbf{e}_2)$ (Section 1.8). Any vector satisfying (6.6-9) can be taken as $\mathbf{e}_1$ or $\mathbf{e}_2$. The vector $\mathbf{e}_2$ or $\mathbf{e}_1$ is then determined from $(6.6\text{-}8b)_1$ or $(6.6\text{-}8b)_2$.

Let $(\mathbf{e}^1 \pm i\mathbf{e}^2)$ and $\mathbf{e}^3$ be the left eigenvectors associated with the eigenvalues $\pm is$ and 0, i.e.,

$$\mathbf{S}^T(\mathbf{e}^1 \pm i\mathbf{e}^2) = \pm is(\mathbf{e}^1 \pm i\mathbf{e}^2), \quad \mathbf{S}^T\mathbf{e}^3 = \mathbf{0}, \tag{6.6-10a}$$

or

$$\mathbf{S}^T\mathbf{e}^1 = -s\mathbf{e}^2, \quad \mathbf{S}^T\mathbf{e}^2 = s\mathbf{e}^1, \quad \mathbf{S}^T\mathbf{e}^3 = \mathbf{0}. \tag{6.6-10b}$$

The first two equations lead to

$$(\mathbf{S}^2)^T\mathbf{e}^i = -s^2\mathbf{e}^i, \quad i = 1, 2. \tag{6.6-11}$$

Again, any vector satisfying (6.6-11) can be taken as $\mathbf{e}^1$ or $\mathbf{e}^2$. The vector $\mathbf{e}^2$ or $\mathbf{e}^1$ is then determined form $(6.6\text{-}10b)_1$ or $(6.6\text{-}10b)_2$. The two independent vectors obtained from (6.6-11) span a left eigenplane.

The left and right eigenvectors associated with different eigenvalues are orthogonal to each other (Section 1.2). If we normalize the eigenvectors we have

$$\begin{gathered}(\mathbf{e}_1 \pm i\mathbf{e}_2)\cdot(\mathbf{e}^1 \pm i\mathbf{e}^2) = 0, \quad (\mathbf{e}_1 \pm i\mathbf{e}_2)\cdot(\mathbf{e}^1 \mp i\mathbf{e}^2) = 1, \\ (\mathbf{e}_1 \pm i\mathbf{e}_2)\cdot\mathbf{e}^3 = 0, \quad (\mathbf{e}^1 \pm i\mathbf{e}^2)\cdot\mathbf{e}_3 = 0, \quad \mathbf{e}^3\cdot\mathbf{e}_3 = 1.\end{gathered} \tag{6.6-12}$$

Equations (6.6-12) are satisfied if

$$\mathbf{e}_i \cdot \mathbf{e}^j = \delta_{ij} = \mathbf{e}_i^T\mathbf{e}^j, \tag{6.6-13}$$

where δ_{ij} is the Kronecker delta. Thus $\mathbf{e}^1$, $\mathbf{e}^2$, $\mathbf{e}^3$ are the *reciprocal vectors* of $\mathbf{e}_1$, $\mathbf{e}_2$, $\mathbf{e}_3$ and vice versa.

Any 3×3 real matrix can be decomposed in terms of one of the following four matrix bases

$$\mathbf{e}^i(\mathbf{e}^j)^T, \quad \mathbf{e}_i(\mathbf{e}_j)^T, \quad \mathbf{e}_i(\mathbf{e}^j)^T, \quad \mathbf{e}^i(\mathbf{e}_j)^T.$$

Thus **S** can be written as

$$\mathbf{S} = S_{ij}\mathbf{e}^i(\mathbf{e}^j)^T = S^{ij}\mathbf{e}_i(\mathbf{e}_j)^T = S^i{}_j\mathbf{e}_i(\mathbf{e}^j)^T = S_i{}^j\mathbf{e}^i(\mathbf{e}_j)^T \tag{6.6-14}$$

in which repeated indices imply summation. Since **S** is a tensor (Section 7.2) the components S_{ij} and S^{ij} are, respectively, the *covariant* and *contravariant* components of **S** while $S^i{}_j$ and $S_i{}^j$ are the *mixed* components. The 3×3 identity matrix **I** can be written as

$$\begin{aligned}\mathbf{I} &= \mathbf{e}_1(\mathbf{e}^1)^T + \mathbf{e}_2(\mathbf{e}^2)^T + \mathbf{e}_3(\mathbf{e}^3)^T \\ &= \mathbf{e}^1(\mathbf{e}_1)^T + \mathbf{e}^2(\mathbf{e}_2)^T + \mathbf{e}^3(\mathbf{e}_3)^T.\end{aligned} \tag{6.6-15}$$

They are verified easily by postmultiplying $(6.6\text{-}15)_1$ and $(6.6\text{-}15)_2$ by $\mathbf{e}_i$ and $\mathbf{e}^i$, respectively, and employing $(6.6\text{-}13)_2$.

In terms of $S^i{}_j$, H^{ij}, and L_{ij}, we will show that

$$S^i{}_j = s\begin{bmatrix}0 & -1 & 0\\ 1 & 0 & 0\\ 0 & 0 & 0\end{bmatrix},\quad H^{ij} = \frac{1}{\mu\kappa}\begin{bmatrix}1-s^2 & 0 & 0\\ 0 & 1-s^2 & 0\\ 0 & 0 & \kappa\end{bmatrix},\quad L_{ij} = \mu\begin{bmatrix}\kappa & 0 & 0\\ 0 & \kappa & 0\\ 0 & 0 & 1\end{bmatrix}, \tag{6.6-16}$$

where s, μ, κ are real parameters. First we write **S** in the form

$$\mathbf{S} = S^i{}_j\mathbf{e}_i(\mathbf{e}^j)^T,\qquad \mathbf{S}^T = S^i{}_j\mathbf{e}^j(\mathbf{e}_i)^T. \tag{6.6-17}$$

Satisfaction of (6.6-8b) means that

$$S^i{}_1\mathbf{e}_i = s\mathbf{e}_2,\qquad S^i{}_2\mathbf{e}_i = -s\mathbf{e}_1,\qquad S^i{}_3\mathbf{e}_i = \mathbf{0}.$$

Therefore the only nonzero $S^i{}_j$ are

$$S^2{}_1 = s,\qquad S^1{}_2 = -s, \tag{6.6-18}$$

which proves $(6.6\text{-}16)_1$. For the matrix **H** and **L** let

$$\mathbf{H} = H^{ij}\mathbf{e}_i(\mathbf{e}_j)^T,\qquad \mathbf{L} = L_{ij}\mathbf{e}^i(\mathbf{e}^j)^T. \tag{6.6-19}$$

Substitution of (6.6-17) and $(6.6\text{-}19)_1$ into $(6.5\text{-}9)_1$ yields

$$S^i{}_kH^{kj} + H^{ik}S^j{}_k = 0 \tag{6.6-20}$$

or, by $(6.6\text{-}16)_1$,

$$\begin{bmatrix}-H^{21} & -H^{22} & -H^{23}\\ H^{11} & H^{12} & H^{13}\\ 0 & 0 & 0\end{bmatrix} + \begin{bmatrix}-H^{12} & H^{11} & 0\\ -H^{22} & H^{21} & 0\\ -H^{32} & H^{31} & 0\end{bmatrix} = \mathbf{0}.$$

Since $H^{ij} = H^{ji}$ the only nonzero elements are

$$H^{11} = H^{22} \quad \text{and} \quad H^{33}.$$

In the same manner, substituting (6.6-17) and $(6.6\text{-}19)_2$ into $(6.5\text{-}9)_2$ and making use of $(6.6\text{-}16)_1$ the only nonzero elements of L_{ij} are

$$L_{11} = L_{22} \quad \text{and} \quad L_{33}.$$

Finally, from the identity $(6.5\text{-}8)_1$ we obtain $(6.6\text{-}16)_{2,3}$.

Although they are in tensor components, the expressions in (6.6-16) are identical to those in (6.4-13) for isotropic materials. For isotropic materials for which $\mathbf{e}_1$, $\mathbf{e}_2$, $\mathbf{e}_3$ are unit vectors in the positive directions of the x_1-, x_2-, x_3- axes, respectively, the covariant, contravariant, and mixed components of a tensor are identical.

In summary, the matrices **S**, **H**, and **L** have the representation

$$\mathbf{S} = s\left\{\mathbf{e}_2(\mathbf{e}^1)^T - \mathbf{e}_1(\mathbf{e}^2)^T\right\}, \tag{6.6-21}$$

$$\mathbf{H} = \frac{1-s^2}{\mu\kappa}\left\{\mathbf{e}_1(\mathbf{e}_1)^T + \mathbf{e}_2(\mathbf{e}_2)^T\right\} + \frac{1}{\mu}\mathbf{e}_3(\mathbf{e}_3)^T, \tag{6.6-22}$$

$$\mathbf{L} = \mu\kappa\left\{\mathbf{e}^1(\mathbf{e}^1)^T + \mathbf{e}^2(\mathbf{e}^2)^T\right\} + \mu\mathbf{e}^3(\mathbf{e}^3)^T. \tag{6.6-23}$$

Noting that **H** and **L** are positive definite and $s>0$, we have

$$1 > s > 0, \qquad \mu > 0, \qquad \kappa > 0. \tag{6.6-24}$$

These were first obtained by Chadwick and Ting (1987). We will show in Section 7.6 that **S** is physically dimensionless, while **L** and **H** have the dimensions of stress and compliance, respectively. Therefore μ has the dimensions of stress, and s and κ are dimensionless. The vectors $\mathbf{e}_1$, $\mathbf{e}_2$, $\mathbf{e}_3$, and $\mathbf{e}^1$, $\mathbf{e}^2$, $\mathbf{e}^3$ have physical interpretations (Section 8.3).

With the orthogonality relations (6.6-13) and the representation (6.6-15) for the identity matrix **I**, it is readily shown that

$$\mathbf{H}^{-1} = \frac{\mu\kappa}{1-s^2}\left\{\mathbf{e}^1(\mathbf{e}^1)^T + \mathbf{e}^2(\mathbf{e}^2)^T\right\} + \mu\mathbf{e}^3(\mathbf{e}^3)^T, \tag{6.6-25}$$

$$\mathbf{L}^{-1} = \frac{1}{\mu\kappa}\left\{\mathbf{e}_1(\mathbf{e}_1)^T + \mathbf{e}_2(\mathbf{e}_2)^T\right\} + \frac{1}{\mu}\mathbf{e}_3(\mathbf{e}_3)^T, \tag{6.6-26}$$

and

$$\mathbf{S}^T = s\left\{\mathbf{e}^1(\mathbf{e}_2)^T - \mathbf{e}^2(\mathbf{e}_1)^T\right\}, \tag{6.6-27}$$

$$\mathbf{S}^2 = -s^2\left\{\mathbf{e}_1(\mathbf{e}^1)^T + \mathbf{e}_2(\mathbf{e}^2)^T\right\}, \tag{6.6-28}$$

$$\mathbf{SL}^{-1} = \frac{s}{\mu\kappa}\left\{\mathbf{e}_2(\mathbf{e}_1)^T - \mathbf{e}_1(\mathbf{e}_2)^T\right\}, \tag{6.6-29}$$

$$\mathbf{SH} = \frac{s(1-s^2)}{\mu\kappa}\left\{\mathbf{e}_2(\mathbf{e}_1)^T - \mathbf{e}_1(\mathbf{e}_2)^T\right\}, \tag{6.6-30}$$

$$\mathbf{H}^{-1}\mathbf{S} = \frac{\mu\kappa s}{1-s^2}\left\{\mathbf{e}^2(\mathbf{e}^1)^T - \mathbf{e}^1(\mathbf{e}^2)^T\right\}, \tag{6.6-31}$$

$$\mathbf{LS} = \mu\kappa s\left\{\mathbf{e}^2(\mathbf{e}^1)^T - \mathbf{e}^1(\mathbf{e}^2)^T\right\}. \tag{6.6-32}$$

Equations (6.6-29)–(6.6-32) tell us that $\mathbf{e}^3$ is the axial vector of $\mathbf{SL}^{-1}$ and $\mathbf{SH}$ while $\mathbf{e}_3$ is the axial vector of $\mathbf{H}^{-1}\mathbf{S}$ and $\mathbf{LS}$. The identities presented in (6.5-14)–(6.5-22) can be verified by employing the representations (6.6-21)–(6.6-32).

The impedance tensor

$$\mathbf{M} = \mathbf{H}^{-1} + i\mathbf{H}^{-1}\mathbf{S} \tag{6.6-33}$$

rewritten from $(5.6\text{-}6)_2$ can now be shown to be a positive definite Hermitian. Indeed, from (6.6-25) and (6.6-31) $\mathbf{M}$ has the structure

$$\mathbf{M} = \frac{\mu\kappa}{1-s^2}(\mathbf{e}^1 + is\mathbf{e}^2)(\mathbf{e}^1 - is\mathbf{e}^2)^T + \mu\kappa\mathbf{e}^2(\mathbf{e}^2)^T + \mu\mathbf{e}^3(\mathbf{e}^3)^T. \tag{6.6-34}$$

The first term is a positive definite Hermitian while the second and the third terms are positive definite real matrices, confirming that $\mathbf{M}$ is positive definite. Likewise the inverse

$$\mathbf{M}^{-1} = \mathbf{L}^{-1} - i\mathbf{SL}^{-1} \tag{6.6-35}$$

defined in $(5.6\text{-}6)_4$ has the representation

$$\mathbf{M}^{-1} = \frac{1}{\mu\kappa}(\mathbf{e}_1 - is\mathbf{e}_2)(\mathbf{e}_1 + is\mathbf{e}_2)^T + \frac{1-s^2}{\mu\kappa}\mathbf{e}_2(\mathbf{e}_2)^T + \frac{1}{\mu}\mathbf{e}_3(\mathbf{e}_3)^T \tag{6.6-36}$$

which is clearly positive definite. The positive definiteness of $\mathbf{M}$ was first proved indirectly by Ingebrigtsen and Tonning (1969) and Lothe and Barnett (1976a). A direct proof was given by Chadwick and Ting (1987).

Bounds can be set for the eigenvalues of $\mathbf{S}$, $\mathbf{H}$, and $\mathbf{L}$ with the aid of the representations presented here. Let

$$H_1 \geq H_2 \geq H_3 > 0 \tag{6.6-37}$$

be the eigenvalues of $\mathbf{H}$ and

$$L_1 \geq L_2 \geq L_3 > 0 \tag{6.6-38}$$

be the eigenvalues of **L**. The eigenvalues of **S** are $\pm is$ and 0, which means that the eigenvalues of **I**+**SS** are $1-s^2$, $1-s^2$, and 1. By taking the determinant on both sides of the equation

$$\mathbf{I}+\mathbf{SS}=\mathbf{HL}$$

obtained from $(6.5\text{-}8)_1$ we have

$$1>(1-s^2)^2=H_1H_2H_3L_1L_2L_3>0. \qquad (6.6\text{-}39)$$

The inequalities come from (6.6-24). The following inequalities are obtained from **H**, **L**, $\mathbf{H}^{-1}$, $\mathbf{L}^{-1}$ given in (6.6-22), (6.6-23), (6.6-25), (6.6-26) with the aid of (1.6-9).

$$H_1\left|\mathbf{e}^3\right|^2\geq\frac{1}{\mu}\geq H_3\left|\mathbf{e}^3\right|^2, \qquad (6.6\text{-}40)$$

$$L_1\left|\mathbf{e}_3\right|^2\geq\mu\geq L_3\left|\mathbf{e}_3\right|^2, \qquad (6.6\text{-}41)$$

$$H_3^{-1}\left|\mathbf{e}_3\right|^2\geq\mu\geq H_1^{-1}\left|\mathbf{e}_3\right|^2, \qquad (6.6\text{-}42)$$

$$L_3^{-1}\left|\mathbf{e}^3\right|^2\geq\frac{1}{\mu}\geq L_1^{-1}\left|\mathbf{e}^3\right|^2, \qquad (6.6\text{-}43)$$

$$H_1\left|\mathbf{e}^1\right|^2\geq\frac{1-s^2}{\mu\kappa}\geq H_3\left|\mathbf{e}^1\right|^2, \qquad (6.6\text{-}44)$$

$$L_1\left|\mathbf{e}_1\right|^2\geq\mu\kappa\geq L_3\left|\mathbf{e}_1\right|^2, \qquad (6.6\text{-}45)$$

$$H_3^{-1}\left|\mathbf{e}_1\right|^2\geq\frac{\mu\kappa}{1-s^2}\geq H_1^{-1}\left|\mathbf{e}_1\right|^2, \qquad (6.6\text{-}46)$$

$$L_3^{-1}\left|\mathbf{e}^1\right|^2\geq\frac{1}{\mu\kappa}\geq L_1^{-1}\left|\mathbf{e}^1\right|^2. \qquad (6.6\text{-}47)$$

When **H** and $\mathbf{L}^{-1}$ are premultiplied by $(\mathbf{e}^3)^T$ and postmultiplied by $\mathbf{e}^3$ we have (6.6-40) and (6.6-43). The same operation with $(\mathbf{e}_3)^T$ and $\mathbf{e}_3$ on **L** and $\mathbf{H}^{-1}$ leads to (6.6-41) and (6.6-42). Inequalities (6.6-44)–(6.6-47) are derived in the same way except that we employ $\mathbf{e}_1$ and $\mathbf{e}^1$ instead of $\mathbf{e}_3$ and $\mathbf{e}^3$. Noting that

$$\left|\mathbf{e}^3\right|^2\left|\mathbf{e}_3\right|^2\geq\left|\mathbf{e}^3\cdot\mathbf{e}_3\right|^2=1, \quad \left|\mathbf{e}^1\right|^2\left|\mathbf{e}_1\right|^2\geq\left|\mathbf{e}^1\cdot\mathbf{e}_1\right|^2=1, \qquad (6.6\text{-}48)$$

inequalities (6.6-42) and (6.6-43), and (6.6-44) and (6.6-45) yield, respectively,

$$H_1L_1\geq 1, \qquad (6.6\text{-}49)$$

$$1 - s^2 \geq H_3 L_3. \tag{6.6-50}$$

From (6.6-39) and the fact that

$$H_1 L_1 \geq H_2 L_2 \geq H_3 L_3, \tag{6.6-51}$$

we have

$$H_3 L_3 < 1, \tag{6.6-52}$$

and the bounds for s can be written as

$$1 > (1 - s^2)^2 \geq \max\{H_2 H_3 L_2 L_3,\ H_3 L_3\} > 0. \tag{6.6-53}$$

We will discuss the bounds for μ and κ in Section 6.7.

In closing this section we present the vectors $\mathbf{e}_1$, $\mathbf{e}_2$, and $\mathbf{e}_3$ explicitly in terms of $\mathbf{S}$ as follows. Let $\mathbf{t}$ be any real vector. It can be decomposed into two components, one along $\mathbf{e}_3$ and the other on the right eigenplane spanned by $\mathbf{e}_1$ and $\mathbf{e}_2$. Since any vector on the eigenplane can be taken as $\mathbf{e}_1$ or $\mathbf{e}_2$, we will call it $\mathbf{e}_1$ and write

$$s^2 \mathbf{t} = \mathbf{e}_1 + \mathbf{e}_3. \tag{6.6-54}$$

There is no need to add coefficients on $\mathbf{e}_1$ and $\mathbf{e}_3$. They are unnormalized. If both sides of (6.6-54) are multiplied by $\mathbf{S}$ and $\mathbf{S}^2$ and use is made of (6.6-8b) and (6.6-9) we have

$$s^2 \mathbf{S}\mathbf{t} = s\mathbf{e}_2, \quad s^2 \mathbf{S}^2 \mathbf{t} = -s^2 \mathbf{e}_1.$$

Together with (6.6-54) one obtains

$$\mathbf{e}_1 = -\mathbf{S}^2 \mathbf{t}, \quad \mathbf{e}_2 = s\mathbf{S}\mathbf{t}, \quad \mathbf{e}_3 = (\mathbf{S}^2 + s^2 \mathbf{I})\mathbf{t}. \tag{6.6-55}$$

These are explicit expressions of $\mathbf{e}_1$, $\mathbf{e}_2$, and $\mathbf{e}_3$ in terms of an arbitrarily given real vector $\mathbf{t}$. The $\mathbf{e}_1$ and $\mathbf{e}_2$ calculated from $(6.6\text{-}55)_{1,2}$ vanish when $\mathbf{t}$ is proportional to $\mathbf{e}_3$. On the other hand, $\mathbf{e}_3$ computed from $(6.6\text{-}55)_3$ vanishes when $\mathbf{t}$ is on the right eigenplane. In both cases a different $\mathbf{t}$ should be assumed. Employing (6.6-6) it is readily shown that (6.6-55) indeed satisfy (6.6-8b) for any real vector $\mathbf{t}$.

For different $\mathbf{t}$'s the complex vectors $\mathbf{e}_1 + i\mathbf{e}_2$ obtained from $(6.6\text{-}55)_{1,2}$ differ by a complex multiplier while the real vector $\mathbf{e}_3$ calculated from $(6.6\text{-}55)_3$ differ by a real multiplier.

6.7 Classification of the Eigenvectors $\mathbf{e}_i$ and $\mathbf{e}^i$

The right null vector $\mathbf{e}_3$ is unique up to an arbitrary real multiplier. The right eigenvectors $\mathbf{e}_1 \mp i\mathbf{e}_2$ are complex vectors. They define an eigenplane (Section 1.8). If $\mathbf{e}_1^{\circ} \mp i\mathbf{e}_2^{\circ}$ are right eigenvec tors, so are

$$\mathbf{e}_1 \mp i\mathbf{e}_2 = e^{-i\psi}(\mathbf{e}_1^{\circ} \mp i\mathbf{e}_2^{\circ}), \tag{6.7-1}$$

where ψ is a real parameter. The real and imaginary parts are

$$\begin{aligned}\mathbf{e}_1 &= \mathbf{e}_1^{\circ}\cos\psi + \mathbf{e}_2^{\circ}\sin\psi,\\ \mathbf{e}_2 &= -\mathbf{e}_1^{\circ}\sin\psi + \mathbf{e}_2^{\circ}\cos\psi.\end{aligned} \tag{6.7-2}$$

The $\mathbf{e}_1$ and $\mathbf{e}_2$ lie on the plane spanned by $\mathbf{e}_1^{\circ}$ and $\mathbf{e}_2^{\circ}$. As ψ varies $\mathbf{e}_1$ and $\mathbf{e}_2$ trace out an ellipse (Fig. 1.1). The plane will be called the Γ_b plane and the ellipse the Γ_b ellipse. The vectors $\mathbf{e}_1$ and $\mathbf{e}_2$ are a pair of conjugate radii of the Γ_b ellipse. Any pair of conjugate radii of the Γ_b ellipse can be taken as $\mathbf{e}_1$ and $\mathbf{e}_2$. When $\mathbf{e}_1 \cdot \mathbf{e}_2 = 0$ they are the principal radii of the ellipse and are denoted by $\hat{\mathbf{e}}_1$ and $\hat{\mathbf{e}}_2$. One may choose ψ in (6.7-2) such that $\mathbf{e}_1 \cdot \mathbf{e}_2 = 0$ (Section 1.8).

The left eigenvectors $\mathbf{e}^1 \pm i\mathbf{e}^2$ also define an eigenplane which will be called the Γ_f plane. If $\mathbf{e}_{\circ}^1 \pm i\mathbf{e}_{\circ}^2$ are left eigenvectors,

$$\mathbf{e}^1 \pm i\mathbf{e}^2 = e^{-i\psi}(\mathbf{e}_{\circ}^1 \pm i\mathbf{e}_{\circ}^2) \tag{6.7-3}$$

where

$$\begin{aligned}\mathbf{e}^1 &= \mathbf{e}_{\circ}^1\cos\psi + \mathbf{e}_{\circ}^2\sin\psi,\\ \mathbf{e}^2 &= -\mathbf{e}_{\circ}^1\sin\psi + \mathbf{e}_{\circ}^2\cos\psi,\end{aligned} \tag{6.7-4}$$

are also left eigenvectors. As ψ varies $\mathbf{e}^1$ and $\mathbf{e}^2$ trace out an ellipse, the Γ_f ellipse. They are a pair of conjugate radii of the Γ_f ellipse. The principal radii of the Γ_f ellipse are denoted by $\hat{\mathbf{e}}^1$, $\hat{\mathbf{e}}^2$.

From (6.7-2) and (6.7-4) it is easily shown that if $(\mathbf{e}_1^{\circ}, \mathbf{e}_2^{\circ}, \mathbf{e}_3)$ and $(\mathbf{e}_{\circ}^1, \mathbf{e}_{\circ}^2, \mathbf{e}^3)$ are orthogonal to each other according to (6.6-13), so are $(\mathbf{e}_1, \mathbf{e}_2, \mathbf{e}_3)$ and $(\mathbf{e}^1, \mathbf{e}^2, \mathbf{e}^3)$. Moreover,

$$\begin{aligned}&\mathbf{e}_2(\mathbf{e}^1)^T - \mathbf{e}_1(\mathbf{e}^2)^T = \mathbf{e}_2^{\circ}(\mathbf{e}_{\circ}^1)^T - \mathbf{e}_1^{\circ}(\mathbf{e}_{\circ}^2)^T,\\ &\mathbf{e}_1(\mathbf{e}_1)^T + \mathbf{e}_2(\mathbf{e}_2)^T = \mathbf{e}_1^{\circ}(\mathbf{e}_1^{\circ})^T + \mathbf{e}_2^{\circ}(\mathbf{e}_2^{\circ})^T,\\ &\mathbf{e}^1(\mathbf{e}^1)^T + \mathbf{e}^2(\mathbf{e}^2)^T = \mathbf{e}_{\circ}^1(\mathbf{e}_{\circ}^1)^T + \mathbf{e}_{\circ}^2(\mathbf{e}_{\circ}^2)^T,\\ &\mathbf{e}_1 \times \mathbf{e}_2 = \mathbf{e}_1^{\circ} \times \mathbf{e}_2^{\circ}, \qquad \mathbf{e}^1 \times \mathbf{e}^2 = \mathbf{e}_{\circ}^1 \times \mathbf{e}_{\circ}^2,\end{aligned} \tag{6.7-5}$$

and hence the expressions on the left sides are independent of ψ. The representations (6.6-21)–(6.6-23) for **S**, **H**, and **L** are therefore invariant with ψ, so are the representations (6.6-25)–(6.6-32). With the exception of the parameter s, the other two parameters μ and κ appearing in these representations are not uniquely determined by **S**, **H**, and **L** even though they do not depend on ψ. For the uniqueness of μ and κ the eigenvectors need to be normalized.

Let

$$|\mathbf{e}_3| = |\mathbf{e}^3|, \quad (\mathbf{e}_1 \times \mathbf{e}_2)\cdot\mathbf{e}_3 = 1. \tag{6.7-6}$$

It is easily shown with the aid of (6.7-6) and the orthogonality relations (6.6-13) that

$$\begin{aligned} &\mathbf{e}^1 = \mathbf{e}_2 \times \mathbf{e}_3, \quad \mathbf{e}^2 = \mathbf{e}_3 \times \mathbf{e}_1, \quad \mathbf{e}^3 = \mathbf{e}_1 \times \mathbf{e}_2, \\ &\mathbf{e}_1 = \mathbf{e}^2 \times \mathbf{e}^3, \quad \mathbf{e}_2 = \mathbf{e}^3 \times \mathbf{e}^1, \quad \mathbf{e}_3 = \mathbf{e}^1 \times \mathbf{e}^2, \end{aligned} \tag{6.7-7}$$

and

$$|\mathbf{e}_3| = |\mathbf{e}^3| = |\mathbf{e}_1 \times \mathbf{e}_2| = |\mathbf{e}^1 \times \mathbf{e}^2| \geq 1, \tag{6.7-8}$$

$$\max|\mathbf{e}_1| \geq 1, \qquad \max|\mathbf{e}^1| \geq 1. \tag{6.7-9}$$

Equations (6.7-8) tell us that the lengths of the null vectors $\mathbf{e}_3$ and $\mathbf{e}^3$ are the same and that the areas of the Γ_b ellipse and the Γ_f ellipse are identical. Equation $(6.7\text{-}6)_2$ implies that the volume of the parallelepiped formed by $(\mathbf{e}_1,\mathbf{e}_2,\mathbf{e}_3)$ is unity, so is the volume of the parallelepiped formed by $(\mathbf{e}^1,\mathbf{e}^2,\mathbf{e}^3)$.

Bounds for μ and κ can now be set using (6.7-8) and (6.7-9). Inequalities (6.6-40) and (6.6-42), and (6.6-41) and (6.6-43) yield,

$$H_1 \geq \frac{1}{\mu} \geq H_3, \tag{6.7-10}$$

$$L_1 \geq \mu \geq L_3, \tag{6.7-11}$$

respectively. They can be combined into one equation as

$$\min(L_1, H_3^{-1}) \geq \mu \geq \max(L_3, H_1^{-1}), \tag{6.7-12a}$$

or

$$\min(H_1, L_3^{-1}) \geq \frac{1}{\mu} \geq \max(H_3, L_1^{-1}). \tag{6.7-12b}$$

Inequalities (6.6-45) and (6.6-47) apply to any $\mathbf{e}_1$ and $\mathbf{e}^1$. When $|\mathbf{e}_1|$ $= \max|\mathbf{e}_1|$ for (6.6-45) and $|\mathbf{e}^1| = \max|\mathbf{e}^1|$ for (6.6-47) we have,

$$L_1 \geq \mu\kappa \geq L_3. \tag{6.7-13}$$

In conjunction with (6.7-12b) the bounds for κ are

$$\min(H_1 L_1,\ L_1/L_3) \geq \kappa \geq \max(H_3 L_3,\ L_3/L_1). \tag{6.7-14}$$

When a pair of conjugate radii $(\mathbf{e}_1,\mathbf{e}_2)$ of the Γ_b ellipse is chosen, the pair of conjugate radii $(\mathbf{e}^1,\mathbf{e}^2)$ on the Γ_f ellipse are determined from the orthogonality relations (6.6-13). This means that if we choose the principal radii of the Γ_b ellipse as $(\mathbf{e}_1,\mathbf{e}_2)$, $(\mathbf{e}^1,\mathbf{e}^2)$ may not be the principal radii of the Γ_f ellipse. From $(6.7\text{-}7)_{1,2}$,

$$\mathbf{e}^1 \cdot \mathbf{e}^2 = (\mathbf{e}_2 \times \mathbf{e}_3)\cdot(\mathbf{e}_3 \times \mathbf{e}_1)$$
$$= (\mathbf{e}_2 \cdot \mathbf{e}_3)(\mathbf{e}_1 \cdot \mathbf{e}_3) - (\mathbf{e}_1 \cdot \mathbf{e}_2)(\mathbf{e}_3 \cdot \mathbf{e}_3).$$

Therefore when $\mathbf{e}_1\cdot\mathbf{e}_2 = 0$, $\mathbf{e}^1\cdot\mathbf{e}^2 = 0$ also if

$$\mathbf{e}_2 \cdot \mathbf{e}_3 = 0 \quad \text{and/or} \quad \mathbf{e}_1 \cdot \mathbf{e}_3 = 0. \tag{6.7-15}$$

Thus $\mathbf{e}_1$ and/or $\mathbf{e}_2$ must be orthogonal to $\mathbf{e}_3$.

The geometry of the Γ_b ellipse, Γ_f ellipse, and the null vectors $\mathbf{e}_3$, $\mathbf{e}^3$ depends on whether $\mathbf{e}_3=\mathbf{e}^3$ or not, and on whether or not $(\mathbf{e}_1,\mathbf{e}_2)$ and $(\mathbf{e}^1,\mathbf{e}^2)$ can be simultaneously the principal radii of their respective ellipses. There are three types of geometry for the vectors $(\mathbf{e}_1,\mathbf{e}_2,\mathbf{e}_3)$ and $(\mathbf{e}^1,\mathbf{e}^2,\mathbf{e}^3)$ which are discussed separately below (Ting, 1990a).

(I) **Type A**: $\mathbf{e}_3=\mathbf{e}^3$. When $\mathbf{e}_3=\mathbf{e}^3$, $(6.7\text{-}15)_{1,2}$ are satisfied due to the orthogonality relations (6.6-13). Hence if $(\mathbf{e}_1,\mathbf{e}_2)$ are the principal radii of the Γ_b ellipse, $(\mathbf{e}^1,\mathbf{e}^2)$ are necessarily the principal radii of the Γ_f ellipse. The Γ_b plane and the Γ_f plane are the same plane, Fig. 6.1. The vectors $(\hat{\mathbf{e}}_1,\hat{\mathbf{e}}_2,\mathbf{e}_3)$ form an orthogonal set, so are the vectors $(\hat{\mathbf{e}}^1,\hat{\mathbf{e}}^2,\mathbf{e}^3)$. They have the relations

$$\begin{gathered} |\hat{\mathbf{e}}_1| = \alpha = |\hat{\mathbf{e}}^2|, \quad |\hat{\mathbf{e}}_2| = \alpha^{-1} = |\hat{\mathbf{e}}^1|, \quad |\mathbf{e}_3| = |\mathbf{e}^3| = 1, \\ \hat{\mathbf{e}}^1 = \alpha^{-2}\hat{\mathbf{e}}_1, \qquad \hat{\mathbf{e}}^2 = \alpha^2\hat{\mathbf{e}}_2, \qquad \mathbf{e}_3 = \mathbf{e}^3. \end{gathered} \tag{6.7-16}$$

In the above $\alpha>1$ or $\alpha<1$ depending on whether $\hat{\mathbf{e}}_1$ is a major or minor axis of the Γ_b ellipse. If $\alpha=1$, both Γ_b and Γ_f ellipses reduce to a circle of unit radius, which is the case for isotropic materials.

Since $\mathbf{e}_3=\mathbf{e}^3$, we have from $(6.6\text{-}8a)_3$ and $(6.6\text{-}10a)_3$

$$(\mathbf{S} - \mathbf{S}^T)\mathbf{e}_3 = \mathbf{0}. \tag{6.7-17}$$

The matrix $\mathbf{S}-\mathbf{S}^T$ is nonzero. If it were zero, $\mathbf{S}$ is symmetric which contradicts the fact that $\mathbf{S}$ has complex eigenvalues. Observing that $\mathbf{S}-\mathbf{S}^T$ is skew-symmetric and nonzero, $\mathbf{e}_3$ is the axial vector of $\mathbf{S}-\mathbf{S}^T$. If we write $\mathbf{S}-\mathbf{S}^T$ in the form

$$\begin{bmatrix} 0 & w_3 & -w_2 \\ -w_3 & 0 & w_1 \\ w_2 & -w_1 & 0 \end{bmatrix}, \tag{6.7-18}$$

$\mathbf{e}_3$ is given by

$$\mathbf{e}_3^T = \pm\frac{1}{|\mathbf{w}|}[w_1, w_2, w_3]. \tag{6.7-19}$$

It is unique to the extent that if $\mathbf{e}_3$ is a solution, so is $-\mathbf{e}_3$.

There is a way to determine if $\mathbf{e}_3=\mathbf{e}^3$ without computing $\mathbf{e}_3$ and $\mathbf{e}^3$. When $\mathbf{e}_3=\mathbf{e}^3$, the Γ_b plane and the Γ_f plane are the same plane. Any vector $\mathbf{x}$ on the plane can be taken as the unnormalized $\mathbf{e}_1$, $\mathbf{e}_2$, $\mathbf{e}^1$, or $\mathbf{e}^2$. From (6.6-9) and (6.6-11),

$$\left(\mathbf{S}^2-(\mathbf{S}^2)^T\right)\mathbf{x}=\mathbf{0}. \tag{6.7-20}$$

The matrix $\mathbf{S}^2-(\mathbf{S}^2)^T$ is skew-symmetric. If $\mathbf{S}^2-(\mathbf{S}^2)^T$ is not identically zero, $\mathbf{x}$ is the axial vector of $\mathbf{S}^2-(\mathbf{S}^2)^T$ and is unique except a multiplier. But $\mathbf{x}$ can be the unnormalized $\mathbf{e}_1$, $\mathbf{e}_2$, $\mathbf{e}^1$, or $\mathbf{e}^2$. Therefore $\mathbf{S}^2=(\mathbf{S}^2)^T$ if $\mathbf{e}_3=\mathbf{e}^3$. The converse is also true. If $\mathbf{S}^2=(\mathbf{S}^2)^T$, (6.6-9) and (6.6-11) define the same eigenplane for the left and right complex eigenvectors, confirming that $\mathbf{e}_3=\mathbf{e}^3$. We therefore conclude that

$$\mathbf{S}^2\begin{cases}=(\mathbf{S}^2)^T & \text{if}\ \ \mathbf{e}_3=\mathbf{e}^3,\\ \neq(\mathbf{S}^2)^T & \text{if}\ \ \mathbf{e}_3\neq\mathbf{e}^3.\end{cases} \tag{6.7-21}$$

Thus we have Type A geometry when $\mathbf{S}^2$ is symmetric, in particular when $\mathbf{S}$ is skew-symmetric.

For monoclinic materials with the symmetry plane at $x_3=0$, $\mathbf{e}_3=\mathbf{e}^3$ and is along the x_3-axis (Section 8.3). Hence we let

$$\begin{gathered}\mathbf{e}_1=\alpha\begin{bmatrix}\cos\Delta\\ \sin\Delta\\ 0\end{bmatrix},\quad \mathbf{e}_2=\alpha^{-1}\begin{bmatrix}-\sin\Delta\\ \cos\Delta\\ 0\end{bmatrix},\quad \mathbf{e}_3=\begin{bmatrix}0\\ 0\\ 1\end{bmatrix},\\ \mathbf{e}^1=\alpha^{-2}\mathbf{e}_1,\qquad \mathbf{e}^2=\alpha^2\mathbf{e}_2,\qquad \mathbf{e}^3=\mathbf{e}_3,\end{gathered} \tag{6.7-22}$$

where α and Δ are real constants. The matrices $\mathbf{S}$, $\mathbf{H}$, and $\mathbf{L}$ obtained from (6.6-21)–(6.6-23) have the structure

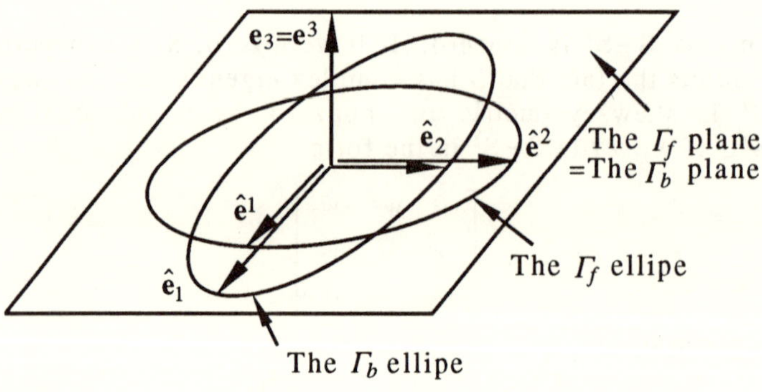

Fig. 6.1 Type A of the eigenvectors $\mathbf{e}_i$ and $\mathbf{e}^i$: $\mathbf{e}_3=\mathbf{e}^3$.

$$\mathbf{S}=\begin{bmatrix} * & * & 0 \\ * & * & 0 \\ 0 & 0 & 0 \end{bmatrix}, \quad \mathbf{H}=\begin{bmatrix} * & * & 0 \\ * & * & 0 \\ 0 & 0 & * \end{bmatrix}, \quad \mathbf{L}=\begin{bmatrix} * & * & 0 \\ * & * & 0 \\ 0 & 0 & * \end{bmatrix}, \tag{6.7-23}$$

in which the asterisk $*$ denotes a possibly nonzero elements. They depend on five constants α, Δ, s, μ, and κ. Equations (6.7-23) agree with the explicit expressions obtained in (6.4-11).

For isotropic materials $\alpha=1$ and (6.7-23) have the expressions

$$\mathbf{S}=s\begin{bmatrix} 0 & -1 & 0 \\ 1 & 0 & 0 \\ 0 & 0 & 0 \end{bmatrix}, \quad \mathbf{H}=\frac{1}{\mu\kappa}\begin{bmatrix} 1-s^2 & 0 & 0 \\ 0 & 1-s^2 & 0 \\ 0 & 0 & \kappa \end{bmatrix}, \quad \mathbf{L}=\mu\begin{bmatrix} \kappa & 0 & 0 \\ 0 & \kappa & 0 \\ 0 & 0 & 1 \end{bmatrix}, \tag{6.7-24}$$

which are independent of Δ. Although it cannot be shown here, s and κ in (6.7-24) are related by (6.4-14).

(II) **Type B**: $\mathbf{e}_3 \neq \mathbf{e}^3$, but $(\mathbf{e}_1,\mathbf{e}_2)$ and $(\mathbf{e}^1,\mathbf{e}^2)$ can be simultaneously the principal radii of their respective ellipses. When $\mathbf{e}_3 \neq \mathbf{e}^3$, the Γ_b and Γ_f planes intersect along a line which can be represented by a vector $\mathbf{k}$, Fig. 6.2. Since $\mathbf{k}$ is on both planes it satisfies (6.6-9) and (6.6-11), and

$$\left(\mathbf{S}^2-(\mathbf{S}^2)^T\right)\mathbf{k}=\mathbf{0}. \tag{6.7-25}$$

The matrix $\mathbf{S}^2-(\mathbf{S}^2)^T$ is skew-symmetric and, by (6.7-21), nonzero. If $\mathbf{S}^2-(\mathbf{S}^2)^T$ is written in the form of (6.7-18) $\mathbf{k}$ is given by (6.7-19). For $(\mathbf{e}_1,\mathbf{e}_2)$ and $(\mathbf{e}^1,\mathbf{e}^2)$ to be simultaneously the principal radii of their respective ellipses, by (6.7-15) $\hat{\mathbf{e}}_2$ or $\hat{\mathbf{e}}_1$ must be orthogonal to $\mathbf{e}_3$. This means that $\hat{\mathbf{e}}_2$ or $\hat{\mathbf{e}}_1$ is along the vector $\mathbf{k}$. Let $\hat{\mathbf{e}}_1$ be along $\mathbf{k}$. It is readily shown that $\hat{\mathbf{e}}^1$ is also along $\mathbf{k}$ as shown in Fig. 6.2. By $(6.6\text{-}8b)_1$,

$$(\hat{\mathbf{e}}^1)^T\mathbf{S}\hat{\mathbf{e}}_1 = s(\hat{\mathbf{e}}^1)^T\hat{\mathbf{e}}_2 = 0,$$

implying that $\mathbf{k}$ satisfies the relation

$$\mathbf{k}^T\mathbf{S}\mathbf{k}=0. \tag{6.7-26}$$

Equation (6.7-26) is the condition for Type B geometry for which $(\mathbf{e}_1,\mathbf{e}_2)$ and $(\mathbf{e}^1,\mathbf{e}^2)$ can be simultaneously the principal radii of their respective ellipses. The vectors $(\hat{\mathbf{e}}_1,\hat{\mathbf{e}}_2,\mathbf{e}_3)$ and $(\hat{\mathbf{e}}^1,\hat{\mathbf{e}}^2,\mathbf{e}^3)$ have the following relations.

$$\begin{aligned} &|\hat{\mathbf{e}}_1|=\alpha, \quad &&|\hat{\mathbf{e}}_2|=\beta\alpha^{-1}, \quad &&|\mathbf{e}_3|=|\mathbf{e}^3|=\beta, \\ &|\hat{\mathbf{e}}^1|=\alpha^{-1}, \quad &&|\hat{\mathbf{e}}^2|=\alpha\beta, \quad &&\hat{\mathbf{e}}^1=\alpha^{-2}\hat{\mathbf{e}}_1. \end{aligned} \tag{6.7-27}$$

For monoclinic materials with the symmetry plane at $x_1=0$, $\mathbf{e}_3 \neq \mathbf{e}^3$

and $\mathbf{k}$ is along the x_1-axis (Section 8.3) which can be taken as the direction of the $\hat{\mathbf{e}}_1$ vector. Hence we can write

$$\mathbf{e}_1 = \begin{bmatrix} \alpha \\ 0 \\ 0 \end{bmatrix}, \quad \mathbf{e}_2 = \beta\alpha^{-1} \begin{bmatrix} 0 \\ \cos\Delta \\ \sin\Delta \end{bmatrix}, \quad \mathbf{e}_3 = \beta \begin{bmatrix} 0 \\ \sin\delta \\ \cos\delta \end{bmatrix},$$
$$\mathbf{e}^1 = \begin{bmatrix} \alpha^{-1} \\ 0 \\ 0 \end{bmatrix}, \quad \mathbf{e}^2 = \alpha\beta \begin{bmatrix} 0 \\ \cos\delta \\ -\sin\delta \end{bmatrix}, \quad \mathbf{e}^3 = \beta \begin{bmatrix} 0 \\ -\sin\Delta \\ \cos\Delta \end{bmatrix}, \tag{6.7-28}$$

where α, δ, and Δ are real constants and

$$\beta = \{\cos(\delta + \Delta)\}^{-1/2}, \quad |\delta + \Delta| < \tfrac{1}{2}\pi. \tag{6.7-29}$$

It can be shown that $\mathbf{S}$, $\mathbf{H}$, and $\mathbf{L}$ obtained from (6.6-21)–(6.6-23) have the structure

$$\mathbf{S} = \begin{bmatrix} 0 & * & * \\ * & 0 & 0 \\ * & 0 & 0 \end{bmatrix}, \quad \mathbf{H} = \begin{bmatrix} * & 0 & 0 \\ 0 & * & * \\ 0 & * & * \end{bmatrix}, \quad \mathbf{L} = \begin{bmatrix} * & 0 & 0 \\ 0 & * & * \\ 0 & * & * \end{bmatrix}, \tag{6.7-30}$$

where the asterisk * denotes a possibly nonzero element. They

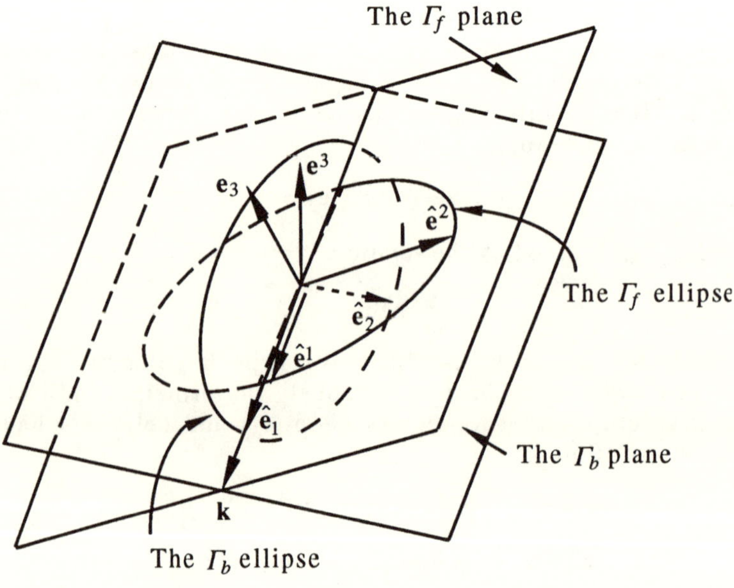

Fig. 6.2 Type B of the eigenvectors $\mathbf{e}_i$ and $\mathbf{e}^i$: $\mathbf{e}_3 \neq \mathbf{e}^3$ and $\mathbf{k}^T\mathbf{S}\mathbf{k}=0$.

depend on six parameters α, δ, Δ, s, μ, and κ. The parameter β is related to δ and Δ through (6.7-29).

For monoclinic materials with the symmetry plane at $x_2=0$, **k** is along the x_2-axis (Section 8.3) which can be designated as the direction for $\hat{\mathbf{e}}_2$. Thus

$$\mathbf{e}_1 = \beta\alpha^{-1}\begin{bmatrix}\cos\Delta\\0\\\sin\Delta\end{bmatrix},\quad \mathbf{e}_2 = \begin{bmatrix}0\\\alpha\\0\end{bmatrix},\quad \mathbf{e}_3 = \beta\begin{bmatrix}\sin\delta\\0\\\cos\delta\end{bmatrix},$$
$$\mathbf{e}^1 = \alpha\beta\begin{bmatrix}\cos\delta\\0\\-\sin\delta\end{bmatrix},\quad \mathbf{e}^2 = \begin{bmatrix}0\\\alpha^{-1}\\0\end{bmatrix},\quad \mathbf{e}^3 = \beta\begin{bmatrix}-\sin\Delta\\0\\\cos\Delta\end{bmatrix}, \tag{6.7-31}$$

where α, δ, and Δ are real constants and β is given in (6.7-29). The matrices **S**, **H**, and **L** obtained from (6.6-21)–(6.6-23) have the structure

$$\mathbf{S} = \begin{bmatrix}0 & * & 0\\ * & 0 & *\\ 0 & * & 0\end{bmatrix},\quad \mathbf{H} = \begin{bmatrix}* & 0 & *\\ 0 & * & 0\\ * & 0 & *\end{bmatrix},\quad \mathbf{L} = \begin{bmatrix}* & 0 & *\\ 0 & * & 0\\ * & 0 & *\end{bmatrix}. \tag{6.7-32}$$

Again, the nonzero elements depends on six parameters.

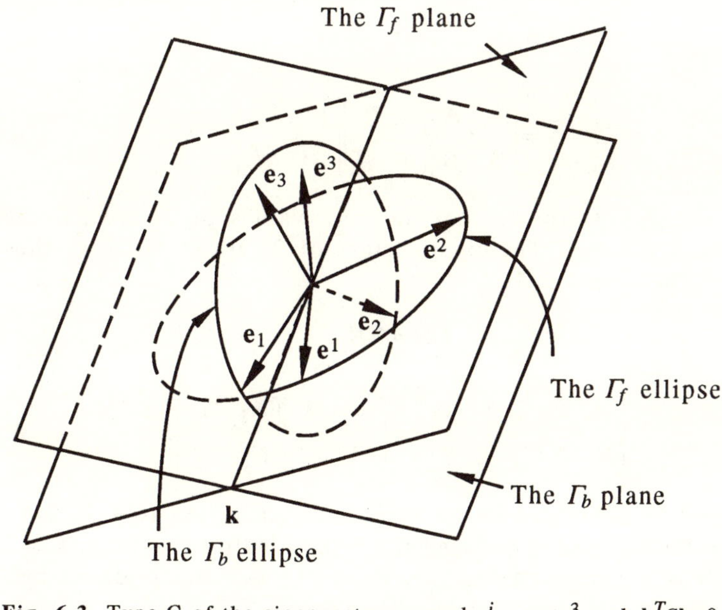

Fig. 6.3 Type C of the eigenvectors $\mathbf{e}_i$ and $\mathbf{e}^i$: $\mathbf{e}_3 \neq \mathbf{e}^3$ and $\mathbf{k}^T\mathbf{S}\mathbf{k} \neq 0$.

(III) **Type C**: $\mathbf{e}_3 \neq \mathbf{e}^3$, and $(\mathbf{e}_1, \mathbf{e}_2)$ and $(\mathbf{e}^1, \mathbf{e}^2)$ cannot be simultaneously the principal radii of their respective ellipses. In this case $\mathbf{S}^2$ is not symmetric and $\mathbf{k}$ determined from (6.7-25) does not satisfy (6.7-26). If $(\mathbf{e}_1, \mathbf{e}_2)$ are the principal radii of the Γ_b ellipse, $(\mathbf{e}^1, \mathbf{e}^2)$ cannot be the principal radii of the Γ_f ellipse, and vice versa. The geometry of Type C is shown in Fig. 6.3.

In summary, Type A, Type B, and Type C can be identified easily as follows. When $\mathbf{S}^2$ is symmetric, and in particular when $\mathbf{S}$ is skew-symmetric, we have Type A. The null vectors $\mathbf{e}_3$ and $\mathbf{e}^3$ are identical and is the axial vector of the skew-symmetric matrix $\mathbf{S}-\mathbf{S}^T$. When $\mathbf{S}^2$ is not symmetric, $\mathbf{e}_3 \neq \mathbf{e}^3$ and the Γ_b and Γ_f planes intersect along the line represented by the vector $\mathbf{k}$. The vector $\mathbf{k}$ is the axial vector of the skew-symmetric matrix $\mathbf{S}^2 - (\mathbf{S}^2)^T$. If $\mathbf{k}^T\mathbf{S}\mathbf{k}$ vanishes we have Type B. Otherwise, we have Type C.

6.8 Commutative 6×6 Matrices

When $\mathbf{N}$ is simple or semisimple, the diagonalization of $\mathbf{N}$ is, by (5.5-20),

$$\mathbf{N} = \begin{bmatrix} \mathbf{A} & \overline{\mathbf{A}} \\ \mathbf{B} & \overline{\mathbf{B}} \end{bmatrix} \begin{bmatrix} \mathbf{P} & \mathbf{0} \\ \mathbf{0} & \overline{\mathbf{P}} \end{bmatrix} \begin{bmatrix} \mathbf{B}^T & \mathbf{A}^T \\ \overline{\mathbf{B}}^T & \overline{\mathbf{A}}^T \end{bmatrix}. \tag{6.8-1}$$

On the other hand the diagonalization of the 6×6 matrix

$$\tilde{\mathbf{N}} = \begin{bmatrix} \mathbf{S} & \mathbf{H} \\ -\mathbf{L} & \mathbf{S}^T \end{bmatrix} \tag{6.8-2}$$

is, by (6.5-3),

$$\tilde{\mathbf{N}} = \begin{bmatrix} \mathbf{A} & \overline{\mathbf{A}} \\ \mathbf{B} & \overline{\mathbf{B}} \end{bmatrix} \begin{bmatrix} i\mathbf{I} & \mathbf{0} \\ \mathbf{0} & -i\mathbf{I} \end{bmatrix} \begin{bmatrix} \mathbf{B}^T & \mathbf{A}^T \\ \overline{\mathbf{B}}^T & \overline{\mathbf{A}}^T \end{bmatrix}. \tag{6.8-3}$$

Equation (6.8-3) shows that the eigenvalues of $\tilde{\mathbf{N}}$ are $\pm i$, each of multiplicity three. The eigenvectors of $\tilde{\mathbf{N}}$ are identical to those of $\mathbf{N}$. By Theorem 1.5–1,

$$\mathbf{N}\tilde{\mathbf{N}} = \tilde{\mathbf{N}}\mathbf{N}, \tag{6.8-4}$$

or

$$\begin{bmatrix} \mathbf{N}_1 & \mathbf{N}_2 \\ \mathbf{N}_3 & \mathbf{N}_1^{\mathrm{T}} \end{bmatrix} \begin{bmatrix} \mathbf{S} & \mathbf{H} \\ -\mathbf{L} & \mathbf{S}^{\mathrm{T}} \end{bmatrix} = \begin{bmatrix} \mathbf{S} & \mathbf{H} \\ -\mathbf{L} & \mathbf{S}^{\mathrm{T}} \end{bmatrix} \begin{bmatrix} \mathbf{N}_1 & \mathbf{N}_2 \\ \mathbf{N}_3 & \mathbf{N}_1^{\mathrm{T}} \end{bmatrix}. \tag{6.8-5}$$

Equation (6.8-4) can also be verified by inserting (6.8-1) and (6.8-3) into (6.8-4) and employing the orthogonality relations (5.5-13).

If $\mathbf{N}$ is nonsemisimple, (6.8-1) and (6.8-3) are replaced by (5.9-20) and (6.5-7), respectively. We have

$$\mathbf{N}=\begin{bmatrix}\mathbf{A} & \overline{\mathbf{A}}\\ \mathbf{B} & \overline{\mathbf{B}}\end{bmatrix}\begin{bmatrix}\mathbf{J} & \mathbf{0}\\ \mathbf{0} & \overline{\mathbf{J}}\end{bmatrix}\begin{bmatrix}\Gamma\mathbf{B}^T & \Gamma\mathbf{A}^T\\ \Gamma\overline{\mathbf{B}}^T & \Gamma\overline{\mathbf{A}}^T\end{bmatrix}. \tag{6.8-6}$$

$$\tilde{\mathbf{N}}=\begin{bmatrix}\mathbf{A} & \overline{\mathbf{A}}\\ \mathbf{B} & \overline{\mathbf{B}}\end{bmatrix}\begin{bmatrix}i\mathbf{I} & \mathbf{0}\\ \mathbf{0} & -i\mathbf{I}\end{bmatrix}\begin{bmatrix}\Gamma\mathbf{B}^T & \Gamma\mathbf{A}^T\\ \Gamma\overline{\mathbf{B}}^T & \Gamma\overline{\mathbf{A}}^T\end{bmatrix}. \tag{6.8-7}$$

A direct substitution of (6.8-6) and (6.8-7) shows that (6.8-4) remains valid by virtue of (5.9-15).

It should be noted that the **A**, **B** in (6.8-1) and (6.8-6) are different. All eigenvectors ξ_α are regular eigenvectors for the **N** in (6.8-1) while two of the eigenvectors, namely ξ_2 and ξ_5, are generalized eigenvectors for the **N** in (6.8-6). Regardless of whether **N** is simple or nonsemisimple, (6.8-3) and (6.8-7) tell us that $\tilde{\mathbf{N}}$ is *semisimple* (see also Section 6.10).

Carrying out the matrix multiplication in (6.8-5) yields

$$\begin{gathered}\mathbf{N}_1\mathbf{S}-\mathbf{N}_2\mathbf{L}=(\mathbf{N}_3\mathbf{H}+\mathbf{N}_1^T\mathbf{S}^T)^T,\\ \mathbf{N}_1\mathbf{H}+\mathbf{N}_2\mathbf{S}^T=(\mathbf{N}_1\mathbf{H}+\mathbf{N}_2\mathbf{S}^T)^T,\\ \mathbf{N}_3\mathbf{S}-\mathbf{N}_1^T\mathbf{L}=(\mathbf{N}_3\mathbf{S}-\mathbf{N}_1^T\mathbf{L})^T.\end{gathered} \tag{6.8-8}$$

The fourth equation is the transpose of the first equation. Equations $(6.8\text{-}8)_{2,3}$ tell us that

$$\mathbf{N}_1\mathbf{H}+\mathbf{N}_2\mathbf{S}^T \quad \text{and} \quad \mathbf{N}_3\mathbf{S}-\mathbf{N}_1^T\mathbf{L}$$

are symmetric. Without the commutative property it would be difficult to prove that these two matrices are symmetric.

The eigenrelation (6.2-2) shows that the eigenvectors of $\mathbf{N}^{-1}$ are also identical to those of **N** and $\tilde{\mathbf{N}}$. Therefore

$$\mathbf{N}^{-1}\tilde{\mathbf{N}}=\tilde{\mathbf{N}}\mathbf{N}^{-1} \tag{6.8-9}$$

and, using the notation (6.2-3) for the submatrices of $\mathbf{N}^{-1}$, we have

$$\begin{gathered}\mathbf{N}_1^{(-1)}\mathbf{S}-\mathbf{N}_2^{(-1)}\mathbf{L}=\left(\mathbf{N}_3^{(-1)}\mathbf{H}+(\mathbf{N}_1^{(-1)})^T\mathbf{S}^T\right)^T,\\ \mathbf{N}_1^{(-1)}\mathbf{H}+\mathbf{N}_2^{(-1)}\mathbf{S}^T=(\mathbf{N}_1^{(-1)}\mathbf{H}+\mathbf{N}_2^{(-1)}\mathbf{S}^T)^T,\\ \mathbf{N}_3^{(-1)}\mathbf{S}-(\mathbf{N}_1^{(-1)})^T\mathbf{L}=\left(\mathbf{N}_3^{(-1)}\mathbf{S}-(\mathbf{N}_1^{(-1)})^T\mathbf{L}\right)^T.\end{gathered} \tag{6.8-10}$$

We may generalize (6.8-4) and (6.8-9) as

$$\mathbf{N}^n\tilde{\mathbf{N}}=\tilde{\mathbf{N}}\mathbf{N}^n \tag{6.8-11}$$

where n is an integer which can be positive or negative. Let

$$\mathbf{N}^n = \begin{bmatrix} \mathbf{N}_1^{(n)} & \mathbf{N}_2^{(n)} \\ \mathbf{N}_3^{(n)} & (\mathbf{N}_1^{(n)})^T \end{bmatrix} \tag{6.8-12}$$

with

$$\mathbf{N}_1^{(n)},\ \mathbf{N}_2^{(n)},\ \mathbf{N}_3^{(n)} = \begin{cases} \mathbf{N}_1,\ \mathbf{N}_2,\ \mathbf{N}_3 & \text{when } n=1, \\ \mathbf{I},\ \ \mathbf{0},\ \ \mathbf{0} & \text{when } n=0. \end{cases} \tag{6.8-13}$$

We then have

$$\begin{aligned} &\mathbf{N}_1^{(n)}\mathbf{S} - \mathbf{N}_2^{(n)}\mathbf{L} = \left(\mathbf{N}_3^{(n)}\mathbf{H} + (\mathbf{N}_1^{(n)})^T\mathbf{S}^T\right)^T, \\ &\mathbf{N}_1^{(n)}\mathbf{H} + \mathbf{N}_2^{(n)}\mathbf{S}^T = (\mathbf{N}_1^{(n)}\mathbf{H} + \mathbf{N}_2^{(n)}\mathbf{S}^T)^T, \\ &\mathbf{N}_3^{(n)}\mathbf{S} - (\mathbf{N}_1^{(n)})^T\mathbf{L} = \left(\mathbf{N}_3^{(n)}\mathbf{S} - (\mathbf{N}_1^{(n)})^T\mathbf{L}\right)^T. \end{aligned} \tag{6.8-14}$$

When n=1 and −1, (6.8-14) reduce to (6.8-8) and (6.8-10), respectively. For n=0, (6.8-14) become trivial identities.

6.9 Identities Connecting p, A, and B to Real Matrices

The eigenrelation

$$\mathbf{N}\boldsymbol{\xi} = p\boldsymbol{\xi}$$

for $p=p_1, p_2, p_3$ can be written as

$$\mathbf{N}\begin{bmatrix} \mathbf{A} \\ \mathbf{B} \end{bmatrix} = \begin{bmatrix} \mathbf{AP} \\ \mathbf{BP} \end{bmatrix} \tag{6.9-1}$$

where

$$\mathbf{P} = \langle p_* \rangle = \mathrm{diag}[p_1,\ p_2,\ p_3].$$

When (6.9-1) is postmultiplied by the 3×6 matrix $[\mathbf{B}^T,\ \mathbf{A}^T]$ we have

$$\begin{bmatrix} \mathbf{N}_1 & \mathbf{N}_2 \\ \mathbf{N}_3 & \mathbf{N}_1^T \end{bmatrix}\begin{bmatrix} \mathbf{AB}^T & \mathbf{AA}^T \\ \mathbf{BB}^T & \mathbf{BA}^T \end{bmatrix} = \begin{bmatrix} \mathbf{APB}^T & \mathbf{APA}^T \\ \mathbf{BPB}^T & \mathbf{BPA}^T \end{bmatrix}.$$

On the other hand, we may write **S**, **H**, and **L** of (6.5-1) as

$$\begin{bmatrix} \mathbf{S} & \mathbf{H} \\ -\mathbf{L} & \mathbf{S}^T \end{bmatrix} = 2i\begin{bmatrix} \mathbf{AB}^T & \mathbf{AA}^T \\ \mathbf{BB}^T & \mathbf{BA}^T \end{bmatrix} - i\,\mathbf{I}. \tag{6.9-2}$$

Therefore

$$2\begin{bmatrix} \mathbf{APB}^T & \mathbf{APA}^T \\ \mathbf{BPB}^T & \mathbf{BPA}^T \end{bmatrix} = \begin{bmatrix} \mathbf{N}_1 & \mathbf{N}_2 \\ \mathbf{N}_3 & \mathbf{N}_1^T \end{bmatrix} - i\begin{bmatrix} \mathbf{N}_1 & \mathbf{N}_2 \\ \mathbf{N}_3 & \mathbf{N}_1^T \end{bmatrix}\begin{bmatrix} \mathbf{S} & \mathbf{H} \\ -\mathbf{L} & \mathbf{S}^T \end{bmatrix} \tag{6.9-3}$$

or

$$\begin{aligned} 2\mathbf{APB}^T &= \mathbf{N}_1 - i(\mathbf{N}_1\mathbf{S} - \mathbf{N}_2\mathbf{L}), \\ 2\mathbf{APA}^T &= \mathbf{N}_2 - i(\mathbf{N}_1\mathbf{H} + \mathbf{N}_2\mathbf{S}^T), \\ 2\mathbf{BPB}^T &= \mathbf{N}_3 - i(\mathbf{N}_3\mathbf{S} - \mathbf{N}_1^T\mathbf{L}), \\ 2\mathbf{BPA}^T &= \mathbf{N}_1^T - i(\mathbf{N}_3\mathbf{H} + \mathbf{N}_1^T\mathbf{S}^T). \end{aligned} \tag{6.9-4}$$

Identities (6.8-8) assure us the $2\mathbf{APA}^T$ and $2\mathbf{BPB}^T$ are indeed symmetric, and that $2\mathbf{APB}^T$ is the transpose of $2\mathbf{BPA}^T$.

Equations (6.9-4) provide an explicit expression for the real and imaginary parts of

$$\mathbf{APB}^T,\ \mathbf{APA}^T,\ \mathbf{BPB}^T,\ \mathbf{BPA}^T, \tag{6.9-5}$$

in terms of the real matrices **S**, **H**, **L**, and $\mathbf{N}_i$. If the Barnett–Lothe integral formalism (Section 7.6) is employed to calculate **S**, **H**, and **L**, the quantities in (6.9-5) can be determined without computing the eigenvalues p and the eigenvectors that form the columns of **A** and **B**. Moreover, even though the left sides of (6.9-4) are not valid for degenerate materials, the right sides do not depend on p, **A**, **B**, and hence remain valid regardless of whether the material is degenerate or not. Thus the problems associated with degenerate materials disappear upon using the identities (6.9-4).

In some applications we need the expressions for

$$\mathbf{APB}^{-1},\ \mathbf{APA}^{-1},\ \mathbf{BPB}^{-1},\ \mathbf{BPA}^{-1}. \tag{6.9-6}$$

Observing that

$$\begin{aligned} \mathbf{B}^{-1} &= \mathbf{B}^T(\mathbf{BB}^T)^{-1} = -2i\mathbf{B}^T\mathbf{L}^{-1}, \\ \mathbf{A}^{-1} &= \mathbf{A}^T(\mathbf{AA}^T)^{-1} = 2i\mathbf{A}^T\mathbf{H}^{-1}, \end{aligned}$$

we obtain from (6.9-4)

$$\begin{aligned} \mathbf{APB}^{-1} &= (\mathbf{N}_2 - \mathbf{N}_1\mathbf{SL}^{-1}) - i\mathbf{N}_1\mathbf{L}^{-1}, \\ \mathbf{APA}^{-1} &= (\mathbf{N}_1 + \mathbf{N}_2\mathbf{S}^T\mathbf{H}^{-1}) + i\mathbf{N}_2\mathbf{H}^{-1}, \\ \mathbf{BPB}^{-1} &= (\mathbf{N}_1^T - \mathbf{N}_3\mathbf{SL}^{-1}) - i\mathbf{N}_3\mathbf{L}^{-1}, \\ \mathbf{BPA}^{-1} &= (\mathbf{N}_3 + \mathbf{N}_1^T\mathbf{S}^T\mathbf{H}^{-1}) + i\mathbf{N}_1^T\mathbf{H}^{-1}. \end{aligned} \tag{6.9-7}$$

The eigenrelation (6.2-2) for $p=p_1, p_2, p_3$ are

$$\begin{bmatrix} \mathbf{N}_1^{(-1)} & \mathbf{N}_2^{(-1)} \\ \mathbf{N}_3^{(-1)} & (\mathbf{N}_1^{(-1)})^T \end{bmatrix} \begin{bmatrix} \mathbf{A} \\ \mathbf{B} \end{bmatrix} = \begin{bmatrix} \mathbf{AP}^{-1} \\ \mathbf{BP}^{-1} \end{bmatrix}. \tag{6.9-8}$$

Following the derivation of (6.9-4) and (6.9-7) we have

$$\begin{aligned} 2\mathbf{A}\mathbf{P}^{-1}\mathbf{B}^T &= \mathbf{N}_1^{(-1)} - i(\mathbf{N}_1^{(-1)}\mathbf{S} - \mathbf{N}_2^{(-1)}\mathbf{L}), \\ 2\mathbf{A}\mathbf{P}^{-1}\mathbf{A}^T &= \mathbf{N}_2^{(-1)} - i(\mathbf{N}_1^{(-1)}\mathbf{H} + \mathbf{N}_2^{(-1)}\mathbf{S}^T), \\ 2\mathbf{B}\mathbf{P}^{-1}\mathbf{B}^T &= \mathbf{N}_3^{(-1)} - i\big(\mathbf{N}_3^{(-1)}\mathbf{S} - (\mathbf{N}_1^{(-1)})^T\mathbf{L}\big), \\ 2\mathbf{B}\mathbf{P}^{-1}\mathbf{A}^T &= (\mathbf{N}_1^{(-1)})^T - i\big(\mathbf{N}_3^{(-1)}\mathbf{H} + (\mathbf{N}_1^{(-1)})^T\mathbf{S}^T\big), \end{aligned} \tag{6.9-9}$$

and

$$\begin{aligned} \mathbf{A}\mathbf{P}^{-1}\mathbf{B}^{-1} &= (\mathbf{N}_2^{(-1)} - \mathbf{N}_1^{(-1)}\mathbf{S}\mathbf{L}^{-1}) - i\mathbf{N}_1^{(-1)}\mathbf{L}^{-1}, \\ \mathbf{A}\mathbf{P}^{-1}\mathbf{A}^{-1} &= (\mathbf{N}_1^{(-1)} + \mathbf{N}_2^{(-1)}\mathbf{S}^T\mathbf{H}^{-1}) + i\mathbf{N}_2^{(-1)}\mathbf{H}^{-1}, \\ \mathbf{B}\mathbf{P}^{-1}\mathbf{B}^{-1} &= \big((\mathbf{N}_1^{(-1)})^T - \mathbf{N}_3^{(-1)}\mathbf{S}\mathbf{L}^{-1}\big) - i\mathbf{N}_3^{(-1)}\mathbf{L}^{-1}, \\ \mathbf{B}\mathbf{P}^{-1}\mathbf{A}^{-1} &= \big(\mathbf{N}_3^{(-1)} + (\mathbf{N}_1^{(-1)})^T\mathbf{S}^T\mathbf{H}^{-1}\big) + i(\mathbf{N}_1^{(-1)})^T\mathbf{H}^{-1}. \end{aligned} \tag{6.9-10}$$

More identities can be obtained from (6.9-7) and (6.9-10) by observing that

$$\begin{aligned} &(\mathbf{A}\mathbf{P}\mathbf{B}^{-1})(\mathbf{B}\mathbf{P}^{-1}\mathbf{A}^{-1}) = \mathbf{I}, \quad (\mathbf{A}\mathbf{P}\mathbf{B}^{-1})(\mathbf{B}\mathbf{P}^{-1}\mathbf{B}^{-1}) = -i\mathbf{M}^{-1}, \\ &(\mathbf{A}\mathbf{P}\mathbf{A}^{-1})(\mathbf{A}\mathbf{P}^{-1}\mathbf{A}^{-1}) = \mathbf{I}, \quad (\mathbf{A}\mathbf{P}\mathbf{A}^{-1})(\mathbf{A}\mathbf{P}^{-1}\mathbf{B}^{-1}) = -i\mathbf{M}^{-1}, \\ &(\mathbf{B}\mathbf{P}\mathbf{B}^{-1})(\mathbf{B}\mathbf{P}^{-1}\mathbf{B}^{-1}) = \mathbf{I}, \quad (\mathbf{B}\mathbf{P}\mathbf{B}^{-1})(\mathbf{B}\mathbf{P}^{-1}\mathbf{A}^{-1}) = i\mathbf{M}, \\ &(\mathbf{B}\mathbf{P}\mathbf{A}^{-1})(\mathbf{A}\mathbf{P}^{-1}\mathbf{B}^{-1}) = \mathbf{I}, \quad (\mathbf{B}\mathbf{P}\mathbf{A}^{-1})(\mathbf{A}\mathbf{P}^{-1}\mathbf{A}^{-1}) = i\mathbf{M}, \end{aligned} \tag{6.9-11}$$

where $\mathbf{M}$ and $\mathbf{M}^{-1}$ are defined in (5.6-6). Thus, for instance, the first equation in (6.9-11) gives

$$\big\{(\mathbf{N}_2 - \mathbf{N}_1\mathbf{S}\mathbf{L}^{-1}) - i\mathbf{N}_1\mathbf{L}^{-1}\big\}\big\{\big(\mathbf{N}_3^{(-1)} + (\mathbf{N}_1^{(-1)})^T\mathbf{S}^T\mathbf{H}^{-1}\big) + i(\mathbf{N}_1^{(-1)})^T\mathbf{H}^{-1}\big\} = \mathbf{I}. \tag{6.9-12}$$

This is equivalent to

$$\begin{aligned} &(\mathbf{N}_2 - \mathbf{N}_1\mathbf{S}\mathbf{L}^{-1})\big(\mathbf{N}_3^{(-1)} + (\mathbf{N}_1^{(-1)})^T\mathbf{S}^T\mathbf{H}^{-1}\big) + \mathbf{N}_1\mathbf{L}^{-1}(\mathbf{N}_1^{(-1)})^T\mathbf{H}^{-1} = \mathbf{I}. \\ &(\mathbf{N}_2 - \mathbf{N}_1\mathbf{S}\mathbf{L}^{-1})(\mathbf{N}_1^{(-1)})^T\mathbf{H}^{-1} - \mathbf{N}_1\mathbf{L}^{-1}\big(\mathbf{N}_3^{(-1)} + (\mathbf{N}_1^{(-1)})^T\mathbf{S}^T\mathbf{H}^{-1}\big) = \mathbf{0}. \end{aligned} \tag{6.9-13}$$

The identity obtained so far can be generalized by considering the eigenrelation

$$\mathbf{N}^n\xi = p^n\xi \tag{6.9-14}$$

where n is an integer which can be positive or negative. Making use of the notation $\mathbf{N}_i^{(n)}$ defined in (6.8-12) and (6.8-13) we obtain

$$
\begin{aligned}
2\mathbf{A}\mathbf{P}^n\mathbf{B}^T &= \mathbf{N}_1^{(n)} - i(\mathbf{N}_1^{(n)}\mathbf{S} - \mathbf{N}_2^{(n)}\mathbf{L}),\\
2\mathbf{A}\mathbf{P}^n\mathbf{A}^T &= \mathbf{N}_2^{(n)} - i(\mathbf{N}_1^{(n)}\mathbf{H} + \mathbf{N}_2^{(n)}\mathbf{S}^T),\\
2\mathbf{B}\mathbf{P}^n\mathbf{B}^T &= \mathbf{N}_3^{(n)} - i\left(\mathbf{N}_3^{(n)}\mathbf{S} - (\mathbf{N}_1^{(n)})^T\mathbf{L}\right),\\
2\mathbf{B}\mathbf{P}^n\mathbf{A}^T &= (\mathbf{N}_1^{(n)})^T - i\left(\mathbf{N}_3^{(n)}\mathbf{H} + (\mathbf{N}_1^{(n)})^T\mathbf{S}^T\right),
\end{aligned}
\tag{6.9-15}
$$

and

$$
\begin{aligned}
\mathbf{A}\mathbf{P}^n\mathbf{B}^{-1} &= (\mathbf{N}_2^{(n)} - \mathbf{N}_1^{(n)}\mathbf{S}\mathbf{L}^{-1}) - i\mathbf{N}_1^{(n)}\mathbf{L}^{-1},\\
\mathbf{A}\mathbf{P}^n\mathbf{A}^{-1} &= (\mathbf{N}_1^{(n)} + \mathbf{N}_2^{(n)}\mathbf{S}^T\mathbf{H}^{-1}) + i\mathbf{N}_2^{(n)}\mathbf{H}^{-1},\\
\mathbf{B}\mathbf{P}^n\mathbf{B}^{-1} &= \left((\mathbf{N}_1^{(n)})^T - \mathbf{N}_3^{(n)}\mathbf{S}\mathbf{L}^{-1}\right) - i\mathbf{N}_3^{(n)}\mathbf{L}^{-1},\\
\mathbf{B}\mathbf{P}^n\mathbf{A}^{-1} &= \left(\mathbf{N}_3^{(n)} + (\mathbf{N}_1^{(n)})^T\mathbf{S}^T\mathbf{H}^{-1}\right) + i(\mathbf{N}_1^{(n)})^T\mathbf{H}^{-1}.
\end{aligned}
\tag{6.9-16}
$$

They reduce to (6.9-4) and (6.9-7) when n=1, and to (6.9-9) and (6.9-10) when n=–1. At n=0, (6.9-15) recover (6.5-1), $(6.9\text{-}16)_{1,4}$ reduce to $(5.6\text{-}6)_{2,4}$, and $(6.9\text{-}16)_{2,3}$ become trivial identities.

Of course, the identities in (6.9-11) can also be generalized by replacing $\mathbf{P}$ and $\mathbf{P}^{-1}$ by $\mathbf{P}^n$ and $\mathbf{P}^{-n}$, respectively. Some of the identities presented here have been obtained by Nishoka and Lothe (1972a,b), Asaro et al. (1973), Bacon, et al. (1978), and Ting (1988a).

6.10 Eigenvectors of the Extraordinary Semisimple Matrix $\tilde{\mathbf{N}}$

We have shown in Section 6.5 that the eigenvalues of $\tilde{\mathbf{N}}$ are $\pm i$, each of multiplicity three, and that the eigenvectors of $\mathbf{N}$ are also the eigenvectors of $\tilde{\mathbf{N}}$. Since $\tilde{\mathbf{N}}$ has three independent eigenvectors associated with the three repeated eigenvalues, it is called *extraordinary semisimple*. The eigenvectors of $\tilde{\mathbf{N}}$ are not unique. We will show that the eigenvectors of $\tilde{\mathbf{N}}$ need not be the eigenvectors of $\mathbf{N}$.

For the eigenvalue i,

$$\tilde{\mathbf{N}}\xi = i\xi \tag{6.10-1a}$$

or

$$\begin{bmatrix} \mathbf{S} & \mathbf{H} \\ -\mathbf{L} & \mathbf{S}^T \end{bmatrix}\begin{bmatrix} \boldsymbol{a} \\ \boldsymbol{b} \end{bmatrix} = i\begin{bmatrix} \boldsymbol{a} \\ \boldsymbol{b} \end{bmatrix}. \tag{6.10-1b}$$

Written in full we have

$$\mathbf{S}\boldsymbol{a} + \mathbf{H}\boldsymbol{b} = i\boldsymbol{a}, \quad -\mathbf{L}\boldsymbol{a} + \mathbf{S}^T\boldsymbol{b} = i\boldsymbol{b}. \tag{6.10-2}$$

Solving for $\boldsymbol{b}$ and $\boldsymbol{a}$ from $(6.10\text{-}2)_{1,2}$, respectively, one has

$$b = i(\mathbf{H}^{-1}+i\mathbf{H}^{-1}\mathbf{S})a = i\mathbf{M}a,$$
$$a = -i(\mathbf{L}^{-1}-i\mathbf{S}\,\mathbf{L}^{-1})b = -i\mathbf{M}^{-1}b, \tag{6.10-3}$$

use being made of (5.6-6). The two equations are equivalent. If we let the vector a be arbitrary, b is given by $(6.10\text{-}3)_1$. If the vector b is arbitrary, a is determined from $(6.10\text{-}3)_2$. The orthogonality relations (5.5-12) can be written as

$$i\,a_\alpha^T(\mathbf{M} + \overline{\mathbf{M}})\,a_\beta = \delta_{\alpha\beta}. \tag{6.10-4}$$

By $(5.6\text{-}6)_2$ this is identical to

$$2i\,a_\alpha^T\mathbf{H}^{-1}\,a_\beta = \delta_{\alpha\beta}. \tag{6.10-5}$$

Equation (6.10-5) may be used to determined a_α (α=1,2,3). The vector a_1 can be chosen arbitrarily subjected to the normalization imposed by (6.10-5) for $\alpha=\beta=1$. The choice of a_2 is not as arbitrary as a_1 since it has to satisfy (6.10-5) for (α,β)=(2,1) and (2,2). The choice of a_3 is the least arbitrary. It has to satisfy (6.10-5) for (α,β)=(3,1), (3,2), and (3,3). The vectors b_α (α=1,2,3) are determined from $(6.10\text{-}3)_1$. It is clear that the eigenvectors so obtained may not be the eigenvectors of **N**.

The three eigenvectors associated with the eigenvalue $-i$ can be computed similarly. They need not be the complex conjugates of the three eigenvectors associated with the eigenvalue i.

A vector a that satisfies the relation

$$a^T\mathbf{H}^{-1}a = 0 \tag{6.10-6}$$

is an *unnormalizable vector*, or an isotropic vector (Section 1.1), because the normalization (6.10-5) for $\alpha=\beta$=1, 2, or 3 cannot be enforced. Since **H** is real and positive definite, a in (6.10-6) must be genuinely complex, i.e., the real and imaginary parts of a are independent vectors. If **H** is diagonalized as (Section 1.3)

$$\mathbf{H} = \mathbf{X}\langle\lambda_*\rangle\mathbf{X}^T$$

where the columns of the matrix **X** are the eigenvectors of **H** and the elements of the diagonal matrix $\langle\lambda_*\rangle$ are the eigenvalues of **H**, (6.10-6) is satisfied by

$$a = \mathbf{X}\langle\sqrt{\lambda_*}\rangle(\mathbf{n}+i\mathbf{m}) \tag{6.10-7}$$

where **n** and **m** are any two mutually orthogonal unit vectors. With a given by (6.10-7) and b by $(6.10\text{-}3)_1$, $\xi=(a,b)$ is an unnormalizable eigenvector for $\tilde{\mathbf{N}}$ (see also Section 12.11).

Chapter 7

TRANSFORMATION OF THE ELASTICITY MATRICES AND DUAL COORDINATE SYSTEMS

When the elasticity matrices are referred to a rotated coordinate system their elements change and assume different values. We will show in this chapter that, under rotations about the x_3-axis, the matrices **A** and **B** are tensors of rank one while **S**, **H**, **L**, and **M** are tensors of rank two. These properties are important in establishing certain invariants that are physically interesting and puzzling. We will also present the amazing Barnett–Lothe integral formalism that allows us to determine **S**, **H**, and **L** without computing the eigenvalues arpd eigenvectors of elastic constants. New tensors $\mathbf{N}_i(\theta)$ (i=1,2,3), $\mathbf{S}(\theta)$, $\mathbf{H}(\theta)$, $\mathbf{L}(\theta)$, and $\mathbf{G}_i(\theta)$ (i=1,3) are introduced, and their properties as well as identities relating them are presented. Also introduced is the idea of dual coordinate systems where the position of a point is referred to one coordinate system while the displacement components are referred to another coordinate system. These will be useful in applications. As in Chapter 6 readers may skip this chapter in the first reading. They can return to this chapter later for specific information.

7.1 Eigenvalues $p(\theta)$

In a fixed coordinate system x_i (i=1,2,3), the displacement **u** for two-dimensional deformations is (Section 5.1)

$$\mathbf{u} = \boldsymbol{a}f(z), \quad z = x_1 + px_2, \tag{7.1-1}$$

where p and the vector $\boldsymbol{a}$ are constants and f is an arbitrary function of z. Consider a new coordinate system x_i^*, Fig. 7.1, obtained by rotating x_i about the x_3-axis an angle θ, i.e.,

$$x_i^* = \Omega_{ij}x_j, \tag{7.1-2}$$

$$\Omega = \begin{bmatrix} \cos\theta & \sin\theta & 0 \\ -\sin\theta & \cos\theta & 0 \\ 0 & 0 & 1 \end{bmatrix}. \tag{7.1-3}$$

In the x_i^* coordinate system (7.1-1) becomes

$$\mathbf{u}^* = \boldsymbol{a}^* f^*(z^*), \qquad z^* = x_1^* + p^* x_2^*. \tag{7.1-4}$$

The displacement $\mathbf{u}$, being a tensor of rank one, has the relation

$$\mathbf{u}^* = \Omega\mathbf{u}. \tag{7.1-5}$$

Thus without loss in generality we may set

$$\boldsymbol{a}^* = \Omega\boldsymbol{a}, \tag{7.1-6}$$

$$f^*(z^*) = f(z). \tag{7.1-7}$$

From $(7.1\text{-}4)_2$ and (7.1-2),

$$\begin{aligned} z^* = x_1^* + p^* x_2^* &= (x_1\cos\theta + x_2\sin\theta) + p^*(-x_1\sin\theta + x_2\cos\theta) \\ &= (\cos\theta - p^*\sin\theta)\left(x_1 + \frac{\sin\theta + p^*\cos\theta}{\cos\theta - p^*\sin\theta}x_2\right). \end{aligned}$$

If the right side of (7.1-7) depends on $x_1 + px_2$, so does the left side of (7.1-7). Hence

$$\frac{\sin\theta + p^*\cos\theta}{\cos\theta - p^*\sin\theta} = p, \tag{7.1-8}$$

or

$$p^* = \frac{p\cos\theta - \sin\theta}{p\sin\theta + \cos\theta} = p(\theta), \quad \text{say}. \tag{7.1-9}$$

$p(\theta)$ is periodic in θ with periodicity π. In the sequel, the dependence of $p(\theta)$ on θ will be given explicitly unless θ=0.

Since p is complex $p(\theta)$ remains complex for all θ. If Imp>0,

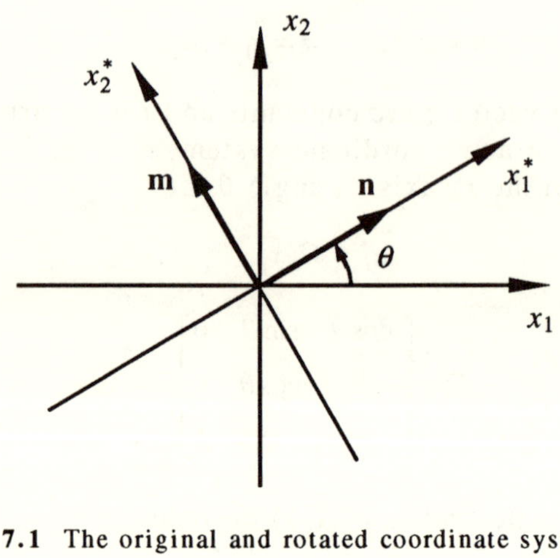

Fig. 7.1 The original and rotated coordinate systems.

$\mathrm{Im}p(\theta)>0$ for all θ. Otherwise $p(\theta)$ becomes real at the θ at which $\mathrm{Im}p(\theta)$ changes the sign. There are six eigenvalues p_α ($\alpha=1,2,...,6$) which consist of three pairs of complex conjugates. Therefore we also have three pairs of complex conjugates for $p_\alpha(\theta)$. As in (5.1-12) we let

$$\mathrm{Im}\, p_\alpha(\theta) > 0, \quad p_{\alpha+3}(\theta) = \bar{p}_\alpha(\theta), \quad (\alpha = 1,2,3). \tag{7.1-10}$$

It is clear that if $p_1=p_2$, $p_1(\theta)=p_2(\theta)$ for all θ. If $p_1 \neq p_2$, $p_1(\theta) \neq p_2(\theta)$ for all θ.

Let

$$p(\theta) = X + iY \tag{7.1-11}$$

where X, Y are real. We will show that, as θ varies from θ to π, $p(\theta)$ on the (X,Y) plane traces out a circle. The following discussions are for

$$\mathrm{Im}\, p_\alpha(\theta) > 0, \quad \text{i.e., } Y > 0.$$

First, we obtain from (7.1-9)

$$p(\theta)\bar{p}(\theta) + 1 = \frac{p\bar{p}+1}{(p\sin\theta+\cos\theta)(\bar{p}\sin\theta+\cos\theta)},$$

$$-i[p(\theta) - \bar{p}(\theta)] = \frac{-i(p-\bar{p})}{(p\sin\theta+\cos\theta)(\bar{p}\sin\theta+\cos\theta)},$$

so that

$$\gamma = \frac{p(\theta)\bar{p}(\theta)+1}{-i[p(\theta)-\bar{p}(\theta)]} = \frac{p\bar{p}+1}{-i[p-\bar{p}]} \geq 0 \tag{7.1-12}$$

is a positive constant independent of θ. It is an invariant. From

$$\rho^2 = \gamma^2 - 1 = \frac{(p^2+1)(\bar{p}^2+1)}{-(p-\bar{p})^2} \geq 0$$

we have

$$\gamma = \frac{p\bar{p}+1}{-i[p-\bar{p}]} \geq 1, \tag{7.1-13}$$

$$\rho = \sqrt{\gamma^2 - 1} = \frac{\sqrt{(p^2+1)(\bar{p}^2+1)}}{-i(p-\bar{p})} \geq 0. \tag{7.1-14}$$

Substitution of (7.1-11) into $(7.1\text{-}12)_1$ yields

$$\frac{X^2+Y^2+1}{2Y} = \gamma$$

or

$$X^2+(Y-\gamma)^2=\gamma^2-1=\rho^2. \tag{7.1-15}$$

This is an equation for a circle with radius ρ and with its center located at $(0,\gamma)$, Fig. 7.2. In view of $(7.1\text{-}15)_2$ the circle intersects orthogonally at points t and t' with a circle of unit radius whose center is located at the origin (Ting, 1982).

We next show that, as θ increases, $p(\theta)$ moves clockwise on the circle. To this end we write (7.1-9) as

$$\begin{aligned} p(\theta) &= \frac{(p\cos\theta-\sin\theta)(\overline{p}\sin\theta+\cos\theta)}{(p\sin\theta+\cos\theta)(\overline{p}\sin\theta+\cos\theta)} \\ &= \frac{(p\overline{p}-1)\sin 2\theta+(p+\overline{p})\cos 2\theta+(p-\overline{p})}{(p\overline{p}+1)-(p\overline{p}-1)\cos 2\theta+(p+\overline{p})\sin 2\theta}. \end{aligned} \tag{7.1-16}$$

From the relation

$$(p\overline{p}-1)^2+(p+\overline{p})^2=(p^2+1)(\overline{p}^2+1)$$

we define a real quantity ψ by

$$\cos 2\psi=\frac{p\overline{p}-1}{\sqrt{(p^2+1)(\overline{p}^2+1)}}, \qquad \sin 2\psi=\frac{-(p+\overline{p})}{\sqrt{(p^2+1)(\overline{p}^2+1)}}, \tag{7.1-17}$$

so that (7.1-16) is written as

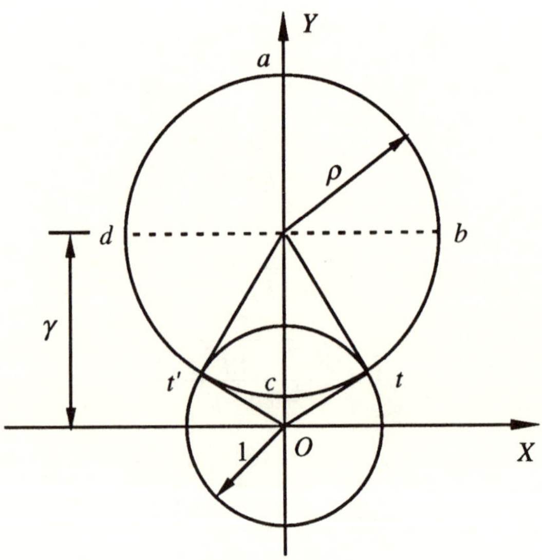

Fig. 7.2 $p(\theta)$=X+iY traces a circle clockwise as θ varies from 0 to π.

$$p(\theta) = \frac{\rho \sin 2(\theta - \psi) + i}{\gamma - \rho \cos 2(\theta - \psi)}. \tag{7.1-18}$$

It can be shown with (7.1-18) that the four extreme points a, b, c, and d on the circle in Fig. 7.2 correspond to

$$p(\theta) = \begin{cases} i\beta & \text{at } \theta = \psi, \\ \rho + i\gamma & \text{at } \theta = \psi + \frac{1}{2}\cot^{-1}\rho, \\ i\beta^{-1} & \text{at } \theta = \psi + \frac{1}{2}\pi, \\ -\rho + i\gamma & \text{at } \theta = \psi - \frac{1}{2}\cot^{-1}\rho, \end{cases} \tag{7.1-19}$$

where

$$\beta = \gamma + \rho = (\gamma - \rho)^{-1}.$$

Therefore $p(\theta)$ traces the circle clockwise once as θ increases from 0 to π. $p(\theta)$ becomes purely imaginary at

$$\theta = \psi, \quad \text{and} \quad \psi + \tfrac{1}{2}\pi,$$
$$\tan 2\psi = \frac{-(p + \bar{p})}{p\bar{p} - 1}. \tag{7.1-20}$$

For Imp<0 the circle $p(\theta)$ is located in the Y<0 plane. As θ varies from 0 to π, $p(\theta)$ traces out the circle counter-clockwise once.

From (7.1-15) circles associated with different γ do not intersect. Therefore if p_1, p_2, p_3 are distinct, there are in general three distinct circles for $p_1(\theta)$, $p_2(\theta)$, and $p_3(\theta)$. However if $p_1 \neq p_2$ but the γ computed from (7.1-13) have the same value, the circles for $p_1(\theta)$ and $p_2(\theta)$ are the same circle. For instance, when

$$\bar{p}_1 = -p_2 \quad \text{or} \quad \bar{p}_1 = p_2^{-1},$$

the circles for $p_1(\theta)$ and $p_2(\theta)$ are the same circle even though $p_1(\theta) \neq p_2(\theta)$ for all θ. For isotropic materials and for certain anisotropic materials for which $p_1 = p_2 = p_3 = i$, all three circles degenerate to a single point.

In closing this section we mention several equations related to $p(\theta)$. Let

$$p = \tan\theta_o \tag{7.1-21}$$

where θ_o is necessarily complex. By writing (7.1-9) as

$$p(\theta) = \frac{p - \tan\theta}{1 + p\tan\theta}$$

we have

$$p(\theta) = \tan(\theta_o - \theta). \tag{7.1-22}$$

Hence $p(\theta)$ satisfies the differential equation

$$\frac{d}{d\theta} p(\theta) = -1 - p^2(\theta) = \frac{-(1+p)^2}{(\cos\theta + p\sin\theta)^2}. \tag{7.1-23}$$

On the other hand, writing (7.1-9) as

$$p(\theta) = \frac{1}{(\cos\theta + p\sin\theta)} \frac{d}{d\theta}(\cos\theta + p\sin\theta) \tag{7.1-24}$$

we have (Chadwick and Smith, 1977)

$$\int_0^\theta p(\eta) d\eta = \ln(\cos\theta + p\sin\theta) \tag{7.1-25}$$

and, by specializing (3.4-8),

$$\int_0^\pi p(\theta) d\theta = \begin{cases} i\pi & \text{if} \quad \operatorname{Im} p > 0, \\ -i\pi & \text{if} \quad \operatorname{Im} p < 0. \end{cases} \tag{7.1-26}$$

Finally, a useful relation that can be deduced from (7.1-9) is

$$(\cos\theta + p\sin\theta)(\cos\theta - p(\theta)\sin\theta) = 1. \tag{7.1-27}$$

This suggests that if we write

$$z = x_1 + p x_2 = r(\cos\theta + p\sin\theta)$$

we have

$$z^{-1} = r^{-1}(\cos\theta - p(\theta)\sin\theta). \tag{7.1-28}$$

A generalization of (7.1-27) is

$$\zeta(\theta, \theta^-)\zeta(\theta^-, \theta_o) = \zeta(\theta, \theta_o) \tag{7.1-29}$$

where θ_o and θ^- are arbitrary constants and

$$\zeta(\theta, \theta_o) = \cos(\theta - \theta_o) + p(\theta_o)\sin(\theta - \theta_o). \tag{7.1-30}$$

In particular,

$$\zeta(\theta, \theta_o)\zeta(\theta_o, \theta) = 1 \tag{7.1-31}$$

which is a generalization of (7.1-27). Equation (7.1-31) can be written as

$$p(\theta) = \frac{p(\theta_o)\cos(\theta - \theta_o) - \sin(\theta - \theta_o)}{p(\theta_o)\sin(\theta - \theta_o) + \cos(\theta - \theta_o)} \tag{7.1-32}$$

which reduces to (7.1-9) when $\theta_o=0$.

7.2 Elasticity Matrices in a Rotated Coordinate System

The eigenvalue p and the eigenvectors $\boldsymbol{a}$, $\boldsymbol{b}$ in the x_i coordinate system are given by (5.1-8) and (5.3-2), namely,

$$\left\{\mathbf{Q}+p(\mathbf{R}+\mathbf{R}^T)+p^2\mathbf{T}\right\}\boldsymbol{a} = \mathbf{0}, \tag{7.2-1}$$

$$\boldsymbol{b} = (\mathbf{R}^T + p\mathbf{T})\boldsymbol{a} = -\frac{1}{p}(\mathbf{Q}+p\mathbf{R})\boldsymbol{a}, \tag{7.2-2}$$

where

$$Q_{ik} = C_{i1k1}, \quad R_{ik} = C_{i1k2}, \quad T_{ik} = C_{i2k2}. \tag{7.2-3}$$

In the rotated coordinate system defined by (7.1-2), they become

$$\left\{\mathbf{Q}^* + p^*\left(\mathbf{R}^* + (\mathbf{R}^*)^T\right) + (p^*)^2\mathbf{T}^*\right\}\boldsymbol{a}^* \tag{7.2-4}$$

$$\boldsymbol{b}^* = \left((\mathbf{R}^*)^T + p^*\mathbf{T}^*\right)\boldsymbol{a}^* = -\frac{1}{p^*}(\mathbf{Q}^* + p^*\mathbf{R}^*)\,\boldsymbol{a}^* \tag{7.2-5}$$

where

$$Q^*_{ik} = C^*_{i1k1}, \quad R^*_{ik} = C^*_{i1k2}, \quad T^*_{ik} = C^*_{i2k2}. \tag{7.2-6}$$

Since C_{ijks} is a tensor of rank four,

$$C^*_{ijks} = \Omega_{ip}\Omega_{jq}\Omega_{kr}\Omega_{st}C_{pqrt}. \tag{7.2-7}$$

Let $\mathbf{n}$, $\mathbf{m}$ be unit vectors along the positive x_1^*- and x_2^*-axes, respectively, Fig. 7.1. We have

$$\mathbf{n}^T = [\cos\theta, \sin\theta, 0], \qquad \mathbf{m}^T = [-\sin\theta, \cos\theta, 0], \tag{7.2-8}$$

and hence by (7.1-3)

$$\Omega_{1j} = n_j, \qquad \Omega_{2j} = m_j. \tag{7.2-9}$$

With (7.2-7) and (7.2-9), (7.2-6) can be written as

$$\begin{aligned} Q^*_{ik} &= \Omega_{ip}(C_{pqrt}n_q n_t)\Omega_{kp}, \\ R^*_{ik} &= \Omega_{ip}(C_{pqrt}n_q m_t)\Omega_{kp}, \\ T^*_{ik} &= \Omega_{ip}(C_{pqrt}m_q m_t)\Omega_{kp}, \end{aligned} \tag{7.2-10}$$

or, in matrix notation,

$$\mathbf{Q}^* = \Omega\mathbf{Q}(\theta)\Omega^T, \quad \mathbf{R}^* = \Omega\mathbf{R}(\theta)\Omega^T, \quad \mathbf{T}^* = \Omega\mathbf{T}(\theta)\Omega^T, \tag{7.2-11}$$

where

$$Q_{ik}(\theta) = C_{ijks}n_j n_s, \quad R_{ik}(\theta) = C_{ijks}n_j m_s, \quad T_{ik}(\theta) = C_{ijks}m_j m_s. \tag{7.2-12}$$

Equations (7.2-12) reduce to (7.2-3) when θ=0. The dependence of $\mathbf{Q}(\theta)$, $\mathbf{R}(\theta)$, $\mathbf{T}(\theta)$ on θ will be given explicitly except when θ=0. The dependence of $\mathbf{Q}^*$, $\mathbf{R}^*$, $\mathbf{T}^*$, and Ω on θ is understood. The matrices $\mathbf{Q}(\theta)$, $\mathbf{T}(\theta)$, $\mathbf{Q}^*$, and $\mathbf{T}^*$ are symmetric and positive definite if the strain energy is positive definite. When the summations over the indices j and s are carried out (7.2-12) yield

$$\begin{aligned}
\mathbf{Q}(\theta) &= \mathbf{Q}\cos^2\theta + (\mathbf{R}+\mathbf{R}^T)\sin\theta\cos\theta + \mathbf{T}\sin^2\theta,\\
\mathbf{R}(\theta) &= \mathbf{R}\cos^2\theta + (\mathbf{T}-\mathbf{Q})\sin\theta\cos\theta - \mathbf{R}^T\sin^2\theta,\\
\mathbf{T}(\theta) &= \mathbf{T}\cos^2\theta - (\mathbf{R}+\mathbf{R}^T)\sin\theta\cos\theta + \mathbf{Q}\sin^2\theta,
\end{aligned} \tag{7.2-13a}$$

which may be written in one equation as

$$\begin{bmatrix} \mathbf{Q}(\theta) & \mathbf{R}(\theta) \\ \mathbf{R}^T(\theta) & \mathbf{T}(\theta) \end{bmatrix} = \begin{bmatrix} \cos\theta\mathbf{I} & \sin\theta\mathbf{I} \\ -\sin\theta\mathbf{I} & \cos\theta\mathbf{I} \end{bmatrix} \begin{bmatrix} \mathbf{Q} & \mathbf{R} \\ \mathbf{R}^T & \mathbf{T} \end{bmatrix} \begin{bmatrix} \cos\theta\mathbf{I} & -\sin\theta\mathbf{I} \\ \sin\theta\mathbf{I} & \cos\theta\mathbf{I} \end{bmatrix}. \tag{7.2-13b}$$

It is clear that $\mathbf{Q}(\theta)$, $\mathbf{R}(\theta)$, $\mathbf{T}(\theta)$ are periodic in θ with periodicity π. Moreover,

$$\mathbf{Q}(\tfrac{\pi}{2}) = \mathbf{T}, \quad \mathbf{R}(\tfrac{\pi}{2}) = -\mathbf{R}^T, \quad \mathbf{T}(\tfrac{\pi}{2}) = \mathbf{Q}, \tag{7.2-14}$$

and

$$\mathbf{Q}(\theta) + \mathbf{T}(\theta) = \mathbf{Q} + \mathbf{T}, \quad \mathbf{R}(\theta) - \mathbf{R}^T(\theta) = \mathbf{R} - \mathbf{R}^T, \tag{7.2-15}$$

are invariants. Another invariant is

$$[\mathbf{Q}(\theta) - \mathbf{T}(\theta)]^2 + [\mathbf{R}(\theta) + \mathbf{R}^T(\theta)]^2 = (\mathbf{Q}-\mathbf{T})^2 + (\mathbf{R}+\mathbf{R}^T)^2. \tag{7.2-16}$$

Inserting (7.2-11) into (7.2-4) we obtain

$$\left\{\mathbf{Q}(\theta) + p^*\left(\mathbf{R}(\theta) + \mathbf{R}^T(\theta)\right) + (p^*)^2\mathbf{T}(\theta)\right\}\Omega^T\boldsymbol{a}^* = \mathbf{0}. \tag{7.2-17}$$

Substitution of (7.2-13a) into the above leads to

$$\left\{\mathbf{Q} + \left(\frac{\sin\theta + p^*\cos\theta}{\cos\theta - p^*\sin\theta}\right)(\mathbf{R}+\mathbf{R}^T) + \left(\frac{\sin\theta + p^*\cos\theta}{\cos\theta - p^*\sin\theta}\right)^2 \mathbf{T}\right\}\Omega^T\boldsymbol{a} = \mathbf{0}.$$

The fact that p^* cannot be real means that the factor $(\cos\theta - p^*\sin\theta)$ is nonzero. Comparison with (7.2-1) indicates that

$$p = \frac{\sin\theta + p^* \cos\theta}{\cos\theta - p^* \sin\theta}, \quad \boldsymbol{a} = \Omega^T \boldsymbol{a}^*,$$

or, solving for p^* and $\boldsymbol{a}^*$ and writing $p(\theta)$ for p^*,

$$p(\theta) = \frac{p\cos\theta - \sin\theta}{p\sin\theta + \cos\theta} = p^*, \tag{7.2-18}$$

$$\boldsymbol{a}^* = \Omega \boldsymbol{a}. \tag{7.2-19}$$

Equations (7.2-18), (7.2-19) were obtained in (7.1-9), (7.1-6) by a different derivation. Finally from $(7.2\text{-}5)_1$, (7.2-11), (7.2-13a), (7.2-18), and (7.2-19),

$$\boldsymbol{b}^* = \frac{1}{p\sin\theta + \cos\theta}\Omega\left\{(\mathbf{R}^T + p\mathbf{T})\cos\theta - (\mathbf{Q} + p\mathbf{R})\sin\theta\right\}\boldsymbol{a}$$

or, by $(7.2\text{-}2)_{1,2}$,

$$\boldsymbol{b}^* = \Omega \boldsymbol{b}. \tag{7.2-20}$$

Equations (7.2-19) and (7.2-20) tells us that $\boldsymbol{a}$ and $\boldsymbol{b}$ are tensors of rank one despite the fact that they are complex-valued. It should be emphasized that Ω here is limited to rotations about the x_3-axis, not any orthogonal transformation. Let $\mathbf{A}^*$, $\mathbf{B}^*$ be 3×3 matrices defined by

$$\mathbf{A}^* = \left[\boldsymbol{a}_1^*, \boldsymbol{a}_2^*, \boldsymbol{a}_3^*\right], \quad \mathbf{B}^* = \left[\boldsymbol{b}_1^*, \boldsymbol{b}_2^*, \boldsymbol{b}_3^*\right].$$

We then have

$$\mathbf{A}^* = \Omega\mathbf{A}, \quad \mathbf{B}^* = \Omega\mathbf{B}, \tag{7.2-21}$$

which are tensors of rank one. The three real matrices **S**, **H**, and **L** of (5.5-17) in the x_i^* coordinate system are

$$\mathbf{S}^* = \Omega\mathbf{S}\Omega^T, \quad \mathbf{H}^* = \Omega\mathbf{H}\Omega^T, \quad \mathbf{L}^* = \Omega\mathbf{L}\Omega^T, \tag{7.2-22}$$

implying that **S**, **H**, and **L** are tensors of rank two. Again, the orthogonal transformation Ω is limited to rotations about the x_3-axis. The matrix **M** defined in (5.6-6) is also a tensor of rank two.

It should be noted that the matrices **Q**, **R**, and **T** are not tensors. If **Q** were a tensor of rank two, it should satisfy the relation

$$\mathbf{Q}^* = \Omega\mathbf{Q}\Omega^T$$

which contradicts $(7.2\text{-}11)_1$. However, $\mathbf{Q}(\theta)$, $\mathbf{R}(\theta)$, and $\mathbf{T}(\theta)$ defined in (7.2-12) are tensors of rank two because C_{ijks}, n_i, and m_i appearing in (7.2-12) are tensors. In fact $\mathbf{Q}(\theta)$ is known as the *acoustic tensor*. The seemingly paradoxical result that $\mathbf{Q}(\theta)$ is a tensor for any θ while $\mathbf{Q}=\mathbf{Q}(0)$ is not a tensor becomes transparent when the transformation law for $\mathbf{Q}(\theta)$ is written explicitly as

$$\mathbf{Q}^*(\theta^*) = \Omega(\psi)\mathbf{Q}(\theta)\Omega^T(\psi), \quad \theta^* = \theta - \psi. \tag{7.2-23}$$

At θ=0 this reduces to

$$\mathbf{Q}^*(-\psi) = \Omega(\psi)\mathbf{Q}(0)\Omega^T(\psi)$$

indicating that $\mathbf{Q}(0)$ is not a tensor of rank two. Notice that (7.2-23) reduces to $(7.2\text{-}11)_1$ when $\psi=\theta$.

Certain components of $\mathbf{Q}(\theta)$, $\mathbf{R}(\theta)$, and $\mathbf{T}(\theta)$ are related. From the symmetry $C_{ijks} = C_{ijsk}$ we have

$$(C_{ijks}n_j n_s)m_k = (C_{ijsk}n_j m_k)n_s, \quad (C_{ijks}m_j n_s)m_k = (C_{ijsk}m_j m_k)n_s,$$

or

$$\mathbf{Q}(\theta)\mathbf{m} = \mathbf{R}(\theta)\mathbf{n}, \quad \mathbf{R}^T(\theta)\mathbf{m} = \mathbf{T}(\theta)\mathbf{n}. \tag{7.2-24}$$

They recover (6.1-3) when θ=0.

By writing (7.2-13b) as

$$\begin{bmatrix} \cos\theta\mathbf{I} & -\sin\theta\mathbf{I} \\ \sin\theta\mathbf{I} & \cos\theta\mathbf{I} \end{bmatrix} \begin{bmatrix} \mathbf{Q}(\theta) & \mathbf{R}(\theta) \\ \mathbf{R}^T(\theta) & \mathbf{T}(\theta) \end{bmatrix} = \begin{bmatrix} \mathbf{Q} & \mathbf{R} \\ \mathbf{R}^T & \mathbf{T} \end{bmatrix} \begin{bmatrix} \cos\theta\mathbf{I} & -\sin\theta\mathbf{I} \\ \sin\theta\mathbf{I} & \cos\theta\mathbf{I} \end{bmatrix}$$

we have

$$\left.\begin{aligned} \cos\theta\mathbf{Q}(\theta) - \sin\theta\mathbf{R}(\theta) &= \cos\theta\mathbf{Q} + \sin\theta\mathbf{R}^T, \\ \cos\theta\mathbf{R}(\theta) + \sin\theta\mathbf{Q}(\theta) &= \cos\theta\mathbf{R} + \sin\theta\mathbf{T}, \\ \cos\theta\mathbf{R}(\theta) - \sin\theta\mathbf{T}(\theta) &= \cos\theta\mathbf{R} - \sin\theta\mathbf{Q}, \\ \cos\theta\mathbf{T}(\theta) + \sin\theta\mathbf{R}(\theta) &= \cos\theta\mathbf{T} - \sin\theta\mathbf{R}^T. \end{aligned}\right. \tag{7.2-25}$$

The $\mathbf{Q}(\theta)$, $\mathbf{R}(\theta)$, $\mathbf{T}(\theta)$ appearing on the left depend on θ, while $\mathbf{Q}$, $\mathbf{R}$, $\mathbf{T}$ appearing on the right are independent on θ.

7.3 Sextic Formalism in Dual Coordinate Systems

The stress-strain laws and the equations of equilibrium in the x_i coordinate system are

$$\sigma_{ij} = C_{ijks}u_{k,s}, \tag{7.3-1}$$

$$C_{ijks}u_{k,s} = 0. \tag{7.3-2}$$

We study the sextic formalism in which the displacement u_i is referred to the x_i coordinate system while the independent variables are referred to a rotated coordinate system x_1^*, x_2^*. The motivation for using dual coordinate systems will be given in Section 7.10.

The derivations presented below follow closely those of Sections 5.1, 5.3, 5.5, and 5.7.

For two-dimensional deformations for which the displacement u_i depends on x_1, x_2 only (7.3-1) is a homogeneous differential equation in two independent variables x_1, x_2. A general solution for u_i is a function of one composite variable that is a linear combination of x_1 and x_2. With a rotation about the x_3-axis the new coordinate system x_i^* as given by (7.1-2) is linear in x_i. In particular, x_1^* and x_2^* are linearly related to x_1 and x_2. Hence u_i can be a function of a linear combination of x_1^* and x_2^*, and we write

$$\mathbf{u} = \boldsymbol{a} f(z^*), \tag{7.3-3}$$

$$z^* = x_1^* + p^* x_2^* = \mathbf{x}\cdot\mathbf{n} + p^*\mathbf{x}\cdot\mathbf{m}, \tag{7.3-4}$$

where $\mathbf{n}$ and $\mathbf{m}$ are unit vectors defined in (7.2-8), Fig. 7.1. Equation (7.3-3) is identical to (7.1-4) when (7.1-5) and (7.1-6) are employed and the asterisk $*$ on f is ignored. It should be pointed out that while p^*, $\mathbf{n}$, and $\mathbf{m}$ depend on θ, the vector $\boldsymbol{a}$ does not. Moreover the components of the vectors $\mathbf{u}$ and $\boldsymbol{a}$ are referred to the x_i, not the x_i^* coordinate system. Differentiation of (7.3-3) with x_s yields

$$u_{k,s} = (n_s + p^* m_s) a_k f'(z^*) \tag{7.3-5}$$

where the prime denotes differentiation with z^*. Differentiating once more with x_j we find that (7.3-2) is satisfied if

$$C_{ijks}(n_j + p^* m_j)(n_s + p^* m_s) a_k = 0$$

or, writing $p(\theta)$ for p^*,

$$\left\{\mathbf{Q}(\theta) + p(\theta)\left(\mathbf{R}(\theta) + \mathbf{R}^T(\theta)\right) + p^2(\theta)\mathbf{T}(\theta)\right\}\boldsymbol{a} = \mathbf{0}. \tag{7.3-6}$$

Thus the eigenvalues $p(\theta)$ are the roots of

$$\left|\mathbf{Q}(\theta) + p(\theta)\left(\mathbf{R}(\theta) + \mathbf{R}^T(\theta)\right) + p^2(\theta)\mathbf{T}(\theta)\right| = 0. \tag{7.3-7}$$

The stress obtained from (7.3-1) and (7.3-5) is

$$\sigma_{ij} = C_{ijks}(n_s + p^* m_s) a_k f'(z^*)$$

from which we have

$$\begin{aligned} \sigma_{ij} n_j &= \{Q_{ik}(\theta) + p(\theta) R_{ik}(\theta)\} a_k f'(z^*), \\ \sigma_{ij} m_j &= \{R_{ki}(\theta) + p(\theta) T_{ik}(\theta)\} a_k f'(z^*). \end{aligned} \tag{7.3-8}$$

If we define the stress function

$$\phi = \boldsymbol{b} f(z^*), \tag{7.3-9}$$

$$\boldsymbol{b} = \{\mathbf{R}^T(\theta) + p(\theta)\mathbf{T}(\theta)\}\boldsymbol{a} = -\frac{1}{p(\theta)}\{\mathbf{Q}(\theta) + p(\theta)\mathbf{R}(\theta)\}\boldsymbol{a}, \tag{7.3-10}$$

where the second equality follows from (7.3-6), (7.3-8) can be written as

$$\sigma_{ij}n_j = -\frac{\partial}{\partial x_2^*}\varphi_i, \qquad \sigma_{ij}m_j = \frac{\partial}{\partial x_1^*}\varphi_i. \tag{7.3-11}$$

These are an application of (5.4-2), and are a generalization of (5.3-4). The solutions $\mathbf{u}$ and ϕ indicated in (7.3-3) and (7.3-9) reduce to those in (5.1-4) and (5.3-3) when θ=0.

The two equations in (7.3-10) can be cast in the form

$$\begin{bmatrix} -\mathbf{Q}(\theta) & \mathbf{0} \\ -\mathbf{R}^T(\theta) & \mathbf{I} \end{bmatrix} \begin{bmatrix} \boldsymbol{a} \\ \boldsymbol{b} \end{bmatrix} = p(\theta) \begin{bmatrix} \mathbf{R}(\theta) & \mathbf{I} \\ \mathbf{T}(\theta) & \mathbf{0} \end{bmatrix} \begin{bmatrix} \boldsymbol{a} \\ \boldsymbol{b} \end{bmatrix} \tag{7.3-12}$$

Making use of the identity

$$\begin{bmatrix} \mathbf{0} & \mathbf{T}^{-1}(\theta) \\ \mathbf{I} & -\mathbf{R}(\theta)\mathbf{T}^{-1}(\theta) \end{bmatrix} \begin{bmatrix} \mathbf{R}(\theta) & \mathbf{I} \\ \mathbf{T}(\theta) & \mathbf{0} \end{bmatrix} = \begin{bmatrix} \mathbf{I} & \mathbf{0} \\ \mathbf{0} & \mathbf{I} \end{bmatrix} \tag{7.3-13}$$

which can be verified easily, (7.3-12) is written as

$$\mathbf{N}(\theta)\xi = p(\theta)\xi, \tag{7.3-14}$$

$$\mathbf{N}(\theta) = \begin{bmatrix} \mathbf{N}_1(\theta) & \mathbf{N}_2(\theta) \\ \mathbf{N}_3(\theta) & \mathbf{N}_1^T(\theta) \end{bmatrix}, \qquad \xi = \begin{bmatrix} \boldsymbol{a} \\ \boldsymbol{b} \end{bmatrix}, \tag{7.3-15}$$

$$\begin{gathered} \mathbf{N}_1(\theta) = -\mathbf{T}^{-1}(\theta)\mathbf{R}^T(\theta), \quad \mathbf{N}_2(\theta) = \mathbf{T}^{-1}(\theta), \\ \mathbf{N}_3(\theta) = \mathbf{R}(\theta)\mathbf{T}^{-1}(\theta)\mathbf{R}^T(\theta) - \mathbf{Q}(\theta). \end{gathered} \tag{7.3-16}$$

They reduce to (5.5-3), (5.5-4), and (5.5-5), respectively, when θ=0. The dependence of $\mathbf{N}(\theta)$ and $\mathbf{N}_i(\theta)$ (i=1,2,3) on θ will be given explicitly unless θ=0. They are periodic in θ with periodicity π. It is clear that $\mathbf{N}_2(\theta)$ and $\mathbf{N}_3(\theta)$ are symmetric, and that $\mathbf{N}_2(\theta)$ is positive definite. We will show in Section 7.5 that $-\mathbf{N}_3(\theta)$ is positive semidefinite. Equation (7.3-14) tells us that $p(\theta)$ is an eigenvalue of $\mathbf{N}(\theta)$ and that ξ, which is an eigenvector of $\mathbf{N}$, is also an eigenvector of $\mathbf{N}(\theta)$. Therefore the diagonalization of $\mathbf{N}(\theta)$ is identical to that of $\mathbf{N}$ in (5.5-20), namely,

$$\mathbf{N}(\theta)=\begin{bmatrix}\mathbf{A} & \overline{\mathbf{A}}\\ \mathbf{B} & \overline{\mathbf{B}}\end{bmatrix}\begin{bmatrix}\mathbf{P}(\theta) & \mathbf{0}\\ \mathbf{0} & \overline{\mathbf{P}}(\theta)\end{bmatrix}\begin{bmatrix}\mathbf{B}^T & \mathbf{A}^T\\ \overline{\mathbf{B}}^T & \overline{\mathbf{A}}^T\end{bmatrix} \tag{7.3-17}$$

where

$$\mathbf{P}(\theta)=\langle p_*(\theta)\rangle=\mathrm{diag}\big[p_1(\theta),\ p_2(\theta),\ p_3(\theta)\big].$$

We next derive the matrix differential equation in the x_i^* coordinate system for two-dimensional deformations of anisotropic elastic bodies. With the aid of the relation

$$\frac{\partial}{\partial x_s}=\Omega_{ps}\frac{\partial}{\partial x_p^*}=n_s\frac{\partial}{\partial x_1^*}+m_s\frac{\partial}{\partial x_2^*},$$

the stress-strain laws (7.3-1) takes the form

$$\sigma_{ij}=C_{ijks}n_s u_{k,1^*}+C_{ijks}m_s u_{k,2^*},$$

where a comma followed by 1* or 2* denotes differentiation with x_1^* and x_2^*, respectively. Equation (7.3-11) can then be written as

$$\begin{aligned}\mathbf{Q}(\theta)\mathbf{u}_{,1^*}+\mathbf{R}(\theta)\mathbf{u}_{,2^*}&=-\phi_{,2^*},\\ \mathbf{R}^T(\theta)\mathbf{u}_{,1^*}+\mathbf{T}(\theta)\mathbf{u}_{,2^*}&=\phi_{,1^*}\end{aligned} \tag{7.3-18a}$$

or

$$\begin{bmatrix}-\mathbf{Q}(\theta) & \mathbf{0}\\ -\mathbf{R}^T(\theta) & \mathbf{I}\end{bmatrix}\begin{bmatrix}\mathbf{u}_{,1^*}\\ \phi_{,1^*}\end{bmatrix}=\begin{bmatrix}\mathbf{R}(\theta) & \mathbf{I}\\ \mathbf{T}(\theta) & \mathbf{0}\end{bmatrix}\begin{bmatrix}\mathbf{u}_{,2^*}\\ \phi_{,2^*}\end{bmatrix}. \tag{7.3-18b}$$

This is similar to (7.3-12). With the aid of (7.3-13) we have

$$\mathbf{N}(\theta)\,\varpi_{,1^*}=\varpi_{,2^*},\quad \varpi=\begin{bmatrix}\mathbf{u}\\ \phi\end{bmatrix}, \tag{7.3-19}$$

or

$$\begin{aligned}\mathbf{u}_{,2^*}&=\mathbf{N}_1(\theta)\mathbf{u}_{,1^*}+\mathbf{N}_2(\theta)\,\phi_{,1^*},\\ \phi_{,2^*}&=\mathbf{N}_3(\theta)\mathbf{u}_{,1^*}+\mathbf{N}_1^T(\theta)\,\phi_{,1^*}.\end{aligned} \tag{7.3-20}$$

Equations (7.3-18)–(7.3-20) reduce to (5.7-3)–(5.7-5) when θ=0. The discussions that follow (5.7-5) apply to the rotated coordinate system. Equations (7.3-20) allow us to compute derivatives of **u** and ϕ in the x_2^* direction in terms of the derivatives in the x_1^* direction. It should be emphasized again that **u** and ϕ are referred to the x_i, not the x_i^* coordinate system.

As in (5.7-4) the derivation of (7.3-19) does not require differentiation of elastic constants C_{ijks}. Hence (7.9-19) applies to inhomogeneous materials for which C_{ijks} implicit in $\mathbf{N}(\theta)$ may depend on x_1^* and x_2^*.

If $x_2^*=0$ is a traction-free surface setting $\boldsymbol{\phi}_{,1^*}=\mathbf{0}$ in $(7.3\text{-}20)_2$ gives

$$\boldsymbol{\phi}_{,2^*}=\mathbf{N}_3(\theta)\mathbf{u}_{,1^*}$$

or, by $(7.3\text{-}11)_1$,

$$\mathbf{t}_n=-\mathbf{N}_3(\theta)\mathbf{u}_{,1^*},\qquad (\mathbf{t}_n)_i=\sigma_{ij}n_j. \tag{7.3-21}$$

$\mathbf{t}_n$ is the *hoop stress vector* on the surface whose normal is $\mathbf{n}$. The surface is perpendicular the surface $x_2^*=0$. The shear stress $\mathbf{t}_n\cdot\mathbf{m}$ on this surface must vanish. This is assured by the identity

$$\mathbf{N}_3(\theta)\mathbf{m}(\theta)=\mathbf{0}. \tag{7.3-22}$$

which will be proved in Section 7.5.

When $x_2^*=0$ is a rigidly clamped surface, setting $\mathbf{u}=\mathbf{0}$ in $(7.3\text{-}20)_2$ and using (7.3-11) yields the hoop stress vector

$$\mathbf{t}_n=-\mathbf{N}_1^T(\theta)\mathbf{t}_m,\qquad (\mathbf{t}_m)_i=\sigma_{ij}m_j, \tag{7.3-23}$$

where $\mathbf{t}_m$ is the traction vector on the surface $x_2^*=0$. If $x_2^*\leq 0$ is a rigid body the displacement $\mathbf{u}$ in $x_2^*\leq 0$ can be a rigid body translation $\mathbf{u}_o$ and a rotation about the x_3-axis, i.e.,

$$\mathbf{u}=\mathbf{u}_o+k\mathbf{i}_3\times\mathbf{x} \tag{7.3-24}$$

where k is a constant and $\mathbf{i}_3$ is the unit vector along the x_3-axis. In this case the displacement $\mathbf{u}$ at $x_2^*=0$ need not vanish. However, from (7.3-24)

$$\mathbf{u}_{,1^*}=k\mathbf{i}_3\times\mathbf{n}=k\mathbf{m}$$

and, by virtue of (7.3-22),

$$\mathbf{N}_3(\theta)\mathbf{u}_{,1^*}=\mathbf{0}. \tag{7.3-25}$$

Thus $(7.3\text{-}23)_1$ remains valid if $x_2^*\leq 0$ is a rigid body which is allowed to move so that the displacement $\mathbf{u}$ at $x_2^*=0$ is nonzero.

In applications we may use the cylindrical coordinate system (r,θ,x_3) where

$$x_1=r\cos\theta,\quad x_2=r\sin\theta.$$

Equation (7.3-19) is written as

$$\mathbf{N}(\theta)\frac{\partial}{\partial r}\boldsymbol{\varpi}=\frac{\partial}{r\partial\theta}\boldsymbol{\varpi},\quad \boldsymbol{\varpi}=\begin{bmatrix}\mathbf{u}\\ \boldsymbol{\phi}\end{bmatrix}. \tag{7.3-26}$$

This is not exactly a differential equation in the cylindrical coordi nate system because the components of $\mathbf{u}$ and $\boldsymbol{\phi}$ are referred to the x_i coordinate system. If u_r, u_θ, u_3 are the displacement components in the cylindrical coordinates,

$$u_r = \mathbf{n}\cdot\mathbf{u}, \quad u_\theta = \mathbf{m}\cdot\mathbf{u}, \quad u_3 = (\mathbf{u})_3.$$

Again we point out that (7.3-26) remains valid for inhomogeneous materials in which the elastic constants C_{ijks} implicit in $\mathbf{N}(\theta)$ may depend on r and θ. By a different approach Kirchner (1989) has shown that C_{ijks} may depend on θ.

Let $\mathbf{t}_r$, $\mathbf{t}_\theta$, be, respectively, the traction vectors on a cylindrical surface r=constant and on a radial plane θ=constant. Their components are

$$(\mathbf{t}_r)_i = \sigma_{ij}n_j, \quad (\mathbf{t}_\theta)_i = \sigma_{ij}m_j \tag{7.3-27}$$

or, by (7.3-11),

$$\mathbf{t}_r = -\frac{\partial}{r\partial\theta}\boldsymbol{\phi}, \quad \mathbf{t}_\theta = \frac{\partial}{\partial r}\boldsymbol{\phi}. \tag{7.3-28}$$

The cylindrical components of the stress are

$$\begin{aligned} &\sigma_{rr} = \mathbf{n}\cdot\mathbf{t}_r, \quad \sigma_{r\theta} = \mathbf{m}\cdot\mathbf{t}_r = \mathbf{n}\cdot\mathbf{t}_\theta, \\ &\sigma_{\theta\theta} = \mathbf{m}\cdot\mathbf{t}_\theta, \quad \sigma_{r3} = (\mathbf{t}_r)_3, \quad \sigma_{\theta 3} = (\mathbf{t}_\theta)_3. \end{aligned} \tag{7.3-29}$$

The stress component σ_{33} is determined from the condition that ε_{33} vanishes.

7.4 Nonsemisimple $\mathbf{N}(\theta)$

The analysis presented in the previous section indicated that the $\boldsymbol{\xi}$ is an eigenvector of $\mathbf{N}$ as well as $\mathbf{N}(\theta)$. Therefore the eigenvectors of $\mathbf{N}(\theta)$ are independent of θ (Barnett and Lothe, 1973; Chadwick and Smith, 1977). When $\mathbf{N}(\theta)$ is nonsemisimple, say $p_1(\theta)=p_2(\theta)$, and there exists only one eigenvector $\boldsymbol{\xi}_1$ for $p_1(\theta) = p_2(\theta)$, the second eigenvector is provided by the generalized eigenvector $\boldsymbol{\xi}_2(\theta)$ which must be presumed to vary with θ. In fact the normalized eigenvector $\boldsymbol{\xi}_1$ may also vary with θ. However, the unnormalized $\boldsymbol{\xi}_1$ can be taken as independent of θ and we have

$$\begin{aligned} &\mathbf{N}(\theta)\boldsymbol{\xi}_1 = p_1(\theta)\boldsymbol{\xi}_1, \\ &\mathbf{N}(\theta)\boldsymbol{\xi}_2(\theta) = p_1(\theta)\boldsymbol{\xi}_2(\theta) + \boldsymbol{\xi}_1. \end{aligned} \tag{7.4-1}$$

Differentiating $(7.4\text{-}1)_2$ with θ and employing $(7.1\text{-}23)_1$ and (7.5-2) to be proved in Section 7.5, we obtain

$$\mathbf{N}(\theta)\boldsymbol{\xi}'_2(\theta) = p_1(\theta)\boldsymbol{\xi}'_2(\theta) + 2p_1(\theta)\boldsymbol{\xi}_1, \tag{7.4-2}$$

where the prime denotes differentiation with θ. With reference to $(7.4\text{-}1)_2$ we conclude that

$$\xi'_2(\theta)=2p_1(\theta)\xi_2(\theta)+\beta(\theta)\xi_1, \qquad (7.4\text{-}3)$$

where $\beta(\theta)$ is an arbitrary function of θ. The solution of (7.4-3) subject to the condition $\xi_2(0)=\xi_2$ is

$$\xi_2(\theta)=g^2(\theta)\xi_2+h(\theta)\xi_1, \qquad (7.4\text{-}4)$$

where

$$\begin{aligned} g(\theta) &= e^{\int_0^\theta p_1(\lambda)d\lambda} = e^{\ln(\cos\theta+p_1\sin\theta)} = (\cos\theta+p_1\sin\theta), \\ h(\theta) &= g^2(\theta)\int_0^\theta g^{-2}(\lambda)\beta(\lambda)d\lambda, \end{aligned} \qquad (7.4\text{-}5)$$

appeal being made to (7.1-25). $h(\theta)$ is a disposable function because it depends on the arbitrary function $\beta(\theta)$. These results are due to Lothe and Barnett (1976a).

It should be noted that the solutions of (7.4-1) for ξ_1 and $\xi_2(\theta)$ are not unique. We will normalize ξ_1 and $\xi_2(\theta)$ so that the orthogonality relations between the right and left eigenvectors are maintained. The following derivation follows closely that of Section 5.9. The left eigenvectors $\eta_1(\theta)$ and η_2 are obtained from (5.9-4) as

$$\eta_1(\theta)=\hat{\mathbf{I}}\xi_2(\theta), \qquad \eta_2=\hat{\mathbf{I}}\xi_1. \qquad (7.4\text{-}6)$$

Assuming that, at $\theta=0$, ξ_1, ξ_2, η_1, and η_2 have been normalized to satisfy the orthogonality relations

$$\eta_\alpha\cdot\xi_\beta=\delta_{\alpha\beta}, \qquad (7.4\text{-}7)$$

the normalized $\xi_1(\theta)$, $\xi_2(\theta)$, $\eta_1(\theta)$, and $\eta_2(\theta)$, can be shown to be

$$\begin{aligned} \xi_1(\theta) &= g^{-1}(\theta)\xi_1, \quad \xi_2(\theta) = g(\theta)\xi_2, \\ \eta_1(\theta) &= g(\theta)\eta_1, \quad \eta_2(\theta) = g^{-1}(\theta)\eta_2. \end{aligned} \qquad (7.4\text{-}8)$$

The eigenvectors $\xi_3(\theta)$ and $\eta_3(\theta)$ associated with $p_3(\theta)$ are independent of θ. The eigenrelation for nonsemisimple $\mathbf{N}(\theta)$ for $p_1(\theta)=p_2(\theta)$ and $p_3(\theta)$ can now be combined into one equation as

$$\mathbf{N}(\theta)[\xi_1(\theta), \xi_2(\theta), \xi_3(\theta)] = [\xi_1(\theta), \xi_2(\theta), \xi_3(\theta)]\mathbf{J}(\theta) \qquad (7.4\text{-}9)$$

where

$$\mathbf{J}(\theta)=\begin{bmatrix} p_1(\theta) & 1 & 0 \\ 0 & p_1(\theta) & 0 \\ 0 & 0 & p_3(\theta) \end{bmatrix} \qquad (7.4\text{-}10)$$

is a Jordan canonical form. By (7.4-8),

$$[\xi_1(\theta),\ \xi_2(\theta),\ \xi_3(\theta)] = [\xi_1,\ \xi_2,\ \xi_3]\mathbf{K}(\theta) = \begin{bmatrix} \mathbf{AK}(\theta) \\ \mathbf{BK}(\theta) \end{bmatrix},$$

$$[\eta_1(\theta),\ \eta_2(\theta),\ \eta_3(\theta)] = [\eta_1,\ \eta_2,\ \eta_3]\,\mathbf{K}^{-1}(\theta) = \begin{bmatrix} \mathbf{B\Gamma K}^{-1}(\theta) \\ \mathbf{A\Gamma K}^{-1}(\theta) \end{bmatrix}, \tag{7.4-11}$$

where use being made of (5.9-14) and

$$\mathbf{K}(\theta) = \mathrm{diag}[g^{-1}(\theta),\ g(\theta),\ 1]. \tag{7.4-12}$$

The orthogonality relations (7.4-7) for all six eigenvectors can be written as

$$\begin{bmatrix} \mathbf{K}^{-1}(\theta)\Gamma\mathbf{B}^T & \mathbf{K}^{-1}(\theta)\Gamma\mathbf{A}^T \\ \overline{\mathbf{K}}^{-1}(\theta)\Gamma\overline{\mathbf{B}}^T & \overline{\mathbf{K}}^{-1}(\theta)\Gamma\overline{\mathbf{A}}^T \end{bmatrix} \begin{bmatrix} \mathbf{AK}(\theta) & \overline{\mathbf{AK}}(\theta) \\ \mathbf{BK}(\theta) & \overline{\mathbf{BK}}(\theta) \end{bmatrix} = \mathbf{I}. \tag{7.4-13}$$

This is equivalent to

$$\begin{bmatrix} \mathbf{K}^{-1}(\theta) & \mathbf{0} \\ \mathbf{0} & \overline{\mathbf{K}}^{-1}(\theta) \end{bmatrix} \begin{bmatrix} \Gamma\mathbf{B}^T & \Gamma\mathbf{A}^T \\ \Gamma\overline{\mathbf{B}}^T & \Gamma\overline{\mathbf{A}}^T \end{bmatrix} \begin{bmatrix} \mathbf{A} & \overline{\mathbf{A}} \\ \mathbf{B} & \overline{\mathbf{B}} \end{bmatrix} \begin{bmatrix} \mathbf{K}(\theta) & \mathbf{0} \\ \mathbf{0} & \overline{\mathbf{K}}(\theta) \end{bmatrix} = \mathbf{I},$$

which is a trivial identity in view of (5.9-15).

With $(7.4\text{-}11)_1$, (7.4-9) and its complex conjugate are

$$\mathbf{N}(\theta) \begin{bmatrix} \mathbf{AK}(\theta) & \overline{\mathbf{AK}}(\theta) \\ \mathbf{BK}(\theta) & \overline{\mathbf{BK}}(\theta) \end{bmatrix} = \begin{bmatrix} \mathbf{AK}(\theta) & \overline{\mathbf{AK}}(\theta) \\ \mathbf{BK}(\theta) & \overline{\mathbf{BK}}(\theta) \end{bmatrix} \begin{bmatrix} \mathbf{J}(\theta) & \mathbf{0} \\ \mathbf{0} & \overline{\mathbf{J}}(\theta) \end{bmatrix}. \tag{7.4-14}$$

Noting that the two 6×6 matrices on the left of (7.4-13) are the inverses of each other and that their products can be interchanged, we have

$$\mathbf{N}(\theta) = \begin{bmatrix} \mathbf{AK}(\theta) & \overline{\mathbf{AK}}(\theta) \\ \mathbf{BK}(\theta) & \overline{\mathbf{BK}}(\theta) \end{bmatrix} \begin{bmatrix} \mathbf{J}(\theta) & \mathbf{0} \\ \mathbf{0} & \overline{\mathbf{J}}(\theta) \end{bmatrix} \begin{bmatrix} \mathbf{K}^{-1}(\theta)\Gamma\mathbf{B}^T & \mathbf{K}^{-1}(\theta)\Gamma\mathbf{A}^T \\ \overline{\mathbf{K}}^{-1}(\theta)\Gamma\overline{\mathbf{B}}^T & \overline{\mathbf{K}}^{-1}(\theta)\Gamma\overline{\mathbf{A}}^T \end{bmatrix}. \tag{7.4-15}$$

This is the diagonalization of a nonsemisimple $\mathbf{N}(\theta)$. At $\theta=0$, $\mathbf{K}(0) = \mathbf{I}$ and (7.4-15) reduces to (5.9-20). It is clear that if $\mathbf{N}(\theta)$ is nonsemisimple at $\theta=0$, it remains so for all θ.

We close this section by noting from $(7.4\text{-}5)_1$ and (7.1-23) that

$$g^{-2}(\theta) = -(1+p_1)^{-2}p_1'(\theta).$$

Since $p_1(0)=p_1(\pi)$,

$$\int_0^{\pi} g^{-2}(\theta)d\theta = 0. \tag{7.4-16}$$

7.5 Properties of $\mathbf{N}(\theta)$ and $\mathbf{N}_i(\theta)$

We first show that (Chadwick and Smith, 1977)

$$\mathbf{N}(\theta)\mathbf{N}(\theta+\tfrac{\pi}{2}) = -\mathbf{I} = \mathbf{N}(\theta+\tfrac{\pi}{2})\mathbf{N}(\theta). \tag{7.5-1}$$

From the relations

$$\mathbf{n}(\theta+\tfrac{\pi}{2}) = \mathbf{m}(\theta), \quad \mathbf{m}(\theta+\tfrac{\pi}{2}) = -\mathbf{n}(\theta),$$

it is easily verified that

$$\mathbf{Q}(\theta+\tfrac{\pi}{2}) = \mathbf{T}(\theta), \quad \mathbf{R}(\theta+\tfrac{\pi}{2}) = -\mathbf{R}^T(\theta), \quad \mathbf{T}(\theta+\tfrac{\pi}{2}) = \mathbf{Q}(\theta).$$

Writing $\mathbf{N}(\theta)$ as

$$\mathbf{N}(\theta) = \begin{bmatrix} \mathbf{0} & \mathbf{T}^{-1}(\theta) \\ \mathbf{I} & -\mathbf{R}(\theta)\mathbf{T}^{-1}(\theta) \end{bmatrix} \begin{bmatrix} -\mathbf{Q}(\theta) & \mathbf{0} \\ -\mathbf{R}^T(\theta) & \mathbf{I} \end{bmatrix}$$

we have

$$\mathbf{N}(\theta+\tfrac{\pi}{2}) = \begin{bmatrix} \mathbf{0} & \mathbf{Q}^{-1}(\theta) \\ \mathbf{I} & \mathbf{R}^T(\theta)\mathbf{Q}^{-1}(\theta) \end{bmatrix} \begin{bmatrix} -\mathbf{T}(\theta) & \mathbf{0} \\ \mathbf{R}(\theta) & \mathbf{I} \end{bmatrix}$$

Equation (7.5-1) is established by carrying out the matrix multiplication of $\mathbf{N}(\theta)$ by $\mathbf{N}(\theta+\tfrac{\pi}{2})$.

We next show that $\mathbf{N}(\theta)$ satisfies the differential equation (Barnett and Lothe, 1973)

$$\mathbf{N}'(\theta) = -\mathbf{I} - \mathbf{N}^2(\theta) \tag{7.5-2}$$

where the prime denotes differentiation with θ. Equation (7.5-2) is equivalent to

$$\begin{aligned} -\mathbf{N}_1'(\theta) &= \mathbf{I} + \mathbf{N}_1^2(\theta) + \mathbf{N}_2(\theta)\mathbf{N}_3(\theta), \\ -\mathbf{N}_2'(\theta) &= \mathbf{N}_1(\theta)\mathbf{N}_2(\theta) + \mathbf{N}_2(\theta)\mathbf{N}_1^T(\theta), \\ -\mathbf{N}_3'(\theta) &= \mathbf{N}_3(\theta)\mathbf{N}_1(\theta) + \mathbf{N}_1^T(\theta)\mathbf{N}_3(\theta). \end{aligned} \tag{7.5-3}$$

Noting that

$$\mathbf{n}' = \mathbf{m}, \quad \mathbf{m}' = -\mathbf{n}, \tag{7.5-4}$$

differentiation of $\mathbf{Q}(\theta)$, $\mathbf{R}(\theta)$, $\mathbf{T}(\theta)$ defined in (7.2-12) yields

$$\begin{aligned} \mathbf{Q}'(\theta) &= \mathbf{R}(\theta) + \mathbf{R}^T(\theta), \\ \mathbf{R}'(\theta) &= \mathbf{T}(\theta) - \mathbf{Q}(\theta), \\ \mathbf{T}'(\theta) &= -\mathbf{R}(\theta) - \mathbf{R}^T(\theta). \end{aligned} \tag{7.5-5}$$

It is interesting to note that $\mathbf{R}'(\theta)$ is symmetric. When (7.3-16)$_{2,1,3}$, in that order, are written as

$$\mathbf{N}_2(\theta)\mathbf{T}(\theta) = \mathbf{I}, \quad \mathbf{N}_1(\theta) = -\mathbf{N}_2(\theta)\mathbf{R}^T(\theta), \quad \mathbf{N}_3(\theta) = -\mathbf{R}(\theta)\mathbf{N}_1(\theta) - \mathbf{Q}(\theta),$$

a direct differentiation with θ and with the aid of (7.5-5) verifies (7.5-3).

From (7.3-16)$_{1,3}$ and (7.2-24) we have

$$\mathbf{N}_1(\theta)\mathbf{m} = -\mathbf{T}^{-1}(\theta)\mathbf{R}^T(\theta)\mathbf{m} = -\mathbf{T}^{-1}(\theta)\mathbf{T}(\theta)\mathbf{n},$$

$$\begin{aligned} \mathbf{N}_3(\theta)\mathbf{m} &= \mathbf{R}(\theta)\mathbf{T}^{-1}(\theta)\mathbf{R}^T(\theta)\mathbf{m} - \mathbf{Q}(\theta)\mathbf{m} \\ &= \mathbf{R}(\theta)\mathbf{T}^{-1}(\theta)\mathbf{T}(\theta)\mathbf{n} - \mathbf{R}(\theta)\mathbf{n}, \end{aligned}$$

so that

$$\mathbf{N}_1(\theta)\mathbf{m}(\theta) = -\mathbf{n}(\theta), \qquad \mathbf{N}_3(\theta)\mathbf{m}(\theta) = \mathbf{0}. \tag{7.5-6}$$

The dependence of $\mathbf{n}$ and $\mathbf{m}$ on θ is now shown explicitly. When θ=0 (7.5-6)$_{1,2}$ confirm the structure of $\mathbf{N}_1$ and $\mathbf{N}_3$ given in (6.1-5) and (6.1-6). It can be proved that the identities

$$\begin{aligned} \cos(\theta - \theta_o)\mathbf{n}(\theta) - \sin(\theta - \theta_o)\mathbf{m}(\theta) &= \mathbf{n}(\theta_o), \\ \sin(\theta - \theta_o)\mathbf{n}(\theta) + \cos(\theta - \theta_o)\mathbf{m}(\theta) &= \mathbf{m}(\theta_o), \end{aligned} \tag{7.5-7}$$

hold for any θ and θ_o. With appeal to (7.5-6) we obtain (Hwu and Ting, 1989)

$$\begin{aligned} \cos(\theta - \theta_o)\mathbf{N}_1(\theta)\mathbf{n}(\theta) &= \mathbf{N}_1(\theta)\mathbf{n}(\theta_o) - \sin(\theta - \theta_o)\mathbf{n}(\theta), \\ \sin(\theta - \theta_o)\mathbf{N}_1(\theta)\mathbf{n}(\theta) &= \mathbf{N}_1(\theta)\mathbf{m}(\theta_o) + \cos(\theta - \theta_o)\mathbf{n}(\theta), \\ \cos(\theta - \theta_o)\mathbf{N}_3(\theta)\mathbf{n}(\theta) &= \mathbf{N}_3(\theta)\mathbf{n}(\theta_o), \\ \sin(\theta - \theta_o)\mathbf{N}_3(\theta)\mathbf{n}(\theta) &= \mathbf{N}_3(\theta)\mathbf{m}(\theta_o). \end{aligned} \tag{7.5-8}$$

When θ=θ_o (7.5-8)$_{2,4}$ reduce to (7.5-6)$_{1,2}$ while (7.5-8)$_{1,3}$ become trivial identities.

From (7.5-4) and (7.1-23)$_1$ it is readily shown that

$$\frac{d}{d\theta}\big(\mathbf{n}(\theta) + p(\theta)\mathbf{m}(\theta)\big) = -p(\theta)\big(\mathbf{n}(\theta) + p(\theta)\mathbf{m}(\theta)\big), \tag{7.5-9}$$

indicating that the direction of the vector $\mathbf{n}(\theta)$+$p(\theta)\mathbf{m}(\theta)$ is invariant with θ (Barnett and Lothe, 1973). Indeed, a direct substitution of $p(\theta)$, $\mathbf{n}(\theta)$, and $\mathbf{m}(\theta)$ from (7.2-18) and (7.2-8) yields

$$\mathbf{n}(\theta) + p(\theta)\mathbf{m}(\theta) = \frac{1}{\cos\theta + p\sin\theta}\begin{bmatrix} 1 \\ p \\ 0 \end{bmatrix} \tag{7.5-10}$$

which is a multiple of a constant vector.

In the x_i^* coordinate system,

$$\mathbf{N}_1^* = -(\mathbf{T}^*)^{-1}(\mathbf{R}^*)^T, \quad \mathbf{N}_2^* = (\mathbf{T}^*)^{-1}, \quad \mathbf{N}_3^* = \mathbf{R}^*(\mathbf{T}^*)^{-1}(\mathbf{R}^*)^T - \mathbf{Q}^*,$$

and, by (7.2-11) and (7.3-16),

$$\mathbf{N}_1^* = \Omega\mathbf{N}_1(\theta)\Omega^T, \quad \mathbf{N}_2^* = \Omega\mathbf{N}_2(\theta)\Omega^T, \quad \mathbf{N}_3^* = \Omega\mathbf{N}_3(\theta)\Omega^T. \tag{7.5-11}$$

They are similar to (7.2-11). The 3×3 matrices $\mathbf{N}_1(\theta)$, $\mathbf{N}_2(\theta)$, and $\mathbf{N}_3(\theta)$ are tensors because $\mathbf{Q}(\theta)$, $\mathbf{R}(\theta)$, and $\mathbf{T}(\theta)$ are (Section 7.2). Like $\mathbf{Q}$, $\mathbf{R}$, $\mathbf{T}$, however, $\mathbf{N}_1$, $\mathbf{N}_2$, $\mathbf{N}_3$ are not tensors (see the discussion of (7.2-23)). In the x_i^* coordinate system $\mathbf{N}_1^*$ and $\mathbf{N}_3^*$ have the same structure as have $\mathbf{N}_1$ and $\mathbf{N}_3$ in the x_i coordinate system shown in (6.1-5) and (6.1-6). Since $-\mathbf{N}_3$ is positive semidefinite (Section 6.1) so is $-\mathbf{N}_3^*$. From $(7.5\text{-}11)_3$,

$$-\mathbf{N}_3(\theta) = \Omega^T(-\mathbf{N}_3^*)\Omega$$

and $-\mathbf{N}_3(\theta)$ is also positive semidefinite. Therefore

$$\mathbf{y}^T\left(-\mathbf{N}_3(\theta)\right)\mathbf{y} \geq 0$$

for any real unit vector $\mathbf{y}$. By $(7.5\text{-}6)_2$ and the fact that $\mathbf{N}_3(\theta)$ is symmetric, the equality holds only when $\mathbf{y}=\mathbf{m}(\theta)$ and we have

$$\mathbf{y}^T\left(-\mathbf{N}_3(\theta)\right)\mathbf{y}\begin{cases} >0 & \text{if } \mathbf{y} \neq \mathbf{m}(\theta), \\ =0 & \text{if } \mathbf{y} = \mathbf{m}(\theta). \end{cases} \tag{7.5-12}$$

We will show that

$$(\cos\theta\mathbf{I} + \sin\theta\mathbf{N})(\cos\theta\mathbf{I} - \sin\theta\mathbf{N}(\theta)) = \mathbf{I}. \tag{7.5-13}$$

When $\mathbf{N}(\theta)$ is simple or semisimple, the identity can be proved easily by observing that

$$\begin{aligned} (\cos\theta\mathbf{I} + \sin\theta\mathbf{N})\boldsymbol{\xi} &= (\cos\theta + p\sin\theta)\boldsymbol{\xi}, \\ (\cos\theta\mathbf{I} - \sin\theta\mathbf{N}(\theta))\boldsymbol{\xi} &= (\cos\theta - p(\theta)\sin\theta)\boldsymbol{\xi}, \end{aligned} \tag{7.5-14}$$

and hence by virtue of (7.1-27),

$$(\cos\theta\mathbf{I} + \sin\theta\mathbf{N})(\cos\theta\mathbf{I} - \sin\theta\mathbf{N}(\theta))\boldsymbol{\xi} = \boldsymbol{\xi}. \tag{7.5-15}$$

The fact that this equation applies to all six eigenvectors ξ tells us that (7.5-13) must hold. If $\mathbf{N}(\theta)$ is nonsemisimple, say $p_1(\theta)=p_2(\theta)$, and there exists only one eigenvector ξ_1 associated with $p_1(\theta)=p_2(\theta)$, the generalized eigenvector $\xi_2(\theta)$ satisfies the eigenrelation $(7.4\text{-}1)_2$, and is given by (7.4-4). Upon multiplication on both sides of (7.4-4) by $\mathbf{N}$ one has

$$\begin{aligned}\mathbf{N}\xi_2(\theta) &= g^2(\theta)\mathbf{N}\xi_2 + h(\theta)\mathbf{N}\xi_1\\ &= g^2(\theta)(p_1\xi_2 + \xi_1) + h(\theta)p_1\xi_1\\ &= p_1(g^2(\theta)\xi_2 + h(\theta)\xi_1) + g^2(\theta)\xi_1,\end{aligned}$$

appeal being made to $(7.4\text{-}1)_{1,2}$ with $\theta=0$. Hence

$$\mathbf{N}\xi_2(\theta)=p_1\xi_2(\theta) + g^2(\theta)\xi_1. \tag{7.5-16}$$

By $(7.4\text{-}1)_2$ and (7.5-16),

$$\begin{aligned}(\cos\theta\mathbf{I}+\sin\theta\mathbf{N})\xi_2(\theta) &= (\cos\theta+p\sin\theta)\xi_2(\theta) + g^2(\theta)\sin\theta\xi_1,\\ (\cos\theta\mathbf{I}-\sin\theta\mathbf{N}(\theta))\xi_2(\theta) &= (\cos\theta-p(\theta)\sin\theta)\xi_2(\theta) - \sin\theta\xi_1.\end{aligned} \tag{7.5-17}$$

With $g(\theta)$ defined in (7.4-5) and using (7.1-27), it can be shown that (7.5-15) remains valid for $\xi=\xi_2(\theta)$ and $\xi_5(\theta)$, the latter is the complex conjugate of $\xi_2(\theta)$. Therefore (7.5-13) holds also for a nonsemisimple $\mathbf{N}(\theta)$ (Ting, 1988d).

When multiplied by r^2 (7.5-13) is

$$(x_1\mathbf{I}+x_2\mathbf{N})(x_1\mathbf{I}-x_2\mathbf{N}(\theta))=r^2\mathbf{I} \tag{7.5-18}$$

which provides an explicit inverse of $x_1+x_2\mathbf{N}$ as

$$(x_1\mathbf{I}+x_2\mathbf{N})^{-1}=r^{-2}(x_1\mathbf{I}-x_2\mathbf{N}(\theta)). \tag{7.5-19}$$

A generalization of (7.5-13) is (Hwu and Ting, 1990)

$$\mathbf{N}(\theta,\theta^-)\mathbf{N}(\theta^-,\theta_o)=\mathbf{N}(\theta,\theta_o)=\mathbf{N}(\theta^-,\theta_o)\mathbf{N}(\theta,\theta^-) \tag{7.5-20}$$

where θ_o and θ^- are arbitrary real constants and

$$\mathbf{N}(\theta,\theta_o)=\cos(\theta-\theta_o)\mathbf{I}+\sin(\theta-\theta_o)\mathbf{N}(\theta_o). \tag{7.5-21}$$

The proof can be accomplished with the aid of the identity (7.1-29). A special case of (7.5-20) is

$$\mathbf{N}(\theta,\theta_o)\mathbf{N}(\theta_o,\theta)=\mathbf{I}=\mathbf{N}(\theta_o,\theta)\mathbf{N}(\theta,\theta_o). \tag{7.5-22}$$

Thus the inverse of $\mathbf{N}(\theta,\theta_o)$ is simply $\mathbf{N}(\theta_o,\theta)$. Since

$$\mathbf{N}(\theta+\tfrac{\pi}{2},\theta)=\mathbf{N}(\theta),\quad \mathbf{N}(\theta,\theta+\tfrac{\pi}{2})=-\mathbf{N}(\theta+\tfrac{\pi}{2}), \tag{7.5-23}$$

identity (7.5-1) is a special case of (7.5-22).

7.6 The Barnett–Lothe Integral Formalism

When $\mathbf{N}(\theta)$ is simple or semisimple the diagonalization of the 6×6 matrix $\mathbf{N}(\theta)$ is, by (7.3-17),

$$\mathbf{N}(\theta)=\begin{bmatrix}\mathbf{A} & \overline{\mathbf{A}}\\ \mathbf{B} & \overline{\mathbf{B}}\end{bmatrix}\begin{bmatrix}\mathbf{P}(\theta) & \mathbf{0}\\ \mathbf{0} & \overline{\mathbf{P}}(\theta)\end{bmatrix}\begin{bmatrix}\mathbf{B}^T & \mathbf{A}^T\\ \overline{\mathbf{B}}^T & \overline{\mathbf{A}}^T\end{bmatrix}, \tag{7.6-1}$$

$$\mathbf{P}(\theta)=\langle p_*(\theta)\rangle=\mathrm{diag}[p_1(\theta),\ p_2(\theta),\ p_3(\theta)].$$

From (7.1-26) we have

$$\frac{1}{\pi}\int_0^{\pi}\mathbf{P}(\theta)d\theta=i\mathbf{I}$$

and integration of (7.6-1) with θ gives

$$\frac{1}{\pi}\int_0^{\pi}\mathbf{N}(\theta)d\theta=\begin{bmatrix}\mathbf{S} & \mathbf{H}\\ -\mathbf{L} & \mathbf{S}^T\end{bmatrix} \tag{7.6-2}$$

where use has been made of (6.5-3). If $\mathbf{N}(\theta)$ is nonsemisimple (7.6-1) is replaced by (7.4-15) which can be rewritten as

$$\mathbf{N}(\theta)=\begin{bmatrix}\mathbf{A} & \overline{\mathbf{A}}\\ \mathbf{B} & \overline{\mathbf{B}}\end{bmatrix}\begin{bmatrix}\mathbf{U}(\theta) & \mathbf{0}\\ \mathbf{0} & \overline{\mathbf{U}}(\theta)\end{bmatrix}\begin{bmatrix}\Gamma\mathbf{B}^T & \Gamma\mathbf{A}^T\\ \Gamma\overline{\mathbf{B}}^T & \Gamma\overline{\mathbf{A}}^T\end{bmatrix} \tag{7.6-3}$$

where, by (7.4-10) and (7.4-12),

$$\mathbf{U}(\theta)=\mathbf{K}(\theta)\mathbf{J}(\theta)\mathbf{K}^{-1}(\theta)=\begin{bmatrix}p_1(\theta) & g^{-2}(\theta) & 0\\ 0 & p_1(\theta) & 0\\ 0 & 0 & p_3(\theta)\end{bmatrix}.$$

With the aid of (7.1-26) and (7.4-16),

$$\frac{1}{\pi}\int_0^{\pi}\mathbf{U}(\theta)d\theta=i\mathbf{I}$$

and integration of (7.6-3) leads to (7.6-2) by virtue of (6.5-7).

Equation (7.6-2) is equivalent to

$$\mathbf{S} = \frac{1}{\pi}\int_0^{\pi} \mathbf{N}_1(\theta)d\theta, \quad \mathbf{H} = \frac{1}{\pi}\int_0^{\pi} \mathbf{N}_2(\theta)d\theta, \quad \mathbf{L} = -\frac{1}{\pi}\int_0^{\pi} \mathbf{N}_3(\theta)d\theta, \quad (7.6\text{-}4)$$

indicating that **S**, **H**, and **L** are, respectively, the average values of $\mathbf{N}_1(\theta)$, $\mathbf{N}_2(\theta)$, and $-\mathbf{N}_3(\theta)$ over the interval $\theta=(0,\pi)$. Since $\mathbf{N}_i(\theta)$ $(i=1,2,3)$ is periodic in θ with periodicity π (Section 7.3), the range of integration $(0,\pi)$ in (7.6-4) can be replaced by $(\alpha,\alpha+\pi)$ where α is any real constant. Equations (7.6-4) were first obtained by Barnett and Lothe (1973). The matrices **S**, **H**, **L**, which are tensors (Section 7.2), are thus called the *Barnett–Lothe tensors*. Equations (7.6-4) allow us to compute **S**, **H**, **L** directly from the elastic stiffnesses C_{ijks} without computing the eigenvalues p_α and the associated eigenvectors $\boldsymbol{a}_\alpha$ and $\boldsymbol{b}_\alpha$ of the fundamental elasticity matrix **N**. The problems associated with degenerate (and almost degenerate) materials are therefore circumvented. This is indeed a remarkable result, and is one of the major contributions to the theory of anisotropic elasticity.

It is obvious that **H** is positive definite because $\mathbf{N}_2(\theta)$ is. To show that **L** is positive definite consider

$$\mathbf{y}^T\mathbf{L}\mathbf{y} = \frac{1}{\pi}\int_0^{\pi} \mathbf{y}^T\left(-\mathbf{N}_3(\theta)\right)\mathbf{y}d\theta \qquad (7.6\text{-}5)$$

where **y** is any real unit vector. It is shown in Section 7.5 that $-\mathbf{N}_3(\theta)$ is positive semidefinite and, by (7.5-12), the integrand in (7.6-5) vanishes only at the θ at which $\mathbf{y}=\mathrm{m}(\theta)$. Thus the integral in (7.6-5) is positive and nonzero for any **y**, confirming that **L** is positive definite. The positive definiteness of **L** was first proved indirectly by Chadwick and Smith (1977). A direct proof was given by Gundersen et al. (1987) and by Ting (1988a). The method of proof of the latter has been employed here.

In general the integral in (7.6-4) have to be computed numerically (Gundersen and Lothe, 1987). There is no need to compute all three integrals. Kirchner and Lothe (1986) has shown that **S** and **L** can be found algebraically in terms of $\mathbf{N}_1$, $\mathbf{N}_2$, and **H**. Similarly, **S** and **H** can be found in terms of $\mathbf{N}_1$, $\mathbf{N}_3$, and **L**.

Related to the Barnett–Lothe tensors are the associated tensors

$$\begin{aligned} \mathbf{S}(\theta) &= \frac{1}{\pi}\int_0^{\theta} \mathbf{N}_1(\lambda)d\lambda, \\ \mathbf{H}(\theta) &= \frac{1}{\pi}\int_0^{\theta} \mathbf{N}_2(\lambda)d\lambda, \qquad (7.6\text{-}6) \\ \mathbf{L}(\theta) &= -\frac{1}{\pi}\int_0^{\theta} \mathbf{N}_3(\lambda)d\lambda, \end{aligned}$$

They reduce to (7.6-4) when $\theta=\pi$. The dependence of $\mathbf{S}(\theta)$, $\mathbf{H}(\theta)$, and $\mathbf{L}(\theta)$ on θ will be given explicitly unless $\theta=\pi$. Following the argument in proving the positive definiteness of $\mathbf{H}$ and $\mathbf{L}$, it is clear that $\mathbf{H}(\theta)$ and $\mathbf{L}(\theta)$ are positive definite. If the integrals in (7.6-4) are called the *complete integrals*, the integrals in (7.6-6) are the *incomplete integrals*. Integrating (7.6-1) with θ and making use of (7.1-25) the algebraic representation of $\mathbf{S}(\theta)$, $\mathbf{H}(\theta)$, and $\mathbf{L}(\theta)$ are

$$\begin{aligned}\mathbf{S}(\theta) &= \frac{2}{\pi}\mathrm{Re}\{\mathbf{A}\langle \ln(\cos\theta + p_* \sin\theta)\rangle \mathbf{B}^T\},\\ \mathbf{H}(\theta) &= \frac{2}{\pi}\mathrm{Re}\{\mathbf{A}\langle \ln(\cos\theta + p_* \sin\theta)\rangle \mathbf{A}^T\},\\ \mathbf{L}(\theta) &= -\frac{2}{\pi}\mathrm{Re}\{\mathbf{B}\langle \ln(\cos\theta + p_* \sin\theta)\rangle \mathbf{B}^T\}.\end{aligned} \tag{7.6-7}$$

They reduce to (6.5-2) when $\theta=\pi$. Explicit expressions of $\mathbf{S}(\theta)$, $\mathbf{H}(\theta)$, and $\mathbf{L}(\theta)$ using (7.6-7) for monoclinic materials with the symmetry plane at $x_3=0$ have been obtained by Ting (1992a).

Equations (7.6-4) and (7.6-6) tell us that

$$\mathbf{S}(\pi+\theta) = \mathbf{S}+\mathbf{S}(\theta), \quad \mathbf{H}(\pi+\theta) = \mathbf{H}+\mathbf{H}(\theta), \quad \mathbf{L}(\pi+\theta) = \mathbf{L}+\mathbf{L}(\theta). \tag{7.6-8}$$

Following (7.2-23) the $\mathbf{N}_i(\theta)$ in a rotated coordinate system is

$$\mathbf{N}_i^*(\theta^*) = \Omega(\psi)\mathbf{N}_i(\theta)\Omega^T(\psi), \quad \theta^* = \theta - \psi.$$

Integration with θ^* leads to (Dongye and Ting, 1989)

$$\begin{aligned}\mathbf{S}^*(\theta^*) &= \Omega(\psi)[\mathbf{S}(\theta) - \mathbf{S}(\psi)]\Omega^T(\psi),\\ \mathbf{H}^*(\theta^*) &= \Omega(\psi)[\mathbf{H}(\theta) - \mathbf{H}(\psi)]\Omega^T(\psi),\\ \mathbf{L}^*(\theta^*) &= \Omega(\psi)[\mathbf{L}(\theta) - \mathbf{L}(\psi)]\Omega^T(\psi).\end{aligned} \tag{7.6-9}$$

At $\theta^*=\pi$ they reduce to (7.2-22) by virtue of (7.6-8).

Since $\mathbf{N}_1(\theta)$ is dimensionless while $\mathbf{N}_2(\theta)$ and $\mathbf{N}_3(\theta)$ have the dimensions of stress and compliance, respectively, $\mathbf{S}$, $\mathbf{S}(\theta)$ are dimensionless. $\mathbf{H}$, $\mathbf{H}(\theta)$ and $\mathbf{L}$, $\mathbf{L}(\theta)$ have the dimensions of stress and compliances, respectively. The tensor $\mathbf{L}(\theta)$ has a physical interpretation. We will show in Section 9.2 that, with the possible exception of the σ_{33} component, $\sigma_{ij} = L_{ij}(\theta)$ represents an admissible stress state in anisotropic elastic materials.

7.7 Invariants of S, H, and L

We have shown in (7.2-22) that under rotations about the x_3-axis $\mathbf{S}$, $\mathbf{H}$, and $\mathbf{L}$ are tensors of rank two, i.e.,

$$\mathbf{S}^* = \Omega\mathbf{S}\Omega^T, \quad \mathbf{H}^* = \Omega\mathbf{H}\Omega^T, \quad \mathbf{L}^* = \Omega\mathbf{L}\Omega^T, \tag{7.7-1}$$

where

$$\Omega = \begin{bmatrix} \cos\theta & \sin\theta & 0 \\ -\sin\theta & \cos\theta & 0 \\ 0 & 0 & 1 \end{bmatrix} = \begin{bmatrix} \Omega_o & \mathbf{0} \\ \mathbf{0} & 1 \end{bmatrix}, \text{ say.} \tag{7.7-2}$$

Consider the invariants of **S** first, since it is in general neither symmetric nor skew-symmetric. By partitioning **S** similar to (7.7-2) as

$$\mathbf{S} = \begin{bmatrix} S_{11} & S_{12} & S_{13} \\ S_{21} & S_{22} & S_{23} \\ S_{31} & S_{32} & S_{33} \end{bmatrix} = \begin{bmatrix} \mathbf{S}_o & \mathbf{s}_1 \\ \mathbf{s}_2^T & S_{33} \end{bmatrix}, \tag{7.7-3}$$

$(7.7\text{-}1)_1$ is decomposed into

$$\mathbf{S}_o^* = \Omega_o\mathbf{S}_o\Omega_o^T, \quad \mathbf{s}_k^* = \Omega_o\mathbf{s}_k \quad (k=1,2), \quad S_{33}^* = S_{33}. \tag{7.7-4}$$

It can be shown from (7.7-4) that

$$\begin{gathered} S_{33}, \quad S_{11}+S_{22}, \quad S_{11}S_{22}-S_{12}S_{21}, \quad S_{13}^2+S_{23}^2, \\ S_{31}^2+S_{32}^2, \quad S_{13}S_{31}+S_{23}S_{32}, \quad S_{21}-S_{12}, \end{gathered} \tag{7.7-5}$$

are invariants. The last invariant is obtained from the relation

$$(\mathbf{S}_o^* - \mathbf{S}_o^{*T}) = \Omega_o(\mathbf{S}_o - \mathbf{S}_o^T)\Omega_o^T$$

and observing that $(\mathbf{S}_o - \mathbf{S}_o^T)$ is skew-symmetric. Of course, more invariants can be obtained by combining the invariants shown above. An example is

$$(S_{13} \pm S_{31})^2 + (S_{23} \pm S_{32})^2.$$

For **H** and **L** which are symmetric the equations analogous to (7.7-4) are

$$\begin{gathered} \mathbf{H}_o^* = \Omega_o\mathbf{H}_o\Omega_o^T, \quad \mathbf{h}^* = \Omega_o\mathbf{h}, \quad H_{33}^* = H_{33}, \\ \mathbf{L}_o^* = \Omega_o\mathbf{L}_o\Omega_o^T, \quad \boldsymbol{l}^* = \Omega_o\boldsymbol{l}, \quad L_{33}^* = L_{33}. \end{gathered} \tag{7.7-6}$$

Hence,

$$\begin{gathered} H_{33}, \quad H_{11}+H_{22}, \quad H_{11}H_{22}-H_{12}^2, \quad H_{33}, \\ L_{33}, \quad L_{11}+L_{22}, \quad L_{11}L_{22}-L_{12}^2, \quad L_{33}, \end{gathered} \tag{7.7-7}$$

are invariants. Moreover from $(7.7\text{-}4)_2$ and $(7.7\text{-}6)_{2,5}$ we have

$$\mathbf{h}^*\cdot\mathbf{s}_k^* = \mathbf{h}\cdot\mathbf{s}_k, \quad \boldsymbol{l}^*\cdot\mathbf{s}_k^* = \boldsymbol{l}\cdot\mathbf{s}_k, \quad \boldsymbol{l}^*\cdot\mathbf{h}^* = \boldsymbol{l}\cdot\mathbf{h}.$$

Thus

$$\begin{gathered} H_{13}S_{13}+H_{23}S_{23}, \quad H_{13}S_{31}+H_{23}S_{32}, \\ L_{13}S_{13}+L_{23}S_{23}, \quad L_{13}S_{31}+L_{23}S_{32}, \\ H_{13}L_{13}+H_{23}L_{23}, \end{gathered} \tag{7.7-8}$$

are invariants.

The tensors $\mathbf{SL}^{-1}$, $\mathbf{LS}$, $\mathbf{H}^{-1}\mathbf{S}$, and $\mathbf{SH}$ are skew-symmetric. A skew-symmetric tensor $\mathbf{W}$ has the structure

$$\mathbf{W} = \begin{bmatrix} 0 & w_3 & -w_2 \\ -w_3 & 0 & w_1 \\ w_2 & -w_1 & 0 \end{bmatrix}$$

which can be written as

$$W_{ij} = \delta_{ijk}w_k, \quad w_k = \tfrac{1}{2}\delta_{kij}W_{ij}. \tag{7.7-9}$$

Since

$$\mathbf{W}^* = \Omega\mathbf{W}\Omega^T, \quad W_{ij}^* = \Omega_{ip}\Omega_{jq}W_{pq},$$

$(7.7\text{-}9)_2$ in the x_i^* coordinate system is

$$w_k^* = \tfrac{1}{2}\delta_{kij}W_{ij}^* = \tfrac{1}{2}\delta_{kij}\Omega_{ip}\Omega_{jq}W_{pq} = (\tfrac{1}{2}\delta_{kij}\delta_{pqs}\Omega_{ip}\Omega_{jq})w_s.$$

The quantity in the parentheses is $\hat{\Omega}_{ks}$, the cofactor of Ω_{ks}. From $(1.1\text{-}20)_2$ and bearing in mind that Ω is an orthogonal tensor,

$$\hat{\Omega} = |\Omega|\Omega = \pm\Omega.$$

For proper orthogonal Ω, $|\Omega| = 1$ and we have the result

$$\mathbf{w}^* = \Omega\mathbf{w}, \quad \mathbf{w} = \begin{bmatrix} w_1 \\ w_2 \\ w_3 \end{bmatrix}. \tag{7.7-10}$$

With the Ω defined in (7.7-2) it is readily shown that

$$w_3, \quad w_1^2 + w_2^2, \tag{7.7-11a}$$

are invariants. In terms of W_{ij} the invariants are

$$W_{12}, \quad W_{13}^2 + W_{23}^2. \tag{7.7-11b}$$

Therefore

$$
\begin{aligned}
&(\mathbf{SL}^{-1})_{12}, \quad \left((\mathbf{SL}^{-1})_{13}\right)^2 + \left((\mathbf{SL}^{-1})_{23}\right)^2, \\
&(\mathbf{LS})_{12}, \quad \left((\mathbf{LS})_{13}\right)^2 + \left((\mathbf{LS})_{23}\right)^2, \\
&(\mathbf{H}^{-1}\mathbf{S})_{12}, \quad \left((\mathbf{H}^{-1}\mathbf{S})_{13}\right)^2 + \left((\mathbf{H}^{-1}\mathbf{S})_{23}\right)^2, \\
&(\mathbf{SH})_{12}, \quad \left((\mathbf{SH})_{13}\right)^2 + \left((\mathbf{SH})_{23}\right)^2,
\end{aligned}
\tag{7.7-12}
$$

are invariants. More invariants involving components of **S**, **H**, and **L** can be found in (Ting, 1991).

Is it possible to have invariant Barnett–Lothe tensors? In other words, what materials give all components of **S**, **H**, and **L** invariant under rotations about the x_3-axis? The obvious choices are isotropic materials and transversely isotropic materials with the axis of symmetry at the x_3-axis. We will show that there are other more general materials for which all components of **S**, **H**, **L** are invariant.

If the symmetric tensor **H** is invariant we must have

$$\mathbf{H}_o^* = \mathbf{H}_o, \quad \mathbf{h}^* = \mathbf{h}.$$

By $(7.7\text{-}6)_{1,2}$ $\mathbf{H}_o$ must be a multiple of the 2×2 identity tensor **I** and **h** vanishes identically. The same statement applies to **L**. Thus **H** and **L** must have the structure

$$
\mathbf{H} = \begin{bmatrix} H_{11} & 0 & 0 \\ 0 & H_{11} & 0 \\ 0 & 0 & H_{33} \end{bmatrix}, \qquad
\mathbf{L} = \begin{bmatrix} L_{11} & 0 & 0 \\ 0 & L_{11} & 0 \\ 0 & 0 & L_{33} \end{bmatrix},
\tag{7.7-13}
$$

when they are invariants. If **S** is an invariant it may have the structure (7.7-13) or

$$
\mathbf{S} = \begin{bmatrix} 0 & -S_{21} & 0 \\ S_{21} & 0 & 0 \\ 0 & 0 & 0 \end{bmatrix}.
\tag{7.7-14}
$$

The fact that $\mathbf{SL}^{-1}$ is skew-symmetric and $\mathbf{L}^{-1}$ is positive definite rules out the possibility of **S** having the structure (7.7-13). This conclusion is independent of whether **L** is an invariant or not. The same argument shows that the S_{33} component in (7.7-14) must vanish. Therefore an invariant **S** must have the structure (7.7-14), regardless of whether or not **H** and/or **L** are invariants.

For monoclinic materials with the symmetry plane at x_3=0, explicit expressions of the components of **S**, **H**, and **L** are given in (6.4-11)–(6.4-13). Comparison with (7.7-13) and (7.7-14) shows that they are invariants if

$$d = 0 \quad \text{and} \quad b = e.$$

From $(6.4\text{-}8)_4$ this means that $c=-1$, and $(6.4\text{-}12)_{1,3,4}$ lead to

$$s'_{16} = -s'_{26}, \quad s'_{11} = s'_{22}. \tag{7.7-15}$$

For monoclinic materials with the symmetry plane at $x_3=0$, it can be shown that (7.7-15) are equivalent to (see (A.12) of Ting, 1992a)

$$C_{16} = -C_{26}, \quad C_{11} = C_{22}. \tag{7.7-16}$$

We therefore have the result that, when (7.7-16) are satisfied, monoclinic materials with the symmetry plane at $x_3=0$ possess invariant Barnett–Lothe tensors. Special materials such as tetragonal and cubic materials whose $C_{\alpha\beta}$ are shown in (2.6-6) and (2.6-8), respectively, satisfy conditions (7.7-16). The Barnett–Lothe tensors for these materials are therefore invariant with rotations about the x_3-axis.

It should be pointed out that conditions (7.7-16) do not imply that C_{11} and C_{16} are invariant with the angle of rotation θ about the x_3-axis. However, it can be shown that for monoclinic materials with the symmetry plane at $x_3=0$ the equalities in (7.7-16) hold for all θ if they hold for $\theta=0$.

7.8 Commutative 6×6 Matrices

When $\mathbf{N}(\theta)$ is simple or semisimple the diagonalization of $\mathbf{N}(\theta)$ given in (7.6-1) suggests that $\mathbf{N}(\theta)$ and $\mathbf{N}(\psi)$ share the same set of eigenvectors. By Theorem 1.5–1 the products of $\mathbf{N}(\theta)$ and $\mathbf{N}(\psi)$ commute (Kirchner and Lothe, 1986),

$$\mathbf{N}(\psi)\mathbf{N}(\theta) = \mathbf{N}(\theta)\mathbf{N}(\psi). \tag{7.8-1}$$

If $\mathbf{N}(\theta)$ is nonsemisimple, the diagonalization of $\mathbf{N}(\theta)$ is given by (7.4-15). Again, $\mathbf{N}(\theta)$ and $\mathbf{N}(\psi)$ share the same set of eigenvectors and the Jordan canonical forms for $\mathbf{N}(\theta)$ and $\mathbf{N}(\psi)$ have the same structure. By Theorem 1.5–3 the products of $\mathbf{N}(\theta)$ and $\mathbf{N}(\psi)$ commute and (7.8-1) holds. Equation (7.8-1) can also be verified directly for nonsemisimple $\mathbf{N}(\theta)$ by inserting (7.4-15) in (7.8-1) and using (5.9-15).

When $\psi=0$ (7.8-1) has the expression

$$\begin{bmatrix} \mathbf{N}_1 & \mathbf{N}_2 \\ \mathbf{N}_3 & \mathbf{N}_1^T \end{bmatrix} \begin{bmatrix} \mathbf{N}_1(\theta) & \mathbf{N}_2(\theta) \\ \mathbf{N}_3(\theta) & \mathbf{N}_1^T(\theta) \end{bmatrix} = \begin{bmatrix} \mathbf{N}_1(\theta) & \mathbf{N}_2(\theta) \\ \mathbf{N}_3(\theta) & \mathbf{N}_1^T(\theta) \end{bmatrix} \begin{bmatrix} \mathbf{N}_1 & \mathbf{N}_2 \\ \mathbf{N}_3 & \mathbf{N}_1^T \end{bmatrix}. \tag{7.8-2}$$

Written explicitly we have

$$\begin{aligned} &\mathbf{N}_1\mathbf{N}_1(\theta)+\mathbf{N}_2\mathbf{N}_3(\theta)=\left(\mathbf{N}_3\mathbf{N}_2(\theta)+\mathbf{N}_1^T\mathbf{N}_1^T(\theta)\right)^T, \\ &\mathbf{N}_1\mathbf{N}_2(\theta)+\mathbf{N}_2\mathbf{N}_1^T(\theta)=\left(\mathbf{N}_1\mathbf{N}_2(\theta)+\mathbf{N}_2\mathbf{N}_1^T(\theta)\right)^T, \\ &\mathbf{N}_3\mathbf{N}_1(\theta)+\mathbf{N}_1^T\mathbf{N}_3(\theta)=\left(\mathbf{N}_3\mathbf{N}_1(\theta)+\mathbf{N}_1^T\mathbf{N}_3(\theta)\right)^T. \end{aligned} \tag{7.8-3}$$

The fourth equation is the transpose of the firs equation. Equations $(7.8\text{-}3)_{2,3}$ tell us that

$$\mathbf{N}_1\mathbf{N}_2(\theta)+\mathbf{N}_2\mathbf{N}_1^T(\theta) \quad \text{and} \quad \mathbf{N}_3\mathbf{N}_1(\theta)+\mathbf{N}_1^T\mathbf{N}_3(\theta)$$

are symmetric. Let

$$\tilde{\mathbf{N}}(\theta)=\frac{1}{\pi}\int_0^\theta \mathbf{N}(\lambda)d\lambda=\begin{bmatrix} \mathbf{S}(\theta) & \mathbf{H}(\theta) \\ -\mathbf{L}(\theta) & \mathbf{S}^T(\theta) \end{bmatrix}, \tag{7.8-4}$$

$$\tilde{\mathbf{N}}=\frac{1}{\pi}\int_0^\pi \mathbf{N}(\theta)d\theta=\begin{bmatrix} \mathbf{S} & \mathbf{H} \\ -\mathbf{L} & \mathbf{S}^T \end{bmatrix}. \tag{7.8-5}$$

Regardless of whether $\mathbf{N}(\theta)$ is simple, semisimple, or nonsemisimple, $\tilde{\mathbf{N}}$ is always semisimple (Sections 6.5 and 6.10). Integration of (7.8-1) with θ yields

$$\mathbf{N}(\psi)\tilde{\mathbf{N}}(\theta)=\tilde{\mathbf{N}}(\theta)\mathbf{N}(\psi) \tag{7.8-6}$$

or

$$\begin{bmatrix} \mathbf{N}_1(\psi) & \mathbf{N}_2(\psi) \\ \mathbf{N}_3(\psi) & \mathbf{N}_1^T(\psi) \end{bmatrix}\begin{bmatrix} \mathbf{S}(\theta) & \mathbf{H}(\theta) \\ -\mathbf{L}(\theta) & \mathbf{S}^T(\theta) \end{bmatrix}=\begin{bmatrix} \mathbf{S}(\theta) & \mathbf{H}(\theta) \\ -\mathbf{L}(\theta) & \mathbf{S}^T(\theta) \end{bmatrix}\begin{bmatrix} \mathbf{N}_1(\psi) & \mathbf{N}_2(\psi) \\ \mathbf{N}_3(\psi) & \mathbf{N}_1^T(\psi) \end{bmatrix}. \tag{7.8-7}$$

This is equivalent to

$$\begin{aligned} &\mathbf{N}_1(\psi)\mathbf{S}(\theta)-\mathbf{N}_2(\psi)\mathbf{L}(\theta)=\left(\mathbf{N}_3(\psi)\mathbf{H}(\theta)+\mathbf{N}_1^T(\psi)\mathbf{S}^T(\theta)\right)^T, \\ &\mathbf{N}_1(\psi)\mathbf{H}(\theta)+\mathbf{N}_2(\psi)\mathbf{S}^T(\theta)=\left(\mathbf{N}_1(\psi)\mathbf{H}(\theta)+\mathbf{N}_2(\psi)\mathbf{S}^T(\theta)\right)^T, \\ &\mathbf{N}_3(\psi)\mathbf{S}(\theta)-\mathbf{N}_1^T(\psi)\mathbf{L}(\theta)=\left(\mathbf{N}_3(\psi)\mathbf{S}(\theta)-\mathbf{N}_1^T(\psi)\mathbf{L}(\theta)\right)^T. \end{aligned} \tag{7.8-8}$$

Equations (7.8-8) reduce to (6.8-8) when $\psi=0$ and $\theta=\pi$. A useful special case is when $\theta=\pi$ and ψ is replaced by θ. We have

$$\mathbf{N}(\theta)\tilde{\mathbf{N}}=\tilde{\mathbf{N}}\mathbf{N}(\theta) \tag{7.8-9}$$

or

$$\begin{aligned}
\mathbf{N}_1(\theta)\mathbf{S} - \mathbf{N}_2(\theta)\mathbf{L} &= \left(\mathbf{N}_3(\theta)\mathbf{H} + \mathbf{N}_1^T(\theta)\mathbf{S}^T\right)^T, \\
\mathbf{N}_1(\theta)\mathbf{H} + \mathbf{N}_2(\theta)\mathbf{S}^T &= \left(\mathbf{N}_1(\theta)\mathbf{H} + \mathbf{N}_2(\theta)\mathbf{S}^T\right)^T, \\
\mathbf{N}_3(\theta)\mathbf{S} - \mathbf{N}_1^T(\theta)\mathbf{L} &= \left(\mathbf{N}_3(\theta)\mathbf{S} - \mathbf{N}_1^T(\theta)\mathbf{L}\right)^T.
\end{aligned} \tag{7.8-10}$$

If we integrate (7.8-6) with ψ we have

$$\tilde{\mathbf{N}}(\psi)\tilde{\mathbf{N}}(\theta) = \tilde{\mathbf{N}}(\theta)\tilde{\mathbf{N}}(\psi) \tag{7.8-11}$$

or

$$\begin{aligned}
\mathbf{S}(\psi)\mathbf{S}(\theta) - \mathbf{H}(\psi)\mathbf{L}(\theta) &= \left(-\mathbf{L}(\psi)\mathbf{H}(\theta) + \mathbf{S}^T(\psi)\mathbf{S}^T(\theta)\right)^T, \\
\mathbf{S}(\psi)\mathbf{H}(\theta) + \mathbf{H}(\psi)\mathbf{S}^T(\theta) &= \left(\mathbf{S}(\psi)\mathbf{H}(\theta) + \mathbf{H}(\psi)\mathbf{S}^T(\theta)\right)^T, \\
\mathbf{L}(\psi)\mathbf{S}(\theta) + \mathbf{S}^T(\psi)\mathbf{L}(\theta) &= \left(\mathbf{L}(\psi)\mathbf{S}(\theta) + \mathbf{S}^T(\psi)\mathbf{L}(\theta)\right)^T.
\end{aligned} \tag{7.8-12}$$

In particular at $\psi=\pi$,

$$\begin{aligned}
\mathbf{S}\mathbf{S}(\theta) - \mathbf{H}\mathbf{L}(\theta) &= \left(-\mathbf{L}\mathbf{H}(\theta) + \mathbf{S}^T\mathbf{S}^T(\theta)\right)^T, \\
\mathbf{S}\mathbf{H}(\theta) + \mathbf{H}\mathbf{S}^T(\theta) &= \left(\mathbf{S}\mathbf{H}(\theta) + \mathbf{H}\mathbf{S}^T(\theta)\right)^T, \\
\mathbf{L}\mathbf{S}(\theta) + \mathbf{S}^T\mathbf{L}(\theta) &= \left(\mathbf{L}\mathbf{S}(\theta) + \mathbf{S}^T\mathbf{L}(\theta)\right)^T.
\end{aligned} \tag{7.8-13}$$

The commutative relations presented in (7.8-1), (7.8-6), and (7.8-10) can be generalized to include to products of

$$\mathbf{N}^m, \quad \mathbf{N}^n(\theta), \quad \tilde{\mathbf{N}}^s, \quad \tilde{\mathbf{N}}^t(\theta)$$

where m, n, s, and t are any nonzero integer, positive or negative.

7.9 Identities Connecting A, B, and $f(z)$ to Real Matrices

In applications the final solution to a problem contains the matrices

$$\mathbf{A}\langle f(z_*)\rangle\mathbf{B}^T, \quad \mathbf{A}\langle f(z_*)\rangle\mathbf{A}^T, \quad \mathbf{B}\langle f(z_*)\rangle\mathbf{B}^T, \quad \mathbf{B}\langle f(z_*)\rangle\mathbf{A}^T, \tag{7.9-1}$$

where

$$\langle f(z_*)\rangle = \text{diag}[f(z_1),\ f(z_2),\ f(z_3)],$$

$$z_\alpha = x_1 + p_\alpha x_2 \quad (\alpha = 1,2,3).$$

We will present in this section explicit expressions of (7.9-1) in terms of the real quantities

$$\mathbf{N}_i,\ \mathbf{N}(\theta),\ \mathbf{S},\ \mathbf{H},\ \mathbf{L},\ \mathbf{S}(\theta),\ \mathbf{H}(\theta),\ \mathbf{L}(\theta),$$

for certain functions $f(z)$. We will consider materials which are not degenerate, i.e., the matrix **N** is not nonsemisimple.

The trivial case is the one when $f(z)$ is a constant. The matrices in (7.9-1) then reduce to a constant times

$$\mathbf{AB}^T,\ \mathbf{AA}^T,\ \mathbf{BB}^T,\ \mathbf{BA}^T,$$

which are, by (5.5-17),

$$\mathbf{AB}^T = \tfrac{1}{2}(\mathbf{I} - i\mathbf{S}), \quad \mathbf{AA}^T = -\tfrac{1}{2}i\mathbf{H}, \quad \mathbf{BB}^T = \tfrac{1}{2}i\mathbf{L}, \tag{7.9-2}$$

and $\mathbf{BA}^T$ is the transpose of $\mathbf{AB}^T$. They can be cast in the form

$$\begin{bmatrix} \mathbf{AB}^T & \mathbf{AA}^T \\ \mathbf{BB}^T & \mathbf{BA}^T \end{bmatrix} = \frac{1}{2}\mathbf{I} - \frac{i}{2}\begin{bmatrix} \mathbf{S} & \mathbf{H} \\ -\mathbf{L} & \mathbf{S}^T \end{bmatrix} = \frac{1}{2}(\mathbf{I} - i\tilde{\mathbf{N}}). \tag{7.9-3}$$

The simplest function is $f(z)=z$. From the eigenrelation

$$\mathbf{N}\xi = p\xi$$

we have

$$(x_1\mathbf{I} + x_2\mathbf{N})\xi = (x_1 + px_2)\xi = z\xi. \tag{7.9-4}$$

Equation (7.9-4) for $p=p_1, p_2, p_3$ can be written in one equation as

$$(x_1\mathbf{I} + x_2\mathbf{N})\begin{bmatrix} \mathbf{A} \\ \mathbf{B} \end{bmatrix} = \begin{bmatrix} \mathbf{A}\langle z_* \rangle \\ \mathbf{B}\langle z_* \rangle \end{bmatrix} \tag{7.9-5}$$

where

$$\langle z_* \rangle = \mathrm{diag}[z_1,\ z_2,\ z_3].$$

When both sides of (7.9-5) are postmultiplied by the 3×6 matrix $[\mathbf{B}^T, \mathbf{A}^T]$ we have

$$(x_1\mathbf{I} + x_2\mathbf{N})\begin{bmatrix} \mathbf{AB}^T & \mathbf{AA}^T \\ \mathbf{BB}^T & \mathbf{BA}^T \end{bmatrix} = \begin{bmatrix} \mathbf{A}\langle z_* \rangle\mathbf{B}^T & \mathbf{A}\langle z_* \rangle\mathbf{A}^T \\ \mathbf{B}\langle z_* \rangle\mathbf{B}^T & \mathbf{B}\langle z_* \rangle\mathbf{A}^T \end{bmatrix}$$

or, by (7.9-3),

$$\begin{bmatrix} \mathbf{A}\langle z_* \rangle\mathbf{B}^T & \mathbf{A}\langle z_* \rangle\mathbf{A}^T \\ \mathbf{B}\langle z_* \rangle\mathbf{B}^T & \mathbf{B}\langle z_* \rangle\mathbf{A}^T \end{bmatrix} = \frac{1}{2}(x_1\mathbf{I} + x_2\mathbf{N})(\mathbf{I} - i\tilde{\mathbf{N}}). \tag{7.9-6}$$

Carrying out the matrix multiplication we obtain

$$
\begin{aligned}
2\mathbf{A}\langle z_*\rangle\mathbf{B}^T &= x_1(\mathbf{I}-i\mathbf{S})+x_2\{\mathbf{N}_1-i(\mathbf{N}_1\mathbf{S}-\mathbf{N}_2\mathbf{L})\},\\
2\mathbf{A}\langle z_*\rangle\mathbf{A}^T &= -ix_1\mathbf{H}+x_2\{\mathbf{N}_2-i(\mathbf{N}_1\mathbf{H}+\mathbf{N}_2\mathbf{S}^T)\},\\
2\mathbf{B}\langle z_*\rangle\mathbf{B}^T &= ix_1\mathbf{L}+x_2\{\mathbf{N}_3-i(\mathbf{N}_3\mathbf{S}-\mathbf{N}_1^T\mathbf{L})\},\\
2\mathbf{B}\langle z_*\rangle\mathbf{A}^T &= x_1(\mathbf{I}-i\mathbf{S}^T)+x_2\{\mathbf{N}_1^T-i(\mathbf{N}_3\mathbf{H}+\mathbf{N}_1^T\mathbf{S}^T)\}.
\end{aligned}
\tag{7.9-7}
$$

Notice that in (7.9-6),

$$
(x_1\mathbf{I}+x_2\mathbf{N})(\mathbf{I}-i\tilde{\mathbf{N}})=(\mathbf{I}-i\tilde{\mathbf{N}})(x_1\mathbf{I}+x_2\mathbf{N})
\tag{7.9-8}
$$

because the matrices $(x_1\mathbf{I}+x_2\mathbf{N})$ and $(\mathbf{I}-i\tilde{\mathbf{N}})$ share the same set of eigenvectors (Section 1.5). With (7.9-8) we are assured that $\mathbf{A}\langle z_*\rangle\mathbf{B}^T$ in (7.9-7) is indeed the transpose of $\mathbf{B}\langle z_*\rangle\mathbf{A}^T$, and $\mathbf{A}\langle z_*\rangle\mathbf{A}^T$ and $\mathbf{B}\langle z_*\rangle\mathbf{B}^T$ are symmetric (see also (6.8-8)).

From the relations

$$
\begin{aligned}
\mathbf{B}^{-1} &= \mathbf{B}^T(\mathbf{B}\mathbf{B}^T)^{-1}=-2i\mathbf{B}^T\mathbf{L}^{-1},\\
\mathbf{A}^{-1} &= \mathbf{A}^T(\mathbf{A}\mathbf{A}^T)^{-1}=2i\mathbf{A}^T\mathbf{H}^{-1},
\end{aligned}
\tag{7.9-9}
$$

(7.9-7) can be transformed into

$$
\begin{aligned}
\mathbf{A}\langle z_*\rangle\mathbf{B}^{-1} &= -x_1(\mathbf{S}+i\mathbf{I})\mathbf{L}^{-1}+x_2\{(\mathbf{N}_2-\mathbf{N}_1\mathbf{S}\mathbf{L}^{-1})-i\mathbf{N}_1\mathbf{L}^{-1}\},\\
\mathbf{A}\langle z_*\rangle\mathbf{A}^{-1} &= x_1\mathbf{I}+x_2\{(\mathbf{N}_1+\mathbf{N}_2\mathbf{S}^T\mathbf{H}^{-1})+i\mathbf{N}_2\mathbf{H}^{-1}\},\\
\mathbf{B}\langle z_*\rangle\mathbf{B}^{-1} &= x_1\mathbf{I}+x_2\{(\mathbf{N}_1^T-\mathbf{N}_3\mathbf{S}\mathbf{L}^{-1})-i\mathbf{N}_3\mathbf{L}^{-1}\},\\
\mathbf{B}\langle z_*\rangle\mathbf{A}^{-1} &= x_1(\mathbf{S}^T+i\mathbf{I})\mathbf{H}^{-1}+x_2\{(\mathbf{N}_3+\mathbf{N}_1^T\mathbf{S}^T\mathbf{H}^{-1})+i\mathbf{N}_1^T\mathbf{H}^{-1}\}.
\end{aligned}
\tag{7.9-10}
$$

The derivation presented above for $f(z)=z$ can be extended for $f(z)=z^n$ where n is a positive integer. From (7.9-4),

$$
(\mathbf{x}_1\mathbf{I}+\mathbf{x}_2\mathbf{N})^n\xi=z^n\xi
$$

and (7.9-6) is generalized to

$$
\begin{bmatrix}
\mathbf{A}\langle z_*^n\rangle\mathbf{B}^T & \mathbf{A}\langle z_*^n\rangle\mathbf{A}^T\\
\mathbf{B}\langle z_*^n\rangle\mathbf{B}^T & \mathbf{B}\langle z_*^n\rangle\mathbf{A}^T
\end{bmatrix}
=\frac{1}{2}(x_1\mathbf{I}+x_2\mathbf{N})^n(\mathbf{I}-i\tilde{\mathbf{N}}).
\tag{7.9-11}
$$

Employment of (7.9-9) allows us to write

$$
\begin{bmatrix}
\mathbf{A}\langle z_*^n\rangle\mathbf{B}^{-1} & \mathbf{A}\langle z_*^n\rangle\mathbf{A}^{-1}\\
\mathbf{B}\langle z_*^n\rangle\mathbf{B}^{-1} & \mathbf{B}\langle z_*^n\rangle\mathbf{A}^{-1}
\end{bmatrix}
=(x_1\mathbf{I}+x_2\mathbf{N})^n(\mathbf{I}-i\tilde{\mathbf{N}})
\begin{bmatrix}
-i\mathbf{L}^{-1} & \mathbf{0}\\
\mathbf{0} & i\mathbf{H}^{-1}
\end{bmatrix}.
\tag{7.9-12}
$$

Alternately, from the relations

$$\begin{aligned}\mathbf{A}\langle z_*^n\rangle\mathbf{B}^{-1} &= \left(\mathbf{A}\langle z_*\rangle\mathbf{A}^{-1}\right)\left(\mathbf{A}\langle z_*\rangle\mathbf{B}^{-1}\right)^{n-1},\\ \mathbf{A}\langle z_*^n\rangle\mathbf{A}^{-1} &= \left(\mathbf{A}\langle z_*\rangle\mathbf{A}^{-1}\right)^n,\\ \mathbf{B}\langle z_*^n\rangle\mathbf{B}^{-1} &= \left(\mathbf{B}\langle z_*\rangle\mathbf{B}^{-1}\right)^n,\\ \mathbf{B}\langle z_*^n\rangle\mathbf{A}^{-1} &= \left(\mathbf{B}\langle z_*\rangle\mathbf{B}^{-1}\right)\left(\mathbf{B}\langle z_*\rangle\mathbf{A}^{-1}\right)^{n-1},\end{aligned} \qquad (7.9\text{-}13)$$

the quantities on the left can be computed in terms of those given in (7.9-10).

Of particular interest in application is the expression

$$\mathbf{V} = \mathbf{B}\left\langle\sqrt{z_*^2 - a^2}\right\rangle\mathbf{B}^{-1} \qquad (7.9\text{-}14)$$

where a is a constant. Since

$$\mathbf{V}^2 = \mathbf{B}\langle z_*^2 - a^2\rangle\mathbf{B}^{-1} = \mathbf{B}\langle z_*^2\rangle\mathbf{B}^{-1} - a^2\mathbf{I},$$

by $(7.9\text{-}13)_3$ and $(7.9\text{-}10)_3$,

$$\mathbf{B}\left\langle\sqrt{z_*^2 - a^2}\right\rangle\mathbf{B}^{-1} = \sqrt{\left[x_1\mathbf{I} + x_2\left\{(\mathbf{N}_1^T - \mathbf{N}_3\mathbf{S}\mathbf{L}^{-1}) - i\mathbf{N}_3\mathbf{L}^{-1}\right\}\right]^2 - a^2\mathbf{I}}\,. \qquad (7.9\text{-}15)$$

The square root of a square matrix is unique if the eigenvalues of $\mathbf{V}$ is known (Section 1.9). From (7.9-14) it is clear that the three eigenvalues of $\mathbf{V}$ are

$$\sqrt{z_\alpha^2 - a^2}, \quad \alpha = 1,2,3,$$

so that, when the Riemann surface with a proper cut is introduced (Section 3.4), the $\mathbf{V}$ in (7.9-15) can be determined uniquely.

Next we investigate the function $f(z) = z^{-m}$ where m is a positive integer. From (7.9-4) we write

$$(\mathbf{x}_1\mathbf{I} + \mathbf{x}_2\mathbf{N})^{-m}\xi = z^{-m}\xi, \qquad (7.9\text{-}16)$$

or, by (7.5-19),

$$r^{-2m}\left(x_1\mathbf{I} - x_2\mathbf{N}(\theta)\right)^m\xi = z^{-m}\xi. \qquad (7.9\text{-}17)$$

Following the derivation of (7.9-6) one obtains

$$\begin{bmatrix}\mathbf{A}\langle z_*^{-m}\rangle\mathbf{B}^T & \mathbf{A}\langle z_*^{-m}\rangle\mathbf{A}^T\\ \mathbf{B}\langle z_*^{-m}\rangle\mathbf{B}^T & \mathbf{B}\langle z_*^{-m}\rangle\mathbf{A}^T\end{bmatrix} = \frac{1}{2}r^{-2m}\left(x_1\mathbf{I} - x_2\mathbf{N}(\theta)\right)^m(\mathbf{I} - i\tilde{\mathbf{N}}). \qquad (7.9\text{-}18)$$

In particular, for $m=1$,

$$\begin{aligned}
2r\mathbf{A}\langle z_*^{-1}\rangle\mathbf{B}^T &= (\cos\theta\mathbf{I}-\sin\theta\mathbf{N}_1(\theta))(\mathbf{I}-i\mathbf{S})-i\sin\theta\mathbf{N}_2(\theta)\mathbf{L},\\
2r\mathbf{A}\langle z_*^{-1}\rangle\mathbf{A}^T &= -i(\cos\theta\mathbf{I}-\sin\theta\mathbf{N}_1(\theta))\mathbf{H}-\sin\theta\mathbf{N}_2(\theta)(\mathbf{I}-i\mathbf{S}^T),\\
2r\mathbf{B}\langle z_*^{-1}\rangle\mathbf{B}^T &= i(\cos\theta\mathbf{I}-\sin\theta\mathbf{N}_1^T(\theta))\mathbf{L}-\sin\theta\mathbf{N}_3(\theta)(\mathbf{I}-i\mathbf{S}),\\
2r\mathbf{B}\langle z_*^{-1}\rangle\mathbf{A}^T &= (\cos\theta\mathbf{I}-\sin\theta\mathbf{N}_1^T(\theta))(\mathbf{I}-i\mathbf{S}^T)+i\sin\theta\mathbf{N}_3(\theta)\mathbf{H}.
\end{aligned} \tag{7.9-19}$$

As in (7.9-8) the products of $(x_1\mathbf{I}-x_2\mathbf{N}(\theta))^m$ and $(\mathbf{I}-i\tilde{\mathbf{N}})$ in (7.9-18) commute. For instance,

$$(x_1\mathbf{I}-x_2\mathbf{N}(\theta))(\mathbf{I}-i\tilde{\mathbf{N}}) = (\mathbf{I}-i\tilde{\mathbf{N}})(x_1\mathbf{I}-x_2\mathbf{N}(\theta)), \tag{7.9-20a}$$

$$\begin{aligned}
(x_1\mathbf{I}-x_2\mathbf{N}(\theta))^2(\mathbf{I}-i\tilde{\mathbf{N}}) &= (x_1\mathbf{I}-x_2\mathbf{N}(\theta))(\mathbf{I}-i\tilde{\mathbf{N}})(x_1\mathbf{I}-x_2\mathbf{N}(\theta))\\
&= (\mathbf{I}-i\tilde{\mathbf{N}})(x_1\mathbf{I}-x_2\mathbf{N}(\theta))^2.
\end{aligned} \tag{7.9-20b}$$

From (7.9-20a) or (7.8-9) the symmetry of $\mathbf{A}\langle z_*^{-1}\rangle\mathbf{A}^T$ and $\mathbf{B}\langle z_*^{-1}\rangle\mathbf{B}^T$ in (7.9-19) is confirmed, so is the fact that $\mathbf{A}\langle z_*^{-1}\rangle\mathbf{B}^T$ is the transpose of $\mathbf{B}\langle z_*^{-1}\rangle\mathbf{A}^T$.

Finally consider the function

$$f(z) = \ln z = \ln r + \ln(\cos\theta + p\sin\theta).$$

On integrating the eigenrelation

$$\mathbf{N}(\theta)\boldsymbol{\xi} = p(\theta)\boldsymbol{\xi}$$

and making use of (7.1-25) and the $\tilde{\mathbf{N}}(\theta)$ defined in (7.8-4) one has

$$\pi\tilde{\mathbf{N}}(\theta)\boldsymbol{\xi} = \ln(\cos\theta + p\sin\theta)\boldsymbol{\xi} = (\ln z - \ln r)\boldsymbol{\xi}. \tag{7.9-21}$$

For $\boldsymbol{\xi} = \boldsymbol{\xi}_1, \boldsymbol{\xi}_2, \boldsymbol{\xi}_3$, (7.9-21) can be written as

$$\left((\ln r)\mathbf{I}+\pi\tilde{\mathbf{N}}(\theta)\right)\begin{bmatrix}\mathbf{A}\\ \mathbf{B}\end{bmatrix} = \begin{bmatrix}\mathbf{A}\langle\ln z_*\rangle\\ \mathbf{B}\langle\ln z_*\rangle\end{bmatrix}$$

or, more generally,

$$\left((\ln r)\mathbf{I}+\pi\tilde{\mathbf{N}}(\theta)\right)^s\begin{bmatrix}\mathbf{A}\\ \mathbf{B}\end{bmatrix} = \begin{bmatrix}\mathbf{A}\langle\ln z_*\rangle^s\\ \mathbf{B}\langle\ln z_*\rangle^s\end{bmatrix}$$

where s is any positive integer. When both sides are Postmultiplied by $[\mathbf{B}^T, \mathbf{A}^T]$ we obtain

$$\begin{bmatrix} \mathbf{A}\langle \ln z_* \rangle^s \mathbf{B}^T & \mathbf{A}\langle \ln z_* \rangle^s \mathbf{A}^T \\ \mathbf{B}\langle \ln z_* \rangle^s \mathbf{B}^T & \mathbf{B}\langle \ln z_* \rangle^s \mathbf{A}^T \end{bmatrix} = \frac{1}{2}\Big((\ln r)\mathbf{I} + \pi\tilde{\mathbf{N}}(\theta)\Big)^s (\mathbf{I} - i\tilde{\mathbf{N}}), \tag{7.9-22}$$

use being made of (7.9-3). For s=1 the result is

$$\begin{aligned} 2\mathbf{A}\langle \ln z_* \rangle \mathbf{B}^T &= ((\ln r)\mathbf{I} + \pi\mathbf{S}(\theta))(\mathbf{I} - i\mathbf{S}) + i\pi\mathbf{H}(\theta)\mathbf{L}, \\ 2\mathbf{A}\langle \ln z_* \rangle \mathbf{A}^T &= -i((\ln r)\mathbf{I} + \pi\mathbf{S}(\theta))\mathbf{H} + \pi\mathbf{H}(\theta)(\mathbf{I} - i\mathbf{S}^T), \\ 2\mathbf{B}\langle \ln z_* \rangle \mathbf{B}^T &= i\Big((\ln r)\mathbf{I} + \pi\mathbf{S}^T(\theta)\Big)\mathbf{L} - \pi\mathbf{L}(\theta)(\mathbf{I} - i\mathbf{S}), \\ 2\mathbf{B}\langle \ln z_* \rangle \mathbf{A}^T &= \Big((\ln r)\mathbf{I} + \pi\mathbf{S}^T(\theta)\Big)(\mathbf{I} - i\mathbf{S}^T) + i\pi\mathbf{L}(\theta)\mathbf{H}. \end{aligned} \tag{7.9-23}$$

Again, the products of the matrices $\Big((\ln r)\mathbf{I} + \tilde{\mathbf{N}}(\theta)\Big)$ and $(\mathbf{I} - i\tilde{\mathbf{N}})$ in (7.9-22) commute, i.e.,

$$\Big((\ln r)\mathbf{I} + \pi\tilde{\mathbf{N}}(\theta)\Big)(\mathbf{I} - i\tilde{\mathbf{N}}) = (\mathbf{I} - i\tilde{\mathbf{N}})\Big((\ln r)\mathbf{I} + \pi\tilde{\mathbf{N}}(\theta)\Big) \tag{7.9-24}$$

so that $\mathbf{A}\langle \ln z_* \rangle \mathbf{A}^T$ and $\mathbf{B}\langle \ln z_* \rangle \mathbf{B}^T$ are indeed symmetric and $\mathbf{A}\langle \ln z_* \rangle \mathbf{B}^T$ is the transpose of $\mathbf{B}\langle \ln z_* \rangle \mathbf{A}^T$ (see also (7.8-9)).

The identities obtained above can be combined into one equation as

$$\begin{bmatrix} \mathbf{A}\langle z_*^n (\ln z_*)^s \rangle \mathbf{B}^T & \mathbf{A}\langle z_*^n (\ln z_*)^s \rangle \mathbf{A}^T \\ \mathbf{B}\langle z_*^n (\ln z_*)^s \rangle \mathbf{B}^T & \mathbf{B}\langle z_*^n (\ln z_*)^s \rangle \mathbf{A}^T \end{bmatrix} = \tfrac{1}{2}(x_1\mathbf{I} + x_2\mathbf{N})^n \Big((\ln r)\mathbf{I} + \pi\tilde{\mathbf{N}}(\theta)\Big)^s (\mathbf{I} - i\tilde{\mathbf{N}}) \tag{7.9-25}$$

in which the products of the matrices

$$(\mathbf{x}_1\mathbf{I} + x_2\mathbf{N}), \quad \Big((\mathrm{lnr})\mathbf{I} + \pi\tilde{\mathbf{N}}(\theta)\Big), \quad (\mathbf{I} - \mathrm{i}\tilde{\mathbf{N}})$$

commute. When n<0, we use (7.5-19) to find the inverse of $(x_1\mathbf{I}+ x_2\mathbf{N})$.

7.10 General Solutions and Identities in Dual Coordinate Systems

While the general solutions **u** and $\boldsymbol{\phi}$ given in (7.3-3) and (7.3-9) for dual coordinate systems have been employed extensively in the literature, their full potential for applications has not been really exploited. Consider for example a straight crack of finite length in an orthotropic material with x_3=0 being one of the symmetry planes of the material. The crack line makes an angle θ_0 with a symmetry plane of the material, Fig. 7.3. If we choose the coordinate system

x_i that coincides with the symmetry planes of the material, the crack line is disoriented with respect to the x_i coordinate system. If we choose the $\hat{x}_i$ coordinate system such that the crack line is on the $\hat{x}_1$-axis, the elastic constants $\hat{C}_{ijks}$ referred to the $\hat{x}_i$ coordinate system no longer have a simple expression. The dual coordinate systems allow us to keep both the x_i and $\hat{x}_i$ coordinate systems. There are other advantages. As we will see in Chapters 8–11 some solutions or certain aspects of the solutions may not depend on θ_o, leading to interesting and puzzling invariant properties.

In the notation of Fig. 7.3 the general solutions (7.3-3) and (7.3-9) for dual coordinate systems are written as.

$$\mathbf{u} = \boldsymbol{a} f(\hat{z}), \quad \boldsymbol{\phi} = \boldsymbol{b} f(\hat{z}), \tag{7.10-1}$$

where the vectors $\mathbf{u}$ and $\boldsymbol{\phi}$ are referred to the x_i coordinate system, while

$$\hat{z} = \hat{x}_1 + p(\theta_o)\hat{x}_2, \tag{7.10-2}$$

$$\begin{aligned} \hat{x}_1 &= \mathbf{x} \cdot \mathbf{n}(\theta_o) = x_1 \cos\theta_o + x_2 \sin\theta_o, \\ \hat{x}_2 &= \mathbf{x} \cdot \mathbf{m}(\theta_o) = -x_1 \sin\theta_o + x_2 \cos\theta_o, \end{aligned} \tag{7.10-3}$$

are referred to the $\hat{x}_i$ coordinate system. Since

$$x_1 = r\cos\theta, \quad x_2 = r\sin\theta,$$

(7.10-3) and (7.10-2) can also be written as

$$\hat{x}_1 = r\cos(\theta - \theta_o), \quad \hat{x}_2 = r\sin(\theta - \theta_o), \tag{7.10-4}$$

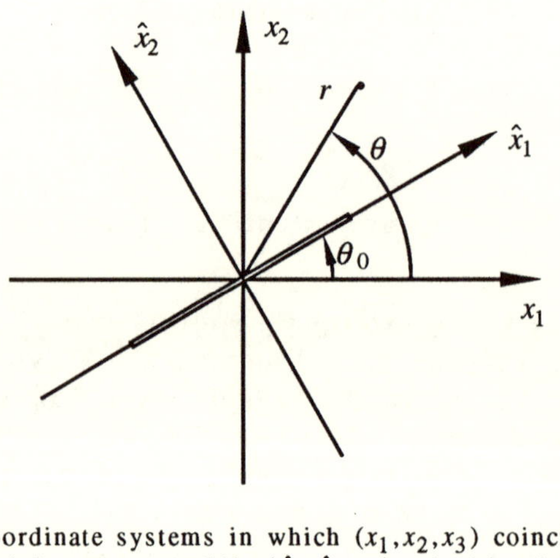

Fig. 7.3 Dual coordinate systems in which (x_1, x_2, x_3) coincide with a plane or planes of material symmetry while $(\hat{x}_1, \hat{x}_2, x_3)$ coincide with the crack.

$$\hat{z} = r\{\cos(\theta - \theta_o) + p(\theta_o)\sin(\theta - \theta_o)\}. \tag{7.10-5}$$

In particular,

$$\hat{x}_1 = r, \quad \hat{x}_2 = 0, \quad \text{at } \theta = \theta_o.$$

The transformation between z and $\hat{z}$ can be written in a concise form by introducing the notation

$$\zeta(\theta, \theta_o) = \cos(\theta - \theta_o) + p(\theta_o)\sin(\theta - \theta_o). \tag{7.10-6}$$

We then have

$$z = r\zeta(\theta, 0), \quad \hat{z} = r\zeta(\theta, \theta_o). \tag{7.10-7}$$

From the identities

$$\begin{aligned} \zeta(\theta,0)\zeta(0,\theta_o) &= \zeta(\theta,\theta_o), \\ \zeta(\theta_o,0)\zeta(0,\theta_o) &= 1, \end{aligned} \tag{7.10-8}$$

deduced from (7.1-29) one obtains the transformation laws

$$\hat{z} = z\zeta(0, \theta_o), \quad z = \hat{z}\zeta(\theta_o, 0). \tag{7.10-9}$$

Superimposing six solutions of the form (7.10-1) associated with six eigenvalues $p_\alpha(\theta_o)$ α=1, 2,..., 6 yields

$$\begin{aligned} \mathbf{u} &= \sum_{\alpha=1}^{3} \{ \boldsymbol{a}_\alpha f_\alpha(\hat{z}_\alpha) + \bar{\boldsymbol{a}}_\alpha f_{\alpha+3}(\hat{\bar{z}}_\alpha) \} \\ \boldsymbol{\phi} &= \sum_{\alpha=1}^{3} \{ \boldsymbol{b}_\alpha f_\alpha(\hat{z}_\alpha) + \bar{\boldsymbol{b}}_\alpha f_{\alpha+3}(\hat{\bar{z}}_\alpha) \} \end{aligned} \tag{7.10-10}$$

which are a generalization of (5.3-7). The counterparts of (5.3-10) and (5.3-13) are

$$\mathbf{u} = 2\,\mathrm{Re}\{\mathbf{A}\langle f(\hat{z}_*)\rangle \mathbf{q}\}, \quad \boldsymbol{\phi} = 2\,\mathrm{Re}\{\mathbf{B}\langle f(\hat{z}_*)\rangle \mathbf{q}\}, \tag{7.10-11}$$

and

$$\mathbf{u} = 2\,\mathrm{Im}\{\mathbf{A}\langle f(\hat{z}_*)\rangle \mathbf{q}\}, \quad \boldsymbol{\phi} = 2\,\mathrm{Im}\{\mathbf{B}\langle f(\hat{z}_*)\rangle \mathbf{q}\}. \tag{7.10-12}$$

It should be noted that the arbitrary constant vector $\mathbf{q}$ in (7.10-11), (7.10-12), (5.3-10), and (5.3-13) are all different.

From the eigenrelation

$$\mathbf{N}(\theta_o)\boldsymbol{\xi} = p(\theta_o)\boldsymbol{\xi}$$

we have

$$(\hat{x}_1\mathbf{I}+\hat{x}_2\mathbf{N}(\theta_o))\xi = (\hat{x}_1+p(\theta_o)\hat{x}_2)\xi = \hat{z}\xi. \tag{7.10-13}$$

Equation (7.10-13) can be written in one equation for $p=p_1,p_2,p_3$ as

$$(\hat{x}_1\mathbf{I}+\hat{x}_2\mathbf{N}(\theta_o))\begin{bmatrix}\mathbf{A}\\ \mathbf{B}\end{bmatrix}=\begin{bmatrix}\mathbf{A}\langle\hat{z}_*\rangle\\ \mathbf{B}\langle\hat{z}_*\rangle\end{bmatrix}. \tag{7.10-14}$$

This is a generalization of (7.9-5). Following the derivation of (7.9-6) one obtains

$$\begin{bmatrix}\mathbf{A}\langle\hat{z}_*\rangle\mathbf{B}^T & \mathbf{A}\langle\hat{z}_*\rangle\mathbf{A}^T\\ \mathbf{B}\langle\hat{z}_*\rangle\mathbf{B}^T & \mathbf{B}\langle\hat{z}_*\rangle\mathbf{A}^T\end{bmatrix}=\frac{1}{2}(\hat{x}_1\mathbf{I}+\hat{x}_2\mathbf{N}(\theta_o))(\mathbf{I}-i\tilde{\mathbf{N}}), \tag{7.10-15}$$

or

$$\begin{aligned}
2\mathbf{A}\langle\hat{z}_*\rangle\mathbf{B}^T &= \hat{x}_1(\mathbf{I}-i\mathbf{S})+\hat{x}_2\{\mathbf{N}_1(\theta_o)-i[\mathbf{N}_1(\theta_o)\mathbf{S}-\mathbf{N}_2(\theta_o)\mathbf{L}]\},\\
2\mathbf{A}\langle\hat{z}_*\rangle\mathbf{A}^T &= -i\hat{x}_1\mathbf{H}+\hat{x}_2\{\mathbf{N}_2(\theta_o)-i[\mathbf{N}_1(\theta_o)\mathbf{H}+\mathbf{N}_2(\theta_o)\mathbf{S}^T]\},\\
2\mathbf{B}\langle\hat{z}_*\rangle\mathbf{B}^T &= i\hat{x}_1\mathbf{L}+\hat{x}_2\{\mathbf{N}_3(\theta_o)-i[\mathbf{N}_3(\theta_o)\mathbf{S}-\mathbf{N}_1^T(\theta_o)\mathbf{L}]\},\\
2\mathbf{B}\langle\hat{z}_*\rangle\mathbf{A}^T &= \hat{x}_1(\mathbf{I}-i\mathbf{S}^T)+\hat{x}_2\{\mathbf{N}_1^T(\theta_o)-i[\mathbf{N}_3(\theta_o)\mathbf{H}+\mathbf{N}_1^T(\theta_o)\mathbf{S}^T]\}.
\end{aligned} \tag{7.10-16}$$

This is a generalization of (7.9-7). It can be shown that generalization of (7.9-10)–(7.9-12) and (7.9-18) are

$$\begin{aligned}
\mathbf{A}\langle\hat{z}_*\rangle\mathbf{B}^{-1} &= -\hat{x}_1(\mathbf{S}+i\mathbf{I})\mathbf{L}^{-1}+\hat{x}_2\{[\mathbf{N}_2(\theta_o)-\mathbf{N}_1(\theta_o)\mathbf{S}\mathbf{L}^{-1}]-i\mathbf{N}_1(\theta_o)\mathbf{L}^{-1}\},\\
\mathbf{A}\langle\hat{z}_*\rangle\mathbf{A}^{-1} &= \hat{x}_1\mathbf{I}+\hat{x}_2\{[\mathbf{N}_1(\theta_o)+\mathbf{N}_2(\theta_o)\mathbf{S}^T\mathbf{H}^{-1}]+i\mathbf{N}_2(\theta_o)\mathbf{H}^{-1}\},\\
\mathbf{B}\langle\hat{z}_*\rangle\mathbf{B}^{-1} &= \hat{x}_1\mathbf{I}+\hat{x}_2\{[\mathbf{N}_1^T(\theta_o)-\mathbf{N}_3(\theta_o)\mathbf{S}\mathbf{L}^{-1}]-i\mathbf{N}_3(\theta_o)\mathbf{L}^{-1}\},\\
\mathbf{B}\langle\hat{z}_*\rangle\mathbf{A}^{-1} &= \hat{x}_1(\mathbf{S}^T+i\mathbf{I})\mathbf{H}^{-1}+\hat{x}_2\{[\mathbf{N}_3(\theta_o)+\mathbf{N}_1^T(\theta_o)\mathbf{S}^T\mathbf{H}^{-1}]+i\mathbf{N}_1^T(\theta_o)\mathbf{H}^{-1}\},
\end{aligned} \tag{7.10-17}$$

$$\begin{bmatrix}\mathbf{A}\langle\hat{z}_*^n\rangle\mathbf{B}^T & \mathbf{A}\langle\hat{z}_*^n\rangle\mathbf{A}^T\\ \mathbf{B}\langle\hat{z}_*^n\rangle\mathbf{B}^T & \mathbf{B}\langle\hat{z}_*^n\rangle\mathbf{A}^T\end{bmatrix}=\frac{1}{2}\{\hat{x}_1\mathbf{I}+\hat{x}_2\mathbf{N}(\theta_o)\}^n(\mathbf{I}-i\tilde{\mathbf{N}}), \tag{7.10-18}$$

$$\begin{bmatrix}\mathbf{A}\langle\hat{z}_*^n\rangle\mathbf{B}^{-1} & \mathbf{A}\langle\hat{z}_*^n\rangle\mathbf{A}^{-1}\\ \mathbf{B}\langle\hat{z}_*^n\rangle\mathbf{B}^{-1} & \mathbf{B}\langle\hat{z}_*^n\rangle\mathbf{A}^{-1}\end{bmatrix}=\{\hat{x}_1\mathbf{I}+\hat{x}_2\mathbf{N}(\theta_o)\}^n(\mathbf{I}-i\tilde{\mathbf{N}})\begin{bmatrix}-i\mathbf{L}^{-1} & \mathbf{0}\\ \mathbf{0} & i\mathbf{H}^{-1}\end{bmatrix}, \tag{7.10-19}$$

and

$$\begin{bmatrix} \mathbf{A}\langle \hat{z}_*^{-m} \rangle \mathbf{B}^T & \mathbf{A}\langle \hat{z}_*^{-m} \rangle \mathbf{A}^T \\ \mathbf{B}\langle \hat{z}_*^{-m} \rangle \mathbf{B}^T & \mathbf{B}\langle \hat{z}_*^{-m} \rangle \mathbf{A}^T \end{bmatrix} = \frac{1}{2} r^{-2m} [\hat{x}_1 \mathbf{I} - \hat{x}_2 \mathbf{N}(\theta)]^m (\mathbf{I} - i\tilde{\mathbf{N}}), \quad (7.10\text{-}20)$$

respectively. In deriving (7.10-20) one needs the identity

$$[\hat{x}_1 \mathbf{I} + \hat{x}_2 \mathbf{N}(\theta_o)][\hat{x}_1 \mathbf{I} - \hat{x}_2 \mathbf{N}(\theta)] = r^2 \mathbf{I} \quad (7.10\text{-}21)$$

which is obtained from (7.5-22) after multiplication on both sides by r^2. It is a generalization of (7.5-18). For $n=1$ (7.10-19) is written explicitly in (7.10-17). For $m=1$ (7.10-20) is

$$\begin{aligned}
2r\mathbf{A}\langle \hat{z}_*^{-1} \rangle \mathbf{B}^T &= [\cos(\theta - \theta_o)\mathbf{I} - \sin(\theta - \theta_o)\mathbf{N}_1(\theta)](\mathbf{I} - i\mathbf{S}) - i\sin(\theta - \theta_o)\mathbf{N}_2(\theta)\mathbf{L}, \\
2r\mathbf{A}\langle \hat{z}_*^{-1} \rangle \mathbf{A}^T &= -i[\cos(\theta - \theta_o)\mathbf{I} - \sin(\theta - \theta_o)\mathbf{N}_1(\theta)]\mathbf{H} - \sin(\theta - \theta_o)\mathbf{N}_2(\theta)(\mathbf{I} - i\mathbf{S}^T), \\
2r\mathbf{B}\langle \hat{z}_*^{-1} \rangle \mathbf{B}^T &= i[\cos(\theta - \theta_o)\mathbf{I} - \sin(\theta - \theta_o)\mathbf{N}_1^T(\theta)]\mathbf{L} - \sin(\theta - \theta_o)\mathbf{N}_3(\theta)(\mathbf{I} - i\mathbf{S}), \\
2r\mathbf{B}\langle \hat{z}_*^{-1} \rangle \mathbf{A}^T &= [\cos(\theta - \theta_o)\mathbf{I} - \sin(\theta - \theta_o)\mathbf{N}_1^T(\theta)](\mathbf{I} - i\mathbf{S}^T) + i\sin(\theta - \theta_o)\mathbf{N}_3(\theta)\mathbf{H}.
\end{aligned} \quad (7.10\text{-}22)$$

With $\zeta(\theta, \theta_o)$ defined in (7.10-6), (7.9-21) is

$$\pi\tilde{\mathbf{N}}(\theta)\boldsymbol{\xi} = \{\ln \zeta(\theta, 0)\}\boldsymbol{\xi}.$$

Hence by $(7.10\text{-}8)_2$,

$$\pi\tilde{\mathbf{N}}(\theta_o)\boldsymbol{\xi} = \{\ln \zeta(\theta_o, 0)\}\boldsymbol{\xi} = -\{\ln \zeta(0, \theta_o)\}\boldsymbol{\xi}.$$

Subtracting the two equations and making use of $(7.10\text{-}8)_1$ and $(7.10\text{-}7)_2$ leads to

$$\pi\tilde{\mathbf{N}}(\theta, \theta_o)\boldsymbol{\xi} = \{\ln \zeta(\theta, \theta_o)\}\boldsymbol{\xi} = (\ln \hat{z} - \ln r)\boldsymbol{\xi} \quad (7.10\text{-}23)$$

where

$$\tilde{\mathbf{N}}(\theta, \theta_o) = \tilde{\mathbf{N}}(\theta) - \tilde{\mathbf{N}}(\theta_o) = \frac{1}{\pi} \int_{\theta_o}^{\theta} \mathbf{N}(\lambda) d\lambda. \quad (7.10\text{-}24)$$

Following the derivation of (7.9-22) from (7.9-21) we obtain from (7.10-23)

$$\begin{bmatrix} \mathbf{A}\langle \ln \hat{z}_* \rangle^s \mathbf{B}^T & \mathbf{A}\langle \ln \hat{z}_* \rangle^s \mathbf{A}^T \\ \mathbf{B}\langle \ln \hat{z}_* \rangle^s \mathbf{B}^T & \mathbf{B}\langle \ln \hat{z}_* \rangle^s \mathbf{A}^T \end{bmatrix} = \frac{1}{2} \left((\ln r)\mathbf{I} + \pi\tilde{\mathbf{N}}(\theta, \theta_o) \right)^s (\mathbf{I} - i\tilde{\mathbf{N}}). \quad (7.10\text{-}25)$$

In particular, for $s=1$,

$$\begin{aligned}
2\mathbf{A}\langle \ln \hat{z}_* \rangle \mathbf{B}^T &= ((\ln r)\mathbf{I} + \pi \mathbf{S}(\theta,\theta_o))(\mathbf{I} - i\mathbf{S}) + i\pi \mathbf{H}(\theta,\theta_o)\mathbf{L},\\
2\mathbf{A}\langle \ln \hat{z}_* \rangle \mathbf{A}^T &= -i((\ln r)\mathbf{I} + \pi \mathbf{S}(\theta,\theta_o))\mathbf{H} + \pi \mathbf{H}(\theta,\theta_o)(\mathbf{I} - i\mathbf{S}^T),\\
2\mathbf{B}\langle \ln \hat{z}_* \rangle \mathbf{B}^T &= i((\ln r)\mathbf{I} + \pi \mathbf{S}^T(\theta,\theta_o))\mathbf{L} - \pi \mathbf{L}(\theta,\theta_o)(\mathbf{I} - i\mathbf{S}),\\
2\mathbf{B}\langle \ln \hat{z}_* \rangle \mathbf{A}^T &= ((\ln r)\mathbf{I} + \pi \mathbf{S}^T(\theta,\theta_o))(\mathbf{I} - i\mathbf{S}^T) + i\pi \mathbf{L}(\theta,\theta_o)\mathbf{H}.
\end{aligned} \tag{7.10-26}$$

As in (7.9-25) we may combine the results in one equation as

$$\begin{bmatrix} \mathbf{A}\langle \hat{z}_*^n (\ln \hat{z}_*)^s \rangle \mathbf{B}^T & \mathbf{A}\langle \hat{z}_*^n (\ln \hat{z}_*)^s \rangle \mathbf{A}^T \\ \mathbf{B}\langle \hat{z}_*^n (\ln \hat{z}_*)^s \rangle \mathbf{B}^T & \mathbf{B}\langle \hat{z}_*^n (\ln \hat{z}_*)^s \rangle \mathbf{A}^T \end{bmatrix} =$$

$$\tfrac{1}{2}(\hat{x}_1\mathbf{I} + \hat{x}_2\mathbf{N}(\theta_o))^n \left((\ln r)\mathbf{I} + \pi \tilde{\mathbf{N}}(\theta,\theta_o)\right)^s (\mathbf{I} - i\tilde{\mathbf{N}}). \tag{7.10-27}$$

With the notation defined in (7.5-21) we have

$$\begin{aligned}
\hat{x}_1\mathbf{I} + \hat{x}_2\mathbf{N}(\theta_o) &= r\mathbf{N}(\theta,\theta_o),\\
\hat{x}_1\mathbf{I} - \hat{x}_2\mathbf{N}(\theta) &= r\mathbf{N}(\theta_o,\theta),
\end{aligned} \tag{7.10-28}$$

and, by (7.5-22),

$$\mathbf{N}^{-1}(\theta,\theta_o) = \mathbf{N}(\theta_o,\theta). \tag{7.10-29}$$

Equation (7.10-29) is useful for computing the inverse of $\mathbf{N}(\theta,\theta_o)$ for $n<0$ in (7.10-27). The products of

$$\mathbf{N}(\theta,\theta_o),\quad \left((\ln r)\mathbf{I} + \pi \tilde{\mathbf{N}}(\theta,\theta_o)\right),\quad (\mathbf{I} - i\tilde{\mathbf{N}})$$

in (7.10-27) commute (Section 7.8).

7.11 The Tensors $\mathbf{G}_1(\theta)$ and $\mathbf{G}_3(\theta)$

If we set $\hat{x}_1=0$ and $\hat{x}_2=1$, the identity $(7.10\text{-}17)_3$ is

$$\mathbf{B}\langle p_*(\theta) \rangle \mathbf{B}^{-1} = \mathbf{G}_1(\theta) + i\mathbf{G}_2(\theta), \tag{7.11-1}$$

$$\mathbf{G}_1(\theta) = \mathbf{N}_1^T(\theta) - \mathbf{N}_3(\theta)\mathbf{S}\mathbf{L}^{-1}, \quad \mathbf{G}_3(\theta) = -\mathbf{N}_3(\theta)\mathbf{L}^{-1}. \tag{7.11-2}$$

The tensors $\mathbf{G}_1^T(\theta)$ and $\mathbf{G}_3^T(\theta)$ are physically dimensionless. They share certain properties with the tensors $\mathbf{N}_1(\theta)$ and $\mathbf{N}_3(\theta)$. They are periodic in θ with periodicity π, and satisfy the same identities (7.5-6) and (7.5-8), namely,

$$\mathbf{G}_1^T(\theta)\mathbf{m}(\theta) = -\mathbf{n}(\theta), \quad \mathbf{G}_3^T(\theta)\mathbf{m}(\theta) = \mathbf{0}, \tag{7.11-3}$$

$$\begin{aligned}
\cos(\theta-\theta_o)\mathbf{G}_1^T(\theta)\mathbf{n}(\theta) &= \mathbf{G}_1^T(\theta)\mathbf{n}(\theta_o)-\sin(\theta-\theta_o)\mathbf{n}(\theta),\\
\sin(\theta-\theta_o)\mathbf{G}_1^T(\theta)\mathbf{n}(\theta) &= \mathbf{G}_1^T(\theta)\mathbf{m}(\theta_o)+\cos(\theta-\theta_o)\mathbf{n}(\theta),\\
\cos(\theta-\theta_o)\mathbf{G}_3^T(\theta)\mathbf{n}(\theta) &= \mathbf{G}_3^T(\theta)\mathbf{n}(\theta_o),\\
\sin(\theta-\theta_o)\mathbf{G}_3^T(\theta)\mathbf{n}(\theta) &= \mathbf{G}_3^T(\theta)\mathbf{m}(\theta_o).
\end{aligned} \tag{7.11-4}$$

By taking the trace on both sides of (7.11-1) one has

$$\mathrm{tr}\langle p_*(\theta)\rangle = \mathrm{tr}\,\mathbf{G}_1(\theta)+i\,\mathrm{tr}\,\mathbf{G}_3(\theta). \tag{7.11-5}$$

Since $p_\alpha(\theta)$ (α=1,2,3) and their complex conjugates are eigenvalues of $\mathbf{N}(\theta)$,

$$\mathrm{tr}\langle p_*(\theta)\rangle + \mathrm{tr}\langle \bar{p}_*(\theta)\rangle = \mathrm{tr}\,\mathbf{N}(\theta) = 2\,\mathrm{tr}\,\mathbf{N}_1(\theta).$$

We therefore have the relations

$$\mathrm{Re}\{\mathrm{tr}\langle p_*(\theta)\rangle\} = \mathrm{tr}\,\mathbf{G}_1(\theta) = \mathrm{tr}\,\mathbf{N}_1(\theta), \tag{7.11-6}$$

$$\mathrm{Im}\{\mathrm{tr}\langle p_*(\theta)\rangle\} = \mathrm{tr}\,\mathbf{G}_3(\theta) > 0, \tag{7.11-7}$$

and, by (7.11-2),

$$\mathrm{tr}\{\mathbf{N}_3(\theta)\mathbf{S}\mathbf{L}^{-1}\} = 0 = \mathrm{tr}\{\mathbf{G}_3(\theta)\mathbf{S}^T\}. \tag{7.11-8}$$

From the structures of $\mathbf{N}_1$ and $\mathbf{N}_3$ given in (6.1-5) and (6.1-6) it is easily shown that $\mathbf{G}_1(0)$ and $\mathbf{G}_3(0)$ have the structures

$$\mathbf{G}_1(0) = \begin{bmatrix} * & * & * \\ -1 & 0 & * \\ * & * & * \end{bmatrix}, \quad \mathbf{G}_3(0) = \begin{bmatrix} * & * & * \\ 0 & 0 & 0 \\ * & * & * \end{bmatrix} \tag{7.11-9}$$

where the asterisk $*$ denotes a possibly nonzero element. By setting θ=0 in (7.5-1) one obtains

$$\mathbf{N}(\tfrac{\pi}{2}) = -\mathbf{N}^{-1}(0).$$

Hence

$$\mathbf{N}_1(\tfrac{\pi}{2}) = -\mathbf{N}_1^{(-1)}(0), \quad \mathbf{N}_3(\tfrac{\pi}{2}) = -\mathbf{N}_3^{(-1)}(0).$$

The structures of $\mathbf{N}_1^{(-1)}(0)$ and $\mathbf{N}_3^{(-1)}(0)$ have been given in (6.2-5) and (6.2-6), and we have

$$\mathbf{G}_1(\tfrac{\pi}{2}) = \begin{bmatrix} 0 & 1 & 0 \\ * & * & * \\ * & * & * \end{bmatrix}, \quad \mathbf{G}_3(\tfrac{\pi}{2}) = \begin{bmatrix} 0 & 0 & 0 \\ * & * & * \\ * & * & * \end{bmatrix}. \tag{7.11-10}$$

For isotropic materials it can be shown that (Ting and Yan, 1991)

$$\mathbf{G}_1(\theta) = \begin{bmatrix} \sin 2\theta & -\cos 2\theta & 0 \\ -\cos 2\theta & -\sin 2\theta & 0 \\ 0 & 0 & 0 \end{bmatrix},$$

$$\mathbf{G}_3(\theta) = \begin{bmatrix} 1+\cos 2\theta & \sin 2\theta & 0 \\ \sin 2\theta & 1-\cos 2\theta & 0 \\ 0 & 0 & 1 \end{bmatrix}. \tag{7.11-11}$$

Chapter 8

GREEN'S FUNCTIONS FOR INFINITE SPACE, HALF-SPACE, AND COMPOSITE SPACE

One of the basic solutions for anisotropic elasticity, and for other subject for that matter, is the Green's function. The importance of Green's functions in constructing solutions to boundary value problems has been well recognized. We will study in this chapter Green's functions for infinite spaces, half-spaces, and bimaterials that consist of two half-spaces of dissimilar materials bonded together. Also studied are a composite space that consists of wedges of dissimilar materials and an angularly inhomogeneous space. Green's functions for the infinite space with the presence of a crack, an elliptic hole, or an elliptic inclusion will be studied in separate chapters. We will be concerned mainly with Green's functions due to a line of concentrated forces and a line dislocation that have the r^{-1} stress singularity where r is the radial distance from the line of forces or the line dislocation. Green's functions due to a concentrated couple, a double force, a center of dilatation, etc., that provide the r^{-2} stress singularities will be discussed in Section 8.12. We will see that most solutions can be expressed in a real form with the identities presented in Chapters 6 and 7.

8.1 Line Force and Line Dislocation in Infinite Space

The Green's function for two-dimensional deformations of an infinite anisotropic elastic material subject to a line dislocation has been obtained by Eshelby et al. (1953), Stroh (1958), Willis (1970), Malen and Lothe (1970), and Malen (1971). Further developments of the Green's function to include a line force was given by Barnett and Lothe (1975a). The solution was in a complex form. A real form solution using an integral representation was derived by Barnett and Swanger (1971) and Asaro et al. (1973) (see also Mura, 1975). Most of the real form solutions obtained by these authors were for the displacement gradient, and hence for the strain. The stress was then obtained indirectly through the stress-strain relations. Chadwick and Smith (1977) did present real form solutions for the displacement as well as the stress directly. The solutions required the inverse of the 6×6 matrix $(x_1\mathbf{I} + x_2\mathbf{N})$ which can be achieved by employing (7.9-17). Kirchner and Lothe (1987) appear to be the first ones to present real form solutions in a simplest form. A comprehensive treatment of dislocations can be found in the books by Hirth and Lothe (1982) and, more recently, Indenbom and Lothe (1992). The problem of continuous distributions of dislocations have been studied by Bilby (1960), Bilby and Smith (1956), Bilby and Gardner (1958), and Bilby et al. (1955, 1958).

The general solutions for the displacement $\mathbf{u}$ and the stress function $\boldsymbol{\phi}$ are (Section 5.3)

$$\mathbf{u} = 2\,\mathrm{Im}\{\mathbf{A}\langle f(z_*)\rangle\mathbf{q}\}, \quad \boldsymbol{\phi} = 2\,\mathrm{Im}\{\mathbf{B}\langle f(z_*)\rangle\mathbf{q}\}, \tag{8.1-1}$$

$$\langle f(z_*)\rangle = \mathrm{diag}[f(z_1),\ f(z_2),\ f(z_3)],$$

$$z_\alpha = x_1 + p_\alpha x_2 \quad (\alpha = 1,2,3).$$

In the above $\mathbf{A}$ and $\mathbf{B}$ are elasticity matrices, $\mathbf{q}$ is an arbitrary constant vector, f is an arbitrary function of z_α, and p_α are the eigenvalues of the fundamental elasticity matrix. The stresses σ_{ij} are determined by

$$\sigma_{i1} = -\varphi_{i,2}, \quad \sigma_{i2} = \varphi_{i,1}, \tag{8.1-2}$$

where a comma stands for differentiation with x_1 or x_2, and φ_i are the components of the vector $\boldsymbol{\phi}$. We record below identities from (6.5-4), (6.5-8), (6.5-9), (5.6-6) and (7.5-6)–(7.5-8) which will be used frequently in the sequel.

$$\tilde{\mathbf{N}}\tilde{\mathbf{N}} = \begin{bmatrix} \mathbf{S} & \mathbf{H} \\ -\mathbf{L} & \mathbf{S}^T \end{bmatrix}\begin{bmatrix} \mathbf{S} & \mathbf{H} \\ -\mathbf{L} & \mathbf{S}^T \end{bmatrix} = -\mathbf{I}, \tag{8.1-3a}$$

$$\begin{aligned} &\mathbf{HL} - \mathbf{SS} = \mathbf{I} = \mathbf{LH} - \mathbf{S}^T\mathbf{S}^T, \\ &\mathbf{SH} + \mathbf{HS}^T = \mathbf{0} = \mathbf{LS} + \mathbf{S}^T\mathbf{L}, \end{aligned} \tag{8.1-3b}$$

$$\begin{aligned} &\mathbf{M} = -i\mathbf{BA}^{-1} = \mathbf{H}^{-1} + i\mathbf{H}^{-1}\mathbf{S}, \\ &\mathbf{M}^{-1} = i\mathbf{AB}^{-1} = \mathbf{L}^{-1} - i\mathbf{SL}^{-1}, \end{aligned} \tag{8.1-4}$$

$$\mathbf{N}_1(\theta)\mathbf{m}(\theta) = -\mathbf{n}(\theta), \qquad \mathbf{N}_3(\theta)\mathbf{m}(\theta) = \mathbf{0}, \tag{8.1-5}$$

$$\begin{aligned} &\cos(\theta - \theta_o)\mathbf{n}(\theta) - \sin(\theta - \theta_o)\mathbf{m}(\theta) = \mathbf{n}(\theta_o), \\ &\sin(\theta - \theta_o)\mathbf{n}(\theta) + \cos(\theta - \theta_o)\mathbf{m}(\theta) = \mathbf{m}(\theta_o), \end{aligned} \tag{8.1-6}$$

$$\begin{aligned} &\cos(\theta - \theta_o)\mathbf{N}_1(\theta)\mathbf{n}(\theta) = \mathbf{N}_1(\theta)\mathbf{n}(\theta_o) - \sin(\theta - \theta_o)\mathbf{n}(\theta), \\ &\sin(\theta - \theta_o)\mathbf{N}_1(\theta)\mathbf{n}(\theta) = \mathbf{N}_1(\theta)\mathbf{m}(\theta_o) + \cos(\theta - \theta_o)\mathbf{n}(\theta), \end{aligned} \tag{8.1-7$_{1,2}$}$$

$$\cos(\theta-\theta_o)\mathbf{N}_3(\theta)\mathbf{n}(\theta)=\mathbf{N}_3(\theta)\mathbf{n}(\theta_o),$$
$$\sin(\theta-\theta_o)\mathbf{N}_3(\theta)\mathbf{n}(\theta)=\mathbf{N}_3(\theta)\mathbf{m}(\theta_o). \qquad (8.1\text{-}7)_{3,4}$$

The 3×3 real matrices **S**, **H**, and **L** are the Barnett–Lothe tensors. **H** and **L** are symmetric and positive definite. **M** is the impedance tensor which is a positive definite Hermitian (section 6.6).

In a cylindrical coordinate system (r,θ,x_3), (8.1-2) are replaced by (Section 7.3)

$$\mathbf{t}_r=-\frac{1}{r}\boldsymbol{\phi}_{,\theta},\quad \mathbf{t}_\theta=\boldsymbol{\phi}_{,r}, \qquad (8.1\text{-}8a)$$

where a comma followed by r or θ denotes differentiation with r or θ and

$$(\mathbf{t}_r)_i=\sigma_{ij}n_j,\quad (\mathbf{t}_\theta)_i=\sigma_{ij}m_j, \qquad (8.1\text{-}8b)$$

$$\mathbf{n}^T=[\cos\theta,\ \sin\theta,\ 0],\quad \mathbf{m}^T=[-\sin\theta,\ \cos\theta,\ 0], \qquad (8.1\text{-}9)$$

$$\mathbf{n}_{,\theta}=\mathbf{m},\qquad \mathbf{m}_{,\theta}=-\mathbf{n}. \qquad (8.1\text{-}10)$$

The vectors $\mathbf{t}_r$ and $\mathbf{t}_\theta$ are, respectively, the traction vectors on the cylindrical surface r=constant and on the radial plane θ=constant. The stress components in the cylindrical coordinate system are

$$\sigma_{rr}=\mathbf{n}\cdot\mathbf{t}_r,\quad \sigma_{r\theta}=\mathbf{m}\cdot\mathbf{t}_r=\mathbf{n}\cdot\mathbf{t}_\theta,$$
$$\sigma_{\theta\theta}=\mathbf{m}\cdot\mathbf{t}_\theta,\quad \sigma_{r3}=\mathbf{i}_3\cdot\mathbf{t}_r,\quad \sigma_{\theta3}=\mathbf{i}_3\cdot\mathbf{t}_\theta, \qquad (8.1\text{-}11)$$

where $\mathbf{i}_3$ is the unit vector in the x_3-direction. Insertion from (8.1-8a) and making use of (8.1-10) we have

$$\sigma_{rr}=-\frac{1}{r}\{(\mathbf{n}\cdot\boldsymbol{\phi})_{,\theta}-(\mathbf{m}\cdot\boldsymbol{\phi})\},\quad \sigma_{\theta\theta}=(\mathbf{m}\cdot\boldsymbol{\phi})_{,r},$$
$$\sigma_{r\theta}=-\frac{1}{r}\{(\mathbf{m}\cdot\boldsymbol{\phi})_{,\theta}+(\mathbf{n}\cdot\boldsymbol{\phi})\}=(\mathbf{n}\cdot\boldsymbol{\phi})_{,r}, \qquad (8.1\text{-}12)$$
$$\sigma_{r3}=-\frac{1}{r}\varphi_{3,\theta},\quad \sigma_{\theta3}=\varphi_{3,r}.$$

Equations (8.1-12) provide the cylindrical components of stress in terms of the stress function $\boldsymbol{\phi}$. In some cases (8.1-12) is more convenient than (8.1-11).

Let a line of uniformly distributed forces **f** per unit length be applied on the x_3-axis and a line dislocation be at the x_3-axis with Burgers vector **b**. Without loss in generality we take the plane $x_2=0$, $x_1<0$ to be the slip plane. Introducing a branch cut along the negative x_1-axis we have

$$r>0,\qquad -\pi<\theta<\pi.$$

The boundary conditions are

$$\sigma_{ij} = 0 \quad \text{at} \quad |\mathbf{x}| = \infty, \tag{8.1-13}$$

$$\mathbf{u}(r,\pi) - \mathbf{u}(r,-\pi) = \mathbf{b}, \quad \boldsymbol{\phi}(r,\pi) - \boldsymbol{\phi}(r,-\pi) = \mathbf{f}. \tag{8.1-14}$$

As discussed in Section 3.4 the complex function $\ln z_\alpha$ is a multiple-valued function of θ which increases its value by $2\pi i$ each time θ is increased by 2π. It assumes the values

$$\ln z_\alpha = \begin{cases} \ln r, & \text{at } \theta = 0, \\ \ln r \pm i\pi, & \text{at } \theta = \pm\pi, \end{cases} \tag{8.1-15}$$

independent of α. Thus $\ln z_\alpha$ is the function for satisfying the boundary conditions (8.1-14). We write (8.1-1) as

$$\mathbf{u} = \frac{1}{\pi}\mathrm{Im}\{\mathbf{A}\langle \ln z_* \rangle \mathbf{q}^\infty\}, \quad \boldsymbol{\phi} = \frac{1}{\pi}\mathrm{Im}\{\mathbf{B}\langle \ln z_* \rangle \mathbf{q}^\infty\}, \tag{8.1-16}$$

where $\mathbf{q}^\infty$ is an arbitrary constant vector to be determined. The boundary condition (8.1-13) is satisfied when we insert $(8.1\text{-}16)_2$ into (8.1-2). Upon substitution of (8.1-16) into (8.1-14) we have

$$2\,\mathrm{Re}(\mathbf{A}\mathbf{q}^\infty) = \mathbf{b}, \quad 2\,\mathrm{Re}(\mathbf{B}\mathbf{q}^\infty) = \mathbf{f}, \tag{8.1-17a}$$

which can be written as

$$\begin{bmatrix} \mathbf{A} & \overline{\mathbf{A}} \\ \mathbf{B} & \overline{\mathbf{B}} \end{bmatrix} \begin{bmatrix} \mathbf{q}^\infty \\ \overline{\mathbf{q}}^\infty \end{bmatrix} = \begin{bmatrix} \mathbf{b} \\ \mathbf{f} \end{bmatrix}. \tag{8.1-17b}$$

Using the orthogonality relation (5.5-13) it leads to

$$\begin{bmatrix} \mathbf{q}^\infty \\ \overline{\mathbf{q}}^\infty \end{bmatrix} = \begin{bmatrix} \mathbf{B}^T & \mathbf{A}^T \\ \overline{\mathbf{B}}^T & \overline{\mathbf{A}}^T \end{bmatrix} \begin{bmatrix} \mathbf{b} \\ \mathbf{f} \end{bmatrix},$$

or

$$\mathbf{q}^\infty = \mathbf{A}^T\mathbf{f} + \mathbf{B}^T\mathbf{b}. \tag{8.1-18}$$

Reversing the procedure, (8.1-17a,b) are the solutions of (8.1-18) for the real vectors $\mathbf{f}$ and $\mathbf{b}$ when the complex vector $\mathbf{q}^\infty$ is given.

We therefore obtain the Green's function for the infinite space given by (8.1-16) and (8.1-18) which are written as

$$\begin{bmatrix} \mathbf{u} \\ \boldsymbol{\phi} \end{bmatrix} = \frac{1}{\pi}\mathrm{Im}\begin{bmatrix} \mathbf{A}\langle \ln z_* \rangle \mathbf{B}^T & \mathbf{A}\langle \ln z_* \rangle \mathbf{A}^T \\ \mathbf{B}\langle \ln z_* \rangle \mathbf{B}^T & \mathbf{B}\langle \ln z_* \rangle \mathbf{A}^T \end{bmatrix} \begin{bmatrix} \mathbf{b} \\ \mathbf{f} \end{bmatrix}.$$

The identity presented in (7.9-22) allows us to convert the solution into a real form

$$\begin{bmatrix} \mathbf{u} \\ \boldsymbol{\phi} \end{bmatrix} = -\frac{1}{2\pi}\Big((\ln r)\mathbf{I} + \pi\tilde{\mathbf{N}}(\theta)\Big)\begin{bmatrix} \mathbf{h} \\ \mathbf{g} \end{bmatrix} \tag{8.1-19}$$

where

$$\tilde{\mathbf{N}}(\theta) = \frac{1}{\pi}\int_0^\theta \mathbf{N}(\lambda)d\lambda = \begin{bmatrix} \mathbf{S}(\theta) & \mathbf{H}(\theta) \\ -\mathbf{L}(\theta) & \mathbf{S}^T(\theta) \end{bmatrix},$$

$$\tilde{\mathbf{N}} = \tilde{\mathbf{N}}(\pi) = \begin{bmatrix} \mathbf{S} & \mathbf{H} \\ -\mathbf{L} & \mathbf{S}^T \end{bmatrix},$$

and

$$\begin{bmatrix} \mathbf{h} \\ \mathbf{g} \end{bmatrix} = \tilde{\mathbf{N}}\begin{bmatrix} \mathbf{b} \\ \mathbf{f} \end{bmatrix} \tag{8.1-20a}$$

or

$$\mathbf{h} = \mathbf{Sb} + \mathbf{Hf}, \quad \mathbf{g} = \mathbf{S}^T\mathbf{f} - \mathbf{Lb}. \tag{8.1-20b}$$

Equation (8.1-19) is equivalent to

$$\begin{aligned} 2\mathbf{u} &= -\frac{1}{\pi}(\ln r)\mathbf{h} - \mathbf{S}(\theta)\mathbf{h} - \mathbf{H}(\theta)\mathbf{g}, \\ 2\boldsymbol{\phi} &= -\frac{1}{\pi}(\ln r)\mathbf{g} + \mathbf{L}(\theta)\mathbf{h} - \mathbf{S}^T(\theta)\mathbf{g}. \end{aligned} \tag{8.1-21}$$

This is the real form solution for the Green's function for the infinite space subject to a line force **f** and a line dislocation **b**. The integral representations of the tensors **S**, **H**, **L**, **S**(θ), **H**(θ), and **L**(θ) are defined in (7.6-4) and (7.6-6). The algebraic representations are given in (5.5-17), (6.5-2), and (7.6-7). We will show in Sections 8.10 and 8.11 that (8.1-21) remains valid for angularly inhomogeneous materials.

The vectors **g** and **h** have physical interpretations. From $(8.1\text{-}21)_1$,

$$\mathbf{u} \to \frac{1}{2\pi}\ln(r^{-1})\mathbf{h}, \quad \text{as } r \to 0,$$

indicating that **h** represents the direction of the infinite displacement at r=0. As to the vector **g**, insertion of $(8.1\text{-}21)_2$ into (8.1-8a) leads to

$$\mathbf{t}_r = \frac{1}{2r}\Big(\mathbf{N}_3(\theta)\mathbf{h} + \mathbf{N}_1^T(\theta)\mathbf{g}\Big), \quad \mathbf{t}_\theta = \frac{-1}{2\pi r}\mathbf{g}. \tag{8.1-22}$$

Thus $-$**g** represents the direction of the traction vector $\mathbf{t}_\theta$. Notice that $\mathbf{t}_\theta$ is independent of θ. Therefore if $\mathbf{t}_\theta$ vanishes at a particular θ, it vanishes for all θ.

From (8.1-22), (8.1-11), and the identities (8.1-5) we obtain

$$\begin{aligned}
\sigma_{rr} &= \frac{1}{2r}\mathbf{n}^T(\theta)\big(\mathbf{N}_3(\theta)\mathbf{h} + \mathbf{N}_1^T(\theta)\mathbf{g}\big),\\
\sigma_{r\theta} &= \frac{-1}{2\pi r}\mathbf{n}^T(\theta)\mathbf{g}, \quad \sigma_{\theta\theta} = \frac{-1}{2\pi r}\mathbf{m}^T(\theta)\mathbf{g},\\
\sigma_{r3} &= \frac{1}{2r}\mathbf{i}_3^T\big(\mathbf{N}_3(\theta)\mathbf{h} + \mathbf{N}_1^T(\theta)\mathbf{g}\big), \quad \sigma_{\theta 3} = \frac{-1}{2\pi r}g_3.
\end{aligned} \tag{8.1-23}$$

The stresses are inversely proportional to r. The only stress component missing is σ_{33} which is determined from (4.6-10), the condition for $\varepsilon_{33}=0$.

The fact that $\mathbf{t}_\theta$ is independent of θ is a consequence of the stress field that is proportional to r^{-1} (Section 8.11). Its validity is independent of the material property. Thus as long as the stress is proportional to r^{-1}, $\mathbf{t}_\theta$ is invariant with θ regardless of whether the material is isotropic or anisotropic, homogeneous or nonhomogeneous, elastic or inelastic, and linear or nonlinear.

The strain energy density is

$$\tfrac{1}{2}\sigma_{ij}\varepsilon_{ij} = \tfrac{1}{2}\sigma_{ij}u_{i,j} = \tfrac{1}{2}(\sigma_{ij}u_i)_{,j} \tag{8.1-24}$$

in which the last equality is reached by employing the equation of equilibrium. Integration over a volume with unit thickness in the x_3-direction and bounded by a curve Γ on the (x_1,x_2) plane can be converted into a line integral along the curve Γ by the divergence theorem. The energy $\mathcal{E}$ in the volume is

$$\mathcal{E} = \tfrac{1}{2}\int_\Gamma \sigma_{ij}u_i v_j ds = \tfrac{1}{2}\int_\Gamma \mathbf{u}\cdot\mathbf{t}ds \tag{8.1-25}$$

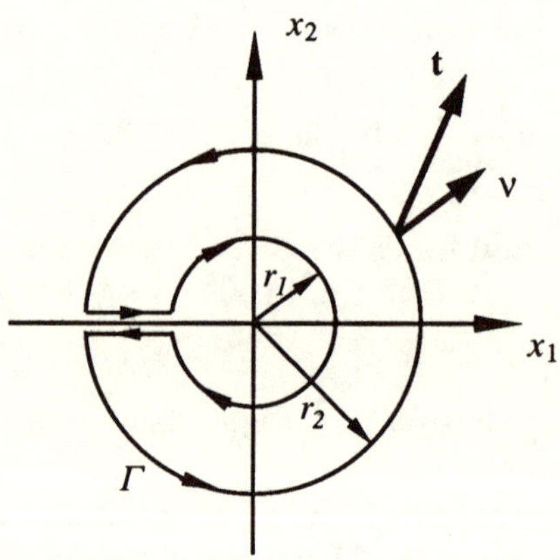

Fig. 8.1 Contour of the line integral in (8.1-25).

where v_j is a unit normal vector on Γ and $\mathbf{t}$ is the surface traction on Γ. For the annulus regions $0<r_1\le r\le r_2$ the Γ consists of two circles of radii r_1 and r_2 and the lines above and below the branch cut on the negative x_1-axis between $r=r_1$ and r_2, Fig. 8.1. The displacement inside Γ is continuous. On the circles $r=r_1$ and r_2,

$$r_2\mathbf{t}(r_2,\theta)=-r_1\mathbf{t}(r_1,\theta),\qquad \mathbf{t}(r_2,\theta)=-\frac{\partial}{r_2\partial\theta}\boldsymbol{\phi}(r_2,\theta),$$

and the integrals on the two circles are

$$\begin{aligned}&-\tfrac{1}{2}\int_{-\pi}^{\pi}\big(\mathbf{u}(r_2,\theta)-\mathbf{u}(r_1,\theta)\big)\cdot\frac{\partial}{\partial\theta}\boldsymbol{\phi}(r_2,\theta)d\theta\\&=\frac{1}{4\pi}\ln\left(\frac{r_2}{r_1}\right)\mathbf{h}\cdot[\boldsymbol{\phi}(r_2,\pi)-\boldsymbol{\phi}(r_2,-\pi)]=\frac{1}{4\pi}\ln\left(\frac{r_2}{r_1}\right)\mathbf{h}\cdot\mathbf{f},\end{aligned}$$

use being made of $(8.1\text{-}21)_1$ and $(8.1\text{-}14)_2$. On the lines above and below the branch cut,

$$\mathbf{t}(r,\pi)=-\mathbf{t}(r,-\pi),\qquad \mathbf{t}(r,\pi)=\frac{\partial}{\partial r}\boldsymbol{\phi}(r,\pi),$$

and the integrals are, after employing $(8.1\text{-}14)_1$ and $(8.1\text{-}21)_2$,

$$\begin{aligned}&\tfrac{1}{2}\int_{r_1}^{r_2}\big(\mathbf{u}(r,\pi)-\mathbf{u}(r,-\pi)\big)\cdot\frac{\partial}{\partial r}\boldsymbol{\phi}(r,\pi)dr\\&=\tfrac{1}{2}\mathbf{b}\cdot[\boldsymbol{\phi}(r_2,\pi)-\boldsymbol{\phi}(r_1,\pi)]=\frac{-1}{4\pi}\ln\left(\frac{r_2}{r_1}\right)\mathbf{b}\cdot\mathbf{g}.\end{aligned}$$

Equation (8.1-25) for the annulus region then reduces to

$$\varepsilon=\frac{1}{4\pi}\ln\left(\frac{r_2}{r_1}\right)(\mathbf{f}\cdot\mathbf{h}-\mathbf{b}\cdot\mathbf{g}). \tag{8.1-26}$$

Substitution from (8.1-20b) leads to the explicit expression

$$\varepsilon=\frac{1}{4\pi}\ln\left(\frac{r_2}{r_1}\right)(\mathbf{f}^T\mathbf{H}\mathbf{f}+\mathbf{b}^T\mathbf{L}\mathbf{b}). \tag{8.1-27}$$

Equation (8.1-27) provides an indirect proof that $\mathbf{H}$ and $\mathbf{L}$ must be positive definite if the strain energy is positive (Chadwick and Smith, 1977). Notice that there is no interaction of strain energy between the line force $\mathbf{f}$ and the line dislocation $\mathbf{b}$ (Eshelby, 1956).

Substituting $\mathbf{h}$ and $\mathbf{g}$ from (8.1-20b) we obtain

$$\mathbf{N}_3(\theta)\mathbf{h}+\mathbf{N}_1^T(\theta)\mathbf{g}=\mathbf{V}(\theta)\mathbf{b}+\mathbf{W}^T(\theta)\mathbf{f}, \tag{8.1-28}$$

where

$$\mathbf{V}(\theta) = \mathbf{N}_3(\theta)\mathbf{S} - \mathbf{N}_1^T(\theta)\mathbf{L} = \left(\mathbf{N}_3(\theta)\mathbf{S} - \mathbf{N}_1^T(\theta)\mathbf{L}\right)^T,$$
$$\mathbf{W}(\theta) = \mathbf{N}_1(\theta)\mathbf{S} - \mathbf{N}_2(\theta)\mathbf{L} = \left(\mathbf{N}_3(\theta)\mathbf{H} + \mathbf{N}_1^T(\theta)\mathbf{S}^T\right)^T. \tag{8.1-29}$$

The second equalities for $\mathbf{V}(\theta)$ and $\mathbf{W}(\theta)$ are taken from (7.8-10). The tensor $\mathbf{V}(\theta)$ is symmetric. The expression on the left of (8.1-28) has appeared in (8.1-22) and (8.1-23).

8.2 One-Component Green's Functions

The solution (8.1-16) for the infinite space can be recast into three independent solutions as

$$\mathbf{u} = \frac{1}{\pi}\mathrm{Im}\sum_{\alpha=1}^{3}(\ln z_\alpha)q_\alpha^\infty \boldsymbol{a}_\alpha, \quad \boldsymbol{\phi} = \frac{1}{\pi}\mathrm{Im}\sum_{\alpha=1}^{3}(\ln z_\alpha)q_\alpha^\infty \boldsymbol{b}_\alpha, \tag{8.2-1}$$

where $\boldsymbol{a}_\alpha$, $\boldsymbol{b}_\alpha$ are, respectively, the columns of $\mathbf{A}$ and $\mathbf{B}$,

$$\mathbf{A} = [\boldsymbol{a}_1,\ \boldsymbol{a}_2,\ \boldsymbol{a}_3], \quad \mathbf{B} = [\boldsymbol{b}_1,\ \boldsymbol{b}_2,\ \boldsymbol{b}_3].$$

The 6×1 vectors $(\boldsymbol{a}_\alpha, \boldsymbol{b}_\alpha)$ $(\alpha=1,2,3)$ are the eigenvectors of the fundamental elasticity matrix $\mathbf{N}$ (Section 5.5). We employ the italic $\boldsymbol{b}_\alpha$ for the columns of $\mathbf{B}$ (Chapter 5) to distinguish it from the Burgers vector $\mathbf{b}$. Each of the three solutions in (8.2-1) is a *one-component* Green's function of the form

$$\mathbf{u} = \frac{1}{\pi}\mathrm{Im}\left\{(\ln z)q^\infty \boldsymbol{a}\right\}, \quad \boldsymbol{\phi} = \frac{1}{\pi}\mathrm{Im}\left\{(\ln z)q^\infty \boldsymbol{b}\right\}, \tag{8.2-2}$$

subjected to a line force $\mathbf{f}$ and a line dislocation with Burgers vector $\mathbf{b}$ which are determined by making use of (8.1-17a) as

$$\mathbf{b} = 2\,\mathrm{Re}(q^\infty \boldsymbol{a}), \quad \mathbf{f} = 2\,\mathrm{Re}(q^\infty \boldsymbol{b}). \tag{8.2-3}$$

Let $\boldsymbol{a}'$, $\boldsymbol{a}''$ be, respectively, the real and imaginary parts of the vector $\boldsymbol{a}$. We assume that $\boldsymbol{a}$ is *genuinely complex* so that $\boldsymbol{a}'$ and $\boldsymbol{a}''$ are independent of each other . Otherwise the unnormalized $\boldsymbol{a}$ can be taken as a real vector multiplied by a complex factor. The vectors $\boldsymbol{a}'$ and $\boldsymbol{a}''$ span a plane in the (x_1,x_2,x_3) space. If $\boldsymbol{a}$ is multiplied by a complex factor, the real and imaginary parts of the new vector $\boldsymbol{a}$ span the same plane (Section 1.8). The plane is called the eigenplane $\boldsymbol{a}$. Likewise the real and imaginary parts of the vector $\boldsymbol{b}$ span an eigenplane $\boldsymbol{b}$. Equations (8.2-2) tell us that the displacement $\mathbf{u}$ is polarized on the eigenplane $\boldsymbol{a}$, while the stress function ϕ is polarized on the eigenplane $\boldsymbol{b}$. This means that the traction vector $\mathbf{t}$ on any curved boundary Γ is polarized on the eigenplane $\boldsymbol{b}$. The line dislocation $\mathbf{b}$ and the line force $\mathbf{f}$ as given by (8.2-3) show that they are polarized on the eigenplanes $\boldsymbol{a}$ and $\boldsymbol{b}$, respectively. Thus the Burgers vector $\mathbf{b}$ and the force $\mathbf{f}$ cannot be prescribed arbitrarily for a one-component Green's function. In fact only one of them, say $\mathbf{b}$, can be prescribed subjected to the condition that it lies on the eigenplane $\boldsymbol{a}$. Equation $(8.2\text{-}3)_1$ can then be solved for q^∞, and $(8.2\text{-}3)_2$ provides the associated $\mathbf{f}$.

To determine q^∞ from $(8.2\text{-}3)_1$ when the Burgers vector $\mathbf{b}$ on the eigenplane $\boldsymbol{a}$ is prescribed, we write $(8.2\text{-}3)_1$ as

$$\mathbf{b} = q^\infty \boldsymbol{a} + \bar{q}^\infty \bar{\boldsymbol{a}}.$$

Application of the scalar products on both sides by $\boldsymbol{a}$ and $\bar{\boldsymbol{a}}$, respectively, yields

$$\boldsymbol{a}\cdot\mathbf{b} = q^\infty \boldsymbol{a}\cdot\boldsymbol{a} + \bar{q}^\infty \boldsymbol{a}\cdot\bar{\boldsymbol{a}},$$

$$\bar{\boldsymbol{a}}\cdot\mathbf{b} = q^\infty \bar{\boldsymbol{a}}\cdot\boldsymbol{a} + \bar{q}^\infty \bar{\boldsymbol{a}}\cdot\bar{\boldsymbol{a}},$$

from which we have

$$q^\infty = \frac{(\bar{\boldsymbol{a}}\cdot\boldsymbol{a})(\bar{\boldsymbol{a}}\cdot\mathbf{b}) - (\bar{\boldsymbol{a}}\cdot\bar{\boldsymbol{a}})(\boldsymbol{a}\cdot\mathbf{b})}{(\bar{\boldsymbol{a}}\cdot\boldsymbol{a})(\bar{\boldsymbol{a}}\cdot\boldsymbol{a}) - (\bar{\boldsymbol{a}}\cdot\bar{\boldsymbol{a}})(\boldsymbol{a}\cdot\boldsymbol{a})}. \tag{8.2-4}$$

The denominator can be expressed in terms of the real and imaginary parts of $\boldsymbol{a}$ as

$$4\left\{(\boldsymbol{a}'\cdot\boldsymbol{a}')(\boldsymbol{a}''\cdot\boldsymbol{a}'') - (\boldsymbol{a}'\cdot\boldsymbol{a}'')^2\right\} > 0.$$

The inequality holds because $\boldsymbol{a}$ is genuinely complex so that neither $\boldsymbol{a}'$ nor $\boldsymbol{a}''$ can be a zero vector.

In summary one can have a one-component Green's function if the Burgers vector $\mathbf{b}$ is on the eigenplane $\boldsymbol{a}$ and $\mathbf{f}$ is on the eigenplane $\boldsymbol{b}$. Moreover $\mathbf{f}$ is related to $\mathbf{b}$ through $(8.2\text{-}3)_2$ and (8.2-4). A one-component Green's function has the feature that the displacement $\mathbf{u}$ is polarized on the eigenplane $\boldsymbol{a}$ while the traction vector $\mathbf{t}$ on any curved surface is polarized on the eigenplane $\boldsymbol{b}$.

It is possible to have a two-component Green's function. For instance, if the Burgers vector $\mathbf{b}$ is normal to the eigenplane $\boldsymbol{b}_1$ and $\mathbf{f}$ is normal to the eigenplane $\boldsymbol{a}_1$, q_1^∞ vanishes according to (8.1-18). We then have a two-component Green's function associated with q_2^∞ and q_3^∞.

When $\mathbf{b}$ and $\mathbf{f}$ are prescribed arbitrarily all three one-component Green's functions are required for the solution.

For degenerate materials for which $(\boldsymbol{a}_1,\boldsymbol{b}_1)$ and $(\boldsymbol{a}_3,\boldsymbol{b}_3)$ are the eigenvectors and $(\boldsymbol{a}_2,\boldsymbol{b}_2)$ is the generalized eigenvector, there are still three one-component Green's functions. However, the displacement **u** and the traction vector **t** of the one-component Green's function associated with the generalized eigenvector $(\boldsymbol{a}_2,\boldsymbol{b}_2)$ are no longer polarized on a single plane (Ting, 1992e).

8.3 Physical Meaning of the Eigenvectors $\mathbf{e}_i$ and $\mathbf{e}^i$

We have seen in the previous section interesting relations between **b**, **f** and **u**, **t**. In this section we investigate relations between **b**, **f** and **g**, **h**, which are given by (8.1-20b), namely,

$$\mathbf{h} = \mathbf{S}\mathbf{b} + \mathbf{H}\mathbf{f}, \quad \mathbf{g} = \mathbf{S}^T\mathbf{f} - \mathbf{L}\mathbf{b}. \tag{8.3-1}$$

As pointed out in Section 8.1 **h** represents the direction of the infinite displacement at r=0 and $-\mathbf{g}$ represents the direction of the surface traction $\mathbf{t}_\theta$ on any radial plane. The vector $\mathbf{t}_\theta$ is invariant with θ. Making use of (8.1-3b) it is readily shown that

$$\mathbf{g}\cdot\mathbf{h} = \mathbf{f}\cdot\mathbf{b}. \tag{8.3-2}$$

Therefore if **f** and **b** are orthogonal to each other, so are **g** and **h**. In particular, if the infinite space is subjected to a line force **f** only, or a line dislocation **b** only, **g** and **h** are orthogonal to each other. We also have from (8.3-1)

$$\left.\begin{aligned} &\mathbf{g} = \mathbf{0}, \quad \text{if} \quad \mathbf{b} = \mathbf{L}^{-1}\mathbf{S}^T\mathbf{f} = -\mathbf{S}\mathbf{L}^{-1}\mathbf{f}, \\ &\mathbf{h} = \mathbf{0}, \quad \text{if} \quad \mathbf{f} = -\mathbf{H}^{-1}\mathbf{S}\mathbf{b}. \end{aligned}\right\} \tag{8.3-3}$$

Noting that $\mathbf{S}\mathbf{L}^{-1}$ and $\mathbf{H}^{-1}\mathbf{S}$ are skew-symmetric they possess axial vectors. If the axial vectors are $\mathbf{e}^3$ and $\mathbf{e}_3$, respectively, special cases of (8.3-3) are

$$\left.\begin{aligned} &\mathbf{g} = \mathbf{0} \text{ and } \mathbf{b} = \mathbf{0}, \quad \text{if} \quad \mathbf{f} = \mathbf{e}^3, \\ &\mathbf{h} = \mathbf{0} \text{ and } \mathbf{f} = \mathbf{0}, \quad \text{if} \quad \mathbf{b} = \mathbf{e}_3. \end{aligned}\right\} \tag{8.3-4}$$

It should be pointed out that **h=0** implies that, by $(8.1\text{-}21)_1$,

$$\mathbf{u} = -\tfrac{1}{2}\mathbf{H}(\theta)\mathbf{g} \tag{8.3-5}$$

which is bounded everywhere including at r=0. In fact **u** is independent of r. Therefore when **h**=0 the displacement at r=0 is bounded. However the displacement at r=0 depends on θ and hence is discontinuous there. Equations (8.3-4) tell us that when the infinite space is subjected to a line force **f** acting in the direction of

$\mathbf{e}^3$, the traction on any radial plane vanishes. If the infinite space is subjected to a line dislocation $\mathbf{b}$ in the direction of $\mathbf{e}_3$, the displacement is given by (8.3-5) which is bounded at r=0.

The structure of $\mathbf{S}$, $\mathbf{H}$, $\mathbf{L}$ has been analyzed in Section 6.6. It was shown there that the eigenvalues of $\mathbf{S}$ are $\pm is$, 0 where

$$1 > s = \left\{-\tfrac{1}{2}\mathrm{tr}(\mathbf{S}^2)\right\}^{1/2} > 0. \tag{8.3-6}$$

Let the associated eigenvectors be $\mathbf{e}_1 \mp i\mathbf{e}_2$, $\mathbf{e}_3$ in which $\mathbf{e}_1$, $\mathbf{e}_2$, $\mathbf{e}_3$, are real vectors and let the reciprocal vectors $\mathbf{e}^1$, $\mathbf{e}^2$, $\mathbf{e}^3$, satisfy the orthogonality relations

$$\mathbf{e}_i \cdot \mathbf{e}^j = \delta_{ij} = \mathbf{e}_i{}^T\mathbf{e}^j. \tag{8.3-7}$$

We then have

$$\begin{aligned}
\mathbf{S} &= s\left\{\mathbf{e}_2(\mathbf{e}^1)^T - \mathbf{e}_1(\mathbf{e}^2)^T\right\},\\
\mathbf{H} &= \frac{1-s^2}{\mu\kappa}\left\{\mathbf{e}_1(\mathbf{e}_1)^T + \mathbf{e}_2(\mathbf{e}_2)^T\right\} + \frac{1}{\mu}\mathbf{e}_3(\mathbf{e}_3)^T,\\
\mathbf{L} &= \mu\kappa\left\{\mathbf{e}^1(\mathbf{e}^1)^T + \mathbf{e}^2(\mathbf{e}^2)^T\right\} + \mu\mathbf{e}^3(\mathbf{e}^3)^T,
\end{aligned} \tag{8.3-8}$$

where μ and κ are positive, nonzero constants. Equations (8.3-1) can now be written as

$$\mathbf{h} = s\left\{\mathbf{e}_2(\mathbf{e}^1\cdot\mathbf{b}) - \mathbf{e}_1(\mathbf{e}^2\cdot\mathbf{b})\right\} + \frac{1-s^2}{\mu\kappa}\left\{\mathbf{e}_1(\mathbf{e}_1\cdot\mathbf{f}) + \mathbf{e}_2(\mathbf{e}_2\cdot\mathbf{f})\right\} + \frac{1}{\mu}\mathbf{e}_3(\mathbf{e}_3\cdot\mathbf{f}),$$

$$\mathbf{g} = s\left\{\mathbf{e}^1(\mathbf{e}_2\cdot\mathbf{f}) - \mathbf{e}^2(\mathbf{e}_1\cdot\mathbf{f})\right\} - \mu\kappa\left\{\mathbf{e}^1(\mathbf{e}^1\cdot\mathbf{b}) + \mathbf{e}^2(\mathbf{e}^2\cdot\mathbf{b})\right\} - \mu\mathbf{e}^3(\mathbf{e}^3\cdot\mathbf{b}),$$

and, by virtue of (8.3-7), the following results emerge.

(1) Let $\mathbf{b}=\mathbf{0}$. Then

$$\begin{aligned}
\mathbf{h} &= \frac{1-s^2}{\mu\kappa}\mathbf{e}_1, &\quad \mathbf{g} &= -s\mathbf{e}^2, &\quad &\text{if } \mathbf{f} = \mathbf{e}^1,\\
\mathbf{h} &= \frac{1-s^2}{\mu\kappa}\mathbf{e}_2, & \mathbf{g} &= s\mathbf{e}^1, & &\text{if } \mathbf{f} = \mathbf{e}^2,\\
\mathbf{h} &= \frac{1}{\mu}\mathbf{e}_3, & \mathbf{g} &= \mathbf{0}, & &\text{if } \mathbf{f} = \mathbf{e}^3.
\end{aligned}$$

(2) Let $\mathbf{f}=\mathbf{0}$. Then

$$\begin{aligned}
\mathbf{h} &= s\mathbf{e}_2, &\quad \mathbf{g} &= -\mu\kappa\mathbf{e}^1, &\quad &\text{if } \mathbf{b} = \mathbf{e}_1.\\
\mathbf{h} &= -s\mathbf{e}_1, & \mathbf{g} &= -\mu\kappa\mathbf{e}^2, & &\text{if } \mathbf{b} = \mathbf{e}_2.\\
\mathbf{h} &= \mathbf{0}, & \mathbf{g} &= -\mu\mathbf{e}^3, & &\text{if } \mathbf{b} = \mathbf{e}_3.
\end{aligned}$$

The plane spanned by the vectors $\mathbf{e}_1$ and $\mathbf{e}_2$ is called the Γ_b plane and the plane spanned by $\mathbf{e}^1$ and $\mathbf{e}^2$ the Γ_f plane (Section 6.7). The vectors $\mathbf{e}_3$ and $\mathbf{e}^3$ are, respectively, the right and left null vectors of $\mathbf{S}$. They are also the axial vectors of $\mathbf{H}^{-1}\mathbf{S}$ and $\mathbf{SL}^{-1}$, respectively. The results presented above state that if $\mathbf{f}$ stays on the Γ_f plane and $\mathbf{b}$ stays on the Γ_b plane, then $\mathbf{g}$ is on the Γ_f plane and $\mathbf{h}$ is on the Γ_b plane. Also, if $\mathbf{f}$ is along the direction of $\mathbf{e}^3$, $\mathbf{h}$ is in the direction of $\mathbf{e}_3$ and $\mathbf{g}=\mathbf{0}$. If $\mathbf{b}$ is along the direction of $\mathbf{e}_3$, $\mathbf{g}$ is in the direction of $\mathbf{e}^3$ and $\mathbf{h}=\mathbf{0}$.

For monoclinic materials with the symmetry plane at $x_3=0$, if the force $\mathbf{f}$ is in the direction of the x_3-axis it is expected that $\mathbf{h}$ is also in the direction of the x_3-axis. Hence $\mathbf{e}_3=\mathbf{e}^3=\mathbf{i}_3$, a unit vector along the x_3-axis. For monoclinic materials with the symmetry plane at $x_1=0$, if $\mathbf{f}$ is in the direction of the x_1-axis so is $\mathbf{h}$. In this case $\mathbf{e}_1$ and $\mathbf{e}^1$ are along the x_1-axis. Likewise, for monoclinic materials with the symmetry plane at $x_2=0$ $\mathbf{e}_2$ and $\mathbf{e}^2$ are along the x_2-axis.

From (8.3-8) it can be shown that (see also (6.6-29) and (6.6-31))

$$\begin{aligned}\mathbf{SL}^{-1} &= \frac{s}{\mu\kappa}\left\{\mathbf{e}_2(\mathbf{e}_1)^T - \mathbf{e}_1(\mathbf{e}_2)^T\right\},\\ \mathbf{H}^{-1}\mathbf{S} &= \frac{\mu\kappa s}{1-s^2}\left\{\mathbf{e}^2(\mathbf{e}^1)^T - \mathbf{e}^1(\mathbf{e}^2)^T\right\},\end{aligned} \tag{8.3-9}$$

and, by (8.3-3), we have the following results.

(3) $\mathbf{g}=\mathbf{0}$,

(i) if $\mathbf{f}=\mathbf{e}^1$ and $\mathbf{b}=-\dfrac{s}{\mu\kappa}\mathbf{e}_2$, (then $\mathbf{h}=\dfrac{1}{\mu\kappa}\mathbf{e}_1$);

(ii) if $\mathbf{f}=\mathbf{e}^2$ and $\mathbf{b}=\dfrac{s}{\mu\kappa}\mathbf{e}_1$, (then $\mathbf{h}=\dfrac{1}{\mu\kappa}\mathbf{e}_2$);

(iii) if $\mathbf{f}=\mathbf{e}^3$ and $\mathbf{b}=\mathbf{0}$, (then $\mathbf{h}=\dfrac{1}{\mu}\mathbf{e}_3$).

(4) $\mathbf{h}=\mathbf{0}$,

(i) if $\mathbf{b}=\mathbf{e}_1$ and $\mathbf{f}=-\dfrac{\mu\kappa s}{1-s^2}\mathbf{e}^2$, (then $\mathbf{g}=-\dfrac{\mu\kappa}{1-s^2}\mathbf{e}^1$);

(ii) if $\mathbf{b}=\mathbf{e}_2$ and $\mathbf{f}=\dfrac{\mu\kappa s}{1-s^2}\mathbf{e}^1$, (then $\mathbf{g}=-\dfrac{\mu\kappa}{1-s^2}\mathbf{e}^2$);

(iii) if $\mathbf{b}=\mathbf{e}_3$ and $\mathbf{f}=\mathbf{0}$, (then $\mathbf{g}=-\mu\mathbf{e}^3$).

Thus it is possible to have Green's functions with nonzero $\mathbf{b}$ and $\mathbf{f}$ for which the traction on all radial planes vanishes or the displacement is bounded everywhere.

8.4 Interaction of Two Line Dislocations

Let a line dislocation with Burgers vector $\mathbf{b}^{(1)}$ be located at the origin $(x_1,x_2) = (0,0)$ and a second line dislocation with Burgers vector $\mathbf{b}^{(2)}$ be at $(x_1,x_2) = (x_1^o,x_2^o)$. When we change the location of the second dislocation slightly, the total strain energy changes. With the postulation that the dislocations tend to move into positions of lower energy, the negative of the derivatives of the total energy with respect to the position (x_1^o,x_2^o) of the second dislocation provide the *image force*, or the Peach–Koehler (1950) force exerted on the second dislocation. The *interaction energy* V is that part of the total energy that depends on the position (x_1^o,x_2^o) of the second dislocation. It can be determined by the formation of the second line dislocation in the presence of the first line dislocation (Cottrell, 1949). The second line dislocation is formed by making a branch cut along the line $x_2 = x_2^o$, $-\infty < x_1 < x_1^o$ and giving the surfaces of the cut a relative displacement $\mathbf{b}^{(2)}$. The work done in this process with the presence of the stress field by the first line dislocation is

$$V = -b_i^{(2)} \int_{-\infty}^{x_1^o} \sigma_{i2}^{(1)}(x_1, x_2^o)dx_1 . \tag{8.4-1a}$$

An alternate is to make a cut along $x_1 = x_1^o$, $x_2^o < x_2 < \infty$. We have

$$V = -b_i^{(2)} \int_{x_2^o}^{\infty} \sigma_{i1}^{(1)}(x_1^o, x_2)dx_2 . \tag{8.4-1b}$$

It is a simple matter to prove that V given by (8.4-1a) represents the interaction energy that depends on the position (x_1^o,x_2^o) of the second line dislocation. From the equation of equilibrium we have

$$0 = \sigma_{ij,j}^{(1)} u_i^{(2)} = (\sigma_{ij}^{(1)} u_i^{(2)})_{,j} - \sigma_{ij}^{(1)} u_{i,j}^{(2)} = (\sigma_{ij}^{(1)} u_i^{(2)})_{,j} - C_{ijks} u_{k,s}^{(1)} u_{i,j}^{(2)} .$$

Interchanging (1) and (2) leads to the *reciprocal theorem*

$$(\sigma_{ij}^{(1)} u_i^{(2)})_{,j} = (\sigma_{ij}^{(2)} u_i^{(1)})_{,j} . \tag{8.4-2}$$

From (8.1-24) and (8.4-2) the strain energy density is

$$\tfrac{1}{2}\left\{(\sigma_{ij}^{(1)} + \sigma_{ij}^{(2)})(u_i^{(1)} + u_i^{(2)})\right\}_{,j} = \tfrac{1}{2}\left\{\sigma_{ij}^{(1)} u_i^{(1)} + \sigma_{ij}^{(2)} u_i^{(2)} + 2\sigma_{ij}^{(1)} u_i^{(2)}\right\}_{,j}$$

and, following the derivation of (8.1-25) we obtain the total energy,

$$\mathcal{E} = \tfrac{1}{2}\int_{\Gamma} \left\{\mathbf{u}^{(1)} \cdot \mathbf{t}^{(1)} + \mathbf{u}^{(2)} \cdot \mathbf{t}^{(2)} + 2\mathbf{u}^{(2)} \cdot \mathbf{t}^{(1)}\right\} ds . \tag{8.4-3}$$

The Γ encircles a region inside of which the displacement and the stress are continuous. For the first term in (8.4-3) the Γ is given by the one shown in Fig. 8.1 with $r_1 \to 0$ and $r_2 \to \infty$. The integral of the first term thus represents the total strain energy due to the dislocation $\mathbf{b}^{(1)}$ alone and hence is independent of (x_1^o, x_2^o). The Γ for the second term in (8.4-3) is similar to that shown in Fig. 8.1 except that the origin is moved to (x_1^o, x_2^o). The integral of the second term also represents the total strain energy due to the line dislocation $\mathbf{b}^{(2)}$ alone and is independent of the location of $\mathbf{b}^{(2)}$. Thus the interaction energy comes from the third term in (8.4-3) which depends on the displacement $\mathbf{u}^{(2)}$ caused by the dislocation $\mathbf{b}^{(2)}$ and the traction $\mathbf{t}^{(1)}$ caused by the dislocation $\mathbf{b}^{(1)}$. This justifies the name *interaction energy*. The Γ of the third term in (8.4-3) consists of one infinitely large circle of radius R encircling both dislocations, two infinitesimal circles of radii R_1, R_2 encircling the dislocations at $(x_1, x_2) = (0,0)$ and (x_1^o, x_2^o), respectively, and two lines each along the upper and lower surfaces of the two branch cuts at $x_2 = 0$, $-\infty < x_1 < 0$, and $x_2 = x_2^o$, $-\infty < x_1 < x_1^o$. As $R \to \infty$, $\mathbf{u}^{(2)}$ is practically independent of (x_1^o, x_2^o) so that the integral on the large circle can be ignored. The integral on the small circle R_1 as $R_1 \to 0$ vanishes because the net force due to tractions on this circle vanishes. The integral on the small circle R_2 as $R_2 \to 0$ also vanishes because $\mathbf{t}^{(1)}$ is bounded at (x_1^o, x_2^o). The integral along the cut $x_2 = 0$, $-\infty < x_1 < 0$ vanishes due to the fact that $\mathbf{u}^{(2)}$ and $\mathbf{t}^{(1)}$ are continuous across the cut. That leaves the integral of the third term in (8.4-3) along the cut at $x_2 = x_2^o$, $-\infty < x_1 < x_1^o$, which is precisely (8.4-1a). If the cut for $\mathbf{b}^{(2)}$ is at $x_1 = x_1^o$, $x_2^o < x_2 < \infty$, we have (8.4-1b).

The components F_1, F_2 of the image force $\mathbf{F}$ obtained from (8.4-1a,b) are

$$\begin{aligned} F_1 &= -\frac{\partial V}{\partial x_1^o} = b_i^{(2)} \sigma_{i2}^{(1)}(x_1^o, x_2^o) = \mathbf{b}^{(2)} \cdot \boldsymbol{\phi}_{,1}^{(1)}(x_1^o, x_2^o), \\ F_2 &= -\frac{\partial V}{\partial x_2^o} = -b_i^{(2)} \sigma_{i1}^{(1)}(x_1^o, x_2^o) = \mathbf{b}^{(2)} \cdot \boldsymbol{\phi}_{,2}^{(1)}(x_1^o, x_2^o). \end{aligned} \tag{8.4-4}$$

Hence,

$$\mathbf{F} = -\nabla V, \quad V = -\mathbf{b}^{(2)} \cdot \boldsymbol{\phi}^{(1)}(x_1^o, x_2^o). \tag{8.4-5}$$

For the cylindrical components, Fig. 8.2, we have

$$\begin{aligned} F_r &= -\frac{\partial V}{\partial r} = \mathbf{b}^{(2)} \cdot \mathbf{t}_\theta^{(1)}(x_1^o, x_2^o), \\ F_\theta &= -\frac{\partial V}{r \partial \theta} = -\mathbf{b}^{(2)} \cdot \mathbf{t}_r^{(1)}(x_1^o, x_2^o), \end{aligned} \tag{8.4-6}$$

use being made of (8.1-8a). The change in strain energy due to a small displacement (δx_1, δx_2) of the dislocation $\mathbf{b}^{(2)}$ is identical to the change in strain energy due to a small displacement ($-\delta x_1$, $-\delta x_2$) of the dislocation $\mathbf{b}^{(1)}$. Therefore the image force exerted on $\mathbf{b}^{(1)}$ is equal but opposite in sign to the image force exerted on $\mathbf{b}^{(2)}$. Thus if F_r is positive (or negative), the image force is repulsive (or attractive). The F_θ component of the image force tends to rotate the two dislocations counter-clockwise (or clockwise) if $F_\theta>0$ (or $F_\theta<0$). Hence the two dislocations are in *equilibrium* if $F_\theta=0$.

Substitution of $\mathbf{t}_r$, $\mathbf{t}_\theta$ from (8.1-22) into (8.4-6) with $\mathbf{g}$ and $\mathbf{h}$ inserted from (8.1-20b) with $\mathbf{f}=\mathbf{0}$ gives us the real form expressions

$$F_r=\frac{1}{2\pi r}(\mathbf{b}^{(2)})^T\mathbf{L}\mathbf{b}^{(1)}, \quad F_\theta=\frac{-1}{2r}(\mathbf{b}^{(2)})^T\mathbf{V}(\theta)\mathbf{b}^{(1)}, \tag{8.4-7}$$

where $\mathbf{V}(\theta)$ is defined in (8.1-29). Since $\mathbf{L}$ is symmetric F_r is invariant when $\mathbf{b}^{(1)}$ and $\mathbf{b}^{(2)}$ are interchanged. So is F_θ because $\mathbf{V}(\theta)$ is symmetric. Therefore the image force remains the same when we interchange $\mathbf{b}^{(1)}$ and $\mathbf{b}^{(2)}$, Fig. 8.2.

If $\mathbf{b}^{(1)}=\mathbf{b}^{(2)}$, F_r is positive by virtue of the positive definiteness of $\mathbf{L}$ and the image force is repulsive. If $\mathbf{b}^{(1)}=-\mathbf{b}^{(2)}$, the image force is attractive. A particularly simple result is obtained when $\mathbf{b}^{(1)}=\pm\mathbf{b}^{(2)}=b\mathbf{m}$ where $\mathbf{m}^T=[-\sin\theta, \cos\theta, 0]$. Making use of (8.1-5) we have

$$F_r=\pm\frac{b^2}{2\pi r}\mathbf{m}^T\mathbf{L}\mathbf{m}, \quad F_\theta=\mp\frac{b^2}{2r}\mathbf{n}^T\mathbf{L}\mathbf{m}, \tag{8.4-8}$$

where $\mathbf{n}^T=[\cos\theta, \sin\theta, 0]$. In particular, when $\theta=0$,

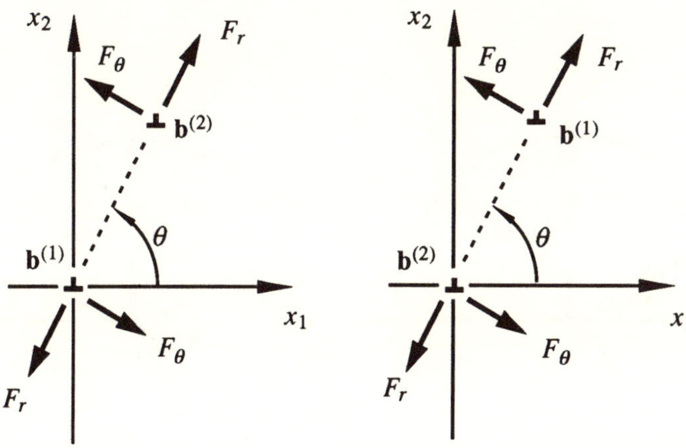

Fig. 8.2 The image force (F_r,f_θ) remains the same if $\mathbf{b}^{(1)}$ and $\mathbf{b}^{(2)}$ are interchanged

$$F_r = \pm\frac{b^2}{2\pi r}L_{22}, \qquad F_\theta = \mp\frac{b^2}{2r}L_{12}.$$

8.5 Surface Green's Function

Let the material occupy the half-space $x_2>0$ and let a line force **f** apply at $r=0$, Fig. 8.3(a). The surface $x_2=0$ is otherwise traction-free. The boundary conditions are

$$\begin{gathered}\boldsymbol{\phi}(r,0)=\text{constant}, \quad \boldsymbol{\phi}(r,\pi)=\text{constant}, \\ \boldsymbol{\phi}(r,\pi) - \boldsymbol{\phi}(r,0) = \mathbf{f}.\end{gathered} \tag{8.5-1}$$

If the solution is given by

$$\mathbf{u} = \frac{1}{\pi}\mathrm{Im}\{\mathbf{A}\langle \ln z_* \rangle \mathbf{q}\}, \qquad \boldsymbol{\phi} = \frac{1}{\pi}\mathrm{Im}\{\mathbf{B}\langle \ln z_* \rangle \mathbf{q}\}, \tag{8.5-2}$$

conditions in (8.5-1) are satisfied when **Bq** is real and

$$\mathbf{Bq} = \mathbf{f}, \quad \text{or} \quad \mathbf{q} = \mathbf{B}^{-1}\mathbf{f} = -2i\mathbf{B}^T\mathbf{L}^{-1}\mathbf{f}. \tag{8.5-3}$$

The real form solution for (8.5-2) with the **q** given by $(8.5\text{-}3)_3$ is, from (7.9-25),

$$\mathbf{u} = -\left(\tfrac{1}{\pi}(\ln r)\mathbf{I} + \mathbf{S}(\theta)\right)\mathbf{L}^{-1}\mathbf{f}, \qquad \boldsymbol{\phi} = \mathbf{L}(\theta)\mathbf{L}^{-1}\mathbf{f}. \tag{8.5-4}$$

From (8.1-8a)

$$\mathbf{t}_r = \frac{1}{\pi r}\mathbf{N}_3(\theta)\mathbf{L}^{-1}\mathbf{f}, \qquad \mathbf{t}_\theta = \mathbf{0}.$$

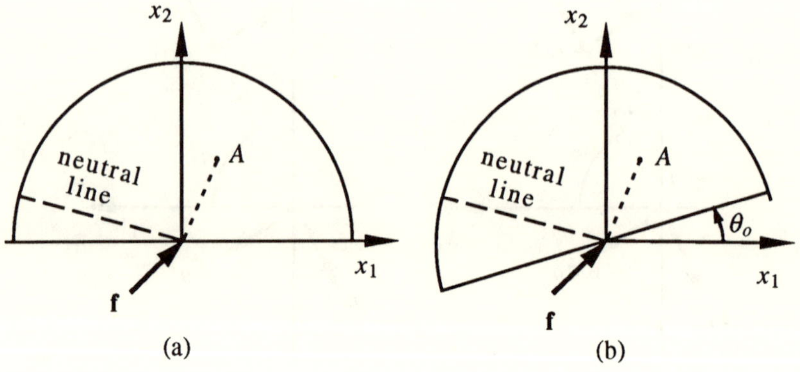

Fig. 8.3 The half-space is subjected to a concentrated force **f**. The stress at any point A is independent of θ_o, the orientation of the free surface.

Thus the traction vector on all radial plane θ=constant vanishes. The solution (8.5-4) could have also been obtained by employing the real form solution (8.1-21) by letting **g=0** and applying the conditions (8.5-1). The only nonzero stresses are σ_{33}, σ_{rr}, and σ_{r3}. The angle θ at which

$$\sigma_{rr} = \mathbf{n}(\theta)\cdot\mathbf{t}_r = 0 \quad \text{or} \quad \mathbf{n}^T(\theta)\mathbf{N}_3(\theta)\mathbf{L}^{-1}\mathbf{f} = 0$$

is called the *neutral line* (Lekhnitskii, 1963).

It should be pointed out that the displacement **u** is unique up to an arbitrary rigid body translation and rotation. Thus we could modify $(8.5\text{-}4)_1$ as

$$\mathbf{u} = -\left(\tfrac{1}{\pi}(\ln r)\mathbf{I} + \mathbf{S}(\theta) - \tfrac{1}{2}\mathbf{S}\right)\mathbf{L}^{-1}\mathbf{f} \tag{8.5-5}$$

so that

$$\mathbf{u} = \begin{cases} -\left(\tfrac{1}{\pi}(\ln r)\mathbf{I} - \tfrac{1}{2}\mathbf{S}\right)\mathbf{L}^{-1}\mathbf{f} & \text{at } \theta = 0, \\ -\left(\tfrac{1}{\pi}(\ln r)\mathbf{I} + \tfrac{1}{2}\mathbf{S}\right)\mathbf{L}^{-1}\mathbf{f} & \text{at } \theta = \pi. \end{cases} \tag{8.5-6}$$

If the free surface x_2=0 is replaced by the surface $\theta=\theta_o$, Fig. 8.3(b), the solution is still given by (8.5-4) because

$$\boldsymbol{\phi} = \begin{cases} \mathbf{L}(\theta_o)\mathbf{L}^{-1}\mathbf{f}, & \text{at } \theta = \theta_o, \\ \mathbf{L}(\theta_o + \pi)\mathbf{L}^{-1}\mathbf{f}, & \text{at } \theta = \theta_o. \end{cases}$$

But

$$\mathbf{L}(\theta_o + \pi)\mathbf{L}^{-1}\mathbf{f} = \left(\mathbf{L} + \mathbf{L}(\theta_o)\right)\mathbf{L}^{-1}\mathbf{f} = \mathbf{f} + \mathbf{L}(\theta_o)\mathbf{L}^{-1}\mathbf{f}$$

so that the boundary condition

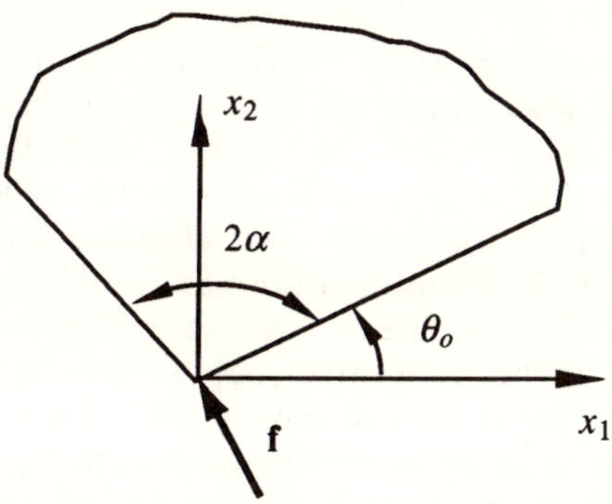

Fig. 8.4 The wedge of wedge angle 2α is subjected to a concentrated force **f**.

$$\phi(r,\ \theta_o+\pi) - \phi(r,\ \theta_o) = \mathbf{f}$$

is satisfied. Thus the solution (8.5-4) is independent of the orientation θ_o of the traction-free surface. One could have reached the same conclusion by applying the dual coordinate systems introduced in Section 7.10. The stress at any point A in Figs 8.3(a) and (b) are identical. The location of the neutral line is independent of θ_o. For isotropic materials the neutral line is perpendicular to the projection of $\mathbf{f}$ on the (x_1,x_2)-plane.

The half-space is a special case of a wedge of wedge angle $2\alpha=\pi$, Fig. 8.4. For the solution of the wedge subjected to a line force $\mathbf{f}$ at the wedge apex we may employ the real form solution (8.1-21) by setting $\mathbf{g=0}$, i.e.,

$$2\mathbf{u} = -\left(\tfrac{1}{\pi}(\ln r)\mathbf{I} + \mathbf{S}(\theta)\right)\mathbf{h}, \quad 2\phi = \mathbf{L}(\theta)\mathbf{h}. \tag{8.5-7}$$

To satisfy the applied force $\mathbf{f}$ we must have

$$\phi(r,\theta_o+2\alpha) - \phi(r,\theta_o) = \mathbf{f} \tag{8.5-8}$$

or

$$\mathbf{h} = 2\left\{\mathbf{L}(\theta_o+2\alpha) - \mathbf{L}(\theta_o)\right\}^{-1}\mathbf{f}. \tag{8.5-9}$$

When $2\alpha=\pi$, (8.5-7) and (8.5-9) recovers (8.5-4).

8.6 Green's Functions for Half-Space

The Green's function for an anisotropic elastic half-space $x_2>0$ subject to a line force and/or a line dislocation inside the half-space has been considered by Head (1953, 1965), Chou (1966), Tucker (1969), Willis (1970), Braekhus and Lothe (1971), Nakahara and Willis (1973), Barnett and Lothe (1974b), Tewary et al. (1989a), Suo (1990), Qu and Li (1991), and Qu (1991). In the earlier work the Green's function for the half-space with the surface $x_2=0$ being traction-free was obtained from the Green's function for the infinite space by adding a distribution of forces along the surface $x_2=0$ so that the net surface traction vanishes. With this approach the solution is not explicit in that the final solution requires integration of the distributed forces at $x_2=0$. An alternate approach is to use the method of analytical continuation (Suo, 1990) so that the need of integration is circumvented. A more direct approach (Ting, 1992b) which will be presented below was motivated by the solutions obtained by Hwu and Yen (1991) for Green's functions for an elliptic hole in an infinite anisotropic elastic space.

Let the half-space $x_2>0$ be occupied by an anisotropic elastic material and a line force $\mathbf{f}$ and a line dislocation with Burgers vector

b be applied at

$$(x_1,x_2)=(0,d), \quad d>0. \tag{8.6-1}$$

Before we present the solution we make the following observations regarding the general solution (8.1-1). First, the general solution remains valid if $f(z_\alpha)$ is replace by

$$f(z_\alpha - z_\alpha^{\mathrm{o}}), \quad z_\alpha^{\mathrm{o}} = x_1^{\mathrm{o}} + p_\alpha x_2^{\mathrm{o}},$$

where $x_1^{\mathrm{o}}, x_2^{\mathrm{o}}$ are real constants. This simply implies that the origin of the coordinate system is moved to $(x_1^{\mathrm{o}}, x_2^{\mathrm{o}})$. Secondly, the general solution also remains valid if $f(z_\alpha)$ is replaced by

$$f(z_\alpha - \gamma)$$

where γ is an arbitrary complex constant which may depend on p. Making use of these two observations we now modify (8.1-16) and consider the solution

$$\begin{aligned}\mathbf{u} &= \frac{1}{\pi}\operatorname{Im}\left\{\mathbf{A}\left\langle \ln(z_* - p_* d)\right\rangle \mathbf{q}^\infty\right\} + \frac{1}{\pi}\operatorname{Im}\sum_{\beta=1}^{3}\left\{\mathbf{A}\left\langle \ln(z_* - \bar{p}_\beta d)\right\rangle \mathbf{q}_\beta\right\},\\ \boldsymbol{\phi} &= \frac{1}{\pi}\operatorname{Im}\left\{\mathbf{B}\left\langle \ln(z_* - p_* d)\right\rangle \mathbf{q}^\infty\right\} + \frac{1}{\pi}\operatorname{Im}\sum_{\beta=1}^{3}\left\{\mathbf{B}\left\langle \ln(z_* - \bar{p}_\beta d)\right\rangle \mathbf{q}_\beta\right\},\end{aligned} \tag{8.6-2}$$

where $\mathbf{q}^\infty$ is given in (8.1-18) and $\mathbf{q}_\beta$ (β = 1,2,3) are unknown vectors to be determined. In the above,

$$\begin{aligned}\left\langle \ln(z_* - p_* d)\right\rangle &= \operatorname{diag}[\ln(z_1 - p_1 d),\ \ln(z_2 - p_2 d),\ \ln(z_3 - p_3 d)],\\ \left\langle \ln(z_* - \bar{p}_\beta d)\right\rangle &= \operatorname{diag}[\ln(z_1 - \bar{p}_\beta d),\ \ln(z_2 - \bar{p}_\beta d),\ \ln(z_3 - \bar{p}_\beta d)].\end{aligned} \tag{8.6-3}$$

Thus the asterisk $*$ is for identifying the element in the diagonal matrix. The first terms on the right of (8.6-2) represent the Green's function for an infinite space with a line force **f** and a line dislocation with Burgers vector **b** applied at the location given by (8.6-1). The second terms in (8.6-2) are the *disturbed* solutions which are added to satisfy the boundary conditions at x_2=0.

Consider first the case in which the surface x_2=0 is traction-free so that

$$\boldsymbol{\phi} = \mathbf{0} \quad \text{at} \quad x_2 = 0. \tag{8.6-4}$$

Substitution of $(8.6\text{-}2)_2$ into (8.6-4) leads to

$$\operatorname{Im}\left\{\mathbf{B}\left\langle \ln(x_1 - p_* d)\right\rangle \mathbf{q}^\infty\right\} + \operatorname{Im}\sum_{\beta=1}^{3}\left\{\ln(x_1 - \bar{p}_\beta d)\mathbf{B}\mathbf{q}_\beta\right\} = \mathbf{0}. \tag{8.6-5}$$

The first term can be replaced by the negative of its complex conjugate, i.e.,

$$\mathrm{Im}\left\{\mathbf{B}\langle \ln(x_1 - p_* d)\rangle \mathbf{q}^\infty\right\} = -\mathrm{Im}\left\{\overline{\mathbf{B}}\langle \ln(x_1 - \bar{p}_* d)\rangle \overline{\mathbf{q}}^\infty\right\} \tag{8.6-6}$$

in which

$$\langle \ln(x_1 - \bar{p}_* d)\rangle = \sum_{\beta=1}^{3} \ln(x_1 - \bar{p}_\beta d)\mathbf{I}_\beta \tag{8.6-7}$$

and

$$\mathbf{I}_1 = \mathrm{diag}[1,\ 0,\ 0],\quad \mathbf{I}_2 = \mathrm{diag}[0,\ 1,\ 0],\quad \mathbf{I}_3 = \mathrm{diag}[0,\ 0,\ 1]. \tag{8.6-8}$$

A trivial identity is

$$\sum_{\beta=1}^{3} \mathbf{I}_\beta = \mathbf{I}. \tag{8.6-9}$$

Equation (8.6-5) is now reduced to

$$\mathrm{Im}\sum_{\beta=1}^{3} \ln(x_1 - \bar{p}_\beta d)\left\{-\overline{\mathbf{B}}\mathbf{I}_\beta \overline{\mathbf{q}}^\infty + \mathbf{B}\mathbf{q}_\beta\right\} = \mathbf{0} \tag{8.6-10}$$

which gives

$$\mathbf{q}_\beta = \mathbf{B}^{-1}\overline{\mathbf{B}}\mathbf{I}_\beta \overline{\mathbf{q}}^\infty. \tag{8.6-11}$$

If the boundary $x_2=0$ is a rigidly clamped surface,

$$\mathbf{u} = \mathbf{0} \quad \text{at} \quad x_2 = 0.$$

The same procedure shows that the solution is given by (8.6-2) with

$$\mathbf{q}_\beta = \mathbf{A}^{-1}\overline{\mathbf{A}}\mathbf{I}_\beta \overline{\mathbf{q}}^\infty. \tag{8.6-12}$$

Using the method of analytical continuation Suo (1990) obtained the solutions which, though not explicitly given, can be shown to be identical to (8.6-2) with $\mathbf{q}_\beta$ given by (8.6-11) or (8.6-12).

It has been shown in Section 5.6 that $i\mathbf{B}^{-1}\overline{\mathbf{B}}$ and $-i\mathbf{A}^{-1}\overline{\mathbf{A}}$ are orthogonal, positive definite Hermitian. Noting that $\mathbf{I}_\beta \overline{\mathbf{q}}^\infty$ is a column matrix whose βth element is $\bar{q}_\beta^\infty$ while other elements are zero, (8.6-11) indicates that $\mathbf{q}_\beta$ is the βth column of $i\mathbf{B}^{-1}\overline{\mathbf{B}}$ multiplied by $-i\bar{q}_\beta^\infty$. The βth element of the βth column of $i\mathbf{B}^{-1}\overline{\mathbf{B}}$ is a diagonal element of $i\mathbf{B}^{-1}\overline{\mathbf{B}}$ which is necessarily positive and nonzero. Thus the βth element of $\mathbf{q}_\beta$ is $-i\bar{q}_\beta^\infty$ multiplied by a positive nonzero quantity. Likewise, from (8.6-12) $\mathbf{q}_\beta$ is the βth column of $-i\mathbf{A}^{-1}\overline{\mathbf{A}}$ multiplied by $i\bar{q}_\beta^\infty$. The βth element of $\mathbf{q}_\beta$ is $i\bar{q}_\beta^\infty$ multiplied by a positive nonzero quantity.

When d=0 (8.6-2) reduce to

$$\mathbf{u}=\frac{1}{\pi}\text{Im}\{\mathbf{A}\langle\ln z_*\rangle\mathbf{q}\},\qquad \boldsymbol{\phi}=\frac{1}{\pi}\text{Im}\{\mathbf{B}\langle\ln z_*\rangle\mathbf{q}\},\tag{8.6-13}$$

where

$$\mathbf{q}=\mathbf{q}^{\infty}+\sum_{\beta=1}^{3}\mathbf{q}_{\beta}.\tag{8.6-14}$$

To see if the solution reproduces the surface Green's function presented in Section 8.5 we set **b**=**0** in (8.1-18) and substitute (8.6-11) into (8.6-14). The result is

$$\mathbf{q}=\mathbf{A}^T\mathbf{f}+\mathbf{B}^{-1}\overline{\mathbf{B}}\overline{\mathbf{A}}^T\mathbf{f}=\mathbf{B}^{-1}\mathbf{f},$$

use being made of $(5.5\text{-}16)_2$. This agrees with $(8.5\text{-}3)_2$.

The *image force* that exerts on the line dislocation in the half-space is the negative of the gradient of the interaction energy. The interaction energy is that part of the strain energy that depends on the location of the line dislocation. For the problem under consideration the strain energy is obviously unchanged if we move the dislocation in the x_1 direction. It will change only if we move it in the x_2 direction. Following the derivation in Section 8.4 for the interaction energy of two dislocations we consider the work done by the formation of the line dislocation at $(x_1,x_2)=(0,d)$ due to the presence of the stress field given by the second terms in $(8.6\text{-}2)_2$. The interaction energy is given by (8.4-1b) and the image force is, by $(8.4\text{-}4)_{5,6}$,

$$F=-b_i\sigma_{i1}^d(0,d)=\mathbf{b}\cdot\boldsymbol{\phi}_{,2}^d(0,d)\tag{8.6-15}$$

where $\boldsymbol{\phi}^d$ represents the second terms in $(8.6\text{-}2)_2$. The image force is normal to the surface $x_2=0$, the boundary of the half-space. It is a repulsive force if $F>0$ and an attractive force if $F<0$. Using (8.6-8) we have

$$\begin{aligned}F&=\frac{1}{\pi d}\text{Im}\sum_{\beta=1}^{3}\mathbf{b}^T\left\{\mathbf{B}\left\langle\frac{p_*}{p_*-\bar{p}_\beta}\right\rangle\mathbf{q}_\beta\right\}\\&=\frac{1}{\pi d}\text{Im}\sum_{\beta=1}^{3}\sum_{\alpha=1}^{3}\mathbf{b}^T\left\{\frac{p_\alpha}{p_\alpha-\bar{p}_\beta}\mathbf{B}\mathbf{I}_\alpha\mathbf{q}_\beta\right\}.\end{aligned}\tag{8.6-16}$$

When the boundary $x_2=0$ is traction-free, substitution of (8.6-11) with $\mathbf{q}^{\infty}=\mathbf{B}^T\mathbf{b}$ into (8.6-16) leads to

$$F=\frac{-1}{2\pi d}\text{Re}\sum_{\beta=1}^{3}\sum_{\alpha=1}^{3}\mathbf{b}^T\left\{\frac{p_\alpha}{p_\alpha-\bar{p}_\beta}(\mathbf{B}\mathbf{I}_\alpha\mathbf{B}^{-1})\mathbf{L}(\overline{\mathbf{B}}\mathbf{I}_\beta\overline{\mathbf{B}}^{-1})^T\right\}\mathbf{b},\tag{8.6-17}$$

where use has been made of the relations $\mathbf{L}=-2i\mathbf{B}\mathbf{B}^T=2i\overline{\mathbf{B}}\overline{\mathbf{B}}^T$. We now perform the following three operations on the right of (8.6-17) which leaves F unchanged. First the α and β are interchanged. Next the right side is replaced by its complex conjugate. Finally we apply the matrix transpose on the right. The result is

$$F=\frac{-1}{2\pi d}\operatorname{Re}\sum_{\beta=1}^{3}\sum_{\alpha=1}^{3}\mathbf{b}^T\left\{\frac{-\bar{p}_\beta}{p_\alpha-\bar{p}_\beta}(\mathbf{B}\mathbf{I}_\alpha\mathbf{B}^{-1})\mathbf{L}(\overline{\mathbf{B}}\mathbf{I}_\beta\overline{\mathbf{B}}^{-1})^T\right\}\mathbf{b}. \tag{8.6-18}$$

Addition of (8.6-17) and (8.6-18) yields, after using (8.6-9),

$$F=\frac{-1}{4\pi d}\mathbf{b}^T\mathbf{L}\mathbf{b}. \tag{8.6-19}$$

Since $\mathbf{L}$ is positive definite, $F<0$ and the image force is attractive to the free-surface.

If the boundary $x_2=0$ is a fixed boundary insertion of (8.6-12) into (8.6-16) with $\mathbf{q}^\infty=\mathbf{B}^T\mathbf{b}$ and making use of the relations $\mathbf{H}=2i\mathbf{A}\mathbf{A}^T=-2i\overline{\mathbf{A}}\overline{\mathbf{A}}^T$ we have

$$F=\frac{1}{2\pi d}\operatorname{Re}\sum_{\beta=1}^{3}\sum_{\alpha=1}^{3}\mathbf{b}^T\left\{\frac{p_\alpha}{p_\alpha-\bar{p}_\beta}(\mathbf{B}\mathbf{I}_\alpha\mathbf{A}^{-1})\mathbf{H}(\overline{\mathbf{B}}\mathbf{I}_\beta\overline{\mathbf{A}}^{-1})^T\right\}\mathbf{b}. \tag{8.6-20}$$

Following the derivation of (8.6-18) from (8.6-17), (8.6-20) can be rewritten as

$$F=\frac{1}{2\pi d}\operatorname{Re}\sum_{\beta=1}^{3}\sum_{\alpha=1}^{3}\mathbf{b}^T\left\{\frac{-\bar{p}_\beta}{p_\alpha-\bar{p}_\beta}(\mathbf{B}\mathbf{I}_\alpha\mathbf{A}^{-1})\mathbf{H}(\overline{\mathbf{B}}\mathbf{I}_\beta\overline{\mathbf{A}}^{-1})^T\right\}\mathbf{b}. \tag{8.6-21}$$

Addition of (8.6-20) and (8.6-21) leads to

$$F=\frac{1}{4\pi d}\mathbf{b}^T\operatorname{Re}\left\{(\mathbf{B}\mathbf{A}^{-1})\mathbf{H}(\overline{\mathbf{B}}\overline{\mathbf{A}}^{-1})^T\right\}\mathbf{b}$$

or, using $(8.1\text{-}4)_2$ and noting that $\mathbf{H}^{-1}\mathbf{S}$ is skew-symmetric,

$$F=\frac{1}{4\pi d}\mathbf{b}^T\left\{\mathbf{H}^{-1}+\mathbf{S}^T\mathbf{H}^{-1}\mathbf{S}\right\}\mathbf{b}. \tag{8.6-22}$$

By writing $(8.1\text{-}3b)_1$ as

$$\mathbf{H}^{-1}-\mathbf{L}=\mathbf{S}^T\mathbf{H}^{-1}\mathbf{S} \tag{8.6-23}$$

we have the alternate form

$$F = \frac{1}{4\pi d}\mathbf{b}^T(2\mathbf{H}^{-1} - \mathbf{L})\mathbf{b}. \tag{8.6-24}$$

Since $\mathbf{S}$ is singular and $\mathbf{H}^{-1}$ is positive definite, the right side of (8.6-23) is positive semidefinite, so is $\mathbf{H}^{-1}$–$\mathbf{L}$. Therefore

$$F = \frac{1}{2\pi d}\mathbf{b}^T(\mathbf{H}^{-1} - \mathbf{L})\mathbf{b} + \frac{1}{4\pi d}\mathbf{b}^T\mathbf{L}\mathbf{b} \geq \frac{1}{4\pi d}\mathbf{b}^T\mathbf{L}\mathbf{b} > 0 \tag{8.6-25}$$

which tells us that the image force for the fixed boundary is repulsive, and that the magnitude of the image force is equal to or larger than the image force for a traction-free boundary (Ting and Barnett, 1993). The equality in (8.6-25) holds only when $\mathbf{b}$ is a right null vector of $\mathbf{S}$, i.e., when

$$\mathbf{S}\mathbf{b} = \mathbf{0}.$$

Specialized to isotropic half-spaces, the results agree with those obtained by Dundurs (1969) and Mura (1987). It should be mentioned that Tucker (1969) has studied a variety of problems involving dislocations interacting with straight boundaries in anisotropic elastic half-spaces and bimaterials. Tucker's method did not make use of the orthogonality relations inherent between the matrices $\mathbf{A}$ and $\mathbf{B}$ and, as a result, his solutions are in complex form and are somewhat less transparent than those presented here.

We now derive the displacement $\mathbf{u}$ and the surface traction $\mathbf{t}$ at the boundary x_2=0 in terms of the displacement and traction at x_2=0 deduced from the Green's function for the infinite space. Let $\mathbf{u}^\infty$, ϕ^∞ stand for the first terms on the right of (8.6-2). They represent the Green's function for the infinite space subjected to a line force $\mathbf{f}$ and a line dislocation with Burger vector $\mathbf{b}$ applied at $(x_1,x_2)=(0,d)$. At x_2=0,

$$\begin{aligned} \mathbf{u}^\infty(x_1,0) &= \frac{1}{\pi}\mathrm{Im}\{\mathbf{A}\langle\ln(x_1 - p_* d)\rangle\mathbf{q}^\infty\}, \\ \phi^\infty(x_1,0) &= \frac{1}{\pi}\mathrm{Im}\{\mathbf{B}\langle\ln(x_1 - p_* d)\rangle\mathbf{q}^\infty\}, \end{aligned} \tag{8.6-26}$$

or, employing the real form solution (8.1-21),

$$\begin{aligned} 2\mathbf{u}^\infty(x_1,0) &= -\frac{1}{\pi}(\ln\hat{r})\mathbf{h} - \mathbf{S}(\hat{\theta})\mathbf{h} - \mathbf{H}(\hat{\theta})\mathbf{g}, \\ 2\phi^\infty(x_1,0) &= -\frac{1}{\pi}(\ln\hat{r})\mathbf{g} + \mathbf{L}(\hat{\theta})\mathbf{h} - \mathbf{S}^T(\hat{\theta})\mathbf{g}, \end{aligned} \tag{8.6-27}$$

where

$$\hat{r}\cos\hat{\theta} = x_1, \quad \hat{r}\sin\hat{\theta} = -d. \tag{8.6-28}$$

The constants **h** and **g** are given in (8.1-20b). Writing

$$\mathbf{A} = (\mathbf{A}\mathbf{B}^{-1})\mathbf{B} = -(\mathbf{S}\mathbf{L}^{-1} + i\mathbf{L}^{-1})\mathbf{B}$$

we have from $(8.6\text{-}26)_1$

$$\begin{aligned}\mathbf{u}^{\infty}(x_1,0) &= -\frac{1}{\pi}\operatorname{Im}\left\{(\mathbf{S}\mathbf{L}^{-1} + i\mathbf{L}^{-1})\mathbf{B}\langle\ln(x_1 - p_* d)\rangle\mathbf{q}^{\infty}\right\} \\ &= -\mathbf{S}\mathbf{L}^{-1}\boldsymbol{\phi}^{\infty}(x_1,0) - \frac{1}{\pi}\mathbf{L}^{-1}\operatorname{Re}\left\{\mathbf{B}\langle\ln(x_1 - p_* d)\rangle\mathbf{q}^{\infty}\right\}.\end{aligned}$$

Noting that $\mathbf{S}\mathbf{L}^{-1} = -\mathbf{L}^{-1}\mathbf{S}^T$ we obtain

$$\frac{1}{\pi}\operatorname{Re}\left\{\mathbf{B}\langle\ln(x_1 - p_* d)\rangle\mathbf{q}^{\infty}\right\} = \mathbf{S}^T\boldsymbol{\phi}^{\infty}(x_1,0) - \mathbf{L}\mathbf{u}^{\infty}(x_1,0). \qquad (8.6\text{-}29)$$

Similarly, setting

$$\mathbf{B} = (\mathbf{B}\mathbf{A}^{-1})\mathbf{A} = (-\mathbf{H}^{-1}\mathbf{S} + i\mathbf{H}^{-1})\mathbf{A}$$

it can be shown from $(8.6\text{-}26)_2$ that

$$\frac{1}{\pi}\operatorname{Re}\left\{\mathbf{A}\langle\ln(x_1 - p_* d)\rangle\mathbf{q}^{\infty}\right\} = \mathbf{S}\mathbf{u}^{\infty}(x_1,0) + \mathbf{H}\boldsymbol{\phi}^{\infty}(x_1,0). \qquad (8.6\text{-}30)$$

When $x_2=0$ is a traction-free boundary, following the derivation from (8.6-6) to (8.6-10) we obtain from $(8.6\text{-}2)_1$

$$\mathbf{u}(x_1,0) = \frac{1}{\pi}\operatorname{Im}\sum_{\beta=1}^{3}\ln(x_1 - \bar{p}_\beta d)\left\{-\overline{\mathbf{A}}\mathbf{I}_\beta\overline{\mathbf{q}}^{\infty} + \mathbf{A}\mathbf{q}_\beta\right\}.$$

Inserting $\mathbf{q}_\beta$ from (8.6-11) one has

$$-\overline{\mathbf{A}}\mathbf{I}_\beta\overline{\mathbf{q}}^{\infty} + \mathbf{A}\mathbf{q}_\beta = (-\overline{\mathbf{A}}\overline{\mathbf{B}}^{-1} + \mathbf{A}\mathbf{B}^{-1})\overline{\mathbf{B}}\mathbf{I}_\beta\overline{\mathbf{q}}^{\infty}$$

and, after employing $(8.1\text{-}4)_4$, (8.6-7), and (8.6-6),

$$\mathbf{u}(x_1,0) = -\frac{2}{\pi}\mathbf{L}^{-1}\operatorname{Re}\left\{\mathbf{B}\langle\ln(x_1 - p_* d)\rangle\mathbf{q}^{\infty}\right\}. \qquad (8.6\text{-}31)$$

In view of (8.6-29), (8.6-31) is

$$\mathbf{u}(x_1,0) = 2\left\{\mathbf{u}^{\infty}(x_1,0) + \mathbf{S}\mathbf{L}^{-1}\boldsymbol{\phi}^{\infty}(x_1,0)\right\}. \qquad (8.6\text{-}32)$$

This can be written in a remarkably simple and explicit form by employing (8.6-27), (8.1-20b), and the identities (7.8-13) as (Kirchner and Lothe, 1987)

$$\mathbf{u}(x_1,0) = -\mathbf{L}^{-1}\left\{\frac{1}{\pi}(\ln\hat{r})\mathbf{f} + \mathbf{S}^T(\hat{\theta})\mathbf{f} - \mathbf{L}(\hat{\theta})\mathbf{b}\right\}. \tag{8.6-33}$$

The $\hat{r}$ and $\hat{\theta}$ are defined in (8.6-28). To find the *hoop stress vector* $\mathbf{t}_1$ at the surface $x_2=0$ we employ (5.7-9) which is

$$\mathbf{t}_1 = -\mathbf{N}_3\mathbf{u}_{,1}(x_1,0), \qquad (\mathbf{t}_1)_i = \sigma_{i1}(x_1,0),$$

and the relation

$$\frac{\partial}{\partial x_1} = \cos\hat{\theta}\frac{\partial}{\partial \hat{r}} - \sin\hat{\theta}\frac{\partial}{\hat{r}\partial\hat{\theta}}. \tag{8.6-34}$$

The result is

$$\mathbf{t}_1 = \frac{1}{\pi\hat{r}}\mathbf{N}_3\mathbf{L}^{-1}\left\{\cos\hat{\theta}\,\mathbf{f} - \sin\hat{\theta}[\mathbf{N}_1^T(\hat{\theta})\mathbf{f} + \mathbf{N}_3(\hat{\theta})\mathbf{b}]\right\}. \tag{8.6-35}$$

If $x_2=0$ is a rigid surface the derivations are similar. In place of (8.6-31)–(8.6-33), it can be shown that

$$\boldsymbol{\phi}(x_1,0) = \frac{2}{\pi}\mathbf{H}^{-1}\,\mathrm{Re}\left\{\mathbf{A}\langle\ln(x_1 - p_* d)\rangle\mathbf{q}^\infty\right\}, \tag{8.6-36}$$

$$\boldsymbol{\phi}(x_1,0) = 2\left\{\boldsymbol{\phi}^\infty(x_1,0) + \mathbf{H}^{-1}\mathbf{S}\mathbf{u}^\infty(x_1,0)\right\}, \tag{8.6-37}$$

$$\boldsymbol{\phi}(x_1,0) = \mathbf{H}^{-1}\left\{\frac{1}{\pi}(\ln\hat{r})\mathbf{b} + \mathbf{S}(\hat{\theta})\mathbf{b} + \mathbf{H}(\hat{\theta})\mathbf{f}\right\}. \tag{8.6-38}$$

The surface traction $\mathbf{t}_2$ at $x_2=0$ is

$$\mathbf{t}_2 = \boldsymbol{\phi}_{,1}(x_1,0), \qquad (\mathbf{t}_2)_i = \sigma_{i1}(x_1,0),$$

or, by (8.6-34) and (8.6-38),

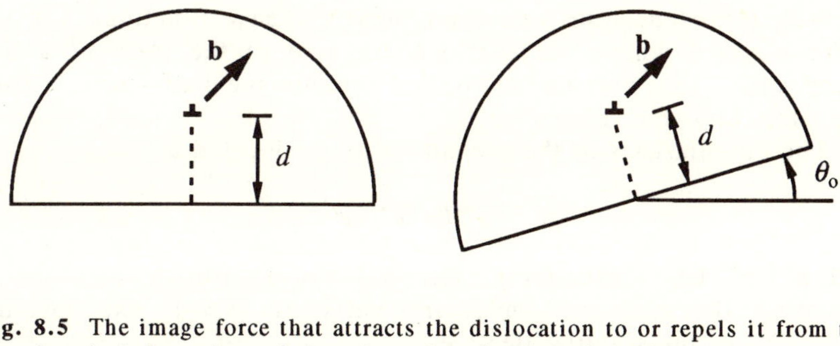

Fig. 8.5 The image force that attracts the dislocation to or repels it from the boundary is independent of the orientation θ_0 of the boundary.

$$\mathbf{t}_2 = \frac{1}{\pi \hat{r}^2}\mathbf{H}^{-1}\left\{x_1\, \mathbf{b} + d[\mathbf{N}_1(\hat{\theta})\mathbf{b} + \mathbf{N}_2(\hat{\theta})\mathbf{f}]\right\}. \tag{8.6-39}$$

The total traction force on the fixed surface $x_2=0$ is

$$\boldsymbol{\phi}(\infty,0) - \boldsymbol{\phi}(-\infty,0) = \mathbf{H}^{-1}\mathbf{S}\mathbf{b} + \mathbf{f}$$

which is not equal to $\mathbf{f}$ unless $\mathbf{b}$ vanishes, or is proportional to the null vector of $\mathbf{H}^{-1}\mathbf{S}$. The *hoop stress vector* $\mathbf{t}_1$ at $x_2=0$ is, from (5.7-10),

$$\mathbf{t}_1 = -\mathbf{N}_1^T \mathbf{t}_2.$$

Under orthogonal transformations the Burgers vector $\mathbf{b}$ transforms as a tensor of rank one. The tensors $\mathbf{S}$, $\mathbf{H}$, and $\mathbf{L}$ are tensors of rank two when the transformations are limited to rotations about the x_3-axis (Section 7.2). Therefore the image force F given in (8.6-19) and (8.6-24) are invariant under rotations about the x_3-axis. The image force is also invariant, obviously, with translation of the coordinate system in the x_1 direction. Thus one reaches the conclusion that the image force for half-spaces remains the same if the boundary $x_2=0$ is replaced by any plane boundary which is parallel to the x_3-axis such that the distance between the line dislocation and the new boundary has the same value d, Fig. 8.5. For isotropic materials and for transversely isotropic materials with the axis of symmetry along the x_3-axis this means that, if the boundary remains at $x_2=0$, the image force is unchanged when we rotate the Burgers vector $\mathbf{b}$ about the dislocation line.

8.7 Image Singularities

The Green's function for the half-space as given by (8.6-2) consists of two parts. The first terms on the right represent the Green's function for the infinite space with singularities in the form of a line force $\mathbf{f}$ and a line dislocation with Burgers vector $\mathbf{b}$ located at $(x_1,x_2) = (0,d)$. We will show that the second terms on the right of (8.6-2) represent nine one-component Green's functions for the infinite space whose singularities are also in the form of a line force and a line dislocation located outside the half-space. They represent *image singularities*.

The singularities of the second terms are located at

$$z_\alpha - \bar{p}_\beta d = x_1 + p_\alpha x_2 - \bar{p}_\beta d = 0. \tag{8.7-1}$$

Let p', p'' be, respectively, the real and imaginary parts of p. Equating the real and imaginary parts of (8.7-1) the location $(x_1^{\alpha\beta}, x_2^{\alpha\beta})$ of the singularities are

$$\frac{x_1^{\alpha\beta}}{d} = \frac{p_\alpha'' p_\beta' + p_\beta'' p_\alpha'}{p_\alpha''}, \qquad \frac{x_2^{\alpha\beta}}{d} = -\frac{p_\beta''}{p_\alpha''}. \tag{8.7-2}$$

Since the imaginary parts of p_α (α=1,2,3) are nonzero and are taken to be positive, p_α'', p_β'' are positive and nonzero. It follows that $x_1^{\alpha\beta}$ and $x_2^{\alpha\beta}$ exist and that $x_2^{\alpha\beta}<0$, confirming that the image singularities are located outside the half-space occupied by the material.

If we write (8.7-1) as

$$\begin{aligned} z_\alpha - \bar{p}_\beta d &= (x_1 - x_1^{\alpha\beta}) + p_\alpha (x_2 - x_2^{\alpha\beta}) = z_\alpha - z^{\alpha\beta}, \\ z^{\alpha\beta} &= x_1^{\alpha\beta} + p_\alpha x_2^{\alpha\beta}, \end{aligned} \tag{8.7-3}$$

we have

$$\left\langle \ln(z_* - \bar{p}_\beta d) \right\rangle = \left\langle \ln(z_* - z^{*\beta}) \right\rangle = \sum_{\alpha=1}^{3} \left\{ \ln(z_\alpha - z^{\alpha\beta}) \right\} \mathbf{I}_\alpha .$$

Equation (8.6-2) is equivalent to (Ting, 1992e)

$$\mathbf{u} = \sum_{\alpha=1}^{3} \mathbf{u}_\alpha^\infty + \sum_{\alpha=1}^{3}\sum_{\beta=1}^{3} \mathbf{u}^{\alpha\beta}, \qquad \boldsymbol{\phi} = \sum_{\alpha=1}^{3} \boldsymbol{\phi}_\alpha^\infty + \sum_{\alpha=1}^{3}\sum_{\beta=1}^{3} \boldsymbol{\phi}^{\alpha\beta},$$

in which

$$\mathbf{u}_\alpha^\infty = \frac{1}{\pi} \operatorname{Im}\left\{ \ln(z_\alpha - p_\alpha d) q_\alpha^\infty \mathbf{a}_\alpha \right\}, \quad \boldsymbol{\phi}_\alpha^\infty = \frac{1}{\pi} \operatorname{Im}\left\{ \ln(z_\alpha - p_\alpha d) q_\alpha^\infty \mathbf{b}_\alpha \right\},$$

and

$$\begin{aligned} \mathbf{u}^{\alpha\beta} &= \frac{1}{\pi} \operatorname{Im}\left\{ \ln(z_\alpha - z^{\alpha\beta}) q_{\alpha\beta} \boldsymbol{a}_\alpha \right\}, \\ \boldsymbol{\phi}^{\alpha\beta} &= \frac{1}{\pi} \operatorname{Im}\left\{ \ln(z_\alpha - z^{\alpha\beta}) q_{\alpha\beta} \boldsymbol{b}_\alpha \right\}. \end{aligned} \tag{8.7-4}$$

In the above $q_{\alpha\beta}$ is the αth component of $\mathbf{q}_\beta$, i.e.,

$$q_{\alpha\beta} = (\mathbf{q}_\beta)_\alpha . \tag{8.7-5}$$

The displacements $\mathbf{u}_\alpha^\infty$ and $\mathbf{u}^{\alpha\beta}$ (β=1,2,3) are polarized on the eigenplane associated with $\boldsymbol{a}_\alpha$ while the surface tractions deduced from $\boldsymbol{\phi}_\alpha^\infty$ and $\boldsymbol{\phi}^{\alpha\beta}$ (β=1,2,3) are polarized on the eigenplane associated with $\boldsymbol{b}_\alpha$ (Section 8.2). ($\mathbf{u}^{\alpha\beta}, \boldsymbol{\phi}^{\alpha\beta}$) represent nine one-component Green's functions with singularities located at $(x_1^{\alpha\beta}, x_2^{\alpha\beta})$. The image singularities consist of a line force $\mathbf{f}^{\alpha\beta}$ and a line dislocation with Burgers vector $\mathbf{b}^{\alpha\beta}$ which are, by (8.2-3),

$$\mathbf{f}^{\alpha\beta} = 2\operatorname{Re}(q_{\alpha\beta} \boldsymbol{b}_\alpha) = 2\operatorname{Re}(\mathbf{B}\mathbf{I}_\alpha \mathbf{q}_\beta), \tag{8.7-6a}$$

$$\mathbf{b}^{\alpha\beta} = 2\,\mathrm{Re}(q_{\alpha\beta}\boldsymbol{a}_\alpha) = 2\,\mathrm{Re}(\mathbf{A}\mathbf{I}_\alpha \mathbf{q}_\beta). \tag{8.7-6b}$$

If $x_2=0$ is a traction-free surface, substituting $\mathbf{q}_\beta$ from (8.6-11) and using (8.1-18) yields

$$\begin{aligned}\mathbf{f}^{\alpha\beta} &= 2\,\mathrm{Re}\left\{\mathbf{B}\mathbf{I}_\alpha \mathbf{B}^{-1}\overline{\mathbf{B}}\mathbf{I}_\beta(\overline{\mathbf{A}}^T\mathbf{f} + \overline{\mathbf{B}}^T\mathbf{b})\right\},\\ \mathbf{b}^{\alpha\beta} &= 2\,\mathrm{Re}\left\{\mathbf{A}\mathbf{I}_\alpha \mathbf{B}^{-1}\overline{\mathbf{B}}\mathbf{I}_\beta(\overline{\mathbf{A}}^T\mathbf{f} + \overline{\mathbf{B}}^T\mathbf{b})\right\}.\end{aligned} \tag{8.7-7}$$

We see that both $\mathbf{f}^{\alpha\beta}$ and $\mathbf{b}^{\alpha\beta}$ are in general nonzero even if $\mathbf{f}$ or $\mathbf{b}$ is zero. Making use of the identities (8.6-9), (5.5-17), and (8.1-4) it is easily shown that

$$\sum_{\alpha=1}^{3}\sum_{\beta=1}^{3}\mathbf{f}^{\alpha\beta} = \mathbf{f}, \qquad \sum_{\alpha=1}^{3}\sum_{\beta=1}^{3}\mathbf{b}^{\alpha\beta} = -\mathbf{b} - 2\mathbf{S}\mathbf{L}^{-1}\mathbf{f}.$$

Thus the total force due to image singularities (not to be confused with the image force studied in Section 8.6) is identical to the applied force $\mathbf{f}$. This explains that, in the limiting case of the surface Green's function for which $\mathbf{b}=\mathbf{0}$ and $d=0$ considered in Section 8.5, the total force from image singularities and the applied force do not form a double-force.

If $x_2=0$ is a rigid surface, inserting (8.6-12) into (8.7-6) yields

$$\begin{aligned}\mathbf{f}^{\alpha\beta} &= 2\,\mathrm{Re}\left\{\mathbf{B}\mathbf{I}_\alpha \mathbf{A}^{-1}\overline{\mathbf{A}}\mathbf{I}_\beta(\overline{\mathbf{A}}^T\mathbf{f} + \overline{\mathbf{B}}^T\mathbf{b})\right\},\\ \mathbf{b}^{\alpha\beta} &= 2\,\mathrm{Re}\left\{\mathbf{A}\mathbf{I}_\alpha \mathbf{A}^{-1}\overline{\mathbf{A}}\mathbf{I}_\beta(\overline{\mathbf{A}}^T\mathbf{f} + \overline{\mathbf{B}}^T\mathbf{b})\right\}.\end{aligned} \tag{8.7-8}$$

The total force and total dislocation from image singularities are

$$\sum_{\alpha=1}^{3}\sum_{\beta=1}^{3}\mathbf{f}^{\alpha\beta} = -\mathbf{f} - 2\mathbf{H}^{-1}\mathbf{S}\mathbf{b}, \qquad \sum_{\alpha=1}^{3}\sum_{\beta=1}^{3}\mathbf{b}^{\alpha\beta} = \mathbf{b}.$$

It should be pointed out that not all nine locations $(x_1^{\alpha\beta}, x_2^{\alpha\beta})$ given in (8.7-2) have image singularities. For monoclinic materials with the plane of symmetry at $x_3=0$ both matrices $\mathbf{A}$ and $\mathbf{B}$ have the structure (Section 6.3)

$$\begin{bmatrix} * & * & 0 \\ * & * & 0 \\ 0 & 0 & * \end{bmatrix}$$

where the asterisks represent possibly nonzero elements. It is readily shown from (8.7-7) and (8.7-8) that $\mathbf{f}^{\alpha\beta}$ and $\mathbf{b}^{\alpha\beta}$ vanish when α or β (but not both) assumes the value 3. Therefore there are only five image singularities for monoclinic materials with the plane of symmetry at $x_3=0$.

It was stated in Section 3.4 that the complex variable

$$z = x_1 + px_2 = (x_1 + p'x_2) + ip''x_2$$

is a mapping from the (x_1,x_2)-plane to the z-plane. The mapping can be regarded as a deformation. The (x_1,x_2)-plane is sheared in the x_1-direction with the shear strain p' and stretched (or contracted) in the x_2-direction with the stretch ratio p''. Hence the x_1-axis does not move. It is easily shown that (i) straight lines deform into straight lines and parallel lines deform into parallel lines. (ii) The point $D=(0,1)$ is moved to $p=(p',p'')$. Any point other than D is moved in the direction parallel to the movement of D to p. The magnitude of the movement is proportional to its distance from the x_1-axis. (iii) The mapping of a straight line is a rotation about its intersection with the x_1-axis accompanied by a stretching or contraction.

Using the properties mentioned above the position $(x_1^{\alpha\beta},x_2^{\alpha\beta})$ of the image singularities can be determined graphically. For simplicity define X_1, X_2 by

$$X_1 = x_1/d, \quad X_2 = x_2/d. \tag{8.7-9}$$

The real singularities as given by the first terms in (8.6-2) are at

$$z_\alpha/d = X_1 + p_\alpha X_2 = p_\alpha, \quad (\alpha=1,2,3). \tag{8.7-10}$$

They are at

$$(X_1,X_2)=(0,1)$$

which is designated as the point D in Fig. 8.6. It is mapped to three points p_α $(\alpha=1,2,3)$ in the z-plane. The image singularities of the second terms in (8.6-2) are, using (8.7-3), at

$$z^{\alpha\beta}/d = X_1^{\alpha\beta} + p_\alpha X_2^{\alpha\beta} = \bar{p}_\beta. \tag{8.7-11}$$

Knowing $\bar{p}_\beta$ $(\beta=1,2,3)$, the problem is to locate $X^{\alpha\beta} = (X_1^{\alpha\beta},X_2^{\alpha\beta})$ in the (X_1,X_2)-plane that would map to $\bar{p}_\beta$ in the z-plane. To illustrate the graphical solution consider $\beta=1$, i.e.,

$$X_1^{\alpha 1} + p_\alpha X_2^{\alpha 1} = \bar{p}_1 \quad (\alpha=1,2,3).$$

Let P_{11} be the point of intersection of the line $p_1\bar{p}_1$ with the X_1-axis. By drawing the line $\bar{p}_1 X^{11}$ parallel to Dp_1 and the line $D\,P_{11}$, X^{11} is at the intersection of these two lines. This is based on properties (ii) and (iii) of the mapping stated earlier. Likewise X^{21} is at the intersection of the line $\bar{p}_1 X^{21}$ parallel to Dp_2 and the line $D\,P_{21}$

where P_{21} is the intersection of the line $p_2\bar{p}_1$ with the X_1-axis. Instead of drawing the line DP_{21} we could draw the line $P_{11}X^{21}$ parallel to DP_{22}. The latter is based on property (i) that the parallel lines $P_{11}X^{21}$ and DP_{22} deform into parallel lines $P_{11}\bar{p}_1$ and p_2P_{22}. With this alternate approach X^{31} is located by drawing the line $\bar{p}_1X^{31}$ parallel to Dp_3 and the line $P_{11}X^{31}$ parallel to DP_{33}.

The graphical solutions of $X^{\alpha 2}$ (α=1,2,3) and $X^{\alpha 3}$ (α=1,2,3) are similar to that for $X^{\alpha 1}$ (α=1,2,3). As shown in Fig. 8.6 the triangles formed by (X^{11}, X^{21}, X^{31}), (X^{12}, X^{22}, X^{32}), and (X^{13}, X^{23}, X^{33}) are similar triangles.

Equation (8.7-10) tells us that the real singularity D=(0,1) in the (X_1,X_2)-plane is mapped to three points p_1, p_2, p_3 in the z-plane while (8.7-11) implies that the image singularities $X^{\alpha 1}$, $X^{\alpha 2}$, $X^{\alpha 3}$ are mapped to $\bar{p}_1$, $\bar{p}_2$, $\bar{p}_3$, respectively. p_1, p_2, p_3 and $\bar{p}_1$, $\bar{p}_2$, $\bar{p}_3$ are *mirror images* of each other in the z-plane. Thus the real singularities and the image singularities are mirror images of each other in the z-plane.

There are in general nine distinct image singularities for $X^{\alpha\beta}$ when p_1, p_2, p_3 are distinct. Different graphical solutions for the nine image singularities and special cases can be found in (Ting, 1992b). If $p_1' = p_2'$, X^{11} and X^{22} coalesce into the same point and X^{11}, X^{12}, X^{21}, and D are collinear. When $p_1'' = p_2''$, X^{12} and X^{21} are the same point and X^{11}, X^{12}, and X^{22} lie on a line parallel to the X_1-axis at the distance −1 from the X_1-axis. If p_1, p_2, p_3 are collinear,

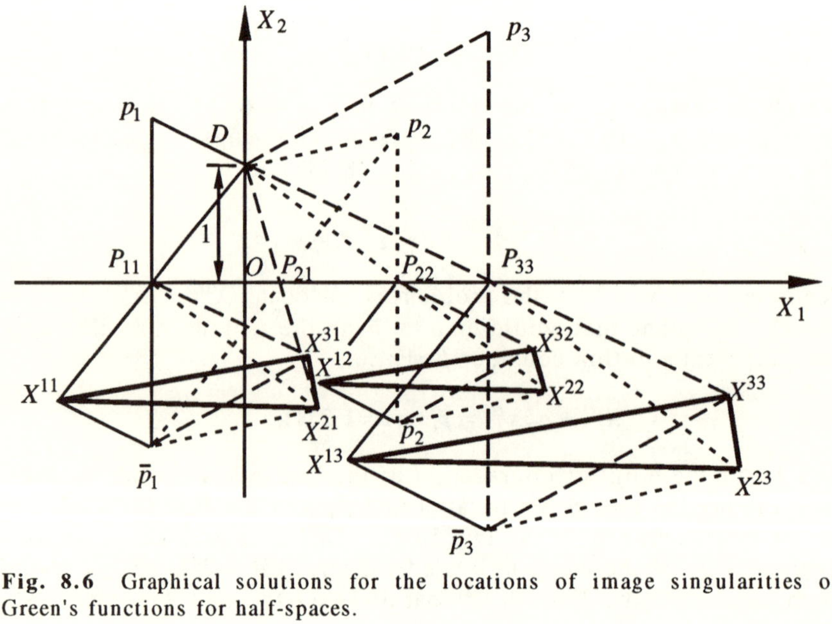

Fig. 8.6 Graphical solutions for the locations of image singularities of Green's functions for half-spaces.

the three points $X^{\alpha\beta}$ for a fixed α or a fixed β are collinear. Moreover the three lines $X^{\alpha\beta}$ (β=1,2,3) are parallel to each other.

When p_α are distinct and some of the locations of the singularities coalesce into one point the singularities at the coalesced point are simply the sum of all individual singularities. The situation is quite different if the coalescence of the locations of the image singularities is due to degeneracy of the materials. For instance, when $p_1=p_2$ there are only four distinct locations of image singularities. They are

$$\begin{gathered} X^{11}=X^{12}=X^{21}=X^{22}, \\ X^{13}=X^{31}, \quad X^{23}=X^{32}, \quad X^{33}. \end{gathered} \tag{8.7-12}$$

The forces $\mathbf{f}^{\alpha\beta}$ and the dislocation $\mathbf{b}^{\alpha\beta}$ all become infinite with the exception of $\mathbf{f}^{33}$ and $\mathbf{b}^{33}$. Although $(\mathbf{f}^{11}+\mathbf{f}^{12}+\mathbf{f}^{21}+\mathbf{f}^{22})$, $(\mathbf{f}^{13}+\mathbf{f}^{23})$, $(\mathbf{f}^{31}+\mathbf{f}^{32})$, and the same expressions for $\mathbf{b}^{\alpha\beta}$ can be shown to be bounded, it does not necessarily mean that these are the only forces and dislocations at the first three locations listed in (8.7-12). As is well-known for isotropic materials the image singularities are not simply concentrated forces and dislocations (Dundurs, 1969; Hirth and Lothe, 1982). There may be double-forces, concentrated couples and other higher-order singularities which will be discussed in Section 8.12.

It is clear from the analyses that the methods presented here for determining the locations and nature of image singularities are general, not limited to Green's functions for line forces and line dislocations. For Green's functions for other types of singularities such as concentrated couples, we simply replace $\langle \ln z_* \rangle$, $\langle \ln(z_*-p_*d) \rangle$, and $\langle \ln(z_*-\bar{p}_\beta d) \rangle$ by $\langle f(z_*) \rangle$, $\langle f(z_*-p_*d) \rangle$, and $\langle f(z_*-\bar{p}_\beta d) \rangle$ where f is the function appropriate for the particular singularity under consideration. The location of an image singularity is independent of the type of singularity concerned. It depends on elastic constants only.

8.8 Interfacial Green's Function

A bimaterial consists of two dissimilar anisotropic elastic half-spaces bonded together. Let the half-space

$$r>0, \quad \theta_o+\pi>\theta>\theta_o$$

be occupied by material 1, and the half-space

$$r>0, \quad \theta_o-\pi<\theta<\theta_o$$

be occupied by material 2. The plane $\theta=\theta_o$ and $\theta_o \pm \pi$ are the interface, Fig. 8.7(a). In particular, $x_2=0$ is the interface when $\theta_o=0$,

Fig. 8.7(b). The bimaterial is subjected to a line force **f** and a line dislocation with Burgers vector **b** at the origin r=0. For the displacement **u** to be single-valued we introduce a branch cut at the interface $\theta=\theta_o\pm\pi$. The conditions at the interface are

$$\mathbf{u}_1(r,\theta_o)=\mathbf{u}_2(r,\theta_o), \qquad \phi_1(r,\theta_o)=\phi_2(r,\theta_o), \tag{8.8-1a}$$

$$\begin{aligned} \mathbf{u}_1(r,\theta_o+\pi)-\mathbf{u}_2(r,\theta_o-\pi)&=\mathbf{b},\\ \phi_1(r,\theta_o+\pi)-\phi_2(r,\theta_o-\pi)&=\mathbf{f}. \end{aligned} \tag{8.8-1b}$$

The subscripts 1 and 2 denote quantities referred to materials 1 and 2, respectively.

The real form solution (8.1-21) with **g** and **h** regarded as arbitrary real vectors is a general solution for stress fields that are proportional to r^{-1}. Since the stresses are unchanged if we add constant terms to the right of (8.1-21), we modify (8.1-21) and write

$$\begin{aligned} 2\mathbf{u}_1&=-\frac{1}{\pi}(\ln r)\mathbf{h}-[\mathbf{S}_1(\theta)-\mathbf{S}_1(\theta_o)]\mathbf{h}-[\mathbf{H}_1(\theta)-\mathbf{H}_1(\theta_o)]\mathbf{g},\\ 2\phi_1&=-\frac{1}{\pi}(\ln r)\mathbf{g}+[\mathbf{L}_1(\theta)-\mathbf{L}_1(\theta_o)]\mathbf{h}-[\mathbf{S}_1^T(\theta)-\mathbf{S}_1^T(\theta_o)]\mathbf{g}, \end{aligned} \tag{8.8-2}$$

for material 1 in $\theta_o+\pi>\theta>\theta_o$ and

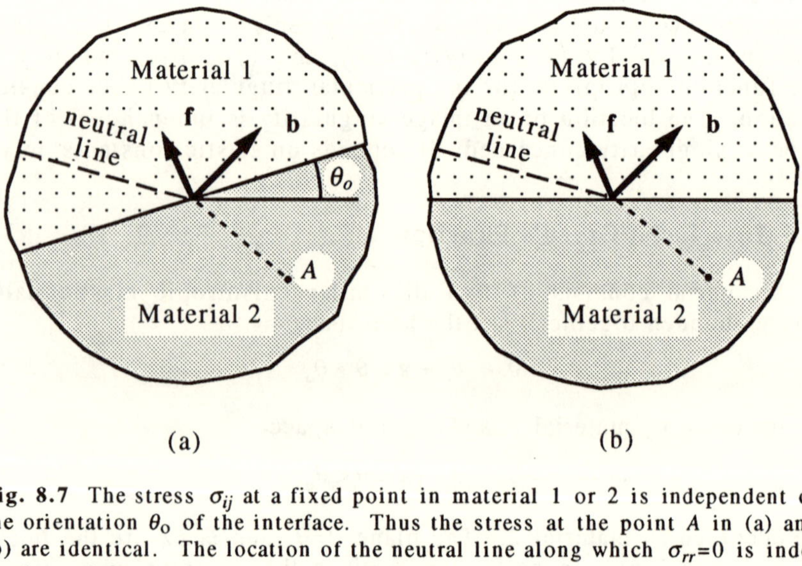

(a) (b)

Fig. 8.7 The stress σ_{ij} at a fixed point in material 1 or 2 is independent of the orientation θ_o of the interface. Thus the stress at the point A in (a) and (b) are identical. The location of the neutral line along which σ_{rr}=0 is independent of θ_o.

$$
\begin{aligned}
2\mathbf{u}_2 &= -\frac{1}{\pi}(\ln r)\mathbf{h} - [\mathbf{S}_2(\theta) - \mathbf{S}_2(\theta_o)]\mathbf{h} - [\mathbf{H}_2(\theta) - \mathbf{H}_2(\theta_o)]\mathbf{g}, \\
2\boldsymbol{\phi}_2 &= -\frac{1}{\pi}(\ln r)\mathbf{g} + [\mathbf{L}_2(\theta) - \mathbf{L}_2(\theta_o)]\mathbf{h} - [\mathbf{S}_2^T(\theta) - \mathbf{S}_2^T(\theta_o)]\mathbf{g},
\end{aligned} \tag{8.8-3}
$$

for material 2 in $\theta_o - \pi < \theta < \theta_o$. The physical meanings of **g** and **h** explained in Section 8.1 for the homogeneous infinite space remains valid here. The vector **h** represents the direction of infinite displacement at $r=0$ while $-\mathbf{g}$ represents the direction of the surface traction on any radial plane θ = constant. In view of this, and for the continuity of surface traction at the interface $\theta=\theta_o$ as required in (8.8-1a), **g** and **h** are identical in both materials as shown in (8.8-2) and (8.8-3). Noting that $\mathbf{S}(\theta)$, $\mathbf{H}(\theta)$, $\mathbf{L}(\theta)$ are periodic in θ with periodicity π and that $\mathbf{S}(\pi)=\mathbf{S}$, $\mathbf{H}(\pi)=\mathbf{H}$, $\mathbf{L}(\pi)=\mathbf{L}$, substitution of (8.8-2) and (8.8-3) into (8.8-1b) leads to

$$
\begin{aligned}
-(\mathbf{S}_1 + \mathbf{S}_2)\mathbf{h} - (\mathbf{H}_1 + \mathbf{H}_2)\mathbf{g} &= 2\mathbf{b}, \\
(\mathbf{L}_1 + \mathbf{L}_2)\mathbf{h} - (\mathbf{S}_1^T + \mathbf{S}_2^T)\mathbf{g} &= 2\mathbf{f}.
\end{aligned} \tag{8.8-4}
$$

Before we solve for **g** and **h** it should be noted that **g** and **h** obtained from (8.8-4) are independent of θ_o, the orientation of the interface. We therefore have the interesting and puzzling result that the stress is *invariant* with the orientation of the interface in the bimaterial. It should be stressed that the vectors **f** and **b** remain fixed, so do the orientation of the two materials. Only the orientation of the interface is changed. The stress at any point in the bimaterial, say the point *A* in Fig. 8.7, is invariant with the orientation of the interface. So is the location of the *neutral line* along which $\sigma_{rr}=0$.

Let $\tilde{\mathbf{S}}$, $\tilde{\mathbf{H}}$, and $\tilde{\mathbf{L}}$ be 3×3 real tensors defined by

$$
\begin{bmatrix} \tilde{\mathbf{S}} & \tilde{\mathbf{H}} \\ -\tilde{\mathbf{L}} & \tilde{\mathbf{S}}^T \end{bmatrix}
\begin{bmatrix} \mathbf{S}_1+\mathbf{S}_2 & \mathbf{H}_1+\mathbf{H}_2 \\ -\mathbf{L}_1-\mathbf{L}_2 & \mathbf{S}_1^T+\mathbf{S}_2^T \end{bmatrix}
= -\mathbf{I} =
\begin{bmatrix} \mathbf{S}_1+\mathbf{S}_2 & \mathbf{H}_1+\mathbf{H}_2 \\ -\mathbf{L}_1-\mathbf{L}_2 & \mathbf{S}_1^T+\mathbf{S}_2^T \end{bmatrix}
\begin{bmatrix} \tilde{\mathbf{S}} & \tilde{\mathbf{H}} \\ -\tilde{\mathbf{L}} & \tilde{\mathbf{S}}^T \end{bmatrix} \tag{8.8-5a}
$$

or

$$
\begin{aligned}
\tilde{\mathbf{H}}(\mathbf{L}_1 + \mathbf{L}_2) - \tilde{\mathbf{S}}(\mathbf{S}_1 + \mathbf{S}_2) = \mathbf{I} &= \tilde{\mathbf{L}}(\mathbf{H}_1 + \mathbf{H}_2) - \tilde{\mathbf{S}}^T(\mathbf{S}_1^T + \mathbf{S}_2^T), \\
\tilde{\mathbf{S}}^T(\mathbf{L}_1 + \mathbf{L}_2) + \tilde{\mathbf{L}}(\mathbf{S}_1 + \mathbf{S}_2) = \mathbf{0} &= \tilde{\mathbf{S}}(\mathbf{H}_1 + \mathbf{H}_2) + \tilde{\mathbf{H}}(\mathbf{S}_1^T + \mathbf{S}_2^T),
\end{aligned} \tag{8.8-5b}
$$

$$
\begin{aligned}
(\mathbf{L}_1 + \mathbf{L}_2)\tilde{\mathbf{H}} - (\mathbf{S}_1^T + \mathbf{S}_2^T)\tilde{\mathbf{S}}^T = \mathbf{I} &= (\mathbf{H}_1 + \mathbf{H}_2)\tilde{\mathbf{L}} - (\mathbf{S}_1 + \mathbf{S}_2)\tilde{\mathbf{S}}, \\
(\mathbf{L}_1 + \mathbf{L}_2)\tilde{\mathbf{S}} + (\mathbf{S}_1^T + \mathbf{S}_2^T)\tilde{\mathbf{L}} = \mathbf{0} &= (\mathbf{H}_1 + \mathbf{H}_2)\tilde{\mathbf{S}}^T + (\mathbf{S}_1 + \mathbf{S}_2)\tilde{\mathbf{H}}.
\end{aligned} \tag{8.8-5c}
$$

It can be shown from $(8.8\text{-}5b)_{3,4}$, $(8.8\text{-}5b)_1$, and $(8.8\text{-}5b)_2$ that

$$\tilde{\mathbf{S}} = -\tilde{\mathbf{H}}(\mathbf{S}_1^T + \mathbf{S}_2^T)(\mathbf{H}_1 + \mathbf{H}_2)^{-1} = -(\mathbf{L}_1 + \mathbf{L}_2)^{-1}(\mathbf{S}_1^T + \mathbf{S}_2^T)\tilde{\mathbf{L}},$$
$$\tilde{\mathbf{H}} = \left\{(\mathbf{L}_1 + \mathbf{L}_2) + (\mathbf{S}_1^T + \mathbf{S}_2^T)(\mathbf{H}_1 + \mathbf{H}_2)^{-1}(\mathbf{S}_1 + \mathbf{S}_2)\right\}^{-1}, \tag{8.8-6}$$
$$\tilde{\mathbf{L}} = \left\{(\mathbf{H}_1 + \mathbf{H}_2) + (\mathbf{S}_1 + \mathbf{S}_2)(\mathbf{L}_1 + \mathbf{L}_2)^{-1}(\mathbf{S}_1^T + \mathbf{S}_2^T)\right\}^{-1}.$$

By writing (8.8-4) as

$$\begin{bmatrix} \mathbf{S}_1+\mathbf{S}_2 & \mathbf{H}_1+\mathbf{H}_2 \\ -\mathbf{L}_1-\mathbf{L}_2 & \mathbf{S}_1^T+\mathbf{S}_2^T \end{bmatrix}\begin{bmatrix} \mathbf{h} \\ \mathbf{g} \end{bmatrix} = -2\begin{bmatrix} \mathbf{b} \\ \mathbf{f} \end{bmatrix}$$

and using (8.8-5a) we obtain

$$\begin{bmatrix} \mathbf{h} \\ \mathbf{g} \end{bmatrix} = 2\begin{bmatrix} \tilde{\mathbf{S}} & \tilde{\mathbf{H}} \\ -\tilde{\mathbf{L}} & \tilde{\mathbf{S}}^T \end{bmatrix}\begin{bmatrix} \mathbf{b} \\ \mathbf{f} \end{bmatrix} \tag{8.8-7a}$$

or

$$\mathbf{h} = 2(\tilde{\mathbf{S}}\mathbf{b} + \tilde{\mathbf{H}}\mathbf{f}), \quad \mathbf{g} = 2(\tilde{\mathbf{S}}^T\mathbf{f} - \tilde{\mathbf{L}}\mathbf{b}). \tag{8.8-7b}$$

Equations (8.8-2), (8.8-3), and (8.8-7b) provide the interfacial Green's function for bimaterials.

From (8.8-6) the tensors $\tilde{\mathbf{H}}$ and $\tilde{\mathbf{L}}$ are positive definite because $\mathbf{L}_1$, $\mathbf{L}_2$, $\mathbf{H}_1$, $\mathbf{H}_2$ are positive definite. When the impedance of the two materials are identical, $\mathbf{S}_1=\mathbf{S}_2$, $\mathbf{H}_1=\mathbf{H}_2$, and $\mathbf{L}_1=\mathbf{L}_2$. Hence

$$\tilde{\mathbf{H}} = \tfrac{1}{2}\mathbf{H}_1 = \tfrac{1}{2}\mathbf{H}_2, \quad \tilde{\mathbf{L}} = \tfrac{1}{2}\mathbf{L}_1 = \tfrac{1}{2}\mathbf{L}_2, \quad \tilde{\mathbf{S}} = \tfrac{1}{2}\mathbf{S}_1 = \tfrac{1}{2}\mathbf{S}_2, \tag{8.8-8}$$

and (8.8-5b), (8.8-7b) reduce to (8.1-3b), (8.1-20b), respectively.

Equation (8.1-26) remains valid for the strain energy ε for the annulus region $0 < r_1 \le r \le r_2$ of the bimaterial under consideration. Substitution of $\mathbf{h}$ and $\mathbf{g}$ from (8.8-7b) yields

$$\varepsilon = \frac{1}{2\pi}\ln\left(\frac{r_2}{r_1}\right)(\mathbf{f}^T\tilde{\mathbf{H}}\mathbf{f} + \mathbf{b}^T\tilde{\mathbf{L}}\mathbf{b}) \tag{8.8-9}$$

which is independent of θ_0, the orientation of the interface (Barnett and Lothe, 1974b). Again there is no interaction of strain energy between the line force $\mathbf{f}$ and the line dislocation $\mathbf{b}$ for the bimaterial. The tensors $\tilde{\mathbf{H}}$ and $\tilde{\mathbf{L}}$ are given by (8.8-6). Alternate expressions of $\tilde{\mathbf{H}}$ and $\tilde{\mathbf{L}}$ will be given in (8.8-29) and (8.8-26).

For isotropic materials Dundurs and Sendeckyj (1965a) have shown that the applied force $\mathbf{f}$ does not transmit in equal portions to materials 1 and 2. Let $\mathbf{f}_1$ be the total traction force on a plane surface in materials 1 which is parallel to the interface, and $\mathbf{f}_2$ be the total traction force on a plane surface in material 2 which is parallel to the interface. They are given by

$$\mathbf{f}_1 = \boldsymbol{\phi}_1(r,\ \theta_o+\pi) - \boldsymbol{\phi}_1(r,\ \theta_o) = \tfrac{1}{2}(\mathbf{L}_1\mathbf{h} - \mathbf{S}_1^T\mathbf{g}),$$

$$\mathbf{f}_2 = \boldsymbol{\phi}_2(r,\ \theta_o) - \boldsymbol{\phi}_2(r,\ \theta_o-\pi) = \tfrac{1}{2}(\mathbf{L}_2\mathbf{h} - \mathbf{S}_2^T\mathbf{g}),$$

Inserting **h** and **g** from (8.8-7b) we have

$$\begin{aligned} \mathbf{f}_1 &= (\mathbf{L}_1\tilde{\mathbf{H}} - \mathbf{S}_1^T\tilde{\mathbf{S}}^T)\mathbf{f} + (\mathbf{L}_1\tilde{\mathbf{S}} + \mathbf{S}_1^T\tilde{\mathbf{L}})\mathbf{b}, \\ \mathbf{f}_2 &= (\mathbf{L}_2\tilde{\mathbf{H}} - \mathbf{S}_2^T\tilde{\mathbf{S}}^T)\mathbf{f} + (\mathbf{L}_2\tilde{\mathbf{S}} + \mathbf{S}_2^T\tilde{\mathbf{L}})\mathbf{b}. \end{aligned} \qquad (8.8\text{-}10)$$

With the identities $(8.8\text{-}5c)_{1,3}$ it is readily shown that

$$\mathbf{f}_1 + \mathbf{f}_2 = \mathbf{f}.$$

It will be shown in (8.8-34a) that the coefficients of **b** in (8.8-10) are skew-symmetric tensors.

The dislocation **b** is also not equally divided between the two materials. If $\mathbf{b}_1$, $\mathbf{b}_2$ are defined as

$$\mathbf{b}_1 = \mathbf{u}_1(r,\ \theta_o+\pi) - \mathbf{u}_1(r,\ \theta_o) = -\tfrac{1}{2}(\mathbf{S}_1\mathbf{h} + \mathbf{H}_1\mathbf{g}),$$

$$\mathbf{b}_2 = \mathbf{u}_2(r,\ \theta_o) - \mathbf{u}_2(r,\ \theta_o-\pi) = -\tfrac{1}{2}(\mathbf{S}_2\mathbf{h} + \mathbf{H}_2\mathbf{g}),$$

a similar derivation leads to

$$\begin{aligned} \mathbf{b}_1 &= (\mathbf{H}_1\tilde{\mathbf{L}} - \mathbf{S}_1\tilde{\mathbf{S}})\mathbf{b} - (\mathbf{H}_1\tilde{\mathbf{S}}^T + \mathbf{S}_1\tilde{\mathbf{H}})\mathbf{f}, \\ \mathbf{b}_2 &= (\mathbf{H}_2\tilde{\mathbf{L}} - \mathbf{S}_2\tilde{\mathbf{S}})\mathbf{b} - (\mathbf{H}_2\tilde{\mathbf{S}}^T + \mathbf{S}_2\tilde{\mathbf{H}})\mathbf{f}. \end{aligned} \qquad (8.8\text{-}11)$$

Again it is easily seen from $(8.8\text{-}5c)_{2,4}$ that

$$\mathbf{b}_1 + \mathbf{b}_2 = \mathbf{b}.$$

We will show in (8.8-34b) that the coefficients of **f** in (8.8-11) are skew-symmetric tensors. We will also show in (8.8-35) that the coefficients of **f** in $(8.8\text{-}10)_{1,2}$ are, respectively, the transpose of the coefficients of **b** in $(8.8\text{-}11)_{2,1}$.

Rearranging (8.8-10) and (8.8-11) in the form

$$\begin{bmatrix} \mathbf{b}_i \\ \mathbf{f}_i \end{bmatrix} = -\begin{bmatrix} \mathbf{S}_i & \mathbf{H}_i \\ -\mathbf{L}_i & \mathbf{S}_i^T \end{bmatrix} \begin{bmatrix} \tilde{\mathbf{S}} & \tilde{\mathbf{H}} \\ -\tilde{\mathbf{L}} & \tilde{\mathbf{S}} \end{bmatrix} \begin{bmatrix} \mathbf{b} \\ \mathbf{f} \end{bmatrix} \quad (i = 1,2) \qquad (8.8\text{-}12)$$

we can write (8.8-7a) as, by virtue of (8.1-3a),

$$\begin{bmatrix} \mathbf{h} \\ \mathbf{g} \end{bmatrix} = 2\begin{bmatrix} \mathbf{S}_i & \mathbf{H}_i \\ -\mathbf{L}_i & \mathbf{S}_i^T \end{bmatrix} \begin{bmatrix} \mathbf{b}_i \\ \mathbf{f}_i \end{bmatrix} \quad (i \text{ not summed}). \qquad (8.8\text{-}13)$$

Comparison with (8.1-20a) suggests that the solution (8.8-2) for material 1 represents the Green's function for the infinite space occupied by material 1 subjected to a line force $2\mathbf{f}_1$ and a line dislocation with Burgers vector $2\mathbf{b}_1$ at $r=0$. Likewise (8.8-3) represents the Green's function for the infinite space occupied by material 2 subjected to a line force $2\mathbf{f}_2$ and a line dislocation with Burgers vector $2\mathbf{b}_2$ at $r=0$ (Kirchner and Lothe, 1987). Rewriting (8.8-5a) as

$$\begin{bmatrix} \mathbf{S}_1 & \mathbf{H}_1 \\ -\mathbf{L}_1 & \mathbf{S}_1^T \end{bmatrix} \begin{bmatrix} \tilde{\mathbf{S}} & \tilde{\mathbf{H}} \\ -\tilde{\mathbf{L}} & \tilde{\mathbf{S}} \end{bmatrix} + \begin{bmatrix} \mathbf{S}_2 & \mathbf{H}_2 \\ -\mathbf{L}_2 & \mathbf{S}_2^T \end{bmatrix} \begin{bmatrix} \tilde{\mathbf{S}} & \tilde{\mathbf{H}} \\ -\tilde{\mathbf{L}} & \tilde{\mathbf{S}} \end{bmatrix} = -\mathbf{I},$$

(8.8-12) has the alternate expression

$$\begin{bmatrix} \mathbf{b}_1 \\ \mathbf{f}_1 \end{bmatrix} = \frac{1}{2} \begin{bmatrix} \mathbf{b} \\ \mathbf{f} \end{bmatrix} - \frac{1}{2} \begin{bmatrix} \mathbf{S}_1 - \mathbf{S}_2 & \mathbf{H}_1 - \mathbf{H}_2 \\ -\mathbf{L}_1 + \mathbf{L}_2 & \mathbf{S}_1^T - \mathbf{S}_2^T \end{bmatrix} \begin{bmatrix} \tilde{\mathbf{S}} & \tilde{\mathbf{H}} \\ -\tilde{\mathbf{L}} & \tilde{\mathbf{S}} \end{bmatrix} \begin{bmatrix} \mathbf{b} \\ \mathbf{f} \end{bmatrix},$$

$$\begin{bmatrix} \mathbf{b}_2 \\ \mathbf{f}_2 \end{bmatrix} = \frac{1}{2} \begin{bmatrix} \mathbf{b} \\ \mathbf{f} \end{bmatrix} + \frac{1}{2} \begin{bmatrix} \mathbf{S}_1 - \mathbf{S}_2 & \mathbf{H}_1 - \mathbf{H}_2 \\ -\mathbf{L}_1 + \mathbf{L}_2 & \mathbf{S}_1^T - \mathbf{S}_2^T \end{bmatrix} \begin{bmatrix} \tilde{\mathbf{S}} & \tilde{\mathbf{H}} \\ -\tilde{\mathbf{L}} & \tilde{\mathbf{S}} \end{bmatrix} \begin{bmatrix} \mathbf{b} \\ \mathbf{f} \end{bmatrix}.$$

The following different derivation of the interfacial Green's function is more transparent in determining $\mathbf{f}_1$, $\mathbf{f}_2$, $\mathbf{b}_1$, and $\mathbf{b}_2$. For simplicity we let $\theta_o=0$ so that the interface is at $x_2=0$. We use the original complex form solution (8.1-16) and let

$$\mathbf{u}_1 = \frac{1}{\pi}\mathrm{Im}\left\{\mathbf{A}_1\left\langle \ln z_*^{(1)}\right\rangle \mathbf{q}_1\right\}, \quad \boldsymbol{\phi}_1 = \frac{1}{\pi}\mathrm{Im}\left\{\mathbf{B}_1\left\langle \ln z_*^{(1)}\right\rangle \mathbf{q}_1\right\}, \tag{8.8-14}$$

for material 1 in $x_2>0$ and

$$\mathbf{u}_2 = \frac{1}{\pi}\mathrm{Im}\left\{\mathbf{A}_2\left\langle \ln z_*^{(2)}\right\rangle \mathbf{q}_2\right\}, \quad \boldsymbol{\phi}_2 = \frac{1}{\pi}\mathrm{Im}\left\{\mathbf{B}_2\left\langle \ln z_*^{(2)}\right\rangle \mathbf{q}_2\right\}, \tag{8.8-15}$$

for material 2 in $x_2<0$. The superscripts (1) and (2) referred to materials 1 and 2, respectively. The constant vectors $\mathbf{q}_1$ and $\mathbf{q}_2$ are determined by substituting (8.8-14) and (8.8-15) into (8.8-1a) and (8.8-1b) with $\theta_o=0$. We have

$$\mathrm{Im}(\mathbf{A}_1\mathbf{q}_1) = \mathrm{Im}(\mathbf{A}_2\mathbf{q}_2), \quad \mathrm{Im}(\mathbf{B}_1\mathbf{q}_1) = \mathrm{Im}(\mathbf{B}_2\mathbf{q}_2), \tag{8.8-16}$$

$$\mathrm{Re}(\mathbf{A}_1\mathbf{q}_1) + \mathrm{Re}(\mathbf{A}_2\mathbf{q}_2) = \mathbf{b}, \quad \mathrm{Re}(\mathbf{B}_1\mathbf{q}_1) + \mathrm{Re}(\mathbf{B}_2\mathbf{q}_2) = \mathbf{f}. \tag{8.8-17}$$

Equations (8.8-16) and (8.8-17) are equivalent to

$$\mathbf{A}_1\mathbf{q}_1 + \overline{\mathbf{A}}_2\overline{\mathbf{q}}_2 = \mathbf{b}, \quad \mathbf{B}_1\mathbf{q}_1 + \overline{\mathbf{B}}_2\overline{\mathbf{q}}_2 = \mathbf{f}, \tag{8.8-18}$$

which can be solved for $\mathbf{q}_1$ and $\mathbf{q}_2$. It should be noted that

$$\mathbf{b}_1 = \mathbf{u}_1(r, \pi) - \mathbf{u}_1(r, 0) = \mathrm{Re}(\mathbf{A}_1\mathbf{q}_1),$$
$$\mathbf{b}_2 = \mathbf{u}_2(r, 0) - \mathbf{u}_2(r, -\pi) = \mathrm{Re}(\mathbf{A}_2\mathbf{q}_2),$$
$$\mathbf{f}_1 = \boldsymbol{\phi}_1(r, \pi) - \boldsymbol{\phi}_1(r, 0) = \mathrm{Re}(\mathbf{B}_1\mathbf{q}_1),$$
$$\mathbf{f}_2 = \boldsymbol{\phi}_2(r, 0) - \boldsymbol{\phi}_2(r, -\pi) = \mathrm{Re}(\mathbf{B}_2\mathbf{q}_2).$$

Moreover,

$$\mathrm{Im}(\mathbf{A}_1\mathbf{q}_1) = \mathrm{Im}(\mathbf{A}_2\mathbf{q}_2) = -\tfrac{1}{2}\mathbf{h},$$
$$\mathrm{Im}(\mathbf{B}_1\mathbf{q}_1) = \mathrm{Im}(\mathbf{B}_2\mathbf{q}_2) = -\tfrac{1}{2}\mathbf{g}.$$

Therefore

$$\begin{aligned} \mathbf{A}_1\mathbf{q}_1 &= \mathbf{b}_1 - \tfrac{i}{2}\mathbf{h}, \quad \overline{\mathbf{A}}_2\overline{\mathbf{q}}_2 = \mathbf{b}_2 + \tfrac{i}{2}\mathbf{h}, \\ \mathbf{B}_1\mathbf{q}_1 &= \mathbf{f}_1 - \tfrac{i}{2}\mathbf{g}, \quad \overline{\mathbf{B}}_2\overline{\mathbf{q}}_2 = \mathbf{f}_2 + \tfrac{i}{2}\mathbf{g}. \end{aligned} \tag{8.8-19}$$

To solve (8.8-18) for $\mathbf{B}_1\mathbf{q}_1$ and $\overline{\mathbf{B}}_2\overline{\mathbf{q}}_2$ we rewrite $(8.8\text{-}18)_1$ in terms of $\mathbf{M}^{-1}$ using $(8.1\text{-}4)_3$. We have

$$\mathbf{M}_1^{-1}(\mathbf{B}_1\mathbf{q}_1) - \overline{\mathbf{M}}_2^{-1}(\overline{\mathbf{B}}_2\overline{\mathbf{q}}_2) = i\mathbf{b}$$

and the solutions are

$$\begin{aligned} \mathbf{B}_1\mathbf{q}_1 &= (\mathbf{M}_1^{-1} + \overline{\mathbf{M}}_2^{-1})^{-1}(\overline{\mathbf{M}}_2^{-1}\mathbf{f} + i\mathbf{b}), \\ \overline{\mathbf{B}}_2\overline{\mathbf{q}}_2 &= (\mathbf{M}_1^{-1} + \overline{\mathbf{M}}_2^{-1})^{-1}(\mathbf{M}_1^{-1}\mathbf{f} - i\mathbf{b}). \end{aligned} \tag{8.8-20}$$

Similarly, to solve (8.8-18) for $\mathbf{A}_1\mathbf{q}_1$ and $\overline{\mathbf{A}}_2\overline{\mathbf{q}}_2$ we rewrite $(8.8\text{-}18)_2$ in terms of $\mathbf{M}$ using $(8.1\text{-}4)_1$ as

$$\mathbf{M}_1(\mathbf{A}_1\mathbf{q}_1) - \overline{\mathbf{M}}_2(\overline{\mathbf{A}}_2\overline{\mathbf{q}}_2) = -i\mathbf{f}.$$

The solutions are

$$\begin{aligned} \mathbf{A}_1\mathbf{q}_1 &= (\mathbf{M}_1 + \overline{\mathbf{M}}_2)^{-1}(\overline{\mathbf{M}}_2\mathbf{b} - i\mathbf{f}), \\ \overline{\mathbf{A}}_2\overline{\mathbf{q}}_2 &= (\mathbf{M}_1 + \overline{\mathbf{M}}_2)^{-1}(\mathbf{M}_1\mathbf{b} + i\mathbf{f}). \end{aligned} \tag{8.8-21}$$

Noting that the impedance tensor $\mathbf{M}$ is a positive definite Hermitian, so are $\mathbf{M}_1 + \overline{\mathbf{M}}_2$ and $\mathbf{M}_1^{-1} + \overline{\mathbf{M}}_2^{-1}$. By virtue of the identities in (8.1-4) let

$$\begin{aligned} \mathbf{M}_1^{-1} + \overline{\mathbf{M}}_2^{-1} &= \mathbf{D} - i\mathbf{W}, \\ \mathbf{M}_1 + \overline{\mathbf{M}}_2 &= \hat{\mathbf{D}} - i\hat{\mathbf{W}}, \end{aligned} \tag{8.8-22}$$

where

$$\begin{aligned} \mathbf{D} &= \mathbf{L}_1^{-1} + \mathbf{L}_2^{-1}, \quad \mathbf{W} = \mathbf{S}_1\mathbf{L}_1^{-1} - \mathbf{S}_2\mathbf{L}_2^{-1}, \\ \hat{\mathbf{D}} &= \mathbf{H}_1^{-1} + \mathbf{H}_2^{-1}, \quad \hat{\mathbf{W}} = \mathbf{S}_1^T\mathbf{H}_1^{-1} - \mathbf{S}_2^T\mathbf{H}_2^{-1}. \end{aligned} \tag{8.8-23}$$

Both $\mathbf{W}$ and $\hat{\mathbf{W}}$ are skew-symmetric. Let

$$(\mathbf{M}_1^{-1}+\overline{\mathbf{M}}_2^{-1})^{-1}=\tilde{\mathbf{L}}+i\tilde{\mathbf{W}}. \tag{8.8-24}$$

From the relations

$$(\mathbf{D}-i\mathbf{W})(\tilde{\mathbf{L}}+i\tilde{\mathbf{W}})=\mathbf{I}=(\tilde{\mathbf{L}}+i\tilde{\mathbf{W}})(\mathbf{D}-i\mathbf{W})$$

we have

$$\begin{aligned}\mathbf{D}\tilde{\mathbf{L}}+\mathbf{W}\tilde{\mathbf{W}}&=\mathbf{I}=\tilde{\mathbf{L}}\mathbf{D}+\tilde{\mathbf{W}}\mathbf{W},\\ \mathbf{W}\tilde{\mathbf{L}}-\mathbf{D}\tilde{\mathbf{W}}&=\mathbf{0}=\tilde{\mathbf{L}}\mathbf{W}-\tilde{\mathbf{W}}\mathbf{D},\end{aligned} \tag{8.8-25}$$

and hence

$$\begin{aligned}\tilde{\mathbf{W}}&=\mathbf{D}^{-1}\mathbf{W}\tilde{\mathbf{L}}=\tilde{\mathbf{L}}\mathbf{W}\mathbf{D}^{-1},\\ \tilde{\mathbf{L}}&=(\mathbf{D}-\mathbf{W}^T\mathbf{D}^{-1}\mathbf{W})^{-1}.\end{aligned} \tag{8.8-26}$$

The employment of the notation $\tilde{\mathbf{L}}$ is intentional because it can be shown directly (Qu and Li, 1991) that the $\tilde{\mathbf{L}}$ in (8.8-26) is equivalent to the $\tilde{\mathbf{L}}$ defined in (8.8-6). In the same manner let

$$(\mathbf{M}_1+\overline{\mathbf{M}}_2)^{-1}=\tilde{\mathbf{H}}+i\breve{\mathbf{W}}. \tag{8.8-27}$$

From the relations

$$(\hat{\mathbf{D}}-i\hat{\mathbf{W}})(\tilde{\mathbf{H}}+i\breve{\mathbf{W}})=\mathbf{I}=(\tilde{\mathbf{H}}+i\breve{\mathbf{W}})(\hat{\mathbf{D}}-i\hat{\mathbf{W}})$$

we have

$$\begin{aligned}\hat{\mathbf{D}}\tilde{\mathbf{H}}+\hat{\mathbf{W}}\breve{\mathbf{W}}&=\mathbf{I}=\tilde{\mathbf{H}}\hat{\mathbf{D}}+\breve{\mathbf{W}}\hat{\mathbf{W}},\\ \hat{\mathbf{W}}\tilde{\mathbf{H}}-\hat{\mathbf{D}}\breve{\mathbf{W}}&=\mathbf{0}=\breve{\mathbf{W}}\hat{\mathbf{D}}-\tilde{\mathbf{H}}\hat{\mathbf{W}},\end{aligned} \tag{8.8-28}$$

and hence

$$\begin{aligned}\breve{\mathbf{W}}&=\hat{\mathbf{D}}^{-1}\hat{\mathbf{W}}\tilde{\mathbf{H}}=\tilde{\mathbf{H}}\hat{\mathbf{W}}\hat{\mathbf{D}}^{-1},\\ \tilde{\mathbf{H}}&=(\hat{\mathbf{D}}-\hat{\mathbf{W}}^T\hat{\mathbf{D}}^{-1}\hat{\mathbf{W}})^{-1}.\end{aligned} \tag{8.8-29}$$

Again it can be shown directly that the $\tilde{\mathbf{H}}$ in (8.8-29) is equivalent to the $\tilde{\mathbf{H}}$ defined in (8.8-6). The tensors $\mathbf{D}$, $\hat{\mathbf{D}}$, $\tilde{\mathbf{L}}$, and $\tilde{\mathbf{H}}$ are symmetric and positive definite while $\mathbf{W}$, $\hat{\mathbf{W}}$, $\tilde{\mathbf{W}}$, and $\breve{\mathbf{W}}$ are skew-symmetric.

With (8.8-24), (8.8-27), and (8.1-4) the $\mathbf{A}_1\mathbf{q}_1$, $\overline{\mathbf{A}}_2\overline{\mathbf{q}}_2$, $\mathbf{B}_1\mathbf{q}_1$, and $\overline{\mathbf{B}}_2\overline{\mathbf{q}}_2$ in (8.8-21) and (8.8-20) can be expressed in terms of real quantities. From (8.8-19) we have the following results:

$$\begin{aligned}\mathbf{h}&=2\left\{-(\tilde{\mathbf{H}}\mathbf{S}_1^T-\breve{\mathbf{W}})\mathbf{H}_1^{-1}\mathbf{b}+\tilde{\mathbf{H}}\mathbf{f}\right\}=2\left\{-(\tilde{\mathbf{H}}\mathbf{S}_2^T+\breve{\mathbf{W}})\mathbf{H}_2^{-1}\mathbf{b}+\tilde{\mathbf{H}}\mathbf{f}\right\},\\ \mathbf{g}&=2\left\{-(\tilde{\mathbf{L}}\mathbf{S}_1-\tilde{\mathbf{W}})\mathbf{L}_1^{-1}\mathbf{f}-\tilde{\mathbf{L}}\mathbf{b}\right\}=2\left\{-(\tilde{\mathbf{L}}\mathbf{S}_2+\tilde{\mathbf{W}})\mathbf{L}_2^{-1}\mathbf{f}-\tilde{\mathbf{L}}\mathbf{b}\right\},\end{aligned} \tag{8.8-30}$$

$$\mathbf{f}_1 = (\tilde{\mathbf{L}} - \tilde{\mathbf{W}}\mathbf{S}_2)\mathbf{L}_2^{-1}\mathbf{f} - \tilde{\mathbf{W}}\mathbf{b},$$
$$\mathbf{f}_2 = (\tilde{\mathbf{L}} + \tilde{\mathbf{W}}\mathbf{S}_1)\mathbf{L}_1^{-1}\mathbf{f} + \tilde{\mathbf{W}}\mathbf{b}, \tag{8.8-31}$$

$$\mathbf{b}_1 = (\tilde{\mathbf{H}} - \breve{\mathbf{W}}\mathbf{S}_2^T)\mathbf{H}_2^{-1}\mathbf{b} + \breve{\mathbf{W}}\mathbf{f},$$
$$\mathbf{b}_2 = (\tilde{\mathbf{H}} + \breve{\mathbf{W}}\mathbf{S}_1^T)\mathbf{H}_1^{-1}\mathbf{b} - \breve{\mathbf{W}}\mathbf{f}. \tag{8.8-32}$$

Comparison of (8.8-30) with (8.8-7b) confirms that the two definitions of $\tilde{\mathbf{H}}$ and $\tilde{\mathbf{L}}$ are indeed equivalent. We also obtain the following identities.

$$\tilde{\mathbf{S}} = -(\tilde{\mathbf{H}}\mathbf{S}_1^T - \breve{\mathbf{W}})\mathbf{H}_1^{-1} = -(\tilde{\mathbf{H}}\mathbf{S}_2^T + \breve{\mathbf{W}})\mathbf{H}_2^{-1}$$
$$= -\mathbf{L}_1^{-1}(\mathbf{S}_1^T\tilde{\mathbf{L}} + \tilde{\mathbf{W}}) = -\mathbf{L}_2^{-1}(\mathbf{S}_2^T\tilde{\mathbf{L}} - \tilde{\mathbf{W}}). \tag{8.8-33}$$

Solving for $\tilde{\mathbf{W}}$ and $\breve{\mathbf{W}}$ one has

$$\breve{\mathbf{W}} = \tilde{\mathbf{S}}\mathbf{H}_1 + \tilde{\mathbf{H}}\mathbf{S}_1^T = -(\tilde{\mathbf{S}}\mathbf{H}_2 + \tilde{\mathbf{H}}\mathbf{S}_2^T),$$
$$\tilde{\mathbf{W}} = -(\mathbf{L}_1\tilde{\mathbf{S}} + \mathbf{S}_1^T\tilde{\mathbf{L}}) = \mathbf{L}_2\tilde{\mathbf{S}} + \mathbf{S}_2^T\tilde{\mathbf{L}} \tag{8.8-34a}$$

and, due to the fact that $\tilde{\mathbf{W}}$ and $\breve{\mathbf{W}}$ are skew-symmetric,

$$\breve{\mathbf{W}} = -(\mathbf{H}_1\tilde{\mathbf{S}}^T + \mathbf{S}_1\tilde{\mathbf{H}}) = \mathbf{H}_2\tilde{\mathbf{S}}^T + \mathbf{S}_2\tilde{\mathbf{H}},$$
$$\tilde{\mathbf{W}} = \tilde{\mathbf{S}}^T\mathbf{L}_1 + \tilde{\mathbf{L}}\mathbf{S}_1 = -(\tilde{\mathbf{S}}^T\mathbf{L}_2 + \tilde{\mathbf{L}}\mathbf{S}_2). \tag{8.8-34b}$$

Additional identities are deduced by comparing (8.8-31), (8.8-32) with (8.8-10), (8.8-11).

$$\mathbf{L}_1\tilde{\mathbf{H}} - \mathbf{S}_1^T\tilde{\mathbf{S}}^T = (\tilde{\mathbf{L}} - \tilde{\mathbf{W}}\mathbf{S}_2)\mathbf{L}_2^{-1} = \tilde{\mathbf{L}}\mathbf{H}_2 - \tilde{\mathbf{S}}^T\mathbf{S}_2^T,$$
$$\mathbf{L}_2\tilde{\mathbf{H}} - \mathbf{S}_2^T\tilde{\mathbf{S}}^T = (\tilde{\mathbf{L}} + \tilde{\mathbf{W}}\mathbf{S}_1)\mathbf{L}_1^{-1} = \tilde{\mathbf{L}}\mathbf{H}_1 - \tilde{\mathbf{S}}^T\mathbf{S}_1^T, \tag{8.8-35}$$

$$\mathbf{H}_1\tilde{\mathbf{L}} - \mathbf{S}_1\tilde{\mathbf{S}} = (\tilde{\mathbf{H}} - \breve{\mathbf{W}}\mathbf{S}_2^T)\mathbf{H}_2^{-1} = \tilde{\mathbf{H}}\mathbf{L}_2 - \tilde{\mathbf{S}}\mathbf{S}_2,$$
$$\mathbf{H}_2\tilde{\mathbf{L}} - \mathbf{S}_2\tilde{\mathbf{S}} = (\tilde{\mathbf{H}} + \breve{\mathbf{W}}\mathbf{S}_1^T)\mathbf{H}_1^{-1} = \tilde{\mathbf{H}}\mathbf{L}_1 - \tilde{\mathbf{S}}\mathbf{S}_1. \tag{8.8-36}$$

Equations $(8.8\text{-}35)_{2,4}$ and $(8.8\text{-}36)_{2,4}$ are deduced by inserting $\tilde{\mathbf{W}}$ from $(8.8\text{-}34b)_{4,3}$, $\breve{\mathbf{W}}$ from $(8.8\text{-}34a)_{2,1}$, and making use of (8.1-3b). By (8.8-35) and (8.8-36) the coefficients of $\mathbf{f}$ in $(8.8\text{-}31)_{1,2}$ are the transpose of the coefficients of $\mathbf{b}$ in $(8.8\text{-}32)_{2,1}$, respectively.

When $\mathbf{f}=\mathbf{0}$ (8.8-31) simplifies to

$$\mathbf{f}_2 = -\mathbf{f}_1 = \tilde{\mathbf{W}}\mathbf{b} \tag{8.8-37}$$

and, due to the skew-symmetry of $\tilde{\mathbf{W}}$,

$$\mathbf{f}_2 \cdot \mathbf{b} = -\mathbf{f}_1 \cdot \mathbf{b} = \mathbf{0}.$$

Thus the traction forces $\mathbf{f}_1$, $\mathbf{f}_2$ induced by the line dislocation are normal to $\mathbf{b}$ (Barnett and Lothe, 1974b). The traction forces vanish if $\tilde{\mathbf{W}}=\mathbf{0}$ or $\mathbf{b}$ is collinear with the axial vector of $\tilde{\mathbf{W}}$. It is easily seen from (8.8-26) that $\tilde{\mathbf{W}}=\mathbf{0}$ if and only if $\mathbf{W}=\mathbf{0}$. Thus when

$$\mathbf{W} = \mathbf{S}_1\mathbf{L}_1^{-1} - \mathbf{S}_2\mathbf{L}_2^{-1} = \mathbf{0} \tag{8.8-38}$$

the traction forces $\mathbf{f}_1$ and $\mathbf{f}_2$ induced by the line dislocation $\mathbf{b}$ vanish. Condition (8.8-38) also have an important physical significance for an interfacial crack to be studied in Chapter 11.

If $\mathbf{b}=\mathbf{0}$ (8.8-32) reduce to

$$\mathbf{b}_1 = -\mathbf{b}_2 = \breve{\mathbf{W}}\mathbf{f} \tag{8.8-39}$$

and, in view of skew-symmetry of $\breve{\mathbf{W}}$,

$$\mathbf{b}_1 \cdot \mathbf{f} = -\mathbf{b}_2 \cdot \mathbf{f} = 0.$$

The Burgers vectors $\mathbf{b}_1$ and $\mathbf{b}_2$ induced by the line force $\mathbf{f}$ are normal to $\mathbf{f}$. The vectors $\mathbf{b}_1$ and $\mathbf{b}_2$ vanish when $\breve{\mathbf{W}}=\mathbf{0}$ or when $\mathbf{f}$ is collinear with the axial vector of $\breve{\mathbf{W}}$. From (8.8-29) $\breve{\mathbf{W}}=\mathbf{0}$ if and only if $\hat{\mathbf{W}}=\mathbf{0}$. Therefore when

$$\hat{\mathbf{W}} = \mathbf{S}_1^T\mathbf{H}_1^{-1} - \mathbf{S}_2^T\mathbf{H}_2^{-1} = \mathbf{0} \tag{8.8-40}$$

the Burgers vectors $\mathbf{b}_1$ and $\mathbf{b}_2$ induced by the line force $\mathbf{f}$ vanish. Condition (8.8-40) also have an important physical significance for an interfacial rigid line inclusion (Section 9.6).

In closing this section we point out that (8.8-25) can be recast in the form

$$\begin{bmatrix} \breve{\mathbf{S}} & \mathbf{D}^{-1} \\ -\tilde{\mathbf{L}}^{-1} & \breve{\mathbf{S}}^T \end{bmatrix} \begin{bmatrix} \breve{\mathbf{S}} & \mathbf{D}^{-1} \\ -\tilde{\mathbf{L}}^{-1} & \breve{\mathbf{S}}^T \end{bmatrix} = -\begin{bmatrix} \mathbf{I} & \mathbf{0} \\ \mathbf{0} & \mathbf{I} \end{bmatrix}, \tag{8.8-41}$$

$$\breve{\mathbf{S}} = \mathbf{D}^{-1}\mathbf{W} = \tilde{\mathbf{W}}\tilde{\mathbf{L}}^{-1}. \tag{8.8-42}$$

Likewise (8.8-28) is equivalent to

$$\begin{bmatrix} \hat{\mathbf{S}} & \tilde{\mathbf{H}}^{-1} \\ -\hat{\mathbf{D}}^{-1} & \hat{\mathbf{S}}^T \end{bmatrix} \begin{bmatrix} \hat{\mathbf{S}} & \tilde{\mathbf{H}}^{-1} \\ -\hat{\mathbf{D}}^{-1} & \hat{\mathbf{S}}^T \end{bmatrix} = -\begin{bmatrix} \mathbf{I} & \mathbf{0} \\ \mathbf{0} & \mathbf{I} \end{bmatrix}, \tag{8.8-43}$$

$$\hat{\mathbf{S}} = \hat{\mathbf{W}}\hat{\mathbf{D}}^{-1} = \tilde{\mathbf{H}}^{-1}\breve{\mathbf{W}}. \tag{8.8-44}$$

Equations (8.8-41) and (8.8-43) are similar to (8.1-3a) for $\hat{\mathbf{S}}$, $\hat{\mathbf{H}}$, and $\mathbf{L}$. Therefore the three real tensors ($\breve{\mathbf{S}}$, $\mathbf{D}^{-1}$, $\tilde{\mathbf{L}}^{-1}$) and ($\hat{\mathbf{S}}$, $\tilde{\mathbf{H}}^{-1}$, $\hat{\mathbf{D}}^{-1}$) have structures similar to those of ($\mathbf{S}$, $\mathbf{H}$, $\mathbf{L}$) discussed in Section 6.6. The identities (8.8-34)–(8.8-36) can be summarized in the following compact forms.

$$\begin{aligned}\begin{bmatrix}\tilde{\mathbf{S}} & \tilde{\mathbf{H}}\\ -\tilde{\mathbf{L}} & \tilde{\mathbf{S}}\end{bmatrix}\begin{bmatrix}\mathbf{S}_1 & \mathbf{H}_1\\ -\mathbf{L}_1 & \mathbf{S}_1^T\end{bmatrix}&=\begin{bmatrix}\mathbf{S}_2 & \mathbf{H}_2\\ -\mathbf{L}_2 & \mathbf{S}_2^T\end{bmatrix}\begin{bmatrix}\tilde{\mathbf{S}} & \tilde{\mathbf{H}}\\ -\tilde{\mathbf{L}} & \tilde{\mathbf{S}}\end{bmatrix}\\ &=-\begin{bmatrix}(\tilde{\mathbf{H}}+\breve{\mathbf{W}}\mathbf{S}_1^T)\mathbf{H}_1^{-1} & -\breve{\mathbf{W}}\\ \tilde{\mathbf{W}} & (\tilde{\mathbf{L}}+\tilde{\mathbf{W}}\mathbf{S}_1)\mathbf{L}_1^{-1}\end{bmatrix}\end{aligned} \tag{8.8-45}$$

$$\begin{aligned}\begin{bmatrix}\tilde{\mathbf{S}} & \tilde{\mathbf{H}}\\ -\tilde{\mathbf{L}} & \tilde{\mathbf{S}}\end{bmatrix}\begin{bmatrix}\mathbf{S}_2 & \mathbf{H}_2\\ -\mathbf{L}_2 & \mathbf{S}_2^T\end{bmatrix}&=\begin{bmatrix}\mathbf{S}_1 & \mathbf{H}_1\\ -\mathbf{L}_1 & \mathbf{S}_1^T\end{bmatrix}\begin{bmatrix}\tilde{\mathbf{S}} & \tilde{\mathbf{H}}\\ -\tilde{\mathbf{L}} & \tilde{\mathbf{S}}\end{bmatrix}\\ &=-\begin{bmatrix}(\tilde{\mathbf{H}}-\breve{\mathbf{W}}\mathbf{S}_2^T)\mathbf{H}_2^{-1} & \breve{\mathbf{W}}\\ -\tilde{\mathbf{W}} & (\tilde{\mathbf{L}}-\tilde{\mathbf{W}}\mathbf{S}_2)\mathbf{L}_2^{-1}\end{bmatrix}\end{aligned} \tag{8.8-46}$$

from which one obtains

$$\begin{bmatrix}\tilde{\mathbf{S}} & \tilde{\mathbf{H}}\\ -\tilde{\mathbf{L}} & \tilde{\mathbf{S}}\end{bmatrix}\begin{bmatrix}\mathbf{S}_i & \mathbf{H}_i\\ -\mathbf{L}_i & \mathbf{S}_i^T\end{bmatrix}+\begin{bmatrix}\mathbf{S}_i & \mathbf{H}_i\\ -\mathbf{L}_i & \mathbf{S}_i^T\end{bmatrix}\begin{bmatrix}\tilde{\mathbf{S}} & \tilde{\mathbf{H}}\\ -\tilde{\mathbf{L}} & \tilde{\mathbf{S}}\end{bmatrix}=-\begin{bmatrix}\mathbf{I} & \mathbf{0}\\ \mathbf{0} & \mathbf{I}\end{bmatrix} \tag{8.8-47}$$

(i=1,2), use being made of $(8.8\text{-}25)_2$ and $(8.8\text{-}28)_2$.

8.9 Green's Function for Bimaterials

The Green's function for anisotropic bimaterials subjected to a line force and/or a line dislocation not located at the interface has been investigated by Barnett and Lothe (1974a), Belov et al. (1983), Kirchner and Lothe (1987), Tewary et al. (1989a), and Ting (1992b). Similar to the derivation of Green's functions for half-spaces we will modify the Green's function for the infinite space to obtain the Green's function for bimaterials. With this approach the solution is explicit for the whole field. No integration with distributed dislocations and line forces along the interface is required. We will also show that a remarkably simple expression can be obtained for the displacement, surface traction, and the hoop stress vector along the interface.

Let the half-space $x_2>0$ be occupied by material 1 and the other half-space $x_2<0$ be occupied by material 2. They are rigidly bonded together along the interface $x_2=0$ so that

$$\mathbf{u}_1=\mathbf{u}_2,\quad \boldsymbol{\phi}_1=\boldsymbol{\phi}_2,\quad \text{at } x_2=0. \tag{8.9-1}$$

The subscripts 1 and 2 denote materials 1 and 2, respectively. Let a line force **f** and a line dislocation with Burgers vector **b** be applied at

$$(x_1, x_2) = (0, d), \quad d > 0. \tag{8.9-2}$$

The solution can be written as

$$\begin{aligned}\mathbf{u}_1 &= \frac{1}{\pi}\mathrm{Im}\{\mathbf{A}_1\langle \ln(z_*^{(1)} - p_*^{(1)} d)\rangle \mathbf{q}^\infty\} + \frac{1}{\pi}\mathrm{Im}\sum_{\beta=1}^{3}\{\mathbf{A}_1\langle \ln(z_*^{(1)} - \bar{p}_\beta^{(1)} d)\rangle \mathbf{q}_\beta^{(1)}\},\\ \boldsymbol{\phi}_1 &= \frac{1}{\pi}\mathrm{Im}\{\mathbf{B}_1\langle \ln(z_*^{(1)} - p_*^{(1)} d)\rangle \mathbf{q}^\infty\} + \frac{1}{\pi}\mathrm{Im}\sum_{\beta=1}^{3}\{\mathbf{B}_1\langle \ln(z_*^{(1)} - \bar{p}_\beta^{(1)} d)\rangle \mathbf{q}_\beta^{(1)}\},\end{aligned} \tag{8.9-3}$$

for material 1 in $x_2 > 0$ and

$$\begin{aligned}\mathbf{u}_2 &= \frac{1}{\pi}\mathrm{Im}\sum_{\beta=1}^{3}\{\mathbf{A}_2\langle \ln(z_*^{(2)} - p_\beta^{(1)} d)\rangle \mathbf{q}_\beta^{(2)}\},\\ \boldsymbol{\phi}_2 &= \frac{1}{\pi}\mathrm{Im}\sum_{\beta=1}^{3}\{\mathbf{B}_2\langle \ln(z_*^{(2)} - p_\beta^{(1)} d)\rangle \mathbf{q}_\beta^{(2)}\},\end{aligned} \tag{8.9-4}$$

for material 2 in $x_2 < 0$. When subscripts are already present, superscripts (1), (2) are employed to label materials 1 and 2, respectively. Equations (8.9-3) are identical to (8.6-2) for Green's functions for half-spaces. The first terms on the right of (8.9-3) represent the Green's function for the infinite space with the line force **f** and the line dislocation **b** applied at the location (8.9-2). Hence the $\mathbf{q}^\infty$ in the first terms is given by (8.1-18), i.e.,

$$\mathbf{q}^\infty = \mathbf{A}_1^T \mathbf{f} + \mathbf{B}_1^T \mathbf{b}. \tag{8.9-5}$$

The constant vectors $\mathbf{q}_\beta^{(1)}$ and $\mathbf{q}_\beta^{(2)}$ (β=1,2,3) are determined by the interface conditions (8.9-1).

Following the derivation of (8.6-10), substitution of (8.9-3) and (8.9-4) into (8.9-1) yields

$$\begin{aligned}\mathbf{A}_1 \mathbf{q}_\beta^{(1)} + \overline{\mathbf{A}}_2 \overline{\mathbf{q}}_\beta^{(2)} &= \overline{\mathbf{A}}_1 \mathbf{I}_\beta \overline{\mathbf{q}}^\infty,\\ \mathbf{B}_1 \mathbf{q}_\beta^{(1)} + \overline{\mathbf{B}}_2 \overline{\mathbf{q}}_\beta^{(2)} &= \overline{\mathbf{B}}_1 \mathbf{I}_\beta \overline{\mathbf{q}}^\infty.\end{aligned} \tag{8.9-6}$$

This is similar to (8.8-18). We can therefore make use of (8.8-20) and (8.8-21) to write the solutions. They are

$$\begin{aligned}\mathbf{B}_1\mathbf{q}_\beta^{(1)} &= (\mathbf{M}_1^{-1}+\overline{\mathbf{M}}_2^{-1})^{-1}(\overline{\mathbf{M}}_2^{-1}-\overline{\mathbf{M}}_1^{-1})\overline{\mathbf{B}}_1\mathbf{I}_\beta\overline{\mathbf{q}}^\infty,\\ \overline{\mathbf{B}}_2\overline{\mathbf{q}}_\beta^{(2)} &= (\mathbf{M}_1^{-1}+\overline{\mathbf{M}}_2^{-1})^{-1}(\mathbf{M}_1^{-1}+\overline{\mathbf{M}}_1^{-1})\overline{\mathbf{B}}_1\mathbf{I}_\beta\overline{\mathbf{q}}^\infty,\end{aligned} \tag{8.9-7}$$

$$\begin{aligned}\mathbf{A}_1\mathbf{q}_\beta^{(1)} &= (\mathbf{M}_1+\overline{\mathbf{M}}_2)^{-1}(\overline{\mathbf{M}}_2-\overline{\mathbf{M}}_1)\overline{\mathbf{A}}_1\mathbf{I}_\beta\overline{\mathbf{q}}^\infty,\\ \overline{\mathbf{A}}_2\overline{\mathbf{q}}_\beta^{(2)} &= (\mathbf{M}_1+\overline{\mathbf{M}}_2)^{-1}(\mathbf{M}_1+\overline{\mathbf{M}}_1)\overline{\mathbf{A}}_1\mathbf{I}_\beta\overline{\mathbf{q}}^\infty,\end{aligned} \tag{8.9-8}$$

appeal being made to (8.1-4).

It is interesting to note that when $\mathbf{M}_1=\mathbf{M}_2$,

$$\mathbf{q}_\beta^{(1)}=0,\quad \mathbf{B}_2\mathbf{q}_\beta^{(2)}=\mathbf{B}_1\mathbf{I}_\beta\mathbf{q}^\infty,\quad \mathbf{A}_2\mathbf{q}_\beta^{(2)}=\mathbf{A}_1\mathbf{I}_\beta\mathbf{q}^\infty.$$

The second terms on the right of (8.9-3) can be deleted while (8.9-4) is replaced by

$$\mathbf{u}_2=\frac{1}{\pi}\mathrm{Im}\sum_{\beta=1}^{3}\left\{\mathbf{A}_2\left\langle \ln(z_*^{(2)}-p_\beta^{(1)}d)\right\rangle\mathbf{A}_2^{-1}\mathbf{A}_1\mathbf{I}_\beta\mathbf{q}^\infty\right\},$$

$$\phi_2=\frac{1}{\pi}\mathrm{Im}\sum_{\beta=1}^{3}\left\{\mathbf{B}_2\left\langle \ln(z_*^{(2)}-p_\beta^{(1)}d)\right\rangle\mathbf{B}_2^{-1}\mathbf{B}_1\mathbf{I}_\beta\mathbf{q}^\infty\right\}.$$

Thus when the impedance tensors of the two materials are identical the Green's function for material 1 is identical to the Green's function for the infinite space with the singularities located at $(x_1,x_2)=(0,d)$.

The image force F that tends to move the dislocation line away from the interface is given by (8.6-15) in which ϕ^d is the second term in (8.9-3) (Barnett and Lothe, 1974b; Rice, 1985). We have

$$\begin{aligned}F &= \frac{1}{\pi d}\mathrm{Im}\sum_{\beta=1}^{3}\mathbf{b}^T\left\{\mathbf{B}_1\left\langle\frac{p_*^{(1)}}{p_*^{(1)}-\bar{p}_\beta^{(1)}}\right\rangle\mathbf{q}_\beta^{(1)}\right\}\\ &= \frac{1}{\pi d}\mathrm{Im}\sum_{\beta=1}^{3}\sum_{\alpha=1}^{3}\mathbf{b}^T\left\{\frac{p_\alpha^{(1)}}{p_\alpha^{(1)}-\bar{p}_\beta^{(1)}}\mathbf{B}_1\mathbf{I}_\alpha\mathbf{q}_\beta^{(1)}\right\}.\end{aligned}$$

From $(8.9\text{-}7)_1$, (8.1-4), and (8.8-24),

$$\begin{aligned}\mathbf{q}_\beta^{(1)} &= \mathbf{B}_1^{-1}(\mathbf{M}_1^{-1}+\overline{\mathbf{M}}_1^{-1})^{-1}(\mathbf{M}_1^{-1}+\overline{\mathbf{M}}_2^{-1}-2\mathbf{L}_1^{-1})\overline{\mathbf{B}}_1\mathbf{I}_\beta\overline{\mathbf{q}}^\infty\\ &= \mathbf{B}_1^{-1}\left\{\mathbf{I}-2(\tilde{\mathbf{L}}+i\tilde{\mathbf{W}})\mathbf{L}_1^{-1}\right\}\overline{\mathbf{B}}_1\mathbf{I}_\beta\overline{\mathbf{q}}^\infty.\end{aligned}$$

Inserting $\overline{\mathbf{q}}^\infty=\overline{\mathbf{B}}^T\mathbf{b}$ obtained from (8.9-5) with $\mathbf{f}=\mathbf{0}$ one obtains

$$F=\frac{-1}{2\pi d}\mathrm{Re}\sum_{\beta=1}^{3}\sum_{\alpha=1}^{3}\mathbf{b}^T\left\{\frac{p_\alpha^{(1)}}{p_\alpha^{(1)}-\bar{p}_\beta^{(1)}}(\mathbf{B}_1\mathbf{I}_\alpha\mathbf{B}_1^{-1})\mathbf{L}_1(\overline{\mathbf{B}}_1\mathbf{I}_\beta\overline{\mathbf{B}}_1^{-1})^T\right\}\mathbf{b}$$

$$+\frac{1}{\pi d}\mathrm{Re}\sum_{\beta=1}^{3}\sum_{\alpha=1}^{3}\mathbf{b}^T\left\{\frac{p_\alpha^{(1)}}{p_\alpha^{(1)}-\bar{p}_\beta^{(1)}}(\mathbf{B}_1\mathbf{I}_\alpha\mathbf{B}_1^{-1})(\tilde{\mathbf{L}}+i\tilde{\mathbf{W}})(\overline{\mathbf{B}}_1\mathbf{I}_\beta\overline{\mathbf{B}}_1^{-1})^T\right\}\mathbf{b}.$$

The first term on the right is identical to that shown in (8.6-17) and can be simplified to the one given in (8.6-19). The second term can be simplified by following the derivation of (8.6-19) to

$$\frac{1}{2\pi d}\mathrm{Re}\left\{\mathbf{b}^T(\tilde{\mathbf{L}}+i\tilde{\mathbf{W}})\mathbf{b}\right\}.$$

We therefore have

$$F=\frac{1}{4\pi d}\mathbf{b}^T(2\tilde{\mathbf{L}}-\mathbf{L}_1)\mathbf{b}. \tag{8.9-9}$$

Since $\mathbf{b}$, $\mathbf{L}_1$, and $\tilde{\mathbf{L}}$ are tensors when the transformation of the coordinate system is a rotation about the x_3-axis, F is invariant with rotations about the x_3-axis. The image force remains the same if the interface $x_2=0$ is replaced by any plane that is parallel to the x_3-axis and the distance between the line dislocation and the new interface remains the same, Fig. 8.8.

For isotropic materials with $\mathbf{b}^T=(b_1,b_2,b_3)$ and $\mathbf{S}$, $\mathbf{H}$, $\mathbf{L}$ given by (6.4-13), (8.9-9) reduces to

$$F=\frac{\mu_1}{4\pi d(1-\nu_1)}\left(\frac{\alpha+\beta^2}{1-\beta^2}\right)\left(b_1^2+b_2^2\right)-\frac{\mu_1}{4\pi d}\left(\frac{\mu_1-\mu_2}{\mu_1+\mu_2}\right)b_3^2$$

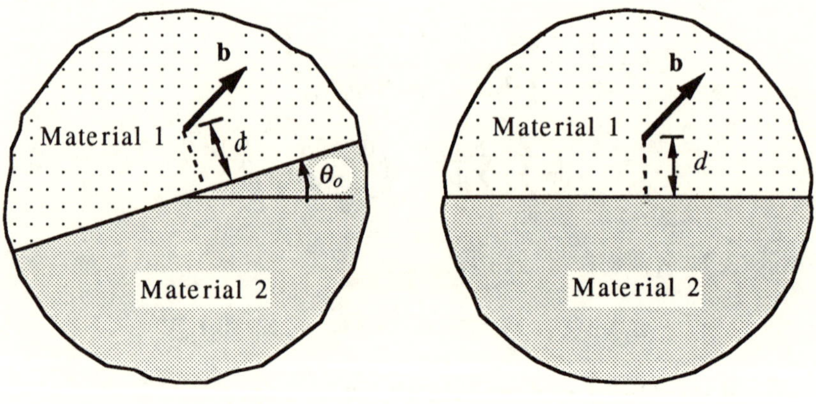

Fig. 8.8 The image force that attracts the dislocation to or repels it from the interface is independent of θ_o, the orientation of the interface.

where α,β are the Dundurs (1970) constants

$$\alpha=\frac{\mu_2(1-\nu_1)-\mu_1(1-\nu_2)}{\mu_2(1-\nu_1)+\mu_1(1-\nu_2)},\quad \beta=\frac{\mu_2(1-2\nu_1)-\mu_1(1-2\nu_2)}{2[\mu_2(1-\nu_1)+\mu_1(1-\nu_2)]},$$

and μ, ν are the shear modulus and Poisson ratio, respectively. The second term gives the image force due to a screw dislocation b_3. It agrees with the result obtained in (3.6-10). The first term gives the image force due to an edge dislocation with Burgers vector (b_1,b_2). It depends on $b_1^2+b_2^2$ only, and hence is invariant with the direction of the vector (b_1,b_2). The image force on an edge dislocation is positive (repulsive from the interface) if $\alpha+\beta^2>0$, and is negative (attractive to the interface) if $\alpha+\beta^2<0$ (Dundurs, 1969; see also Dundurs and Sendeckyj, 1965a). If the dislocation is located in material 2, we change the sign of α and β. This implies that when $-\beta^2<\alpha<\beta^2$ the image force is mutual repulsive regardless of whether the dislocation is located in material 1 or 2. For a combined edge and screw dislocation Barnett and Lothe (1995) have shown that mutual attraction is possible.

Consider the special bimaterial for which

$$C^{(2)}_{ijks}=\lambda C^{(1)}_{ijks},\quad \lambda\geq 0, \tag{8.9-10}$$

where λ is an arbitrary real positive constant. $\lambda=0$ when material 2 is a vacuum, $\lambda=1$ if the two materials are identical, and $\lambda=\infty$ when material 2 is a rigid body. From (5.1-8) the unnormalized eigenvectors $\boldsymbol{a}$ for materials 1 and 2 can be taken identical and, by (5.3-11)$_1$, $\mathbf{A}_2=\mathbf{A}_1$. When normalized, however, $\mathbf{A}_2$ should be taken as proportional to $\mathbf{A}_1$, i.e.,

$$\mathbf{A}_2=k\mathbf{A}_1$$

where k is the proportionality factor. From (5.3-2) and (5.3-11)$_2$,

$$\mathbf{B}_2=\lambda k\mathbf{B}_1$$

and the orthogonality relation (5.5-14)$_1$ yields

$$k=\lambda^{-1/2}.$$

We therefore have the result that, when (8.9-10) holds,

$$\mathbf{A}_2=\lambda^{-1/2}\mathbf{A}_1,\quad \mathbf{B}_2=\lambda^{1/2}\mathbf{B}_1. \tag{8.9-11}$$

From (5.5-17) and (8.1-4),

$$\begin{aligned}&\mathbf{S}_2=\mathbf{S}_1,\quad \mathbf{H}_2=\tfrac{1}{\lambda}\mathbf{H}_1,\quad \mathbf{L}_2=\lambda\mathbf{L}_1,\\ &\mathbf{M}_2=\lambda\mathbf{M}_1,\quad \mathbf{M}_2^{-1}=\tfrac{1}{\lambda}\mathbf{M}_1^{-1}.\end{aligned} \tag{8.9-12}$$

The tensors $\mathbf{D}$, $\mathbf{W}$, $\hat{\mathbf{D}}$, $\hat{\mathbf{W}}$, $\tilde{\mathbf{L}}$, and $\tilde{\mathbf{H}}$ defined in (8.8-23), (8.8-22), (8.8-26), and (8.8-29) are

$$\begin{aligned} \mathbf{D} &= \tfrac{\lambda+1}{\lambda}\mathbf{L}_1^{-1}, \quad & \mathbf{W} &= \tfrac{\lambda-1}{\lambda}\mathbf{S}_1\mathbf{L}_1^{-1}, \\ \hat{\mathbf{D}} &= (\lambda+1)\mathbf{H}_1^{-1}, \quad & \hat{\mathbf{W}} &= (\lambda-1)\mathbf{H}_1^{-1}\mathbf{S}_1, \end{aligned} \tag{8.9-13}$$

$$\tilde{\mathbf{L}} = \tfrac{\lambda}{\lambda+1}\mathbf{L}_1\left\{\mathbf{I} + \left(\tfrac{\lambda-1}{\lambda+1}\right)^2\mathbf{S}_1^2\right\}^{-1}, \quad \tilde{\mathbf{H}} = \tfrac{1}{\lambda+1}\left\{\mathbf{I} + \left(\tfrac{\lambda-1}{\lambda+1}\right)^2\mathbf{S}_1^2\right\}^{-1}\mathbf{H}_1 \tag{8.9-14}$$

and, by (8.8-42) and (8.8-44),

$$\begin{aligned} \breve{\mathbf{S}} &= \mathbf{D}^{-1}\mathbf{W} = \tilde{\mathbf{W}}\tilde{\mathbf{L}}^{-1} = \tfrac{1-\lambda}{1+\lambda}\mathbf{S}_1^T, \\ \hat{\mathbf{S}} &= \hat{\mathbf{W}}\hat{\mathbf{D}}^{-1} = \tilde{\mathbf{H}}^{-1}\breve{\mathbf{W}} = \tfrac{1-\lambda}{1+\lambda}\mathbf{S}_1^T. \end{aligned} \tag{8.9-15}$$

It is interesting to note that $\mathbf{S}_2$ is independent of λ and that $\tilde{\mathbf{L}}$, $\tilde{\mathbf{H}}$, and $\breve{\mathbf{S}}=\hat{\mathbf{S}}$ are bounded for all positive λ. Equations (8.9-11)–(8.9-15) are useful for specializing bimaterial problems to problems involving only one material whose boundary can be either traction-free or rigidly clamped.

As an illustration we will deduce the image forces in (8.6-19) and (8.6-24) for half-spaces from the image force in (8.9-9) for bimaterials. When the boundary $x_2=0$ is traction-free material 2 is a vacuum. Assuming that the elastic stiffnesses in the two materials are related by (8.9-10), setting $\lambda=0$ in $(8.9\text{-}14)_1$ yields $\tilde{\mathbf{L}}=\mathbf{0}$ so that (8.9-9) reduces to (8.6-19). If the boundary $x_2=0$ is a fixed boundary, material 2 can be considered as infinitely rigid, whereupon setting $\lambda=\infty$ in $(8.9\text{-}14)_1$ leads to

$$\tilde{\mathbf{L}} = \mathbf{L}_1(\mathbf{I}+\mathbf{S}_1^2)^{-1} = \mathbf{H}_1^{-1}$$

by $(8.1\text{-}3\text{b})_1$. Equation (8.9-9) reduces to (8.6-24).

The solution (8.9-3) for material 1 is identical in expression to the solution (8.6-2) for a half-space. Analogous to the image singularities for half-spaces discussed in Section 8.7, the second terms in (8.9-3) represent nine one-component Green's functions for the infinite space occupied by material 1 whose singularities are located in $x_2<0$. The locations $(x_1^{\alpha\beta}, x_2^{\alpha\beta})$ of the nine image singularities are given by (8.7-2) where p'_α and p''_α are the real and imaginary parts of $p_\alpha^{(1)}$, respectively. Thus the locations of the image singularities for material 1 are independent of the property of material 2. The solution (8.9-4) for material 2 represents nine one-component Green's functions for the infinite space occupied by material 2 whose singularities are located in $x_2>0$. The locations $(x_1^{\alpha\beta}, x_2^{\alpha\beta})$ of the nine image singularities for material 2 are

$$\frac{x_1^{\alpha\beta}}{d} = \frac{p_\alpha''^{(2)} p_\beta'^{(1)} - p_\beta''^{(1)} p_\alpha'^{(2)}}{p_\alpha''^{(2)}}, \qquad \frac{x_2^{\alpha\beta}}{d} = \frac{p_\beta''^{(1)}}{p_\alpha''^{(2)}}, \tag{8.9-16}$$

which depend on both materials. They can be determined graphically (Ting, 1992b).

We will next determine the displacement, surface traction, and the hoop stress vector along the interface $x_2=0$. At $x_2=0$ (8.9-4) reduce to

$$\begin{aligned} \mathbf{u}_2(x_1,0) &= \frac{1}{\pi}\mathrm{Im}\sum_{\beta=1}^{3}\left\{\ln(x_1 - p_\beta^{(1)}d)\mathbf{A}_2\mathbf{q}_\beta^{(2)}\right\}, \\ \boldsymbol{\phi}_2(x_1,0) &= \frac{1}{\pi}\mathrm{Im}\sum_{\beta=1}^{3}\left\{\ln(x_1 - p_\beta^{(1)}d)\mathbf{B}_2\mathbf{q}_\beta^{(2)}\right\}. \end{aligned} \tag{8.9-17}$$

Making use of (8.8-24), (8.8-27), and (8.1-4) to write $(8.9\text{-}8)_2$ and $(8.9\text{-}7)_2$ as

$$\begin{aligned} \overline{\mathbf{A}}_2\overline{\mathbf{q}}_\beta^{(2)} &= 2(\tilde{\mathbf{H}} + i\breve{\mathbf{W}})\mathbf{H}^{-1}\overline{\mathbf{A}}_1\mathbf{I}_\beta\overline{\mathbf{q}}^\infty, \\ \overline{\mathbf{B}}_2\overline{\mathbf{q}}_\beta^{(2)} &= 2(\tilde{\mathbf{L}} + i\tilde{\mathbf{W}})\mathbf{L}^{-1}\overline{\mathbf{B}}_1\mathbf{I}_\beta\overline{\mathbf{q}}^\infty, \end{aligned} \tag{8.9-18}$$

we have

$$\begin{aligned} \mathbf{u}_2(x_1,0) &= \frac{2}{\pi}\mathrm{Im}\left\{(\tilde{\mathbf{H}} - i\breve{\mathbf{W}})\mathbf{H}_1^{-1}\mathbf{A}_1\left\langle \ln(x_1 - p_*^{(1)}d)\right\rangle\mathbf{q}^\infty\right\}, \\ \boldsymbol{\phi}_2(x_1,0) &= \frac{2}{\pi}\mathrm{Im}\left\{(\tilde{\mathbf{L}} - i\tilde{\mathbf{W}})\mathbf{L}_1^{-1}\mathbf{B}_1\left\langle \ln(x_1 - p_*^{(1)}d)\right\rangle\mathbf{q}^\infty\right\}. \end{aligned}$$

From (8.6-26), (8.6-29), and (8.6-30) this is equivalent to

$$\begin{aligned} \begin{bmatrix} \mathbf{u}_2(x_1,0) \\ \boldsymbol{\phi}_2(x_1,0) \end{bmatrix} &= 2\begin{bmatrix} (\tilde{\mathbf{H}} + \breve{\mathbf{W}}\mathbf{S}_1^T)\mathbf{H}_1^{-1} & -\breve{\mathbf{W}} \\ \tilde{\mathbf{W}} & (\tilde{\mathbf{L}} + \tilde{\mathbf{W}}\mathbf{S}_1)\mathbf{L}_1^{-1} \end{bmatrix}\begin{bmatrix} \mathbf{u}_1^\infty(x_1,0) \\ \boldsymbol{\phi}_1^\infty(x_1,0) \end{bmatrix} \\ &= -2\begin{bmatrix} \tilde{\mathbf{S}} & \tilde{\mathbf{H}} \\ -\tilde{\mathbf{L}} & \tilde{\mathbf{S}} \end{bmatrix}\begin{bmatrix} \mathbf{S}_1 & \mathbf{H}_1 \\ -\mathbf{L}_1 & \mathbf{S}_1^T \end{bmatrix}\begin{bmatrix} \mathbf{u}_1^\infty(x_1,0) \\ \boldsymbol{\phi}_1^\infty(x_1,0) \end{bmatrix} \end{aligned}$$

by (8.8-45). The $\mathbf{u}_1^\infty(x_1,0)$ and $\boldsymbol{\phi}_1^\infty(x_1,0)$ shown in (8.6-27) with $\mathbf{h}$ and $\mathbf{g}$ given in (8.1-20a) can be written as

$$2\begin{bmatrix} \mathbf{u}_1^\infty(x_1,0) \\ \boldsymbol{\phi}_1^\infty(x_1,0) \end{bmatrix} = -\frac{1}{\pi}\tilde{\mathbf{N}}^{(1)}\left((\ln\hat{r})\mathbf{I} + \pi\tilde{\mathbf{N}}^{(1)}(\hat{\theta})\right)\begin{bmatrix} \mathbf{b} \\ \mathbf{f} \end{bmatrix} \tag{8.9-19}$$

where use has been made of the commutative relation

$$\tilde{\mathbf{N}}(\theta)\tilde{\mathbf{N}} = \tilde{\mathbf{N}}\tilde{\mathbf{N}}(\theta) \tag{8.9-20}$$

proved in (7.8-11). In (8.9-19) $\hat{r}$ and $\hat{\theta}$ are defined in (8.6-28) and

$$\tilde{\mathbf{N}}^{(1)}(\hat{\theta}) = \begin{bmatrix} \mathbf{S}_1(\hat{\theta}) & \mathbf{H}_1(\hat{\theta}) \\ -\mathbf{L}_1(\hat{\theta}) & \mathbf{S}_1^T(\hat{\theta}) \end{bmatrix}. \tag{8.9-21}$$

We have thus the following remarkably simple and explicit expression for the displacement and the stress function at the interface (Kirchner and Lothe, 1987)

$$\begin{bmatrix} \mathbf{u}(x_1,0) \\ \phi(x_1,0) \end{bmatrix} = -\frac{1}{\pi}\begin{bmatrix} \tilde{\mathbf{S}} & \tilde{\mathbf{H}} \\ -\tilde{\mathbf{L}} & \tilde{\mathbf{S}}^T \end{bmatrix}\Big((\ln\hat{r})\mathbf{I} + \pi\tilde{\mathbf{N}}^{(1)}(\hat{\theta})\Big)\begin{bmatrix} \mathbf{b} \\ \mathbf{f} \end{bmatrix}. \tag{8.9-22}$$

The subscript 2 for $\mathbf{u}$ and ϕ in (8.9-22) have been deleted in view of the continuity conditions $\mathbf{u}_1 = \mathbf{u}_2$, $\phi_1 = \phi_2$ at the interface $x_2=0$.

The Green's function for the infinite space as given by (8.1-19) and (8.1-20a) are

$$\begin{bmatrix} \mathbf{u}(x_1,x_2) \\ \phi(x_1,x_2) \end{bmatrix} = -\frac{1}{2\pi}\tilde{\mathbf{N}}\Big((\ln r)\mathbf{I} + \pi\tilde{\mathbf{N}}(\theta)\Big)\begin{bmatrix} \mathbf{b} \\ \mathbf{f} \end{bmatrix}. \tag{8.9-23}$$

use being made of (8.9-20). This is for a line force $\mathbf{f}$ and a line dislocation $\mathbf{b}$ applied at $(x_1, x_2) = (0, 0)$. If they are applied at $(x_1, x_2) = (0, d)$, $\mathbf{u}$ and ϕ at $x_2=0$ are

$$\begin{bmatrix} \mathbf{u}(x_1,0) \\ \phi(x_1,0) \end{bmatrix} = -\frac{1}{2\pi}\tilde{\mathbf{N}}\Big((\ln\hat{r})\mathbf{I} + \pi\tilde{\mathbf{N}}(\hat{\theta})\Big)\begin{bmatrix} \mathbf{b} \\ \mathbf{f} \end{bmatrix}. \tag{8.9-24}$$

When the impedances of the two materials are identical, by (8.8-8) $\tilde{\mathbf{S}}$, $\tilde{\mathbf{H}}$, $\tilde{\mathbf{L}}$ reduce to one-halves of $\mathbf{S}_1$, $\mathbf{H}_1$, $\mathbf{L}_1$, respectively, so that (8.9-22) recovers (8.9-24).

With the help of (8.6-34) differentiation of (8.9-22) with x_1 is

$$\begin{bmatrix} \mathbf{u}_{,1}(x_1,0) \\ \phi_{,1}(x_1,0) \end{bmatrix} = -\frac{1}{\pi\hat{r}}\begin{bmatrix} \tilde{\mathbf{S}} & \tilde{\mathbf{H}} \\ -\tilde{\mathbf{L}} & \tilde{\mathbf{S}}^T \end{bmatrix}\Big(\cos\hat{\theta}\,\mathbf{I} - \sin\hat{\theta}\,\mathbf{N}^{(1)}(\hat{\theta})\Big)\begin{bmatrix} \mathbf{b} \\ \mathbf{f} \end{bmatrix} \tag{8.9-25}$$

where

$$\mathbf{N}^{(1)}(\hat{\theta}) = \begin{bmatrix} \mathbf{N}_1^{(1)}(\hat{\theta}) & \mathbf{N}_2^{(1)}(\hat{\theta}) \\ \mathbf{N}_3^{(1)}(\hat{\theta}) & \mathbf{N}_1^{(1)T}(\hat{\theta}) \end{bmatrix}.$$

Equation (8.9-25) provides the surface tractions

$$\mathbf{t}_2^{(1)} = -\phi_{,1}(x_1, 0), \quad \mathbf{t}_2^{(2)} = \phi_{,1}(x_1, 0) \tag{8.9-26}$$

at the interface $x_2=0$. The hoop stress vector $\mathbf{t}_1^{(1)}$ at $x_2=0$ on the side of material 1 and $\mathbf{t}_1^{(2)}$ at $x_2=0$ on the side of material 2 are determined by employing $(5.7\text{-}7)_2$ and (8.9-25). We have

$$\mathbf{t}_1^{(i)} = -\mathbf{N}_3^{(i)}\mathbf{u}_{,1}(x_1,0) - \mathbf{N}_1^{(i)T}\boldsymbol{\phi}_{,1}(x_1,0) \qquad (i=1,2). \tag{8.9-27}$$

The total surface traction $\mathbf{f}_2$ transmitted to material 2 is

$$\mathbf{f}_2 = \boldsymbol{\phi}(\infty,\ 0) - \boldsymbol{\phi}(-\infty,\ 0)$$

or, by (8.9-22),

$$\mathbf{f}_2 = (\tilde{\mathbf{L}}\mathbf{H}_1 - \tilde{\mathbf{S}}^T\mathbf{S}_1^T)\mathbf{f} + (\tilde{\mathbf{L}}\mathbf{S}_1 + \tilde{\mathbf{S}}^T\mathbf{L}_1)\mathbf{b}. \tag{8.9-28}$$

The right side is independent of the location $(x_1,x_2) = (0,d)$ of the singularities and hence $\mathbf{f}_2$ in (8.9-28) should be identical to the $\mathbf{f}_2$ in $(8.8\text{-}10)_2$ when the singularities are located at the interface. The two expressions are indeed equivalent in view of the identities (8.8-35) and $(8.8\text{-}34\text{b})_4$. One can also show that the total traction force $\mathbf{f}_1$ on a plane $x_1>d$ in material 1 is given by $(8.8\text{-}10)_1$.

8.10 Composite Space and Composite Wedge

Consider a composite space that consists of n wedges of different anisotropic elastic materials, Fig. 8.9(a). In the cylindrical coordinate system (r,θ,x_3) each wedge occupies the region

$$\begin{gathered}\theta_{k-1} \le \theta \le \theta_k, \quad k=1,2,\ldots,n,\\ 0 \le \theta_0 < \theta_1 < \cdots < \theta_n = 2\pi + \theta_0.\end{gathered} \tag{8.10-1}$$

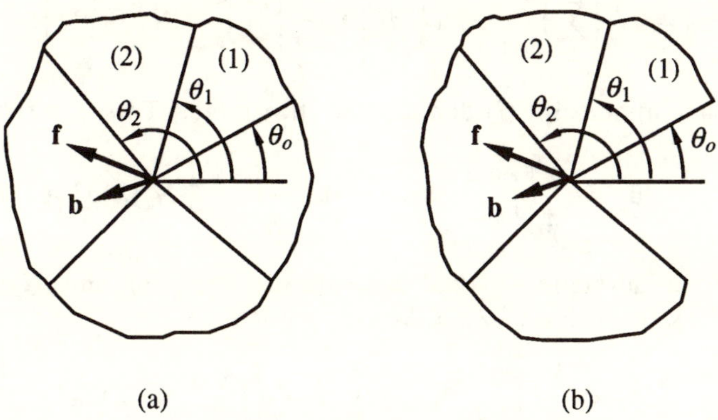

Fig. 8.9 Geometry of (a) a composite space and (b) a composite wedge.

If $\theta_n<2\pi+\theta_o$ we have a composite wedge, Fig. 8.9(b). Along each interface $\theta=\theta_k$ $(k=1,2,\ldots,n-1)$ the displacement and the traction are continuous. Let a line force **f** and a line dislocation with Burgers vector **b** be applied at $r=0$. The real form solution (8.1-21) can be applied to each wedge. Following (8.8-2) we let

$$
\begin{aligned}
2\mathbf{u}_k &= -\frac{1}{\pi}(\ln r)\mathbf{h} - [\mathbf{S}_k(\theta) - \mathbf{S}_k(\theta_{k-1})]\mathbf{h} - [\mathbf{H}_k(\theta) - \mathbf{H}_k(\theta_{k-1})]\mathbf{g} + 2\hat{\mathbf{u}}_k, \\
2\boldsymbol{\phi}_k &= -\frac{1}{\pi}(\ln r)\mathbf{g} + [\mathbf{L}_k(\theta) - \mathbf{L}_k(\theta_{k-1})]\mathbf{h} - [\mathbf{S}_k^T(\theta) - \mathbf{S}_k^T(\theta_{k-1})]\mathbf{g} + 2\hat{\boldsymbol{\phi}}_k.
\end{aligned} \tag{8.10-2}
$$

The subscript k referred to the kth wedge and $\hat{\mathbf{u}}_k$, $\hat{\boldsymbol{\phi}}_k$ are constant vectors which do not alter the stresses. For the continuity of displacement across an interface between two wedges it is necessary that the $(\ln r)$ term be continuous. Hence the constant vector **h** must be the same for all wedges. By a similar argument, continuity of the surface traction across an interface dictates that **g** must be the same for all wedges. In fact the surface traction $\mathbf{t}_\theta$ on any radial plane θ = constant is

$$
\mathbf{t}_\theta = \frac{-1}{2\pi r}\mathbf{g} \tag{8.10-3}
$$

which is independent of θ.

To satisfy the continuity of displacement **u** and the stress function $\boldsymbol{\phi}$ at interfaces except at $\theta=\theta_o$ we let $\hat{\mathbf{u}}_1=\mathbf{0}$ and, for $k>1$,

$$
\begin{aligned}
2\hat{\mathbf{u}}_k &= -\sum_{i=1}^{k-1}\{\mathbf{S}_i(\theta_i) - \mathbf{S}_i(\theta_{i-1})\}\mathbf{h} - \sum_{i=1}^{k-1}\{\mathbf{H}_i(\theta_i) - \mathbf{H}_i(\theta_{i-1})\}\mathbf{g} \\
&= -\frac{1}{\pi}\left\{\sum_{i=1}^{k-1}\int_{\theta_{i-1}}^{\theta_i} \mathbf{N}_1^{(i)}(\lambda)d\lambda\right\}\mathbf{h} - \frac{1}{\pi}\left\{\sum_{i=1}^{k-1}\int_{\theta_{i-1}}^{\theta_i} \mathbf{N}_2^{(i)}(\lambda)d\lambda\right\}\mathbf{g}
\end{aligned}
$$

where the superscript (i) denotes the ith wedge. This is written as

$$
2\hat{\mathbf{u}}_k = -\frac{1}{\pi}\left\{\int_{\theta_o}^{\theta_{k-1}} \mathbf{N}_1(\lambda)d\lambda\right\}\mathbf{h} - \frac{1}{\pi}\left\{\int_{\theta_o}^{\theta_{k-1}} \mathbf{N}_2(\lambda)d\lambda\right\}\mathbf{g} \tag{8.10-4}
$$

with the understanding that the values of $\mathbf{N}_1(\lambda)$ and $\mathbf{N}_2(\lambda)$ are dictated by the value of λ. Likewise we let

$$
2\hat{\boldsymbol{\phi}}_k = -\frac{1}{\pi}\left\{\int_{\theta_o}^{\theta_{k-1}} \mathbf{N}_3(\lambda)d\lambda\right\}\mathbf{h} - \frac{1}{\pi}\left\{\int_{\theta_o}^{\theta_{k-1}} \mathbf{N}_1^T(\lambda)d\lambda\right\}\mathbf{g} \tag{8.10-5}
$$

Equation (8.10-2) can then be written as

$$2\mathbf{u} = -\frac{1}{\pi}(\ln r)\mathbf{h} - [\mathbf{S}(\theta) - \mathbf{S}(\theta_o)]\mathbf{h} - [\mathbf{H}(\theta) - \mathbf{H}(\theta_o)]\mathbf{g},$$
$$2\boldsymbol{\phi} = -\frac{1}{\pi}(\ln r)\mathbf{g} + [\mathbf{L}(\theta) - \mathbf{L}(\theta_o)]\mathbf{h} - [\mathbf{S}^T(\theta) - \mathbf{S}^T(\theta_o)]\mathbf{g}, \tag{8.10-6}$$

where

$$\mathbf{S}(\theta) = \frac{1}{\pi}\int_0^\theta \mathbf{N}_1(\lambda)d\lambda,$$
$$\mathbf{H}(\theta) = \frac{1}{\pi}\int_0^\theta \mathbf{N}_2(\lambda)d\lambda, \tag{8.10-7}$$
$$\mathbf{L}(\theta) = -\frac{1}{\pi}\int_0^\theta \mathbf{N}_3(\lambda)d\lambda.$$

We have omitted the subscript k from $\mathbf{u}_k$ and $\boldsymbol{\phi}_k$ in (8.10-6) because the right sides of (8.10-6) are independent of k. Therefore (8.10-6) apply to any wedge in the composite space. The continuity of $\mathbf{u}$ and $\boldsymbol{\phi}$ at interfaces are automatically satisfied for $\theta=\theta_1, \theta_2, \ldots \theta_{n-1}$. It should be pointed out that the real matrices $\mathbf{S}(\theta)$, $\mathbf{H}(\theta)$, and $\mathbf{L}(\theta)$ defined in (8.10-7) are a generalization of their counterparts in (7.6-6) for homogeneous materials. $\mathbf{S}(\theta)$, $\mathbf{H}(\theta)$, and $\mathbf{L}(\theta)$ here are periodic in θ with periodicity 2π, not π which is the case for homogeneous materials.

For a line force $\mathbf{f}$ and a line dislocation with Burgers vector $\mathbf{b}$ applied at $r=0$ (8.10-6) must satisfy the conditions

$$\mathbf{u}(r, 2\pi+\theta_o) - \mathbf{u}(r, \theta_o) = \mathbf{b},$$
$$\boldsymbol{\phi}(r, 2\pi+\theta_o) - \boldsymbol{\phi}(r, \theta_o) = \mathbf{f},$$

or

$$-\mathbf{S}(2\pi)\mathbf{h} - \mathbf{H}(2\pi)\mathbf{g} = 2\mathbf{b},$$
$$\mathbf{L}(2\pi)\mathbf{h} - \mathbf{S}^T(2\pi)\mathbf{g} = 2\mathbf{f}. \tag{8.10-8}$$

They reduce to (8.8-4) when the composite space consists of two half-spaces. Following the derivation of (8.8-7b) let $\tilde{\mathbf{S}}$, $\tilde{\mathbf{H}}$, and $\tilde{\mathbf{L}}$ be 3×3 real matrices defined by

$$\begin{bmatrix} \tilde{\mathbf{S}} & \tilde{\mathbf{H}} \\ -\tilde{\mathbf{L}} & \tilde{\mathbf{S}}^T \end{bmatrix} \begin{bmatrix} \mathbf{S}(2\pi) & \mathbf{H}(2\pi) \\ -\mathbf{L}(2\pi) & \mathbf{S}^T(2\pi) \end{bmatrix} = -\mathbf{I} \tag{8.10-9a}$$

or

$$\tilde{\mathbf{H}}\mathbf{L}(2\pi) - \tilde{\mathbf{S}}\mathbf{S}(2\pi) = \mathbf{I} = \tilde{\mathbf{L}}\mathbf{H}(2\pi) - \tilde{\mathbf{S}}^T\mathbf{S}^T(2\pi),$$
$$\tilde{\mathbf{S}}^T\mathbf{L}(2\pi) + \tilde{\mathbf{L}}\mathbf{S}(2\pi) = \mathbf{0} = \tilde{\mathbf{S}}\mathbf{H}(2\pi) + \tilde{\mathbf{H}}\mathbf{S}^T(2\pi). \tag{8.10-9b}$$

From $(8.10\text{-}9b)_{3,4}$, $(8.10\text{-}9b)_1$, and $(8.10\text{-}9b)_2$,

$$\begin{aligned} \tilde{\mathbf{S}} &= -\tilde{\mathbf{H}}\mathbf{S}^T(2\pi)\mathbf{H}^{-1}(2\pi) = -\mathbf{L}^{-1}(2\pi)\mathbf{S}^T(2\pi)\tilde{\mathbf{L}}, \\ \tilde{\mathbf{H}} &= \left\{\mathbf{L}(2\pi) + \mathbf{S}^T(2\pi)\mathbf{H}^{-1}(2\pi)\mathbf{S}(2\pi)\right\}^{-1}, \\ \tilde{\mathbf{L}} &= \left\{\mathbf{H}(2\pi) + \mathbf{S}(2\pi)\mathbf{L}^{-1}(2\pi)\mathbf{S}^T(2\pi)\right\}^{-1}. \end{aligned} \tag{8.10-10}$$

By writing (8.10-8) as

$$\begin{bmatrix} \mathbf{S}(2\pi) & \mathbf{H}(2\pi) \\ -\mathbf{L}(2\pi) & \mathbf{S}^T(2\pi) \end{bmatrix} \begin{bmatrix} \mathbf{h} \\ \mathbf{g} \end{bmatrix} = -2 \begin{bmatrix} \mathbf{b} \\ \mathbf{f} \end{bmatrix}$$

and using (8.10-9a) we have (Ting, 1988b)

$$\begin{bmatrix} \mathbf{h} \\ \mathbf{g} \end{bmatrix} = 2 \begin{bmatrix} \tilde{\mathbf{S}} & \tilde{\mathbf{H}} \\ -\tilde{\mathbf{L}} & \tilde{\mathbf{S}}^T \end{bmatrix} \begin{bmatrix} \mathbf{b} \\ \mathbf{f} \end{bmatrix} \tag{8.10-11a}$$

or

$$\mathbf{h} = 2(\tilde{\mathbf{S}}\mathbf{b} + \tilde{\mathbf{H}}\mathbf{f}), \quad \mathbf{g} = 2(\tilde{\mathbf{S}}^T\mathbf{f} - \tilde{\mathbf{L}}\mathbf{b}). \tag{8.10-11b}$$

Equations (8.10-6) and (8.10-11b) provide the solution for the composite space subjected to a line force **f** and a line dislocation with Burgers vector **b** applied at r=0.

The strain energy $\mathcal{E}$ given in (8.1-26) for the annulus region $0<r_1 \le r \le r_2$ in the homogeneous space remains valid for the composite space. Substitution of (8.10-11b) into (8.1-26) leads to (8.8-9).

If $\theta_n < 2\pi + \theta_o$ we have a composite wedge, Fig. 8.9(b). The solution is given by (8.10-6) with $\mathbf{g}=\mathbf{0}$, i.e.,

$$\begin{aligned} 2\mathbf{u} &= -\left(\tfrac{1}{\pi}(\ln r)\mathbf{I} + \mathbf{S}(\theta) - \mathbf{S}(\theta_o)\right)\mathbf{h}, \\ 2\boldsymbol{\phi} &= \left(\mathbf{L}(\theta) - \mathbf{L}(\theta_o)\right)\mathbf{h}. \end{aligned} \tag{8.10-12}$$

With the line force **f** applied at r=0, $\boldsymbol{\phi}$ must satisfy the condition

$$\boldsymbol{\phi}(r, \theta_n) - \boldsymbol{\phi}(r, \theta_o) = \mathbf{f}$$

or, by $(8.10\text{-}12)_2$,

$$\mathbf{h} = 2\{\mathbf{L}(\theta_n) - \mathbf{L}(\theta_o)\}^{-1}\mathbf{f}. \tag{8.10-13}$$

Equations (8.10-12) and (8.10-13) provide the solution for the composite wedge subject to a line force **f** at the wedge apex.

We note that the expressions for the stress components listed in (8.1-23) remain applicable for the composite space and composite wedge studied in this section.

Extension of the problem to piezoelectric composite space and composite wedge has been considered by Chung and Ting (1995a).

8.11 Angularly Inhomogeneous Space and Wedge

The composite spaces and wedges studied in the previous section are special cases of angularly inhomogeneous spaces and wedges. The Stroh formalism does not apply to inhomogeneous materials. We will employ a different approach in treating the problem of angularly inhomogeneous spaces that are subjected to a line force and a line dislocation at r=0. With the new approach (Ting, 1989), the fact that the surface traction on any radial plane is invariant with θ applies to any material.

The equations of equilibrium are

$$\sigma_{ij,j} = 0 \tag{8.11-1}$$

where a comma denotes differentiation. For two-dimensional deformations for which the infinite space is subjected to a line force **f**, dimensional analyses show that the stress must be proportional to r^{-1} in order to balance the applied force **f**. Let

$$\sigma_{ij} = r^{-1}\tau_{ij}(\theta) \tag{8.11-2}$$

where τ_{ij} depends on θ only. Substituting (8.11-2) into (8.11-1) and employing the relations

$$r_{,j} = n_j, \quad \theta_{,j} = r^{-1}m_j, \quad n_{j,\theta} = m_j, \quad m_{j,\theta} = -n_j \tag{8.11-3}$$

where the unit vectors **n** and **m** are

$$\mathbf{n}^T = (\cos\theta,\ \sin\theta,\ 0), \qquad \mathbf{m}^T = (-\sin\theta,\ \cos\theta,\ 0), \tag{8.11-4}$$

we obtain

$$(\tau_{ij}m_j)_{,\theta} = 0.$$

Hence

$$\tau_{ij}m_j = \frac{-1}{2\pi}g_i \tag{8.11-5}$$

where **g** is a constant vector.

The surface traction vector $\mathbf{t}_\theta$ on any radial plane θ=constant is

$$(\mathbf{t}_\theta)_i = \sigma_{ij}m_j = r^{-1}\tau_{ij}m_j = \frac{-1}{2\pi r}g_i,$$

or

$$\mathbf{t}_\theta = \frac{-1}{2\pi r}\mathbf{g}. \tag{8.11-6}$$

Therefore the surface traction vector $\mathbf{t}_\theta$ is independent of θ. This conclusion is reached based on the equations of equilibrium alone. It is independent of the material property.

Let the stress-strain law of an angularly inhomogeneous anisotropic material be written as

$$\sigma_{ij} = C_{ijks}(\theta)u_{k,s} \tag{8.11-7}$$

where the elastic stiffnesses C_{ijks} depend on θ. The C_{ijks} need not be continuous in θ. For the composite spaces considered in Section 8.10 C_{ijks} are piece-wise continuous in θ. Assuming that the displacement $\mathbf{u}$ has the form

$$\mathbf{u} = -\frac{1}{2\pi}(\ln r)\mathbf{h} + \mathbf{d}(\theta) \tag{8.11-8}$$

where $\mathbf{h}$ is a constant vector while the vector $\mathbf{d}$ depends on θ, substitution of (8.11-8) into (8.11-7) and with the use of (8.11-3) lead to

$$\sigma_{ij} = r^{-1}\left\{-\frac{1}{2\pi}C_{ijks}(\theta)n_s h_k + C_{ijks}(\theta)m_s d_{k,\theta}\right\} \tag{8.11-9}$$

which provides the expression for τ_{ij} in (8.11-2). Inserting into (8.11-5) we obtain the differential equation for $\mathbf{d}(\theta)$,

$$-\frac{1}{2\pi}C_{ijks}(\theta)n_s m_j h_k + C_{ijks}(\theta)m_s m_s d_{k,\theta} = -\frac{1}{2\pi}g_i.$$

In matrix notation this is

$$-\frac{1}{2\pi}\mathbf{R}^T(\theta)\mathbf{h} + \mathbf{T}(\theta)\mathbf{d}_{,\theta} = -\frac{1}{2\pi}\mathbf{g} \tag{8.11-10}$$

where the superscript T denotes the transpose and

$$\begin{aligned} Q_{ik}(\theta) &= C_{ijks}(\theta)n_j n_s, \\ R_{ik}(\theta) &= C_{ijks}(\theta)n_j m_s, \\ T_{ik}(\theta) &= C_{ijks}(\theta)m_j m_s. \end{aligned} \tag{8.11-11}$$

The matrix $\mathbf{Q}(\theta)$ did not appear in (8.11-10) but will be needed below. The matrices $\mathbf{Q}(\theta)$ and $\mathbf{T}(\theta)$ are symmetric and, subject to positive definiteness of strain energy, positive definite. We rewrite (8.11-10) as

$$\mathbf{d}_{,\theta} = -\frac{1}{2\pi}\{\mathbf{N}_1(\theta)\mathbf{h} + \mathbf{N}_2(\theta)\mathbf{g}\} \tag{8.11-12}$$

where

$$\mathbf{N}_1(\theta) = -\mathbf{T}^{-1}(\theta)\mathbf{R}^T(\theta), \quad \mathbf{N}_2(\theta) = \mathbf{T}^{-1}(\theta),$$
$$\mathbf{N}_3(\theta) = \mathbf{R}(\theta)\mathbf{T}^{-1}(\theta)\mathbf{R}^T(\theta) - \mathbf{Q}(\theta). \tag{8.11-13}$$

The matrix $\mathbf{N}_3(\theta)$ did not appear in (8.11-12) but will be needed later. The matrices $\mathbf{N}_2(\theta)$ and $\mathbf{N}_3(\theta)$ are symmetric and $\mathbf{N}_2(\theta)$ is positive definite. As in the case of a homogeneous medium it can be shown that $\mathbf{N}_3(\theta)$ is positive semidefinite. However $\mathbf{Q}(\theta)$, $\mathbf{R}(\theta)$, $\mathbf{T}(\theta)$, and $\mathbf{N}_i(\theta)$ (i=1,2,3) here are periodic in θ with periodicity 2π, not π, which is the case for a homogeneous medium.

If we define $\mathbf{S}(\theta)$, $\mathbf{H}(\theta)$, and $\mathbf{L}(\theta)$ according to (8.10-7), integration of (8.11-12) yields

$$\mathbf{d}(\theta) = -\tfrac{1}{2}\{[\mathbf{S}(\theta) - \mathbf{S}(\theta_o)]\mathbf{h} + [\mathbf{H}(\theta) - \mathbf{H}(\theta_o)]\mathbf{g}\}, \tag{8.11-14}$$

where θ_o is a constant. Again $\mathbf{S}(\theta)$, $\mathbf{H}(\theta)$, and $\mathbf{L}(\theta)$ here are periodic in θ with periodicity 2π, not π. The displacement given by (8.11-8) can now be written as (Ting, 1989)

$$2\mathbf{u} = -\frac{1}{\pi}(\ln r)\mathbf{h} - [\mathbf{S}(\theta) - \mathbf{S}(\theta_o)]\mathbf{h} - [\mathbf{H}(\theta) - \mathbf{H}(\theta_o)]\mathbf{g}. \tag{8.11-15}$$

The surface traction $\mathbf{t}_r$ on a cylindrical surface r = constant is

$$(\mathbf{t}_r)_i = \sigma_{ij} n_j$$

or, by (8.11-9), (8.11-12), and (8.11-13),

$$\mathbf{t}_r = \tfrac{1}{2\pi r}\left\{\mathbf{N}_3(\theta)\mathbf{h} + \mathbf{N}_1^T(\theta)\mathbf{g}\right\}. \tag{8.11-16}$$

In terms of the stress function ϕ the relations (8.1-8a) hold since their validity is independent of the material property. Hence (8.11-6) and (8.11- 16) suggest that

$$2\phi = -\tfrac{1}{\pi}(\ln r)\mathbf{g} + [\mathbf{L}(\theta) - \mathbf{L}(\theta_o)]\mathbf{h} - [\mathbf{S}^T(\theta) - \mathbf{S}^T(\theta_o)]\mathbf{g}. \tag{8.11-17}$$

Equations (8.11-15) and (8.11-17) are the general solution for angularly inhomogeneous anisotropic elastic materials for which the stress is proportional to r^{-1}. They are identical in expression to (8.10-6) for composite spaces except that $\mathbf{S}(\theta)$, $\mathbf{H}(\theta)$, and $\mathbf{L}(\theta)$ defined here are more general. The C_{ijks} required in the integration for $\mathbf{S}(\theta)$, $\mathbf{H}(\theta)$, and $\mathbf{L}(\theta)$ depend on θ. The dependence on θ needs not be continuous. Thus the analyses after (8.10-7) in Section 8.10 apply to the present problem. In particular, for the infinite space subjected to a line force $\mathbf{f}$ and a line dislocation with Burgers vector $\mathbf{b}$ at r=0, the constant vectors $\mathbf{h}$ and $\mathbf{g}$ that appeared in (8.11-15) and (8.11-17) are given by (8.10-11b). This solution was also obtained by Kirchner (1987, 1989) by a different derivation. For an angularly inhomogeneous wedge with $\theta_0 \le \theta \le \theta_n \ne 2\pi + \theta_0$ subjected to a line force **f** at the wedge apex the solution is given by (8.10-12) and (8.10-13).

Extension of the problem to angularly inhomogeneous piezoelectric space and wedge has been studied by Chung and Ting (1995b).

8.12 Green's Functions for Concentrated Couple, Double Force, and Other Higher Order Singularities

In this section we study Green's functions for the infinite anisotropic elastic space due to a double force with or without moment, a concentrated couple, a center of dilatation, etc., that produce a stress field proportional to r^{-2}. Before we present the general solution for this type of Green's functions we derive certain relations between the stress components when the stress field is proportional to r^{-2} or, more generally, r^{δ} where δ is an arbitrary constant. δ need not be real (Section 9.2). These relations are obtained by the equations of equilibrium alone, and hence are applicable to any material.

In the cylindrical coordinate system (r,θ,x_3) the stress σ_{ij} depends on r and θ only for two-dimensional deformations. Employing (8.11-3) the equations of equilibrium are

$$\sigma_{ij,j} = \sigma_{ij,r}n_j + r^{-1}\sigma_{ij,\theta}m_j = 0$$

or

$$\sigma_{ij,\theta}m_j = -r\sigma_{ij,r}n_j.$$

This can be written as

$$(\sigma_{ij}m_j)_{,\theta} = -(\delta+1)\sigma_{ij}n_j - r^{\delta+1}(r^{-\delta}\sigma_{ij})_{,r}n_j$$

in which δ is an arbitrary constant. If σ_{ij} is proportional to r^{δ}, $r^{-\delta}\sigma_{ij}$ is independent of r and the last term vanishes. We have

$$\mathbf{t}_{\theta,\theta} = -(\delta+1)\mathbf{t}_r \tag{8.12-1}$$

where

$$(\mathbf{t}_\theta)_i = \sigma_{ij}m_j, \quad (\mathbf{t}_r)_i = \sigma_{ij}n_j.$$

Equation (8.12-1) could also be obtained by the argument that if σ_{ij} is proportional to r^{δ} the stress function ϕ must be proportional to $r^{\delta+1}$, and (8.12-1) emerges from the relations (8.1-8a). Employing (8.1-11) and (8.12-1) it is easily shown that

$$\sigma_{r\theta,\theta} = -(\delta+1)\sigma_{rr} + \sigma_{\theta\theta}, \quad \sigma_{\theta\theta,\theta} = -(\delta+2)\sigma_{r\theta},$$
$$\sigma_{\theta 3,\theta} = -(\delta+1)\sigma_{r3}. \tag{8.12-2}$$

Equations (8.12-1) and (8.12-2) are universal in that they are independent of the material property. They are valid as long as the stress field is proportional to r^{δ} regardless of whether the material is isotropic or anisotropic, elastic or inelastic, homogeneous or inhomogeneous, linear or nonlinear. For $\delta=-1$, $\mathbf{t}_{\theta,\theta}=0$ implies that $\mathbf{t}_{\theta}$ is independent of θ, a result we have established many times through this chapter.

In this section we consider the case $\delta=-2$. Equations (8.12-1) and (8.12-2) reduce to

$$\mathbf{t}_{\theta,\theta} = \mathbf{t}_r, \quad \sigma_{r\theta,\theta} = \sigma_{rr} + \sigma_{\theta\theta}, \quad \sigma_{\theta\theta,\theta} = 0, \quad \sigma_{\theta 3,\theta} = \sigma_{r3}. \tag{8.12-3}$$

By $(8.12\text{-}3)_3$ the stress $\sigma_{\theta\theta}$ is independent of θ. Therefore if $\sigma_{\theta\theta}$ vanishes at a radial plane, it vanishes everywhere. For a wedge subjected to a concentrated couple at the wedge apex with the sides of the wedge being traction-free, $\sigma_{\theta\theta}$ vanishes everywhere. This conclusion in completely independent of the material property.

When the stress field is proportional to r^{-2} the displacement $\mathbf{u}$ and the stress function ϕ are proportional to r^{-1}. The singularities can be viewed as the result of the coalesce of two line forces and/or two line dislocations. Let a line force $\mathbf{f}$ and a line dislocation with Burgers vector $\mathbf{b}$ be located at

$$(\hat{x}_1, \hat{x}_2) = (-\tfrac{1}{2}\varepsilon, 0), \qquad \varepsilon > 0$$

where $(\hat{x}_1, \hat{x}_2)$ is a rotated coordinate system, Fig. 8.10(a). The angle between the $\hat{x}_1$-axis and the x_1-axis is θ_o. Another line force $-\mathbf{f}$ and line dislocation with Burgers vector $-\mathbf{b}$ are located at

$$(\hat{x}_1, \hat{x}_2) = (\tfrac{1}{2}\varepsilon, 0), \qquad \varepsilon > 0.$$

As $\varepsilon \to 0$ both $\mathbf{f}$ and $\mathbf{b}$ become infinite but the products

$$\hat{\mathbf{f}} = \varepsilon\mathbf{f}, \quad \hat{\mathbf{b}} = \varepsilon\mathbf{b}, \tag{8.12-4}$$

remain bounded. The solution is obtained by superposing two Green's functions for the infinite space and by taking the limit $\varepsilon \to 0$. With the dual coordinate systems (Section 7.10) the solution for the two Green's functions are

$$\mathbf{u} = \frac{1}{\pi}\mathrm{Im}\left\{\mathbf{A}\left\langle \ln(\hat{z}_* + \tfrac{1}{2}\varepsilon)\right\rangle \mathbf{q}^{\infty}\right\} - \frac{1}{\pi}\mathrm{Im}\left\{\mathbf{A}\left\langle \ln(\hat{z}_* - \tfrac{1}{2}\varepsilon)\right\rangle \mathbf{q}^{\infty}\right\},$$

$$\phi = \frac{1}{\pi}\mathrm{Im}\left\{\mathbf{B}\left\langle \ln(\hat{z}_* + \tfrac{1}{2}\varepsilon)\right\rangle \mathbf{q}^\infty\right\} - \frac{1}{\pi}\mathrm{Im}\left\{\mathbf{B}\left\langle \ln(\hat{z}_* - \tfrac{1}{2}\varepsilon)\right\rangle \mathbf{q}^\infty\right\},$$

where (Section 7.10)

$$\hat{z}_\alpha = \hat{x}_1 + p_\alpha(\theta_o)\hat{x}_2, \tag{8.12-5}$$

$$\begin{aligned} \hat{x}_1 &= \mathbf{x}\cdot\mathbf{n}(\theta_o) = r\cos(\theta - \theta_o), \\ \hat{x}_2 &= \mathbf{x}\cdot\mathbf{m}(\theta_o) = r\sin(\theta - \theta_o), \end{aligned} \tag{8.12-6}$$

and

$$\mathbf{q}^\infty = \mathbf{A}^T\mathbf{f} + \mathbf{B}^T\mathbf{b}.$$

In the limit as $\varepsilon \to 0$,

$$\mathbf{u} = \frac{1}{\pi}\mathrm{Im}\left\{\mathbf{A}\left\langle \hat{z}_*^{-1}\right\rangle \mathbf{q}\right\}, \qquad \phi = \frac{1}{\pi}\mathrm{Im}\left\{\mathbf{B}\left\langle \hat{z}_*^{-1}\right\rangle \mathbf{q}\right\}, \tag{8.12-7}$$

$$\mathbf{q} = \mathbf{A}^T\widehat{\mathbf{f}} + \mathbf{B}^T\widehat{\mathbf{b}}. \tag{8.12-8}$$

Equation (8.12-7) is the general solution of Green's functions for a stress field that is proportional to r^{-2}. The complex vector $\mathbf{q}$ or the real vectors $\widehat{\mathbf{f}}$ and $\widehat{\mathbf{b}}$ are arbitrary. One could also obtain (8.12-7) by differentiating the Green's function for the infinite space (8.1-16) with $\hat{x}_1$ using the relations

$$\frac{\partial}{\partial \hat{x}_1} = \cos\theta_o \frac{\partial}{\partial x_1} + \sin\theta_o \frac{\partial}{\partial x_2},$$

$$\hat{z}_\alpha = (\cos\theta_o + p_\alpha \sin\theta_o)^{-1} z_\alpha .$$

The second equation is taken from (7.10-9) and $(7.10\text{-}8)_2$.

Equation (8.12-7) together with (8.12-8) can be converted into a real form by the identity (7.10-20) as

$$\begin{bmatrix}\mathbf{u}\\ \phi\end{bmatrix} = \frac{-1}{2\pi r}\{\cos(\theta - \theta_o)\mathbf{I} - \sin(\theta - \theta_o)\mathbf{N}(\theta)\}\tilde{\mathbf{N}}\begin{bmatrix}\widehat{\mathbf{b}}\\ \widehat{\mathbf{f}}\end{bmatrix}, \tag{8.12-9}$$

$$\tilde{\mathbf{N}} = \begin{bmatrix}\mathbf{S} & \mathbf{H}\\ -\mathbf{L} & \mathbf{S}^T\end{bmatrix}.$$

Again, (8.12-9) could be obtained by differentiating the real form Green's function for the infinite space given by (8.1-19) and (8.1-20a) with $\hat{x}_1$ and employing the relation

$$\frac{\partial}{\partial \hat{x}_1} = \cos(\theta - \theta_o)\frac{\partial}{\partial r} - \sin(\theta - \theta_o)\frac{\partial}{r\partial\theta}.$$

The image singularities for the Green's function for the half-space subjected to a line force and a line dislocation are in general in the form of a line force and a line dislocation (Section 8.7). For degenerate materials some of the image singularities coalesce into one singularity of higher order. The image forces and dislocations are not necessarily of equal strength and opposite in sign before they coalesce. Let $\mathbf{f}^{(1)}$, $\mathbf{b}^{(1)}$ be the line force and line dislocation located at $(\hat{x}_1, \hat{x}_2) = (-\frac{1}{2}\varepsilon, 0)$ and $\mathbf{f}^{(2)}$, $\mathbf{b}^{(2)}$ be the second line force and line dislocation located at $(\frac{1}{2}\varepsilon, 0)$. If they are not equal and opposite in sign, and if they become infinite as $\varepsilon \to 0$, consider the symmetric parts

$$\tfrac{1}{2}(\mathbf{f}^{(1)} + \mathbf{f}^{(2)}), \qquad \tfrac{1}{2}(\mathbf{b}^{(1)} + \mathbf{b}^{(2)}),$$

and the skew-symmetric parts

$$\mathbf{f} = \tfrac{1}{2}(\mathbf{f}^{(1)} - \mathbf{f}^{(2)}), \qquad \mathbf{b} = \tfrac{1}{2}(\mathbf{b}^{(1)} - \mathbf{b}^{(2)}),$$

separately. When the symmetric parts are applied at $(-\frac{1}{2}\varepsilon, 0)$ and $(\frac{1}{2}\varepsilon, 0)$, they should remain bounded as $\varepsilon \to 0$. Otherwise the problem is physically meaningless. When the skew-symmetric parts $\mathbf{f}$ and $\mathbf{b}$ are applied at $(-\frac{1}{2}\varepsilon, 0)$ and $-\mathbf{f}$, $-\mathbf{b}$ are applied at $(\frac{1}{2}\varepsilon, 0)$, we have the problem shown in Fig. 8.10(a).

Equation (8.12-9) shows that the solution depends on θ_o. The line $\theta = \theta_o$ is called the *axis of limit* because the two line forces and the two line dislocations coalesce into one in the limit $\varepsilon \to 0$ along this axis. A given Green's function with a specified axis of limit can be replaced by a different axis of limit with the line force $\mathbf{f}$ and line dislocation $\mathbf{b}$ properly modified. Let $\theta = \theta'$ be a different axis of limit with $\mathbf{f}'$ and $\mathbf{b}'$ being the corresponding force and dislocation. Equation (8.12-9) for the new axis of limit is

$$\begin{bmatrix} \mathbf{u} \\ \boldsymbol{\phi} \end{bmatrix} = \frac{-1}{2\pi r}\{\cos(\theta - \theta')\mathbf{I} - \sin(\theta - \theta')\mathbf{N}(\theta)\}\tilde{\mathbf{N}} \begin{bmatrix} \hat{\mathbf{b}}' \\ \hat{\mathbf{f}}' \end{bmatrix}. \qquad (8.12\text{-}10)$$

The identity (7.5-20) can be written as

$$\mathbf{N}(\theta_o, \theta')\mathbf{N}(\theta', \theta) = \mathbf{N}(\theta_o, \theta) = \mathbf{N}(\theta', \theta)\mathbf{N}(\theta_o, \theta') \qquad (8.12\text{-}11)$$

where

$$\mathbf{N}(\theta_o, \theta) = \cos(\theta - \theta_o)\mathbf{I} - \sin(\theta - \theta_o)\mathbf{N}(\theta). \qquad (8.12\text{-}12)$$

Noting that the products of $\mathbf{N}(\theta)$ and $\tilde{\mathbf{N}}$ commute, it is easily seen that (8.12-9) and (8.12-10) are equivalent if

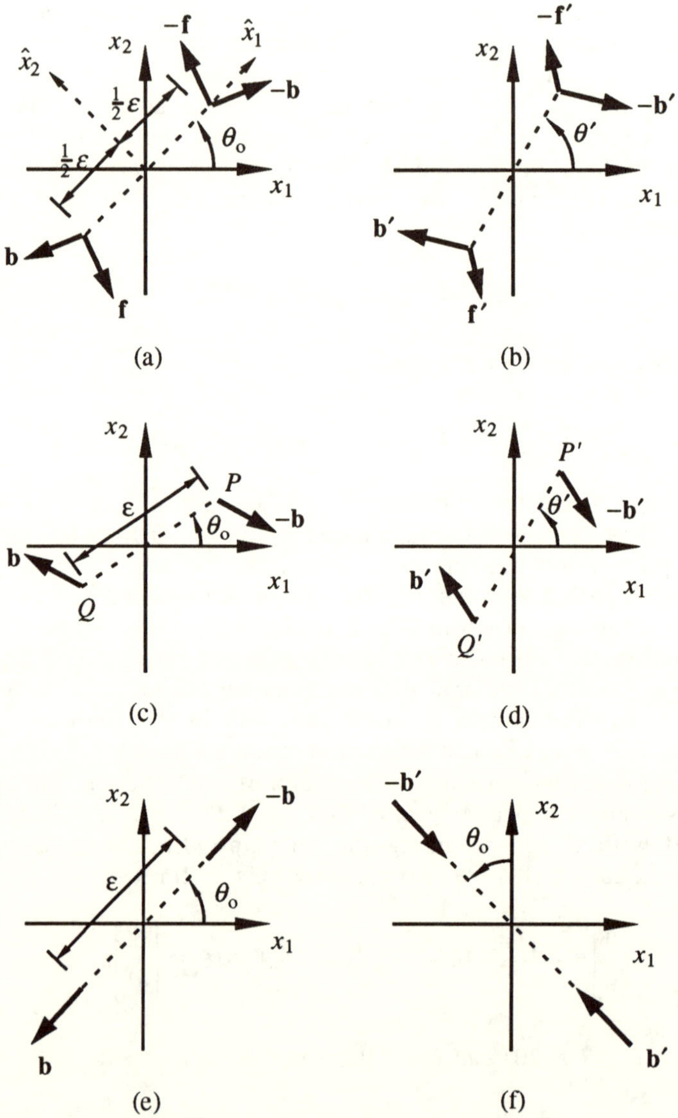

Fig. 8.10 As $\varepsilon \to 0$ the stress fields generated in (a) and (b) are identical if **f**, **b**, **f**′, **b**′ are related by (8.12-13). In particular the stress fields generated by the two dislocations in (c) and (d) are identical as $\varepsilon \to 0$. The Burgers vector **b**′ in (d) is perpendicular to the line PQ in (c), and the Burgers vector **b** in (c) is perpendicular the line $P'Q'$ in (d). When $\theta_o' = \theta_o + \pi/2$ we have (e) and (f).

$$\begin{bmatrix} \mathbf{b}' \\ \mathbf{f}' \end{bmatrix} = \mathbf{N}(\theta_o, \theta') \begin{bmatrix} \mathbf{b} \\ \mathbf{f} \end{bmatrix}, \tag{8.12-13a}$$

or

$$\begin{bmatrix} \mathbf{b} \\ \mathbf{f} \end{bmatrix} = \mathbf{N}(\theta', \theta_o) \begin{bmatrix} \mathbf{b}' \\ \mathbf{f}' \end{bmatrix}. \tag{8.12-13b}$$

In view of (8.12-4), we have replaced $\hat{\mathbf{f}}$, $\hat{\mathbf{b}}$, $\hat{\mathbf{f}}'$, $\hat{\mathbf{b}}'$, by $\mathbf{f}$, $\mathbf{b}$, $\mathbf{f}'$, $\mathbf{b}'$, respectively. When we multiply both sides of (8.12-13b) by $\mathbf{N}(\theta, \theta')$ we have a symmetric form

$$\mathbf{N}(\theta, \theta') \begin{bmatrix} \mathbf{b} \\ \mathbf{f} \end{bmatrix} = \mathbf{N}(\theta, \theta_o) \begin{bmatrix} \mathbf{b}' \\ \mathbf{f}' \end{bmatrix}. \tag{8.12-13c}$$

Therefore if $\mathbf{f}$, $\mathbf{b}$, $\mathbf{f}'$, and $\mathbf{b}'$ are related by (8.12-13) the stress field given by (8.12-9) and (8.12-10) are identical, Fig. 8.10(a), (b). As a special case it is readily shown with the aid of (8.1-5) that (8.12-13c) is satisfied by

$$\mathbf{b} = \mathbf{m}(\theta'), \quad \mathbf{b}' = \mathbf{m}(\theta_o), \quad \mathbf{f} = \mathbf{f}' = \mathbf{0}. \tag{8.12-14}$$

This is shown in Figs. 8.10(c), (d). The vector $\mathbf{b}$ in (c) is perpendicular to the axis of limit $P'Q'$ in (d), and the vector $\mathbf{b}'$ in (d) is perpendicular to the axis of limit PQ in (c). In particular, when $\theta' = \theta_o + \pi/2$ we have the situation shown in Figs. 8.10(e), (f). The stress field generated by the dislocations in Figs. 8.10(e) and (f) are identical in the limit $\varepsilon \rightarrow 0$.

Let the real vectors $\mathbf{h}$ and $\mathbf{g}$ be given by

$$\begin{bmatrix} \mathbf{h} \\ \mathbf{g} \end{bmatrix} = \tilde{\mathbf{N}} \begin{bmatrix} \hat{\mathbf{b}} \\ \hat{\mathbf{f}} \end{bmatrix} = \begin{bmatrix} \mathbf{S} & \mathbf{H} \\ -\mathbf{L} & \mathbf{S}^T \end{bmatrix} \begin{bmatrix} \hat{\mathbf{b}} \\ \hat{\mathbf{f}} \end{bmatrix}, \tag{8.12-15}$$

so that (8.12-9) is

$$\begin{aligned} \mathbf{u} &= \frac{-1}{2\pi r}\{\cos(\theta - \theta_o)\mathbf{h} - \sin(\theta - \theta_o)[\mathbf{N}_1(\theta)\mathbf{h} + \mathbf{N}_2(\theta)\mathbf{g}]\}, \\ \boldsymbol{\phi} &= \frac{-1}{2\pi r}\{\cos(\theta - \theta_o)\mathbf{g} - \sin(\theta - \theta_o)[\mathbf{N}_3(\theta)\mathbf{h} + \mathbf{N}_1^T(\theta)\mathbf{g}]\}. \end{aligned} \tag{8.12-16}$$

We have

$$\mathbf{u} = \frac{-1}{2\pi r}\mathbf{h}, \quad \boldsymbol{\phi} = \frac{-1}{2\pi r}\mathbf{g}, \quad \text{at } \theta = \theta_o. \tag{8.12-17}$$

Thus the constant vector $\mathbf{h}$ represents the direction of the displacement at the axis of limit, while $\mathbf{g}$ represents the direction of the sur face traction $\mathbf{t}_\theta$ on the radial plane $\theta=\theta_o$. Making use of the identities (8.1-5)–(8.1-7),

$$\mathbf{n}(\theta)\cdot\boldsymbol{\phi} = \frac{1}{2\pi r}\mathbf{m}^T(\theta_o)\{\mathbf{N}_3(\theta)\mathbf{h} + \mathbf{N}_1^T(\theta)\mathbf{g}\},$$
$$\mathbf{m}(\theta)\cdot\boldsymbol{\phi} = \frac{-1}{2\pi r}\mathbf{m}^T(\theta_o)\mathbf{g}. \tag{8.12-18}$$

The stresses are then determined from (8.1-12) and (8.12-3) as

$$\sigma_{r\theta} = \{\mathbf{n}(\theta)\cdot\boldsymbol{\phi}\}_{,r}, \quad \sigma_{\theta\theta} = \{\mathbf{m}(\theta)\cdot\boldsymbol{\phi}\}_{,r}, \quad \sigma_{\theta 3} = \varphi_{3,r},$$
$$\sigma_{rr} = \sigma_{r\theta,\theta} - \sigma_{\theta\theta}, \quad \sigma_{r3} = \sigma_{\theta 3,\theta}, \tag{8.12-19}$$

in which φ_3 is the third component of $\boldsymbol{\phi}$.

From (7.8-9) the product $\mathbf{N}(\theta)\tilde{\mathbf{N}}$ in (8.12-10) commutes, i.e.,

$$\begin{bmatrix}\mathbf{W}(\theta) & \mathbf{U}(\theta)\\ \mathbf{V}(\theta) & \mathbf{W}^T(\theta)\end{bmatrix} = \mathbf{N}(\theta)\tilde{\mathbf{N}} = \tilde{\mathbf{N}}\mathbf{N}(\theta), \tag{8.12-20a}$$

or

$$\mathbf{W}(\theta) = \mathbf{N}_1(\theta)\mathbf{S} - \mathbf{N}_2(\theta)\mathbf{L} = \mathbf{S}\mathbf{N}_1(\theta) + \mathbf{H}\mathbf{N}_3(\theta),$$
$$\mathbf{U}(\theta) = \mathbf{N}_1(\theta)\mathbf{H} + \mathbf{N}_2(\theta)\mathbf{S}^T = \mathbf{H}\mathbf{N}_1^T(\theta) + \mathbf{S}\mathbf{N}_2(\theta), \tag{8.12-20b}$$
$$\mathbf{V}(\theta) = \mathbf{N}_3(\theta)\mathbf{S} - \mathbf{N}_1^T(\theta)\mathbf{L} = \mathbf{S}^T\mathbf{N}_3(\theta) - \mathbf{L}\mathbf{N}_1(\theta).$$

The second equalities for $\mathbf{U}(\theta)$ and $\mathbf{V}(\theta)$ show that they are symmetric tensors. By (8.1-5) and (8.1-7),

$$\mathbf{W}(\theta)\mathbf{m}(\theta) = -\mathbf{S}\mathbf{n}(\theta), \quad \mathbf{V}(\theta)\mathbf{m}(\theta) = \mathbf{L}\mathbf{n}(\theta), \tag{8.12-21}$$

$$\cos(\theta-\theta_o)\mathbf{W}(\theta)\mathbf{n}(\theta) = \mathbf{W}(\theta)\mathbf{n}(\theta_o) - \sin(\theta-\theta_o)\mathbf{S}\mathbf{n}(\theta),$$
$$\sin(\theta-\theta_o)\mathbf{W}(\theta)\mathbf{n}(\theta) = \mathbf{W}(\theta)\mathbf{m}(\theta_o) + \cos(\theta-\theta_o)\mathbf{S}\mathbf{n}(\theta),$$
$$\cos(\theta-\theta_o)\mathbf{V}(\theta)\mathbf{n}(\theta) = \mathbf{V}(\theta)\mathbf{n}(\theta_o) + \sin(\theta-\theta_o)\mathbf{L}\mathbf{n}(\theta), \tag{8.12-22}$$
$$\sin(\theta-\theta_o)\mathbf{V}(\theta)\mathbf{n}(\theta) = \mathbf{V}(\theta)\mathbf{m}(\theta_o) - \cos(\theta-\theta_o)\mathbf{L}\mathbf{n}(\theta).$$

We have now an alternate to (8.12-16) and (8.12-18) the following expressions

$$\mathbf{u} = \frac{-1}{2\pi r}\{\cos(\theta-\theta_o)[\mathbf{S}\hat{\mathbf{b}} + \mathbf{H}\hat{\mathbf{f}}] - \sin(\theta-\theta_o)[\mathbf{W}(\theta)\hat{\mathbf{b}} + \mathbf{U}(\theta)\hat{\mathbf{f}}]\}, \tag{8.12-23$_1$}$$

$$\phi = \frac{-1}{2\pi r}\left\{\cos(\theta-\theta_o)[-\mathbf{L}\hat{\mathbf{b}}+\mathbf{S}^T\hat{\mathbf{f}}]-\sin(\theta-\theta_o)[\mathbf{V}(\theta)\hat{\mathbf{b}}+\mathbf{W}^T(\theta)\hat{\mathbf{f}}]\right\}, \quad 8.12\text{-}23)_2$$

$$\mathbf{n}(\theta)\cdot\phi = \frac{1}{2\pi r}\mathbf{m}^T(\theta_o)\left\{\mathbf{V}(\theta)\hat{\mathbf{b}}+\mathbf{W}^T(\theta)\hat{\mathbf{f}}\right\},$$

$$\mathbf{m}(\theta)\cdot\phi = \frac{1}{2\pi r}\mathbf{m}^T(\theta_o)\left\{\mathbf{L}\hat{\mathbf{b}}-\mathbf{S}^T\hat{\mathbf{f}}\right\}. \quad (8.12\text{-}24)$$

The stresses are determined by (8.12-19).

The strength D of dilatation is defined as

$$D = -\int_0^{2\pi}\sigma_{rr}r^2 d\theta = r^2\int_0^{2\pi}\sigma_{\theta\theta}d\theta = 2\pi r^2\sigma_{\theta\theta}, \quad (8.12\text{-}25a)$$

use being made of $(8.12\text{-}19)_4$ and the fact that $\sigma_{r\theta}$ at θ=0 and 2π are identical and $\sigma_{\theta\theta}$ is independent of θ. By $(8.12\text{-}19)_2$

$$D = 2\pi r^2\{\mathbf{m}(\theta)\cdot\phi\}_{,r}. \quad (8.12\text{-}25b)$$

The counterclockwise moment or torque T about the x_3-axis generated by the singularity is

$$T = -\int_0^{2\pi}\sigma_{r\theta}r^2 d\theta = -\int_0^{2\pi} r^2\{\mathbf{n}(\theta)\cdot\phi\}_{,r}d\theta. \quad (8.12\text{-}26)$$

The following relations are useful in the sequel. From (8.12-20b) and $(8.1\text{-}3b)_2$,

$$\frac{1}{\pi}\int_0^{\pi}\mathbf{W}(\theta)d\theta = \mathbf{SS}-\mathbf{LH} = -\mathbf{I}, \quad (8.12\text{-}27)$$

$$\frac{1}{\pi}\int_0^{\pi}\mathbf{V}(\theta)d\theta = -\mathbf{LS}-\mathbf{S}^T\mathbf{L} = \mathbf{0}. \quad (8.12\text{-}28)$$

We now discuss Green's functions due to various combinations of $\hat{\mathbf{f}}$ and $\hat{\mathbf{b}}$ below.

(1) **Double Force Without Moment**. Consider the coalesce of two equal and opposite line forces shown in Fig. 8.11(a). The two forces are acting away from each other and hence no moment is generated (Love, 1927). As $\varepsilon\to 0$ $\mathbf{f}$ becomes infinity but $\hat{\mathbf{f}}=\varepsilon\mathbf{f}$ is bounded. The solution is given by (8.12-23) and (8.12-24) with

$$\hat{\mathbf{b}} = \mathbf{0}, \quad \hat{\mathbf{f}} = -P\mathbf{n}(\theta_o), \quad (8.12\text{-}29)$$

where P is the magnitude of $\hat{\mathbf{f}}_o$. Hence

$$\mathbf{u} = \frac{P}{2\pi r}\{\cos(\theta-\theta_o)\mathbf{H} - \sin(\theta-\theta_o)\mathbf{U}(\theta)\}\mathbf{n}(\theta_o),$$
$$\phi = \frac{P}{2\pi r}\{\cos(\theta-\theta_o)\mathbf{S}^T - \sin(\theta-\theta_o)\mathbf{W}^T(\theta)\}\mathbf{n}(\theta_o), \tag{8.12-30}$$

$$\mathbf{n}(\theta)\cdot\phi = \frac{-P}{2\pi r}\mathbf{n}^T(\theta_o)\mathbf{W}(\theta)\mathbf{m}(\theta_o),$$
$$\mathbf{m}(\theta)\cdot\phi = \frac{P}{2\pi r}\mathbf{n}^T(\theta_o)\mathbf{S}\mathbf{m}(\theta_o). \tag{8.12-31}$$

We could derive (8.12-31) directly from $(8.12\text{-}30)_2$ with the aid of (8.12-21) and (8.12-22). It is seen that the solution depends on θ_o, the axis of limit. By (8.12-26) and (8.12-27) T=0 so that no concentrated force is generated. For isotropic materials it can be shown that

$$\phi = \frac{-P}{4\pi r(1-\nu)}\{(1-2\nu)\mathbf{m}(\theta) - \sin 2(\theta-\theta_o)\mathbf{n}(\theta)\}$$

where ν is the Poisson's ratio.

(2) **Double Force With Moment.** In this case the two line forces which are equal and opposite in direction are perpendicular to the axis of limit connecting the two forces before they coalesce, Fig. 8.11(b). A counterclockwise couple C about the x_3-axis is generated. For this case

$$\hat{\mathbf{b}} = \mathbf{0}, \quad \hat{\mathbf{f}} = -C\mathbf{m}(\theta_o), \tag{8.12-32}$$

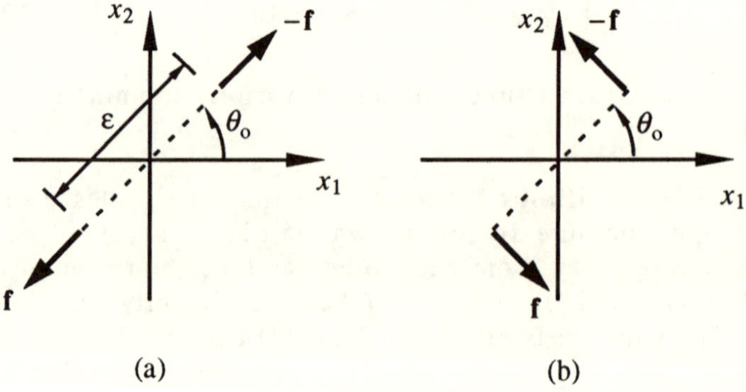

Fig. 8.11 As $\varepsilon \to 0$ the stress field produces (a) a double force without moment and (b) a double force with moment.

$$\mathbf{u}=\frac{C}{2\pi r}\left\{\cos(\theta-\theta_o)\mathbf{H}-\sin(\theta-\theta_o)\mathbf{U}(\theta)\right\}\mathbf{m}(\theta_o),$$
$$\boldsymbol{\phi}=\frac{C}{2\pi r}\left\{\cos(\theta-\theta_o)\mathbf{S}^T-\sin(\theta-\theta_o)\mathbf{W}^T(\theta)\right\}\mathbf{m}(\theta_o), \tag{8.12-33}$$

$$\mathbf{n}(\theta)\cdot\boldsymbol{\phi}=\frac{-C}{2\pi r}\mathbf{m}^T(\theta_o)\mathbf{W}(\theta)\mathbf{m}(\theta_o),$$
$$\mathbf{m}(\theta)\cdot\boldsymbol{\phi}=\frac{C}{2\pi r}\mathbf{m}^T(\theta_o)\mathbf{S}\mathbf{m}(\theta_o). \tag{8.12-34}$$

Again the solution depends on θ_o. It is readily shown from (8.12-25), (8.12-26), and (8.12-27) that $T=C$ while D may be nonzero. For isotropic materials $D=0$ and

$$\boldsymbol{\phi}=\frac{-C}{4\pi r(1-\nu)}\{2(1-\nu)-\cos2(\theta-\theta_o)\}\mathbf{n}(\theta).$$

(3) **Center of Dilatation**. A center of dilatation is obtained when the axes of limit for two double forces without moment are at θ_o and $\theta_o+\pi/2$, Fig. 8.12(a). The solution for a double force without moment with the axis of limit at $\theta_o+\pi/2$ is obtained from (8.12-30) and (8.12-31) by replacing θ_o by $\theta_o+\pi/2$. We have

$$\mathbf{u}=\frac{P}{2\pi r}\{\sin(\theta-\theta_o)\mathbf{H}+\cos(\theta-\theta_o)\mathbf{U}(\theta)\}\mathbf{m}(\theta_o),$$
$$\boldsymbol{\phi}=\frac{P}{2\pi r}\left\{\sin(\theta-\theta_o)\mathbf{S}^T+\cos(\theta-\theta_o)\mathbf{W}^T(\theta)\right\}\mathbf{m}(\theta_o), \tag{8.12-35}$$

$$\mathbf{n}(\theta)\cdot\boldsymbol{\phi}=\frac{P}{2\pi r}\mathbf{m}^T(\theta_o)\mathbf{W}(\theta)\mathbf{n}(\theta_o),$$
$$\mathbf{m}(\theta)\cdot\boldsymbol{\phi}=\frac{-P}{2\pi r}\mathbf{m}^T(\theta_o)\mathbf{S}\mathbf{n}(\theta_o). \tag{8.12-36}$$

Superposition of (8.12-35), (8.12-36) on (8.12-30), (8.12-31), and employing the identities (8.1-6) leads to

$$\mathbf{u}=\frac{P}{2\pi r}\{\mathbf{H}\mathbf{n}(\theta)+\mathbf{U}(\theta)\mathbf{m}(\theta)\},\quad \boldsymbol{\phi}=\frac{P}{2\pi r}\left\{\mathbf{S}^T\mathbf{n}(\theta)+\mathbf{W}^T(\theta)\mathbf{m}(\theta)\right\}, \tag{8.12-37}$$

$$\mathbf{n}(\theta)\cdot\boldsymbol{\phi}=\frac{P}{2\pi r}\mathbf{m}^T(\theta_o)\left\{\mathbf{W}(\theta)-\mathbf{W}^T(\theta)\right\}\mathbf{n}(\theta_o)$$
$$=\frac{P}{2\pi r}\{W_{21}(\theta)-W_{12}(\theta)\}, \tag*{(8.12-38)$_{1,2}$}$$

$$\mathbf{m}(\theta)\cdot\boldsymbol{\phi} = \frac{P}{2\pi r}\mathbf{n}^T(\theta_o)\{\mathbf{S}-\mathbf{S}^T\}\mathbf{m}(\theta_o)$$
$$= \frac{P}{2\pi r}\{S_{12} - S_{21}\}. \qquad (8.12\text{-}38)_{3,4}$$

The solution (8.12-37) is independent of θ_o. By differentiating $(8.12\text{-}38)_1$ with θ_o it can be shown that $\mathbf{n}(\theta)\cdot\boldsymbol{\phi}$ and $\mathbf{m}(\theta)\cdot\boldsymbol{\phi}$ are independent of θ_o. Therefore setting $\theta_o=0$ we obtain the remarkably simple and explicit expressions in the second equalities. We could also replace θ_o by θ to obtain an alternate expression that can be derived directly from $(8.12\text{-}37)_2$ and (8.12-21). For isotropic materials $(8.12\text{-}33)_2$ reduces to

$$\boldsymbol{\phi} = \frac{-(1-2\nu)}{2\pi r(1-\nu)}P\mathbf{m}(\theta).$$

The stresses $\sigma_{r\theta}$ and $\sigma_{\theta\theta}$ from (8.12-19) and (8.12-38) are

$$\sigma_{r\theta} = \frac{P}{2\pi r^2}\{W_{12}(\theta) - W_{21}(\theta)\}, \quad \sigma_{\theta\theta} = \frac{P}{2\pi r^2}\{S_{21} - S_{12}\}. \qquad (8.12\text{-}39)$$

$\sigma_{\theta\theta}$ is independent of θ as it should be. So is $\sigma_{r\theta}$ if the material is isotropic.

By (8.12-26) and (8.12-27) we have T=0. The strength D of dilatation obtained from (8.12-25a) is

$$D = P(S_{21} - S_{12}). \qquad (8.12\text{-}40)$$

It is reasonable to expect that D and P have the same sign. Hence

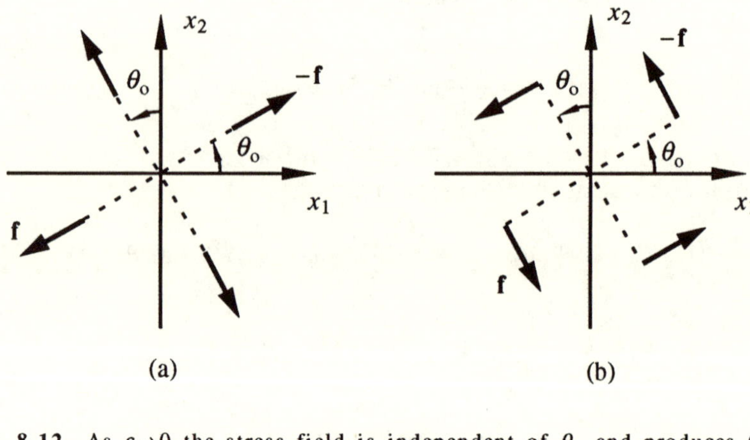

Fig. 8.12 As $\varepsilon \to 0$ the stress field is independent of θ_o and produces (a) a center of dilatation and (b) a center of rotation or a concentrated couple.

$$S_{21} > S_{12}. \tag{8.12-41}$$

(4) **Concentrated Couple.** A concentrated couple is also called a *center of rotation* (Love, 1927). It is obtained by a double force with moment that have the axis of limit at $\theta=\theta_o$ and an identical double force with moment with the axis of limit at $\theta=\theta_o+\pi/2$, Fig. 8.12(b). The solution is obtained by superimposing on (8.12-33), (8.12-34) the solution obtained from them with θ_o replaced by $\theta_o+\pi/2$. Following the derivation of (8.12-37) and (8.12-38) it can be shown that

$$\mathbf{u} = \frac{C}{2\pi r}\{\mathbf{Hm}(\theta) - \mathbf{U}(\theta)\mathbf{n}(\theta)\}, \quad \boldsymbol{\phi} = \frac{C}{2\pi r}\left\{\mathbf{S}^T\mathbf{m}(\theta) - \mathbf{W}^T(\theta)\mathbf{n}(\theta)\right\}, \tag{8.12-42}$$

$$\sigma_{r\theta} = \frac{C}{2\pi r^2}\{W_{11}(\theta) + W_{22}(\theta)\}, \quad \sigma_{\theta\theta} = \frac{C}{2\pi r^2}S_{33}. \tag{8.12-43}$$

In the above we have employed the fact that tr$\mathbf{S}$ = $S_{11}+S_{22}+S_{33}=0$. Equations (8.12-42) show that the solution is independent of θ_o. Therefore we may set $\theta_o=0$ in Fig. 8.12(b) in taking the limit $\varepsilon \to 0$. As expected $\sigma_{\theta\theta}$ is independent of θ. So is $\sigma_{r\theta}$ if the material is isotropic. For isotropic materials

$$\boldsymbol{\phi} = \frac{C}{\pi r}\mathbf{n}(\theta).$$

The total moment or torque T about the x_3-axis is

$$T = 2C. \tag{8.12-44}$$

Substitution of $\sigma_{\theta\theta}$ from (8.12-43) into (8.12-25a) leads to

$$D = CS_{33}$$

so that the Green's function for a concentrated couple may be accompanied by a center of dilatation unless $S_{33}=0$, which is the case for monoclinic materials with the symmetry plane at $x_3=0$. To get rid of the center of dilatation we set

$$P = \frac{-S_{33}}{S_{21} - S_{12}}C$$

in (8.12-37), (8.12-39), and superimpose them onto (8.12-42), (8.12-43), respectively.

The following three cases are counterparts of cases (1), (2), and (4), respectively, with **f** replaced by **b**. The counterpart of (3) leads to zero stress everywhere (See Figs. 8.10(e), (f)). We list only the results below.

(5) **Center of Shear**. For this case, Fig. 8.10(e),

$$\hat{\mathbf{b}} = -\gamma \mathbf{n}(\theta_o), \quad \hat{\mathbf{f}} = 0,$$

where γ is a constant. The solution is

$$\begin{aligned} \mathbf{u} &= \frac{\gamma}{2\pi r}\{\cos(\theta-\theta_o)\mathbf{S} - \sin(\theta-\theta_o)\mathbf{W}(\theta)\}\mathbf{n}(\theta_o), \\ \boldsymbol{\phi} &= \frac{-\gamma}{2\pi r}\{\cos(\theta-\theta_o)\mathbf{L} + \sin(\theta-\theta_o)\mathbf{V}(\theta)\}\mathbf{n}(\theta_o), \end{aligned} \tag{8.12-45}$$

$$\begin{aligned} \mathbf{n}(\theta)\cdot\boldsymbol{\phi} &= \frac{-\gamma}{2\pi r}\mathbf{n}^T(\theta_o)\mathbf{V}(\theta)\mathbf{m}(\theta_o), \\ \mathbf{m}(\theta)\cdot\boldsymbol{\phi} &= \frac{-\gamma}{2\pi r}\mathbf{n}^T(\theta_o)\mathbf{L}\mathbf{m}(\theta_o). \end{aligned} \tag{8.12-46}$$

It can be shown that T=0 but D may be nonzero. For isotropic materials D=0 and

$$\boldsymbol{\phi} = \frac{-\mu\gamma}{2\pi r(1-\nu)}\cos 2(\theta-\theta_o)\mathbf{n}(\theta)$$

where μ is the shear modulus.

(6) **Center of Extension**. This case is shown in Fig. 8.11(b) with **f** replaced by **b**. Hence

$$\hat{\mathbf{b}} = -\kappa \mathbf{m}(\theta_o), \quad \hat{\mathbf{f}} = 0$$

$$\begin{aligned} \mathbf{u} &= \frac{\kappa}{2\pi r}\{\cos(\theta-\theta_o)\mathbf{S} - \sin(\theta-\theta_o)\mathbf{W}(\theta)\}\mathbf{m}(\theta_o), \\ \boldsymbol{\phi} &= \frac{-\kappa}{2\pi r}\{\cos(\theta-\theta_o)\mathbf{L} + \sin(\theta-\theta_o)\mathbf{V}(\theta)\}\mathbf{m}(\theta_o), \end{aligned} \tag{8.12-47}$$

$$\begin{aligned} \mathbf{n}(\theta)\cdot\boldsymbol{\phi} &= \frac{-\kappa}{2\pi r}\mathbf{m}^T(\theta_o)\mathbf{V}(\theta)\mathbf{m}(\theta_o), \\ \mathbf{m}(\theta)\cdot\boldsymbol{\phi} &= \frac{-\kappa}{2\pi r}\mathbf{m}^T(\theta_o)\mathbf{L}\mathbf{m}(\theta_o). \end{aligned} \tag{8.12-48}$$

It can be shown that T=0 but D≠0. For isotropic materials

$$\boldsymbol{\phi} = \frac{-mk}{2\pi r(1-\nu)}\{\mathbf{m}(\theta) + \sin 2(\theta-\theta_o)\mathbf{n}(\theta)\}.$$

(7) **Center of Expansion.** This case is shown in Fig. 8.12(b) with **f** replaced by **b**. The solution is a superposition of (8.12-47), (8.12-48) with the solution obtained from them by replacing θ_o by $\theta_o+\pi/2$. We have

$$\mathbf{u}=\frac{\kappa}{2\pi r}\{\mathbf{S}\mathbf{m}(\theta)-\mathbf{W}(\theta)\mathbf{n}(\theta)\},\quad \boldsymbol{\phi}=\frac{-\kappa}{2\pi r}\{\mathbf{L}\mathbf{m}(\theta)+\mathbf{V}(\theta)\mathbf{n}(\theta)\},\quad (8.12\text{-}49)$$

$$\sigma_{r\theta}=\frac{\kappa}{2\pi r^2}\{V_{11}(\theta)+V_{22}(\theta)\},\quad \sigma_{\theta\theta}=\frac{\kappa}{2\pi r^2}(L_{11}+L_{22}).\quad (8.12\text{-}50)$$

The solution is independent of θ_o. The moment T=0 but D≠0. For isotropic materials

$$\boldsymbol{\phi}=\frac{-\mu\kappa}{\pi r(1-\nu)}\mathbf{m}(\theta)$$

so that it is identical to the solution for a center of dilatation except a constant multiplier.

8.13 Wedge Under a Concentrated Couple

The classical problem of an isotropic elastic wedge of wedge angle 2α subjected to a concentrated couple of magnitude T at the wedge apex has been studied by Carothers (1912), Inglis (1922), Sternberg and Koiter (1958), and Neuber (1963). The corresponding problem for anisotropic elastic wedge was investigated by Ting (1988d) and Hwu and Ting (1990). Let an anisotropic elastic wedge of wedge angle 2α occupy the region

$$\theta^-\le\theta\le\theta^+,\quad r>0,\quad (8.13\text{-}1)$$

where

$$\theta^+=\theta_c+\alpha,\quad \theta^-=\theta_c-\alpha.\quad (8.13\text{-}2)$$

The line $\theta=\theta_c$ is the center line of the wedge, Fig. 8.13. The surfaces $\theta=\theta^-,\theta^+$ are traction free so that

$$\boldsymbol{\phi}=\mathbf{0},\quad \text{at } \theta=\theta^-,\theta^+.\quad (8.13\text{-}3)$$

The couple or the torque T applied at the wedge apex must be balanced by the moment about the wedge apex due to tractions at any cylindrical surface r=constant, i. e.,

$$T=-\int_{\theta^-}^{\theta^+}\sigma_{r\theta}r^2d\theta.\quad (8.13\text{-}4)$$

T is considered positive if it is counterclockwise. Since the

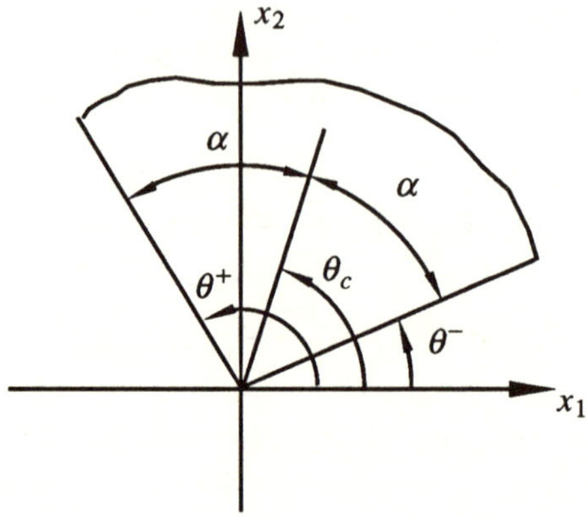

Fig. 8.13 The geometry of a wedge

stresses are proportional to r^{-2} the general solution presented in (8.12-16) applies here. If we replace θ_o in (8.12-16) by θ^- and set $\mathbf{g}=\mathbf{0}$ we have

$$\mathbf{u} = \frac{-1}{2\pi r}[\cos(\theta-\theta^-)\mathbf{I} - \sin(\theta-\theta^-)\mathbf{N}_1(\theta)]\mathbf{h},$$
$$\boldsymbol{\phi} = \frac{1}{2\pi r}\sin(\theta-\theta^-)\mathbf{N}_3(\theta)\mathbf{h}. \tag{8.13-5}$$

This satisfies the boundary condition $\boldsymbol{\phi}=\mathbf{0}$ at $\theta=\theta^-$. The other boundary condition $\boldsymbol{\phi}=\mathbf{0}$ at $\theta=\theta^+$ is satisfied if

$$\sin 2\alpha \mathbf{N}_3(\theta^+)\mathbf{h} = \mathbf{0}. \tag{8.13-6}$$

When $2\alpha \neq \pi$ or 2π,

$$\mathbf{h} = k\mathbf{m}(\theta^+)$$

where k is an arbitrary constant and (8.13-5) becomes

$$\mathbf{u} = \frac{-k}{2\pi r}[\cos(\theta-\theta^-)\mathbf{I} - \sin(\theta-\theta^-)\mathbf{N}_1(\theta)]\mathbf{m}(\theta^+),$$
$$\boldsymbol{\phi} = \frac{k}{2\pi r}\sin(\theta-\theta^-)\mathbf{N}_3(\theta)\mathbf{m}(\theta^+). \tag{8.13-7}$$

From the identities $(8.1\text{-}7)_4$ and $(8.1\text{-}5)_2$,

$$\mathbf{n}(\theta)\cdot\boldsymbol{\phi} = \frac{k}{2\pi r}\mathbf{m}^T(\theta^-)\mathbf{N}_3(\theta)\mathbf{m}(\theta^+), \quad \mathbf{m}(\theta)\cdot\boldsymbol{\phi} = 0, \tag{8.13-8}$$

so that, by (8.12-19),

$$\sigma_{r\theta} = \frac{-k}{2\pi r^2}\mathbf{m}^T(\theta^-)\mathbf{N}_3(\theta)\mathbf{m}(\theta^+),$$
$$\sigma_{\theta\theta} = 0, \quad \sigma_{rr} = \sigma_{r\theta,\theta}. \tag{8.13-9}$$

Equation (8.13-4) now yields

$$k = 2T/\Delta \tag{8.13-10}$$

where

$$\Delta = -\mathbf{m}^T(\theta^-)\mathbf{L}(\theta^+,\theta^-)\mathbf{m}(\theta^+), \tag{8.13-11}$$

$$\mathbf{L}(\theta^+,\theta^-) = -\frac{1}{\pi}\int_{\theta^-}^{\theta^+}\mathbf{N}_3(\lambda)d\lambda = \mathbf{L}(\theta^+) - \mathbf{L}(\theta^-). \tag{8.13-12}$$

Since $-\mathbf{N}_3(\theta)$ is symmetric and positive semidefinite, it can be shown that $\mathbf{L}(\theta^+,\theta^-)$ is symmetric and positive definite (Section 7.6). It is interesting to note from $(8.13\text{-}7)_1$ and the identities (8.1-6) and $(8.1\text{-}5)_1$ that

$$2\pi r\mathbf{u} = \begin{cases} -k\,\mathbf{m}(\theta^-) & \text{at } \theta = \theta^+, \\ -k\,\mathbf{m}(\theta^+) & \text{at } \theta = \theta^-. \end{cases} \tag{8.13-13}$$

The displacement at the surface $\theta = \theta^+$ is normal to the surface $\theta = \theta^-$, and vice versa.

The following identities are easily established.

$$\mathbf{n}(\theta^\pm) = \cos\alpha\,\mathbf{n}(\theta_c) \pm \sin\alpha\,\mathbf{m}(\theta_c),$$
$$\mathbf{m}(\theta^\pm) = \cos\alpha\,\mathbf{m}(\theta_c) \mp \sin\alpha\,\mathbf{n}(\theta_c). \tag{8.13-14}$$

With $(8.13\text{-}14)_2$ and noting that $\mathbf{L}(\theta^+,\theta^-)$ is symmetric, (8.13-11) can be written as

$$\Delta = \sin^2\alpha\, L_{11}^*(\theta^+,\theta^-) - \cos^2\alpha\, L_{22}^*(\theta^+,\theta^-), \tag{8.13-15}$$

where

$$L_{11}^*(\theta^+,\theta^-) = \mathbf{n}^T(\theta_c)\mathbf{L}(\theta^+,\theta^-)\mathbf{n}(\theta_c),$$
$$L_{22}^*(\theta^+,\theta^-) = \mathbf{m}^T(\theta_c)\mathbf{L}(\theta^+,\theta^-)\mathbf{m}(\theta_c).$$

$L_{ij}^*(\theta^+,\theta^-)$ are the components of $\mathbf{L}(\theta^+,\theta^-)$ referred to a coordinate system x_i^* obtained by rotating the x_i coordinate system about the x_3-axis counterclockwise an angle θ_c. Due to the positive definiteness of $\mathbf{L}(\theta^+,\theta^-)$, $L_{11}^*(\theta^+,\theta^-)$ and $L_{22}^*(\theta^+,\theta^-)$ are positive. If the wedge occupies the region $-\alpha \le \theta \le \alpha$, (8.13-15) is simplified to

$$\Delta = \sin^2\alpha\, L_{11}(\alpha,-\alpha) - \cos^2\alpha\, L_{22}(\alpha,-\alpha). \tag{8.13-16}$$

For isotropic materials (8.13-15) and (8.13-16) both reduce to

$$\Delta = \eta(\sin 2\alpha - 2\alpha\cos 2\alpha),$$

where η is a material constant. Using the same technique the stress $\sigma_{r\theta}$ given in (8.13-9) can be rewritten in a simple form.

Equation (8.13-10) for k is valid provided $\Delta \neq 0$. The wedge angle $2\alpha = 2\alpha_*$ that satisfies $\Delta = 0$ is a *critical angle* of the wedge. For isotropic materials $2\alpha_*$ is given by

$$\sin 2\alpha_* - 2\alpha_*\cos 2\alpha_* = 0, \quad (\text{or } 2\alpha_* \cong 257.4534 \text{ degrees}),$$

and is independent of θ_c. For anisotropic materials $2\alpha_*$ in general depends on θ_c. Regardless of θ_c we will show that

$$\pi < 2\alpha_* < 2\pi. \tag{8.13-17}$$

Differentiation of (8.13-11) with α using (8.13-2) and $(8.1\text{-}5)_2$ yields

$$\begin{aligned}\frac{\partial\Delta}{\partial\alpha} &= \mathbf{m}^T(\theta^-)\mathbf{L}(\theta^+,\theta^-)\mathbf{n}(\theta^+) - \mathbf{n}^T(\theta^-)\mathbf{L}(\theta^+,\theta^-)\mathbf{m}(\theta^+) \\ &= \sin 2\alpha\left\{L^*_{11}(\theta^+,\theta^-) + L^*_{22}(\theta^+,\theta^-)\right\}.\end{aligned} \tag{8.13-18}$$

The second equality is reached by employing (8.13-14) and noting that $\mathbf{L}(\theta^+,\theta^-)$ is symmetric. Thus Δ is a monotonically increasing function of α for $0<2\alpha<\pi$ and a decreasing function of α for $\pi<2\alpha<2\pi$. In conjunction with the fact deduced from (8.13-15) that $\Delta=0$, >0, and <0 at $2\alpha=0$, π, and 2π, respectively, established (8.13-17). We could get the following sharper results.

$$\begin{aligned}&\pi < 2\alpha_* < \tfrac{3}{2}\pi, \text{ if } L^*_{22}(\theta_c+\tfrac{3}{4}\pi,\ \theta_c-\tfrac{3}{4}\pi) > L^*_{11}(\theta_c+\tfrac{3}{4}\pi,\ \theta_c-\tfrac{3}{4}\pi),\\ &\tfrac{3}{2}\pi < 2\alpha_* < 2\pi, \text{ if } L^*_{22}(\theta_c+\tfrac{3}{4}\pi,\ \theta_c-\tfrac{3}{4}\pi) < L^*_{11}(\theta_c+\tfrac{3}{4}\pi,\ \theta_c-\tfrac{3}{4}\pi).\end{aligned} \tag{8.13-19}$$

At the critical wedge angle $2\alpha_*$, k becomes infinity so that the stresses given by (8.13-9) are infinite everywhere. This is the classical *paradox* that has been discussed extensively by Sternberg and Koiter (1958) for isotropic materials. The pathological behavior remains when the nonlinearity of the elasticity is taken into consideration (Klingbeil and Shield, 1966). The same pathological phenomenon also occurs in the Jeffery–Hammel viscous flow in which the fluid is injected into a wedge region from the apex. The solution for the flow becomes infinity everywhere when the wedge angle is $2\alpha_*$ (Moffatt and Duffy, 1980). A solution for the critical wedge angle $2\alpha_*$ that remains bounded everywhere except at the wedge apex has been obtained by Ting (1985a) for isotropic materials and by Hwu and Ting (1990) for anisotropic materials. The same problem for a rotationally inhomogeneous wedge has been studied by Belov and Kirchner (1995). Hansen (1995) discovered that the same critical angle exists for the three-dimensional problem of a wedge cylinder subjected to a concentrated couple at a point on the line of the wedge apex.

Finally we discuss the case $2\alpha=\pi$ and 2π. According to (8.13-6) **h** is arbitrary and the solution is given (8.13-5). Since the vector **h** has three components, (8.13-5) corresponds to three independent solutions. As we will see in Section 9.5 they are eigenfunctions associated with the eigenvalue $\delta=-2$ of multiplicity three. For general anisotropic materials all three eigenfunctions produce a nonzero couple about the x_3-axis and hence are admissible solutions. For isotropic materials only one of them produces a nonzero couple. The other two can therefore be ignored. It should be pointed out that even for $2\alpha\neq\pi$, 2π, and $2\alpha_*$, the solutions presented here are not unique. One can always add eigenfunctions that leave the sides $\theta=\theta^-, \theta^+$ traction free.

Chapter 9

PARTICULAR SOLUTIONS, STRESS SINGULARITIES, AND STRESS DECAY

Not all boundary value problems are amenable to a simple analytical solution. This is particularly the case when the geometry of the boundary on which the boundary conditions are specified contains a corner. A crack tip is a special corner. When an analytical solution for the entire region is not available, asymptotic solutions near the corner can be obtained which provide useful information on the nature of stress singularities at the corner. They also provide more accurate numerical solutions by a finite element scheme in which the asymptotic solution at the corner is employed in a special element at the corner with regular elements elsewhere (see, for example, Stolarski and Chiang, 1989). Other information that can be obtained by an asymptotic analysis is the decay factor of stress at a large distance from a point at which a self-equilibrated load is applied (Crafter, et al., 1993). For singularities that arise in non-linear elastic materials the reader is referred to the book by Antman (1995). As in Chapter 8 most solutions can be expressed in a real form with the aid of identities presented in Chapters 6 and 7.

9.1 Wedge Subjected to Tractions

Consider a two-dimensional body with a hole of arbitrary shape subjected to prescribed tractions at the surface of the hole, Fig. 9.1. Let the point O be a corner of the hole. To study the stress near the corner O we take O as the origin of a rectangular coordinate system. Asymptotically, the points near the origin are in a wedge of infinite extent bounded by

$$\theta^- \le \theta \le \theta^+, \quad r \ge 0, \tag{9.1-1}$$

$$\theta^- = \theta_c - \alpha, \quad \theta^+ = \theta_c + \alpha. \tag{9.1-2}$$

The wedge angle is $2\alpha = \theta^+ - \theta^-$ and $\theta = \theta_c$ is the center line bisecting the wedge into two equal angles (see Fig. 8.13). If the traction vector on the hole surface can be expanded into a Taylor series about the point O, the first term is a constant vector. We therefore consider the boundary conditions

$$\mathbf{t}_\theta = \phi_{,r} = -\mathbf{t}^- \quad \text{at } \theta = \theta^-,$$
$$\mathbf{t}_\theta = \phi_{,r} = \mathbf{t}^+ \quad \text{at } \theta = \theta^+, \tag{9.1-3}$$

where $\mathbf{t}^-$ and $\mathbf{t}^+$ are the constant traction vectors applied at the wedge surfaces $\theta = \theta^-$ and θ^+, respectively. The classical solution of the problem for isotropic materials was obtained by Levy (1899).

In the literature the coordinate system for the wedge of wedge angle 2α is often chosen such that $\theta^+ = \pi + \alpha$ and $\theta^- = \pi - \alpha$. This makes the geometry of the wedge symmetric with the x_1-axis. It is convenient and desirable when the material is isotropic. For anisotropic materials the coordinate system so chosen may not coincide with a symmetry plane of the material when it possesses a symmetry plane. If a coordinate system x_i is chosen to coincide with a symmetry plane of the material, the geometry of the wedge may not be symmetric with the x_1-axis. This is where the dual coordinate systems and the identities presented in Section 7.10 become useful. We will choose the coordinate system x_i to coincide with a symmetry plane of the material when such a plane exists. If not, the x_i coordinate system can be arbitrary. The angle θ_o in Section 7.10 will be taken to be θ^-, not θ_c. The wedge is not symmetric with the second coordinate system $\hat{x}_i$, but the choice of $\theta = \theta^-$ will simplify the application of the boundary condition at $\theta = \theta^-$. This is more general than the treatment of the wedge problem by Hwu and Ting (1990). We will see that the analysis is simpler for the general geometry when the dual coordinate systems are employed.

Equations (8.1-1)–(8.1-12) will be employed in this chapter. We

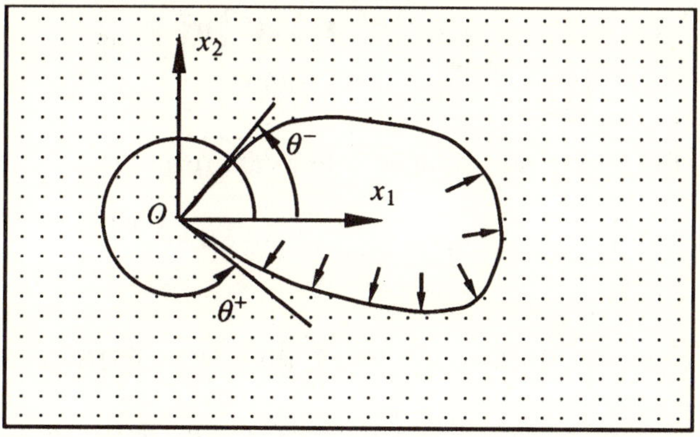

Fig. 9.1 The geometry of the hole near the corner O can be approximated by a wedge of infinite extent.

also need the identities

$$\begin{aligned}&\left\{\cos(\theta-\theta^-)\mathbf{I}+\sin(\theta-\theta^-)\mathbf{N}_1(\theta^-)\right\}\mathbf{n}(\theta^-)=-\mathbf{N}_1(\theta^-)\mathbf{m}(\theta),\\&\left\{\cos(\theta-\theta^-)\mathbf{I}+\sin(\theta-\theta^-)\mathbf{N}_1(\theta^-)\right\}\mathbf{m}(\theta^-)=\mathbf{m}(\theta).\end{aligned}\tag{9.1-4}$$

Equation $(9.1\text{-}4)_1$ is equivalent to $(8.1\text{-}7)_2$ while $(9.1\text{-}4)_2$ is obtained from $(8.1\text{-}5)_1$ and $(8.1\text{-}6)_2$. When $\theta=\theta^+$ $(9.1\text{-}4)_2$ is reduced to

$$\left\{\cos 2\alpha\mathbf{I}+\sin 2\alpha\mathbf{N}_1(\theta^-)\right\}\mathbf{m}(\theta^-)=\mathbf{m}(\theta^+)\tag{9.1-5}$$

or, by $(8.1\text{-}5)_1$,

$$\cos 2\alpha\,\mathbf{m}(\theta^-)-\sin 2\alpha\,\mathbf{n}(\theta^-)=\mathbf{m}(\theta^+).\tag{9.1-6}$$

Since the boundary conditions (9.1-3) are constant let us try a uniform stress solution by choosing $f(\hat{z})=\hat{z}$ in the general solution (7.10-11) and write

$$\mathbf{u}=2\mathrm{Re}\left\{\mathbf{A}\langle\hat{z}_*\rangle\mathbf{q}\right\},\quad \boldsymbol{\phi}=2\mathrm{Re}\left\{\mathbf{B}\langle\hat{z}_*\rangle\mathbf{q}\right\},\tag{9.1-7}$$

where

$$\hat{z}=\hat{x}_1+p(\theta^-)\hat{x}_2,\tag{9.1-8}$$

$$\begin{aligned}\hat{x}_1&=\mathbf{x}\cdot\mathbf{n}(\theta^-)=r\cos(\theta-\theta^-),\\\hat{x}_2&=\mathbf{x}\cdot\mathbf{m}(\theta^-)=r\sin(\theta-\theta^-).\end{aligned}\tag{9.1-9}$$

Setting

$$\mathbf{q}=\mathbf{A}^T\mathbf{g}+\mathbf{B}^T\mathbf{h}$$

in which $\mathbf{g}$ and $\mathbf{h}$ are real constant vectors, (9.1-7) can be written as

$$\begin{bmatrix}\mathbf{u}\\\boldsymbol{\phi}\end{bmatrix}=2\mathrm{Re}\begin{bmatrix}\mathbf{A}\langle\hat{z}_*\rangle\mathbf{B}^T & \mathbf{A}\langle\hat{z}_*\rangle\mathbf{A}^T\\\mathbf{B}\langle\hat{z}_*\rangle\mathbf{B}^T & \mathbf{B}\langle\hat{z}_*\rangle\mathbf{A}^T\end{bmatrix}\begin{bmatrix}\mathbf{h}\\\mathbf{g}\end{bmatrix}.\tag{9.1-10}$$

By (7.10-15) this is transformed into a real form as

$$\begin{bmatrix}\mathbf{u}\\\boldsymbol{\phi}\end{bmatrix}=r\left\{\cos(\theta-\theta^-)\mathbf{I}+\sin(\theta-\theta^-)\mathbf{N}(\theta^-)\right\}\begin{bmatrix}\mathbf{h}\\\mathbf{g}\end{bmatrix}\tag{9.1-11a}$$

or

$$\begin{aligned}\mathbf{u}/r&=\left\{\cos(\theta-\theta^-)\mathbf{I}+\sin(\theta-\theta^-)\mathbf{N}_1(\theta^-)\right\}\mathbf{h}+\sin(\theta-\theta^-)\mathbf{N}_2(\theta^-)\mathbf{g},\\\boldsymbol{\phi}/r&=\sin(\theta-\theta^-)\mathbf{N}_3(\theta^-)\mathbf{h}+\left\{\cos(\theta-\theta^-)\mathbf{I}+\sin(\theta-\theta^-)\mathbf{N}_1^T(\theta^-)\right\}\mathbf{g}.\end{aligned}\tag{9.1-11b}$$

Satisfaction of the boundary conditions (9.1-3) means that

$$\mathbf{g} = -\mathbf{t}^-,$$
$$\sin 2\alpha \mathbf{N}_3(\theta^-)\mathbf{h} = \mathbf{t}^+ + \left\{\cos 2\alpha \mathbf{I} + \sin 2\alpha \mathbf{N}_1^T(\theta^-)\right\}\mathbf{t}^-. \tag{9.1-12}$$

Thus $-\mathbf{g}$ represents the traction $\mathbf{t}^-$. By $(9.1\text{-}11b)_1$, $\mathbf{h}=\mathbf{u}/r$ at $\theta=\theta^-$.

There are two problems with $(9.1\text{-}12)_2$. If $2\alpha\neq\pi$ or 2π, multiplication of $(9.1\text{-}12)_2$ by $\mathbf{m}^T(\theta^-)$ and making use of $(8.1\text{-}5)_2$ and (9.1-5) yields

$$\mathbf{0} = \mathbf{m}^T(\theta^-)\mathbf{t}^+ + \mathbf{m}^T(\theta^+)\mathbf{t}^-. \tag{9.1-13}$$

When (9.1-13) is satisfied, $(9.1\text{-}12)_2$ can be solved for $\mathbf{h}$ but it is not unique. If $\mathbf{h}$ is a solution so is $\mathbf{h}+k\mathbf{m}(\theta^-)$ where k is an arbitrary constant. However, $\mathbf{h}=k\mathbf{m}(\theta^-)$ contributes nothing to $\boldsymbol{\phi}$, and hence to stresses. It represents a rigid body motion and can be ignored. Omitting $k\mathbf{m}(\theta^-)$, a solution for $\mathbf{h}$ exists because $-\mathbf{N}_3(\theta^-)$ is positive semidefinite (Section 7.5). In particular, when $\theta^-=0$ the *pseudo inverse* of $-\mathbf{N}_3$ is $\mathbf{q}_2$ according to (6.1-15) where $\mathbf{q}_2$ is defined in (6.1-9). For the stress function $\boldsymbol{\phi}$ one can eliminate $\mathbf{N}_3(\theta^-)\mathbf{h}$ between $(9.1\text{-}11b)_2$ and $(9.1\text{-}12)_2$ to obtain

$$\boldsymbol{\phi}/r = \frac{1}{\sin 2\alpha}\left\{\sin(\theta-\theta^-)\mathbf{t}^+ + \sin(\theta-\theta^+)\mathbf{t}^-\right\}. \tag{9.1-14}$$

This is a uniform stress solution when $2\alpha\neq\pi$, 2π, and (9.1-13) holds. It is independent of material property.

If the wedge angle $2\alpha=\pi$ or 2π, $\sin 2\alpha=0$ so that the right side of $(9.1\text{-}12)_2$ must vanish, i.e.,

$$\mathbf{t}^+ = \mathbf{t}^-, \quad \text{if } 2\alpha=\pi,$$
$$\mathbf{t}^+ = -\mathbf{t}^-, \quad \text{if } 2\alpha=2\pi. \tag{9.1-15}$$

When (9.1-15) holds $\mathbf{h}$ in $(9.1\text{-}12)_2$ is arbitrary. But the $\mathbf{h}$ term in $(9.1\text{-}11b)_2$ vanishes at $\theta=\theta^-$ and θ^+, where $\theta^+=\theta^-+\pi$ or $\theta^-+2\pi$. Therefore it contributes no tractions at $\theta=\theta^-$ and θ^+. It is an eigenfunction for wedge angle $2\alpha=\pi$ or 2π (Section 9.5) which can always be added to a particular solution without changing the boundary conditions. Setting $\mathbf{h}=\mathbf{0}$ in $(9.1\text{-}11b)_2$,

$$\boldsymbol{\phi}/r = -\left\{\cos(\theta-\theta^-)\mathbf{I} + \sin(\theta-\theta^-)\mathbf{N}_1^T(\theta^-)\right\}\mathbf{t}^- \tag{9.1-16}$$

is a uniform stress solution when $2\alpha=\pi$ or 2π and (9.1-15) holds. The second term depends on material property, but it vanishes at $\theta=\theta^-$ and θ^+. However we cannot simply omit the second term and keep the first term because the second term by itself is not an eigenfunction. It can be shown that

$$\phi / r = -\{\cos(\theta - \theta^-)\mathbf{I} - \sin(\theta - \theta^-)\mathbf{m}(\theta^-)\mathbf{n}^T(\theta^-)\}\mathbf{t}^- \qquad (9.1\text{-}17)$$

is a uniform stress solution when $2\alpha=\pi$ or 2π and (9.1-15) holds. The difference between (9.1-16) and (9.1-17) represents an eigenfunction for $2\alpha=\pi$ or 2π.

The stress is not uniform when $2\alpha\neq\pi$ or 2π and (9.1-13) is violated, or when $2\alpha=\pi$ or 2π and (9.1-15) is violated. This is studied in the next section.

9.2 Nonuniform Stress Solutions in a Wedge

In this section we assume that $2\alpha\neq\pi$ or 2π and (9.1-13) does not hold or, if $2\alpha=\pi$ or 2π , (9.1-13) does not hold. The reason (9.1-11b) fails to produce solutions for arbitrarily prescribed tractions $\mathbf{t}^+$ and $\mathbf{t}^-$ is that the arbitrary constant vector $\mathbf{h}$ that appeared in the equation does not really consist of three arbitrary components. If $\mathbf{h}$ is referred to the bases $\mathbf{n}(\theta^-)$, $\mathbf{m}(\theta^-)$, and $\mathbf{i}_3$ where $\mathbf{i}_3$ is a unit vector in the x_3-direction it was pointed out in the previous section that the $\mathbf{m}(\theta^-)$ component of $\mathbf{h}$ is redundant because it contributes only to a rigid body motion. Therefore it is necessary to restore the function of the $\mathbf{m}(\theta^-)$ component of $\mathbf{h}$.

Following Hwu and Ting (1990) we replace $\hat{z}$ in (9.1-10) by

$$\hat{z}^{1+\delta} = \hat{z} + \delta\, \hat{z}(\ln \hat{z}) + \tfrac{1}{2}\delta^2\, \hat{z}(\ln \hat{z})^2 + \cdots \qquad (9.2\text{-}1)$$

where δ is an arbitrary constant and $\mathbf{g}$ and $\mathbf{h}$ in (9.1-10) are allowed to depend on δ. Let

$$\begin{aligned} \mathbf{g}(\delta) &= \mathbf{g}^{(0)} + \delta\mathbf{g}^{(1)} + \cdots, \\ \mathbf{h}(\delta) &= \frac{k}{\delta}\mathbf{m}(\theta^-) + \mathbf{h}^{(0)} + \delta\mathbf{h}^{(1)} + \cdots, \end{aligned} \qquad (9.2\text{-}2)$$

in which k is an arbitrary constant and $\mathbf{g}^{(0)}$, $\mathbf{g}^{(1)}$, $\mathbf{h}^{(0)}$, $\mathbf{h}^{(1)}$ are arbitrary constant vectors. Making use of identities (7.10-28), the modified (9.1-10) is written in a real form as

$$\begin{bmatrix} \mathbf{u} \\ \phi \end{bmatrix} = \frac{k}{\delta}\begin{bmatrix} \mathbf{u}_R \\ \phi_R \end{bmatrix} + \begin{bmatrix} \mathbf{u}^{(0)} \\ \phi^{(0)} \end{bmatrix} + \delta\begin{bmatrix} \mathbf{u}^{(1)} \\ \phi^{(1)} \end{bmatrix} + \cdots,$$

where

$$\begin{bmatrix} \mathbf{u}_R \\ \phi_R \end{bmatrix} = r\mathbf{N}(\theta,\theta^-)\begin{bmatrix} \mathbf{m}(\theta^-) \\ \mathbf{0} \end{bmatrix} = r\begin{bmatrix} \mathbf{m}(\theta) \\ \mathbf{0} \end{bmatrix}, \qquad (9.2\text{-}3)$$

$$\begin{bmatrix} \mathbf{u}^{(0)} \\ \boldsymbol{\phi}^{(0)} \end{bmatrix} = r\mathbf{N}(\theta,\theta^-)\begin{bmatrix} \mathbf{h}^{(0)} \\ \mathbf{g}^{(0)} \end{bmatrix} + rk\left\{(\ln r)\mathbf{I} + \pi\tilde{\mathbf{N}}(\theta,\theta^-)\right\}\begin{bmatrix} \mathbf{m}(\theta) \\ \mathbf{0} \end{bmatrix}, \tag{9.2-4}$$

$$\begin{bmatrix} \mathbf{u}^{(1)} \\ \boldsymbol{\phi}^{(1)} \end{bmatrix} = r\mathbf{N}(\theta,\theta^-)\begin{bmatrix} \mathbf{h}^{(1)} \\ \mathbf{g}^{(1)} \end{bmatrix} + r\left\{(\ln r)\mathbf{I} + \pi\tilde{\mathbf{N}}(\theta,\theta^-)\right\}\begin{bmatrix} \mathbf{h}^{(0)} \\ \mathbf{g}^{(0)} \end{bmatrix}$$

$$+\tfrac{1}{2}rk\left\{(\ln r)\mathbf{I} + \pi\tilde{\mathbf{N}}(\theta,\theta^-)\right\}^2\begin{bmatrix} \mathbf{m}(\theta) \\ \mathbf{0} \end{bmatrix}. \tag{9.2-5}$$

In the above $\mathbf{N}(\theta,\theta^-)$ and $\tilde{\mathbf{N}}(\theta,\theta^-)$ are both 6×6 matrices defined in (7.5-21) and (7.10-24), respectively. The matrix $\mathbf{N}(\theta,\theta^-)$ is

$$\mathbf{N}(\theta,\theta^-) = \cos(\theta-\theta^-)\mathbf{I} + \sin(\theta-\theta^-)\mathbf{N}(\theta^-). \tag{9.2-6}$$

If the four 3×3 submatrices are defined as

$$\mathbf{N}(\theta,\theta^-) = \begin{bmatrix} \mathbf{N}_1(\theta,\theta^-) & \mathbf{N}_2(\theta,\theta^-) \\ \mathbf{N}_3(\theta,\theta^-) & \mathbf{N}_1^T(\theta,\theta^-) \end{bmatrix} \tag{9.2-7a}$$

we have

$$\begin{aligned} &\mathbf{N}_1(\theta,\theta^-) = \cos(\theta-\theta^-)\mathbf{I} + \sin(\theta-\theta^-)\mathbf{N}_1(\theta^-), \\ &\mathbf{N}_2(\theta,\theta^-) = \sin(\theta-\theta^-)\mathbf{N}_2(\theta^-), \\ &\mathbf{N}_3(\theta,\theta^-) = \sin(\theta-\theta^-)\mathbf{N}_3(\theta^-). \end{aligned} \tag{9.2-7b}$$

There is not much saving in employing the notations $\mathbf{N}_2(\theta,\theta^-)$ and $\mathbf{N}_3(\theta,\theta^-)$ so that we will not use them. We will employ $\mathbf{N}_1(\theta,\theta^-)$ for simplicity. Equation $(9.2\text{-}3)_2$, which can be verified by $(9.1\text{-}4)_2$ and $(8.1\text{-}5)_2$, has been used in (9.2-4) and (9.2-5). We have also used the fact that products of $\tilde{\mathbf{N}}(\theta,\theta^-)$ and $\mathbf{N}(\theta,\theta^-)$ commute (Section 7.8). The 6×6 matrix $\tilde{\mathbf{N}}(\theta,\theta^-)$ is

$$\tilde{\mathbf{N}}(\theta,\theta^-) = \tilde{\mathbf{N}}(\theta) - \tilde{\mathbf{N}}(\theta^-) = \frac{1}{\pi}\int_{\theta^-}^{\theta} \mathbf{N}(\lambda)d\lambda. \tag{9.2-8}$$

Let the four 3×3 submatrices be denoted by

$$\tilde{\mathbf{N}}(\theta,\theta^-) = \begin{bmatrix} \mathbf{S}(\theta,\theta^-) & \mathbf{H}(\theta,\theta^-) \\ -\mathbf{L}(\theta,\theta^-) & \mathbf{S}^T(\theta,\theta^-) \end{bmatrix}. \tag{9.2-9a}$$

We then have

$$\mathbf{S}(\theta,\theta^-) = \mathbf{S}(\theta) - \mathbf{S}(\theta^-) = \frac{1}{\pi}\int_{\theta^-}^{\theta} \mathbf{N}_1(\lambda)d\lambda, \tag{9.2-9b$_1$}$$

$$\mathbf{H}(\theta,\theta^-)=\mathbf{H}(\theta)-\mathbf{H}(\theta^-)=\frac{1}{\pi}\int_{\theta^-}^{\theta}\mathbf{N}_2(\lambda)d\lambda\,, \qquad (9.2\text{-}9b)_2$$

$$\mathbf{L}(\theta,\theta^-)=\mathbf{L}(\theta)-\mathbf{L}(\theta^-)=-\frac{1}{\pi}\int_{\theta^-}^{\theta}\mathbf{N}_3(\lambda)d\lambda\,. \qquad (9.2\text{-}9b)_3$$

Equation (9.2-3), (9.2-4), or (9.2-5) taken individually is an admissible solution for an anisotropic elastic body. As we pointed out in the previous section the solution given by (9.2-3) represents a rigid body motion. The solution given by (9.2-4) is identical to (9.1-11a) for uniform stress solution except the k term. Written explicitly (9.2-4) is

$$\mathbf{u}/r=\mathbf{N}_1(\theta,\theta^-)\mathbf{h}^{(0)}-\sin(\theta-\theta^-)\mathbf{N}_2(\theta^-)\mathbf{t}^-+k\{(\ln r)\mathbf{I}+\pi\mathbf{S}(\theta,\theta^-)\}\mathbf{m}(\theta), \qquad (9.2\text{-}10)$$

$$\boldsymbol{\phi}/r=\sin(\theta-\theta^-)\mathbf{N}_3(\theta^-)\mathbf{h}^{(0)}-\mathbf{N}_1^T(\theta,\theta^-)\mathbf{t}^- -k\pi\mathbf{L}(\theta,\theta^-)\mathbf{m}(\theta),$$

in which we have set $\mathbf{g}^{(0)}=-\mathbf{t}^-$ so that the boundary condition $(9.1\text{-}3)_1$ at $\theta=\theta^-$ is satisfied. At $\theta=\theta^-$ and as $r\to 0$, $\mathbf{u}\to r\mathbf{h}^{(0)}$. It should be noted that

$$\mathbf{N}_1(\theta^+,\theta^-)=\cos 2\alpha\,\mathbf{I}+\sin 2\alpha\,\mathbf{N}_1(\theta^-) \qquad (9.2\text{-}11)$$

so that, by (9.1-5),

$$\mathbf{N}_1(\theta^+,\theta^-)\mathbf{m}(\theta^-)=\mathbf{m}(\theta^+). \qquad (9.2\text{-}12)$$

Satisfaction of the boundary condition $(9.1\text{-}3)_2$ at $\theta=\theta^+$ yields

$$\sin 2\alpha\,\mathbf{N}_3(\theta^-)\mathbf{h}^{(0)}=\mathbf{t}^+ +\mathbf{N}_1^T(\theta^+,\theta^-)\mathbf{t}^- +k\pi\mathbf{L}(\theta^+,\theta^-)\mathbf{m}(\theta^+). \qquad (9.2\text{-}13)$$

This is identical to $(9.1\text{-}12)_2$ except the k term. If (9.2-13) is multiplied by $\mathbf{m}^T(\theta^-)$ and use is made of (9.2-12) we obtain

$$k\pi\Delta=\mathbf{m}^T(\theta^-)\mathbf{t}^+ +\mathbf{m}^T(\theta^+)\mathbf{t}^- \qquad (9.2\text{-}14)$$

where

$$\Delta=-\mathbf{m}^T(\theta^-)\mathbf{L}(\theta^+,\theta^-)\mathbf{m}(\theta^+), \qquad (9.2\text{-}15)$$

$$\mathbf{L}(\theta^+,\theta^-)=-\frac{1}{\pi}\int_{\theta^-}^{\theta^+}\mathbf{N}_3(\lambda)d\lambda=\mathbf{L}(\theta^+)-\mathbf{L}(\theta^-). \qquad (9.2\text{-}16)$$

Assuming that $\Delta\neq 0$ and that $2\alpha\neq\pi$ or 2π, k and $\mathbf{h}^{(0)}$ are obtained from (9.2-14) and (9.2-13). $\mathbf{h}^{(0)}$ determined from (9.2-13) is unique up to an additional term that is a multiple of $\mathbf{m}(\theta^-)$. This term contributes nothing to the stress function $\boldsymbol{\phi}$, and hence to stresses so that it can be ignored. If we substitute $\mathbf{N}_3(\theta^-)\mathbf{h}^{(0)}$ obtained from (9.2-13) into $(9.2\text{-}10)_2$ we have

$$\phi / r = \frac{1}{\sin 2\alpha}\Big\{\sin(\theta - \theta^-)\mathbf{t}^+ + \sin(\theta - \theta^+)\mathbf{t}^-$$

$$+ k\pi[\sin(\theta - \theta^-)\mathbf{L}(\theta^+, \theta^-)\mathbf{m}(\theta^+) - \sin 2\alpha\, \mathbf{L}(\theta, \theta^-)\mathbf{m}(\theta)]\Big\} \quad (9.2\text{-}17)$$

where k is given by (9.2-14).

If $2\alpha = \pi$ or 2π, $\mathbf{m}(\theta^+) = \cos 2\alpha \mathbf{m}(\theta^-)$ and (9.2-13) becomes

$$-k\pi \cos 2\alpha\, \mathbf{L}(\theta^+, \theta^-)\mathbf{m}(\theta^-) = \mathbf{t}^+ + \cos 2\alpha\, \mathbf{t}^-. \quad (9.2\text{-}18)$$

Noting that

$$\mathbf{L}(\theta^+, \theta^-) = \begin{cases} \mathbf{L} & \text{if} \quad 2\alpha = \pi, \\ 2\mathbf{L} & \text{if} \quad 2\alpha = 2\pi, \end{cases}$$

(9.2-18) yields

$$\begin{aligned} k\pi \mathbf{L}\mathbf{m}(\theta^-) &= \mathbf{t}^+ - \mathbf{t}^-, \quad \text{if} \quad 2\alpha = \pi, \\ -2k\pi \mathbf{L}\mathbf{m}(\theta^-) &= \mathbf{t}^+ + \mathbf{t}^-, \quad \text{if} \quad 2\alpha = 2\pi. \end{aligned} \quad (9.2\text{-}19)$$

Equation (9.2-19) imposes the condition that $\mathbf{t}^+ - \mathbf{t}^-$ be co-directional with $\mathbf{Lm}(\theta^-)$ if $2\alpha = \pi$, and $\mathbf{t}^+ + \mathbf{t}^-$ be co-directional with $\mathbf{Lm}(\theta^-)$ if $2\alpha = 2\pi$. When the condition is satisfied, k is uniquely determined from (9.2-19). The solution (9.2-10) then applies to wedge angles $2\alpha = \pi$ or 2π in which $\mathbf{h}^{(0)}$ is arbitrary. However the $\mathbf{h}^{(0)}$ term provides $\phi = \mathbf{0}$ at $\theta = \theta^-$, θ^+ so that the boundary conditions are not affected. It represents an eigenfunction for $2\alpha = \pi$ or 2π that can always be added to the wedge solution without altering the boundary conditions.

It can be shown that the stress derived from $(9.2\text{-}10)_2$, and in particular from (9.2-17), depends on θ only, not on r. The k terms in (9.2-10) with $\theta^- = 0$,

$$\begin{aligned} \mathbf{u} / r &= \left\{\tfrac{1}{\pi}(\ln r)\mathbf{I} + \mathbf{S}(\theta)\right\}\mathbf{m}(\theta), \\ \phi / r &= -\mathbf{L}(\theta)\mathbf{m}(\theta), \end{aligned} \quad (9.2\text{-}20)$$

are an admissible elasticity solution. Using (8.11-3) the stress obtained by inserting $(9.2\text{-}20)_2$ into (8.1-2) can be shown to be

$$\sigma_{i1} = L_{i1}(\theta), \qquad \sigma_{i2} = L_{i2}(\theta),$$

or

$$\sigma_{ij} = L_{ij}(\theta), \quad (ij) \neq (33). \quad (9.2\text{-}21)$$

Thus with the exception of the σ_{33} component, $\sigma_{ij} = L_{ij}(\theta)$ represents a stress state in elastic materials (Ting, 1988c). This provides a physical interpretation of the real tensor $\mathbf{L}(\theta)$.

We have excluded in this section wedges with wedge angles $2\alpha=2\alpha_*$ which is a root of $\Delta=0$ and $2\alpha=\pi$ or 2π unless (9.2-19) is satisfied. These are the *critical angles* of the elastic wedge subjected to uniform tractions. Equation (9.2-15) is identical to (8.13-11) for a wedge subjected to a concentrated couple. It was shown in (8.13-15) that Δ has the alternate expression

$$\Delta = \sin^2\alpha\, L^*_{11}(\theta^+,\theta^-) - \cos^2\alpha\, L^*_{22}(\theta^+,\theta^-), \tag{9.2-22}$$

where $L^*_{ij}(\theta^+,\theta^-)$ are the components of $\mathbf{L}(\theta^+,\theta^-)$ referred to a coordinate system x^*_i obtained by rotating the x_i coordinate system about the x_3-axis counterclockwise an angle θ_c. From (9.1-2) the dependence of $\mathbf{L}(\theta^+,\theta^-)$ on θ^+ and θ^- is equivalent to dependence on α and θ_c. Thus the critical angle $2\alpha_*$ which is a root of $\Delta=0$ depends on θ_c. Regardless of the value of θ_c it was shown in (8.13-17) that

$$\pi < 2\alpha_* < 2\pi. \tag{9.2-23}$$

Sharper inequalities are presented in (8.13-19). For isotropic materials $2\alpha_*$ satisfies the equation

$$\sin 2\alpha_* - 2\alpha_* \cos 2\alpha_* = 0, \quad (\text{or } 2\alpha_* \cong 257.4534 \text{ degrees}).$$

As an illustration of the solution presented here consider a wedge of wedge angle $2\alpha_*=\pi/2$ with $\theta^-=0$ and $\theta^+=\pi/2$ subjected to a uniform traction $\mathbf{t}^+$ on the surface $\theta=\pi/2$, Fig. 9.2. This problem was first considered by Reissner (1944) for an isotropic wedge with the traction $\mathbf{t}^+=-\mathbf{m}(0)$, which is a shear loading of unit magni-

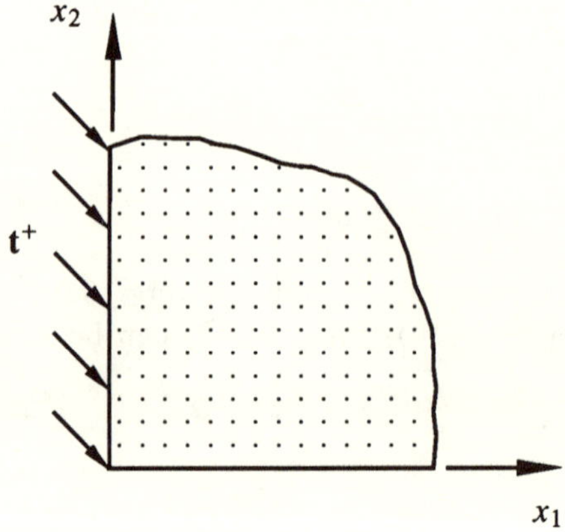

Fig. 9.2 The quarter space is subjected to a uniform traction on one side of the wedge and traction-free on the other side. The solution is shown in (9.2-25) in which k and $\mathbf{h}^{(o)}$ are given in $(9.2\text{-}24)_1$ and (9.2-26).

tude. We will consider general anisotropic materials with $\mathbf{t}^+$ being arbitrary. For this case

$$\mathbf{m}(\theta^-) = \mathbf{m}(0), \quad \mathbf{m}(\theta^+) = -\mathbf{n}(0), \quad \mathbf{t}^- = \mathbf{0},$$

$$\mathbf{L}(\theta^+,\theta^-) = -\frac{1}{\pi}\int_0^{\frac{1}{2}\pi} \mathbf{N}_3(\lambda)d\lambda = \mathbf{L}(\tfrac{\pi}{2}).$$

Equations (9.2-14) and (9.2-13) reduce to

$$\begin{aligned} k\pi L_{21}(\tfrac{\pi}{2}) &= \mathbf{m}^T(0)\mathbf{t}^+ = t_2^+, \\ \mathbf{N}_3\mathbf{h}^{(0)} &= \mathbf{t}^+ - k\pi\mathbf{L}(\tfrac{\pi}{2})\mathbf{n}(0), \end{aligned} \tag{9.2-24}$$

and the solution deduced from (9.2-10) is

$$\begin{aligned} \mathbf{u}/r &= (\cos\theta\mathbf{I} + \sin\theta\mathbf{N}_1)\mathbf{h}^{(0)} + k\{(\ln r)\mathbf{I} + \pi\mathbf{S}(\theta)\}\mathbf{m}(\theta), \\ \boldsymbol{\phi}/r &= \sin\theta\,\mathbf{t}^+ - k\pi\{\mathbf{L}(\theta)\mathbf{m}(\theta) + \sin\theta\,\mathbf{L}(\tfrac{\pi}{2})\mathbf{n}(0)\}. \end{aligned} \tag{9.2-25}$$

We have employed $(9.2\text{-}24)_2$ in writing $(9.2\text{-}25)_2$. The constant k in (9.2-25) is given by $(9.2\text{-}24)_1$. The stress is obtained from (8.1-12). It depends on θ only, not on r. Although the inverse of $\mathbf{N}_3$ does not exist we may employ the pseudo-inverse introduced in (6.1-15) to solve $\mathbf{h}^{(0)}$ in $(9.2\text{-}24)_2$. We have

$$\mathbf{h}^{(0)} = -\mathbf{q}_2\left\{\mathbf{t}^+ - k\pi\mathbf{L}(\tfrac{\pi}{2})\mathbf{n}(0)\right\} \tag{9.2-26}$$

where the 3×3 matrix $\mathbf{q}_2$ is defined in $(6.1\text{-}9)_1$. $\mathbf{h}^{(0)}$ given by (9.2-26) is unique up to an additional term that is a multiple of $\mathbf{m}(0)$. This term represents a rigid body motion and can be ignored. In the special case $k\pi=1$ and $\mathbf{t}^+ = \mathbf{L}(\frac{\pi}{2})\mathbf{n}(0)$, (9.2-25) reduce to (9.2-20). As discussed in (9.2-21) the stress σ_{ij} for this special case is given by $L_{ij}(\theta)$ except the σ_{33} component.

9.3 Solutions for Critical Wedge Angles

When the wedge angle $2\alpha = \pi$ or 2π and (9.2-19) does not hold or when $2\alpha = 2\alpha_*$, the nonuniform stress solutions presented in the previous section do not apply. We then employ the higher order solution $\mathbf{u}^{(1)}$, $\boldsymbol{\phi}^{(1)}$ given in (9.2-5). Written explicitly they are

$$\begin{aligned} \mathbf{u}/r = {} & \mathbf{N}_1(\theta,\theta^-)\mathbf{h}^{(1)} - \sin(\theta-\theta^-)\mathbf{N}_2(\theta^-)\mathbf{t}^- \\ & + \left\{[(\ln r)\mathbf{I} + \pi\mathbf{S}(\theta,\theta^-)]\mathbf{N}_1(\theta,\theta^-) + \pi\sin(\theta-\theta^-)\mathbf{H}(\theta,\theta^-)\mathbf{N}_3(\theta^-)\right\}\mathbf{h}^{(0)} \\ & + \tfrac{1}{2}k\left\{[(\ln r)\mathbf{I} + \pi\mathbf{S}(\theta,\theta^-)]^2 - \pi^2\mathbf{H}(\theta,\theta^-)\mathbf{L}(\theta,\theta^-)\right\}\mathbf{m}(\theta), \end{aligned} \tag{9.3-1a}$$

$$\phi / r = \sin(\theta-\theta^-)\mathbf{N}_3(\theta^-)\mathbf{h}^{(1)} - \mathbf{N}_1^T(\theta,\theta^-)\mathbf{t}^-$$
$$+\left\{\sin(\theta-\theta^-)[(\ln r)\mathbf{I}+\pi\mathbf{S}^T(\theta,\theta^-)]\mathbf{N}_3(\theta^-) - \pi\mathbf{L}(\theta,\theta^-)\mathbf{N}_1(\theta,\theta^-)\right\}\mathbf{h}^{(0)}$$
$$-k\pi\left\{(\ln r)\mathbf{L}(\theta,\theta^-) + \tfrac{1}{2}\pi[\mathbf{L}(\theta,\theta^-)\mathbf{S}(\theta,\theta^-) + \mathbf{S}^T(\theta,\theta^-)\mathbf{L}(\theta,\theta^-)\right\}\mathbf{m}(\theta). \tag{9.3-1b}$$

The stress now has a ln$r$ singularity. We have set $\mathbf{g}^{(0)}=\mathbf{0}$ and $\mathbf{g}^{(1)}=\mathbf{t}^-$ so that the boundary condition $(9.1\text{-}3)_1$ at $\theta=\theta^-$ is satisfied. Satisfaction of the boundary condition $(9.1\text{-}3)_2$ at $\theta=\theta^+$ yields

$$\sin 2\alpha\,\mathbf{N}_3(\theta^-)\mathbf{h}^{(0)} = k\pi\mathbf{L}(\theta^+,\theta^-)\mathbf{m}(\theta^+), \tag{9.3-2}$$

$$\sin 2\alpha\,\mathbf{N}_3(\theta^-)\mathbf{h}^{(1)} - \pi\mathbf{L}(\theta^+,\theta^-)\mathbf{N}_1(\theta^+,\theta^-)\mathbf{h}^{(0)} = \mathbf{t}^+ + \mathbf{N}_1^T(\theta^+,\theta^-)\mathbf{t}^-$$
$$+\tfrac{1}{2}k\pi^2[\mathbf{L}(\theta^+,\theta^-)\mathbf{S}(\theta^+,\theta^-) - \mathbf{S}^T(\theta^+,\theta^-)\mathbf{L}(\theta^+,\theta^-)]\mathbf{m}(\theta^+). \tag{9.3-3}$$

Equation (9.3-2) represents the vanishing of the coefficients of the ln$r$ terms of ϕ at $\theta=\theta^+$. Equation (9.3-2) has been used in deriving (9.3-3). Notice that (9.3-2) is identical to (9.2-13) when $\mathbf{t}^+=\mathbf{t}^-=\mathbf{0}$. By (9.2-14)

$$k=0 \quad \text{if} \quad 2\alpha \neq 2\alpha_*. \tag{9.3-4}$$

We now discuss $2\alpha=\pi$, 2π, and $2\alpha_*$ separately.

(1) $2\alpha=\pi$ or 2π. With k=0 and sin2α=0 (9.3-2) is automatically satisfied. Equation (9.3-3) becomes

$$-\pi\cos 2\alpha\,\mathbf{L}(\theta^+,\theta^-)\mathbf{h}^{(0)} = \mathbf{t}^+ + \cos 2\alpha\,\mathbf{t}^-. \tag{9.3-5}$$

This is similar to (9.2-18) and hence, by (9.2-19),

$$\begin{aligned} \pi\mathbf{L}\mathbf{h}^{(0)} &= \mathbf{t}^+ - \mathbf{t}^-, \quad \text{if} \quad 2\alpha=\pi, \\ -2\pi\mathbf{L}\mathbf{h}^{(0)} &= \mathbf{t}^+ + \mathbf{t}^-, \quad \text{if} \quad 2\alpha=2\pi. \end{aligned} \tag{9.3-6}$$

With $\mathbf{h}^{(0)}$ given by (9.3-6) $\mathbf{h}^{(1)}$ is arbitrary. However the $\mathbf{h}^{(1)}$ term provides no tractions at $\theta=\theta^-$, θ^+ and represents an eigenfunction. Ignoring the $\mathbf{h}^{(1)}$ term the final solution deduced from (9.3-1) is

$$\mathbf{u}/r = -\sin(\theta-\theta^-)\mathbf{N}_2(\theta^-)\mathbf{t}^- + \left\{[(\ln r)\mathbf{I} + \pi\mathbf{S}(\theta,\theta^-)]\mathbf{N}_1(\theta,\theta^-)\right.$$
$$\left.+\pi\sin(\theta-\theta^-)\mathbf{H}(\theta,\theta^-)\mathbf{N}_3(\theta^-)\right\}\mathbf{h}^{(0)}, \tag{9.3-7a}$$

$$\phi/r = -\mathbf{N}_1^T(\theta,\theta^-)\mathbf{t}^- + \left\{\sin(\theta-\theta^-)[(\ln r)\mathbf{I} + \pi\mathbf{S}^T(\theta,\theta^-)]\mathbf{N}_3(\theta^-)\right.$$
$$\left.-\pi\mathbf{L}(\theta,\theta^-)\mathbf{N}_1(\theta,\theta^-)\right\}\mathbf{h}^{(0)}. \tag{9.3-7b}$$

The stress has the ln$r$ singularity at r=0 unless

$$\mathbf{N}_3(\theta^-)\mathbf{h}^{(0)} = \mathbf{0} \quad \text{or} \quad \mathbf{h}^{(0)} = \gamma \mathbf{m}(\theta^-)$$

where γ is an arbitrary constant. By (9.3-6) this means that

$$\begin{aligned} \gamma\pi \mathbf{L}\mathbf{m}(\theta^-) &= \mathbf{t}^+ - \mathbf{t}^-, \quad \text{if} \quad 2\alpha = \pi, \\ -2\gamma\pi \mathbf{L}\mathbf{m}(\theta^-) &= \mathbf{t}^+ + \mathbf{t}^-, \quad \text{if} \quad 2\alpha = 2\pi. \end{aligned} \tag{9.3-8}$$

These agree with (9.2-19).

As an illustration consider the half-space $x_2>0$ subjected to a uniform traction $\mathbf{t}^+$ at the surface $x_2=0$, $x_1<0$, Fig. 9.3(a). The solution is

$$\begin{aligned} \mathbf{u}/r &= \left\{[\tfrac{1}{\pi}(\ln r)\mathbf{I} + \mathbf{S}(\theta)][\cos\theta\, \mathbf{I} + \sin\theta\, \mathbf{N}_1] + \sin\theta\, \mathbf{H}(\theta)\mathbf{N}_3\right\}\mathbf{L}^{-1}\mathbf{t}^+, \\ \boldsymbol{\phi}/r &= \left\{\sin\theta\, [\tfrac{1}{\pi}(\ln r)\mathbf{I} + \mathbf{S}^T(\theta)]\mathbf{N}_3 - \mathbf{L}(\theta)[\cos\theta\, \mathbf{I} + \sin\theta\, \mathbf{N}_1]\right\}\mathbf{L}^{-1}\mathbf{t}^+. \end{aligned} \tag{9.3-9}$$

For the special case $\mathbf{t}^+ = \tau \mathbf{L}\mathbf{n}(0)$ where τ is a constant we have, from $(9.1\text{-}4)_1$ and $(8.1\text{-}7)_4$,

$$\begin{aligned} \mathbf{u}/r &= -\tau\left\{\tfrac{1}{\pi}(\ln r)\mathbf{N}_1 + [\mathbf{S}(\theta)\mathbf{N}_1 + \mathbf{H}(\theta)\mathbf{N}_3]\right\}\mathbf{m}(\theta), \\ \boldsymbol{\phi}/r &= -\tau\left\{\tfrac{1}{\pi}(\ln r)\mathbf{N}_3 + [\mathbf{S}^T(\theta)\mathbf{N}_3 - \mathbf{L}(\theta)\mathbf{N}_1]\right\}\mathbf{m}(\theta). \end{aligned} \tag{9.3-10}$$

By (7.8-8) the quantities in the square brackets can be replaced by

$$\mathbf{S}(\theta)\mathbf{N}_1 + \mathbf{H}(\theta)\mathbf{N}_3 = \mathbf{N}_1\mathbf{S}(\theta) - \mathbf{N}_2\mathbf{L}(\theta),$$

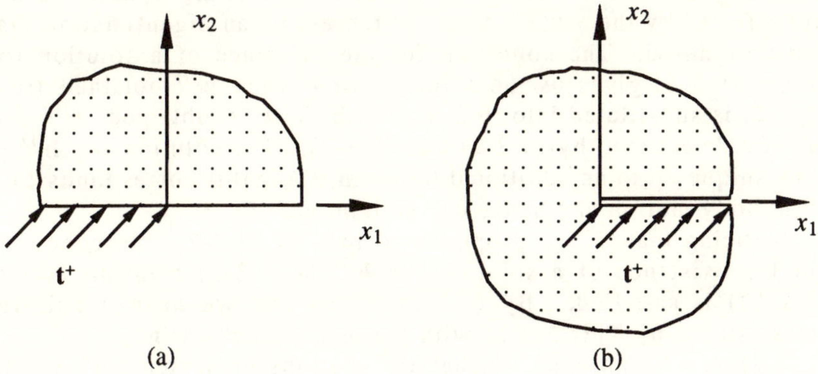

Fig. 9.3 An elastic wedge of wedge angle π or 2π is subjected to a uniform traction on one side of the wedge. The solution for (a) is given in (9.3-9). The solution for (b) is also given by (9.3-9) with $\mathbf{u}$ and $\boldsymbol{\phi}$ replaced by $-2\mathbf{u}$ and $-2\boldsymbol{\phi}$, respectively.

$$\mathbf{S}^T(\theta)\mathbf{N}_3 - \mathbf{L}(\theta)\mathbf{N}_1 = \mathbf{N}_3\mathbf{S}(\theta) - \mathbf{N}_1^T\mathbf{L}(\theta).$$

The second equation implies that $\mathbf{S}^T(\theta)\mathbf{N}_3 - \mathbf{L}(\theta)\mathbf{N}_1$ is a symmetric matrix. There is a $\ln r$ stress singularity. If $\mathbf{t}^+=p\mathbf{Lm}(0)$ where p is a constant, by $(9.1\text{-}4)_2$ and $(8.1\text{-}5)_2$,

$$\mathbf{u}/r = p\left[\tfrac{1}{\pi}(\ln r)\mathbf{I} + \mathbf{S}(\theta)\right]\mathbf{m}(\theta),$$
$$\boldsymbol{\phi}/r = -p\mathbf{L}(\theta)\mathbf{m}(\theta). \tag{9.3-11}$$

This is identical to (9.2-20) when $p=1$ According to (9.2-21) the stress σ_{ij} deduced from $(9.3\text{-}11)_2$ is equal to $pL_{ij}(\theta)$ except the σ_{33} component. There is no $\ln r$ stress singularity. The stress depends on θ only, not on r. For materials such as isotropic materials for which $\mathbf{L}$ is a diagonal matrix, $\mathbf{t}^+=\tau\mathbf{Ln}(0)$ and $p\mathbf{Lm}(0)$ represent, respectively, a shear stress and a normal pressure applied at the wedge surface θ^+. For these materials there is a $\ln r$ stress singularity due to a shear loading but not due to a pressure loading.

The situation is the same for the wedge angle $2\alpha=2\pi$, Fig. 9.3(b). The solutions are given by (9.3-9)–(9.3-11) with $\mathbf{u}$ and $\boldsymbol{\phi}$ replaced by $-2\mathbf{u}$ and $-2\boldsymbol{\phi}$, respectively.

(2) $2\alpha=2\alpha_*$. Equations (9.3-2) and (9.3-3) can be solved for $\mathbf{h}^{(0)}$, $\mathbf{h}^{(1)}$, and k as follows. Since $\mathbf{m}^T(\theta^-)\mathbf{N}_3(\theta^-)=\mathbf{0}$, the condition for the existence of a solution for $\mathbf{h}^{(0)}$ in (9.3-2) is the vanishing of the right side when it is multiplied by $\mathbf{m}^T(\theta^-)$. This leads to $\Delta=0$ which is the case for $2\alpha=2\alpha_*$. Therefore $\mathbf{h}^{(0)}$ exists and is unique up to an additional term $\gamma\mathbf{m}(\theta^-)$ where γ is an arbitrary constant. However, if $\mathbf{h}^{(0)}$ is modified by $\gamma\mathbf{m}(\theta^-)$ (9.3-3) can remain valid by adjusting $\mathbf{h}^{(1)}$ and k. This means that the boundary conditions are not affected by the $\gamma\mathbf{m}(\theta^-)$ term. It represents an eigenfunction that can be ignored. The condition for the existence of a solution for $\mathbf{h}^{(1)}$ in (9.3-3) gives us the solution for k when $\mathbf{h}^{(0)}$ obtained from (9.3-2) is substituted into (9.3-3). With the k so obtained, $\mathbf{h}^{(0)}$ and $\mathbf{h}^{(1)}$ are determined by (9.3-2) and (9.3-3). The solution for $\mathbf{h}^{(1)}$ is also unique up to an additional term $\gamma\mathbf{m}(\theta^-)$, but it corresponds to a rigid body motion and hence can be ignored.

Noticing that $\mathbf{h}^{(0)}$ is proportional to k by (9.3-2), the condition for the existence of a solution for $\mathbf{h}^{(1)}$ in (9.3-3) leads to $k=0$ if (9.1-13) is satisfied. By (9.3-2) $\mathbf{h}^{(0)}=\mathbf{0}$ and we have a uniform stress solution. This agrees with the result stated earlier.

It should be pointed out that all solutions presented here and in the previous two sections are *particular solutions*. As we will discuss in the next section there are infinitely many eigenfunctions that satisfy traction-free conditions at the wedge surfaces $\theta=\theta^-$, θ^+. These eigenfunctions can be added to the particular solution.

A unique solution can be expected when the boundary conditions at the far field $r=\infty$, or at $r=R$ for a finite wedge $0\le r\le R$ are specified. An example is the isotropic elastic wedge of finite dimension $0\le r\le R$ subjected to antiplane shears on the surfaces of the wedge (Ting, 1985b). The critical wedge angles for this case are $2\alpha=\pi$ and 2π.

The analysis presented in this section follows essentially the work of Hwu and Ting (1990) except that the present analysis applies to arbitrary orientation of the wedge. The corresponding solutions for isotropic wedges were obtained by Dempsey (1981) and Ting (1984).

The critical wedge angles and the solutions at the critical wedge angles for a rotationally inhomogeneous anisotropic wedge have been studied by Belov and Kirchner (1995). They showed that the equations for the critical angles of the wedge subjected to uniform tractions and a concentrated couple are different.

A question may arise regarding the solutions for a wedge whose wedge angle is nearly equal to one of the critical wedge angles. A solution that is uniformly valid for the wedge angle near or equal to a critical wedge angle can be obtained by superimposing to the particular solution an eigenfunction that is proportional to the particular solution when the wedge angle is at the critical wedge angle. By judicially choosing the coefficient of the eigenfunction a uniformly valid solution was obtained for isotropic elastic wedges at critical wedge angles (Ting, 1984). The same technique can be employed for anisotropic elastic wedges.

9.4 Eigenfunctions for a Wedge

The solutions presented in the previous sections are particular solutions. They satisfy the prescribed tractions at the wedge surfaces. There are infinitely many *homogeneous solutions* that satisfy traction-free conditions at the wedge surfaces. These are the *eigenfunctions* of the wedge. A general solution can be obtained when all eigenfunctions are superimposed on the particular solution. To find the eigenfunctions for the wedge one assumes that the displacement is proportional to $r^{\delta+1}$ where δ is an eigenvalue to be determined. The stress is then proportional to r^{δ}. If $\delta<0$ the stress is singular at the wedge apex $r=0$.

The problem of finding stress singularities in an isotropic wedge was first considered by Knein (1926) and Williams (1952). It was shown that $\delta<0$ when the wedge angle is larger than π so that there is a stress singularity for an isotropic wedge when the wedge angle is larger than π (see also England, 1971; Ting, 1984). The technique was applied to a crack along (Williams, 1959; England,

1965) and normal to (Zak and Williams, 1963; Cook and Erdogan, 1972) an interface and to other geometry of isotropic composites (Bogy, 1970, 1971a,b; Bogy and Wang, 1971, Erdogan and Gupta, 1972; Delale et al., 1984). A systematic presentation of the equations for finding the singularities was given by Dempsey and Sinclair (1979, 1981) who also provided a large number of references on the subject. Investigation of the associated problems for anisotropic materials was started by Sih et al. (1965) and Sih and Liebowitz (1968), and was followed by Bogy (1972, 1974), Kuo and Bogy (1974), Delale and Erdogan (1979), Sih and Chen (1981), Hoenig (1982), and Ting (1986).

One of the most studied problems in layered composites is the stress distribution at a *free-edge*, i.e., at the point where the interface of two materials meets perpendicularly with a free surface. The order of *free-edge stress singularity* in anisotropic composites has been studied by Ting and Chou (1981), Wang and Choi (1982), Zwiers et al. (1982), Wang (1984), and Tewary and Kritz (1991). The associated problem where the free surface is replaced by a contact surface was considered by Zwiers and Ting (1983). Ting and Hoang (1984) studied the problem of a bimaterial with a crack terminating normal to an interface. The associated problem for axisymmetric deformations of transversely isotropic materials was investigated by Ting et al. (1985) and Jin and Ting (1988).

The eigenvalue δ can be complex. Let

$$\delta = \delta' + i\delta''$$

where δ' and δ'' are real. From

$$\begin{aligned} r^{\delta} &= r^{\delta'} r^{i\delta''} = r^{\delta'} e^{i\delta'' \ln r} \\ &= r^{\delta'}[\cos(\delta'' \ln r) + i \sin(\delta'' \ln r)] \end{aligned} \tag{9.4-1}$$

it is seen that, when $\delta'' \neq 0$, both real and imaginary parts of r^{δ} oscillate with increasing frequency as $r \to 0$. Thus the displacement and stress oscillate with increasing frequency as one approaches the wedge apex. If $\delta' = \mathrm{Re}(\delta) < 0$ the stress oscillates violently with increasing magnitude as one approaches the wedge apex.

Consider the strain energy $\mathcal{E}$ in the region $r_1 \le r \le r_2$ of the wedge,

$$\mathcal{E} = \frac{1}{2} \int_{\theta^-}^{\theta^+} \int_{r_1}^{r_2} r \sigma_{ij} \varepsilon_{ij} dr d\theta .$$

If σ_{ij} is proportional to r^{δ} so is ε_{ij} and the strain energy density is

$$\tfrac{1}{2} \sigma_{ij} \varepsilon_{ij} = r^{2\delta} f(\theta),$$

f being a function of θ. If δ is complex, we superimpose one solu tion associated with δ and one associated with its complex conjugate $\overline{\delta}$ so that the solution is real. The strain energy density is then

$$\tfrac{1}{2}(\sigma_{ij}+\overline{\sigma}_{ij})(\varepsilon_{ij}+\overline{\varepsilon}_{ij}) = r^{2\delta'} f(\theta).$$

Regardless of δ being real or complex, we have

$$\varepsilon = \frac{1}{2(\delta'+1)}[r_2^{2(\delta'+1)} - r_1^{2(\delta'+1)}]\int_{\theta^-}^{\theta^+} f(\theta)d\theta.$$

As $r_1 \to 0$ the strain energy is bounded if

$$-1 < \mathrm{Re}(\delta). \tag{9.4-2}$$

There are exceptions to (9.4-2) when the wedge apex is subjected to a loading. As seen in Chapter 8, $\delta=-1$ when a concentrated force is applied at $r=0$, and $\delta=-2$ when a concentrated couple or a double force is applied at $r=0$.

For *stress singularities* in the wedge we will be interested in δ's in the region

$$-1 < \mathrm{Re}(\delta) < 0. \tag{9.4-3}$$

In many applications however δ's with a positive real part are also needed (Tong et al., 1973; Lin and Mar, 1976; Wang and Choi, 1982). This means that one has to search numerically for all δ's in the region given by (9.4-2). For special wedge angles and special composite wedges we will see that an explicit expression of δ can be obtained.

When the stress is proportional to r^δ, it was shown in (8.12-1) and (8.12-2) that

$$\mathbf{t}_{\theta,\theta} = -(\delta+1)\mathbf{t}_r,$$

$$\sigma_{r\theta,\theta} = -(\delta+1)\sigma_{rr} + \sigma_{\theta\theta}, \quad \sigma_{\theta\theta,\theta} = -(\delta+2)\sigma_{r\theta},$$

$$\sigma_{\theta 3,\theta} = -(\delta+1)\sigma_{r3},$$

where

$$(\mathbf{t}_\theta)_i = \sigma_{ij} m_j, \quad (\mathbf{t}_r)_i = \sigma_{ij} n_j.$$

These results are based on equations of equilibrium alone, and hence are independent of material property.

To find the eigenfunctions for an anisotropic wedge we employ the original general solution (5.3-7) with $f(z)$ being proportional to $z^{\delta+1}$. As in Section 9.1 we use dual coordinate systems with z replaced by $\hat{z}$ so that

$$\mathbf{u} = \mathbf{A}\left\langle \hat{z}_*^{\delta+1} \right\rangle \mathbf{q} + \overline{\mathbf{A}}\left\langle \bar{\hat{z}}_*^{\delta+1} \right\rangle \tilde{\mathbf{q}}, \quad \boldsymbol{\phi} = \mathbf{B}\left\langle \hat{z}_*^{\delta+1} \right\rangle \mathbf{q} + \overline{\mathbf{B}}\left\langle \bar{\hat{z}}_*^{\delta+1} \right\rangle \tilde{\mathbf{q}}, \tag{9.4-4}$$

where $\mathbf{q}$ and $\tilde{\mathbf{q}}$ are arbitrary constant vectors. If δ is real, $\tilde{\mathbf{q}}$ becomes the complex conjugate of $\mathbf{q}$, and $\mathbf{u}$ and $\boldsymbol{\phi}$ are real. If δ is complex, $\tilde{\mathbf{q}}$ is not necessarily the complex conjugate of $\mathbf{q}$, and $\mathbf{u}$ and $\boldsymbol{\phi}$ may not be real. We will see later that if a complex δ is an eigenvalue, so is its complex conjugate $\overline{\delta}$. One can then superimpose two solutions of the form (9.4-4) associated with δ and $\overline{\delta}$ to obtain real values for $\mathbf{u}$ and $\boldsymbol{\phi}$.

From (9.1-8) and (9.1-9) $\hat{z}$ can be written as

$$\hat{z} = r\zeta(\theta,\theta^-) \tag{9.4-5}$$

where

$$\zeta(\theta,\theta^-) = \cos(\theta-\theta^-) + p(\theta^-)\sin(\theta-\theta^-). \tag{9.4-6}$$

In particular,

$$\zeta(\theta^-,\theta^-) = 1, \quad \zeta(\theta^+,\theta^-) = \cos 2\alpha + p(\theta^-)\sin 2\alpha. \tag{9.4-7}$$

Equation (9.4-4) has the alternate expression

$$\begin{aligned}\mathbf{u} &= r^{\delta+1}\left\{\mathbf{A}\left\langle \zeta_*^{\delta+1}(\theta,\theta^-)\right\rangle\mathbf{q} + \overline{\mathbf{A}}\left\langle \overline{\zeta}_*^{\delta+1}(\theta,\theta^-)\right\rangle\tilde{\mathbf{q}}\right\},\\ \boldsymbol{\phi} &= r^{\delta+1}\left\{\mathbf{B}\left\langle \zeta_*^{\delta+1}(\theta,\theta^-)\right\rangle\mathbf{q} + \overline{\mathbf{B}}\left\langle \overline{\zeta}_*^{\delta+1}(\theta,\theta^-)\right\rangle\tilde{\mathbf{q}}\right\}.\end{aligned} \tag{9.4-8}$$

We will consider the following three types of boundary conditions:

$$\text{(a) } \boldsymbol{\phi}=\mathbf{0} \quad \text{at} \quad \theta=\theta^-,\ \theta^+, \tag{9.4-9}$$

$$\text{(b) } \mathbf{u}=\mathbf{0} \quad \text{at} \quad \theta=\theta^-,\ \theta^+, \tag{9.4-10}$$

$$\text{(c) } \boldsymbol{\phi}=\mathbf{0} \quad \text{at} \quad \theta=\theta^-, \text{ and } \mathbf{u}=\mathbf{0} \quad \text{at} \quad \theta=\theta^+. \tag{9.4-11}$$

We call the wedge (a) free-free, (b) fixed-fixed, and (c) free-fixed, respectively. One could also consider boundary conditions for a slippery surface (see Section 14.1).

(a) **The Free-Free Wedge.** Insertion of (9.4-8) into (9.4-9) yields

$$\begin{aligned}&\mathbf{B}\mathbf{q} + \overline{\mathbf{B}}\tilde{\mathbf{q}} = \mathbf{0},\\ &\mathbf{B}\left\langle \zeta_*^{\delta+1}(\theta^+,\theta^-)\right\rangle\mathbf{q} + \overline{\mathbf{B}}\left\langle \overline{\zeta}_*^{\delta+1}(\theta^+,\theta^-)\right\rangle\tilde{\mathbf{q}} = \mathbf{0},\end{aligned} \tag{9.4-12}$$

or

$$\mathbf{q} = \mathbf{B}^{-1}\mathbf{g}, \quad \tilde{\mathbf{q}} = -\overline{\mathbf{B}}^{-1}\mathbf{g}, \tag{9.4-13a}$$

$$\mathbf{K}\mathbf{g} = \mathbf{0}, \tag{9.4-13b}$$

where $\mathbf{g}$ is an arbitrary constant vector and

$$\begin{aligned}
\mathbf{K} &= \mathbf{B}\left\langle \zeta_*^{\delta+1}(\theta^+,\theta^-)\right\rangle \mathbf{B}^{-1} - \overline{\mathbf{B}}\left\langle \overline{\zeta}_*^{\delta+1}(\theta^+,\theta^-)\right\rangle \overline{\mathbf{B}}^{-1} \\
&= \mathbf{B}\left\{\left\langle \zeta_*^{\delta+1}(\theta^+,\theta^-)\right\rangle \mathbf{B}^{-1}\overline{\mathbf{B}} - \mathbf{B}^{-1}\overline{\mathbf{B}}\left\langle \overline{\zeta}_*^{\delta+1}(\theta^+,\theta^-)\right\rangle\right\} \overline{\mathbf{B}}^{-1} \qquad (9.4\text{-}14) \\
&= \left\{\mathbf{B}\left\langle \zeta_*^{\delta+1}(\theta^+,\theta^-)\right\rangle \mathbf{B}^{T} + \overline{\mathbf{B}}\left\langle \overline{\zeta}_*^{\delta+1}(\theta^+,\theta^-)\right\rangle \overline{\mathbf{B}}^{T}\right\}(-2i\mathbf{L}^{-1}).
\end{aligned}$$

For a nontrivial solution of $\mathbf{g}$ from (9.4-13b) δ must be a root of

$$\det\mathbf{K} = |\mathbf{K}| = 0. \qquad (9.4\text{-}15)$$

When δ is a root of (9.4-15) it is an eigenvalue of the elastic wedge and $\mathbf{g}$ of (9.4-13b) is the associated eigenvector. With $\mathbf{q}$ and $\tilde{\mathbf{q}}$ determined from (9.4-13a), (9.4-8) provides the corresponding eigenfunction. If δ is a simple root, the eigenfunction is unique up to a constant multiplier. The case of repeated eigenvalues will be studied in Section 9.7. When we take the complex conjugate of $(9.4\text{-}14)_1$ it is easily shown that if δ is a root of (9.4-15) so is its complex conjugate $\overline{\delta}$. Therefore complex eigenvalues always appear in pairs of complex conjugates.

Insertion of (9.4-14) into (9.4-15) yields

$$\left|\mathbf{B}\left\langle \zeta_*^{\delta+1}(\theta^+,\theta^-)\right\rangle \mathbf{B}^{-1} - \overline{\mathbf{B}}\left\langle \overline{\zeta}_*^{\delta+1}(\theta^+,\theta^-)\right\rangle \overline{\mathbf{B}}^{-1}\right| = 0, \qquad (9.4\text{-}16a)$$

$$\left|\left\langle \zeta_*^{\delta+1}(\theta^+,\theta^-)\right\rangle \mathbf{B}^{-1}\overline{\mathbf{B}} - \mathbf{B}^{-1}\overline{\mathbf{B}}\left\langle \overline{\zeta}_*^{\delta+1}(\theta^+,\theta^-)\right\rangle\right| = 0, \qquad (9.4\text{-}16b)$$

$$\left|\mathbf{B}\left\langle \zeta_*^{\delta+1}(\theta^+,\theta^-)\right\rangle \mathbf{B}^{T} + \overline{\mathbf{B}}\left\langle \overline{\zeta}_*^{\delta+1}(\theta^+,\theta^-)\right\rangle \overline{\mathbf{B}}^{T}\right| = 0. \qquad (9.4\text{-}16c)$$

Each of (9.4-16) has its own special features. Explicit expressions of the matrix $\mathbf{B}$ and its inverse $\mathbf{B}^{-1}$ are given in (6.3-2) and (6.3-4) in which the k_i (i=1,2,3) are the normalization factors for the columns of $\mathbf{B}$. Since $\left\langle \zeta_*^{\delta+1}(\theta^+,\theta^-)\right\rangle$ is a diagonal matrix it is easily shown that the k_i cancel out when we employ (9.4-16a). We may also omit the k_i in computing (9.4-16b) because the deduction of $(9.4\text{-}14)_2$ from $(9.4\text{-}14)_1$ does not require normalization of $\mathbf{B}$. In most cases (9.4-16b) is the simplest in computing the determinant. The matrix product $\mathbf{B}^{-1}\overline{\mathbf{B}}$ appearing in (9.4-16b) has an interesting property. It was shown in Section 5.6 that $i\mathbf{B}^{-1}\overline{\mathbf{B}}$ is an orthogonal, positive definite Hermitian. Equation (9.4-16c) is the least convenient in computing the determinant because the normalization factors k_i cannot be omitted. However it has the following interesting relation. From the eigenequation

$$\mathbf{N}(\theta^-)\boldsymbol{\xi} = p(\theta^-)\boldsymbol{\xi}$$

we have

$$\mathbf{N}(\theta,\theta^-)\xi = \zeta(\theta,\theta^-)\xi \tag{9.4-17}$$

where

$$\mathbf{N}(\theta,\theta^-) = \cos(\theta-\theta^-)\mathbf{I} + p(\theta^-)\sin(\theta-\theta^-)\mathbf{N}(\theta^-). \tag{9.4-18}$$

Hence

$$\mathbf{N}^{\delta+1}(\theta,\theta^-)\xi = \zeta^{\delta+1}(\theta,\theta^-)\xi. \tag{9.4-19}$$

Equation (9.4-19) applies to integer δ, but we will allow δ to be non-integer and complex. If eigenequation (9.4-19) is written for the six eigenvectors ξ_1, ξ_2,..., we have

$$\mathbf{N}^{\delta+1}(\theta,\theta^-)\begin{bmatrix}\mathbf{A} & \overline{\mathbf{A}}\\ \mathbf{B} & \overline{\mathbf{B}}\end{bmatrix} = \begin{bmatrix}\mathbf{A} & \overline{\mathbf{A}}\\ \mathbf{B} & \overline{\mathbf{B}}\end{bmatrix}\begin{bmatrix}\langle\zeta_*^{\delta+1}(\theta,\theta^-)\rangle & \mathbf{0}\\ \mathbf{0} & \langle\overline{\zeta}_*^{\delta+1}(\theta,\theta^-)\rangle\end{bmatrix}$$

or, by the closure relation (5.5-15),

$$\mathbf{N}^{\delta+1}(\theta,\theta^-) = \begin{bmatrix}\mathbf{A} & \overline{\mathbf{A}}\\ \mathbf{B} & \overline{\mathbf{B}}\end{bmatrix}\begin{bmatrix}\langle\zeta_*^{\delta+1}(\theta,\theta^-)\rangle & \mathbf{0}\\ \mathbf{0} & \langle\overline{\zeta}_*^{\delta+1}(\theta,\theta^-)\rangle\end{bmatrix}\begin{bmatrix}\mathbf{B}^T & \mathbf{A}^T\\ \overline{\mathbf{B}}^T & \overline{\mathbf{A}}^T\end{bmatrix}.$$

Let the four 3×3 submatrices of $\mathbf{N}^{\delta+1}(\theta,\theta^-)$ be denoted by

$$\mathbf{N}^{\delta+1}(\theta,\theta^-) = \begin{bmatrix}\mathbf{N}_1^{(\delta+1)}(\theta,\theta^-) & \mathbf{N}_2^{(\delta+1)}(\theta,\theta^-)\\ \mathbf{N}_3^{(\delta+1)}(\theta,\theta^-) & \mathbf{N}_1^{(\delta+1)^T}(\theta,\theta^-)\end{bmatrix}. \tag{9.4-20}$$

One then has

$$\mathbf{N}_1^{(\delta+1)}(\theta,\theta^-) = \mathbf{A}\langle\zeta_*^{\delta+1}(\theta,\theta^-)\rangle\mathbf{B}^T + \overline{\mathbf{A}}\langle\overline{\zeta}_*^{\delta+1}(\theta,\theta^-)\rangle\overline{\mathbf{B}}^T, \tag{9.4-21a}$$

$$\mathbf{N}_2^{(\delta+1)}(\theta,\theta^-) = \mathbf{A}\langle\zeta_*^{\delta+1}(\theta,\theta^-)\rangle\mathbf{A}^T + \overline{\mathbf{A}}\langle\overline{\zeta}_*^{\delta+1}(\theta,\theta^-)\rangle\overline{\mathbf{A}}^T, \tag{9.4-21b}$$

$$\mathbf{N}_3^{(\delta+1)}(\theta,\theta^-) = \mathbf{B}\langle\zeta_*^{\delta+1}(\theta,\theta^-)\rangle\mathbf{B}^T + \overline{\mathbf{B}}\langle\overline{\zeta}_*^{\delta+1}(\theta,\theta^-)\rangle\overline{\mathbf{B}}^T. \tag{9.4-21c}$$

Therefore (9.4-16c) is equivalent to

$$\left|\mathbf{N}_3^{(\delta+1)}(\theta^+,\theta^-)\right| = 0. \tag{9.4-22}$$

With (9.4-13a) the eigenfunction (9.4-8) is

$$\begin{aligned}\mathbf{u} &= r^{\delta+1}\left\{\mathbf{A}\langle\zeta_*^{\delta+1}(\theta,\theta^-)\rangle\mathbf{B}^{-1} - \overline{\mathbf{A}}\langle\overline{\zeta}_*^{\delta+1}(\theta,\theta^-)\rangle\overline{\mathbf{B}}^{-1}\right\}\mathbf{g},\\ \phi &= r^{\delta+1}\left\{\mathbf{B}\langle\zeta_*^{\delta+1}(\theta,\theta^-)\rangle\mathbf{B}^{-1} - \overline{\mathbf{B}}\langle\overline{\zeta}_*^{\delta+1}(\theta,\theta^-)\rangle\overline{\mathbf{B}}^{-1}\right\}\mathbf{g},\end{aligned} \tag{9.4-23a}$$

or, by (9.4-21a,c),

$$\mathbf{u} = r^{\delta+1}\mathbf{N}_1^{(\delta+1)}(\theta,\theta^-)\mathbf{h},$$
$$\phi = r^{\delta+1}\mathbf{N}_3^{(\delta+1)}(\theta,\theta^-)\mathbf{h}, \tag{9.4-23b}$$

where

$$\mathbf{h} = -2i\mathbf{L}^{-1}\mathbf{g}. \tag{9.4-24}$$

At $\theta = \theta^-$, $\mathbf{u} = r^{\delta+1}\mathbf{h}$. Equation (9.4-23b) is useful when δ is an integer.

We will now employ (9.4-16b) to compute the determinant $|\mathbf{K}|$ for monoclinic materials with the symmetry at $x_3=0$. The $\mathbf{B}$ and $\mathbf{B}^{-1}$ are given by (6.3-8) and (6.3-9) so that

$$\mathbf{B}^{-1}\overline{\mathbf{B}} = \frac{1}{p_1-p_2}\begin{bmatrix} k_1^{-1}\bar{k}_1(\bar{p}_1-p_2) & k_1^{-1}\bar{k}_2(\bar{p}_2-p_2) & 0 \\ k_2^{-1}\bar{k}_1(p_1-\bar{p}_1) & k_2^{-1}\bar{k}_2(p_1-\bar{p}_2) & 0 \\ 0 & 0 & (p_1-p_2)k_3^{-1}\bar{k}_3 \end{bmatrix}. \tag{9.4-25}$$

Equation (4.4-16b) now yields

$$\zeta_3^{\delta+1}(\theta^+,\theta^-) - \bar{\zeta}_3^{\delta+1}(\theta^+,\theta^-) = 0 \tag{9.4-26}$$

and

$$\begin{aligned}(\bar{p}_1-p_2)(p_1-\bar{p}_2)&\left\{\zeta_1^{\delta+1}(\theta^+,\theta^-) - \bar{\zeta}_1^{\delta+1}(\theta^+,\theta^-)\right\}\left\{\zeta_2^{\delta+1}(\theta^+,\theta^-) - \bar{\zeta}_2^{\delta+1}(\theta^+,\theta^-)\right\} \\ -(p_1-\bar{p}_1)(\bar{p}_2-p_2)&\left\{\zeta_1^{\delta+1}(\theta^+,\theta^-) - \bar{\zeta}_2^{\delta+1}(\theta^+,\theta^-)\right\}\left\{\zeta_2^{\delta+1}(\theta^+,\theta^-) - \bar{\zeta}_1^{\delta+1}(\theta^+,\theta^-)\right\} \\ =0.\end{aligned} \tag{9.4-27}$$

The normalization factors k_i disappeared in (9.4-26) and (9.4-27), confirming the statement made earlier that the k_i can be omitted in employing (9.4-16b). Equation (9.4-26) applies to antiplane deformations while (9.4-27) applies to inplane deformations.

Let p'_v and p''_v be, respectively, the real and imaginary parts of p_v. Then

$$\begin{aligned}\zeta_v(\theta^+,\theta^-) &= \cos 2\alpha + p_v(\theta^-)\sin 2\alpha \\ &= \cos 2\alpha + p'_v(\theta^-)\sin 2\alpha + ip''_v(\theta^-)\sin 2\alpha \\ &= \rho_v e^{i\psi_v}\end{aligned} \tag{9.4-28}$$

where

$$\begin{aligned}\rho_v \cos\psi_v &= \cos 2\alpha + p'_v(\theta^-)\sin 2\alpha, \\ \rho_v \sin\psi_v &= p''_v(\theta^-)\sin 2\alpha.\end{aligned} \tag{9.4-29}$$

Equation (9.4-26) reduces to

$$\sin(\delta+1)\psi_3=0, \quad \text{or} \quad (\delta+1)\psi_3=n\pi, \tag{9.4-30}$$

where n is any integer. For isotropic materials $\psi_3=2\alpha$ and (9.4-30) recovers the results obtained in (Ting, 1985b). Equation (9.4-27) reduces to

$$[(p_1'-p_2')^2+(p_1''-p_2'')^2](\rho_1\rho_1)^{\delta+1}\sin^2[\tfrac{1}{2}(\delta+1)(\psi_1+\psi_2)]=$$
$$[(p_1'-p_2')^2+(p_1''+p_2'')^2](\rho_1\rho_2)^{\delta+1}\sin^2[\tfrac{1}{2}(\delta+1)(\psi_1-\psi_2)]+p_1''p_2''(\rho_1^{\delta+1}-\rho_2^{\delta+1})^2. \tag{9.4-31}$$

For isotropic materials we let $\theta^-=0$, $p_1'=p_2'=0$, and divide (9.4-31) by $p_1''-p_2''$. In the limit $p_1''=p_2''=1$ we obtain

$$\sin^2[2\alpha(\delta+1)]=(\delta+1)^2\sin^2 2\alpha, \tag{9.4-32}$$

in agreement with the result first obtained by Williams (1952).

(b) **The Fixed-Fixed Wedge**. For this case the boundary conditions are given by (9.4-10). The results obtained in (9.4-13a)–(9.4-16) remain valid if **B** is replaced by **A**, and **L** in $(9.4\text{-}14)_3$ is replaced by $-$**H**. Thus

$$\mathbf{q}=\mathbf{A}^{-1}\mathbf{g}, \quad \tilde{\mathbf{q}}=-\overline{\mathbf{A}}^{-1}\mathbf{g}, \tag{9.4-33}$$

$$\mathbf{K}\mathbf{g}=\mathbf{0}, \tag{9.4-34}$$

where

$$\begin{aligned}\mathbf{K}&=\mathbf{A}\langle\zeta_*^{\delta+1}(\theta^+,\theta^-)\rangle\mathbf{A}^{-1}-\overline{\mathbf{A}}\langle\overline{\zeta}_*^{\delta+1}(\theta^+,\theta^-)\rangle\overline{\mathbf{A}}^{-1}\\&=\mathbf{A}\{\langle\zeta_*^{\delta+1}(\theta^+,\theta^-)\rangle\mathbf{A}^{-1}\overline{\mathbf{A}}-\mathbf{A}^{-1}\overline{\mathbf{A}}\langle\overline{\zeta}_*^{\delta+1}(\theta^+,\theta^-)\rangle\}\overline{\mathbf{A}}^{-1}\\&=\{\mathbf{A}\langle\zeta_*^{\delta+1}(\theta^+,\theta^-)\rangle\mathbf{A}^T+\overline{\mathbf{A}}\langle\overline{\zeta}_*^{\delta+1}(\theta^+,\theta^-)\rangle\overline{\mathbf{A}}^T\}(2i\mathbf{H}^{-1}).\end{aligned} \tag{9.4-35}$$

It is shown in Section 5.6 that $-i\mathbf{A}^{-1}\overline{\mathbf{A}}$ is an orthogonal, positive definite Hermitian. The eigenvalue δ is a root of $|\mathbf{K}|=0$ which is, by (9.4-21b), equivalent to

$$\left|\mathbf{N}_2^{(\delta+1)}(\theta^+,\theta^-)\right|=0. \tag{9.4-36}$$

The eigenfunction is

$$\begin{aligned}\mathbf{u}&=r^{\delta+1}\{\mathbf{A}\langle\zeta_*^{\delta+1}(\theta,\theta^-)\rangle\mathbf{A}^{-1}-\overline{\mathbf{A}}\langle\overline{\zeta}_*^{\delta+1}(\theta,\theta^-)\rangle\overline{\mathbf{A}}^{-1}\}\mathbf{g},\\ \boldsymbol{\phi}&=r^{\delta+1}\{\mathbf{B}\langle\zeta_*^{\delta+1}(\theta,\theta^-)\rangle\mathbf{A}^{-1}-\overline{\mathbf{B}}\langle\overline{\zeta}_*^{\delta+1}(\theta,\theta^-)\rangle\overline{\mathbf{A}}^{-1}\}\mathbf{g},\end{aligned} \tag{9.4-37a}$$

or

$$\mathbf{u} = r^{\delta+1}\mathbf{N}_2^{(\delta+1)}(\theta,\theta^-)\mathbf{h},$$

$$\boldsymbol{\phi} = r^{\delta+1}\mathbf{N}_1^{(\delta+1)^T}(\theta,\theta^-)\mathbf{h}, \tag{9.4-37b}$$

where

$$\mathbf{h} = 2i\mathbf{H}^{-1}\mathbf{g}. \tag{9.4-38}$$

(c) **The Free-Fixed Wedge.** The boundary conditions for this case are given by (9.4-11). The analysis is similar to that for the free-free wedge so that we only record the results below. We have

$$\mathbf{q} = \mathbf{B}^{-1}\mathbf{g}, \quad \tilde{\mathbf{q}} = -\overline{\mathbf{B}}^{-1}\mathbf{g}, \tag{9.4-39}$$

$$\mathbf{K}\mathbf{g} = \mathbf{0}, \tag{9.4-40}$$

where

$$\begin{aligned}\mathbf{K} &= \mathbf{A}\langle\zeta_*^{\delta+1}(\theta^+,\theta^-)\rangle\mathbf{B}^{-1} - \overline{\mathbf{A}}\langle\overline{\zeta}_*^{\delta+1}(\theta^+,\theta^-)\rangle\overline{\mathbf{B}}^{-1} \\ &= \mathbf{A}\{\langle\zeta_*^{\delta+1}(\theta^+,\theta^-)\rangle\mathbf{B}^{-1}\overline{\mathbf{B}} - \mathbf{A}^{-1}\overline{\mathbf{A}}\langle\overline{\zeta}_*^{\delta+1}(\theta^+,\theta^-)\rangle\}\overline{\mathbf{B}}^{-1} \\ &= \{\mathbf{A}\langle\zeta_*^{\delta+1}(\theta^+,\theta^-)\rangle\mathbf{B}^T + \overline{\mathbf{A}}\langle\overline{\zeta}_*^{\delta+1}(\theta^+,\theta^-)\rangle\overline{\mathbf{B}}^T\}(-2i\mathbf{L}^{-1}).\end{aligned} \tag{9.4-41}$$

The eigenvalue δ is a root of $|\mathbf{K}|=0$, which is equivalent to

$$\left|\mathbf{N}_1^{(\delta+1)}(\theta^+,\theta^-)\right| = 0. \tag{9.4-42}$$

The eigenfunction is

$$\mathbf{u} = r^{\delta+1}\{\mathbf{A}\langle\zeta_*^{\delta+1}(\theta,\theta^-)\rangle\mathbf{B}^{-1} - \overline{\mathbf{A}}\langle\overline{\zeta}_*^{\delta+1}(\theta,\theta^-)\rangle\overline{\mathbf{B}}^{-1}\}\mathbf{g},$$

$$\boldsymbol{\phi} = r^{\delta+1}\{\mathbf{B}\langle\zeta_*^{\delta+1}(\theta,\theta^-)\rangle\mathbf{B}^{-1} - \overline{\mathbf{B}}\langle\overline{\zeta}_*^{\delta+1}(\theta,\theta^-)\rangle\overline{\mathbf{B}}^{-1}\}\mathbf{g}, \tag{9.4-43a}$$

or

$$\mathbf{u} = r^{\delta+1}\mathbf{N}_1^{(\delta+1)}(\theta,\theta^-)\mathbf{h},$$

$$\boldsymbol{\phi} = r^{\delta+1}\mathbf{N}_3^{(\delta+1)}(\theta,\theta^-)\mathbf{h}, \tag{9.4-43b}$$

where

$$\mathbf{h} = -2i\mathbf{L}^{-1}\mathbf{g}. \tag{9.4-44}$$

9.5 Stress Singularities for Wedge Angles π or 2π

The results presented in the previous section are simplified substantially for wedge angles $2\alpha=\pi$ or 2π. The wedge angle $2\alpha=\pi$ corresponds to a half-space and $2\alpha=2\pi$ corresponds to a crack in the infinite space. In general an eigenvalue of the wedge depends not only on the wedge angle 2α but also on θ^-. When $2\alpha=\pi$ or 2π the dependence of δ on θ^- disappears. This is so because, in the general solution (9.4-8), the only term that contains δ that is relevant to the determination of δ is $\zeta^{\delta+1}(\theta,\theta^-)$. At $\theta=\theta^-$ and θ^+,

$$\zeta(\theta^-,\theta^-)=1, \quad \zeta(\theta^+,\theta^-)=\cos 2\alpha+p(\theta^-)\sin 2\alpha. \tag{9.5-1}$$

For $2\alpha=\pi$ or 2π,

$$\zeta(\theta^+,\theta^-)=\begin{cases} e^{i\pi} & \text{when } 2\alpha=\pi, \\ e^{2i\pi} & \text{when } 2\alpha=2\pi, \end{cases}$$

so that

$$\zeta^{\delta+1}(\theta^+,\theta^-)=e^{i(\delta+1)\pi}=-e^{i\delta\pi}, \quad \text{when } 2\alpha=\pi, \tag{9.5-2a}$$

$$\zeta^{\delta+1}(\theta^+,\theta^-)=e^{2i(\delta+1)\pi}=e^{2i\delta\pi}, \quad \text{when } 2\alpha=2\pi. \tag{9.5-2b}$$

Since (9.5-1) and (9.5-2) are independent of θ^-, the eigenvalue δ is independent of θ^- when $2\alpha=\pi$ or 2π.

Equation (9.5-2a) remains unchanged when δ is replaced by $\delta+n$ where n is an integer. Hence if δ is an eigenvalue so is $\delta+n$. It suffices to consider

$$-1<\mathrm{Re}(\delta)\leq 0 \tag{9.5-3}$$

when $2\alpha=\pi$. Similarly, (9.5-2b) tells us that if δ is an eigenvalue, so is $\delta+\frac{1}{2}n$ and we need only study

$$-1<\mathrm{Re}(\delta)\leq \tfrac{1}{2} \tag{9.5-4}$$

when $2\alpha=2\pi$. In fact comparison of (9.5-2a,b) shows that if δ is an eigenvalue for $2\alpha=\pi$, $\frac{1}{2}\delta$ is an eigenvalue for $2\alpha=2\pi$. Thus it is sufficient to consider the wedge angle $2\alpha=\pi$.

In summary we have the following results when $2\alpha=\pi$ or 2π:
(i) The eigenvalue δ is independent of θ^-.
(ii) If δ is an eigenvalue so is $\delta+n$.
(iii) If δ is an eigenvalue for $2\alpha=\pi$, $\frac{1}{2}\delta$ is an eigenvalue for $2\alpha=2\pi$.
These conclusions are independent of the types of boundary conditions at $\theta=\theta^-$ and θ^+. For (iii), it remains valid as long as the boundary condition at $\theta=\theta^-$ and θ^+ for $2\alpha=\pi$ and 2π are of the same type. The boundary conditions given in (9.4-9)–(9.4-11) are simple boundary conditions. More general boundary conditions including those for a slippery boundary will be discussed in Section 14.1.

In the following we discuss separately three types of boundary conditions given in (9.4-9)–(9.4-11) (Ting, 1986).

Case I. Wedges with $2\alpha=\pi$.

As stated earlier, if δ is an eigenvalue so is $\delta+n$ where n is an integer. We will therefore find δ in the range given by (9.5-3).

(Ia) **The Free-Free Wedge.** Substitution of (9.5-2a) into $(9.4\text{-}14)_1$ leads to

$$\mathbf{K} = -2i\sin\delta\pi\,\mathbf{I}. \tag{9.5-5}$$

The roots of $|\mathbf{K}|=0$ are

$$\delta = 0 \quad (3). \tag{9.5-6}$$

The number 3 in the parentheses implies that the root is of multiplicity three. At $\delta=0$, $\mathbf{K}=\mathbf{0}$ so that by (9.4-13b) the vector $\mathbf{g}$ is arbitrary. The eigenfunction associated with $\delta=0$ is obtained from (9.4-23a) as

$$\mathbf{u}/r = \left\{\mathbf{A}\langle\zeta_*(\theta,\theta^-)\rangle\mathbf{B}^{-1} - \overline{\mathbf{A}}\langle\overline{\zeta}_*(\theta,\theta^-)\rangle\overline{\mathbf{B}}^{-1}\right\}\mathbf{g},$$

$$\boldsymbol{\phi}/r = \left\{\mathbf{B}\langle\zeta_*(\theta,\theta^-)\rangle\mathbf{B}^{-1} - \overline{\mathbf{B}}\langle\overline{\zeta}_*(\theta,\theta^-)\rangle\overline{\mathbf{B}}^{-1}\right\}\mathbf{g},$$

or, from (9.4-23b),

$$\begin{aligned}\mathbf{u}/r &= \left\{\cos(\theta-\theta^-)\mathbf{I} + \sin(\theta-\theta^-)\mathbf{N}_1(\theta^-)\right\}\mathbf{h},\\ \boldsymbol{\phi}/r &= \sin(\theta-\theta^-)\mathbf{N}_3(\theta^-)\mathbf{h}.\end{aligned} \tag{9.5-7}$$

The constant vector $\mathbf{h}$ is arbitrary, and is related to $\mathbf{g}$ by (9.4-24). Since $\mathbf{h}$ is a 3-vector we have three independent eigenfunctions associated with the triple eigenvalues $\delta=0$. Referring to the bases $\mathbf{n}(\theta^-)$, $\mathbf{m}(\theta^-)$ and $\mathbf{i}_3$ where $\mathbf{i}_3$ is a unit vector in the direction of the x_3-axis, the $\mathbf{m}(\theta^-)$ component of $\mathbf{h}$ furnishes $\boldsymbol{\phi}=\mathbf{0}$ so that it represents a rigid body motion.

(Ib) **The Fixed-Fixed Wedge.** In this case the matrix $\mathbf{K}$ is given by $(9.4\text{-}35)_1$. It is easily shown that insertion of (9.5-2a) into $(9.4\text{-}35)_1$ leads to (9.5-5) so that the eigenvalues are again

$$\delta = 0 \quad (3) \tag{9.5-8}$$

while the eigenvector $\mathbf{g}$ is arbitrary. The eigenfunction for $\delta=0$ is obtained from (9.4-37). For the real form we have

$$\begin{aligned}\mathbf{u}/r &= \sin(\theta-\theta^-)\mathbf{N}_2(\theta^-)\mathbf{h},\\ \boldsymbol{\phi}/r &= \left\{\cos(\theta-\theta^-)\mathbf{I} + \sin(\theta-\theta^-)\mathbf{N}_1^T(\theta^-)\right\}\mathbf{h},\end{aligned} \tag{9.5-9}$$

where $\mathbf{h}$ is an arbitrary constant vector.

(Ic) **The Free-Fixed Wedge**. From (9.5-2a), $(9.4\text{-}41)_3$, and (9.4-44) the eigenrelation (9.4-40) for this case reduces to

$$(e^{i\delta\pi}\mathbf{A}\mathbf{B}^T + e^{-i\delta\pi}\overline{\mathbf{A}\mathbf{B}}^T)\mathbf{h} = \mathbf{0}.$$

This is written in a real form with the aid of $(5.5\text{-}17)_1$ as

$$\{(\sin\delta\pi)\mathbf{S} + (\cos\delta\pi)\mathbf{I}\}\mathbf{h} = \mathbf{0}.$$

When δ=0, $\mathbf{h}=\mathbf{0}$ is the only solution so that δ=0 is not an eigenvalue. Hence the above equation can be written as

$$\{\mathbf{S} + (\cot\delta\pi)\mathbf{I}\}\mathbf{h} = \mathbf{0}, \tag{9.5-10}$$

implying that $-\cot\delta\pi$ and $\mathbf{h}$ are, respectively, an eigenvalue and eigenvector of the tensor $\mathbf{S}$. It was shown in Section 6.6 that the eigenvalues of $\mathbf{S}$ are 0 and $\pm is$ where

$$0 < s = \left\{-\tfrac{1}{2}\mathrm{tr}(\mathbf{S}^2)\right\}^{1/2} < 1. \tag{9.5-11}$$

We therefore have

$$\delta = -\tfrac{1}{2} \quad \text{and} \quad -\tfrac{1}{2} \pm i\gamma,$$
$$\gamma = \tfrac{1}{\pi}\tanh^{-1}s = \tfrac{1}{2\pi}\ln\frac{1+s}{1-s}. \tag{9.5-12}$$

The inequalities for s in (9.5-11) assure us that γ is real and nonzero. Thus two of the three δ's in (9.5-12) are complex. There are three stress singularities. For monoclinic materials with the symmetry plane at x_3=0, $\delta=-\frac{1}{2}$ is associated with the antiplane deformations while $\delta=-\frac{1}{2}\pm i\gamma$ are associated with the inplane deformations.

Case II. Wedges with $2\alpha=2\pi$.

We stated earlier in this section that if δ is an eigenvalue for wedge angle $2\alpha=\pi$, $\frac{1}{2}\delta$ is an eigenvalue for wedge angle $2\alpha=2\pi$. Moreover, if δ is an eigenvalue for $2\alpha=2\pi$, so is $\delta+\frac{1}{2}n$ where n is any integer. Therefore δ's for the present case can be deduced from those in the previous case for $2\alpha=\pi$. In the range indicated by (9.5-4) we have the following results.

(IIa) **The Free-Free Wedge**. The eigenvalues are

$$\delta = 0 \quad (3), \quad \text{and } -\tfrac{1}{2} \quad (3), \tag{9.5-13}$$

each of multiplicity three. The eigenfunction for δ=0 is identical to the one given by (9.5-7).

(IIb) **The Fixed-Fixed Wedge**. The eigenvalues are again

$$\delta = 0 \quad (3), \quad \text{and } -\tfrac{1}{2} \quad (3). \tag{9.5-14}$$

The eigenfunction for δ=0 is identical to the one given by (9.5-9).

(IIc) **The Free-Fixed Wedge**. Equation (9.5-10) is replaced by

$$\{\mathbf{S} + (\cot 2\delta\pi)\mathbf{I}\}\mathbf{h} = \mathbf{0}. \tag{9.5-15}$$

The eigenvalues are

$$\delta = -\tfrac{1}{4}, \quad -\tfrac{3}{4}, \quad -\tfrac{1}{4} \pm \tfrac{1}{2} i\gamma, \quad -\tfrac{3}{4} \pm \tfrac{1}{2} i\gamma, \tag{9.5-16}$$

in which the γ is given by (9.5-12). There are six stress singularities. For monoclinic materials with the symmetry plane at x_3=0, $\delta = -\frac{1}{4}$ and $-\frac{3}{4}$ are associated with the antiplane deformations while the remaining four complex δ's are associated with the inplane deformations.

9.6 Stress Singularities at an Interfacial Crack

In this section we consider a bimaterial that consists of two dissimilar anisotropic half-spaces. Let the half-space

$$r > 0, \quad \theta_o \le \theta \le \theta_o + \pi$$

be occupied by material 1, and the half-space

$$r > 0, \quad \theta_o - \pi \le \theta \le \theta_o$$

be occupied by material 2, Fig. 9.4. The two materials are rigidly bonded at the interface $\theta = \theta_o$. The surfaces $\theta = \theta_o \pm \pi$ can be both traction-free, both fixed, or one surface is traction-free while the other is fixed. We will study stress singularities at the crack tip r=0.

Employing (9.4-8) for materials 1 and 2 we write

$$\begin{aligned} \mathbf{u}_i &= r^{\delta+1}\left\{\mathbf{A}_i\left\langle \zeta_*^{(i)\delta+1}(\theta,\theta_o)\right\rangle \mathbf{q}_i + \overline{\mathbf{A}}_i\left\langle \overline{\zeta}_*^{(i)\delta+1}(\theta,\theta_o)\right\rangle \tilde{\mathbf{q}}_i\right\}, \\ \boldsymbol{\phi}_i &= r^{\delta+1}\left\{\mathbf{B}_i\left\langle \zeta_*^{(i)\delta+1}(\theta,\theta_o)\right\rangle \mathbf{q}_i + \overline{\mathbf{B}}_i\left\langle \overline{\zeta}_*^{(i)\delta+1}(\theta,\theta_o)\right\rangle \tilde{\mathbf{q}}_i\right\}, \end{aligned} \tag{9.6-1}$$

in which i=1 for material 1 and i=2 for material 2 and

$$\zeta^{(i)}(\theta,\theta_o) = \cos(\theta - \theta_o) + p^{(i)}(\theta_o)\sin(\theta - \theta_o).$$

At $\theta = \theta_o$, $\theta_o + \pi$, and $\theta_o - \pi$,

$$\begin{aligned} \zeta^{(i)\delta+1}(\theta_o, \theta_o) &= 1, \quad (i = 1,2), \\ \zeta^{(1)\delta+1}(\theta_o + \pi, \theta_o) &= -e^{i\delta\pi}, \\ \zeta^{(2)\delta+1}(\theta_o - \pi, \theta_o) &= -e^{-i\delta\pi}. \end{aligned} \tag{9.6-2}$$

They are independent of θ_o. These quantities are the only ones that contain δ that are needed in satisfying the interface conditions at $\theta = \theta_o$ and the boundary conditions at $\theta = \theta_o + \pi$ and $\theta_o - \pi$. Therefore δ is *invariant* with θ_o, the orientation of the crack-interface line (Ting, 1986). Moreover, the right sides of (9.6-2) remain the same if δ is replaced by $\delta + n$, where n is any integer. Thus if δ is an eigenvalue, so is $\delta + n$. It suffices to determine δ in the region

$$-1 < \text{Re}(\delta) \le 0. \tag{9.6-3}$$

The above conclusions are independent of the interfacial conditions at $\theta = \theta_o$ and the boundary conditions at $\theta = \theta_o + \pi$ and $\theta_o - \pi$. The two materials need not be rigidly bonded at the interface. The crack surface can be a slippery surface (see the discussions in the paragraph following (9.5-4)).

In this section we assume that the two materials are rigidly bonded at the interface $\theta = \theta_o$. The continuity of displacement and surface traction at the interface means that

$$\mathbf{u}_1 = \mathbf{u}_2, \quad \boldsymbol{\phi}_1 = \boldsymbol{\phi}_2, \quad \text{at } \theta = \theta_o, \tag{9.6-4}$$

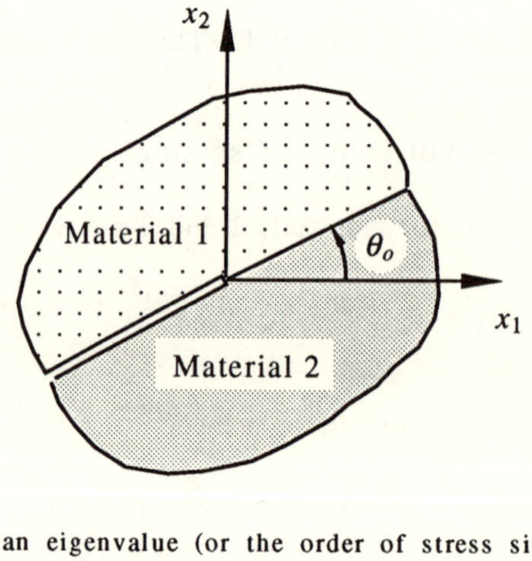

Fig. 9.4 If δ is an eigenvalue (or the order of stress singularity) for the interfacial crack, so is $\delta + n$ where n is an integer. δ is independent of θ_o, the orientation of the crack-interface.

or, by (9.6-1) and $(9.6\text{-}2)_1$,

$$\mathbf{A}_1\mathbf{q}_1 + \overline{\mathbf{A}}_1\tilde{\mathbf{q}}_1 = \mathbf{A}_2\mathbf{q}_2 + \overline{\mathbf{A}}_2\tilde{\mathbf{q}}_2 = \mathbf{h}, \text{ say},$$

$$\mathbf{B}_1\mathbf{q}_1 + \overline{\mathbf{B}}_1\tilde{\mathbf{q}}_1 = \mathbf{B}_2\mathbf{q}_2 + \overline{\mathbf{B}}_2\tilde{\mathbf{q}}_2 = \mathbf{g}, \text{ say}.$$

They can be written as

$$\begin{bmatrix} \mathbf{A}_i & \overline{\mathbf{A}}_i \\ \mathbf{B}_i & \overline{\mathbf{B}}_i \end{bmatrix} \begin{bmatrix} \mathbf{q}_i \\ \tilde{\mathbf{q}}_i \end{bmatrix} = \begin{bmatrix} \mathbf{h} \\ \mathbf{g} \end{bmatrix} \quad (i=1,2).$$

From the orthogonality relation (5.5-13) one obtains

$$\begin{bmatrix} \mathbf{q}_i \\ \tilde{\mathbf{q}}_i \end{bmatrix} = \begin{bmatrix} \mathbf{B}_i^T & \mathbf{A}_i^T \\ \overline{\mathbf{B}}_i^T & \overline{\mathbf{A}}_i^T \end{bmatrix} \begin{bmatrix} \mathbf{h} \\ \mathbf{g} \end{bmatrix}$$

or

$$\mathbf{q}_i = \mathbf{B}_i^T\mathbf{h} + \mathbf{A}_i^T\mathbf{g}, \quad \tilde{\mathbf{q}}_i = \overline{\mathbf{B}}_i^T\mathbf{h} + \overline{\mathbf{A}}_i^T\mathbf{g}. \tag{9.6-5}$$

Equations (9.6-1) become

$$\begin{aligned} \mathbf{u}_i &= r^{\delta+1}\left\{\mathbf{A}_i\left\langle \zeta_*^{(i)\delta+1}(\theta,\theta_o)\right\rangle \mathbf{B}_i^T + \overline{\mathbf{A}}_i\left\langle \overline{\zeta}_*^{(i)\delta+1}(\theta,\theta_o)\right\rangle \overline{\mathbf{B}}_i^T\right\}\mathbf{h} \\ &\quad + r^{\delta+1}\left\{\mathbf{A}_i\left\langle \zeta_*^{(i)\delta+1}(\theta,\theta_o)\right\rangle \mathbf{A}_i^T + \overline{\mathbf{A}}_i\left\langle \overline{\zeta}_*^{(i)\delta+1}(\theta,\theta_o)\right\rangle \overline{\mathbf{A}}_i^T\right\}\mathbf{g}, \\ \boldsymbol{\phi}_i &= r^{\delta+1}\left\{\mathbf{B}_i\left\langle \zeta_*^{(i)\delta+1}(\theta,\theta_o)\right\rangle \mathbf{B}_i^T + \overline{\mathbf{B}}_i\left\langle \overline{\zeta}_*^{(i)\delta+1}(\theta,\theta_o)\right\rangle \overline{\mathbf{B}}_i^T\right\}\mathbf{h} \\ &\quad + r^{\delta+1}\left\{\mathbf{B}_i\left\langle \zeta_*^{(i)\delta+1}(\theta,\theta_o)\right\rangle \mathbf{A}_i^T + \overline{\mathbf{B}}_i\left\langle \overline{\zeta}_*^{(i)\delta+1}(\theta,\theta_o)\right\rangle \overline{\mathbf{A}}_i^T\right\}\mathbf{g}. \end{aligned} \tag{9.6-6}$$

By (9.4-21) they can be written symbolically as

$$\begin{aligned} \mathbf{u}_i &= r^{\delta+1}\left\{\mathbf{N}_1^{(i)(\delta+1)}(\theta,\theta_o)\mathbf{h} + \mathbf{N}_2^{(i)(\delta+1)}(\theta,\theta_o)\mathbf{g}\right\}, \\ \boldsymbol{\phi} &= r^{\delta+1}\left\{\mathbf{N}_3^{(i)(\delta+1)}(\theta,\theta_o)\mathbf{h} + \mathbf{N}_1^{(i)(\delta+1)^T}(\theta,\theta_o)\mathbf{g}\right\}. \end{aligned} \tag{9.6-7}$$

Equations (9.6-7) are useful when δ is an integer.

At $\theta = \theta_o$ (9.6-6) in conjunction with (5.5-16) reduce to

$$\mathbf{u}_i(r,\theta_o) = r^{\delta+1}\mathbf{h}, \quad \boldsymbol{\phi}_i(r,\theta_o) = r^{\delta+1}\mathbf{g}, \quad (i=1,2)$$

so that **h** and **g** represent, respectively, the direction of the displacement and surface traction at the interface $\theta = \theta_o$.

At $\theta = \theta_o \pm \pi$ (9.6-6) reduce to

$$\begin{aligned} \mathbf{u}_1(t,\theta_o+\pi) &= -r^{\delta+1}\big\{(\cos\delta\pi)\mathbf{h} + (\sin\delta\pi)(\mathbf{H}_1\mathbf{g} + \mathbf{S}_1\mathbf{h})\big\}, \\ \boldsymbol{\phi}_1(t,\theta_o+\pi) &= -r^{\delta+1}\big\{(\cos\delta\pi)\mathbf{g} + (\sin\delta\pi)(\mathbf{S}_1^T\mathbf{g} - \mathbf{L}_1\mathbf{h})\big\}, \end{aligned} \tag{9.6-8}$$

and

$$\begin{aligned} \mathbf{u}_2(t,\theta_o-\pi) &= -r^{\delta+1}\big\{(\cos\delta\pi)\mathbf{h} - (\sin\delta\pi)(\mathbf{H}_2\mathbf{g} + \mathbf{S}_2\mathbf{h})\big\}, \\ \boldsymbol{\phi}_2(t,\theta_o-\pi) &= -r^{\delta+1}\big\{(\cos\delta\pi)\mathbf{g} - (\sin\delta\pi)(\mathbf{S}_2^T\mathbf{g} - \mathbf{L}_2\mathbf{h})\big\}. \end{aligned} \tag{9.6-9}$$

We now discuss various types of boundary conditions at the crack surfaces $\theta = \theta_o \pm \pi$.

(a) **The Free-Free Crack.** In this case

$$\boldsymbol{\phi}_1(r,\ \theta_o+\pi) = \boldsymbol{\phi}_2(r,\ \theta_o-\pi) = \mathbf{0}$$

so that, by $(9.6\text{-}8)_2$ and $(9.6\text{-}9)_2$,

$$\begin{aligned} (\cos\delta\pi)\mathbf{L}_1^{-1}\mathbf{g} - (\sin\delta\pi)(\mathbf{S}_1\mathbf{L}_1^{-1}\mathbf{g} + \mathbf{h}) &= \mathbf{0}, \\ (\cos\delta\pi)\mathbf{L}_2^{-1}\mathbf{g} + (\sin\delta\pi)(\mathbf{S}_2\mathbf{L}_2^{-1}\mathbf{g} + \mathbf{h}) &= \mathbf{0}. \end{aligned} \tag{9.6-10}$$

If we set $\mathbf{g}=\mathbf{0}$ we have

$$(\sin\delta\pi)\mathbf{h} = \mathbf{0}$$

which means that

$$\delta = 0 \ \ (3) \tag{9.6-11}$$

is an eigenvalue of multiplicity three and $\mathbf{h}$ is arbitrary. When $\delta \neq 0$, elimination of $\mathbf{h}$ in (9.6-10) yields

$$\big\{\breve{\mathbf{S}} - (\cot\delta\pi)\mathbf{I}\big\}\mathbf{g} = \mathbf{0} \tag{9.6-12}$$

where

$$\breve{\mathbf{S}} = \mathbf{D}^{-1}\mathbf{W}, \tag{9.6-13}$$

$$\mathbf{D} = \mathbf{L}_1^{-1} + \mathbf{L}_2^{-1}, \quad \mathbf{W} = \mathbf{S}_1\mathbf{L}_1^{-1} - \mathbf{S}_2\mathbf{L}_2^{-1}. \tag{9.6-14}$$

Therefore $\cot\delta\pi$ and $\mathbf{g}$ are, respectively, an eigenvalue and eigenvector of the tensor $\breve{\mathbf{S}}$. Like the tensor $\mathbf{S}$, $\breve{\mathbf{S}}$ is the product of a positive definite tensor $\mathbf{D}^{-1}$ and a skew-symmetric tensor $\mathbf{W}$. It was shown in (8.8-41) that $\breve{\mathbf{S}}$, $\mathbf{D}^{-1}$, $\tilde{\mathbf{L}}^{-1}$ have the same mathematical structure as have the three Barnett–Lothe tensors $\mathbf{S}$, $\mathbf{H}$, $\mathbf{L}$. Thus the eigenvalues of $\breve{\mathbf{S}}$ are 0 and $\pm i\beta$ where

$$0 \le \beta = \big\{-\tfrac{1}{2}\mathrm{tr}(\breve{\mathbf{S}}^2)\big\}^{1/2} < 1. \tag{9.6-15}$$

The inequalities can be verified following the proof of (6.6-5) except that β can be zero because $\breve{\mathbf{S}}$ can vanish. We therefore have

$$\delta = -\tfrac{1}{2} \quad \text{and} \quad -\tfrac{1}{2} \mp i\gamma,$$
$$\gamma = \tfrac{1}{\pi}\tanh^{-1}\beta = \tfrac{1}{2\pi}\ln\frac{1+\beta}{1-\beta}. \tag{9.6-16}$$

The inequalities in (9.6-15) assure us that γ in (9.6-16) is real and nonzero unless $\beta=0$.

Equation $(9.6\text{-}16)_2$ gives an explicit expression of γ for an interfacial crack with traction-free crack surfaces. The fact that the singularities at an interfacial crack tip of an anisotropic bimaterial can be written as $-\tfrac{1}{2}\pm i\gamma$ was stated in (Wang, 1984) but no proofs were given. Nor was the expression for γ (or β) given except for an orthotropic material whose planes of symmetry coincide with the coordinate planes. For isotropic bimaterials $\delta=-\tfrac{1}{2}$ is associated with antiplane deformation and $\delta=-\tfrac{1}{2}\pm i\gamma$ are associated with inplane deformations. The β for isotropic bimaterials was first obtained by Williams (1959) and reproduced by Bogy (1971a) in a compact form. Using **S** and **L** in (6.4-13) for isotropic materials it can be shown that

$$\beta = \frac{\mu_2(1-2\nu_1)-\mu_1(1-2\nu_2)}{2[\mu_2(1-\nu_1)+\mu_1(1-\nu_2)]}. \tag{9.6-17}$$

The β shown above is one of the two Dundurs constants for isotropic bimaterials (Dundurs, 1970). Extension of the interfacial singularities to piezoelectric bimaterials has been given by Kuo and Barnett (1991). They showed that there are four δ in which the real δ may not be $-\tfrac{1}{2}$.

(b) **The Fixed-Fixed Crack**. This is equivalent to a rigid line inclusion along the crack. For this case,

$$\mathbf{u}_1(r,\ \theta_o+\pi) = \mathbf{u}_2(r,\ \theta_o-\pi) = \mathbf{0}$$

so that, from $(9.6\text{-}8)_1$ and $(9.6\text{-}9)_1$,

$$(\cos\delta\pi)\mathbf{H}_1^{-1}\mathbf{h} + (\sin\delta\pi)(\mathbf{g} - \mathbf{S}_1^T\mathbf{H}_1^{-1}\mathbf{h}) = \mathbf{0},$$
$$(\cos\delta\pi)\mathbf{H}_2^{-1}\mathbf{h} - (\sin\delta\pi)(\mathbf{g} - \mathbf{S}_2^T\mathbf{H}_2^{-1}\mathbf{h}) = \mathbf{0}. \tag{9.6-18}$$

By setting $\mathbf{h}=\mathbf{0}$ one obtains

$$(\sin\delta\pi)\mathbf{g} = \mathbf{0}$$

implying that

$$\delta = 0 \quad (3) \tag{9.6-19}$$

is an eigenvalue of multiplicity three and $\mathbf{g}$ is arbitrary. When $\delta\neq 0$, elimination of $\mathbf{g}$ in (9.6-18) leads to

$$\left\{\hat{\mathbf{S}}^T + (\cot\delta\pi)\mathbf{I}\right\}\mathbf{h} = \mathbf{0} \tag{9.6-20}$$

where

$$\hat{\mathbf{S}} = \hat{\mathbf{W}}\hat{\mathbf{D}}^{-1}, \tag{9.6-21}$$

$$\hat{\mathbf{D}} = \mathbf{H}_1^{-1} + \mathbf{H}_2^{-1}, \quad \hat{\mathbf{W}} = \mathbf{S}_1^T\mathbf{H}_1^{-1} - \mathbf{S}_2^T\mathbf{H}_2^{-1}. \tag{9.6-22}$$

Thus $-\cot\delta\pi$ and $\mathbf{h}$ are, respectively, an eigenvalue and left eigenvector of $\hat{\mathbf{S}}$. Like the tensor $\mathbf{S}$, $\hat{\mathbf{S}}$ is the product of a positive definite tensor $\hat{\mathbf{D}}^{-1}$ and a skew-symmetric tensor $\hat{\mathbf{W}}$. It was shown in (8.8-43) that $\hat{\mathbf{S}}$, $\tilde{\mathbf{H}}^{-1}$, $\hat{\mathbf{D}}^{-1}$ have the same mathematical structure as have the three Barnett–Lothe tensors $\mathbf{S}$, $\mathbf{H}$, $\mathbf{L}$. Therefore the eigenvalues of $\hat{\mathbf{S}}$ are 0 and $\pm i\hat{\beta}$ where

$$0 \le \hat{\beta} = \left\{-\tfrac{1}{2}\mathrm{tr}(\hat{\mathbf{S}}^2)\right\}^{1/2} < 1. \tag{9.6-23}$$

The inequalities can be verified following the proof of (6.6-5) except that $\hat{\beta}$ can be zero because $\hat{\mathbf{S}}$ can vanish. Hence we have

$$\begin{gathered}\delta = -\tfrac{1}{2} \quad \text{and} \quad -\tfrac{1}{2} \pm i\gamma, \\ \gamma = \tfrac{1}{\pi}\tanh^{-1}\hat{\beta} = \tfrac{1}{2\pi}\ln\frac{1+\hat{\beta}}{1-\hat{\beta}}.\end{gathered} \tag{9.6-24}$$

The inequalities in (9.6-23) assure us that γ in (9.6-24) is real and nonzero unless $\hat{\beta}=0$.

The corresponding problem for isotropic materials was investigated by Erdogan and Gupta (1972). In Case (a) and Case (b) studied here, $\delta=-\frac{1}{2}$ is associated with antiplane deformations and $\delta=-\frac{1}{2}\pm i\gamma$ are associated with inplane deformations for monoclinic materials with the symmetry plane at $x_3=0$.

(c) **The Free-Fixed Crack**. For this case,

$$\boldsymbol{\phi}_2(r, \theta_o-\pi) = \mathbf{u}_1(r, \theta_o+\pi) = \mathbf{0}$$

so that, by $(9.6\text{-}9)_2$ and $(9.6\text{-}8)_1$,

$$(\cos\delta\pi)\mathbf{g} - (\sin\delta\pi)(\mathbf{S}_2^T\mathbf{g} - \mathbf{L}_2\mathbf{h}) = \mathbf{0},$$

$$(\cos\delta\pi)\mathbf{h} + (\sin\delta\pi)(\mathbf{H}_1\mathbf{g} + \mathbf{S}_1\mathbf{h}) = \mathbf{0}.$$

Since $\delta=0$ leads to vanishing of $\mathbf{g}$ and $\mathbf{h}$, $\delta=0$ is not an eigenvalue. The two equations can then be written as

$$\begin{aligned}\left\{\mathbf{S}_2^T - (\cot\delta\pi)\mathbf{I}\right\}\mathbf{g} &= \mathbf{L}_2\mathbf{h}, \\ \left\{\mathbf{S}_1 + (\cot\delta\pi)\mathbf{I}\right\}\mathbf{h} &= -\mathbf{H}_1\mathbf{g},\end{aligned} \tag{9.6-25a}$$

or

$$\begin{bmatrix} \mathbf{S}_1 & \mathbf{H}_1 \\ \mathbf{L}_2 & -\mathbf{S}_2^T \end{bmatrix} \begin{bmatrix} \mathbf{h} \\ \mathbf{g} \end{bmatrix} = -(\cot \delta\pi) \begin{bmatrix} \mathbf{h} \\ \mathbf{g} \end{bmatrix} \tag{9.6-25b}$$

which tells us that $-\cot\delta\pi$ and $[\mathbf{h}, \mathbf{g}]$ are, respectively, an eigenvalue and eigenvector of the 6×6 matrix on the left. The eigenequation is a sextic equation in $\cot\delta\pi$. We will show that it is a cubic equation in $(\cot\delta\pi)^2$.

To this end, we solve $\mathbf{h}$ from $(9.6\text{-}25a)_1$,

$$\mathbf{h} = -\{\mathbf{S}_2 + (\cot \delta\pi)\mathbf{I}\}\mathbf{L}_2^{-1}\mathbf{g},$$

and substitute it into $(9.6\text{-}25a)_2$ to obtain

$$\{[\mathbf{S}_1 + (\cot \delta\pi)\mathbf{I}][\mathbf{S}_2 + (\cot \delta\pi)\mathbf{I}] - \mathbf{H}_1\mathbf{L}_2\}\mathbf{L}_2^{-1}\mathbf{g} = \mathbf{0}. \tag{9.6-26a}$$

If we insert $\mathbf{g}$ obtained from $(9.6\text{-}25)_2$ into $(9.6\text{-}25)_1$ we have

$$\{[\mathbf{S}_2^T - (\cot \delta\pi)\mathbf{I}][\mathbf{S}_1^T - (\cot \delta\pi)\mathbf{I}] - \mathbf{L}_2\mathbf{H}_1\}\mathbf{H}_1^{-1}\mathbf{h} = \mathbf{0}. \tag{9.6-26b}$$

Let

$$\mathbf{K}(\delta) = [\mathbf{S}_1 + (\cot \delta\pi)\mathbf{I}][\mathbf{S}_2 + (\cot \delta\pi)\mathbf{I}] - \mathbf{H}_1\mathbf{L}_2 \tag{9.6-27}$$

so that (9.6-26a,b) are written as

$$\mathbf{K}(\delta)[\mathbf{L}_2^{-1}\mathbf{g}] = \mathbf{0}, \quad \mathbf{K}^T(-\delta)[\mathbf{H}_1^{-1}\mathbf{h}] = \mathbf{0}.$$

Thus $\mathbf{L}_2^{-1}\mathbf{g}$ and $\mathbf{H}_1^{-1}\mathbf{h}$ are, respectively, a right eigenvector of $\mathbf{K}(\delta)$ and a left eigenvector of $\mathbf{K}(-\delta)$. The roots of

$$|\mathbf{K}(\delta)| = 0 \quad \text{and} \quad |\mathbf{K}(-\delta)| = 0$$

must be identical, implying that $-\delta$ is an eigenvalue if δ is. Hence the eigenequation is a cubic equation in $(\cot\delta\pi)^2$. The three pairs of eigenvalues have the form

$$\delta = -\tfrac{1}{2} \pm \eta_1, \quad -\tfrac{1}{2} + (\varepsilon \pm i\gamma), \quad -\tfrac{1}{2} - (\varepsilon \pm i\gamma),$$

if only one root of the cubic equation is real and

$$\delta = -\tfrac{1}{2} \pm \eta_1, \quad -\tfrac{1}{2} \pm i\eta_2, \quad -\tfrac{1}{2} \pm i\eta_3,$$

if all three roots are real. ε and γ are real while η_k (k=1,2,3) is real (or purely imaginary) when the real root $(\cot\delta_k\pi)^2$ is positive (or negative). Explicit expressions of the six δ have been obtained by Ting (1986) and Ballarini (1995) for isotropic bimaterials.

To see how (9.6-26a) reduces to (9.5-10) when the two materials are identical, let

$$\mathbf{Y} = \mathbf{H}_1\mathbf{L}_2 - \mathbf{S}_1\mathbf{S}_2 - \mathbf{I}$$

which vanishes when the two materials are identical. Equation (9.6-26a) can be written as

$$\left\{\tfrac{1}{2}(\mathbf{S}_1 + \mathbf{S}_2) + (\cot\delta\pi)\mathbf{I} - (2\cot\delta\pi)^{-1}\mathbf{Y}\right\}\mathbf{H}_1^{-1}\mathbf{h} = \mathbf{0}.$$

When $\mathbf{Y}=\mathbf{0}$, and in particular when the two materials are identical, this reduces to (9.5-10). It should be added that the $\mathbf{h}$ here is different from the $\mathbf{h}$ in (9.5-10). It would be interesting to see what bimaterials yield $\mathbf{Y}=\mathbf{0}$ besides the special case when the two materials are identical.

9.7 Singularities of the Form $r^{\delta}(\ln r)^{n}$

The stress singularities generated from an eigenfunction presented in the previous sections have the form r^{δ}, $\mathrm{Re}(\delta)<0$. The displacement $\mathbf{u}$ and the stress function $\boldsymbol{\phi}$ are assumed to have the form given by (9.4-8). Upon substitution into the homogeneous boundary conditions (9.4-9), (9.4-10), or (9.4-11) one obtains

$$\mathbf{K}(\delta)\mathbf{g} = \mathbf{0} \tag{9.7-1}$$

in which $\mathbf{K}(\delta)$ is an $n\times n$ matrix that is a function of δ and $\mathbf{g}$ is an eigenvector. We also obtain (9.7-1) for more general homogeneous boundary conditions and for bimaterials and composite wedges. The roots of the determinant

$$|\mathbf{K}(\delta)| = 0 \tag{9.7-2}$$

provides the eigenvalues δ. If $\delta=\delta_o$ is a simple root, the associated eigenvector $\mathbf{g}$ is unique up to a constant multiplier. If $\delta=\delta_o$ is a root of multiplicity m, (9.7-1) may or may not provide m independent eigenvectors $\mathbf{g}$. When (9.7-1) provides m independent eigenvectors we have m eigenfunctions with the stress singularities in the form of r^{δ}. In this section we consider the case in which (9.7-1) provides less than m independent eigenvectors.

When the number of independent eigenvectors is less than the multiplicity of the eigenvalues, there are less than m eigenfunctions associated with the m eigenvalues. The missing eigenfunctions can be determined as follows. Take (9.4-8) as an example. First we observe that (9.4-8) remains an admissible solutions if $\mathbf{q}$ and $\tilde{\mathbf{q}}$ are allowed to depend on δ. Secondly, if we differentiate the right sides of (9.4-8) with δ it remains an admissible solution. Therefore we consider

$$
\begin{aligned}
\mathbf{u} &= \frac{d}{d\delta}\Big[r^{\delta+1}\Big\{\mathbf{A}\big\langle\zeta_*^{\delta+1}(\theta,\theta^-)\big\rangle\mathbf{q}(\delta)+\overline{\mathbf{A}}\big\langle\overline{\zeta}_*^{\delta+1}(\theta,\theta^-)\big\rangle\tilde{\mathbf{q}}(\delta)\Big\}\Big],\\
\boldsymbol{\phi} &= \frac{d}{d\delta}\Big[r^{\delta+1}\Big\{\mathbf{B}\big\langle\zeta_*^{\delta+1}(\theta,\theta^-)\big\rangle\mathbf{q}(\delta)+\overline{\mathbf{B}}\big\langle\overline{\zeta}_*^{\delta+1}(\theta,\theta^-)\big\rangle\tilde{\mathbf{q}}(\delta)\Big\}\Big].
\end{aligned}
\tag{9.7-3}
$$

Upon differentiation the stress function is

$$
\begin{aligned}
\boldsymbol{\phi} &= r^{\delta+1}(\ln r)\Big\{\mathbf{B}\big\langle\zeta_*^{\delta+1}(\theta,\theta^-)\big\rangle\mathbf{q}(\delta)+\overline{\mathbf{B}}\big\langle\overline{\zeta}_*^{\delta+1}(\theta,\theta^-)\big\rangle\tilde{\mathbf{q}}(\delta)\Big\}\\
&\quad + r^{\delta+1}\frac{d}{d\delta}\Big\{\mathbf{B}\big\langle\zeta_*^{\delta+1}(\theta,\theta^-)\big\rangle\mathbf{q}(\delta)+\overline{\mathbf{B}}\big\langle\overline{\zeta}_*^{\delta+1}(\theta,\theta^-)\big\rangle\tilde{\mathbf{q}}(\delta)\Big\}.
\end{aligned}
\tag{9.7-4}
$$

Thus if $\mathbf{q}(\delta)$ and $\tilde{\mathbf{q}}(\delta)$ are nonzero the stress has a $r^{\delta}(\ln r)$ singularity. Equation (9.7-4) can be written as

$$
\boldsymbol{\phi} = r^{\delta+1}\Big((\ln r)+\frac{d}{d\delta}\Big)\Big\{\mathbf{B}\big\langle\zeta_*^{\delta+1}(\theta,\theta^-)\big\rangle\mathbf{q}(\delta)+\overline{\mathbf{B}}\big\langle\overline{\zeta}_*^{\delta+1}(\theta,\theta^-)\big\rangle\tilde{\mathbf{q}}(\delta)\Big\}. \tag{9.7-5}
$$

Substitution of (9.7-3) into prescribed boundary conditions (and interfacial conditions if a composite wedge is considered) yields an equation of the form

$$
\Big((\ln r)+\frac{d}{d\delta}\Big)\{\mathbf{K}(\delta)\mathbf{g}(\delta)\}=\mathbf{0}. \tag{9.7-6}
$$

This means that

$$
\mathbf{K}\mathbf{g}=\mathbf{0},\quad \mathbf{K}'\mathbf{g}+\mathbf{k}\mathbf{g}'=\mathbf{0}, \tag{9.7-7}
$$

where the dependence of $\mathbf{K}$ and $\mathbf{g}$ on δ is understood and a prime denotes differentiation with δ. The vector $\mathbf{g}$ contains $\mathbf{q}(\delta)$ and $\tilde{\mathbf{q}}(\delta)$ while $\mathbf{g}'$ contains $\mathbf{q}'(\delta)$ and $\tilde{\mathbf{q}}'(\delta)$. By (9.7-4) a nonzero $\mathbf{g}$ provides the $r^{\delta}(\ln r)$ singularity.

Let k be the number of independent eigenvectors $\mathbf{g}$ at $\delta=\delta_0$. The rank of $\mathbf{K}$ at $\delta=\delta_0$ is then $n-k$., i.e., the largest submatrix of $\mathbf{K}$ at $\delta=\delta_0$ with nonzero determinant has the size $(n-k)\times(n-k)$. We may rearrange the rows and columns of $\mathbf{K}$ such that

$$
\mathbf{K}=\begin{bmatrix}\mathbf{K}_1 & \mathbf{K}_2\\ \mathbf{K}_3 & \mathbf{K}_4\end{bmatrix},\quad \mathbf{g}=\begin{bmatrix}\mathbf{g}_1\\ \mathbf{g}_2\end{bmatrix} \tag{9.7-8}
$$

in which $\mathbf{K}_1$ is an $(n-k)\times(n-k)$ nonsingular matrix. From

$$
\mathbf{K}=\begin{bmatrix}\mathbf{K}_1 & \mathbf{0}\\ \mathbf{K}_3 & \mathbf{I}\end{bmatrix}\begin{bmatrix}\mathbf{I} & \mathbf{K}_1^{-1}\mathbf{K}_2\\ \mathbf{0} & \mathbf{U}\end{bmatrix}, \tag{9.7-9}
$$

$$
\mathbf{U}=\mathbf{K}_4-\mathbf{K}_3\mathbf{K}_1^{-1}\mathbf{K}_2, \tag{9.7-10}
$$

$(9.7\text{-}7)_1$ is equivalent to

$$\mathbf{g}_1 = -\mathbf{K}_1^{-1}\mathbf{K}_2\mathbf{g}_2, \quad \mathbf{U}\mathbf{g}_2 = \mathbf{0}. \tag{9.7-11}$$

The matrix $\mathbf{U}$ is a $k\times k$ matrix. Equations (9.7-11) are equivalent to $(9.7\text{-}7)_1$ which is supposed to possess k independent eigenvectors $\mathbf{g}$. Therefore the k-vector $\mathbf{g}_2$ must be arbitrary and we have

$$\mathbf{U} = \mathbf{0} \quad \text{at} \quad \delta = \delta_o. \tag{9.7-12}$$

It remains to find a nonzero $\mathbf{g}_2$ that, together with $\mathbf{g}_1$ determined from $(9.7\text{-}11)_1$, satisfies $(9.7\text{-}7)_2$.

If we multiply $(9.7\text{-}7)_2$ by the matrix $[-\mathbf{K}_3\mathbf{K}_1^{-1}, \ \mathbf{I}]$, the $\mathbf{g}'$ is eliminated due to (9.7-12) and we obtain

$$\mathbf{U}'\mathbf{g}_2 = \mathbf{0} \quad \text{at} \quad \delta = \delta_o. \tag{9.7-13}$$

In deducing (9.7-13) we have made use of $(9.7\text{-}11)_1$, (9.7-10), and the relation

$$\mathbf{K}_1'\mathbf{K}_1^{-1} = -\mathbf{K}_1(\mathbf{K}_1^{-1})'$$

obtained by differentiating $\mathbf{K}_1\mathbf{K}_1^{-1}=\mathbf{I}$. For a nonzero solution of $\mathbf{g}_2$ one must have

$$|\mathbf{U}'| = 0 \quad \text{at} \quad \delta = \delta_o. \tag{9.7-14}$$

It is readily shown from (9.7-12) that

$$\begin{aligned} \frac{d^i}{d\delta^i}|\mathbf{U}| &= 0 \quad \text{at} \quad \delta = \delta_o, \quad \text{for } i < k, \\ \frac{d^k}{d\delta^k}|\mathbf{U}| &= k!\,|\mathbf{U}'| \quad \text{at} \quad \delta = \delta_o. \end{aligned} \tag{9.7-15}$$

By (9.7-9) and (9.7-15)

$$\frac{d^k}{d\delta^k}|\mathbf{K}| = k!\,|\mathbf{K}_1||\mathbf{U}'| \quad \text{at} \quad \delta = \delta_o.$$

In conjunction with (9.7-14) we conclude that (9.7-7) has a nonzero solution for $\mathbf{g}$ if

$$\frac{d^k}{d\delta^k}|\mathbf{K}| = 0 \quad \text{at} \quad \delta = \delta_o. \tag{9.7-16}$$

When $\delta=\delta_o$ is an eigenvalue of multiplicity m,

$$\frac{d^i}{d\delta^i}|\mathbf{K}| = 0 \quad \text{at} \quad \delta = \delta_o, \quad \text{for } i < m. \tag{9.7-17}$$

If $k<m$, which is the case under consideration, (9.7-16) is satisfied and we have a nonzero solution for $\mathbf{g}$ (Dempsey and Sinclair, 1979).

The derivation presented above for the existence of a nonzero $\mathbf{g}$ is not the best way to determine $\mathbf{g}$. In applications it is simpler to assume $\mathbf{g}$ in the form

$$\mathbf{g} = c_1\mathbf{g}_1 + c_2\mathbf{g}_2 + \cdots + c_k\mathbf{g}_k$$

where $\mathbf{g}_i$ $(i=1,2,...k)$ are the independent eigenvectors of $\mathbf{K}$ at $\delta=\delta_o$ and c_i are arbitrary constants to be determined. Let $\mathbf{h}_i$ $(i=1,2,...k)$ be the independent left eigenvectors of $\mathbf{K}$ at $\delta=\delta_o$. Multiplication of $(9.7\text{-}7)_2$ by $\mathbf{h}_i^T$ leads to

$$V_{ij}c_j = 0, \tag{9.7-18}$$

where

$$V_{ij} = \mathbf{h}_i^T\mathbf{K}'\mathbf{g}_j.$$

The existence of a solution for $\mathbf{g}$ assures us that the $k\times k$ matrix V_{ij} must be singular. A solution for c_i $(i=1,2,...k)$ can be obtained from (9.7-18).

If $\mathbf{g}$ so obtained has $m-k$ independent solutions we have m eigenfunctions; k of which have the r^δ singularity and $m-k$ of which have the $r^\delta(\ln r)$ singularity. If there are less than $m-k$ independent solutions for $\mathbf{g}$, the missing eigenfunction can be obtained by differentiating (9.7-3) once more with δ. The stress function (9.7-5) is then replaced by

$$\phi = r^{\delta+1}\left((\ln r)+\frac{d}{d\delta}\right)^2\left\{\mathbf{B}\left\langle\zeta_*^{\delta+1}(\theta,\theta^-)\right\rangle\mathbf{q}(\delta)+\overline{\mathbf{B}}\left\langle\overline{\zeta}_*^{\delta+1}(\theta,\theta^-)\right\rangle\tilde{\mathbf{q}}(\delta)\right\}.$$

The stress now has the $r^\delta(\ln r)^2$ singularities. One may continue this process until there are m eigenfunctions associated with the m repeated eigenvalues. The net result is that the stress singularities have the form $r^\delta(\ln r)^n$ $(n=1,2,...)$.

The logarithmic singularities occur also in nonhomogeneous boundary value problems as we have seen in Sections 9.3. In this case (9.7-1) is replaced by

$$\mathbf{K}(0)\mathbf{g} = \mathbf{d} \tag{9.7-19}$$

where $\mathbf{d}$ depends on the traction or the displacement prescribed at the boundaries. Equation (9.7-19) has a unique solution if $\mathbf{K}(0)$ is nonsingular. If $\mathbf{K}(0)$ is singular, it has a solution only if the vector $\mathbf{d}$ is orthogonal to all left eigenvectors of $\mathbf{K}(0)$ (Hildebrand, 1954). If not, a modified solution can be obtained by employing (9.7-3) which has a $\ln r$ singularity. Satisfaction of the boundary conditions with $\delta=0$ leads to

$$\left((\ln r)+\frac{d}{d\delta}\right)\{\mathbf{K}(\delta)\mathbf{g}(\delta)\}=\mathbf{d} \quad \text{at} \quad \delta=0. \tag{9.7-20}$$

This means that

$$\mathbf{Kg}=\mathbf{0}, \quad \mathbf{K}'\mathbf{g}+\mathbf{kg}'=\mathbf{d}. \tag{9.7-21}$$

The stress has a $\ln r$ singularity when a nonzero $\mathbf{g}$ exists. One may follow the proof of the existence of a solution for $\mathbf{g}$ of (9.7-7) to investigate the existence of a solution for $\mathbf{g}$ in (9.7-21) If a nonzero $\mathbf{g}$ does not exist, we differentiate (9.7-3) once more with δ. The stress then has the $(\ln r)^2$ singularity. Continuing this process until one finds a solution, the stress has the $(\ln r)^n$ singularity where n is an integer.

9.8 Singularities Due to Curved Boundaries

The eigenfunctions for a wedge presented in Section 9.4 assume that the boundaries of the wedge are straight lines. The eigenfunctions are proportional to r^δ where δ is an eigenvalue. If δ_1, δ_2 are two eigenvalues with negative real parts and if

$$-1<\mathrm{Re}(\delta_1)<\mathrm{Re}(\delta_2)<0 \tag{9.8-1}$$

the eigenfunction associated with δ_1 provides the strongest stress singularity. The eigenfunction associated with δ_2 gives the second strongest stress singularity. When the boundaries of the wedge are curved, the second strongest singularity may not be the one given by δ_2. For the specimen shown in Fig. 9.5 the corner Q represents a wedge of wedge angle $3\pi/2$ with one curved boundary. The corner M represents a wedge of wedge angle π with one curved boundary. The crack tip R represents a wedge of wedge angle 2π with two curved boundaries. The corresponding problem for isotropic wedges with curved boundaries were investigated by Ting (1985c).

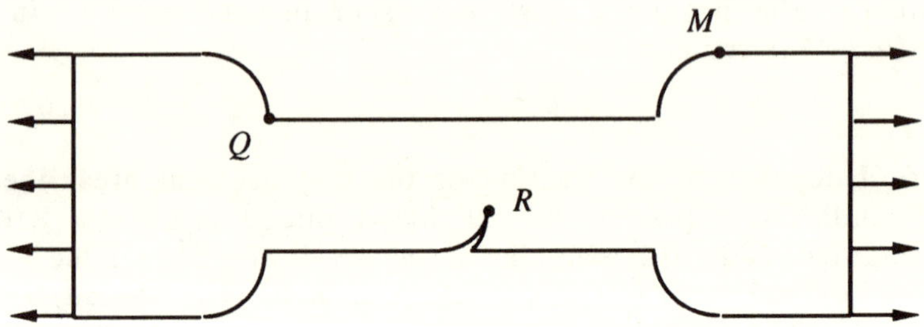

Fig. 9.5 The points M, Q, R represent apexes of wedges of wedge angles π, $3\pi/2$, 2π, respectively, with curved boundaries.

The problem of stress distribution in an infinite isotropic material containing a cusped crack of finite length has been studied by C. H. Wu (1982a,b).

Consider a wedge with two curved boundaries Γ_+ and Γ_-, Fig. 9.6. Let Γ be a curve in the (x_1,x_2) plane defined by

$$\theta = \theta_\Gamma + \kappa_\Gamma r + \cdots \tag{9.8-2}$$

where θ_Γ and k_Γ are constants and the dots denote terms of order higher than r. At r=0 the tangent to the curve Γ makes an angle θ_Γ with the x_1-axis and the curvature of the curve at r=0 is k_Γ. We will modify the eigenfunction given by (9.4-8) for a wedge with curved boundaries. Substitution from (9.8-2) into (9.4-6) leads to

$$\zeta(\theta,\theta^-) = \left\{\cos(\theta_\Gamma - \theta^-) + p(\theta^-)\sin(\theta_\Gamma - \theta^-)\right\}$$
$$+\left\{-\sin(\theta_\Gamma - \theta^-) + p(\theta^-)\cos(\theta_\Gamma - \theta^-)\right\}\kappa_\Gamma r + \cdots$$

or, with the aid of (7.1-31),

$$\zeta(\theta,\theta^-) = \zeta(\theta_\Gamma,\theta^-) + \zeta(\theta_\Gamma,\theta^-)p(\theta_\Gamma)\kappa_\Gamma r + \cdots.$$

Hence

$$\zeta^{\delta+1}(\theta,\theta^-) = \zeta^{\delta+1}(\theta_\Gamma,\theta^-)\{1 + (\delta+1)\kappa_\Gamma p(\theta_\Gamma)r + \cdots\}.$$

Let the equations for the boundaries Γ_+ and Γ_- be given by

$$\theta = \theta^- + \kappa^- r + \cdots, \quad \text{for } \Gamma_-,$$

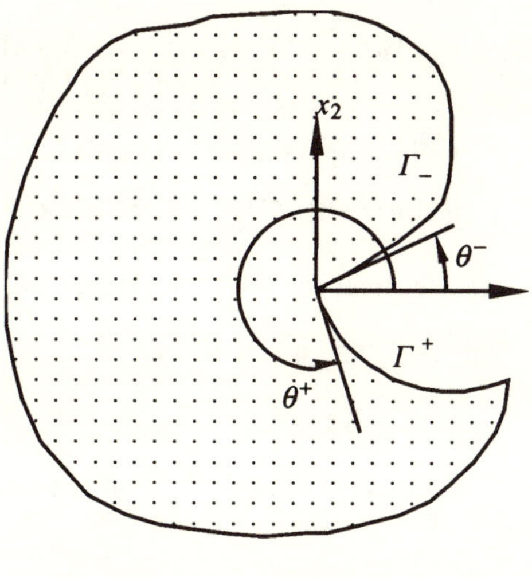

Fig. 9.6 Anisotropic elastic wedge with curved boundaries.

$$\theta = \theta^+ + \kappa^+ r + \cdots, \quad \text{for } \Gamma_+.$$

We then have

$$\zeta^{\delta+1}(\theta,\theta^-) = 1 + (\delta+1)\kappa^- p(\theta^-) r + \cdots, \quad \text{for } \Gamma_-, \tag{9.8-3}$$

$$\zeta^{\delta+1}(\theta,\theta^-) = \zeta^{\delta+1}(\theta^+,\theta^-)\left\{1 + (\delta+1)\kappa^+ p(\theta^+) r + \cdots\right\}, \quad \text{for } \Gamma_+. \tag{9.8-4}$$

This suggests that the eigenfunction (9.4-8) be modified as

$$\begin{aligned} \mathbf{u} = {} & r^{\delta+1}\left\{\mathbf{A}\left\langle\zeta_*^{\delta+1}(\theta,\theta^-)\right\rangle\mathbf{q}^{(0)} + \overline{\mathbf{A}}\left\langle\overline{\zeta}_*^{\delta+1}(\theta,\theta^-)\right\rangle\tilde{\mathbf{q}}^{(0)}\right\} \\ & + r^{\delta+2}\left\{\mathbf{A}\left\langle\zeta_*^{\delta+2}(\theta,\theta^-)\right\rangle\mathbf{q}^{(1)} + \overline{\mathbf{A}}\left\langle\overline{\zeta}_*^{\delta+2}(\theta,\theta^-)\right\rangle\tilde{\mathbf{q}}^{(1)}\right\} + \cdots, \end{aligned} \tag{9.8-5a}$$

$$\begin{aligned} \boldsymbol{\phi} = {} & r^{\delta+1}\left\{\mathbf{B}\left\langle\zeta_*^{\delta+1}(\theta,\theta^-)\right\rangle\mathbf{q}^{(0)} + \overline{\mathbf{B}}\left\langle\overline{\zeta}_*^{\delta+1}(\theta,\theta^-)\right\rangle\tilde{\mathbf{q}}^{(0)}\right\}. \\ & + r^{\delta+2}\left\{\mathbf{B}\left\langle\zeta_*^{\delta+2}(\theta,\theta^-)\right\rangle\mathbf{q}^{(1)} + \overline{\mathbf{B}}\left\langle\overline{\zeta}_*^{\delta+2}(\theta,\theta^-)\right\rangle\tilde{\mathbf{q}}^{(1)}\right\} + \cdots, \end{aligned} \tag{9.8-5b}$$

where $\mathbf{q}^{(0)}$, $\tilde{\mathbf{q}}^{(0)}$, $\mathbf{q}^{(1)}$, and $\tilde{\mathbf{q}}^{(1)}$ are constant vectors to be determined.

Consider a free-free wedge so that

$$\boldsymbol{\phi} = \mathbf{0} \quad \text{at} \quad \Gamma_- \text{ and } \Gamma_+.$$

Insertion of (9.8-3), (9.8-4), and (9.8-3) and (9.8-4) with $\delta+1$ replaced by $\delta+2$ into the above yields

$$\begin{aligned} & \mathbf{B}\mathbf{q}^{(0)} + \overline{\mathbf{B}}\tilde{\mathbf{q}}^{(0)} = \mathbf{0}, \\ & \mathbf{B}\left\langle\zeta_*^{\delta+1}(\theta^+,\theta^-)\right\rangle\mathbf{q}^{(0)} + \overline{\mathbf{B}}\left\langle\overline{\zeta}_*^{\delta+1}(\theta^+,\theta^-)\right\rangle\tilde{\mathbf{q}}^{(0)} = \mathbf{0}, \end{aligned} \tag{9.8-6}$$

$$\mathbf{B}\mathbf{q}^{(1)} + \overline{\mathbf{B}}\tilde{\mathbf{q}}^{(1)} + (\delta+1)\kappa^-\left\{\mathbf{B}\left\langle p_*(\theta^-)\right\rangle\mathbf{q}^{(0)} + \overline{\mathbf{B}}\left\langle\overline{p}_*(\theta^-)\right\rangle\tilde{\mathbf{q}}^{(0)}\right\} = \mathbf{0}, \tag{9.8-7a}$$

$$\begin{aligned} & \mathbf{B}\left\langle\zeta_*^{\delta+1}(\theta^+,\theta^-)\right\rangle\mathbf{q}^{(1)} + \overline{\mathbf{B}}\left\langle\overline{\zeta}_*^{\delta+1}(\theta^+,\theta^-)\right\rangle\tilde{\mathbf{q}}^{(1)} + (\delta+1)\kappa^+ \\ & \left\{\mathbf{B}\left\langle\zeta_*^{\delta+1}(\theta^+,\theta^-)p_*(\theta^+)\right\rangle\mathbf{q}^{(0)} + \overline{\mathbf{B}}\left\langle\overline{\zeta}_*^{\delta+1}(\theta^+,\theta^-)\overline{p}_*(\theta^+)\right\rangle\mathbf{q}^{(0)}\right\} = \mathbf{0}. \end{aligned} \tag{9.8-7b}$$

Equations (9.8-6) are identical to (9.4-12). Hence we may write

$$\mathbf{q}^{(0)} = \mathbf{B}^{-1}\mathbf{g}, \quad \tilde{\mathbf{q}}^{(0)} = -\overline{\mathbf{B}}^{-1}\mathbf{g}, \tag{9.8-8a}$$

$$\mathbf{K}\mathbf{g} = \mathbf{0}, \tag{9.8-8b}$$

where the matrix $\mathbf{K}$ is given in (9.4-14). For a nontrivial solution of $\mathbf{g}$ δ must be a root of

$$\det\mathbf{K} = |\mathbf{K}| = 0$$

which is identical to (9.4-15). Hence the eigenvalues for a wedge with curved boundaries and with straight boundaries are identical. After δ, $\mathbf{q}^{(0)}$, and $\tilde{\mathbf{q}}^{(0)}$ are obtained, we determine $\mathbf{q}^{(1)}$ and $\tilde{\mathbf{q}}^{(1)}$ from (9.8-7). The eigenfunction for the stress associated with the eigenvalue δ has the terms

$$r^{\delta},\ r^{\delta+1},\cdots.$$

Equations (9.8-6) and (9.8-7) can be written, respectively, as

$$\hat{\mathbf{K}}\hat{\mathbf{g}}^{(0)} = \mathbf{0}, \tag{9.8-9}$$

$$\hat{\mathbf{K}}\hat{\mathbf{g}}^{(1)} = \mathbf{d}, \tag{9.8-10}$$

where

$$\hat{\mathbf{K}} = \begin{bmatrix} \mathbf{B} & \overline{\mathbf{B}} \\ \mathbf{B}\left\langle \zeta_*^{\delta+1}(\theta^+,\theta^-)\right\rangle & \overline{\mathbf{B}}\left\langle \overline{\zeta}_*^{\delta+1}(\theta^+,\theta^-)\right\rangle \end{bmatrix},$$

$$\hat{\mathbf{g}}^{(i)} = \begin{bmatrix} \mathbf{q}^{(i)} \\ \tilde{\mathbf{q}}^{(i)} \end{bmatrix}, \quad (i=0,1),$$

$$\mathbf{d} = -(\delta+1)\begin{bmatrix} \kappa^-\mathbf{B}\left\langle p_*(\theta^-)\right\rangle & \kappa^-\overline{\mathbf{B}}\left\langle \overline{p}_*(\theta^-)\right\rangle \\ \kappa^+\mathbf{B}\left\langle \zeta_*^{\delta+1}(\theta^+,\theta^-)p_*(\theta^+)\right\rangle & \kappa^+\overline{\mathbf{B}}\left\langle \overline{\zeta}_*^{\delta+1}(\theta^+,\theta^-)\overline{p}_*(\theta^+)\right\rangle \end{bmatrix}\hat{\mathbf{g}}^{(0)}.$$

The matrix $\hat{\mathbf{K}}$, which depends on δ, is singular when δ is an eigenvalue. Equation (9.8-10) is similar to (9.7-19). It may or may not have a solution for $\hat{\mathbf{g}}^{(1)}$ unless the vector $\mathbf{d}$ is orthogonal to left eigenvectors of $\hat{\mathbf{K}}$. The discussions following (9.7-19) tell us that if (9.8-10) fails to provide a solution we may differentiate the $r^{\delta+2}$ terms in (9.8-5) with δ as many times as we need to obtain a solution. The stress then has the $r^{\delta+1}(\ln r)^n$ singularity where n is an integer. In this case the eigenfunction for the stress associated with the eigenvalue δ has the terms

$$r^{\delta},\ r^{\delta+1}(\ln r)^n,\cdots.$$

Consider a curved boundary more general than (9.8-2) such as

$$\theta = \theta_\Gamma + \kappa_\Gamma r^{\varepsilon} + \cdots,\quad \varepsilon > 0.$$

The eigenvalues are still identical to the ones for wedges with straight boundaries but the eigenfunction for stresses associated with, say $\delta=\delta_1$, has the terms

$$r^{\delta_1},\ r^{\delta_1+\varepsilon},\cdots$$

or

$$r^{\delta_1}, \quad r^{\delta_1+\varepsilon}(\ln r)^n, \cdots$$

depending on whether a solution to an equation similar to (9.8-10) exists. If another eigenvalue δ_2 satisfies the relation (9.8-1) and if

$$\mathrm{Re}(\delta_1) + \varepsilon \mathrm{Re}(\delta_2) < 0,$$

the $r^{\delta_1+\varepsilon}$ or $r^{\delta_1+\varepsilon}(\ln r)^n$ term provides a stronger singularity than the r^{δ_2} term.

9.9 Existence of a Singularity

The eigenfunctions studied in Sections 9.4 through 9.8 are unique up to an arbitrary constant multiplier. In other words, the coefficients of the eigenfunctions are arbitrary. They depend on the conditions at far field $r>>0$. A stress singularity inherited in an eigenfunction may not appear in the solution to a problem if the conditions at far field are such that the coefficient of the eigenfunction vanishes. Therefore the presence of a singularity in an eigenfunction does not assure us the existence of that particular singularity. It does tell us that, if a stress singularity exists it is one of the forms in the eigenfunctions, assuming that the boundary conditions near $r=0$ of the wedge are homogeneous. The form of singularities in the eigenfunctions depends on the material property and the geometry of the wedge near the wedge apex $r=0$.

A stress singularity may also occur when the boundary conditions near $r=0$ of the wedge are not homogeneous. As we have seen in Section 9.3 where the surfaces of the wedge are subjected to a uniform traction, the stress may have a $\ln r$ singularity. This kind of singularity depends on the material property, geometry of the wedge, and the boundary conditions near the wedge apex $r=0$. It is independent of the conditions at far field. Another example is that of a composite wedge consisted of two wedges of dissimilar isotropic materials subjected to an antiplane extension while the surfaces of the composite wedge are traction-free. A logarithmic stress singularity may exist at the wedge apex (Ting and Chou, 1985). A special case of this is that of a layered composite in which the interfaces are normal to the $x_3=0$ plane and to the free-edge $x_1=0$. At the free-edge of an interface the geometry is that of two wedges of equal wedge angles $\pi/2$ joined together. When the layered composite is subjected to an extension in the x_3-direction a logarithmic stress singularity may occur at the free-edge (Zwiers et al., 1982). The existence of a logarithmic stress singularity in this case depends only on the property of the two materials adjacent to the interface. It does not depend on the stacking sequence of the layered composite nor on the boundary conditions at the far field. In other words, if a logarithmic stress singularity exists at the free-edge due to an antiplane extension, the singularity cannot be alleviated by the stacking sequence of the layers nor by the far field boundary conditions. This is in contrast with the singularity arisen from an eigenfunction.

Stress singularities of course exist at a three-dimensional corner. In an excellent review article on a cracked plate Sih (1971) discussed at length the difficulties associated with the task of obtaining an analytical solution for three-dimensional singularities. In most studies either special materials or special geometry are considered. Using the results of Thompson and Little (1970), Bazant and Keer (1974) studied the problem of conical notches and inclusions for isotropic materials. A full three-dimensional analysis was given by Benthem (1977, 1980) and Kawai et al. (1977) who studied stress singularities for an isotropic elastic half space $x_3>0$ with the quarter plane $x_2=0$, $x_1>0$, $x_3>0$ being the crack surface. A collection of three-dimensional singularity analyses is given by Kassir and Sih (1975) and further discussions on three-dimensional analyses can be found in Erdogan (1983). Analytical solutions are largely limited to isotropic materials. Except for special cases (Elliott, 1948, 1949; Somaratna and Ting, 1986a) very few problems are solved analytically for three-dimensional singularities for general anisotropic materials. A numerical procedure for determining three-dimensional singularities was developed by Bazant (1974) for harmonic functions and by Bazant and Estenssoro (1977, 1979) for isotropic materials. Their technique was extended by Somaratna and Ting (1986b) to general anisotropic materials. It should be noted that the bounds for the order of stress singularities δ for three-dimensional problems are different from that presented in (9.4-3) for two-dimensional problems. Following the derivation of (9.4-3) it can be shown that the bounds for three-dimensional stress singularities are

$$-\tfrac{3}{2} < \mathrm{Re}(\delta) < 0. \tag{9.9-1}$$

9.10 Stress Decay in an Anisotropic Elastic Strip

The classical Saint-Venant (1855) principle states that when a self-equilibrated load is applied at a small region of a body, the stress is localized near the region where the load is applied. The stress at far field is negligible. This is a general statement. The precise decay factor of the stress with distance from the region of applied load depends on the geometry of the body, the boundary conditions, the applied load and the material property. In this section we study a semi-infinite anisotropic elastic strip occupying the region

$$x_1 \geq 0, \quad -1 \leq x_2 \leq 1,$$

subjected to a self-equilibrated load at the end $x_1=0$. The related problems for an isotropic elastic strip have been studied by Papkovich (1941), Fadle (1941), and Johnson and Little (1965). Extensions to certain anisotropic strips were carried out by Horgan (1972a,b), Choi and Horgan (1977,1978), Horgan and Simmonds (1991), M. Z. Wang et al. (1993), and Crafter et al. (1993). The analysis we will present here essentially follows that of Wang et al. (1993). They studied much more general boundary conditions and end conditions than the one we will present here. For certain boundary conditions they showed that the decay factors do not require the applied load to be self-equilibrated.

As in the case of wedge we employ the original general solution (5.3-7). We choose $f(z)$ to be proportional to $e^{-\lambda z}$ where λ is an eigenvalue. Therefore consider the eigenfunction

$$\begin{aligned} \mathbf{u} &= \mathbf{A}\left\langle e^{-\lambda z_*}\right\rangle \mathbf{q} + \overline{\mathbf{A}}\left\langle e^{-\lambda \bar{z}_*}\right\rangle \tilde{\mathbf{q}}, \\ \boldsymbol{\phi} &= \mathbf{B}\left\langle e^{-\lambda z_*}\right\rangle \mathbf{q} + \overline{\mathbf{B}}\left\langle e^{-\lambda \bar{z}_*}\right\rangle \tilde{\mathbf{q}}, \end{aligned} \tag{9.10-1}$$

which can be written as

$$\begin{aligned} \mathbf{u} &= e^{-\lambda x_1}\left\{\mathbf{A}\left\langle e^{-\lambda p_* x_2}\right\rangle \mathbf{q} + \overline{\mathbf{A}}\left\langle e^{-\lambda \bar{p}_* x_2}\right\rangle \tilde{\mathbf{q}}\right\}, \\ \boldsymbol{\phi} &= e^{-\lambda x_1}\left\{\mathbf{B}\left\langle e^{-\lambda p_* x_2}\right\rangle \mathbf{q} + \overline{\mathbf{B}}\left\langle e^{-\lambda \bar{p}_* x_2}\right\rangle \tilde{\mathbf{q}}\right\}, \end{aligned} \tag{9.10-2}$$

where $\mathbf{q}$ and $\tilde{\mathbf{q}}$ are arbitrary constant vectors. If λ is real $\tilde{\mathbf{q}}$ is the complex conjugate of $\mathbf{q}$, and $\mathbf{u}$ and $\boldsymbol{\phi}$ are real. If λ is complex $\tilde{\mathbf{q}}$ may not be the complex conjugate of $\mathbf{q}$, and $\mathbf{u}$ and $\boldsymbol{\phi}$ may not be real. We will show later that if a complex λ is an eigenvalue, so is its complex conjugate $\bar{\lambda}$. One can then superimpose two solutions associated with λ and $\bar{\lambda}$ to obtain real solutions for $\mathbf{u}$ and $\boldsymbol{\phi}$. Equation $(9.10\text{-}2)_2$ tells us that, when $\mathrm{Re}\lambda>0$, the stress decays exponentially as $x_1 \to \infty$. The smallest positive $\mathrm{Re}\lambda$ gives the slowest decay.

Let the sides $x_2=\pm 1$ of the strip be traction-free. This means that the vector $\boldsymbol{\phi}$ is a constant at $x_2=\pm 1$. We may set

$$\boldsymbol{\phi} = \mathbf{0} \quad \text{at} \quad x_2 = \pm 1. \tag{9.10-3}$$

This is so because the total traction at the end $x_1=0$ is assumed to vanish, i.e.,

$$\int_{-1}^{1} \sigma_{i1}(0,x_2)dx_2 = 0,$$

or

$$\int_{-1}^{1} \boldsymbol{\phi}_{,2}(0,x_2)dx_2 = \boldsymbol{\phi}(0,1) - \boldsymbol{\phi}(0,-1) = \mathbf{0}.$$

Hence there is no loss in generality in setting $\boldsymbol{\phi}=\mathbf{0}$ at $x_2=\pm 1$. Substitution of $(9.10\text{-}2)_2$ into (9.10-3) yields

$$\begin{aligned} \mathbf{B}\langle e^{-\lambda p_*}\rangle\mathbf{q} + \overline{\mathbf{B}}\langle e^{-\lambda \bar{p}_*}\rangle\tilde{\mathbf{q}} = 0, \\ \mathbf{B}\langle e^{\lambda p_*}\rangle\mathbf{q} + \overline{\mathbf{B}}\langle e^{\lambda \bar{p}_*}\rangle\tilde{\mathbf{q}} = 0. \end{aligned} \tag{9.10-4}$$

Equations (9.10-4) remain the same if λ is replaced by $-\lambda$. Thus if λ is an eigenvalue, so is $-\lambda$. By taking the complex conjugate of (9.10-4), it is easily shown that $\bar{\lambda}$ is an eigenvalue if λ is. The first equation of (9.10-4) is

$$\mathbf{B}\langle e^{-\lambda p_*}\rangle\mathbf{q} = -\overline{\mathbf{B}}\langle e^{-\lambda \bar{p}_*}\rangle\tilde{\mathbf{q}} = \mathbf{g}, \text{ say.}$$

Hence (9.10-4) are equivalent to

$$\mathbf{q} = \langle e^{\lambda p_*}\rangle\mathbf{B}^{-1}\mathbf{g}, \quad \tilde{\mathbf{q}} = -\langle e^{\lambda \bar{p}_*}\rangle\overline{\mathbf{B}}^{-1}\mathbf{g}, \tag{9.10-5a}$$

$$\mathbf{K}\mathbf{g} = \mathbf{0}, \tag{9.10-5b}$$

where $\mathbf{g}$ is a constant vector and

$$\begin{aligned} \mathbf{K} &= \mathbf{B}\langle e^{2\lambda p_*}\rangle\mathbf{B}^{-1} - \overline{\mathbf{B}}\langle e^{2\lambda \bar{p}_*}\rangle\overline{\mathbf{B}}^{-1} \\ &= \mathbf{B}\left\{\langle e^{2\lambda p_*}\rangle\mathbf{B}^{-1}\overline{\mathbf{B}} - \mathbf{B}^{-1}\overline{\mathbf{B}}\langle e^{2\lambda \bar{p}_*}\rangle\right\}\overline{\mathbf{B}}^{-1} \\ &= \left\{\mathbf{B}\langle e^{2\lambda p_*}\rangle\mathbf{B}^T + \overline{\mathbf{B}}\langle e^{2\lambda \bar{p}_*}\rangle\overline{\mathbf{B}}^T\right\}(-2i\mathbf{L}^{-1}). \end{aligned} \tag{9.10-6}$$

It is interesting to compare (9.10-6) with (9.4-14). By (9.10-5b) the eigenvalue λ is a root of

$$\det\mathbf{K} = |\mathbf{K}| = 0 \tag{9.10-7}$$

and $\mathbf{g}$ is the associated eigenvector. As in (9.4-16) we have three different expressions for (9.10-7). The discussions following (9.4-16) tell us that

$$\left|\langle e^{2\lambda p_*}\rangle\mathbf{B}^{-1}\overline{\mathbf{B}} - \mathbf{B}^{-1}\overline{\mathbf{B}}\langle e^{2\lambda \bar{p}_*}\rangle\right| = 0 \tag{9.10-8}$$

is the most convenient for computing λ.

Let the eigenfunction (9.10-2) associated with λ be denoted by $\varpi(\lambda)$ where

$$\varpi = \begin{bmatrix} \mathbf{u} \\ \boldsymbol{\phi} \end{bmatrix}.$$

Since ϖ satisfies the differential equation (Section 5.7)

$$\varpi_{,2} = \mathbf{N}\varpi_{,1}$$

we have

$$\varpi_{,2} = -\lambda \mathbf{N}\,\varpi. \tag{9.10-9}$$

Let λ_k, λ_m be any two eigenvalues. Employing the relation (Section 5.5)

$$\hat{\mathbf{I}}\mathbf{N} = (\hat{\mathbf{I}}\mathbf{N})^T = \mathbf{N}^T\hat{\mathbf{I}}, \quad \hat{\mathbf{I}} = \begin{bmatrix} \mathbf{0} & \mathbf{I} \\ \mathbf{I} & \mathbf{0} \end{bmatrix}$$

and (9.10-9), it can be shown that

$$\{\,\varpi^T(-\lambda_m)\,\hat{\mathbf{I}}\,\varpi(\lambda_k)\}_{,2} = (\lambda_m - \lambda_k)\,\varpi^T(-\lambda_m)\,\hat{\mathbf{I}}\mathbf{N}\,\varpi(\lambda_k). \tag{9.10-10}$$

Noting that

$$\varpi^T(-\lambda_m)\,\hat{\mathbf{I}}\,\varpi(\lambda_k) = \mathbf{u}(-\lambda_m)\cdot\boldsymbol{\phi}(\lambda_k) + \boldsymbol{\phi}(-\lambda_m)\cdot\mathbf{u}(\lambda_k),$$

if the side conditions at $x_2 = \pm 1$ are of the type

$$\mathbf{u}\cdot\boldsymbol{\phi} = 0$$

of which $\mathbf{u}=\mathbf{0}$ or $\boldsymbol{\phi}=\mathbf{0}$ is a special case, integration of (9.10-10) with x_2 yields

$$\int_{-1}^{1} \varpi^T(-\lambda_m)\hat{\mathbf{I}}\mathbf{N}\,\varpi(\lambda_k)dx_2 = \begin{cases} 0, & \text{when } \lambda_k \neq \lambda_m, \\ J_k, & \text{when } \lambda_k = \lambda_m, \end{cases} \tag{9.10-11}$$

where

$$J_k = 2\left\{\mathbf{q}^T(-\lambda_k)\langle p_*\rangle\mathbf{q}(\lambda_k) + \tilde{\mathbf{q}}^T(-\lambda_k)\langle \bar{p}_*\rangle\tilde{\mathbf{q}}(\lambda_k)\right\}. \tag{9.10-12}$$

To deduce (9.10-12) we write the eigenfunction (9.10-2) as

$$\varpi^T(\lambda) = e^{-\lambda x_1}\begin{bmatrix} \mathbf{A} & \bar{\mathbf{A}} \\ \mathbf{B} & \bar{\mathbf{B}} \end{bmatrix}\begin{bmatrix} \langle e^{-\lambda p_* x_2}\rangle & \mathbf{0} \\ \mathbf{0} & \langle e^{-\lambda \bar{p}_* x_2}\rangle \end{bmatrix}\begin{bmatrix} \mathbf{q}(\lambda) \\ \tilde{\mathbf{q}}(\lambda) \end{bmatrix}.$$

With the aid of (5.5-13) and (5.5-18) it is easily shown that

$$\varpi^T(-\lambda)\,\hat{\mathbf{I}}\mathbf{N}\,\varpi(\lambda) = \mathbf{q}^T(-\lambda)\langle p_*\rangle\mathbf{q}(\lambda) + \tilde{\mathbf{q}}^T(-\lambda)\langle \bar{p}_*\rangle\tilde{\mathbf{q}}(\lambda).$$

The right hand side is a constant so that (9.10-12) follows immediately upon integration. With the orthogonality relation (9.10-11) a complete solution can be obtained for the strip when a self-equilibrated traction is prescribed at $x_1=0$. (M. Z. Wang et al., 1993).

For monoclinic materials with the symmetry plane at $x_3=0$ substitution of $\mathbf{B}^{-1}\overline{\mathbf{B}}$ from (9.4-25) into (9.10-8) leads to

$$e^{2\lambda p_3} - e^{2\lambda \bar{p}_3} = 0 \tag{9.10-13}$$

and

$$\begin{vmatrix} (\bar{p}_1-p_2)(e^{2\lambda p_1}-e^{2\lambda\bar{p}_1}) & (\bar{p}_2-p_2)(e^{2\lambda p_1}-e^{2\lambda\bar{p}_2}) \\ (p_1-\bar{p}_1)(e^{2\lambda p_2}-e^{2\lambda\bar{p}_1}) & (p_1-\bar{p}_2)(e^{2\lambda p_2}-e^{2\lambda\bar{p}_2}) \end{vmatrix} = 0. \tag{9.10-14}$$

It can be shown that (9.10-13) applies to antiplane displacements while (9.10-14) applies to in-plane displacements. From (9.10-13) we have

$$\lambda = \frac{n\pi}{2p_3''}, \quad n \text{ integer,}$$

where p_3'' is the imaginary part of p_3. For isotropic materials $p_3''=1$. The eigenvalues are real for antiplane displacements. The slowest decay factor is obtained with $n=1$.

Equation (9.10-14) can be written more concisely as follows. Let p' and p'' be the real and imaginary parts of p, respectively. By

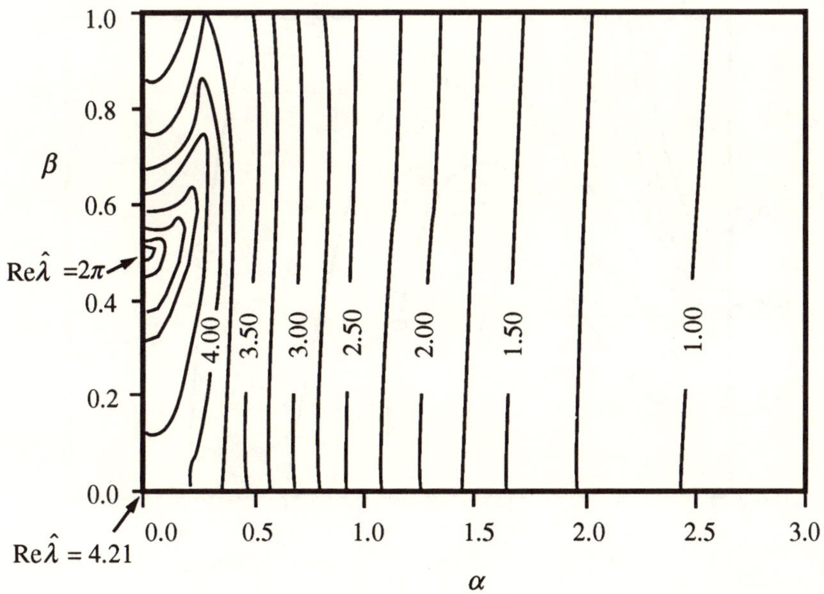

Fig. 9.7 Contour lines for constant Re $\hat{\lambda}$

writing

$$e^{2\lambda p} = e^{2\lambda p'}\{\cos(2\lambda p'') + i\sin(2\lambda p'')\},$$

expansion of the determinant in (9.10-14) leads to

$$\left[(p_1'-p_2')^2+(p_1''-p_2'')^2\right]\sin^2[\lambda(p_1''+p_2'')] = \left[(p_1'-p_2')^2+(p_1''+p_2'')^2\right]$$
$$\sin^2[\lambda(p_1''-p_2'')] + \left[(p_1''+p_2'')^2-(p_1''-p_2'')^2\right]\sinh^2[\lambda(p_1'-p_2')]$$

or (Wang et al., 1993)

$$(\alpha^2+\beta^2)\sin^2\hat{\lambda} = (1+\alpha^2)\sin^2(\hat{\lambda}\beta) + (1-\beta^2)\sinh^2(\hat{\lambda}\alpha) \qquad (9.10\text{-}15)$$

where

$$\hat{\lambda} = \lambda(p_1''+p_2''), \quad \alpha = \frac{p_1'-p_2'}{p_1''+p_2''}, \quad \beta = \frac{p_1''-p_2''}{p_1''+p_2''}.$$

The normalized eigenvalue $\hat{\lambda}$ depends on two material parameters α and β. While α can assume any value, β is limited to $-1<\beta<1$ because p_1'' and p_2'' are positive and nonzero. Equation (9.10-15) remains the same if α is replaced by $-\alpha$ or β is replaced by $-\beta$. It suffices therefore to study $\hat{\lambda}$ for

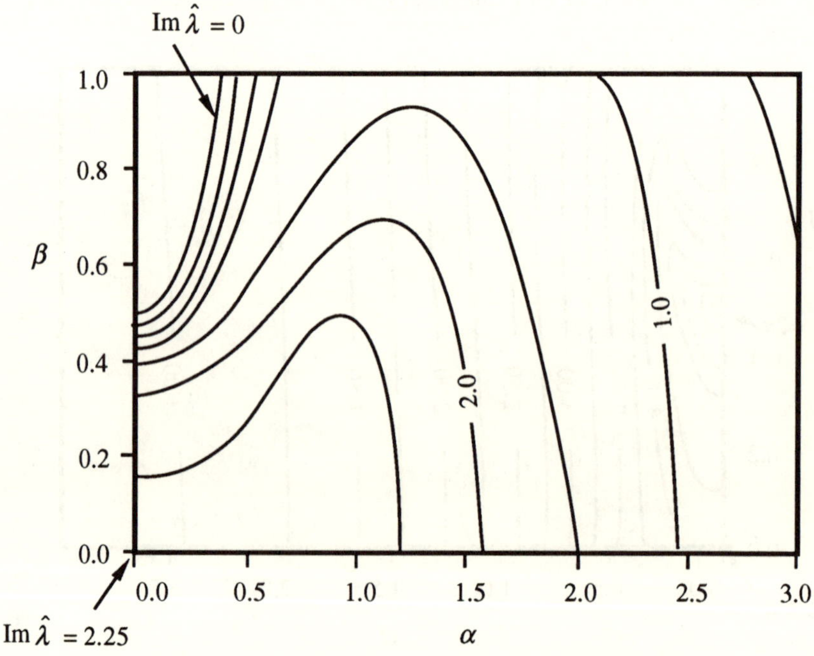

Fig. 9.8 Contour lines for constant Im $\hat{\lambda}$.

$$\alpha \geq 0, \quad 0 \leq \beta < 1.$$

The decay factors are the eigenvalues with positive real parts. Numerical solutions of $\hat{\lambda}$ for the smallest $\mathrm{Re}\,\hat{\lambda}$ are shown in Fig. 9.7 for $\mathrm{Re}\,\hat{\lambda}$ and in Fig.9.8 for the corresponding $\mathrm{Im}\,\hat{\lambda}$. It is seen that the largest $\mathrm{Re}\,\hat{\lambda}$ is 2π which occurs at $\alpha=0$, $\beta=\frac{1}{2}$, or at $p_1'=p_2'$, $p_1''=3\,p_2''$. The special cases when $\alpha=0$ and/or $\beta=0$ are discussed separately below.

(i) $\beta=0$ (or $p_1''=p_2''$). Equation (9.10-15) reduces to

$$\alpha \sin\hat{\lambda} = \pm\sinh(\hat{\lambda}\alpha). \tag{9.10-16}$$

Numerical calculations show that the minus sign should be used for the smallest $\mathrm{Re}\,\hat{\lambda}$. The results for the smallest $\mathrm{Re}\,\hat{\lambda}$ and the corresponding $\mathrm{Im}\,\hat{\lambda}$ are shown in Fig. 9.9. It can be shown that

$$\frac{d\hat{\lambda}}{d\alpha} = 0 \quad \text{at} \quad \alpha=0.$$

The curve $\mathrm{Re}\,\hat{\lambda}$ is monotonic while $\mathrm{Im}\,\hat{\lambda}$ has a maximum at $\alpha=0.61$.

(ii) $\alpha=0$ (or $p_1'=p_2'$). In this case (9.10-15) reduces to

$$\beta \sin\hat{\lambda} = \pm\sin(\hat{\lambda}\beta). \tag{9.10-17}$$

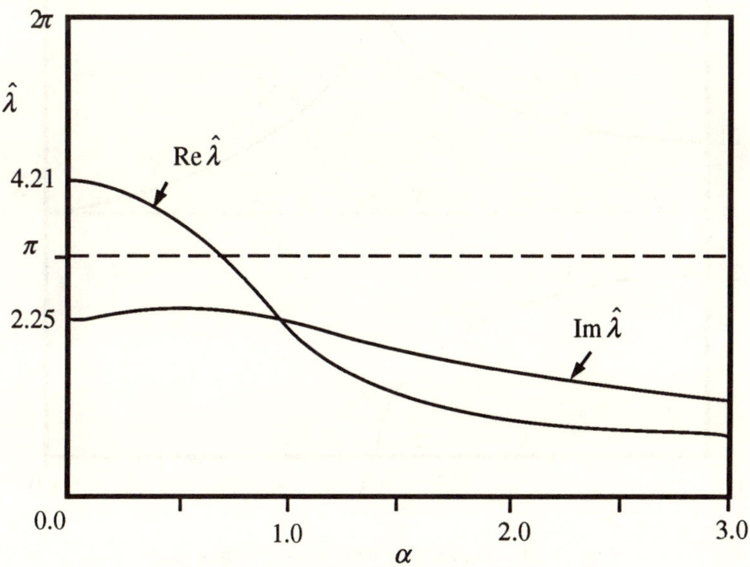

Fig. 9.9 Dependence of $\hat{\lambda}$ on α when $\beta=0$.

Again, the minus sign should be used for the smallest Re$\hat{\lambda}$. Figure 9.10 shows the smallest Re$\hat{\lambda}$ and the corresponding Im$\hat{\lambda}$. The decay factor $\hat{\lambda}$ is real for $\beta \geq \frac{1}{2}$. It can be shown that

$$\hat{\lambda} = \begin{cases} 2\pi \text{ at } \beta = \frac{1}{2}, \\ \pi \text{ at } \beta = 1, \end{cases} \qquad \frac{d\hat{\lambda}}{d\beta} = \begin{cases} 0 \text{ at } \beta = 0, \\ \infty \text{ at } \beta = \frac{1}{2}, \\ -\frac{\pi}{2} \text{ at } \beta = 1. \end{cases}$$

(iii) $\alpha=\beta=0$ (or $p_1=p_2$). By taking the limit $\alpha=0$ in (9.10-16) or $\beta=0$ in (9.10-17) we have

$$\sin\hat{\lambda} = \pm\hat{\lambda}. \tag{9.10-18}$$

This case applies whenever $p_1=p_2$. Isotropic materials are a special case for which $p_1=p_2=i$. Taking the minus sign in (9.10-18), the two $\hat{\lambda}$'s with the smallest Re($\hat{\lambda}$) are (Fadle, 1941)

$$4.2122+2.2508i, \qquad 10.7126+3.1032i.$$

Therefore for an isotropic elastic strip of width two units subjected to a self-equilibrated load at $x_1=0$, the stress decay is equal to, or faster than, $e^{-4.2122x_1}$.

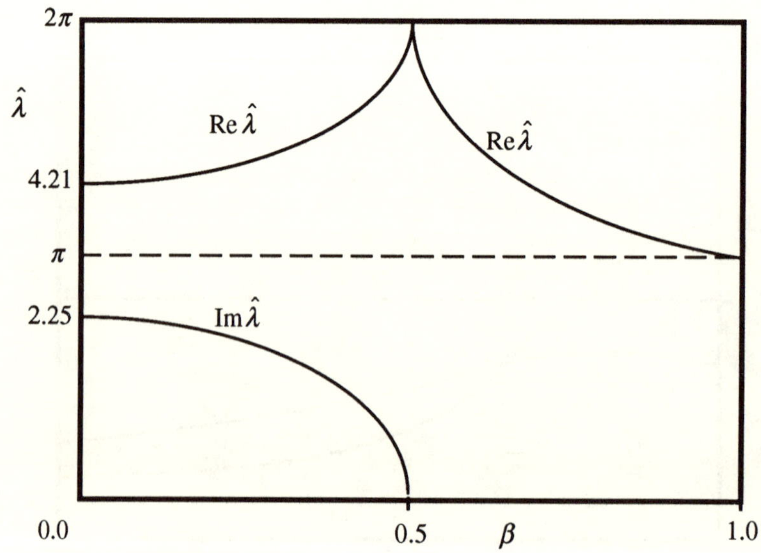

Fig. 9.10 Dependence of $\hat{\lambda}$ on β when $\alpha=0$.

Chapter 10

ANISOTROPIC MATERIALS WITH AN ELLIPTIC BOUNDARY

The determination of stress distribution in a solid with the presence of a hole or an inclusion has been a mathematically interesting and challenging problem. It is also an important problem in applications. The simplest geometry of the hole is a circle. For isotropic materials a hole of arbitrary shape can be transformed, in theory, to a circle by a conformal mapping (Muskhelishvili, 1953; see also Section 3.12). Therefore a circular hole is all one needs to study for isotropic materials. For anisotropic materials there are three complex variables $z_\alpha = x_1 + p_\alpha x_2$ (α=1,2,3). It is in general not possible to transform a hole of given shape to the same circle for all three complex variables. An exception is the ellipse (Lekhnitskii, 1950; Savin, 1961). In this chapter we study various problems involving an elliptic boundary. The ellipse can be a hole, a rigid body, or an inclusion of different anisotropic materials. We will also consider an elliptic body subjected to external forces. For anisotropic elastic materials even the circular hole needs a transformation (Section 10.1). There is practically no difference in the analysis if the ellipse is replaced by a circle. We may employ dual coordinate systems. One coordinate system is chosen to coincide with a symmetry plane of the material when such a plane exists. The other coordinate system is to coincide with the principal axes of the ellipse. The analysis is no more complicated than when a single coordinate system is employed. For some problems employment of dual coordinate systems reveal that certain aspects of the solutions are invariant with the orientation of the ellipse in the material.

10.1 A Force Applied to an Elliptic Rigid Inclusion.

Let an anisotropic elastic material occupy the infinite space with an elliptic rigid inclusion. The principal axes $\hat{x}_1$, $\hat{x}_2$ of the ellipse do not necessarily coincide with the coordinate system x_1, x_2 chosen for the anisotropic material, Fig. 10.1. Let a, b with $a \geq b > 0$ be the principal radii of the ellipse Γ and let the equation of the ellipse be given by

$$\hat{x}_1^\Gamma = a\cos\psi, \quad \hat{x}_2^\Gamma = b\sin\psi, \quad \text{on } \Gamma \qquad (10.1\text{-}1)$$

where ψ is a real parameter. The major axis of the ellipse makes an angle θ_o with the x_1-axis. Consider the transformation

$$\hat{z}_\alpha = c_\alpha \hat{\zeta}_\alpha + d_\alpha \hat{\zeta}_\alpha^{-1}, \quad (\alpha=1,2,3) \tag{10.1-2}$$

where c_α and d_α are constants and (Section 7.10)

$$\hat{z}_\alpha = \hat{x}_1 + p_\alpha(\theta_o)\hat{x}_2, \tag{10.1-3}$$

$$\hat{x}_1 = r\cos(\theta - \theta_o), \quad \hat{x}_2 = r\sin(\theta - \theta_o). \tag{10.1-4}$$

The objective of the transformation (10.1-2) is to choose c_α and d_α such that, when $\hat{x}_1$, $\hat{x}_2$ is on the ellipse Γ, the new complex variable $\hat{\zeta}_\alpha$ is on a unit circle, i.e.,

$$\hat{\zeta}_\alpha = e^{i\psi} \quad \text{on } \Gamma. \tag{10.1-5}$$

Hence

$$a\cos\psi + p_\alpha(\theta_o)b\sin\psi = c_\alpha(\cos\psi + i\sin\psi) + d_\alpha(\cos\psi - i\sin\psi)$$

from which we obtain (Section 3.7)

$$c_\alpha = \tfrac{1}{2}[a - ip_\alpha(\theta_o)b], \quad d_\alpha = \tfrac{1}{2}[a + ip_\alpha(\theta_o)b]. \tag{10.1-6}$$

Notice that when the ellipse is a circle ($a=b$) we still need the transformation (10.1-2) unless $p_\alpha(\theta_o)=i$ ($\alpha=1,2,3$). The transformation is unnecessary when $p_\alpha(\theta_o)=ia/b$ ($\alpha=1,2,3$) so that $c_\alpha=a$ and $d_\alpha=0$. It is important to notice from (10.1-5) that ζ_α assumes the value $e^{i\psi}$ on Γ regardless of whether $\alpha=1$, 2, or 3. With (10.1-6) one of the

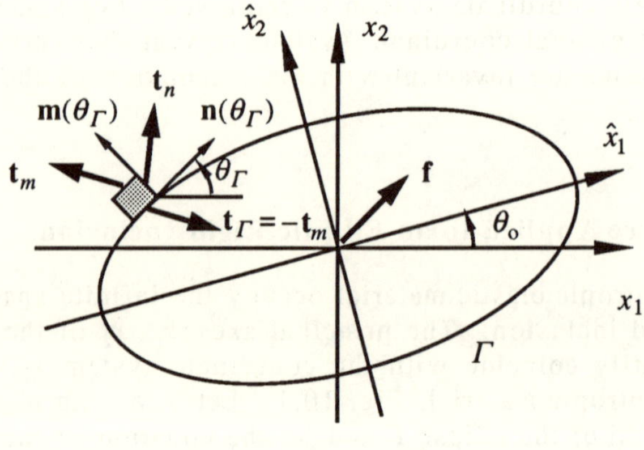

Fig. 10.1 An elliptic rigid inclusion is subjected to a line force **f** at the center of the inclusion. The surface traction $\mathbf{t}_\Gamma$ on the elliptic surface is independent of material property, and is parallel to the applied force **f**.

solutions for ζ_α in (10.1-2) is

$$\hat{\zeta}_\alpha = \frac{\hat{z}_\alpha + \sqrt{\hat{z}_\alpha^2 - [a^2 + p_\alpha^2(\theta_0)b^2]}}{a - ip_\alpha(\theta_0)b}. \tag{10.1-7}$$

This solution has the property that $|\hat{\zeta}_\alpha| \to \infty$ as $|\hat{z}_\alpha| \to \infty$. The other solution gives $|\hat{\zeta}_\alpha| \to 0$ as $|\hat{z}_\alpha| \to \infty$. The inverse of (10.1-7) is

$$\hat{\zeta}_\alpha^{-1} = \frac{\hat{z}_\alpha - \sqrt{\hat{z}_\alpha^2 - [a^2 + p_\alpha^2(\theta_o)b^2]}}{a + ip_\alpha(\theta_o)b}. \tag{10.1-8}$$

We have shown in Section 3.7 that the branch cut required for the square root appearing in (10.1-7) and (10.1-8) is inside the ellipse. Hence the transformation (10.1-2) is one-to-one for points outside the ellipse. However, the transformation is *not* a conformal mapping because the angle between any two infinitesimal line elements in the (x_1,x_2) plane is in general different from the angle between the corresponding line elements in the ζ-plane.

The infinitesimal arclength of the ellipse is $\rho d\psi$ where

$$\rho = \sqrt{a^2 \sin^2 \psi + b^2 \cos^2 \psi}. \tag{10.1-9}$$

At a point on Γ let θ_Γ be the angle the tangent to the ellipse makes with the x_1-axis. The unit vectors tangential and normal to the ellipse are, Fig. 10.1,

$$\mathbf{n}^T(\theta_\Gamma) = (\cos\theta_\Gamma,\ \sin\theta_\Gamma,\ 0),$$
$$\mathbf{m}^T(\theta_\Gamma) = (-\sin\theta_\Gamma,\ \cos\theta_\Gamma,\ 0).$$

It is easily shown that

$$\rho\cos(\theta_\Gamma - \theta_o) = a\sin\psi, \quad \rho\sin(\theta_\Gamma - \theta_o) = -b\cos\psi. \tag{10.1-10}$$

Equation $(10.1\text{-}10)_1$ or $(10.1\text{-}10)_2$ provides the relation between ψ and θ_Γ.

Let a line force $\mathbf{f}$ per unit length of the x_3-axis be applied on the elliptic rigid inclusion. We employ the general solution (7.10-12) for dual coordinate systems where $f(\hat{z}_\alpha)$ is replaced by $\hat{\zeta}_\alpha$, i.e.,

$$\mathbf{u} = \frac{1}{\pi}\mathrm{Im}\left\{\mathbf{A}\left\langle \ln\hat{\zeta}_* \right\rangle \mathbf{q}^\infty\right\}, \quad \boldsymbol{\phi} = \frac{1}{\pi}\mathrm{Im}\left\{\mathbf{B}\left\langle \ln\hat{\zeta}_* \right\rangle \mathbf{q}^\infty\right\}. \tag{10.1-11}$$

Along the elliptic boundary Γ, $\hat{\zeta} = e^{i\psi}$ so that

$$\mathbf{u}_\Gamma = \frac{\psi}{\pi}\mathrm{Re}\{\mathbf{A}\mathbf{q}^\infty\}, \quad \boldsymbol{\phi}_\Gamma = \frac{\psi}{\pi}\mathrm{Re}\{\mathbf{B}\mathbf{q}^\infty\}. \tag{10.1-12}$$

If there were a line dislocation with Burgers vector **b** located on the x_3-axis we must have

$$2\,\mathrm{Re}\{\mathbf{A}\mathbf{q}^{\infty}\}=\mathbf{b},\qquad 2\,\mathrm{Re}\{\mathbf{B}\mathbf{q}^{\infty}\}=\mathbf{f}.$$

These are identical to (8.1-17a) and hence, from (8.1-18),

$$\mathbf{q}^{\infty}=\mathbf{A}^{T}\mathbf{f}+\mathbf{B}^{T}\mathbf{b}. \tag{10.1-13}$$

In the present problem the elliptic inclusion is a rigid body so that $\mathbf{b}=\mathbf{0}$. Equations (10.1-11) and (10.1-12) become

$$\mathbf{u}=\frac{1}{\pi}\mathrm{Im}\left\{\mathbf{A}\left\langle\ln\hat{\zeta}_*\right\rangle\mathbf{A}^{T}\right\}\mathbf{f},\qquad \boldsymbol{\phi}=\frac{1}{\pi}\mathrm{Im}\left\{\mathbf{B}\left\langle\ln\hat{\zeta}_*\right\rangle\mathbf{A}^{T}\right\}\mathbf{f}, \tag{10.1-14}$$

$$\mathbf{u}_{\Gamma}=\mathbf{0},\qquad \boldsymbol{\phi}_{\Gamma}=\frac{\psi}{2\pi}\mathbf{f}. \tag{10.1-15}$$

That $\mathbf{u}_{\Gamma}$ vanishes is not unreasonable because the displacement is unique up to an arbitrary rigid body translation and rotation.

The traction vector $\mathbf{t}_{\Gamma}$ on the elliptic boundary Γ is

$$\mathbf{t}_{\Gamma}=-\mathbf{t}_{m}=\frac{\partial}{\rho\partial\psi}\boldsymbol{\phi}_{\Gamma}. \tag{10.1-16}$$

It is shown in Section 7.3 that the *hoop stress vector* $\mathbf{t}_n$ on the elliptic boundary Γ is

$$\mathbf{t}_{n}=\mathbf{N}_3(\theta_{\Gamma})\frac{\partial}{\rho\partial\psi}\mathbf{u}_{\Gamma} \tag{10.1-17}$$

if the ellipse is a void, and

$$\mathbf{t}_{n}=\mathbf{N}_1^{T}(\theta_{\Gamma})\mathbf{t}_{\Gamma}=\mathbf{N}_1^{T}(\theta_{\Gamma})\frac{\partial}{\rho\partial\psi}\boldsymbol{\phi}_{\Gamma} \tag{10.1-18}$$

if the ellipse is a rigid body. The rigid body need not be stationary. It can have a translation and a rotation. Equation (10.1-18) remains valid (Section 7.3). For the problem under consideration substitution of (10.1-15) into (10.1-16) and (10.1-18) yields

$$\mathbf{t}_{\Gamma}=\frac{1}{2\pi\rho}\mathbf{f},\qquad \mathbf{t}_{n}=\frac{1}{2\pi\rho}\mathbf{N}_1^{T}(\theta_{\Gamma})\mathbf{f}. \tag{10.1-19}$$

The traction vector $\mathbf{t}_{\Gamma}$ is independent of the material property. This is rather unexpected. Also unexpected is the fact that $\mathbf{t}_{\Gamma}$ is parallel to the applied force $\mathbf{f}$. It is independent of the orientation θ_0 of the elliptic inclusion. At the points of the major and minor axes on the elliptic surface (Ting, 1991),

$$\mathbf{t}_\Gamma = \begin{cases} \dfrac{1}{2\pi b}\mathbf{f} & \text{at } \psi = 0, \\ \dfrac{1}{2\pi a}\mathbf{f} & \text{at } \psi = \dfrac{\pi}{2}. \end{cases}$$

Therefore the maximum and minimum surface tractions occur at the major and minor axes, respectively.

10.2 A Moment Applied to an Elliptic Rigid Inclusion

When a moment of magnitude T is applied to the elliptic rigid inclusion, the stress at a large distance from the inclusion must be proportional to r^{-2}. This means that the stress function ϕ is proportional to $\hat{z}^{-1}$. In view of (10.1-2) we take ϕ to be proportional to $\hat{\zeta}^{-1}$ and consider

$$\mathbf{u} = \mathrm{Im}\left\{\mathbf{A}\left\langle\hat{\zeta}_*^{-1}\right\rangle\mathbf{q}\right\}, \qquad \phi = \mathrm{Im}\left\{\mathbf{B}\left\langle\hat{\zeta}_*^{-1}\right\rangle\mathbf{q}\right\}, \tag{10.2-1}$$

where $\mathbf{q}$ is a constant vector to be determined. At the elliptic boundary Γ (10.2-1) reduces to

$$\mathbf{u}_\Gamma = \mathrm{Im}\left\{e^{-i\psi}\mathbf{A}\mathbf{q}\right\}, \qquad \phi_\Gamma = \mathrm{Im}\left\{e^{-i\psi}\mathbf{B}\mathbf{q}\right\}. \tag{10.2-2}$$

Let ω be the counterclockwise rotation of the elliptic rigid inclusion. The displacement $\mathbf{u}_\Gamma$ is

$$\begin{aligned}\mathbf{u}_\Gamma &= \omega\left\{\hat{x}_1^\Gamma\mathbf{m}(\theta_0) - \hat{x}_2^\Gamma\mathbf{n}(\theta_0)\right\} \\ &= \omega\left\{a\cos\psi\,\mathbf{m}(\theta_0) - b\sin\psi\,\mathbf{n}(\theta_0)\right\} \\ &= \omega\,\mathrm{Im}\left\{e^{-i\psi}\mathbf{y}(\theta_0)\right\}\end{aligned} \tag{10.2-3}$$

where

$$\mathbf{y}(\theta_0) = b\mathbf{n}(\theta_0) + ia\mathbf{m}(\theta_0). \tag{10.2-4}$$

Comparison with $(10.2\text{-}2)_1$ leads to

$$\mathbf{q} = \omega\mathbf{A}^{-1}\mathbf{y}(\theta_0). \tag{10.2-5}$$

To determine the rotation ω, the moment about the x_3-axis due to the surface traction $\mathbf{t}_\Gamma$ on the elliptic boundary should be equal to the applied moment T, i.e.,

$$\begin{aligned}T &= \int_0^{2\pi}\{\hat{x}_1\mathbf{m}(\theta_0)\cdot\mathbf{t}_\Gamma - \hat{x}_2\mathbf{n}(\theta_0)\cdot\mathbf{t}_\Gamma\}\rho\,d\psi \\ &= \int_0^{2\pi}\{a\cos\psi\,\mathbf{m}(\theta_0) - b\sin\psi\,\mathbf{n}(\theta_0)\}\cdot\mathbf{t}_\Gamma\rho\,d\psi.\end{aligned} \tag{10.2-6}$$

From (10.1-16), (10.2-2), and (10.2-5),

$$\rho \mathbf{t}_\Gamma = \omega \operatorname{Im}\left\{ e^{-i\psi} \mathbf{M}\mathbf{y}(\theta_o) \right\} \tag{10.2-7}$$

where $\mathbf{M}$ is the impedance tensor (Section 5.6)

$$\begin{aligned} \mathbf{M} &= -i\mathbf{B}\mathbf{A}^{-1} = \mathbf{H}^{-1} + i\mathbf{H}^{-1}\mathbf{S}, \\ \mathbf{M}^{-1} &= i\mathbf{A}\mathbf{B}^{-1} = \mathbf{L}^{-1} - i\mathbf{S}\mathbf{L}^{-1}. \end{aligned} \tag{10.2-8}$$

Using (10.2-7) integration in (10.2-6) leads to

$$T = \omega \pi \bar{\mathbf{y}}^T(\theta_0)\mathbf{M}\mathbf{y}(\theta_0). \tag{10.2-9}$$

The impedance tensor is a positive definite Hermitian, assuring us that the ω in (10.2-9) is positive and nonzero.

It can be shown that

$$\bar{\mathbf{y}}^T(\theta_0)\mathbf{M}\mathbf{y}(\theta_0) = b^2\mathbf{n}^T(\theta_o)\mathbf{H}^{-1}\mathbf{n}(\theta_o) + a^2\mathbf{m}^T(\theta_o)\mathbf{H}^{-1}\mathbf{m}(\theta_o)$$

$$+2ab\mathbf{m}^T(\theta_o)\mathbf{H}^{-1}\mathbf{S}\mathbf{n}(\theta_o). \tag{10.2-10}$$

If $\hat{\mathbf{H}}^{-1}$ denotes the tensor $\mathbf{H}^{-1}$ referred to the $\hat{x}_i$-coordinate system,

$$\bar{\mathbf{y}}^T(\theta_0)\mathbf{M}\mathbf{y}(\theta_0) = b^2(\hat{\mathbf{H}}^{-1})_{11} + a^2(\hat{\mathbf{H}}^{-1})_{22} + 2ab(\hat{\mathbf{H}}^{-1}\hat{\mathbf{S}})_{21}. \tag{10.2-11}$$

It has been shown in Section 7.7 that the components

$$(\hat{\mathbf{H}}^{-1})_{11} + (\hat{\mathbf{H}}^{-1})_{22} \quad \text{and} \quad (\hat{\mathbf{H}}^{-1}\hat{\mathbf{S}})_{21}$$

are invariant with rotations of the coordinate system about the x_3-axis. Hence the right side of (10.2-11) can be written as

$$b^2[(\mathbf{H}^{-1})_{11} + (\mathbf{H}^{-1})_{22}] + 2ab(\mathbf{H}^{-1}\mathbf{S})_{21} + (a^2 - b^2)(\hat{\mathbf{H}}^{-1})_{22}$$

or

$$a^2[(\mathbf{H}^{-1})_{11} + (\mathbf{H}^{-1})_{22}] + 2ab(\mathbf{H}^{-1}\mathbf{S})_{21} - (a^2 - b^2)(\hat{\mathbf{H}}^{-1})_{11}.$$

The parameter θ_o appears only through the last term which vanishes when a=b. Since

$$(\hat{\mathbf{H}}^{-1})_{22} = \mathbf{m}^T(\theta_o)\mathbf{H}^{-1}\mathbf{m}(\theta_o),$$

it can be shown that the maximum or minimum of $(\hat{\mathbf{H}}^{-1})_{22}$ occurs at

$$\tan 2\theta_o = \frac{2(\mathbf{H}^{-1})_{21}}{(\mathbf{H}^{-1})_{11} - (\mathbf{H}^{-1})_{22}}. \tag{10.2-12}$$

The same result is obtained if we take $(\hat{\mathbf{H}}^{-1})_{11}$. Equation (10.2-12) is independent of a and b, the geometry of the ellipse. When θ_o is a solution, so is $\theta_o+\pi/2$. We therefore have the result that, for a given moment T, the orientation θ_o of the elliptic rigid inclusion at which the maximum or minimum rotation ω of the inclusion occurs depends on $\mathbf{H}^{-1}$, not on the geometry of the ellipse. The angles θ_o at which the maximum and minimum of ω occur differ by $\pi/2$.

With (10.1-10), (10.2-7) can be written explicitly as (Ting and Yan, 1991)

$$\mathbf{t}_\Gamma = \omega\mathbf{H}^{-1}\Big\{\mathbf{Sm}(\theta_\Gamma) - \big[\tfrac{b}{a}\cos(\theta_\Gamma-\theta_o)\mathbf{n}(\theta_o) + \tfrac{a}{b}\sin(\theta_\Gamma-\theta_o)\mathbf{m}(\theta_o)\big]\Big\}. \quad (10.2\text{-}13)$$

The hoop stress vector $\mathbf{t}_n$ is obtained from $(10.1\text{-}18)_1$. When the ellipse is a circle, $a=b$ and (10.2-13) reduces to

$$\mathbf{t}_\Gamma = \omega\mathbf{H}^{-1}[\mathbf{Sm}(\theta_\Gamma) - \mathbf{n}(\theta_\Gamma)].$$

It is independent of θ_o as expected.

10.3 Green's Functions for an Elliptic Hole or Rigid Inclusion

The problem of an edge dislocation in an infinite medium with the presence of an elliptic rigid inclusion was solved in a closed form by Santare and Keer (1986) for isotropic materials. The same problem but with a line force and with the presence of an elliptic hole was studied by Hwu and Yen (1991) for general anisotropic elastic materials. Approximate solutions of Green's functions for general anisotropic materials with a hole or a rigid inclusion that is not an ellipse have been studied by Wang and Tarn (1993). In this section we consider a line force $\mathbf{f}$ and a line dislocation with Burgers vector $\mathbf{b}$ located at

$$\hat{x}_1 = \hat{x}_1^o, \quad \hat{x}_2 = \hat{x}_2^o, \quad (10.3\text{-}1)$$

which is outside the ellipse, Fig, 10.2. The ellipse can be a hole or a rigid inclusion. The case of an elliptic inclusion of different materials will be discussed in Section 10.10. Following (8.6-2) for Green's functions for half-spaces and (3.8-2) for Green's functions for antiplane deformations with an elliptic hole or rigid inclusion we let

$$\begin{aligned} \mathbf{u} &= \frac{1}{\pi}\mathrm{Im}\Big\{\mathbf{A}\Big\langle \ln(\hat{\zeta}_* - \hat{\zeta}_{*o})\Big\rangle\mathbf{q}^\infty\Big\} + \frac{1}{\pi}\mathrm{Im}\sum_{\beta=1}^{3}\Big\{\mathbf{A}\Big\langle \ln(\hat{\zeta}_*^{-1} - \bar{\hat{\zeta}}_{\beta o})\Big\rangle\mathbf{q}_\beta\Big\}, \\ \boldsymbol{\phi} &= \frac{1}{\pi}\mathrm{Im}\Big\{\mathbf{B}\Big\langle \ln(\hat{\zeta}_* - \hat{\zeta}_{*o})\Big\rangle\mathbf{q}^\infty\Big\} + \frac{1}{\pi}\mathrm{Im}\sum_{\beta=1}^{3}\Big\{\mathbf{B}\Big\langle \ln(\hat{\zeta}_*^{-1} - \bar{\hat{\zeta}}_{\beta o})\Big\rangle\mathbf{q}_\beta\Big\}, \end{aligned} \quad (10.3\text{-}2)$$

where $\mathbf{q}_\beta$ (β=1,2,3) are arbitrary constant vectors and

$$\mathbf{q}^\infty = \mathbf{A}^T\mathbf{f} + \mathbf{B}^T\mathbf{b}, \tag{10.3-3}$$

$$\hat{\zeta}_{\beta o} = \frac{\hat{z}_{\beta o} + \sqrt{\hat{z}^2_{\beta o} - [a^2 + p^2_\beta(\theta_o)b^2]}}{a - ip_\beta(\theta_o)b}, \tag{10.3-4}$$

$$\hat{z}_{\beta o} = \hat{x}^o_1 + p_\beta(\theta_o)\hat{x}^o_2. \tag{10.3-5}$$

The first terms on the right of (10.3-2) with $\mathbf{q}^\infty$ given by (10.3-3) satisfy the applied force $\mathbf{f}$ and the dislocation $\mathbf{b}$ (Section 10.1). The second terms are added to satisfy the condition at the elliptic boundary Γ.

The procedure of determining the constant vectors $\mathbf{q}_\beta$ is identical to that for Green's functions for half-spaces in Section 8.6. If the ellipse is a hole the surface traction vanishes on Γ, i.e.,

$$\boldsymbol{\phi} = \mathbf{0} \quad \text{on } \Gamma.$$

It can be shown that

$$\mathbf{q}_\beta = \mathbf{B}^{-1}\overline{\mathbf{B}}\mathbf{I}_\beta\overline{\mathbf{q}}^\infty \tag{10.3-6}$$

where

$$\mathbf{I}_1 = \text{diag}[1,\ 0,\ 0],\quad \mathbf{I}_2 = \text{diag}[0,\ 1,\ 0],\quad \mathbf{I}_3 = \text{diag}[0,\ 0,\ 1]. \tag{10.3-7}$$

A trivial identity is

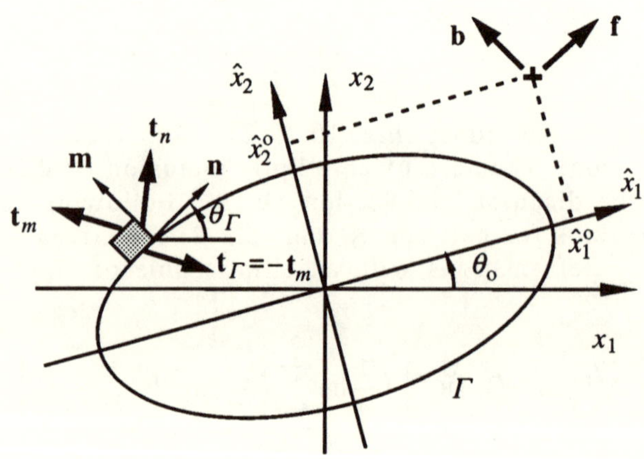

Fig. 10.2 An elliptic hoe or an elliptic rigid inclusion is subjected to a line force $\mathbf{f}$ and a line dislocation with Burgers vector $\mathbf{b}$ at $(\hat{x}_1, \hat{x}_2) = (\hat{x}^o_1, \hat{x}^o_2)$.

$$\sum_{\beta=1}^{3} \mathbf{I}_\beta = \mathbf{I}. \tag{10.3-8}$$

If the ellipse is a rigid inclusion,

$$\mathbf{u} = \mathbf{0} \quad \text{on } \Gamma,$$

and one obtains

$$\mathbf{q}_\beta = \mathbf{A}^{-1}\overline{\mathbf{A}}\mathbf{I}_\beta\overline{\mathbf{q}}^\infty. \tag{10.3-9}$$

Thus the Green's function for an elliptic hole or an elliptic rigid inclusion is given by (10.3-2) and (10.3-6) or (10.3-9).

The singularity of the first terms on the right of (10.3-2) is located at the position indicated in (10.3-1) which is outside the ellipse. Due to the logarithmic term which is a multiple-valued function, a branch cut extended from the singular point to infinity is needed for the function to be single-valued. The second terms on the right of (10.3-2) represent nine one-component Green's functions (Section 8.7). Their singularities are located inside the ellipse, or on the second sheet of the Riemann surface (Section 3.8). From each image singularity we also need a branch cut extended to infinity. These branch cuts from image singularities go through the second Riemann sheet to infinity. They do not appear in the region outside of the ellipse in the first sheet, and hence can be ignored.

The displacement $\mathbf{u}_\Gamma$ and the stress function $\boldsymbol{\phi}_\Gamma$ on the elliptic boundary Γ can be expressed in terms of $\mathbf{u}_\Gamma^\infty$ and $\boldsymbol{\phi}_\Gamma^\infty$ which are the first terms on the right of (10.3-2) evaluated on Γ,

$$\begin{aligned} \mathbf{u}_\Gamma^\infty &= \frac{1}{\pi}\mathrm{Im}\left\{\mathbf{A}\left\langle \ln(e^{i\psi} - \hat{\zeta}_{*o})\right\rangle \mathbf{q}^\infty\right\}, \\ \boldsymbol{\phi}_\Gamma^\infty &= \frac{1}{\pi}\mathrm{Im}\left\{\mathbf{B}\left\langle \ln(e^{i\psi} - \hat{\zeta}_{*o})\right\rangle \mathbf{q}^\infty\right\}. \end{aligned} \tag{10.3-10}$$

The derivation is again similar to that presented in Section 8.6 for determining the displacement and stress function at the boundary on the half-space from the Green's function for the infinite space. When the ellipse is a hole, it can be shown that

$$\mathbf{u}_\Gamma = 2\left\{\mathbf{u}_\Gamma^\infty + \mathbf{S}\mathbf{L}^{-1}\boldsymbol{\phi}_\Gamma^\infty\right\} \tag{10.3-11}$$

which is similar to (8.6-32). If the ellipse is a rigid inclusion,

$$\boldsymbol{\phi}_\Gamma = 2\left\{\boldsymbol{\phi}_\Gamma^\infty + \mathbf{H}^{-1}\mathbf{S}\mathbf{u}_\Gamma^\infty\right\} \tag{10.3-12}$$

which is a counterpart of (8.6-37).

When the ellipse is a hole the hoop stress vector $\mathbf{t}_n$ is determined by (10.1-17). If the ellipse is a rigid inclusion the surface traction $\mathbf{t}_\Gamma$ and the hoop stress vector $\mathbf{t}_n$ are obtained by using (10.1-16) and (10.1-18), respectively.

The Green's function for an elliptic rigid inclusion may generate a nonzero moment T on the ellipse. When $T \neq 0$, and if the ellipse is allowed to rotate freely, we superimpose the Green's function with the solution for the rigid inclusion subjected to a moment $-T$ about the x_3-axis (Section 10.2).

The locations of the nine image singularities can be determined following the special case of antiplane deformation studied in Section 3.8. For simplicity we let $\theta_o=0$ so that the applied force $\mathbf{f}$ and the dislocation $\mathbf{b}$ are at

$$(x_1,x_2)=(x_1^o,x_2^o)$$

denoted by the point D in Fig. 10.3(a). It is mapped to $\zeta_{\beta o}$ (β=1, 2,3) in the ζ-plane through

$$x_1^o + p_\beta x_2^o = z_{\beta o} = c_\beta \zeta_{\beta o} + d_\beta \zeta_{\beta o}^{-1}. \tag{10.3-13}$$

The locations of the image singularities according to the second terms in (10.3-2) are at

$$\zeta_\alpha = \bar{\zeta}_{\beta o}^{-1}$$

where we have omitted the *hat* over the notations. $\bar{\zeta}_{\beta o}^{-1}$ and $\zeta_{\beta o}$ are along the same radial line, Fig. 10.3(b). Their distances from the center of the unit circle are the inverses of each other. Thus $\bar{\zeta}_{\beta o}^{-1}$ and $\zeta_{\beta o}$ are *mirror images* of each other with respect to the unit circle in the ζ-plane. The image singularities in the ζ-plane are now mapped back to the (x_1,x_2)-plane by

$$x_1^{\alpha\beta} + p_\alpha x_2^{\alpha\beta} = z_{\alpha\beta} = c_\alpha \bar{\zeta}_{\beta o}^{-1} + d_\alpha \bar{\zeta}_{\beta o}. \tag{10.3-14}$$

In Fig. 10.3(a) $X^{\alpha\beta}=(x_1^{\alpha\beta},x_2^{\alpha\beta})$ is the location of an image singularity. It lies inside the ellipse if it is on the first sheet of the Riemann surface. If it is located on the second sheet of the Riemann surface, it may be inside or outside of the ellipse. The radial lines in the ζ-plane are hyperbolas in the (x_1,x_2)-plane (Section 3.7). Thus the radial line connecting ζ_{1o} and $\bar{\zeta}_{1o}^{-1}$ in the ζ-plane corresponds to three hyperbolas in the (x_1,x_2)-plane that go through $X^{\alpha 1}$ (α=1,2,3). They intersect each other at the ellipse because the mapping of a point on the ellipse and the unit circle is the same for p_α (α=1,2,3). The three hyperbolas on which X^{11}, X^{22}, X^{33} are located all lead to the point D.

As in the case of Green's function for a half-space (Section 8.7), the three one-component Green's functions associated with $X^{\alpha 1}$, $X^{\alpha 2}$, $X^{\alpha 3}$ have their displacement polarized on the eigenplane associated with $\boldsymbol{a}_\alpha$ and their traction vector polarized on the eigenplane associated with $\boldsymbol{b}_\alpha$.

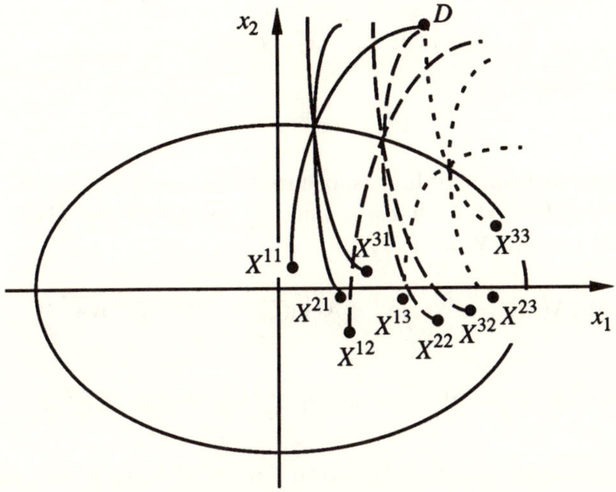

(a) The (x_1, x_2)–plane

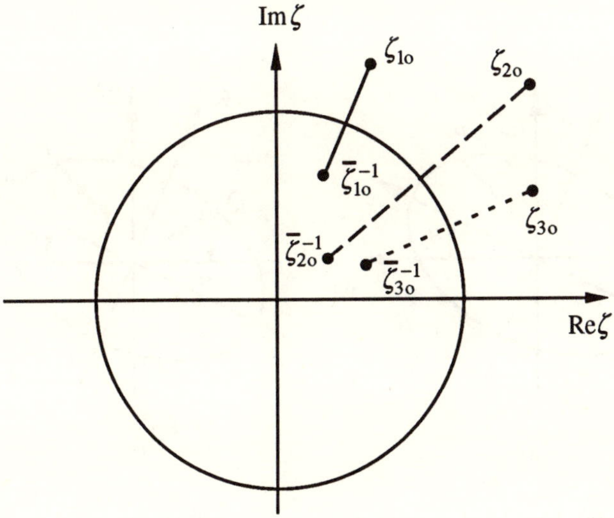

(b) The ζ –plane

Fig. 10-3 The image singularities of an elliptic hole or an elliptic rigid inclusion subjected to a line force and a line dislocation at the point D.

10.4 Surface Green's Function for an Elliptic Hole.

Consider an infinite anisotropic elastic medium with an elliptic hole subjected to a line force **f** which is applied at a point P on the surface of the hole, Fig. 10.4(a). The surface of the elliptic hole is otherwise traction-free. The point P is given by

$$\hat{x}_1^{\rm o} = a\cos\psi_{\rm o}, \quad \hat{x}_2^{\rm o} = b\sin\psi_{\rm o}. \tag{10.4-1}$$

This means that

$$\hat{\zeta}_{\alpha{\rm o}} = e^{i\psi_{\rm o}}. \tag{10.4-2}$$

The solution to this problem is obtained from (10.3-2) in which $\mathbf{q}^\infty$ is given by (10.3-3) with $\mathbf{b}=\mathbf{0}$, and $\mathbf{q}_\beta$ is given by (10.3-6). The result is (Hwu and Yen, 1992)

$$\begin{aligned}\mathbf{u} &= \frac{1}{\pi}\mathrm{Im}\Big\{\mathbf{A}\Big\langle \ln(\hat{\zeta}_* - e^{i\psi_{\rm o}})\Big\rangle\mathbf{A}^T + \mathbf{A}\Big\langle \ln(\hat{\zeta}_*^{-1} - e^{-i\psi_{\rm o}})\Big\rangle\mathbf{B}^{-1}\overline{\mathbf{B}\mathbf{A}}^T\Big\}\mathbf{f},\\ \boldsymbol{\phi} &= \frac{1}{\pi}\mathrm{Im}\Big\{\mathbf{B}\Big\langle \ln(\hat{\zeta}_* - e^{i\psi_{\rm o}})\Big\rangle\mathbf{A}^T + \mathbf{B}\Big\langle \ln(\hat{\zeta}_*^{-1} - e^{-i\psi_{\rm o}})\Big\rangle\mathbf{B}^{-1}\overline{\mathbf{B}\mathbf{A}}^T\Big\}\mathbf{f}.\end{aligned} \tag{10.4-3}$$

This is the surface Green's function for an elliptic hole subjected to a line force on the surface of the hole. For the solution to be single-valued due to the presence of the logarithmic terms, a branch cut starting from the point P where the force **f** is applied to infinity must be introduced. The discontinuity in ϕ across the branch cut is

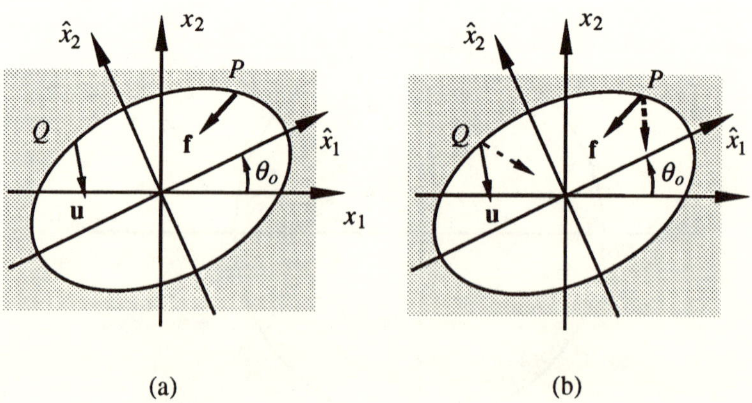

Fig. 10.4 (a) The displacement **u** at any point Q due to the concentrated force **f** at a point P on the elliptical hole is independent of $\theta_{\rm o}$, the orientation of the elliptic hole in the material. (b) If the material is isotropic, this means that when the force **f** is rotated an angle, the displacement vector **u** at Q is also rotated the same angle.

equal to the applied force $\mathbf{f}$.

Equation (104-3) can be simplified by observing that

$$\hat{\zeta}_\alpha - e^{i\psi_o} = \hat{\zeta}_\alpha(1 - e^{i\psi_o}\hat{\zeta}_\alpha^{-1}).$$

On dropping the constant terms that contribute nothing to stresses and replacing $\overline{\mathbf{B}}\overline{\mathbf{A}}^T$ by $\mathbf{I}-\mathbf{B}\mathbf{A}^T$ according to $(5.5\text{-}16)_2$ we have

$$\mathbf{u} = \frac{1}{\pi}\mathrm{Im}\left\{\mathbf{A}\left\langle \ln\left(1 - e^{i\psi_o}\hat{\zeta}_*^{-1}\right)\right\rangle \mathbf{B}^{-1} + \mathbf{A}\left\langle \ln\hat{\zeta}_*\right\rangle \mathbf{A}^T\right\}\mathbf{f},$$

$$\boldsymbol{\phi} = \frac{1}{\pi}\mathrm{Im}\left\{\mathbf{B}\left\langle \ln\left(1 - e^{i\psi_o}\hat{\zeta}_*^{-1}\right)\right\rangle \mathbf{B}^{-1} + \mathbf{B}\left\langle \ln\hat{\zeta}_*\right\rangle \mathbf{A}^T\right\}\mathbf{f}. \tag{10.4-4}$$

On the hole Γ, $\hat{\zeta}_\alpha = e^{i\psi}$ so that the displacement $\mathbf{u}_\Gamma$ is

$$\mathbf{u}_\Gamma = \frac{-1}{\pi}\mathrm{Im}\left\{\ln(1 - e^{-i(\psi-\psi_o)})(\mathbf{S}\mathbf{L}^{-1} + i\mathbf{L}^{-1})\right\}\mathbf{f},$$

use being made of $(10.2\text{-}8)_4$. It is readily shown that

$$1 - e^{-i(\psi-\psi_o)} = 1 - \cos(\psi - \psi_o) + i\sin(\psi - \psi_o) = 2i\sin\Theta e^{-i\Theta},$$

$$\Theta = \tfrac{1}{2}(\psi - \psi_o). \tag{10.4-5}$$

Hence, omitting constant terms that represent a rigid body motion,

$$\mathbf{u}_\Gamma = \frac{1}{\pi}\left\{\Theta\mathbf{S}\mathbf{L}^{-1} - \ln(\sin\Theta)\mathbf{L}^{-1}\right\}\mathbf{f}. \tag{10.4-6}$$

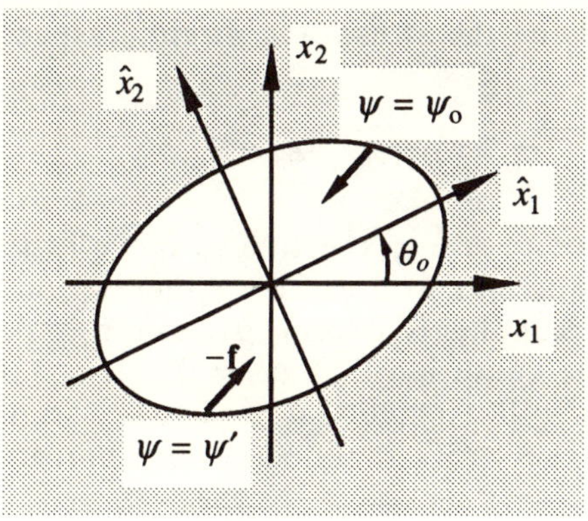

Fig. 10.5 The elliptic hole is subjected to a pair of equal and opposite forces acting on the hole surface. The solution is given in (10.4-8).

Equation (10.4-6) gives an explicit real form expression for $\mathbf{u}_\Gamma$. It should be noted that $\mathbf{u}_\Gamma$ is independent of θ_0, which means that $\mathbf{u}_\Gamma$ is invariant with the orientation of the ellipse. For isotropic materials it implies that if we rotate the applied force $\mathbf{f}$ an angle at the point P, the displacement $\mathbf{u}_\Gamma$ at a point Q on the elliptic surface is also rotated the same angle, Fig. 10.4(b). The hoop stress vector $\mathbf{t}_n$ is, by (10.1-17) and (10.4-6),

$$\mathbf{t}_n = \frac{1}{2\pi\rho}\mathbf{N}_3(\theta_\Gamma)\left\{\mathbf{S}\mathbf{L}^{-1} - (\cot\Theta)\mathbf{L}^{-1}\right\}\mathbf{f}. \tag{10.4-7}$$

The second terms in (10.4-4) do not depend on ψ_0, the location of the applied force $\mathbf{f}$. If there is another force $-\mathbf{f}$ applied at $\psi=\psi'$ on the hole, Fig. 10.5, the solution is

$$\mathbf{u} = \frac{1}{\pi}\mathrm{Im}\left\{\mathbf{A}\left\langle \ln\frac{\hat{\zeta}_* - e^{i\psi_0}}{\hat{\zeta}_* - e^{i\psi'}}\right\rangle\mathbf{B}^{-1}\right\}\mathbf{f},$$

$$\boldsymbol{\phi} = \frac{1}{\pi}\mathrm{Im}\left\{\mathbf{B}\left\langle \ln\frac{\hat{\zeta}_* - e^{i\psi_0}}{\hat{\zeta}_* - e^{i\psi'}}\right\rangle\mathbf{B}^{-1}\right\}\mathbf{f}. \tag{10.4-8}$$

10.5 Representations of a Uniform Stress Solution

In the next two sections we will study an elliptic inclusion in the infinite anisotropic elastic medium subjected to a uniform loading at infinity. In this section we present representations of a uniform stress solution.

Let σ_{ij} and ε_{ij} be the stress and strain, respectively. They are related by

$$\sigma_{ij} = C_{ijks}\varepsilon_{ks}, \quad \varepsilon_{ij} = S_{ijks}\sigma_{ks},$$

where C_{ijks} and S_{ijks} are, respectively, the elastic stiffnesses and compliances. It suffices to prescribe either σ_{ij} or ε_{ij}. When σ_{ij} is prescribed, all components except σ_{33} can be prescribed. σ_{33} is determined by the condition that $\varepsilon_{33} = 0$. If ε_{ij} is prescribed, ε_{33} must vanish. We will assume that σ_{ij} and ε_{ij} are known. The displacement $\mathbf{u}$ and the stress function $\boldsymbol{\phi}$ for a uniform stress solution can be written as

$$\mathbf{u} = x_1\boldsymbol{\varepsilon}_1 + x_2\boldsymbol{\varepsilon}_2, \quad \boldsymbol{\phi} = x_1\mathbf{t}_2 - x_2\mathbf{t}_1, \tag{10.5-1}$$

where

$$\boldsymbol{\varepsilon}_1 = \begin{bmatrix} \varepsilon_{11} \\ 0 \\ 2\varepsilon_{31} \end{bmatrix} = \mathbf{u}_{,1}, \quad \boldsymbol{\varepsilon}_2 = \begin{bmatrix} 2\varepsilon_{12} \\ \varepsilon_{22} \\ 2\varepsilon_{32} \end{bmatrix} = \mathbf{u}_{,2},$$

$$\mathbf{t}_1 = \begin{bmatrix} \sigma_{11} \\ \sigma_{21} \\ \sigma_{31} \end{bmatrix} = -\boldsymbol{\phi}_{,2}, \quad \mathbf{t}_2 = \begin{bmatrix} \sigma_{12} \\ \sigma_{22} \\ \sigma_{32} \end{bmatrix} = \boldsymbol{\phi}_{,1}. \tag{10.5-2}$$

The displacement $\mathbf{u}$ is unique up to a rigid body translation and rotation. In the above we assumed no translation and no rotation of the x_1-axis, justifying the zero element in $\boldsymbol{\varepsilon}_1$. Equation $(10.5\text{-}1)_1$ satisfies the displacement-strain relations (3.1-1) while $(10.5\text{-}1)_2$ satisfies the alternate equilibrium equations (4.1-2).

The uniform stress solution can also be represented by

$$\mathbf{u} = 2\,\mathrm{Re}\{\mathbf{A}\langle z_*\rangle\mathbf{q}\}, \quad \boldsymbol{\phi} = 2\,\mathrm{Re}\{\mathbf{B}\langle z_*\rangle\mathbf{q}\}, \tag{10.5-3}$$

$$\mathbf{q} = \mathbf{A}^T\mathbf{t}_2 + \mathbf{B}^T\boldsymbol{\varepsilon}_1. \tag{10.5-4}$$

To show that (10.5-3) with the $\mathbf{q}$ given by (10.5-4) is equivalent to (10.5-1), we let

$$\mathbf{q} = \mathbf{A}^T\mathbf{g} + \mathbf{B}^T\mathbf{h}$$

where $\mathbf{g}$ and $\mathbf{h}$ are constant vectors. By virtue of the identity (7.9-7), (10.5-3) is transformed into a real form as

$$\mathbf{u} = x_1\mathbf{h} + x_2(\mathbf{N}_1\mathbf{h} + \mathbf{N}_2\mathbf{g}),$$

$$\boldsymbol{\phi} = x_1\mathbf{g} + x_2(\mathbf{N}_3\mathbf{h} + \mathbf{N}_1^T\mathbf{g}).$$

Comparison with (10.5-1) yields

$$\begin{bmatrix} \mathbf{h} \\ \mathbf{g} \end{bmatrix} = \begin{bmatrix} \boldsymbol{\varepsilon}_1 \\ \mathbf{t}_2 \end{bmatrix}, \quad \begin{bmatrix} \boldsymbol{\varepsilon}_2 \\ -\mathbf{t}_1 \end{bmatrix} = \mathbf{N}\begin{bmatrix} \boldsymbol{\varepsilon}_1 \\ \mathbf{t}_2 \end{bmatrix}, \tag{10.5-5}$$

where

$$\mathbf{N} = \begin{bmatrix} \mathbf{N}_1 & \mathbf{N}_2 \\ \mathbf{N}_3 & \mathbf{N}_1^T \end{bmatrix}.$$

Conversely, when the complex vector $\mathbf{q}$ is known, $\mathbf{t}_2$ and $\boldsymbol{\varepsilon}_1$ can be determined by writing (10.5-4) and its complex conjugate as

$$\begin{bmatrix} \mathbf{q} \\ \overline{\mathbf{q}} \end{bmatrix} = \begin{bmatrix} \mathbf{B}^T & \mathbf{A}^T \\ \overline{\mathbf{B}}^T & \overline{\mathbf{A}}^T \end{bmatrix} \begin{bmatrix} \boldsymbol{\varepsilon}_1 \\ \mathbf{t}_2 \end{bmatrix}.$$

By the closure relation (5.5-15) this is equivalent to

$$\begin{bmatrix} \boldsymbol{\varepsilon}_1 \\ \mathbf{t}_2 \end{bmatrix} = \begin{bmatrix} \mathbf{A} & \overline{\mathbf{A}} \\ \mathbf{B} & \overline{\mathbf{B}} \end{bmatrix} \begin{bmatrix} \mathbf{q} \\ \overline{\mathbf{q}} \end{bmatrix}.$$

from which we have

$$\boldsymbol{\varepsilon}_1 = 2\mathrm{Re}(\mathbf{A}\mathbf{q}), \quad \mathbf{t}_2 = 2\mathrm{Re}(\mathbf{B}\mathbf{q}). \tag{10.5-6}$$

$\boldsymbol{\varepsilon}_2$ and $\mathbf{t}_1$ are determined from $(10.5\text{-}5)_2$ as

$$\boldsymbol{\varepsilon}_2 = \mathbf{N}_1\boldsymbol{\varepsilon}_1 + \mathbf{N}_2\mathbf{t}_2, \quad -\mathbf{t}_1 = \mathbf{N}_3\boldsymbol{\varepsilon}_1 + \mathbf{N}_1^T\mathbf{t}_2. \tag{10.5-7}$$

Therefore a uniform stress solution can be represented by either (10.5-1) or (10.5-3) with the $\mathbf{q}$ given by (10.5-4).

10.6 Elliptic Hole or Rigid Inclusion Subjected to a Uniform Loading at Infinity

It is well known that the maximum hoop stress around a circular hole in an infinite isotropic elastic material under a uniform simple tension at infinity is three times the tension applied at infinity. The *stress concentration factor* is three. For an elliptic hole with $a \geq b > 0$ being the principal radii, the stress concentration factor is $1+2(a/b)$ when the applied tension is normal to the major axis (Timoshenko and Goodier, 1970, p.92). Yang and Chou (1978) considered an elliptic rigid inclusion in an orthotropic medium subjected to a uniform loading at infinity. The problem of an elliptic hole in a general anisotropic material was studied by Hwu and Ting (1989) who presented an explicit real expression for the stress concentration factor. Approximate solutions for a hole of geometry other than ellipse in a general anisotropic material were given by Hwu (1990a). In this section we study an elliptic hole or an elliptic rigid inclusion in a general anisotropic elastic medium subjected to a uniform loading at infinity. For the remaining of this chapter we employ a single coordinate system. Equations (10.1-1)–(10.1-10) are specialized by setting $\theta_o = 0$. In particular,

$$z_\alpha = c_\alpha \zeta_\alpha + d_\alpha \zeta_\alpha^{-1}, \quad (\alpha = 1,2,3), \tag{10.6-1}$$

$$c_\alpha = \tfrac{1}{2}(a - ip_\alpha b), \quad d_\alpha = \tfrac{1}{2}(a + ip_\alpha b), \tag{10.6-2}$$

$$\zeta_\alpha = \frac{z_\alpha + \sqrt{z_\alpha^2 - (a^2 + p_\alpha^2 b^2)}}{a - ip_\alpha b}, \tag{10.6-3a}$$

$$\zeta_\alpha^{-1} = \frac{z_\alpha - \sqrt{z_\alpha^2 - (a^2 + p_\alpha^2 b^2)}}{a + ip_\alpha b}, \tag{10.6-3b}$$

$$\rho = \sqrt{a^2 \sin^2 \psi + b^2 \cos^2 \psi}, \tag{10.6-4}$$

$$\rho \cos \theta_\Gamma = a \sin \psi, \quad \rho \sin \theta_\Gamma = -b \cos \psi. \tag{10.6-5}$$

Let σ_{ij}^∞ and ε_{ij}^∞ be the stress and strain at infinity. They are related by the stress-strain laws. As we have discussed in the previous section we can either prescribe σ_{ij}^∞ or ε_{ij}^∞ with the condition that $\varepsilon_{33}^\infty=0$. We assume that both σ_{ij}^∞ and ε_{ij}^∞ are known. Consider the solution in the form

$$\begin{aligned} \mathbf{u} &= x_1 \boldsymbol{\varepsilon}_1^\infty + x_2 \boldsymbol{\varepsilon}_2^\infty + 2\,\mathrm{Re}\{\mathbf{A}\langle \zeta_*^{-1}\rangle \mathbf{q}\}, \\ \boldsymbol{\phi} &= x_1 \mathbf{t}_2^\infty - x_2 \mathbf{t}_1^\infty + 2\,\mathrm{Re}\{\mathbf{B}\langle \zeta_*^{-1}\rangle \mathbf{q}\}, \end{aligned} \tag{10.6-6}$$

where $\mathbf{q}$ is a constant vector to be determined. This is equivalent to

$$\begin{aligned} \mathbf{u} &= 2\,\mathrm{Re}\{\mathbf{A}\langle z_*\rangle \mathbf{q}_o^\infty\} + 2\,\mathrm{Re}\{\mathbf{A}\langle \zeta_*^{-1}\rangle \mathbf{q}\}, \\ \boldsymbol{\phi} &= 2\,\mathrm{Re}\{\mathbf{B}\langle z_*\rangle \mathbf{q}_o^\infty\} + 2\,\mathrm{Re}\{\mathbf{B}\langle \zeta_*^{-1}\rangle \mathbf{q}\}, \end{aligned} \tag{10.6-7}$$

in which $\mathbf{q}_o^\infty$ is, by (10.5-4),

$$\mathbf{q}_o^\infty = \mathbf{A}^T \mathbf{t}_2^\infty + \mathbf{B}^T \boldsymbol{\varepsilon}_1^\infty.$$

Substitution of (10.6-1) into (10.6-7) yields an alternate form (Gao, 1992; Hwu, 1993)

$$\begin{aligned} \mathbf{u} &= 2\,\mathrm{Re}\{\mathbf{A}\langle \zeta_*\rangle \tilde{\mathbf{q}}_o^\infty\} + 2\,\mathrm{Re}\{\mathbf{A}\langle \zeta_*^{-1}\rangle \tilde{\mathbf{q}}\}, \\ \boldsymbol{\phi} &= 2\,\mathrm{Re}\{\mathbf{B}\langle \zeta_*\rangle \tilde{\mathbf{q}}_o^\infty\} + 2\,\mathrm{Re}\{\mathbf{B}\langle \zeta_*^{-1}\rangle \tilde{\mathbf{q}}\}, \end{aligned} \tag{10.6-8}$$

where

$$\tilde{\mathbf{q}}_o^\infty = \langle c_*\rangle \mathbf{q}_o^\infty, \quad \tilde{\mathbf{q}} = \langle d_*\rangle \mathbf{q}_o^\infty + \mathbf{q}.$$

We also have

$$\mathbf{q}_o^\infty = \langle c_*^{-1}\rangle \tilde{\mathbf{q}}_o^\infty, \quad \mathbf{q} = \tilde{\mathbf{q}} - \langle t_*\rangle \tilde{\mathbf{q}}_o^\infty,$$

$$t_\alpha = \frac{d_\alpha}{c_\alpha} = \frac{a + ip_\alpha b}{a - ip_\alpha b}, \qquad |t_\alpha| < 1. \tag{10.6-9}$$

The inequality in (10.6-9) was proved in (3.7-9). We will employ (10.6-6) in the sequel because it is expressed more directly in terms of $\boldsymbol{\varepsilon}_1^\infty$, $\boldsymbol{\varepsilon}_2^\infty$, $\mathbf{t}_1^\infty$, and $\mathbf{t}_2^\infty$.

At the elliptic boundary Γ, $x_1 = a\cos\psi$ and $x_2 = b\sin\psi$ so that

$$\begin{aligned} x_1\boldsymbol{\varepsilon}_1^\infty + x_2\boldsymbol{\varepsilon}_2^\infty &= \mathrm{Re}[e^{-i\psi}(a\boldsymbol{\varepsilon}_1^\infty + ib\boldsymbol{\varepsilon}_2^\infty)], \\ x_1\mathbf{t}_2^\infty - x_2\mathbf{t}_1^\infty &= \mathrm{Re}[e^{-i\psi}(a\mathbf{t}_2^\infty - ib\mathbf{t}_1^\infty)]. \end{aligned} \tag{10.6-10}$$

From (10.6-6) the displacement $\mathbf{u}_\Gamma$ and the stress function $\boldsymbol{\phi}_\Gamma$ at the elliptic boundary Γ are

$$\begin{aligned} \mathbf{u}_\Gamma &= \mathrm{Re}[e^{-i\psi}(a\boldsymbol{\varepsilon}_1^\infty + ib\boldsymbol{\varepsilon}_2^\infty + 2\mathbf{A}\mathbf{q})], \\ \boldsymbol{\phi}_\Gamma &= \mathrm{Re}[e^{-i\psi}(a\mathbf{t}_2^\infty - ib\mathbf{t}_1^\infty + 2\mathbf{B}\mathbf{q})]. \end{aligned} \tag{10.6-11}$$

We now discuss the elliptic hole and elliptic rigid inclusion separately below.

(I) The Elliptic Hole.

The traction on the surface of an elliptic hole vanishes. Setting $\boldsymbol{\phi}_\Gamma = \mathbf{0}$ in $(10.6\text{-}11)_2$ we obtain

$$\mathbf{q} = -\tfrac{1}{2}\mathbf{B}^{-1}(a\mathbf{t}_2^\infty - ib\mathbf{t}_1^\infty) \tag{10.6-12}$$

and $(10.6\text{-}11)_1$ gives

$$\begin{aligned} \mathbf{u}_\Gamma = {} & \cos\psi\,(a\boldsymbol{\varepsilon}_1^\infty + a\mathbf{S}\mathbf{L}^{-1}\mathbf{t}_2^\infty + b\mathbf{L}^{-1}\mathbf{t}_1^\infty) \\ & + \sin\psi\,(b\boldsymbol{\varepsilon}_2^\infty - b\mathbf{S}\mathbf{L}^{-1}\mathbf{t}_1^\infty + a\mathbf{L}^{-1}\mathbf{t}_2^\infty), \end{aligned} \tag{10.6-13}$$

use being made of $(10.2\text{-}8)_4$. From (10.1-17) and (10.6-5) the hoop stress vector $\mathbf{t}_n$ is

$$\begin{aligned} \mathbf{t}_n = -\mathbf{N}_3(\theta_\Gamma)[& \cos\theta_\Gamma(\boldsymbol{\varepsilon}_1^\infty + \mathbf{S}\mathbf{L}^{-1}\mathbf{t}_2^\infty + \tfrac{b}{a}\mathbf{L}^{-1}\mathbf{t}_1^\infty) \\ & + \sin\theta_\Gamma(\boldsymbol{\varepsilon}_2^\infty - \mathbf{S}\mathbf{L}^{-1}\mathbf{t}_1^\infty + \tfrac{a}{b}\mathbf{L}^{-1}\mathbf{t}_2^\infty)]. \end{aligned} \tag{10.6-14}$$

The terms containing $\boldsymbol{\varepsilon}_1^\infty$ and $\boldsymbol{\varepsilon}_2^\infty$ can be replaced by an expression containing $\mathbf{t}_1^\infty$ and $\mathbf{t}_2^\infty$ as follows. Equation (7.3-19) applied to the present problem is

$$\mathbf{N}(\theta)\begin{bmatrix} \mathbf{u}_{,n} \\ \boldsymbol{\phi}_{,n} \end{bmatrix} = \begin{bmatrix} \mathbf{u}_{,m} \\ \boldsymbol{\phi}_{,m} \end{bmatrix}$$

where the subscripts n and m followed by a comma denotes differentiation in the direction of $\mathbf{n}$ and $\mathbf{m}$, respectively. Hence

$$\mathbf{u}_{,m} = \mathbf{m}(\theta)\cdot\nabla\mathbf{u} = -\sin\theta\,\boldsymbol{\varepsilon}_1 + \cos\theta\,\boldsymbol{\varepsilon}_2,$$

$$\mathbf{u}_{,n} = \mathbf{n}(\theta)\cdot\nabla\mathbf{u} = \cos\theta\,\boldsymbol{\varepsilon}_1 + \sin\theta\,\boldsymbol{\varepsilon}_2,$$

$$\boldsymbol{\phi}_{,m} = \mathbf{m}(\theta)\cdot\nabla\boldsymbol{\phi} = -(\sin\theta\,\mathbf{t}_2 + \cos\theta\,\mathbf{t}_1),$$

$$\boldsymbol{\phi}_{,n} = \mathbf{n}(\theta)\cdot\nabla\boldsymbol{\phi} = \cos\theta\,\mathbf{t}_2 - \sin\theta\,\mathbf{t}_1.$$

Therefore

$$\mathbf{N}_3(\theta)(\cos\theta\,\boldsymbol{\varepsilon}_1 + \sin\theta\,\boldsymbol{\varepsilon}_2) + \mathbf{N}_1^T(\theta)(\cos\theta\,\mathbf{t}_2 - \sin\theta\,\mathbf{t}_1)$$
$$= -(\sin\theta\,\mathbf{t}_2 + \cos\theta\,\mathbf{t}_1),$$

and (10.6-14) is replaced by

$$\mathbf{t}_n = \cos\theta_\Gamma[\mathbf{t}_1^\infty + \mathbf{G}_1(\theta_\Gamma)\mathbf{t}_2^\infty + \tfrac{b}{a}\mathbf{G}_3(\theta_\Gamma)\mathbf{t}_1^\infty)]$$
$$+\sin\theta_\Gamma[\mathbf{t}_2^\infty - \mathbf{G}_1(\theta_\Gamma)\mathbf{t}_1^\infty + \tfrac{a}{b}\mathbf{G}_3(\theta_\Gamma)\mathbf{t}_2^\infty)]. \qquad (10.6\text{-}15)$$

where

$$\mathbf{G}_1(\theta) = \mathbf{N}_1^T(\theta) - \mathbf{N}_3(\theta)\mathbf{S}\mathbf{L}^{-1}, \quad \mathbf{G}_3(\theta) = -\mathbf{N}_3(\theta)\mathbf{L}^{-1}. \qquad (10.6\text{-}16)$$

The tensors $\mathbf{G}_1^T(\theta)$ and $\mathbf{G}_3^T(\theta)$ share some properties with $\mathbf{N}_1(\theta)$ and $\mathbf{N}_3(\theta)$ (Section 7.11). They are periodic in θ with periodicity π, and satisfy the same identities (7.11-3) and (7.11-4), i.e.,

$$\mathbf{G}_1^T(\theta)\mathbf{m}(\theta) = -\mathbf{n}(\theta), \qquad \mathbf{G}_3^T(\theta)\mathbf{m}(\theta) = \mathbf{0}, \qquad (10.6\text{-}17)$$

$$\begin{aligned}
\cos(\theta-\theta_o)\mathbf{G}_1^T(\theta)\mathbf{n}(\theta) &= \mathbf{G}_1^T(\theta)\mathbf{n}(\theta_o) - \sin(\theta-\theta_o)\mathbf{n}(\theta),\\
\sin(\theta-\theta_o)\mathbf{G}_1^T(\theta)\mathbf{n}(\theta) &= \mathbf{G}_1^T(\theta)\mathbf{m}(\theta_o) + \cos(\theta-\theta_o)\mathbf{n}(\theta),\\
\cos(\theta-\theta_o)\mathbf{G}_3^T(\theta)\mathbf{n}(\theta) &= \mathbf{G}_3^T(\theta)\mathbf{n}(\theta_o),\\
\sin(\theta-\theta_o)\mathbf{G}_3^T(\theta)\mathbf{n}(\theta) &= \mathbf{G}_3^T(\theta)\mathbf{m}(\theta_o),
\end{aligned} \qquad (10.6\text{-}18)$$

The *hoop stress* $\sigma_{nn} = \mathbf{n}(\theta_\Gamma)\cdot\mathbf{t}_n$ is, after employing (10.6-18),

$$\sigma_{nn} = \mathbf{n}^T(0)\left\{\mathbf{G}_1(\theta_\Gamma)\mathbf{t}_2^\infty + \tfrac{b}{a}\mathbf{G}_3(\theta_\Gamma)\mathbf{t}_1^\infty\right\}$$
$$-\mathbf{m}^T(0)\left\{\mathbf{G}_1(\theta_\Gamma)\mathbf{t}_1^\infty - \tfrac{a}{b}\mathbf{G}_3(\theta_\Gamma)\mathbf{t}_2^\infty\right\}. \qquad (10.6\text{-}19)$$

Consider the special case in which a uniform tension σ_{22}^∞ is applied at infinity. We have

$$\mathbf{t}_1^\infty = \mathbf{0}, \quad \mathbf{t}_2^\infty = \sigma_{22}^\infty\mathbf{m}(0),$$

and (10.6-19) reduces to

$$\sigma_{nn} / \sigma_{22}^{\infty} = [\mathbf{G}_1(\theta_\Gamma)]_{12} + \tfrac{a}{b}[\mathbf{G}_3(\theta_\Gamma)]_{22}. \qquad (10.6\text{-}20)$$

In the above $[*]_{ij}$ denotes the (ij) element of the matrix inside the square brackets. At the point A, Fig. 10.6, of the hole on the major axis of the ellipse $\theta_\Gamma = -\frac{1}{2}\pi$. The structure of $\mathbf{G}_i(\frac{1}{2}\pi)$ which is identical to $\mathbf{G}_i(-\frac{1}{2}\pi)$ $(i=1,3)$ have been given in (7.11-10). Hence

$$\sigma_{nn} / \sigma_{22}^{\infty} = 1 + \tfrac{a}{b}[\mathbf{G}_3(\tfrac{1}{2}\pi)]_{22}, \quad \text{at } \theta_\Gamma = -\tfrac{1}{2}\pi. \qquad (10.6\text{-}21)$$

Employing the explicit expressions of $\mathbf{G}_1(\theta)$, $\mathbf{G}_3(\theta)$ presented in (7.11-11) for isotropic materials, (10.6-21) is reduced to the well known result

$$\sigma_{nn} / \sigma_{22}^{\infty} = 1 + 2(a/b), \quad \text{at } \theta_\Gamma = -\tfrac{1}{2}\pi$$

for isotropic materials. The hoop stress at the point B of the hole on the minor axis of the ellipse is obtained by setting $\theta_\Gamma=0$ in (10.6-20). From the structure of $\mathbf{G}_1(0)$ and $\mathbf{G}_3(0)$ given in (7.11-9) we have

$$\sigma_{nn} / \sigma_{22}^{\infty} = [\mathbf{G}_1(0)]_{12}, \quad \text{at } \theta_\Gamma = 0, \qquad (10.6\text{-}22)$$

which is independent of the shape of the elliptic hole. For isotropic materials this is reduced to

$$\sigma_{nn} / \sigma_{22}^{\infty} = -1, \quad \text{at } \theta_\Gamma = 0.$$

The displacement $\mathbf{u}_\Gamma$ obtained from (10.6-13) is

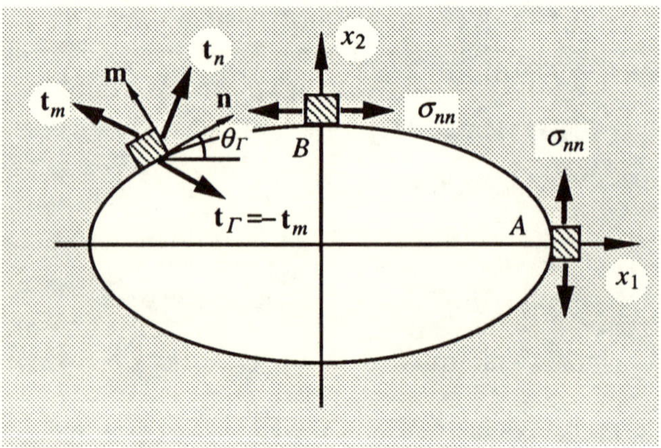

Fig. 10.6 The surface traction $\mathbf{t}_m = -\mathbf{t}_\Gamma$, the hoop stress vector $\mathbf{t}_n$, and the hoop stress σ_{nn}.

$$\mathbf{u}_\Gamma = \cos\psi\,[a\boldsymbol{\varepsilon}_1^\infty + a\sigma_{22}^\infty \mathbf{S}\mathbf{L}^{-1}\mathbf{m}(0)] + \sin\psi\,[b\boldsymbol{\varepsilon}_2^\infty + a\sigma_{22}^\infty \mathbf{L}^{-1}\mathbf{m}(0)].$$

At the point A (ψ=0) the x_2-component of $\mathbf{u}_\Gamma$ vanishes. This implies that the major axis of the elliptic hole is not rotated due to the uniform tension σ_{22}^∞ at infinity. This is a surprising result since no material symmetry has been assumed in the problem.

(II) The Elliptic Rigid Inclusion.

When the elliptic inclusion is a rigid body it may rotate about the x_3-axis unless the rigid inclusion is constrained to rotate. Let ω be the counterclockwise rotation of the elliptic rigid inclusion about the x_3-axis. The displacement $\mathbf{u}_\Gamma$ can be written as

$$\mathbf{u}_\Gamma = \omega[a\cos\psi\,\mathbf{m}(0) - b\sin\psi\,\mathbf{n}(0)] = \omega\,\mathrm{Re}[-ie^{-i\psi}\mathbf{y}]$$

where

$$\mathbf{y} = b\mathbf{n}(0) + ia\mathbf{m}(0).$$

Comparison with the $\mathbf{u}_\Gamma$ in $(10.6\text{-}11)_1$ indicates that

$$\mathbf{q} = -\tfrac{1}{2}\mathbf{A}^{-1}(i\omega\mathbf{y} + a\boldsymbol{\varepsilon}_1^\infty + ib\boldsymbol{\varepsilon}_2^\infty). \qquad (10.6\text{-}23)$$

The surface traction $\mathbf{t}_\Gamma$ along the elliptic boundary is, by (10.1-16) and $(10.6\text{-}11)_2$,

$$\rho\mathbf{t}_\Gamma = \mathrm{Re}[e^{-i\psi}\mathbf{k}] = \cos\psi\,\mathbf{k}' + \sin\psi\,\mathbf{k}'' \qquad (10.6\text{-}24a)$$

where $\mathbf{k}'$ and $\mathbf{k}''$ are, respectively, the real and imaginary parts of

$$\mathbf{k} = -i(a\mathbf{t}_2^\infty - ib\mathbf{t}_1^\infty) - \mathbf{M}(i\omega\mathbf{y} + a\boldsymbol{\varepsilon}_1^\infty + ib\boldsymbol{\varepsilon}_2^\infty). \qquad (10.6\text{-}24b)$$

To determine ω, we impose the condition that the resultant moment about the x_3-axis due to the traction $\mathbf{t}_\Gamma$ on the elliptic boundary vanishes, i.e.,

$$\int_0^{2\pi} [a\cos\psi\,\mathbf{m}(0) - b\sin\psi\,\mathbf{n}(0)]\cdot\mathbf{t}_\Gamma\rho\,d\psi = 0.$$

Substitution of (10.6-24a) into the above leads to

$$a\mathbf{m}(0)\cdot\mathbf{k}' - b\mathbf{n}(0)\cdot\mathbf{k}'' = \mathrm{Re}[i\bar{\mathbf{y}}\cdot\mathbf{k}] = 0,$$

or

$$\mathrm{Re}[\bar{\mathbf{y}}\cdot(a\mathbf{t}_2^\infty - ib\mathbf{t}_1^\infty)] - \mathrm{Re}[i\bar{\mathbf{y}}^T\mathbf{M}(i\omega\mathbf{y} + a\boldsymbol{\varepsilon}_1^\infty + ib\boldsymbol{\varepsilon}_2^\infty)\,] = 0.$$

Since

$$\mathrm{Re}[\bar{\mathbf{y}}\cdot(a\mathbf{t}_2^\infty - ib\mathbf{t}_1^\infty)] = ab[\mathbf{n}(0)\cdot\mathbf{t}_2^\infty - \mathbf{m}(0)\cdot\mathbf{t}_1^\infty] = ab(\sigma_{12}^\infty - \sigma_{21}^\infty) = 0$$

we have

$$\omega = -\frac{\mathrm{Im}[\bar{\mathbf{y}}^T \mathbf{M}(a\boldsymbol{\varepsilon}_1^\infty + ib\boldsymbol{\varepsilon}_2^\infty)]}{\bar{\mathbf{y}}^T \mathbf{M}\mathbf{y}}. \tag{10.6-25}$$

The denominator is real, positive, and nonzero because $\mathbf{M}$ is a positive definite Hermitian. It is identical to (10.2-11) with $\theta_o=0$. Hence

$$\bar{\mathbf{y}}^T \mathbf{M}\mathbf{y} = b^2(\mathbf{H}^{-1})_{11} + a^2(\mathbf{H}^{-1})_{22} + 2ab(\mathbf{H}^{-1}\mathbf{S})_{21}.$$

The surface traction $\mathbf{t}_\Gamma$ and the hoop stress vector $\mathbf{t}_n$ at the elliptic boundary are determined by (10.1-16) and (10.1-18).

10.7 Elliptic Elastic Inclusion Subjected to a Uniform Loading at Infinity

The problem of determining stress distribution in an infinite elastic medium induced by an inclusion has aroused considerable interest among the researchers. It was shown by Hardiman (1954) that, subjected to a uniform loading at infinity, the stress field inside an elliptic elastic inclusion in an infinite elastic plate is uniform. More generally, Eshelby (1957, 1959, 1961) showed that the stress inside an anisotropic ellipsoidal inclusion in an infinite medium of dissimilar anisotropic elastic material is uniform when subjected to a uniform loading at infinity. Most of the work related to the inclusion problems can be found in the book by Mura (1987) and in his review article (Mura, 1988). Although the stress inside the inclusion was shown to be uniform the exact value of the stress inside the inclusion remained to be determined. Jaswon and Bhargava (1961) applied Eshelby's concept to the isotropic plane problem with an elliptic inclusion. Their analysis was extended by Willis (1964) to cubic materials. Subsequent analyses were given by Bhargava and Radhakrishna (1964) for orthotropic materials and by W. T. Chen (1967b) for monoclinic materials. Yang and Chou (1976, 1977, 1978) presented an analysis that is valid for general anisotropic elastic materials. The solutions are in a complex form. An explicit real form solution inside the inclusion was found for orthotropic materials. The following presentation follows essentially that of Hwu and Ting (1989) whose analysis provides an explicit real form solution for the stress inside the inclusion for general anisotropic materials.

We employ (10.6-6) for the solution outside the elliptic inclusion with the complex vector $\mathbf{q}$ replaced by $\mathbf{A}^T\mathbf{g}+\mathbf{B}^T\mathbf{h}$, where $\mathbf{g}$ and $\mathbf{h}$ are real vectors. Hence

$$\mathbf{u} = x_1\boldsymbol{\varepsilon}_1^\infty + x_2\boldsymbol{\varepsilon}_2^\infty + 2\mathrm{Re}\left\{\mathbf{A}\left\langle \zeta_*^{-1}\right\rangle(\mathbf{A}^T\mathbf{g}+\mathbf{B}^T\mathbf{h})\right\}, \tag{10.7-1$_1$}$$

$$\phi = x_1\mathbf{t}_2^\infty - x_2\mathbf{t}_1^\infty + 2\mathrm{Re}\left\{\mathbf{B}\left\langle \zeta_*^{-1}\right\rangle(\mathbf{A}^T\mathbf{g}+\mathbf{B}^T\mathbf{h})\right\}, \qquad (10.7\text{-}1)_2$$

in which ε_1^∞, ε_2^∞, $\mathbf{t}_1^\infty$, and $\mathbf{t}_2^\infty$ are defined in (10.5-2). Since the stress inside the inclusion is uniform we employ (10.5-1) and let

$$\mathbf{u}^\mathrm{o} = x_1\varepsilon_1^\mathrm{o} + x_2\varepsilon_2^\mathrm{o}, \quad \phi^\mathrm{o} = x_1\mathbf{t}_2^\mathrm{o} - x_2\mathbf{t}_1^\mathrm{o}. \qquad (10.7\text{-}2)$$

ε_1^o and ε_2^o have the expressions

$$\varepsilon_1^\mathrm{o} = \begin{bmatrix} \varepsilon_{11} \\ \omega \\ 2\varepsilon_{31} \end{bmatrix} = \mathbf{u}_{,1}, \quad \varepsilon_2^\mathrm{o} = \begin{bmatrix} 2\varepsilon_{12}-\omega \\ \varepsilon_{22} \\ 2\varepsilon_{32} \end{bmatrix} = \mathbf{u}_{,2}, \qquad (10.7\text{-}3)$$

where ω is the rotation of the major axis of the elliptic inclusion. From $(10.5\text{-}5)_2$ we also have the relations

$$\begin{bmatrix} \varepsilon_2^\infty \\ -\mathbf{t}_1^\infty \end{bmatrix} = \mathbf{N}\begin{bmatrix} \varepsilon_1^\infty \\ \mathbf{t}_2^\infty \end{bmatrix}, \quad \begin{bmatrix} \varepsilon_2^\mathrm{o} \\ -\mathbf{t}_1^\mathrm{o} \end{bmatrix} = \mathbf{N}^\mathrm{o}\begin{bmatrix} \varepsilon_1^\mathrm{o} \\ \mathbf{t}_2^\mathrm{o} \end{bmatrix}. \qquad (10.7\text{-}4)$$

Our objective is to determine the real vectors $\mathbf{g}$, $\mathbf{h}$, ε_1^o, and $\mathbf{t}_2^\mathrm{o}$ (or ε_2^o and $\mathbf{t}_1^\mathrm{o}$) from the continuity in displacement and surface traction along the elliptic boundary.

At the elliptic boundary Γ,

$$x_1 = a\cos\psi, \quad x_2 = b\sin\psi, \quad \zeta_\alpha^{-1} = e^{-i\psi}.$$

Making use of the identity (6.9-2), (10.7-1) and (10.7-2) on the elliptic boundary Γ have the expressions

$$\begin{bmatrix} \mathbf{u}_\Gamma \\ \phi_\Gamma \end{bmatrix} = a\cos\psi\begin{bmatrix} \varepsilon_1^\infty \\ \mathbf{t}_2^\infty \end{bmatrix} + b\sin\psi\begin{bmatrix} \varepsilon_2^\infty \\ -\mathbf{t}_1^\infty \end{bmatrix} + \cos\psi\begin{bmatrix} \mathbf{h} \\ \mathbf{g} \end{bmatrix} - \sin\psi\,\tilde{\mathbf{N}}\begin{bmatrix} \mathbf{h} \\ \mathbf{g} \end{bmatrix}, \qquad (10.7\text{-}5)$$

$$\begin{bmatrix} \mathbf{u}_\Gamma^\mathrm{o} \\ \phi_\Gamma^\mathrm{o} \end{bmatrix} = a\cos\psi\begin{bmatrix} \varepsilon_1^\mathrm{o} \\ \mathbf{t}_2^\mathrm{o} \end{bmatrix} + b\sin\psi\begin{bmatrix} \varepsilon_2^\mathrm{o} \\ -\mathbf{t}_1^\mathrm{o} \end{bmatrix}, \qquad (10.7\text{-}6)$$

where (Section 6.5)

$$\tilde{\mathbf{N}} = \begin{bmatrix} \mathbf{S} & \mathbf{H} \\ -\mathbf{L} & \mathbf{S}^T \end{bmatrix}, \qquad \tilde{\mathbf{N}}\tilde{\mathbf{N}} = -\mathbf{I}.$$

The continuity in displacement and surface traction along the elliptic boundary demands that

$$\begin{bmatrix} \mathbf{u}_\Gamma \\ \phi_\Gamma \end{bmatrix} = \begin{bmatrix} \mathbf{u}^o_\Gamma \\ \phi^o_\Gamma \end{bmatrix},$$

which leads to

$$\begin{aligned} a\begin{bmatrix} \varepsilon_1^\infty \\ \mathbf{t}_2^\infty \end{bmatrix} + \begin{bmatrix} \mathbf{h} \\ \mathbf{g} \end{bmatrix} &= a\begin{bmatrix} \varepsilon_1^o \\ \mathbf{t}_2^o \end{bmatrix}, \\ b\begin{bmatrix} \varepsilon_2^\infty \\ -\mathbf{t}_1^\infty \end{bmatrix} - \tilde{\mathbf{N}}\begin{bmatrix} \mathbf{h} \\ \mathbf{g} \end{bmatrix} &= b\begin{bmatrix} \varepsilon_2^o \\ -\mathbf{t}_1^o \end{bmatrix}. \end{aligned} \tag{10.7-7}$$

With the aid of (10.7-4) we obtain

$$\begin{bmatrix} \varepsilon_1^o \\ \mathbf{t}_2^o \end{bmatrix} = \begin{bmatrix} \varepsilon_1^\infty \\ \mathbf{t}_2^\infty \end{bmatrix} + \frac{1}{a}\begin{bmatrix} \mathbf{h} \\ \mathbf{g} \end{bmatrix}, \tag{10.7-8}$$

$$(\tilde{\mathbf{N}} + \tfrac{b}{a}\mathbf{N}^o)\begin{bmatrix} \mathbf{h} \\ \mathbf{g} \end{bmatrix} = b\begin{bmatrix} \xi \\ \eta \end{bmatrix}, \tag{10.7-9}$$

where

$$\begin{bmatrix} \xi \\ \eta \end{bmatrix} = (\mathbf{N} - \mathbf{N}^o)\begin{bmatrix} \varepsilon_1^\infty \\ \mathbf{t}_2^\infty \end{bmatrix}. \tag{10.7-10}$$

Thus **h** and **g** are determined from (10.7-9), ε_1^o and $\mathbf{t}_2^o$ from (10.7-8), and ε_2^o and $\mathbf{t}_1^o$ from $(10.7\text{-}4)_2$. By $(10.7\text{-}3)_1$ and (10.7-8) it is seen that ω, the rotation of the major axis of the elliptic inclusion, is related to the second component of the vector **h** by

$$\omega = h_2 / a. \tag{10.7-11}$$

We now present the solution for **h** and **g** in (10.7-9). Let

$$(\tilde{\mathbf{N}} + \tfrac{b}{a}\mathbf{N}^o) = \begin{bmatrix} \mathbf{D}_1 & \mathbf{D}_2 \\ -\mathbf{D}_3 & \mathbf{D}_1^T \end{bmatrix}$$

where

$$\mathbf{D}_1 = \mathbf{S} + \tfrac{b}{a}\mathbf{N}_1^o, \quad \mathbf{D}_2 = \mathbf{H} + \tfrac{b}{a}\mathbf{N}_2^o, \quad \mathbf{D}_3 = \mathbf{L} - \tfrac{b}{a}\mathbf{N}_3^o.$$

Since **H**, **L**, and $\mathbf{N}_2^o$ are symmetric and positive definite while $-\mathbf{N}_3^o$ is positive semidefinite (Sections 6.1, 7.5), $\mathbf{D}_2$ and $\mathbf{D}_3$ are both symmetric and positive definite. From

$$\begin{bmatrix} \mathbf{D}_1^T\mathbf{D}_2^{-1} & -\mathbf{I} \\ \mathbf{I} & \mathbf{D}_1\mathbf{D}_3^{-1} \end{bmatrix}\begin{bmatrix} \mathbf{D}_1 & \mathbf{D}_2 \\ -\mathbf{D}_3 & \mathbf{D}_1^T \end{bmatrix} = \begin{bmatrix} \mathbf{U} & \mathbf{0} \\ \mathbf{0} & \mathbf{V} \end{bmatrix}$$

where

$$\mathbf{U} = \mathbf{D}_3 + \mathbf{D}_1^T \mathbf{D}_2^{-1} \mathbf{D}_1, \quad \mathbf{V} = \mathbf{D}_2 + \mathbf{D}_1 \mathbf{D}_3^{-1} \mathbf{D}_1^T,$$

the solution is

$$\mathbf{h} = b\mathbf{U}^{-1}(\mathbf{D}_1^T \mathbf{D}_2^{-1} \xi - \eta), \quad \mathbf{g} = b\mathbf{V}^{-1}(\xi + \mathbf{D}_1 \mathbf{D}_3^{-1} \eta). \tag{10.7-12}$$

It should be noticed that **U** and **V** are both positive definite.

The displacement $\mathbf{u}_\Gamma$ and the stress function ϕ_Γ at the elliptic boundary are obtained from (10.7-6). By virtue of (10.6-5),

$$\frac{\partial}{\rho \partial \psi} \mathbf{u}_\Gamma = -(\cos\theta_\Gamma\, \varepsilon_1^{\circ} + \sin\theta_\Gamma\, \varepsilon_2^{\circ}),$$

$$\frac{\partial}{\rho \partial \psi} \phi_\Gamma = -(\cos\theta_\Gamma\, \mathbf{t}_2^{\circ} - \sin\theta_\Gamma\, \mathbf{t}_1^{\circ}).$$

The surface traction vector $\mathbf{t}_\Gamma$ on the elliptic surface of the matrix is

$$\mathbf{t}_\Gamma = \frac{\partial}{\rho \partial \psi} \phi_\Gamma. \tag{10.7-13}$$

The hoop stress vectors $\mathbf{t}_n$ on the side of the matrix and $\mathbf{t}_n^{\circ}$ on the side of the inclusion are obtained by employing $(7.3\text{-}20)_2$. We have

$$\begin{aligned} \mathbf{t}_n &= \mathbf{N}_3(\theta_\Gamma) \frac{\partial}{\rho \partial \psi} \mathbf{u}_\Gamma + \mathbf{N}_1^T(\theta_\Gamma) \frac{\partial}{\rho \partial \psi} \phi_\Gamma, \\ \mathbf{t}_n^{\circ} &= \mathbf{N}_3^{\circ}(\theta_\Gamma) \frac{\partial}{\rho \partial \psi} \mathbf{u}_\Gamma + \mathbf{N}_1^{\circ T}(\theta_\Gamma) \frac{\partial}{\rho \partial \psi} \phi_\Gamma. \end{aligned} \tag{10.7-14}$$

It is clear that the stress outside the elliptic inclusion is uniform if **g**=**h**=**0**. By (10.7-7), $\varepsilon_1^{\circ}, \varepsilon_2^{\circ}$, $\mathbf{t}_1^{\circ}$, $\mathbf{t}_2^{\circ}$, are identical to $\varepsilon_1^{\infty}, \varepsilon_2^{\infty}$, $\mathbf{t}_1^{\infty}$, $\mathbf{t}_2^{\infty}$, respectively. Thus the stress and strain assume the same value inside and outside the inclusion. To see what material combination would lead to a uniform stress everywhere we write the stress-strain laws using the contracted notation as (Section 2.3)

$$\varepsilon_\alpha^{\infty} = s'_{\alpha\beta} \sigma_\beta^{\infty}$$

where $s'_{\alpha\beta}$ are the reduced elastic compliances. For the material inside the inclusion we have

$$\varepsilon_\alpha^{\infty} = s'^{\circ}_{\alpha\beta} \sigma_\beta^{\infty}.$$

Hence

$$(s'_{\alpha\beta} - s'^{\circ}_{\alpha\beta}) \sigma_\beta^{\infty} = 0. \tag{10.7-15}$$

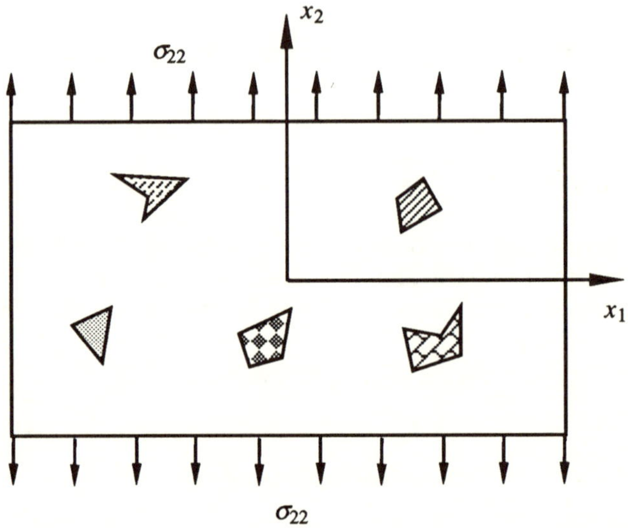

Fig. 10.7 The infinite anisotropic elastic material with any number of dissimilar anisotropic elastic inclusions of arbitrary shape is subjected to a uniform tension in the x_2-direction at infinity. The stress and strain are uniform everywhere in the material and the inclusions if the reduced elastic compliances $s'_{\alpha 2}$ (α=1,2,4,5,6) of the material and inclusions are identical.

Thus when the matrix $(s'_{\alpha\beta} - s'^{o}_{\alpha\beta})$ is singular and σ^{∞}_{β} is a null vector, the stress is uniform everywhere. This conclusion remains valid for any number of inclusions and any shape of the inclusions as long as $s'^{o}_{\alpha\beta}$ of each inclusion satisfies (10.7-15).

Consider the particular case in which σ^{∞}_{2} is the only stress applied at infinity. Equation (10.7-15) is satisfied if

$$s'^{o}_{\alpha 2} = s'_{\alpha 2}, \quad (\alpha=1,2,4,5,6).$$

When $s'^{o}_{\alpha 2}$ of the inclusions are all identical to $s'_{\alpha 2}$ of the matrix, the stress is uniform everywhere, Fig. 10.7. Notice that of the 15 elastic constants $s'_{\alpha\beta}$, only s'_{12}, s'_{22}, s'_{42}, s'_{52}, s'_{62} in the matrix and the inclusions need to be identical. The remaining $s'_{\alpha\beta}$ can be arbitrary. A more general case that involves transformation strains and eigenstrains in the inclusions has been studied by Dvorak and Benveniste (1992).

10.8 Elliptic Hole Subjected to Tractions at the Hole

Let the infinite anisotropic elastic medium with an elliptic hole be subjected to an arbitrarily prescribed traction $\mathbf{t}_{\Gamma}$ on the surface of the hole. If there is a concentrated force applied at the hole surface it can be treated separately (Section 10.4). Henceforth we assume that $\mathbf{t}_\Gamma$ is bounded, and that $\rho\,\mathbf{t}_\Gamma$ can be represented by a Fourier series as (Sneddon, 1951)

$$\rho\mathbf{t}_\Gamma = \mathbf{g}_o + \sum_{k=1}^{\infty}(\cos k\psi\,\mathbf{g}_k + \sin k\psi\,\mathbf{h}_k), \tag{10.8-1}$$

where ρ is defined in (10.6-4) and

$$\mathbf{g}_o = \frac{1}{2\pi}\int_0^{2\pi}\rho\mathbf{t}_\Gamma d\psi,$$

$$\mathbf{g}_k = \frac{1}{\pi}\int_0^{2\pi}\rho\mathbf{t}_\Gamma \cos k\psi\, d\psi,$$

$$\mathbf{h}_k = \frac{1}{\pi}\int_0^{2\pi}\rho\mathbf{t}_\Gamma \sin k\psi\, d\psi,$$

(k=1,2,...). If the tractions on the elliptic surface produce no net force, $\mathbf{g}_o=\mathbf{0}$. The solution to the problem at hand is given by

$$\begin{aligned}\mathbf{u} &= 2\,\mathrm{Im}\left\{\mathbf{A}\langle\ln\zeta_*\rangle\mathbf{A}^T\right\}\mathbf{g}_o - \mathrm{Im}\sum_{k=1}^{\infty}\frac{1}{k}\left\{\mathbf{A}\langle\zeta_*^{-k}\rangle\mathbf{B}^{-1}(\mathbf{g}_k + i\mathbf{h}_k)\right\},\\ \boldsymbol{\phi} &= 2\,\mathrm{Im}\left\{\mathbf{B}\langle\ln\zeta_*\rangle\mathbf{A}^T\right\}\mathbf{g}_o - \mathrm{Im}\sum_{k=1}^{\infty}\frac{1}{k}\left\{\mathbf{B}\langle\zeta_*^{-k}\rangle\mathbf{B}^{-1}(\mathbf{g}_k + i\mathbf{h}_k)\right\}.\end{aligned} \tag{10.8-2}$$

Indeed, the stress function $\boldsymbol{\phi}_\Gamma$ on the elliptic boundary Γ is

$$\boldsymbol{\phi}_\Gamma = \psi\,\mathbf{g}_o + \sum_{k=1}^{\infty}\frac{1}{k}\left\{\sin k\psi\,\mathbf{g}_k - \cos k\psi\,\mathbf{h}_k\right\}. \tag{10.8-3}$$

Substitution of (10.8-3) into (10.1-16) confirms (10.8-1).

The displacement $\mathbf{u}_\Gamma$ obtained from $(10.8\text{-}2)_1$ can be written as

$$\mathbf{u}_\Gamma = \mathbf{S}\mathbf{L}^{-1}(\psi\,\mathbf{g}_o - \boldsymbol{\phi}_\Gamma) + \mathbf{L}^{-1}\sum_{k=1}^{\infty}\frac{1}{k}\left\{\cos k\psi\,\mathbf{g}_k + \sin k\psi\,\mathbf{h}_k\right\}. \tag{10.8-4}$$

Inserting (10.8-3) and (10.8-4) into $(10.7\text{-}14)_1$ the hoop stress vector $\mathbf{t}_n$ can be shown to be (Ting and Yan, 1991)

$$\mathbf{t}_n = \mathbf{G}_1(\theta_\Gamma)\mathbf{t}_\Gamma + \frac{1}{\rho}\mathbf{G}_3(\theta_\Gamma)\left(\mathbf{S}^T\mathbf{g}_o + \sum_{k=1}^{\infty}(\sin k\psi\,\mathbf{g}_k - \cos k\psi\,\mathbf{h}_k)\right) \tag{10.8-5}$$

in which $\mathbf{G}_1(\theta)$ and $\mathbf{G}_3(\theta)$ are defined in (10.6-16). The infinite series in (10.8-5) is the *conjugate function* of $\rho\mathbf{t}_\Gamma$ in (10.8-1). The conjugate function can be written in terms of $\rho\mathbf{t}_\Gamma$ as (Bary, 1964)

$$\sum_{k=1}^{\infty}(\sin k\psi\,\mathbf{g}_k - \cos k\psi\,\mathbf{h}_k) = -\frac{1}{\pi}\int_0^{\pi}\frac{\mathbf{F}(\psi+t)-\mathbf{F}(\psi-t)}{2\tan(t/2)}dt,$$

$$\mathbf{F}(\psi) = \rho(\psi)\mathbf{t}_\Gamma(\psi).$$

Consider the special case in which $\mathbf{t}_\Gamma$ is a uniform pressure, i.e.,

$$\mathbf{t}_\Gamma = p\mathbf{m}(\theta_\Gamma) = p[\cos\theta_\Gamma\,\mathbf{m}(0) - \sin\theta_\Gamma\,\mathbf{n}(0)]$$

where p is the constant pressure. By (10.6-5) this is

$$\rho\mathbf{t}_\Gamma = p[a\sin\psi\,\mathbf{m}(0) + b\cos\psi\,\mathbf{n}(0)].$$

Comparison with (10.8-1) shows that $\mathbf{g}_o=\mathbf{0}$, and

$$\mathbf{g}_1 = pb\mathbf{n}(0), \quad \mathbf{h}_1 = pa\mathbf{m}(0)$$

are the only nonzero $\mathbf{g}_k$ and $\mathbf{h}_k$. The hoop stress vector obtained from (10.8-5) and (10.6-5) is

$$\mathbf{t}_n / p = \mathbf{G}_1(\theta_\Gamma)\mathbf{m}(\theta_\Gamma) + \mathbf{G}_3(\theta_\Gamma)\left(\tfrac{b}{a}\cos\theta_\Gamma\,\mathbf{n}(0) + \tfrac{a}{b}\sin\theta_\Gamma\,\mathbf{m}(0)\right).$$

The hoop stress $\sigma_{nn} = \mathbf{n}(\theta_\Gamma)\cdot\mathbf{t}_n$ is, making use of (10.6-18),

$$\sigma_{nn} / p = \mathbf{n}^T(\theta_\Gamma)\mathbf{G}_1(\theta_\Gamma)\mathbf{m}(\theta_\Gamma) + \tfrac{b}{a}[\mathbf{G}_3(\theta_\Gamma)]_{11} + \tfrac{a}{b}[\mathbf{G}_3(\theta_\Gamma)]_{22}$$

where $[*]_{ij}$ denotes the (ij) component of the matrix inside the square brackets (Ting and Yan, 1991).

If $\mathbf{t}_\Gamma$ is a uniform shear,

$$\mathbf{t}_\Gamma = \tau\mathbf{n}(\theta_\Gamma) = \tau[\cos\theta_\Gamma\,\mathbf{n}(0) + \sin\theta_\Gamma\,\mathbf{m}(0)],$$

where τ is the constant shear. A similar analysis shows that

$$\mathbf{t}_n / \tau = \mathbf{G}_1(\theta_\Gamma)\mathbf{n}(\theta_\Gamma) + \mathbf{G}_3(\theta_\Gamma)\left(\tfrac{a}{b}\sin\theta_\Gamma\,\mathbf{n}(0) - \tfrac{b}{a}\cos\theta_\Gamma\,\mathbf{m}(0)\right),$$

$$t_{nn} / \tau = \mathbf{n}^T(\theta_\Gamma)\mathbf{G}_1(\theta_\Gamma)\mathbf{n}(\theta_\Gamma) + \tfrac{a}{b}[\mathbf{G}_3(\theta_\Gamma)]_{21} - \tfrac{b}{a}[\mathbf{G}_3(\theta_\Gamma)]_{12}.$$

10.9 Anisotropic Elliptic Body

The mapping (10.6-1) or (10.6-3a) is not valid for the region inside the ellipse because there is a branch cut inside the ellipse. Consequently the function ζ_α is not continuous across the branch cut.

However, the function (Lekhnitskii, 1950)

$$f_{k\alpha}(\zeta_\alpha) = (\zeta_\alpha)^k + (t_\alpha)^k (\zeta_\alpha^{-1})^k \quad (\alpha=1,2,3) \tag{10.9-1}$$

where ζ_α, ζ_α^{-1}, and $|t_\alpha|<1$ are defined in (10.6-3a,b) and (10.6-9) is continuous inside the ellipse because the discontinuities at the branch cuts from the first term and the second term cancel each other. In fact the right side of (10.9-1) is a polynomial in z_α of degree k (Section 3.9). At the elliptic boundary Γ (10.9-1) reduces to

$$f_{k\alpha}(e^{i\psi}) = e^{ik\psi} + (t_\alpha)^k e^{-ik\psi} \quad (\alpha=1,2,3). \tag{10.9-2}$$

We study three special problems separately below.

(i) The Ellipse Subjected to Prescribed Tractions

Let $\mathbf{t}_m$ be the traction prescribed on the surface of the elliptic boundary, Fig. 10.8. We assume that $\rho\mathbf{t}_m$ can be expressed in a Fourier series

$$\rho\mathbf{t}_m = \mathbf{g}_o + \sum_{k=1}^{\infty} (\cos k\psi\, \mathbf{g}_k + \sin k\psi\, \mathbf{h}_k), \tag{10.9-3}$$

where ρ is defined in (10.6-4) and

$$\mathbf{g}_o = \frac{1}{2\pi}\int_0^{2\pi} \rho\mathbf{t}_m d\psi, \quad \mathbf{g}_k = \frac{1}{\pi}\int_0^{2\pi} \rho\mathbf{t}_m \cos k\psi\, d\psi,$$

$$\mathbf{h}_k = \frac{1}{\pi}\int_0^{2\pi} \rho\mathbf{t}_m \sin k\psi\, d\psi.$$

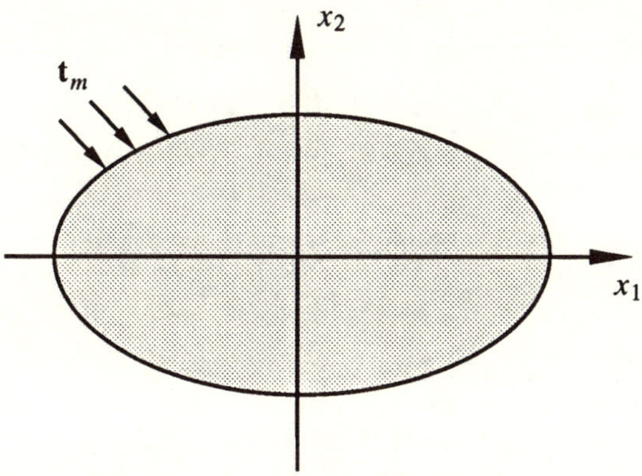

Fig. 10.8 The elliptic body subjected to tractions.

The total force and the total moment about the x_3-axis due to the prescribed traction $\mathbf{t}_m$ must vanish, i.e.,

$$\int_0^{2\pi} \rho \mathbf{t}_m d\psi = \mathbf{0},$$

$$\int_0^{2\pi} (a \cos\psi \mathbf{i}_1 + b \sin\psi \mathbf{i}_2) \times \rho \mathbf{t}_m d\psi = \mathbf{0},$$

$\mathbf{i}_k$ (k=1,2) being the unit vectors in the x_k-direction. Insertion of (10.9-3) into the above leads to

$$\mathbf{g}_o = \mathbf{0}, \quad a\mathbf{i}_1 \times \mathbf{g}_1 + b\mathbf{i}_2 \times \mathbf{h}_1 = \mathbf{0}. \tag{10.9-4}$$

The second equation is equivalent to

$$(\mathbf{g}_1)_3 = (\mathbf{h}_1)_3 = 0, \quad a(\mathbf{g}_1)_2 = b(\mathbf{h}_1)_1. \tag{10.9-5}$$

With $(10.9\text{-}4)_1$ we write (10.9-3) as

$$\rho \mathbf{t}_m = \mathrm{Re} \sum_{k=1}^{\infty} e^{-ik\psi} \mathbf{y}_k, \tag{10.9-6}$$

$$\mathbf{y}_k = \mathbf{g}_k + i\mathbf{h}_k. \tag{10.9-7}$$

To find the solution that satisfies the prescribed traction given by (10.9-6) let

$$\mathbf{u} = \mathrm{Im} \sum_{k=1}^{\infty} \frac{1}{k} \{ \mathbf{A} \langle f_{k*}(\zeta_*) \rangle \mathbf{q}_k \}, \quad \boldsymbol{\phi} = \mathrm{Im} \sum_{k=1}^{\infty} \frac{1}{k} \{ \mathbf{B} \langle f_{k*}(\zeta_*) \rangle \mathbf{q}_k \}, \tag{10.9-8}$$

where $\mathbf{q}_k$ are constant vectors to be determined. At the elliptic boundary Γ,

$$\begin{aligned} \mathbf{u}_\Gamma &= \mathrm{Im} \sum_{k=1}^{\infty} \frac{1}{k} \left\{ e^{-ik\psi} \left(-\overline{\mathbf{A}} \overline{\mathbf{q}}_k + \mathbf{A} \langle t_*^k \rangle \mathbf{q}_k \right) \right\}, \\ \boldsymbol{\phi}_\Gamma &= \mathrm{Im} \sum_{k=1}^{\infty} \frac{1}{k} \left\{ e^{-ik\psi} \left(-\overline{\mathbf{B}} \overline{\mathbf{q}}_k + \mathbf{B} \langle t_*^k \rangle \mathbf{q}_k \right) \right\}. \end{aligned} \tag{10.9-9}$$

We have made used of the fact that $\mathrm{Im}(F) = -\mathrm{Im}(\overline{F})$ for any complex quantity F. Hence

$$\rho \mathbf{t}_m = -\frac{\partial}{\partial \psi} \boldsymbol{\phi}_\Gamma = \mathrm{Re} \sum_{k=1}^{\infty} \left\{ e^{-ik\psi} \left(-\overline{\mathbf{B}} \overline{\mathbf{q}}_k + \mathbf{B} \langle t_*^k \rangle \mathbf{q}_k \right) \right\}$$

and comparison with (10.9-6) leads to

$$\mathbf{B}\langle t_*^k \rangle \mathbf{q}_k - \overline{\mathbf{B}}\overline{\mathbf{q}}_k = \mathbf{y}_k. \tag{10.9-10}$$

If this equation and its complex conjugate are written as

$$\mathbf{U}_k(\mathbf{B}\mathbf{q}_k) - \overline{\mathbf{B}}\overline{\mathbf{q}}_k = \mathbf{y}_k, \quad \overline{\mathbf{U}}_k(\overline{\mathbf{B}}\overline{\mathbf{q}}_k) - \mathbf{B}\mathbf{q}_k = \overline{\mathbf{y}}_k,$$

where

$$\mathbf{U}_k = \mathbf{B}\langle t_*^k \rangle \mathbf{B}^{-1}, \tag{10.9-11}$$

elimination of $\overline{\mathbf{B}}\overline{\mathbf{q}}_k$ yields

$$(\mathbf{I} - \overline{\mathbf{U}}_k \mathbf{U}_k)(\mathbf{B}\mathbf{q}_k) = -(\overline{\mathbf{U}}_k \mathbf{y}_k + \overline{\mathbf{y}}_k). \tag{10.9-12}$$

Equation (10.9-12) provides the solution for $\mathbf{q}_k$.

The special case when the elliptic body is a monoclinic material with the symmetry plane at $x_3=0$ was studied by Lekhnitskii (1957). He obtained a system of equations similar to (10.9-12), and stated that it has a unique solution for $k \neq 1$. The coefficient matrix of $\mathbf{q}_k$ for $k=1$ is singular, but conditions (10.9-5) render the existence of a solution for $\mathbf{q}_1$. There is no need to solve $\mathbf{q}_1$ from (10.9-12) because the $\mathbf{q}_1$ term in (10.9-8) is linear in z_α, and corresponds to a uniform stress solution that can be obtained independently (Section 10.5).

(ii) The Green's Function for the Elliptic Body

Let an anisotropic elliptic body be subjected to a line force $\mathbf{f}$ at

$$(x_1, x_2) = (x_1^o, x_2^o)$$

inside the ellipse while the displacement at the elliptic boundary Γ vanishes, Fig. 10.9. The special case of antiplane deformation has been studied in Section 3.9. Following (3.9-7) let

$$\mathbf{u} - \mathbf{u}_o = \frac{1}{\pi}\mathrm{Im}\left\{\mathbf{A}\langle \ln(z_* - z_{*o}) \rangle \mathbf{q}^\infty\right\} + \frac{1}{\pi}\mathrm{Im}\sum_{\beta=1}^{3}\left\{\mathbf{A}\langle \ln(z_* - z_{*\beta}) \rangle \mathbf{q}_\beta\right\}$$
$$+\frac{1}{\pi}\mathrm{Im}\sum_{k=1}^{\infty}\frac{1}{k}\left\{\mathbf{A}\langle \zeta_*^k + (t_* \zeta_*^{-1})^k \rangle \mathbf{h}_k\right\} \tag{10.9-13}$$

in which $\mathbf{u}_o$, $\mathbf{q}_\beta$, and $\mathbf{h}_k$ are constant vectors and, by (10.3-3),

$$\mathbf{q}^\infty = \mathbf{A}^T \mathbf{f}. \tag{10.9-14}$$

The equation for the stress function ϕ is identical to (10.9-13) when $\mathbf{A}$ and $\mathbf{u}_o$ are replaced by $\mathbf{B}$ and ϕ_o, respectively. The constant vector ϕ_o can be ignored since it does not affect the stress. In (10.9-13),

$$x_1^o + p_\alpha x_2^o = z_{\alpha o} = c_\alpha \zeta_{\alpha o} + d_\alpha \zeta_{\alpha o}^{-1}, \tag{10.9-15}$$

$$x_1^{\alpha\beta} + p_\alpha x_2^{\alpha\beta} = z_{\alpha\beta} = c_\alpha \overline{\zeta}_{\beta o}^{-1} + d_\alpha \overline{\zeta}_{\beta o}, \tag{10.9-16}$$

where

$$\zeta_{\alpha o} = \frac{z_{\alpha o} + \sqrt{z_{\alpha o}^2 - (a^2 + p_\alpha^2 b^2)}}{a - i p_\alpha b}.$$

Equation (10.9-16) determines $(x_1^{\alpha\beta}, x_2^{\alpha\beta})$. The points $(x_1^{\alpha\beta}, x_2^{\alpha\beta})$ represent the positions of nine image singularities that is located outside the ellipse (see Fig. 10.3). By writing

$$z_\alpha - z_{\alpha o} = c_\alpha \zeta_\alpha (1 - \zeta_{\alpha o}\zeta_\alpha^{-1})(1 - \tau_{\alpha o}\zeta_\alpha^{-1}), \quad \tau_{\alpha o} = t_\alpha \zeta_{\alpha o}^{-1}, \tag{10.9-17}$$

$$z_\alpha - z_{\alpha\beta} = -c_\alpha \overline{\zeta}_{\beta o}^{-1}(1 - \overline{\zeta}_{\beta o}\zeta_\alpha)(1 - \tau_{\alpha\beta}\zeta_\alpha^{-1}), \quad \tau_{\alpha\beta} = t_\alpha \overline{\zeta}_{\beta o}, \tag{10.9-18}$$

the displacement $\mathbf{u}_\Gamma$ at the elliptic boundary Γ is

$$\mathbf{u}_\Gamma - \mathbf{u}_o = \frac{1}{\pi}\mathrm{Im}\left\{i\psi \mathbf{A}\mathbf{A}^T + \mathbf{A}\langle \ln c_* \rangle \mathbf{q}^\infty - \overline{\mathbf{A}} \ln\langle 1 - \overline{\zeta}_{*o} e^{i\psi} \rangle \overline{\mathbf{q}}^\infty + \mathbf{A} \ln\langle 1 - \tau_{*o} e^{-i\psi} \rangle \mathbf{q}^\infty \right\}$$

$$+ \frac{1}{\pi}\mathrm{Im}\sum_{\beta=1}^{3}\left\{\mathbf{A}\langle \ln c_* \rangle \mathbf{q}_\beta - \ln(-\overline{\zeta}_{\beta o})\mathbf{A}\mathbf{q}_\beta + \ln(1 - \overline{\zeta}_{\beta o} e^{i\psi})\mathbf{A}\mathbf{q}_\beta + \mathbf{A}\ln\langle 1 - \tau_{*\beta} e^{-i\psi} \rangle \mathbf{q}_\beta \right\}$$

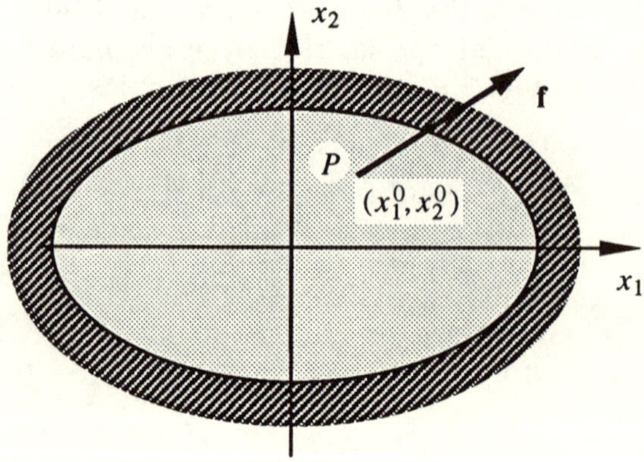

Fig. 10.9 The elliptic body is subjected to a line force with its boundary being fixed. The Green's function is given in (10.9-13).

$$+\frac{1}{\pi}\mathrm{Im}\sum_{k=1}^{\infty}\frac{1}{k}e^{-ik\psi}\left\{-\overline{\mathbf{A}}\overline{\mathbf{h}}_k+\mathbf{A}\left\langle t_*^k\right\rangle\mathbf{h}_k\right\}.$$

With the use of the infinite series

$$\ln(1-x)=-\sum_{k=1}^{\infty}\frac{1}{k}x^k \tag{10.9-19}$$

and noting that $\mathbf{AA}^T$ is purely imaginary, the condition $\mathbf{u}_\Gamma=\mathbf{0}$ yields

$$\mathbf{u}_o=-\frac{1}{\pi}\mathrm{Im}\left\{\mathbf{A}\left\langle\ln c_*\right\rangle(\mathbf{q}^\infty+\mathbf{q})\right\}+\frac{1}{\pi}\mathrm{Im}\sum_{\beta=1}^{3}\left\{\ln(-\bar{\zeta}_{\beta o})\mathbf{A}\mathbf{q}_\beta\right\},$$

$$-\overline{\mathbf{A}}\mathbf{I}_\beta\overline{\mathbf{q}}^\infty+\mathbf{A}\mathbf{q}_\beta=\mathbf{0}, \tag{10.9-20}$$

$$\mathbf{A}\left\langle t_*^k\right\rangle\mathbf{h}_k-\overline{\mathbf{A}}\overline{\mathbf{h}}_k=\mathbf{y}_k, \tag{10.9-21}$$

where $\mathbf{I}_\beta$ (β=1,2,3) are defined in (10.3-7) and

$$\mathbf{q}=\sum_{\beta=1}^{3}\mathbf{q}_\beta,\quad \mathbf{y}_k=\mathbf{A}\left\langle\tau_{*o}^k\right\rangle\mathbf{q}^\infty+\sum_{\beta=1}^{3}\mathbf{A}\left\langle\tau_{*\beta}^k\right\rangle\mathbf{q}_\beta.$$

Equation (10.9-20) gives

$$\mathbf{q}_\beta=\mathbf{A}^{-1}\overline{\mathbf{A}}\mathbf{I}_\beta\overline{\mathbf{q}}^\infty.$$

Equation (10.9-21) is similar to (10.9-10). Hence $\mathbf{h}_k$ is determined by following the derivation of (10.9-12).

For certain special anisotropic materials the infinite series vanishes, and we have a closed form solution. This is discussed at the end of Section 10.10.

(iii) The Elliptic Body Subjected to a Pair of Forces

Consider the case in which a line force $\mathbf{f}$ is applied at

$$(x_1,x_2)=(x_1^+,x_2^+)$$

and another force $-\mathbf{f}$ is applied at

$$(x_1,x_2)=(x_1^-,x_2^-).$$

The elliptic boundary Γ is traction-free, i.e., $\boldsymbol{\phi}_\Gamma=\mathbf{0}$. The direction of $\mathbf{f}$ is such that its projection on the (x_1,x_2)-plane lies on the line connecting (x_1^+,x_2^+) to (x_1^-,x_2^-), Fig. 10.10. This assures us that the net force and net moment about the x_3-axis vanish.

The solution for the problem at hand is obtained by superposition of two solutions of the form (10.9-13). For the stress function ϕ we have

$$\phi - \phi_o = \frac{1}{\pi}\mathrm{Im}\left\{\mathbf{B}\left\langle \ln\frac{z_* - z_{*o}^+}{z_* - z_{*o}^-}\right\rangle \mathbf{q}^\infty\right\} + \frac{1}{\pi}\mathrm{Im}\sum_{\beta=1}^{3}\left\{\mathbf{B}\left\langle \ln\frac{z_* - z_{*\beta}^+}{z_* - z_{*\beta}^-}\right\rangle \mathbf{q}_\beta\right\}$$

$$+\frac{1}{\pi}\mathrm{Im}\sum_{k=1}^{\infty}\frac{1}{k}\left\{\mathbf{B}\left\langle \zeta_*^k + (t_*\zeta_*^{-1})^k\right\rangle \mathbf{h}_k\right\} \tag{10.9-22}$$

in which, following (10.9-15) and (10.9-16),

$$x_1^\pm + p_\alpha x_2^\pm = z_{\alpha o}^\pm = c_\alpha \zeta_{\alpha o}^\pm + d_\alpha(\zeta_{\alpha o}^\pm)^{-1},$$

$$(x_1^{\alpha\beta})^\pm + p_\alpha (x_2^{\alpha\beta})^\pm = z_{\alpha\beta}^\pm = c_\alpha(\overline{\zeta}_{\beta o}^\pm)^{-1} + d_\alpha \overline{\zeta}_{\beta o}^\pm.$$

It is not difficult to show that the condition $\phi=\mathbf{0}$ at the elliptic boundary Γ yields

$$\mathbf{q}_\beta = \mathbf{B}^{-1}\overline{\mathbf{B}}\mathbf{I}_\beta \overline{\mathbf{q}}^\infty, \tag{10.9-23}$$

$$\mathbf{B}\left\langle t_*^k\right\rangle \mathbf{h}_k - \overline{\mathbf{B}}\overline{\mathbf{h}}_k = \mathbf{y}_k, \tag{10.9-24}$$

where

$$\mathbf{y}_k = \mathbf{B}\left\langle (\tau_{*o}^+)^k - (\tau_{*o}^-)^k\right\rangle \mathbf{q}^\infty + \sum_{\beta=1}^{3}\mathbf{B}\left\langle (\tau_{*\beta}^+)^k - (\tau_{*\beta}^-)^k\right\rangle \mathbf{q}_\beta,$$

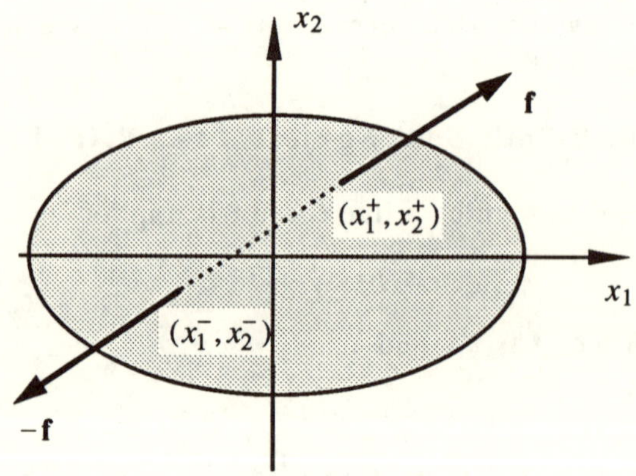

Fig. 10.10 The elliptic body is subjected to a pair of equal and opposite forces with its boundary being traction-free.

$$\tau_{\alpha o}^{\pm} = t_\alpha (\zeta_{\alpha o}^{\pm})^{-1}, \quad \tau_{\alpha\beta}^{\pm} = t_\alpha \overline{\zeta}_{\beta o}^{\pm}.$$

The constant vector ϕ_o can also be determined, but it produces no stresses. Equation (10.9-24) is identical to (10.9-10) if $\mathbf{q}_k$ is replaced by $\mathbf{h}_k$. Hence $\mathbf{h}_k$ is given by (10.9-12).

In the special case when (x_1^+, x_2^+) and (x_1^-, x_2^-) are diametrically located, we let

$$(x_1^+, x_2^+) = (x_1^o, x_2^o) = -(x_1^-, x_2^-),$$

$$z_{\alpha o}^+ = z_{\alpha o} = -z_{\alpha o}^-.$$

Ignoring the constant vector ϕ_o (10.9-22) becomes

$$\phi = \frac{1}{\pi}\mathrm{Im}\left\{\mathbf{B}\left\langle \ln\frac{z_* - z_{*o}}{z_* + z_{*o}}\right\rangle \mathbf{q}^\infty\right\} + \frac{1}{\pi}\mathrm{Im}\sum_{\beta=1}^{3}\left\{\mathbf{B}\left\langle \ln\frac{z_* - z_{*\beta}}{z_* + z_{*\beta}}\right\rangle \mathbf{q}_\beta\right\}$$

$$+\frac{1}{\pi}\mathrm{Im}\sum_{k=1}^{\infty}\frac{1}{k}\left\{\mathbf{B}\left\langle \zeta_*^k + (t_*\zeta_*^{-1})^k\right\rangle \mathbf{h}_k\right\}. \qquad (10.9\text{-}25)$$

If (x_1^o, x_2^o) is on the elliptic boundary,

$$x_1^o = a\cos\psi_o, \quad x_2^o = a\sin\psi_o, \quad \zeta_{\alpha o} = e^{i\psi_o}.$$

The special case of orthotropic circular body subjected to a pair of diametrically loaded forces at the boundary of the circle has been solved numerically by Okubo (1952).

As in Case (ii), the infinite series vanishes for certain special anisotropic materials (see discussion at the end of Section 10.10).

10.10 Green's Functions for an Elliptic Inclusion

The problem of a circular isotropic elastic inclusion subjected to an edge dislocation was solved by Dundurs and Mura (1964). The related problem where the dislocation is inside the circular inclusion was studied by Dundurs and Sendeckyj (1965b). For an inclusion of elliptic shape the problem was solved by Warren (1983) for a dislocation located inside the inclusion in the form of an infinite series. The same technique was employed by Stagni (1982) and Stagni and Lizzio (1983) for a dislocation located outside the inclusion. Hwu and Yen (1993) and Yen and Hwu (1994) investigated the interaction of a dislocation located outside an elliptic inclusion for general anisotropic elastic materials. In this section we study the inclusion problem for a line force $\mathbf{f}$ and a line dislocation with Burgers vector **b** located at

$$(x_1, x_2) = (x_1^{\rm o}, x_2^{\rm o}).$$

We consider separately below the cases when $(x_1^{\rm o}, x_2^{\rm o})$ is outside, inside, and on the interface of the elliptic inclusion. The same problem has been studied by Yen et al. (1995) for a line dislocation. While their solutions are correct, it is not obvious from the expressions of their solutions if the solutions agree with each other in the limit the dislocation is at the interface. Moreover, their solutions did not account for all image singularities. As a result, when the dislocation is very near the interface, the series solution converges very slowly for points near the image singularity that was missing in the solution. With the Green's functions for an elliptic hole and an elliptic body studied in (10.3-2) and (10.9-13), the construction of the solutions for the present problem to account for all image singularities becomes fairly straight forward (Ting, 1996a). The reader is also advised to consult the special case of antiplane deformation presented in Section 3.10.

(I) $(x_1^{\rm o}, x_2^{\rm o})$ **outside the inclusion.**

When the line force **f** and the line dislocation with Burgers vector **b** are located outside the inclusion, Fig. 10.11(a), we let

$$\mathbf{u}_1 = \frac{1}{\pi}\mathrm{Im}\{\mathbf{A}_1\langle \ln(\zeta_*^{(1)} - \zeta_{*\rm o}^{(1)})\rangle \mathbf{q}_1^\infty\} + \frac{1}{\pi}\mathrm{Im}\sum_{\beta=1}^{3}\{\mathbf{A}_1\langle \ln(\zeta_*^{(1)^{-1}} - \bar{\zeta}_{\beta\rm o}^{(1)})\rangle \mathbf{q}_\beta^{(1)}\}$$

$$+\frac{1}{\pi}\mathrm{Im}\sum_{k=1}^{\infty}\frac{1}{k}\{\mathbf{A}_1\langle(\zeta_*^{(1)})^{-k}\rangle \mathbf{g}_k\} \tag{10.10-1}$$

for material 1 outside the inclusion, and

$$\mathbf{u}_2 - \mathbf{u}_2^{\rm o} = \frac{1}{\pi}\mathrm{Im}\sum_{\beta=1}^{3}\{\mathbf{A}_2\langle \ln(z_*^{(2)} - \hat{z}_{*\beta}^{(2)})\rangle \mathbf{q}_\beta^{(2)}\} + \frac{1}{\pi}\mathrm{Im}\sum_{k=1}^{\infty}\frac{1}{k}\{\mathbf{A}_2\langle f_{k*}^{(2)}\rangle \mathbf{h}_k\} \tag{10.10-2}$$

for material 2 inside the inclusion. The stress function $\boldsymbol{\phi}_1$ is identical to $\mathbf{u}_1$ when $\mathbf{A}_1$ is replaced by $\mathbf{B}_1$. The stress function $\boldsymbol{\phi}_2$ is the same as $\mathbf{u}_2$ when $\mathbf{A}_2$ and $\mathbf{u}_2^{\rm o}$ are replaced by $\mathbf{B}_2$ and $\boldsymbol{\phi}_2^{\rm o}$, respectively. $\mathbf{u}_2^{\rm o}$, $\boldsymbol{\phi}_2^{\rm o}$, $\mathbf{q}_\beta^{(1)}$, $\mathbf{q}_\beta^{(2)}$, $\mathbf{g}_k$, and $\mathbf{h}_k$ are constant vectors to be determined, and

$$\mathbf{q}_1^\infty = \mathbf{A}_1^T\mathbf{f} + \mathbf{B}_1^T\mathbf{b}, \quad f_{k\alpha}^{(2)} = (\zeta_\alpha^{(2)})^k + (t_\alpha^{(2)})^k(\zeta_\alpha^{(2)})^{-k},$$

$$x_1^{\rm o} + p_\alpha^{(1)}x_2^{\rm o} = z_{\alpha\rm o}^{(1)} = c_\alpha^{(1)}\zeta_{\alpha\rm o}^{(1)} + d_\alpha^{(1)}(\zeta_{\alpha\rm o}^{(1)})^{-1}, \tag{10.10-3a}$$

$$\hat{x}_1^{\alpha\beta} + p_\alpha^{(2)}\hat{x}_2^{\alpha\beta} = \hat{z}_{\alpha\beta}^{(2)} = c_\alpha^{(2)}\zeta_{\beta o}^{(1)} + d_\alpha^{(2)}(\zeta_{\beta o}^{(1)})^{-1}, \qquad (10.10\text{-}3b)$$

The first term on the right of (10.10-1) takes care of the applied force **f** and the dislocation **b**. The second term represents nine image singularities for material 1 while the first term in (10.10-2) represents nine image singularities for material 2. The solution associated with each image singularity is a one-component Green's function (Section 8.2). The infinite series are added to satisfy the continuity conditions at the elliptic boundary.

The image singularities given by the second term in (10.10-1) are located at

$$\zeta_\alpha^{(1)} = (\bar{\zeta}_{\beta o}^{(1)})^{-1}, \quad \left|(\bar{\zeta}_{\beta o}^{(1)})^{-1}\right| < 1.$$

If $(x_1^{\alpha\beta}, x_2^{\alpha\beta})$ is the location of an image singularity in the (x_1,x_2)-plane it is determined from (10.3-14), i.e.,

$$x_1^{\alpha\beta} + p_\alpha^{(1)} x_2^{\alpha\beta} = z_{\alpha\beta}^{(1)} = c_\alpha^{(1)}(\bar{\zeta}_{\beta o}^{(1)})^{-1} + d_\alpha^{(1)}\bar{\zeta}_{\beta o}^{(1)}. \qquad (10.10\text{-}4)$$

The nine image singularities for material 1 are inside the unit circle in the ζ-plane (see Fig. 10.3). In the (x_1,x_2)-plane they are inside the ellipse or, if they are on the second sheet of the Riemann surface, can be outside the ellipse (Section 3.7). Their locations depend on elastic constants of material 1 only. The first term in (10.10-2) represents nine image singularities for material 2. They are located at $(\hat{x}_1^{\alpha\beta}, \hat{x}_2^{\alpha\beta})$ defined in (10.10-3b). They are outside the ellipse. Their locations depend on elastic constants of both materials.

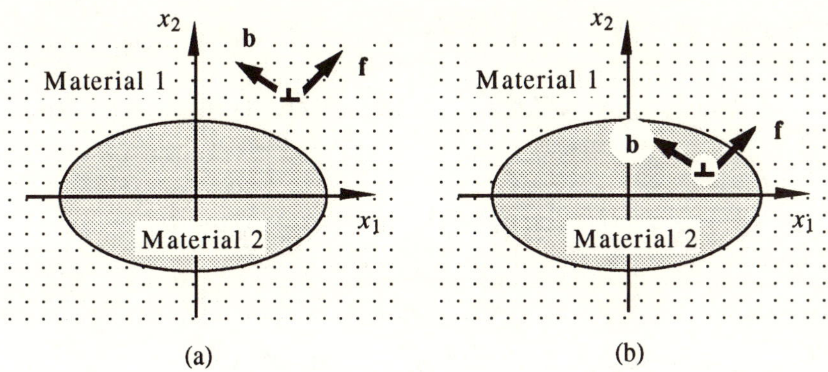

Fig. 10.11 An elliptic inclusion of dissimilar anisotropic material is subjected to a line force **f** and a line dislocation with Burger vector **b** applied at (a) outside the ellipse and (b) inside the ellipse.

By writing

$$z_\alpha^{(2)} - \hat{z}_{\alpha\beta}^{(2)} = c_\alpha^{(2)}\left(\zeta_\alpha^{(2)} - \zeta_{\beta o}^{(1)}\right)\left(1 - \hat{\tau}_{\alpha\beta}^{(2)}\zeta_\alpha^{(2)^{-1}}\right), \quad \hat{\tau}_{\alpha\beta}^{(2)} = t_\alpha^{(2)}\zeta_{\beta o}^{(1)^{-1}}, \tag{10.10-5}$$

(10.10-1) for (x_1,x_2) at the elliptic boundary Γ is

$$\mathbf{u}_1 = \frac{1}{\pi}\mathrm{Im}\left\{-\overline{\mathbf{A}}_1\left\langle \ln(e^{-i\psi} - \overline{\zeta}_{*o}^{(1)})\right\rangle \overline{\mathbf{q}}_1^\infty + \sum_{\beta=1}^{3}\left\{\ln(e^{-i\psi} - \overline{\zeta}_{\beta o}^{(1)})\mathbf{A}_1\mathbf{q}_\beta^{(1)}\right\}\right.$$
$$\left. + \sum_{k=1}^{\infty}\frac{1}{k}e^{-ik\psi}\mathbf{A}_1\mathbf{g}_k\right\},$$

use being made of the relation $\mathrm{Im}(F) = -\mathrm{Im}(\overline{F})$. Replacing $\mathbf{A}_1$ by $\mathbf{B}_1$ we have $\boldsymbol{\phi}_1$ at the elliptic boundary. Equation (10.10-2) for (x_1,x_2) at the elliptic boundary Γ is

$$\mathbf{u}_2 - \mathbf{u}_2^o = \frac{1}{\pi}\mathrm{Im}\left\{\sum_{\beta=1}^{3}\left\{\mathbf{A}_2\left\langle \ln c_*^{(2)}\right\rangle \mathbf{q}_\beta^{(2)} - \ln(e^{-i\psi} - \overline{\zeta}_{\beta o}^{(1)})\overline{\mathbf{A}}_2\overline{\mathbf{q}}_\beta^{(2)}\right.\right.$$
$$\left.\left. + \mathbf{A}_2\left\langle \ln(1 - \hat{\tau}_{*\beta}^{(2)}e^{-i\psi})\right\rangle \mathbf{q}_\beta^{(2)}\right\} + \sum_{k=1}^{\infty}\frac{1}{k}e^{-ik\psi}\left\{-\overline{\mathbf{A}}_2\overline{\mathbf{h}}_k + \mathbf{A}_2\left\langle (t_*^{(2)})^k\right\rangle \mathbf{h}_k\right\}\right\}.$$

When $\mathbf{A}_2$ and $\mathbf{u}_2^o$ are replaced by $\mathbf{B}_2$ and $\boldsymbol{\phi}_2^o$ we obtain $\boldsymbol{\phi}_2$ on Γ. With the use of the series representation (10.9-19) for $\ln(1-x)$ the continuity conditions $\mathbf{u}_1 = \mathbf{u}_2$ and $\boldsymbol{\phi}_1 = \boldsymbol{\phi}_2$ on Γ give

$$\mathbf{u}_2^o = \frac{-1}{\pi}\mathrm{Im}\left\{\mathbf{A}_2\left\langle \ln c_*^{(2)}\right\rangle \mathbf{q}^{(2)}\right\}, \quad \boldsymbol{\phi}_2^o = \frac{-1}{\pi}\mathrm{Im}\left\{\mathbf{B}_2\left\langle \ln c_*^{(2)}\right\rangle \mathbf{q}^{(2)}\right\}, \tag{10.10-6}$$

$$\mathbf{q}^{(2)} = \sum_{\beta=1}^{3}\mathbf{q}_\beta^{(2)},$$

$$\mathbf{A}_1\mathbf{q}_\beta^{(1)} + \overline{\mathbf{A}}_2\overline{\mathbf{q}}_\beta^{(2)} = \overline{\mathbf{A}}_1\mathbf{I}_\beta\overline{\mathbf{q}}_1^\infty, \quad \mathbf{B}_1\mathbf{q}_\beta^{(1)} + \overline{\mathbf{B}}_2\overline{\mathbf{q}}_\beta^{(2)} = \overline{\mathbf{B}}_1\mathbf{I}_\beta\overline{\mathbf{q}}_1^\infty, \tag{10.10-7}$$

and

$$\left.\begin{aligned}\mathbf{A}_1\mathbf{g}_k + \overline{\mathbf{A}}_2\overline{\mathbf{h}}_k &= \mathbf{A}_2\left\{\left\langle (t_*^{(2)})^k\right\rangle \mathbf{h}_k - \sum_{\beta=1}^{3}\left\langle (\hat{\tau}_{*\beta}^{(2)})^k\right\rangle \mathbf{q}_\beta^{(2)}\right\},\\ \mathbf{B}_1\mathbf{g}_k + \overline{\mathbf{B}}_2\overline{\mathbf{h}}_k &= \mathbf{B}_2\left\{\left\langle (t_*^{(2)})^k\right\rangle \mathbf{h}_k - \sum_{\beta=1}^{3}\left\langle (\hat{\tau}_{*\beta}^{(2)})^k\right\rangle \mathbf{q}_\beta^{(2)}\right\}.\end{aligned}\right. \tag{10.10-8}$$

Equations (10.10-7) are identical to (8.9-6) whose solutions are, by (8.9-7),

$$\begin{aligned}\mathbf{B}_1\mathbf{q}_\beta^{(1)} &= (\mathbf{M}_1^{-1}+\overline{\mathbf{M}}_2^{-1})^{-1}(\overline{\mathbf{M}}_2^{-1}-\overline{\mathbf{M}}_1^{-1})\overline{\mathbf{B}}_1\mathbf{I}_\beta\overline{\mathbf{q}}^\infty,\\ \overline{\mathbf{B}}_2\overline{\mathbf{q}}_\beta^{(2)} &= (\mathbf{M}_1^{-1}+\overline{\mathbf{M}}_2^{-1})^{-1}(\mathbf{M}_1^{-1}+\overline{\mathbf{M}}_1^{-1})\overline{\mathbf{B}}_1\mathbf{I}_\beta\overline{\mathbf{q}}^\infty.\end{aligned} \tag{10.10-9}$$

Equations (10.10-8) can be solved for $\mathbf{g}_k$ and $\mathbf{h}_k$ as follows. First with the use of $(10.2\text{-}8)_3$ we write

$$\mathbf{A}_1 = -i\mathbf{M}_1^{-1}\mathbf{B}_1, \quad \mathbf{A}_2 = -i\mathbf{M}_2^{-1}\mathbf{B}_2.$$

The right sides of $(10.10\text{-}8)_{1,2}$ can be eliminated to yield

$$\mathbf{M}_1^{-1}(\mathbf{B}_1\mathbf{g}_k) - \overline{\mathbf{M}}_2^{-1}(\overline{\mathbf{B}}_2\overline{\mathbf{h}}_k) = \mathbf{M}_2^{-1}(\mathbf{B}_1\mathbf{g}_k + \overline{\mathbf{B}}_2\overline{\mathbf{h}}_k).$$

Hence

$$\begin{aligned}\overline{\mathbf{B}}_2\overline{\mathbf{h}}_k &= \frac{1}{2}\mathbf{L}_2(\mathbf{M}_1^{-1}-\mathbf{M}_2^{-1})(\mathbf{B}_1\mathbf{g}_k)\\ &= i\overline{\mathbf{B}}_2\overline{\mathbf{B}}_2^T(\mathbf{M}_1^{-1}-\mathbf{M}_2^{-1})(\mathbf{B}_1\mathbf{g}_k)\end{aligned} \tag{10.10-10}$$

in which the second equality follows from $(5.5\text{-}17)_3$. Equation $(10.10\text{-}8)_2$ now takes the form

$$\mathbf{X}(\mathbf{B}_1\mathbf{g}_k) = i\mathbf{Y}_k(\overline{\mathbf{B}}_1\overline{\mathbf{g}}_k) - \mathbf{y}_k \tag{10.10-11}$$

in which

$$\mathbf{X} = \mathbf{I} + \frac{1}{2}\mathbf{L}_2(\mathbf{M}_1^{-1}-\mathbf{M}_2^{-1}) = \frac{1}{2}\mathbf{L}_2(\mathbf{M}_1^{-1}+\overline{\mathbf{M}}_2^{-1}),$$

$$\mathbf{Y}_k = \mathbf{B}_2\left\langle (t_*^{(2)})^k\right\rangle \mathbf{B}_2^T(\overline{\mathbf{M}}_1^{-1}-\overline{\mathbf{M}}_2^{-1}),$$

$$\mathbf{y}_k = \sum_{\beta=1}^{3}\mathbf{B}_2\left\langle (\overline{\tau}_{*\beta}^{(2)})^k\right\rangle \mathbf{q}_\beta^{(2)}.$$

Elimination of $(\overline{\mathbf{B}}_k\overline{\mathbf{g}}_k)$ between (10.10-11) and its complex conjugate

$$\overline{\mathbf{X}}(\overline{\mathbf{B}}_1\overline{\mathbf{g}}_k) = -i\overline{\mathbf{Y}}_k(\mathbf{B}_1\mathbf{g}_k) - \overline{\mathbf{y}}_k$$

leads to

$$(\mathbf{X} - \mathbf{Y}_k\overline{\mathbf{X}}^{-1}\overline{\mathbf{Y}}_k)(\mathbf{B}_1\mathbf{g}_k) = i\mathbf{Y}_k\overline{\mathbf{X}}^{-1}\overline{\mathbf{y}}_k - \mathbf{y}_k. \tag{10.10-12}$$

This provides the solution for $\mathbf{g}_k$. $\mathbf{h}_k$ is then determined from (10.10-10). The above solutions are in terms of $\mathbf{B}_1\mathbf{g}_k$ and $\mathbf{B}_2\mathbf{h}_k$. We could also derive the solutions in terms of $\mathbf{A}_1\mathbf{g}_k$ and $\mathbf{A}_2\mathbf{h}_k$.

It is interesting to note that when the impedance tensors $\mathbf{M}_1$ and $\mathbf{M}_2$ are identical, i.e., when

$$\mathbf{B}_1\mathbf{A}_1^{-1} = \mathbf{B}_2\mathbf{A}_2^{-1} \quad \text{or} \quad \mathbf{B}_2^{-1}\mathbf{B}_1 = \mathbf{A}_2^{-1}\mathbf{A}_1, \tag{10.10-13}$$

$\mathbf{q}_\beta^{(1)}$ and $\mathbf{h}_k$ vanish according to $(10.10\text{-}9)_1$ and (10.10-10). Thus the second term in (10.10-1) and the infinite series in (10.10-2) can be deleted. By $(10.10\text{-}7)_2$ and $(10.10\text{-}8)_2$,

$$\mathbf{q}_\beta^{(2)} = \mathbf{B}_2^{-1}\mathbf{B}_1\mathbf{I}_\beta\mathbf{q}_1^\infty = \mathbf{A}_2^{-1}\mathbf{A}_1\mathbf{I}_\beta\mathbf{q}_1^\infty,$$

$$\mathbf{g}_k = -\sum_{\beta=1}^{3}\mathbf{B}_1^{-1}\mathbf{B}_2\left\langle(\hat{\tau}_{*\beta}^{(2)})^k\right\rangle\mathbf{q}_\beta^{(2)} = -\sum_{\beta=1}^{3}\mathbf{A}_1^{-1}\mathbf{A}_2\left\langle(\hat{\tau}_{*\beta}^{(2)})^k\right\rangle\mathbf{q}_\beta^{(2)}.$$

By virtue of (10.9-19) the infinite series in (10.10-1) can be expressed in a closed form as

$$\frac{1}{\pi}\mathrm{Im}\sum_{\alpha=1}^{3}\sum_{\beta=1}^{3}\mathbf{A}_1\left\langle\ln\left(1-\hat{\tau}_{\alpha\beta}^{(2)}(\zeta_*^{(1)})^{-1}\right)\right\rangle\mathbf{A}_1^{-1}\mathbf{A}_2\mathbf{I}_\alpha\mathbf{q}_\beta^{(2)}.$$

We therefore have a closed form solution when $\mathbf{M}_1=\mathbf{M}_2$.

(II) $(x_1^{\mathrm{o}},x_2^{\mathrm{o}})$ **inside the inclusion.**

In this case the line force $\mathbf{f}$ and the line dislocation $\mathbf{b}$ are located inside the ellipse, Fig. 10.11(b). We employ the solution (Ting, 1996a)

$$\mathbf{u}_1 = \frac{1}{\pi}\mathrm{Im}\left\{\mathbf{A}_1\left\langle\ln(\zeta_*^{(1)}-\zeta_{*\mathrm{o}}^{(2)})\right\rangle\mathbf{q}_1^\infty\right\} + \frac{1}{\pi}\mathrm{Im}\sum_{\beta=1}^{3}\left\{\mathbf{A}_1\left\langle\ln\left(\zeta_*^{(1)^{-1}}-\zeta_{\beta\mathrm{o}}^{(2)^{-1}}\right)\right\rangle\mathbf{q}_\beta^{(1)}\right\}$$

$$+\frac{1}{\pi}\mathrm{Im}\sum_{k=1}^{\infty}\frac{1}{k}\left\{\mathbf{A}_1\left\langle(\zeta_*^{(1)})^{-k}\right\rangle\mathbf{g}_k\right\} \qquad (10.10\text{-}14)$$

for material 1 outside the inclusion, and

$$\mathbf{u}_2-\mathbf{u}_2^{\mathrm{o}} = \frac{1}{\pi}\mathrm{Im}\left\{\mathbf{A}_2\left\langle\ln(z_*^{(2)}-z_{*\mathrm{o}}^{(2)})\right\rangle\mathbf{q}_2^\infty\right\} + \frac{1}{\pi}\mathrm{Im}\sum_{\beta=1}^{3}\left\{\mathbf{A}_2\left\langle\ln(z_*^{(2)}-\hat{z}_{*\beta}^{(2)})\right\rangle(\mathbf{q}_\beta^{(2)}-\mathbf{I}_\beta\mathbf{q}_2^\infty)\right\}$$

$$+\frac{1}{\pi}\mathrm{Im}\sum_{k=1}^{\infty}\frac{1}{k}\left\{\mathbf{A}_2\left\langle f_{k*}^{(2)}\right\rangle\mathbf{h}_k\right\} \qquad (10.10\text{-}15)$$

for material 2 inside the inclusion. This form of solution is motivated by the special case of antiplane deformation studied in Section 3.10. With this form of solution, (10.10-14), (10.10-15) and (10.10-1), (10.10-2) are identical in the limit $(x_1^{\mathrm{o}},x_2^{\mathrm{o}})$ is on the elliptic boundary. In the above,

$$\mathbf{q}_1^\infty = \mathbf{A}_1^T\mathbf{f} + \mathbf{B}_1^T\mathbf{b}, \quad \mathbf{q}_2^\infty = \mathbf{A}_2^T\mathbf{f} + \mathbf{B}_2^T\mathbf{b}, \qquad (10.10\text{-}16)$$

$$x_1^o + p_\alpha^{(2)} x_2^o = z_{\alpha o}^{(2)} = c_\alpha^{(2)} \zeta_{\alpha o}^{(2)} + d_\alpha^{(2)} (\zeta_{\alpha o}^{(2)})^{-1}, \tag{10.10-17}$$

$$\hat{x}_1^{\alpha\beta} + p_\alpha^{(2)} \hat{x}_2^{\alpha\beta} = \hat{z}_{\alpha\beta}^{(2)} = c_\alpha^{(2)} (\bar{\zeta}_{\beta o}^{(2)})^{-1} + d_\alpha^{(2)} \bar{\zeta}_{\beta o}^{(2)}. \tag{10.10-18}$$

When (x_1^o, x_2^o) is inside the ellipse there is a branch cut extending from (x_1^o, x_2^o) to infinity. The branch cut goes through material 1 outside the ellipse. The first terms in (10.10-14) and (10.10-15) take care of the discontinuity across this branch cut. The second term in (10.10-14) represents nine image singularities for material 1. They are located at

$$\zeta_\alpha^{(1)} = \zeta_{\beta o}^{(2)}, \quad \left|\zeta_{\beta o}^{(2)}\right| < 1.$$

In the (x_1, x_2)-plane they are at $(\hat{x}_1^{\alpha\beta}, \hat{x}_2^{\alpha\beta})$ determined by

$$\hat{x}_1^{\alpha\beta} + p_\alpha^{(1)} \hat{x}_2^{\alpha\beta} = \hat{z}_{\alpha\beta} = c_\alpha^{(1)} \zeta_{\beta o}^{(2)} + d_\alpha^{(1)} (\zeta_{\beta o}^{(2)})^{-1}.$$

They are inside the ellipse or, if they are on the second sheet of the Riemann surface, can be outside the ellipse (Section 3.7). The second term in (10.10-15) represents nine image singularities for material 2. Their locations $(\hat{x}_1^{\alpha\beta}, \hat{x}_2^{\alpha\beta})$ are determined from (10.10-18). They are outside the ellipse, and depend only on elastic constants of material 2.

Making use of (10.10-17) and (10.10-18) we write

$$\zeta_\alpha^{(1)} - \zeta_{\alpha o}^{(2)} = \zeta_\alpha^{(1)} \left(1 - \zeta_{\alpha o}^{(2)} \zeta_\alpha^{(1)^{-1}}\right),$$

$$\zeta_\alpha^{(1)^{-1}} - \zeta_{\beta o}^{(2)^{-1}} = -\zeta_{\beta o}^{(2)^{-1}} \left(1 - \zeta_{\beta o}^{(2)} \zeta_\alpha^{(1)^{-1}}\right),$$

$$z_\alpha^{(2)} - z_{\alpha o}^{(2)} = c_\alpha^{(2)} \zeta_\alpha^{(2)} \left(1 - \zeta_{\alpha o}^{(2)} \zeta_\alpha^{(2)^{-1}}\right)\left(1 - \tau_{\alpha o}^{(2)} \zeta_\alpha^{(2)^{-1}}\right), \quad \tau_{\alpha o}^{(2)} = t_\alpha^{(2)} \zeta_{\alpha o}^{(2)^{-1}},$$

$$z_\alpha^{(2)} - \hat{z}_{\alpha\beta}^{(2)} = -c_\alpha^{(2)} \bar{\zeta}_{\beta o}^{(2)^{-1}} \left(1 - \bar{\zeta}_{\beta o}^{(2)} \zeta_\alpha^{(2)}\right)\left(1 - \hat{\tau}_{\alpha\beta}^{(2)} \zeta_\alpha^{(2)^{-1}}\right), \quad \hat{\tau}_{\alpha\beta}^{(2)} = t_\alpha^{(2)} \bar{\zeta}_{\beta o}^{(2)}.$$

Imposition of the continuity conditions $\mathbf{u}_1 = \mathbf{u}_2$ and $\phi_1 = \phi_2$ at the elliptic boundary leads to (see Sections 3.10 and 10.9)

$$\mathbf{u}_2^o = \frac{-1}{\pi} \mathrm{Im} \sum_{\beta=1}^{3} \left\{ \mathbf{A}_2 \left\langle \ln c_*^{(2)} \right\rangle \mathbf{q}_\beta^{(2)} + \ln(-\zeta_{\beta o}^{(2)}) [\mathbf{A}_2 \mathbf{I}_\beta \mathbf{q}_2^\infty - \mathbf{A}_1 \mathbf{I}_\beta \mathbf{q}_1^\infty] \right\},$$

$$\tag{10.10-19}$$

$$\phi_2^o = \frac{-1}{\pi} \mathrm{Im} \sum_{\beta=1}^{3} \left\{ \mathbf{B}_2 \left\langle \ln c_*^{(2)} \right\rangle \mathbf{q}_\beta^{(2)} + \ln(-\zeta_{\beta o}^{(2)}) [\mathbf{B}_2 \mathbf{I}_\beta \mathbf{q}_2^\infty - \mathbf{B}_1 \mathbf{I}_\beta \mathbf{q}_1^\infty] \right\},$$

$$\mathbf{A}_1\mathbf{q}_\beta^{(1)} + \overline{\mathbf{A}}_2\overline{\mathbf{q}}_\beta^{(2)} = \overline{\mathbf{A}}_1\mathbf{I}_\beta\overline{\mathbf{q}}_1^\infty + 2\,\mathrm{Re}[\mathbf{A}_2\mathbf{I}_\beta\mathbf{q}_2^\infty - \mathbf{A}_1\mathbf{I}_\beta\mathbf{q}_1^\infty],$$
$$\mathbf{B}_1\mathbf{q}_\beta^{(1)} + \overline{\mathbf{B}}_2\overline{\mathbf{q}}_\beta^{(2)} = \overline{\mathbf{B}}_1\mathbf{I}_\beta\overline{\mathbf{q}}_1^\infty + 2\,\mathrm{Re}[\mathbf{B}_2\mathbf{I}_\beta\mathbf{q}_2^\infty - \mathbf{B}_1\mathbf{I}_\beta\mathbf{q}_1^\infty], \qquad (10.10\text{-}20)$$

and

$$\mathbf{A}_1\mathbf{g}_k + \overline{\mathbf{A}}_2\overline{\mathbf{h}}_k = \mathbf{A}_2\left\{\left\langle (t_*^{(2)})^k\right\rangle\mathbf{h}_k - \left\langle (\tau_{*\mathrm{o}}^{(2)})^k\right\rangle\mathbf{q}_2^\infty - \sum_{\beta=1}^{3}\left\langle (\hat{\tau}_{*\beta}^{(2)})^k\right\rangle\left(\mathbf{q}_\beta^{(2)} - \mathbf{I}_\beta\mathbf{q}_2^\infty\right)\right\},$$
$$\mathbf{B}_1\mathbf{g}_k + \overline{\mathbf{B}}_2\overline{\mathbf{h}}_k = \mathbf{B}_2\left\{\left\langle (t_*^{(2)})^k\right\rangle\mathbf{h}_k - \left\langle (\tau_{*\mathrm{o}}^{(2)})^k\right\rangle\mathbf{q}_2^\infty - \sum_{\beta=1}^{3}\left\langle (\hat{\tau}_{*\beta}^{(2)})^k\right\rangle\left(\mathbf{q}_\beta^{(2)} - \mathbf{I}_\beta\mathbf{q}_2^\infty\right)\right\}. \qquad (10.10\text{-}21)$$

In deriving (10.10-19) use has been made of (10.10-20). Equations (10.10-20) and (10.10-21) have the same mathematical structure as have (10.10-7) and (10.10-8). Hence the solutions for $\mathbf{q}_\beta^{(1)}$, $\mathbf{q}_\beta^{(2)}$, $\mathbf{g}_k$, and $\mathbf{h}_k$ can be obtained.

As in Case (I) much simplification is achieved when $\mathbf{M}_1 = \mathbf{M}_2$. It is easily shown that $\mathbf{h}_k$ vanishes while

$$\mathbf{q}_\beta^{(1)} = \mathbf{A}_1^{-1}\mathbf{A}_2\mathbf{I}_\beta\mathbf{q}_2^\infty - \mathbf{I}_\beta\mathbf{q}_1^\infty, \quad \mathbf{q}_\beta^{(2)} = \mathbf{I}_\beta\mathbf{q}_2^\infty,$$

$$\mathbf{g}_k = -\mathbf{A}_1^{-1}\mathbf{A}_2\left\langle (\tau_{*\mathrm{o}}^{(2)})^k\right\rangle\mathbf{q}_2^\infty.$$

The infinite series in (10.10-14) has the closed form expression

$$\frac{1}{\pi}\mathrm{Im}\sum_{\alpha=1}^{3}\mathbf{A}_1\left\langle \ln\left(1 - \tau_{\alpha\mathrm{o}}^{(2)}(\zeta_*^{(1)})^{-1}\right)\right\rangle\mathbf{A}_1^{-1}\mathbf{A}_2\mathbf{I}_\alpha\mathbf{q}_2^\infty. \qquad (10.10\text{-}22)$$

Thus we obtain a closed form solution when $\mathbf{M}_1 = \mathbf{M}_2$.

(III) $(x_1^{\mathrm{o}}, x_2^{\mathrm{o}})$ at the elliptic interface.

When $(x_1^{\mathrm{o}}, x_2^{\mathrm{o}})$ is on the elliptic boundary, say,

$$x_1^{\mathrm{o}} = a\cos\psi_{\mathrm{o}}, \quad x_2^{\mathrm{o}} = b\sin\psi_{\mathrm{o}},$$

we have

$$\zeta_{\alpha\mathrm{o}}^{(1)} = \zeta_{\alpha\mathrm{o}}^{(2)} = e^{i\psi_{\mathrm{o}}}, \quad \ln(-\zeta_{\beta\mathrm{o}}^{(2)}) = i(\psi_{\mathrm{o}} + \pi),$$

$$\hat{\tau}_{\alpha\beta}^{(2)} = \hat{\tau}_{\alpha\beta}^{(2)} = \tau_{\alpha\mathrm{o}}^{(2)} = t_\alpha^{(2)}e^{-i\psi_{\mathrm{o}}},$$

$$\hat{z}_{\alpha\beta}^{(2)} = \hat{z}_{\alpha\beta}^{(2)} = z_{\alpha\mathrm{o}}^{(2)} = x_1^{\mathrm{o}} + p_\alpha^{(2)}x_2^{\mathrm{o}}.$$

Equations (10.10-1) and (10.10-2) reduce to, respectively,

$$\mathbf{u}_1 = \frac{1}{\pi}\text{Im}\left\{\mathbf{A}_1\left\langle \ln(\zeta_*^{(1)} - e^{i\psi_o})\right\rangle \mathbf{q}_1^{\infty}\right\} + \frac{1}{\pi}\text{Im}\,\mathbf{A}_1\left\langle \ln(\zeta_*^{(1)^{-1}} - e^{-i\psi_o})\right\rangle \mathbf{q}^{(1)}$$

$$+\frac{1}{\pi}\text{Im}\sum_{k=1}^{\infty}\frac{1}{k}\left\{\mathbf{A}_1\left\langle (\zeta_*^{(1)})^{-k}\right\rangle \mathbf{g}_k\right\}, \qquad (10.10\text{-}23)$$

$$\mathbf{u}_2 - \mathbf{u}_2^{o} = \frac{1}{\pi}\text{Im}\left\{\mathbf{A}_2\left\langle \ln(z_*^{(2)} - z_{*o}^{(2)})\right\rangle \mathbf{q}^{(2)}\right\} + \frac{1}{\pi}\text{Im}\sum_{k=1}^{\infty}\frac{1}{k}\left\{\mathbf{A}_2\left\langle f_{k*}^{(2)}\right\rangle \mathbf{h}_k\right\}, \quad (10.10\text{-}24)$$

where, by (10.10-9),

$$\mathbf{q}^{(1)} = \sum_{\beta=1}^{3}\mathbf{q}_{\beta}^{(1)} = \mathbf{B}_1^{-1}(\mathbf{M}_1^{-1} + \overline{\mathbf{M}}_2^{-1})^{-1}(\overline{\mathbf{M}}_2^{-1} - \overline{\mathbf{M}}_1^{-1})\overline{\mathbf{B}}_1\overline{\mathbf{q}}_1^{\infty},$$

$$\mathbf{q}^{(2)} = \sum_{\beta=1}^{3}\mathbf{q}_{\beta}^{(2)} = \mathbf{B}_2^{-1}(\overline{\mathbf{M}}_1^{-1} + \mathbf{M}_2^{-1})^{-1}(\overline{\mathbf{M}}_2^{-1} + \mathbf{M}_1^{-1})\mathbf{B}_1\mathbf{q}_1^{\infty}. \qquad (10.10\text{-}25)$$

$\mathbf{g}_k$ and $\mathbf{h}_k$ are determined from (10.10-12) and (10.10-10).

Equations (10.10-14) and (10.10-15) also reduce to (10.10-23) and (10.10-24), respectively. From (10.10-16) and (5.5-17) it is easily established that, for k=1,2,

$$2\sum_{\beta=1}^{3}\mathbf{A}_k\mathbf{I}_{\beta}\mathbf{q}_k^{\infty} = 2\mathbf{A}_k\mathbf{q}_k^{\infty} = \mathbf{b} - i(\mathbf{H}_k\mathbf{f} + \mathbf{S}_k\mathbf{b}),$$

$$2\sum_{\beta=1}^{3}\mathbf{B}_k\mathbf{I}_{\beta}\mathbf{q}_k^{\infty} = 2\mathbf{B}_k\mathbf{q}_k^{\infty} = \mathbf{f} - i(\mathbf{S}_k^T\mathbf{f} - \mathbf{L}_k\mathbf{b}). \qquad (10.10\text{-}26)$$

One can then show that $\mathbf{u}_2^o$, $\boldsymbol{\phi}_2^o$ of (10.10-19) reduce to (10.10-6) when (x_1^o, x_2^o) is on the elliptic boundary. Likewise, (10.10-20) and (10.10-7) are identical when the operation of summation with β=1,2,3 is carried out. Equation (10.10-21) also reduces to (10.10-8). Thus the two solutions in Cases (I) and (II) are identical in the limit (x_1^o, x_2^o) is on the elliptic boundary.

As in Cases (I) and (II), a closed form solution is obtained when $\mathbf{M}_1 = \mathbf{M}_2$. It should be pointed out that when $\mathbf{M}_1 = \mathbf{M}_2$, we have $\mathbf{S}_1 = \mathbf{S}_2$, $\mathbf{H}_1 = \mathbf{H}_2$, and $\mathbf{L}_1 = \mathbf{L}_2$. By (10.10-26), $\mathbf{A}_1\mathbf{q}_1^{\infty} = \mathbf{A}_2\mathbf{q}_2^{\infty}$ and $\mathbf{B}_1\mathbf{q}_1^{\infty} = \mathbf{B}_2\mathbf{q}_2^{\infty}$. Hence

$$\mathbf{h}_k = \mathbf{0}, \quad \mathbf{q}^{(1)} = \mathbf{0}, \quad \mathbf{q}^{(2)} = \mathbf{q}_2^{\infty},$$

and the infinite series in (10.10-23) is given by (10.10-22).

In all three cases closed form solutions are also possible when the three eigenvalues $p_{\alpha}^{(2)}$ for material 2 are identical and given by

$$p_\alpha^{(2)} = ia/b, \quad (\alpha=1,2,3). \tag{10.10-27}$$

Material 1 can be arbitrary. A circular inclusion of isotropic material is an example. When we have (10.10-27), it can be shown that

$$t_\alpha^{(2)}, \quad \tilde{\tau}_{\alpha\beta}^{(2)}, \quad \tau_{\alpha o}^{(2)}, \quad \hat{\tau}_{\alpha\beta}^{(2)}$$

all vanish, as do $\mathbf{g}_k$ and $\mathbf{h}_k$. Consequently the infinite series in the solutions disappear, and the solutions are in a closed form. It should be noted that when $p_\alpha^{(2)}$ are identical and material 2 is degenerate, the solutions for material 2 presented here are not valid. We will show in Section 13.4 how to modify the solutions so that they can be applied to degenerate materials.

Chapter 11

ANISOTROPIC MEDIA WITH A CRACK OR A RIGID LINE INCLUSION

A crack, or cracks, in a material is perhaps one of the most studied problems in solid mechanics. This is due to the fact that many structural failures are related to the presence of a crack in the material. The knowledge of stress distribution near a crack tip is indispensable in a fracture mechanics analysis (Rice, 1968; Sih and Liebowitz, 1968; Sih and Chen, 1981; Kanninen and Popelar, 1985; K. C. Wu, 1989a). A crack is represented by a slit cut whose surfaces are assumed traction-free. This is a mathematical idealization. For a composite material that consists of stiff short fibers or whiskers in the matrix, we have rigid line inclusions. A rigid line inclusion is the counterpart of a crack. It is sometimes called an *anticrack*. The displacement at a rigid line inclusion either vanishes or has a rigid body translation and rotation. One of the puzzling problems for a crack is the one when it is located on the x_1-axis that is an interface between two dissimilar materials. The displacement of the crack surfaces near the crack tips may oscillate, creating a physically unacceptable phenomenon of interpenetration of two materials. The bimaterial tensor $\breve{\mathbf{S}}$ introduced in Section 8.8 plays a key role in the analysis. If $\breve{\mathbf{S}}$ vanishes identically, there is no oscillation. If $\breve{\mathbf{S}}$ is nonzero, we may decompose the tractions applied on the crack surfaces into two components, one along the right null vector of $\breve{\mathbf{S}}$ denoted by $\mathbf{t}_o$ and the other on the right eigenplane of $\breve{\mathbf{S}}$ denoted by $\mathbf{t}_\gamma$. The solution associated with $\mathbf{t}_o$ is not oscillatory. It has the property that the traction on the interface $x_2=0$ is polarized along the right null vector of $\breve{\mathbf{S}}$ while the crack opening displacement is polarized along the left null vector of $\breve{\mathbf{S}}$. The solution associated with $\mathbf{t}_\gamma$ is oscillatory. It has the property that the traction on the interface $x_2=0$ is polarized on the right eigenplane of $\breve{\mathbf{S}}$ while the crack opening displacement is polarized on the left eigenplane of $\breve{\mathbf{S}}$. We will see that these results remain valid for non uniform tractions on the crack surfaces and for multiple interfacial cracks. Again, with the aid of the identities presented in Chapters 6 and 7 real form expressions for the displacement, the surface traction, and the hoop stress vector along the x_1-axis on which the crack is located can be obtained. In the last section we discuss the Comninou (1977b, 1978) crack which, within the linear elasticity theory, solves the interfacial crack problem correctly, both in the mathematical and physical sense.

11.1 A Rigid Line Inclusion

Let an anisotropic elastic medium be referred to a fixed rectangular coordinate system x_i (i=1,2,3). A rigid line of length $2a$ is centrally located on the $\hat{x}_1$-axis that makes an angle θ_o with the x_1-axis, Fig. 11.1. The rigid line is assumed to be inextensible and undeformable. It remains a straight line and is rigidly bonded to the surrounding anisotropic medium during the deformation of the latter. We will study separately the cases when the rigid line is subjected to (a) a concentrated force **f** and (b) a concentrated moment or couple T. The following identities taken from (5.5-17) and (5.6-6) will be employed often in this chapter.

$$\mathbf{S} = i(2\mathbf{A}\mathbf{B}^T - \mathbf{I}), \quad \mathbf{H} = 2i\mathbf{A}\mathbf{A}^T, \quad \mathbf{L} = -2i\mathbf{B}\mathbf{B}^T, \tag{11.1-1}$$

$$\begin{aligned} \mathbf{M} &= -i\mathbf{B}\mathbf{A}^{-1} = \mathbf{H}^{-1} + i\mathbf{H}^{-1}\mathbf{S}, \\ \mathbf{M}^{-1} &= i\mathbf{A}\mathbf{B}^{-1} = \mathbf{L}^{-1} - i\mathbf{S}\mathbf{L}^{-1}. \end{aligned} \tag{11.1-2}$$

Another useful relations taken from (3.4-14) are

$$\sqrt{\hat{z}_\alpha^2 - a^2} = \begin{cases} \pm\sqrt{\hat{x}_1^2 - a^2}\,, & \text{at } \hat{x}_2 = 0, \ \pm\hat{x}_1 > a, \\ \pm i\sqrt{a^2 - \hat{x}_1^2}\,, & \text{at } \hat{x}_2 = \pm 0, \ |\hat{x}_1| < a. \end{cases} \tag{11.1-3}$$

(a) A Force Applied to the Rigid Line Inclusion

Consider a line of force **f** per unit length of the x_3-axis be applied to a rigid line inclusion. This is a special case of a line force

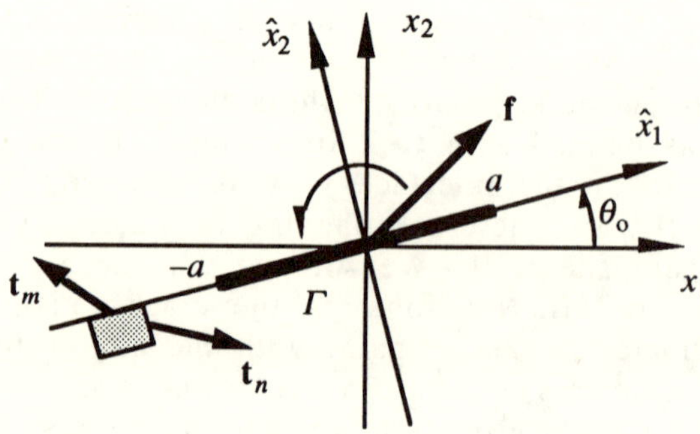

Fig. 11.1 Anisotropic elastic medium with a rigid line inclusion is subjected to a concentrated force **f** at its center and a couple T. If T=0 the surface traction $\mathbf{t}_m$ along the $\hat{x}_1$-axis is independent of θ_o, the orientation of the rigid line in the material. Moreover $\mathbf{t}_m$ is co-directional with **f**.

applied to an elliptic rigid inclusion studied in Section 10.1 if we let the minor axis b of the ellipse vanish. Therefore from (10.1-14) and (10.1-7) the solution is

$$\mathbf{u} = \frac{1}{\pi}\mathrm{Im}\left\{\mathbf{A}\left\langle \ln\left(\hat{z}_* + \sqrt{\hat{z}_*^2 - a^2}\right)\right\rangle \mathbf{A}^T\right\}\mathbf{f},$$

$$\phi = \frac{1}{\pi}\mathrm{Im}\left\{\mathbf{B}\left\langle \ln\left(\hat{z}_* + \sqrt{\hat{z}_*^2 - a^2}\right)\right\rangle \mathbf{A}^T\right\}\mathbf{f}, \tag{11.1-4}$$

where

$$\hat{z}_\alpha = \hat{x}_1 + p_\alpha(\theta_o)\hat{x}_2, \tag{11.1-5}$$

$$\hat{x}_1 = r\cos(\theta - \theta_o), \quad \hat{x}_2 = r\sin(\theta - \theta_o), \tag{11.1-6}$$

r and θ being the polar coordinates referred to (x_1, x_2). A constant term has been omitted in (11.1-4) since it does not affect the stress. The surface traction $\mathbf{t}_m$ and the hoop stress vector $\mathbf{t}_n$ along the $\hat{x}_1$-axis are

$$\mathbf{t}_m = \frac{\partial}{\partial \hat{x}_1}\phi(\hat{x}_1),$$

$$\mathbf{t}_n = -\frac{\partial}{\partial \hat{x}_2}\phi(\hat{x}_2) = -\mathbf{N}_3(\theta_o)\frac{\partial}{\partial \hat{x}_1}\mathbf{u}(\hat{x}_1) - \mathbf{N}_1^T(\theta_o)\frac{\partial}{\partial \hat{x}_1}\phi(\hat{x}_1). \tag{11.1-7}$$

The second equality for $\mathbf{t}_n$ is from $(7.3\text{-}20)_2$. By (11.1-4) and

$$\frac{\partial}{\partial \hat{x}_1}\left\{\hat{z}_* + \sqrt{\hat{z}_*^2 - a^2}\right\} = \left(\hat{z}_*^2 - a^2\right)^{-1/2}$$

it is easily shown that

$$\mathbf{u} = -\frac{1}{2\pi}\ln\left\{\hat{x}_1 + \sqrt{\hat{x}_1^2 - a^2}\right\}\mathbf{H}\mathbf{f},$$

$$\mathbf{t}_m = \frac{\mp 1}{2\pi\sqrt{\hat{x}_1^2 - a^2}}\mathbf{S}^T\mathbf{f}, \quad \mathbf{t}_n = -\mathbf{N}_1^T(\theta_o)\mathbf{t}_m, \tag{11.1-8}$$

for $\hat{x}_2 = 0$, $\pm\hat{x}_1 > a$ and

$$\mathbf{u} = \mathbf{0}, \quad \mathbf{t}_m = \frac{\mp 1}{2\pi\sqrt{a^2 - \hat{x}_1^2}}\mathbf{f}, \quad \mathbf{t}_n = -\mathbf{N}_1^T(\theta_o)\mathbf{t}_m, \tag{11.1-9}$$

for $\hat{x}_2 = \pm 0$, $|\hat{x}_1| < a$. That $\mathbf{u} = \mathbf{0}$ on the rigid line is not unreasonable because the displacement is unique up to an arbitrary rigid body translation and rotation. The traction $\mathbf{t}_m$ is independent of θ_o, the orientation of the rigid line with respect to the material. As expected, $\mathbf{t}_m$ and $\mathbf{t}_n$ have a square root singularity at $\hat{x}_1=\pm a$. $\mathbf{t}_m$ on the rigid line is in the direction of the applied force $\mathbf{f}$.

(b) A Moment Applied to the Rigid Line Inclusion

Again we may specialize the solution presented in Section 10.2 where the problem of a moment applied to an elliptic rigid inclusion was investigated. From (10.2-1), (10.1-8), (10.2-5), and (10.2-4) with b=0 we have

$$\mathbf{u} = \omega \,\mathrm{Re}\left\{\mathbf{A}\left\langle \hat{z}_\alpha - \sqrt{\hat{z}_\alpha^2 - a^2}\right\rangle \mathbf{A}^{-1}\right\}\mathbf{m}(\theta_o),$$

$$\boldsymbol{\phi} = \omega \,\mathrm{Re}\left\{\mathbf{B}\left\langle \hat{z}_\alpha - \sqrt{\hat{z}_\alpha^2 - a^2}\right\rangle \mathbf{A}^{-1}\right\}\mathbf{m}(\theta_o), \qquad (11.1\text{-}10)$$

where ω is the rotation of the rigid line inclusion. By (10.2-9) and (10.2-10) it is related to the applied moment T by

$$T = \omega a^2 \mathbf{m}^T(\theta_o)\mathbf{H}^{-1}\mathbf{m}(\theta_o). \qquad (11.1\text{-}11)$$

From (11.1-10) and (11.1-7) it can be shown that

$$\mathbf{u} = \pm\omega\left\{|\hat{x}_1| - \sqrt{\hat{x}_1^2 - a^2}\right\}\mathbf{m}(\theta_o),$$

$$\mathbf{t}_m = -\omega\left\{1 - \frac{|\hat{x}_1|}{\sqrt{\hat{x}_1^2 - a^2}}\right\}\mathbf{H}^{-1}\mathbf{S}\mathbf{m}(\theta_o),$$

$$\mathbf{t}_n = -\omega\left\{1 - \frac{|\hat{x}_1|}{\sqrt{\hat{x}_1^2 - a^2}}\right\}\left\{\mathbf{N}_3(\theta_o) - \mathbf{N}_1^T(\theta_o)\mathbf{H}^{-1}\mathbf{S}\right\}\mathbf{m}(\theta_o),$$

for $\hat{x}_2=0$, $\pm\hat{x}_1>a$ and

$$\mathbf{u} = \omega\hat{x}_1\mathbf{m}(\theta_o),$$

$$\mathbf{t}_m = -\omega\left\{\mathbf{H}^{-1}\mathbf{S} \pm \frac{\hat{x}_1}{\sqrt{a^2 - \hat{x}_1^2}}\mathbf{H}^{-1}\right\}\mathbf{m}(\theta_o),$$

$$\mathbf{t}_n = -\mathbf{N}_1^T(\theta_o)\mathbf{t}_m,$$

for $\hat{x}_2=\pm0$, $|\hat{x}_1|<a$. Again the stress has a square root singularity at $\hat{x}_1=\pm a$.

11.2 Green's Functions for a Rigid Line or a Crack

In this section we study Green's functions for the infinite space with the presence of a rigid line inclusion or a crack of length $2a$ centrally located along the $\hat{x}_1$-axis. A line force **f** and a line dislocation with Burgers vector **b** are applied at, Fig. 11.2,

$$(\hat{x}_1,\hat{x}_2)=(\hat{x}_1^{\circ},\hat{x}_2^{\circ}).$$

The related problem of the interaction between a dislocation and a crack was investigated by Lo (1978) for a flat crack and by Vitek and Hirth (1977) for an elliptic crack. The solution to this problem can be deduced from the Green's function for the infinite space with the presence of an elliptic rigid inclusion or an elliptic hole studied in Section 10.3. The solution is given by (10.3-2) in which

$$\hat{\zeta}_\alpha=\frac{1}{a}\left\{\hat{z}_\alpha+\sqrt{\hat{z}_\alpha^2-a^2}\right\}, \tag{11.2-1}$$

$$\hat{\zeta}_{\beta\circ}=\frac{1}{a}\left\{\hat{z}_{\beta\circ}+\sqrt{\hat{z}_{\beta\circ}^2-a^2}\right\},$$

$$\hat{z}_{\beta\circ}=\hat{x}_1^{\circ}+p_\beta(\theta_\circ)\hat{x}_2^{\circ}.$$

The constant vectors $\mathbf{q}_\beta$ are given by (10.3-6) if the line is a crack and by (10.3-9) if it is a rigid line inclusion.

The special case when $\mathbf{b}=\mathbf{0}$ and the applied force **f** is on the crack surface, Fig. 11.3(a),

$$(\hat{x}_1^{\circ},\hat{x}_2^{\circ})=(\xi,+0)$$

can also be deduced from the allied problem for an elliptic hole.

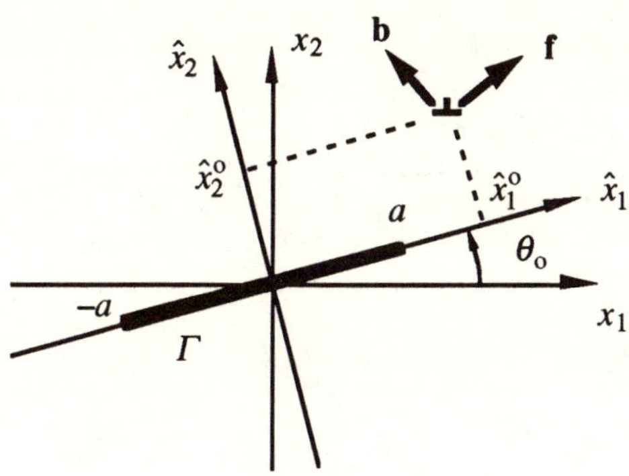

Fig. 11.2 Anisotropic elastic medium with a rigid line inclusion or a crack is subjected to a line force **f** and a line dislocation with Burgers vector **b**.

The solution is given by (10.4-4) in which

$$e^{\pm i\psi_o} = \frac{1}{a}\left\{\xi \pm i\sqrt{a^2-\xi^2}\right\}$$

and $\hat{\zeta}_\alpha$ is given by (11.2-1). It is seen that $\hat{\zeta}_\alpha$ is independent of θ_o along the $\hat{x}_1$-axis,. Therefore the displacement **u** and the surface traction $\mathbf{t}_m$ along the $\hat{x}_1$-axis are invariant with the orientation of the crack with respect to the material.

If, in addition to the force **f** at $(\hat{x}_1^o, \hat{x}_2^o) = (\xi, +0)$, a force $-\mathbf{f}$ is applied at $(\hat{x}_1^o, \hat{x}_2^o) = (\xi, -0)$, Fig. 11.3(b), the solution is obtained by setting $\psi' = -\psi_o$ in (10.4-8), i.e.,

$$\mathbf{u} = \frac{1}{\pi}\mathrm{Im}\left\{\mathbf{A}\langle g(\hat{z}_*)\rangle\mathbf{B}^{-1}\right\}\mathbf{f}, \qquad \boldsymbol{\phi} = \frac{1}{\pi}\mathrm{Im}\left\{\mathbf{B}\langle g(\hat{z}_*)\rangle\mathbf{B}^{-1}\right\}\mathbf{f}, \tag{11.2-2}$$

where

$$g(z) = \ln\frac{\zeta - e^{i\psi_o}}{\zeta - e^{-i\psi_o}} = \ln\frac{z+\sqrt{z^2-a^2}-\xi-i\sqrt{a^2-\xi^2}}{z+\sqrt{z^2-a^2}-\xi+i\sqrt{a^2-\xi^2}}. \tag{11.2-3}$$

The function $g(z)$ can be replaced by other alternate forms. If we add to $g(z)$ the constant term

$$\ln\frac{\sqrt{a+\xi}-i\sqrt{a-\xi}}{\sqrt{a+\xi}+i\sqrt{a-\xi}},$$

the displacements have a rigid body translation while the stresses remain the same. The new $g(z)$ can be written as

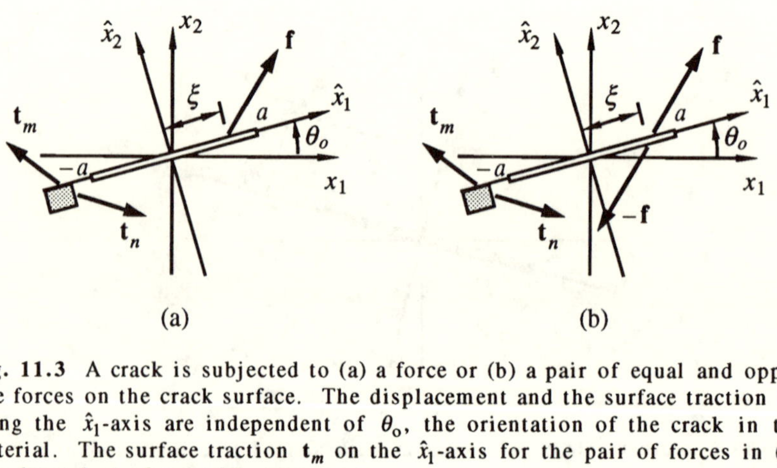

Fig. 11.3 A crack is subjected to (a) a force or (b) a pair of equal and opposite forces on the crack surface. The displacement and the surface traction $\mathbf{t}_m$ along the $\hat{x}_1$-axis are independent of θ_o, the orientation of the crack in the material. The surface traction $\mathbf{t}_m$ on the $\hat{x}_1$-axis for the pair of forces in (b) is independent of material property.

$$g(z) = \ln \frac{\sqrt{(z-a)(a+\xi)} - i\sqrt{(z+a)(a-\xi)}}{\sqrt{(z-a)(a+\xi)} + i\sqrt{(z+a)(a-\xi)}} \tag{11.2-4}$$

or

$$g(z) = \ln\left\{X - i\sqrt{1-X^2}\right\} = -i\Theta \tag{11.2-5a}$$

where

$$X = \cos\Theta = \frac{z\xi - a^2}{a(z-\xi)}, \quad \sqrt{1-X^2} = \frac{\sqrt{z^2-a^2}\sqrt{a^2-\xi^2}}{a(z-\xi)}.$$

Hence we have the remarkably simple form

$$g(z) = -i\cos^{-1}\left\{\frac{z\xi - a^2}{a(z-\xi)}\right\} \tag{11.2-5b}$$

first obtained by Stroh (1958) using a different approach.

Differentiation of $g(\hat{z}_\alpha)$ with respect to $\hat{x}_1$ yields

$$\frac{\partial}{\partial \hat{x}_1} g(\hat{z}_\alpha) = \frac{i\sqrt{a^2-\xi^2}}{\sqrt{\hat{z}_\alpha^2 - a^2}\,(\hat{z}_\alpha - \xi)}.$$

The same result is obtained regardless of whether we employ (11.2-3), (11.2-4), or (11.2-5a,b). Along the $\hat{x}_1$-axis,

$$\frac{\partial}{\partial \hat{x}_1} g(\hat{x}_1) = \frac{\pm i\sqrt{a^2-\xi^2}}{\sqrt{\hat{x}_1^2 - a^2}\,(\hat{x}_1 - \xi)} \quad \text{for } \hat{x}_2=0, \pm\hat{x}_1>a.$$

$$\frac{\partial}{\partial \hat{x}_1} g(\hat{x}_1) = -i\pi\delta(\hat{x}_1 - \xi) \pm \frac{\sqrt{a^2-\xi^2}}{\sqrt{a^2 - \hat{x}_1^2}\,(\hat{x}_1 - \xi)} \quad \text{for } \hat{x}_2 = \pm 0, \ |\hat{x}_1| < a,$$

where $\delta(x)$ is the *delta* function which vanishes everywhere except at $x=0$ and has the property

$$\int_{-\infty}^{\infty} \delta(x)dx = 1. \tag{11.2-6}$$

One can then show from (11.1-7) that

$$\mathbf{t}_m = \frac{\pm\sqrt{a^2-\xi^2}}{\pi\sqrt{\hat{x}_1^2 - a^2}\,(\hat{x}_1 - \xi)}\mathbf{f}, \quad \mathbf{t}_n = -\mathbf{G}_1(\theta_o)\mathbf{t}_m, \tag{11.2-7}$$

for $\hat{x}_2=0$, $\pm\hat{x}_1>a$ and

$$\mathbf{t}_n = \delta(\hat{x}_1 - \xi)\mathbf{G}_1(\theta_o)\mathbf{f} \mp \frac{\sqrt{a^2 - \xi^2}}{\pi\sqrt{a^2 - \hat{x}_1^2}(\hat{x}_1 - \xi)}\mathbf{G}_3(\theta_o)\mathbf{f}, \qquad (11.2\text{-}8)$$

for $\hat{x}_2=\pm 0$, $|\hat{x}_1|<a$. In the above

$$\mathbf{G}_1(\theta) = \mathbf{N}_1^T(\theta) - \mathbf{N}_3(\theta)\mathbf{S}\mathbf{L}^{-1}, \quad \mathbf{G}_3(\theta) = -\mathbf{N}_3(\theta)\mathbf{L}^{-1}. \qquad (11.2\text{-}9)$$

The tensors $\mathbf{G}_1^T(\theta)$ and $\mathbf{G}_3^T(\theta)$ share some properties with $\mathbf{N}_1(\theta)$ and $\mathbf{N}_3(\theta)$. They are periodic in θ with periodicity π. They satisfy the identities listed in (7.11-3) and (7.11-4). For $\theta=0$ and $\pi/2$ they have the structure given by (7.11-9) and (7.11-10).

The *stress intensity factor* $\mathbf{k}$ at the crack tip $\hat{x}_1=a$ is defined by

$$\mathbf{k} = \lim_{\hat{x}_1 \to a} \sqrt{2\pi(\hat{x}_1 - a)}\mathbf{t}_m.$$

We have from $(11.2\text{-}7)_1$

$$\mathbf{k} = \frac{\sqrt{a+\xi}}{\sqrt{\pi a(a-\xi)}} \quad \text{at } \hat{x}_1=a.$$

Likewise the stress intensity factor at $\hat{x}_1=-a$ can be shown to be

$$\mathbf{k} = \frac{\sqrt{a-\xi}}{\sqrt{\pi a(a+\xi)}} \quad \text{at } \hat{x}_1=-a.$$

The displacement along the $\hat{x}_1$-axis can also be expressed in a real form. Let

$$X(\hat{x}_1) = \frac{\hat{x}_1\xi - a^2}{a(\hat{x}_1 - \xi)}. \qquad (11.2\text{-}10)$$

It is easily shown that $X(\hat{x}_1)$ is a monotonically increasing function of $\hat{x}_1$ and that

$$\xi/a < X(\hat{x}_1) < 1 \quad \text{for} \quad -\infty < \hat{x}_1 < -a,$$

$$1 < X(\hat{x}_1) < \infty \quad \text{for} \quad -a < \hat{x}_1 < \xi,$$

$$-\infty < X(\hat{x}_1) < -1 \quad \text{for} \quad \xi < \hat{x}_1 < a,$$

$$-1 < X(\hat{x}_1) < \xi/a \quad \text{for} \quad a < \hat{x}_1 < \infty.$$

Hence by (11.2-5a,b)

$$g(\hat{x}_1) = -i\cos^{-1}\{X(\hat{x}_1)\} \quad \text{for } \hat{x}_2=0, \ |\hat{x}_1|>a,$$

$$g(\hat{x}_1) = \mp\cosh^{-1}\{X(\hat{x}_1)\} \quad \text{for } \hat{x}_2=\pm 0, \ -a<\hat{x}_1<\xi,$$

$$g(\hat{x}_1) = -i\pi \mp \cosh^{-1}\{|X(\hat{x}_1)|\} \quad \text{for } \hat{x}_2 = \pm 0, \ \xi < \hat{x}_1 < a,$$

where

$$0 \le \cos^{-1}\{X(\hat{x}_1)\} \le \pi. \tag{11.2-11}$$

From $(11.2\text{-}2)_1$ the displacement along the $\hat{x}_1$-axis is

$$\mathbf{u} = \frac{1}{\pi}\cos^{-1}\{X(\hat{x}_1)\}\mathbf{S}\mathbf{L}^{-1}\mathbf{f} \quad \text{for } \hat{x}_2 = 0, \ |\hat{x}_1| > a,$$

$$\mathbf{u} = \frac{\pm 1}{\pi}\cosh^{-1}\{X(\hat{x}_1)\}\mathbf{L}^{-1}\mathbf{f} \quad \text{for } \hat{x}_2 = \pm 0, \ -a < \hat{x}_1 < \xi,$$

$$\mathbf{u} = \mathbf{S}\mathbf{L}^{-1}\mathbf{f} \pm \cosh^{-1}\{|X(\hat{x}_1)|\}\mathbf{L}^{-1}\mathbf{f} \quad \text{for } \hat{x}_2 = \pm 0, \ \xi < \hat{x}_1 < a.$$

Noting that $X(\hat{x}_1) > 0$ for $-a < \hat{x}_1 < \xi$, the crack opening displacement $\Delta\mathbf{u}$ is

$$\Delta\mathbf{u} = \mathbf{u}(\hat{x}_1, +0) - \mathbf{u}(\hat{x}_1, -0) = \frac{2}{\pi}\cosh^{-1}\{|X(\hat{x}_1)|\}\mathbf{L}^{-1}\mathbf{f} \quad \text{for } |\hat{x}_1| < a.$$

It should be noted that the traction $\mathbf{t}_m$ as given by (11.2-7) is independent of material property and is in the same direction as the applied force $\mathbf{f}$. This is a more general result than the one reported by Barnett and Asaro (1972) who assumed a uniform distribution of traction on the crack surfaces. As we have pointed out earlier the traction $\mathbf{t}_m$ and the displacement $\mathbf{u}$ on the $\hat{x}_1$-axis are independent of θ_o, the orientation of the crack with respect to the material. The hoop stress vector $\mathbf{t}_n$ on the crack surface as given by (11.2-8) becomes infinite at $\hat{x}_1 = \xi$. However, if $\mathbf{f}$ is proportional to $\mathbf{L}\mathbf{m}(\theta_o)$, $\mathbf{G}_3(\theta_o)\mathbf{f} = \mathbf{0}$ and $\mathbf{t}_n$ vanishes everywhere on the crack surfaces except at $\hat{x}_1 = \xi$. For isotropic materials $\mathbf{L}$ is a diagonal matrix so that $\mathbf{f}$ would be in the direction of $\mathbf{m}(\theta_o)$, which is perpendicular to the crack.

11.3 A Rigid Line Subjected to Uniform Loading

Using the methods of Eshelby (1957, 1959) and Muskhelishvili (1953), Z. Y. Wang et al. (1985, 1986) studied a rigid line inclusion in an isotropic plane elastic body subjected to a uniform loading at infinity. The same problem was also considered by Ballarini (1987) using an integral transform. The more general case in which the line inclusion is an isotropic elastic material was investigated by Sendeckyj (1970), Atkinson (1973), and Selvadurai (1980). Erdogan and Gupta (1972) studied the problem of bonded materials containing a flat inclusion which may be rigid or elastic with negligible bending rigidity. For anisotropic materials Yang and Chou (1982) studied the elliptic inclusion in a cubic material. They obtained explicit solutions for stresses around the inclusion which include the special case of a rigid line inclusion. A rigid line inclusion in a general anisotropic material subjected to a uniform loading at infinity was considered by Li and Ting (1989). They also studied the general case when the line inclusion is an anisotropic material.

For the rest of this chapter we employ a single coordinate system x_i with the rigid line of length $2a$ centrally located on the x_1-axis. Let σ_{ij}^∞ be the stress applied at infinity. The associated strain is denoted by ε_{ij}^∞ with $\varepsilon_{33}^\infty=0$. Within a rigid body translation and rotation the displacement $\mathbf{u}^\infty$ can be written as

$$\mathbf{u}^\infty(x_1,x_2)=\begin{bmatrix} u_1^\infty \\ u_2^\infty \\ u_3^\infty \end{bmatrix}=\begin{bmatrix} \varepsilon_{11}^\infty & 2\varepsilon_{12}^\infty & 0 \\ 0 & \varepsilon_{22}^\infty & 0 \\ 2\varepsilon_{13}^\infty & 2\varepsilon_{23}^\infty & 0 \end{bmatrix}\begin{bmatrix} x_1 \\ x_2 \\ x_3 \end{bmatrix} \tag{11.3-1}$$

which satisfies the condition that $\mathbf{u}^\infty$ is linear in x_1 and x_2. The x_1-axis is not rotated because

$$\mathbf{u}^\infty(x_1,0)=\begin{bmatrix} \varepsilon_{11}^\infty \\ 0 \\ 2\varepsilon_{13}^\infty \end{bmatrix}x_1.$$

If $\varepsilon_{11}^\infty=\varepsilon_{13}^\infty=0$, $\mathbf{u}^\infty(x_1,0)=0$ for all x_1 so that the presence of a rigid line inclusion does not disturb the uniform solution (11.3-1). We will assume that the uniform loading at infinity is such that ε_{11}^∞ and ε_{13}^∞ do not both vanish. We will seek the *disturbed solution* due to nonzero ε_{11}^∞ and ε_{13}^∞.

The disturbed solution satisfies the conditions

$$\sigma_{ij}=0 \quad \text{at} \quad |\mathbf{x}|=\infty, \tag{11.3-2}$$

$$\mathbf{u}(x_1,0)=-\mathbf{q}x_1 \quad \text{for} \quad |x_1|<a, \tag{11.3-3}$$

where

$$\mathbf{q}=\begin{bmatrix} \varepsilon_{11}^\infty \\ -\omega \\ 2\varepsilon_{13}^\infty \end{bmatrix}. \tag{11.3-4}$$

The constant ω represents the counterclockwise rotation of the rigid line. It is determined by the condition that the total moment about the x_3-axis exerted on the rigid line vanish, i.e.,

$$\int_{-a}^{a}\{\sigma_{22}(x_1,+0)-\sigma_{22}(x_1,-0)\}x_1dx_1=0. \tag{11.3-5}$$

If the rigid line is constrained to rotate, ω=0.

It is easily shown that

$$\begin{aligned}\mathbf{u}&=\mathrm{Re}\left\{\mathbf{A}\left\langle\sqrt{z_*^2-a^2}-z_*\right\rangle\mathbf{A}^{-1}\right\}\mathbf{q},\\ \boldsymbol{\phi}&=\mathrm{Re}\left\{\mathbf{B}\left\langle\sqrt{z_*^2-a^2}-z_*\right\rangle\mathbf{A}^{-1}\right\}\mathbf{q},\end{aligned} \tag{11.3-6}$$

where Re stands for the real part satisfies (11.3-2) and (11.3-3). Along the x_1-axis the surface traction vector $(\mathbf{t}_2)_i=\sigma_{i2}$ and the hoop stress vector $(\mathbf{t}_1)_i=\sigma_{i1}$ are

$$\mathbf{t}_2=\boldsymbol{\phi}_{,1},\quad \mathbf{t}_1=-\boldsymbol{\phi}_{,2}. \tag{11.3-7}$$

Making use of (11.1-2), (11.1-3), and the identities

$$\begin{aligned}\mathbf{A}\langle p_*\rangle\mathbf{A}^{-1}&=(\mathbf{N}_1-\mathbf{N}_2\mathbf{H}^{-1}\mathbf{S})+i\mathbf{N}_2\mathbf{H}^{-1},\\ \mathbf{B}\langle p_*\rangle\mathbf{A}^{-1}&=(\mathbf{N}_3-\mathbf{N}_1^T\mathbf{H}^{-1}\mathbf{S})+i\mathbf{N}_1^T\mathbf{H}^{-1},\end{aligned} \tag{11.3-8}$$

taken from (6.9-7) it can be shown that

$$\begin{aligned}\mathbf{t}_2&=-\left\{\frac{|x_1|}{\sqrt{x_1^2-a^2}}-1\right\}\mathbf{H}^{-1}\mathbf{S}\mathbf{q},\\ \mathbf{t}_1&=-\left\{\frac{|x_1|}{\sqrt{x_1^2-a^2}}-1\right\}\left(\mathbf{N}_3-\mathbf{N}_1^T\mathbf{H}^{-1}\mathbf{S}\right)\mathbf{q},\end{aligned} \tag{11.3-9}$$

for x_2=0, $\pm x_1>a$ and

$$\begin{aligned}\mathbf{t}_2&=\mathbf{H}^{-1}\mathbf{S}\mathbf{q}\pm\frac{x_1}{\sqrt{a^2-x_1^2}}\mathbf{H}^{-1}\mathbf{q},\\ \mathbf{t}_1&=\left(\mathbf{N}_3-\mathbf{N}_1^T\mathbf{H}^{-1}\mathbf{S}\right)\mathbf{q}\mp\frac{x_1}{\sqrt{a^2-x_1^2}}\mathbf{N}_1^T\mathbf{H}^{-1}\mathbf{q},\end{aligned} \tag{11.3-10}$$

for $x_2=\pm 0$, $|x_1|<a$. Equations (11.3-9) and (11.3-10) show that there is a square root stress singularity at the tips of the rigid line.

From $(11.3\text{-}10)_1$,

$$\sigma_{22}(x_1,+0)-\sigma_{22}(x_1,-0)=\frac{2x_1}{\sqrt{x_1^2-a^2}}(\mathbf{H}^{-1}\mathbf{q})_2$$

where the subscript 2 outside the parentheses indicates the second element of the matrix. The condition (11.3-5) is satisfied if

$$(\mathbf{H}^{-1}\mathbf{q})_2 = 0$$

which leads to

$$\omega = \left\{(\mathbf{H}^{-1})_{21}\varepsilon_{11}^{\infty} + 2(\mathbf{H}^{-1})_{23}\varepsilon_{13}^{\infty}\right\} \Big/ (\mathbf{H}^{-1})_{22}. \qquad (11.3\text{-}11)$$

Since $\mathbf{H}^{-1}$ is positive definite, $(\mathbf{H}^{-1})_{22}$ is positive and nonzero. Equation (11.3-11) provides the solution for the rotation of the rigid line inclusion.

The displacement $\mathbf{u}$ along the x_1-axis obtained from $(11.3\text{-}6)_1$ is

$$\mathbf{u} = \mp\left(|x_1| - \sqrt{x_1^2 - a^2}\right)\mathbf{q}$$

for $x_2=0$, $\pm x_1>a$. It is in the same direction as the vector $\mathbf{q}$. The displacement components u_1 and u_3 depend on $\varepsilon_{11}^{\infty}$ and $\varepsilon_{13}^{\infty}$, but otherwise independent of material property. The displacement component u_2 depends on ω. By (11.3-11) it depends on material property.

11.4 A Crack in a Homogeneous Medium

Consider a crack of length $2a$ centrally located at $x_2=0$, $|x_1|<a$ in a homogeneous anisotropic elastic medium. A uniform stress σ_{ij}^{∞} is applied at infinity. The crack surfaces are traction-free. The solution consists of a uniform stress solution due to the uniform loading σ_{ij}^{∞} and a disturbed solution due to the presence of the crack. For the disturbed solution a uniform traction $\mathbf{t}_\Gamma$ is applied at the upper crack surface and $-\mathbf{t}_\Gamma$ is applied at the lower crack surface, Fig. 11.4(a). The stress vanishes at infinity. Thus the boundary conditions for the stress function ϕ are

$$\phi = \mathbf{0} \quad \text{for} \quad |\mathbf{x}| = \infty, \qquad (11.4\text{-}1)$$

$$\phi = -x_1\,\mathbf{t}_\Gamma \quad \text{for} \quad x_2=\pm 0,\ |x_1|<a. \qquad (11.4\text{-}2)$$

It is easily shown that the solution given by

$$\mathbf{u} = \mathrm{Re}\left\{\mathbf{A}\langle f_o(z_*)\rangle\mathbf{B}^{-1}\right\}\mathbf{t}_\Gamma, \quad \phi = \mathrm{Re}\left\{\mathbf{B}\langle f_o(z_*)\rangle\mathbf{B}^{-1}\right\}\mathbf{t}_\Gamma, \qquad (11.4\text{-}3)$$

satisfies (11.4-1) and (11.4-2) where

$$f_o(z) = \sqrt{z^2 - a^2} - z. \qquad (11.4\text{-}4)$$

The solution for $\mathbf{u}$ and ϕ given in (11.4-3) can be expressed in terms of real quantities. For ϕ, it has the expression after using $(7.9\text{-}10)_3$ and (7.9-15)

$$\phi = \mathrm{Re}\left\{\sqrt{[x_1\mathbf{I} + x_2(\mathbf{G}_1 + i\mathbf{G}_3)]^2 - a^2\mathbf{I}}\right\}\mathbf{t}_\Gamma - (x_1\mathbf{I} + x_2\mathbf{G}_1)\mathbf{t}_\Gamma \tag{11.4-5}$$

in which

$$\mathbf{G}_1 = \mathbf{G}_1(0), \quad \mathbf{G}_3 = \mathbf{G}_3(0) \tag{11.4-6}$$

and $\mathbf{G}_1(\theta)$, $\mathbf{G}_3(\theta)$ are defined in (11.2-9). The representation (11.4-5) is interesting academically. It is not very useful in computing ϕ because in computing the square root of a matrix (Section 1.9) we still have to introduce the eigenvalues of the matrix which are in this case complex.

With the use of (11.1-2), (11.1-3), (11.3-7), and the identities

$$\mathbf{B}\langle p_*\rangle\mathbf{B}^{-1} = \mathbf{G}_1 + i\mathbf{G}_3 \tag{11.4-7}$$

taken from (6.9-7), it can be shown that

$$\begin{aligned} \mathbf{u} &= \pm\left\{|x_1| - \sqrt{x_1^2 - a^2}\right\}\mathbf{S}\mathbf{L}^{-1}\mathbf{t}_\Gamma, \\ \mathbf{t}_2 &= \left\{\frac{|x_1|}{\sqrt{x_1^2 - a^2}} - 1\right\}\mathbf{t}_\Gamma, \\ \mathbf{t}_1 &= -\left\{\frac{|x_1|}{\sqrt{x_1^2 - a^2}} - 1\right\}\mathbf{G}_1\mathbf{t}_\Gamma, \end{aligned} \tag{11.4-8}$$

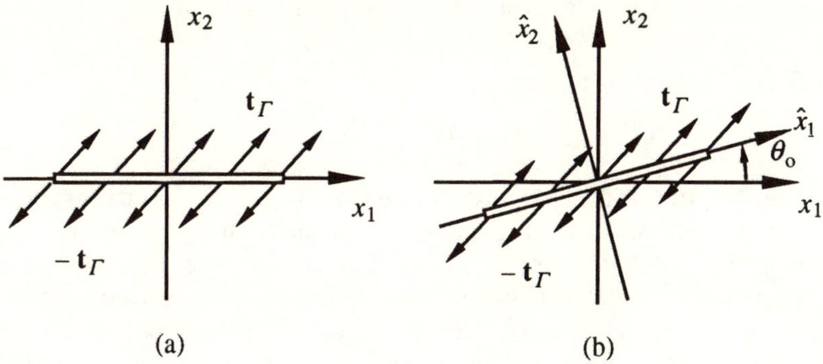

Fig. 11.4 The crack surface is subjected to a uniform traction $\mathbf{t}_\Gamma$ on the upper surface and $-\mathbf{t}_\Gamma$ on the lower surface. The displacement and the surface traction along the x_1-axis in (a) are identical to those along the $\hat{x}_1$-axis in (b). They are independent of θ_o, the orientation of the crack.

for $x_2=0$, $\pm x_1>a$ and

$$\mathbf{u} = \left\{x_1\mathbf{S}\mathbf{L}^{-1} \pm \sqrt{a^2 - x_1^2}\,\mathbf{L}^{-1}\right\}\mathbf{t}_\Gamma,$$

$$\mathbf{t}_2 = -\mathbf{t}_\Gamma, \tag{11.4-9}$$

$$\mathbf{t}_1 = \left\{\mathbf{G}_1 \mp \frac{x_1}{\sqrt{a^2 - x_1^2}}\mathbf{G}_3\right\}\mathbf{t}_\Gamma,$$

for $x_2=\pm 0$, $|x_1|<a$. Equations (11.4-8) tell us that $\mathbf{u}$, $\mathbf{t}_2$, and $\mathbf{t}_1$ along the x_1-axis outside the crack are monotonic in x_1. $\mathbf{t}_2$ is independent of material property (Barnett and Asaro, 1972) and is in the direction of the applied traction $\mathbf{t}_\Gamma$. The displacement of the crack surface is not monotonic, but possesses a single maximum. The crack opening displacement $\Delta\mathbf{u}$ is, by $(11.4\text{-}9)_1$,

$$\Delta\mathbf{u}(x_1) = \mathbf{u}(x_1,+0) - \mathbf{u}(x_1,-0) = 2\sqrt{a^2 - x_1^2}\,\mathbf{L}^{-1}\mathbf{t}_\Gamma.$$

It is in general not in the direction of $\mathbf{t}_\Gamma$.

The stress intensity factor $\mathbf{k}$ at the crack tips $|x_1|=a$ is

$$\mathbf{k} = \lim_{|x_1|\to a} \sqrt{2\pi(|x_1| - a)}\,\mathbf{t}_2$$

in which $\mathbf{t}_2$ is given by $(11.4\text{-}8)_2$. Hence

$$\mathbf{k} = \sqrt{\pi a}\,\mathbf{t}_\Gamma.$$

The *strain energy release rate* G is (Griffith, 1920; Irwin, 1957)

$$G = \frac{\partial}{\partial a}\int_{-a}^{a} \tfrac{1}{2}\mathbf{t}_\Gamma \cdot \Delta\mathbf{u}(x_1)dx_1$$

which can be shown to be

$$G = \mathbf{k}^T\mathbf{L}^{-1}\mathbf{k}.$$

From the fact that $\mathbf{S}$, $\mathbf{L}$ and $\mathbf{t}_\Gamma$ are tensors under rotations of the coordinate system about the x_3-axis, it can be deduced from (11.4-8) and (11.4-9) that $\mathbf{u}$ and $\mathbf{t}_2$ along the x_1-axis are independent of the orientation of the crack. If we rotate the x_1-axis together with the crack an angle θ_o, Fig. 11.4(b), the displacement $\mathbf{u}$ and the surface traction $\mathbf{t}_2$ on the new $\hat{x}_1$-axis remain the same (Qu and Li, 1991). It should be pointed out that $\mathbf{G}_1$ and $\mathbf{G}_3$ are not tensors although $\mathbf{G}_1(\theta)$ and $\mathbf{G}_3(\theta)$ are. They are similar to $\mathbf{N}_1$ and $\mathbf{N}_3$ in that $\mathbf{N}_1(\theta)$ and $\mathbf{N}_3(\theta)$ are tensors but not $\mathbf{N}_1$ and $\mathbf{N}_3$ (Section 7.5). Thus $\mathbf{t}_1$ in (11.4-8) and (11.4-9) are not invariant with θ_o. Other interesting results that can be deduced from (11.4-8) and (11.4-9) are listed below (Ting, 1991).

(i) As we have pointed out, the surface traction $\mathbf{t}_2$ outside the crack is independent of material property and is in the direction of the applied traction $\mathbf{t}_\Gamma$.

(ii) If σ_{22}^∞ is the only loading at infinity,

$$\mathbf{t}_\Gamma = \sigma_{22}^\infty \mathbf{m}(0), \quad \mathbf{m}^T(0) = (0,\ 1,\ 0).$$

The loading is symmetric with respect to the x_1- and x_2- axes. The displacement of the crack surface obtained from (11.4-9) is

$$u_1(x_1,0) = x_1(\mathbf{SL}^{-1})_{12}\,\sigma_{22}^\infty \pm \sqrt{a^2 - x_1^2}\,(\mathbf{L}^{-1})_{12}\,\sigma_{22}^\infty,$$

$$u_2(x_1,0) = \pm\sqrt{a^2 - x_1^2}\,(\mathbf{L}^{-1})_{22}\,\sigma_{22}^\infty.$$

Unless $(\mathbf{L}^{-1})_{12}=0$ the crack opening is not symmetric with the x_1- or the x_2- axis, Fig. 11.5. However,

$$u_2(x_1,0) = 0 \quad \text{at } |x_1| = a,$$

implying that the crack axis is not rotated. This is rather unexpected. Moreover, by $(11.4\text{-}8)_1$ and noting that $\mathbf{SL}^{-1}$ is skew-symmetric,

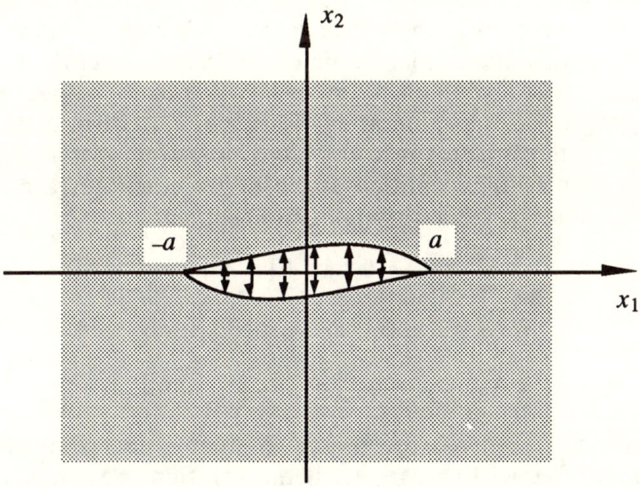

Fig. 11.5 For the uniform loading $-\sigma_{22}^\infty$ at the crack surface the crack opening is in general not symmetric with the x_1- or x_2–axis for anisotropic materials. However, the u_2 component of the displacement vanishes for all points on the x_1-axis outside the crack. This means that the crack is not rotated, and that the x_1-axis outside the crack remains a straight line.

$$u_2(x_1,0)=0 \quad \text{for } |x_1|\geq a.$$

The x_1-axis remains a straight line after application of the load σ_{22}^{∞}. In general, if

$$\mathbf{S}\mathbf{L}^{-1}\mathbf{t}_{\Gamma}=\mathbf{0},$$

i.e., if $\mathbf{t}_{\Gamma}$ is in the direction of the axial vector of $\mathbf{S}\mathbf{L}^{-1}$, u_1, u_2, and u_3 vanish for $x_2=0$, $|x_1|>a$.

(iii) The hoop stress vector $\mathbf{t}_1$ on the crack surface as given by $(11.4\text{-}9)_3$ depends on x_1. It becomes infinity at $|x_1| = a$. If $\mathbf{t}_{\Gamma}$ is proportional to $\mathbf{Lm}(0)$,

$$\mathbf{G}_3\mathbf{t}_{\Gamma}=\mathbf{0} \quad \text{and} \quad \mathbf{t}_1=\mathbf{G}_1\mathbf{t}_{\Gamma}.$$

$\mathbf{t}_1$ is then independent of x_1 and remains bounded at $|x_1|=a$. For isotropic materials $\mathbf{L}$ is a diagonal matrix so that $\mathbf{t}_{\Gamma}$ would be in the direction of $\mathbf{m}(0)$, which is perpendicular to the crack.

The properties (i), (ii), (iii) listed above also apply to the two-force loading shown in Fig. 11.3(b) and to non uniform tractions applied at the crack surface if the statement in (ii) regarding the symmetry about the x_2-axis is deleted.

11.5 Non-Oscillatory Solution for an Interfacial Crack

In this section we study the same problem considered in the previous section except that the crack is now located on the interface between two dissimilar anisotropic elastic media. Let the half-space $x_2>0$ and $x_2<0$ be occupied by materials 1 and 2, respectively. A crack of length $2a$ is located at $x_2=0$, $|x_1|<a$. A uniform traction $\mathbf{t}_{\Gamma}$ is applied at the upper surface of the crack on material 1 and a uniform traction $-\mathbf{t}_{\Gamma}$ is applied at the lower crack surface on material 2. The boundary conditions of the problem are

$$\boldsymbol{\phi}_1 = \boldsymbol{\phi}_2 = \mathbf{0} \quad \text{at } |\mathbf{x}| = \infty, \tag{11.5-1}$$

$$\boldsymbol{\phi}_1 = \boldsymbol{\phi}_2 = -x_1\,\mathbf{t}_{\Gamma} \quad \text{at } x_2=\pm 0,\ |x_1|<a, \tag{11.5-2}$$

$$\mathbf{u}_1 = \mathbf{u}_2, \quad \boldsymbol{\phi}_1 = \boldsymbol{\phi}_2 \quad \text{at } x_2=0,\ |x_1|>a. \tag{11.5-3}$$

The subscripts 1 and 2 for $\mathbf{u}$ and $\boldsymbol{\phi}$ denote materials 1 and 2, respectively. We will investigate if the solution presented in (11.4-3) for homogeneous materials applies to materials 1 and 2. It was shown there that the displacement along the x_1-axis is monotonic outside the crack and possesses a single maximum at the crack. The displacement is not oscillatory.

Let

$$\mathbf{u}_1 = \mathrm{Re}\left\{\mathbf{A}_1\left\langle f_o(z_*^{(1)})\right\rangle\mathbf{B}_1^{-1}\right\}\mathbf{t}_\Gamma,$$
$$\boldsymbol{\phi}_1 = \mathrm{Re}\left\{\mathbf{B}_1\left\langle f_o(z_*^{(1)})\right\rangle\mathbf{B}_1^{-1}\right\}\mathbf{t}_\Gamma, \tag{11.5-4}$$

for material 1 in $x_1>0$ and

$$\mathbf{u}_2 = \mathrm{Re}\left\{\mathbf{A}_2\left\langle f_o(z_*^{(2)})\right\rangle\mathbf{B}_2^{-1}\right\}\mathbf{t}_\Gamma,$$
$$\boldsymbol{\phi}_2 = \mathrm{Re}\left\{\mathbf{B}_2\left\langle f_o(z_*^{(2)})\right\rangle\mathbf{B}_2^{-1}\right\}\mathbf{t}_\Gamma, \tag{11.5-5}$$

for material 2 in $x_2<0$. The superscripts (1) and (2) for z_α refer to materials 1 and 2, respectively. It is readily shown that conditions (11.5-1)–(11.5-3) are satisfied except $(11.5\text{-}3)_1$ which yields

$$(\mathbf{A}_1\mathbf{B}_1^{-1} + \overline{\mathbf{A}}_1\overline{\mathbf{B}}_1^{-1})\mathbf{t}_\Gamma = (\mathbf{A}_2\mathbf{B}_2^{-1} + \overline{\mathbf{A}}_2\overline{\mathbf{B}}_2^{-1})\mathbf{t}_\Gamma.$$

From the identity (11.1-2) this is rewritten as

$$\mathbf{W}\mathbf{t}_\Gamma = \mathbf{0}, \tag{11.5-6}$$

$$\mathbf{W} = \mathbf{S}_1\mathbf{L}_1^{-1} - \mathbf{S}_2\mathbf{L}_2^{-1}. \tag{11.5-7}$$

The tensor $\mathbf{W}$ is skew-symmetric. There are two possibilities for (11.5-6) to be satisfied. If $\mathbf{W}=\mathbf{0}$, (11.5-6) is satisfied for any applied traction $\mathbf{t}_\Gamma$. If $\mathbf{W}\neq\mathbf{0}$, (11.5-6) can still be satisfied if $\mathbf{t}_\Gamma$ is in the direction of the null vector of $\mathbf{W}$.

The two homogeneous materials are said to be *mismatched* when $\mathbf{W}\neq\mathbf{0}$. It is seen that the non-oscillatory solution for the homogeneous medium with a crack applies to the interfacial crack in a dissimilar bimaterial when the two materials are not mismatched, or when they are mismatched but the applied traction $\mathbf{t}_\Gamma$ is in the direction of the null vector of $\mathbf{W}$.

When (11.5-6) holds the displacement $\mathbf{u}$, the surface traction $\mathbf{t}_2$, and the hoop stress vector $\mathbf{t}_1$ along the x_1-axis are given by (11.4-8) and (11.4-9). In particular,

$$\mathbf{t}_2 = \left\{\frac{|x_1|}{\sqrt{x_1^2 - a^2}} - 1\right\}\mathbf{t}_\Gamma \quad \text{for } x_2=0,\ |x_1| > a. \tag{11.5-8}$$

The traction $\mathbf{t}_2$ is independent of material property. The crack opening displacement $\Delta\mathbf{u}$ obtained from $(11.4\text{-}8)_1$ and $(11.4\text{-}9)_1$ can be shown to be

$$\Delta\mathbf{u} = \mathbf{u}(x_1,+0) - \mathbf{u}(x_1,-0) = \sqrt{a^2 - x_1^2}\,\mathbf{D}\mathbf{t}_\Gamma, \tag{11.5-9}$$

$$\mathbf{D} = \mathbf{L}_1^{-1} + \mathbf{L}_2^{-1}. \tag{11.5-10}$$

The tensor **D** is positive definite.

When $\mathbf{W} \neq \mathbf{0}$ and if $\mathbf{t}_\Gamma$ is not in the direction of the null vector of **W**, one could decompose $\mathbf{t}_\Gamma$ into two components. One of which is along the null vector of **W**. The solution associated with this component is given by (11.5-4) and (11.5-5). In the next section we define the other component and the corresponding solution which is characterized by an oscillatory displacement near the crack tips.

11.6 Oscillatory Solution for an Interfacial Crack

The problem of an interfacial crack in isotropic bimaterials was first studied by Williams (1959) and Erdogan (1963) for a semi-infinite crack and by England (1965), Rice and Sih (1965), and Erdogan (1965) for a finite crack. For an interfacial crack in anisotropic bimaterials Gotoh (1967) studied the problem under the condition of plane stress which applied to monoclinic materials with the symmetry plane at $x_3=0$. The problem of a finite interfacial crack in general anisotropic bimaterials was first investigated by Clements (1971) and Willis (1971). The research on composite materials in recent years has rekindled interests on the subject and many work have appeared such as Bassani and Qu (1989), Qu and Bassani (1989, 1993), Ting (1986, 1990b, 1992c), Tewary et al. (1989b), Suo (1990), K. C. Wu (1990, 1991a,b, 1992), Qu and Li (1991), Hwu (1991, 1992b, 1993), Ni and Nemat-Nasser (1991, 1992), and Gao et al. (1992).

Assuming that the stress singularities near the tip of an interfacial crack is proportional to r^δ where r is the radial distance from the crack tip and δ is a constant signifying the order of stress singularity, it was shown in Section 9.6 that

$$\delta = -\tfrac{1}{2} \quad \text{and} \quad -\tfrac{1}{2} \mp i\gamma, \tag{11.6-1}$$

$$\gamma = \tfrac{1}{\pi}\tanh^{-1}\beta = \tfrac{1}{2\pi}\ln\frac{1+\beta}{1-\beta}, \tag{11.6-2}$$

where

$$0 \le \beta = \left\{-\tfrac{1}{2}\mathrm{tr}(\breve{\mathbf{S}}^2)\right\}^{1/2} < 1, \tag{11.6-3}$$

$$\breve{\mathbf{S}} = \mathbf{D}^{-1}\mathbf{W}. \tag{11.6-4}$$

In the above **D** and **W** are defined in (11.5-10) and (11.5-7). The tensor $\breve{\mathbf{S}}$ is the product of a symmetric positive definite tensor $\mathbf{D}^{-1}$ and a skew-symmetric tensor **W**. It was shown in (8.8-41) that $\breve{\mathbf{S}}$, $\mathbf{D}^{-1}$, and $\tilde{\mathbf{L}}^{-1}$ have the similar mathematical properties as have the three Barnett–Lothe tensors **S**, **H**, and **L**. In particular, $\breve{\mathbf{S}}$ has properties similar to those of **S**. The eigenvalues of $\breve{\mathbf{S}}$ are $-i\beta$, $i\beta$, and 0 where β is given by (11.6-3). The inequalities in (11.6-3) are taken from (6.6-5) except that β can vanish here. If the associated eigenvectors are denoted by $\mathbf{d}$, $\bar{\mathbf{d}}$, and $\mathbf{d}_o$, we have

$$\breve{\mathbf{S}}\mathbf{d} = -i\beta\mathbf{d}, \quad \breve{\mathbf{S}}\bar{\mathbf{d}} = i\beta\bar{\mathbf{d}}, \quad \breve{\mathbf{S}}\mathbf{d}_o = \mathbf{0}. \tag{11.6-5}$$

$\mathbf{d}_o$ is the right null vector of $\breve{\mathbf{S}}$ and is a real vector. $\mathbf{d}$ on the other hand is a complex vector or a bivector (Gibbs, 1961; Boulanger and Hayes, 1991). Writing

$$\mathbf{d} = \mathbf{d}_1 + i\mathbf{d}_2 \tag{11.6-6}$$

where $\mathbf{d}_1$ and $\mathbf{d}_2$ are real vectors and equating the real and imaginary parts of $(11.6\text{-}5)_{1,2}$ we have

$$\breve{\mathbf{S}}\mathbf{d}_1 = \beta\mathbf{d}_2, \quad \breve{\mathbf{S}}\mathbf{d}_2 = -\beta\mathbf{d}_1. \tag{11.6-7}$$

Hence

$$\breve{\mathbf{S}}^2\mathbf{d}_i = -\beta^2\mathbf{d}_i \quad (i = 1,2). \tag{11.6-8}$$

The right null vector $\mathbf{d}_o$ is unique up to an arbitrary real constant multiplier. The eigenvectors $\mathbf{d}$ or $\bar{\mathbf{d}}$, on the other hand, is unique up to an arbitrary complex constant multiplier. Regardless of the complex constant multiplier, the real and imaginary parts of the new eigenvector $\mathbf{d}$ or $\bar{\mathbf{d}}$ always lie on the plane spanned by $\mathbf{d}_1$ and $\mathbf{d}_2$. This plane is the *right eigenplane* of $\breve{\mathbf{S}}$. In fact if $e^{i\psi}$ is the complex multiplier, the real and imaginary parts of $e^{i\psi}\mathbf{d}$ form a pair of conjugate radii of an ellipse (Section 1.8).

Consider now the left eigenvectors of $\breve{\mathbf{S}}$. From (11.6-4) and the skew-symmetry property of **W** we have

$$\mathbf{D}\breve{\mathbf{S}} = \mathbf{W} = -\breve{\mathbf{S}}^T\mathbf{D}.$$

When (11.6-5) are multiplied by **D** one obtains

$$\breve{\mathbf{S}}^T(\mathbf{D}\mathbf{d}) = i\beta(\mathbf{D}\mathbf{d}), \quad \breve{\mathbf{S}}^T(\mathbf{D}\bar{\mathbf{d}}) = -\beta(\mathbf{D}\bar{\mathbf{d}}), \quad \breve{\mathbf{S}}^T(\mathbf{D}\mathbf{d}_o) = \mathbf{0}. \tag{11.6-9}$$

Hence $\mathbf{D}\mathbf{d}$, $\mathbf{D}\bar{\mathbf{d}}$, and $\mathbf{D}\mathbf{d}_o$ are the left eigenvectors associated with the eigenvalues $i\beta$, $-i\beta$, and 0, respectively. $\mathbf{D}\mathbf{d}_o$ is the left null vector of $\breve{\mathbf{S}}$, and the plane spanned by $\mathbf{D}\mathbf{d}_1$ and $\mathbf{D}\mathbf{d}_2$ are the left eigenplane of $\breve{\mathbf{S}}$. The left and right eigenvectors associated with different eigenvalues are orthogonal to each other, i.e.,

$$\mathbf{d}^T\mathbf{D}\mathbf{d} = 0 = \bar{\mathbf{d}}^T\mathbf{D}\bar{\mathbf{d}}, \quad \mathbf{d}^T\mathbf{D}\mathbf{d}_o = 0 = \bar{\mathbf{d}}^T\mathbf{D}\mathbf{d}_o. \tag{11.6-10}$$

The last equation tells us that the right null vector $\mathbf{d}_o$ is normal to the left eigenplane, and the left null vector $\mathbf{D}\mathbf{d}_o$ is normal to the right eigenplane.

From (11.6-4) it is clear that $\breve{\mathbf{S}}=\mathbf{0}$ if and only if $\mathbf{W}=\mathbf{0}$. Analogous to the derivation of the equation following (6.6-7), it can be shown that $\beta=0$ if and only if $\mathbf{W}=\mathbf{0}$ (Qu and Bassani, 1989). In conjunction with (11.6-2) we conclude that $\mathbf{W}$, $\breve{\mathbf{S}}$, β, and γ are either all zero or all nonzero. From (11.6-4) and $(11.6\text{-}5)_3$, $\mathbf{d}_o$ is the right null vector of $\breve{\mathbf{S}}$ as well as the null vector of $\mathbf{W}$.

If $\mathbf{W}\neq\mathbf{0}$ and if the applied traction $\mathbf{t}_\Gamma$ is not in the direction of the right null vector $\mathbf{d}_o$ of $\breve{\mathbf{S}}$, we decompose $\mathbf{t}_\Gamma$ into two components

$$\mathbf{t}_\Gamma = \mathbf{t}_o + \mathbf{t}_\gamma \tag{11.6-11}$$

where $\mathbf{t}_o$ is in the direction of $\mathbf{d}_o$ while $\mathbf{t}_\gamma$ is on the right eigenplane. From $(11.6\text{-}5)_3$ and (11.6-8),

$$\breve{\mathbf{S}}\mathbf{t}_o = \mathbf{0}, \quad \breve{\mathbf{S}}^2\mathbf{t}_\gamma = -\beta^2\mathbf{t}_\gamma. \tag{11.6-12}$$

Multiplying (11.6-11) by $\breve{\mathbf{S}}^2$ and using (11.6-12) leads to

$$\mathbf{t}_\gamma = -\beta^{-2}\breve{\mathbf{S}}^2\mathbf{t}_\Gamma, \quad \mathbf{t}_o = \mathbf{t}_\Gamma + \beta^{-2}\breve{\mathbf{S}}^2\mathbf{t}_\Gamma. \tag{11.6-13}$$

Equation (11.6-13) provide an explicit expression for $\mathbf{t}_o$ and $\mathbf{t}_\gamma$ in terms of the applied traction $\mathbf{t}_\Gamma$. The solution associated with $\mathbf{t}_o$ is given in the previous section if $\mathbf{t}_\Gamma$ there is replaced by $\mathbf{t}_o$. $\mathbf{t}_o$ is in the direction of the right null vector of $\breve{\mathbf{S}}$. By (11.5-8) and (11.5-9) the surface traction $\mathbf{t}_2$ on the x_1-axis is also in the direction of the right null vector of $\breve{\mathbf{S}}$ while the crack opening displacement $\Delta\mathbf{u}$ is in the direction of the left null vector of $\breve{\mathbf{S}}$.

It remains to study the solution for the applied traction $\mathbf{t}_\gamma$ which lies on the right eigenplane of $\breve{\mathbf{S}}$. Of the three stress singularities at the crack tips listed in (11.6-1) the singularity $\delta=-\frac{1}{2}$ is represented by the non-oscillatory solution (11.5-4) and (11.5-5) in which the function $f_o(z)$ is defined in (11.4-4). The other two singularities in (11.6-1) can be taken care of by considering the function

$$f(z,\gamma) = (z-a)^{\frac{1}{2}+i\gamma}(z+a)^{\frac{1}{2}-i\gamma} - z \tag{11.6-14}$$

which has the property

$$\bar{f}(z,\gamma) = f(z,-\gamma).$$

Along the x_1-axis,

$$f(x_1,\gamma) = \begin{cases} \pm\sqrt{x_1^2-a^2}\,e^{i\gamma X} - x_1 & \text{for } x_2=0, \ \pm x_1 > a, \\ \pm i\sqrt{a^2-x_1^2}\,e^{\mp\gamma\pi}e^{i\gamma X} - x_1 & \text{for } x_2=\pm 0, \ |x_1|<a, \end{cases} \tag{11.6-15}$$

where

$$X = \ln\left|\frac{x_1 - a}{x_1 + a}\right|.$$

Observing the factor $e^{\mp\gamma\pi}$ in (11.6-15), consider the solution

$$\mathbf{u}_1 = \mathrm{Re}\left\{e^{\gamma\pi}\mathbf{A}_1\left\langle f(z^{(1)},\gamma)\right\rangle\mathbf{B}_1^{-1}\mathbf{d} + e^{-\gamma\pi}\mathbf{A}_1\left\langle \bar{f}(z^{(1)},\gamma)\right\rangle\mathbf{B}_1^{-1}\bar{\mathbf{d}}\right\},$$
$$\boldsymbol{\phi}_1 = \mathrm{Re}\left\{e^{\gamma\pi}\mathbf{B}_1\left\langle f(z^{(1)},\gamma)\right\rangle\mathbf{B}_1^{-1}\mathbf{d} + e^{-\gamma\pi}\mathbf{B}_1\left\langle \bar{f}(z^{(1)},\gamma)\right\rangle\mathbf{B}_1^{-1}\bar{\mathbf{d}}\right\}, \tag{11.6-16}$$

for material 1 in $x_2>0$ and

$$\mathbf{u}_2 = \mathrm{Re}\left\{e^{-\gamma\pi}\mathbf{A}_2\left\langle f(z^{(2)},\gamma)\right\rangle\mathbf{B}_2^{-1}\mathbf{d} + e^{\gamma\pi}\mathbf{A}_2\left\langle \bar{f}(z^{(2)},\gamma)\right\rangle\mathbf{B}_2^{-1}\bar{\mathbf{d}}\right\},$$
$$\boldsymbol{\phi}_2 = \mathrm{Re}\left\{e^{-\gamma\pi}\mathbf{B}_2\left\langle f(z^{(2)},\gamma)\right\rangle\mathbf{B}_2^{-1}\mathbf{d} + e^{\gamma\pi}\mathbf{B}_2\left\langle \bar{f}(z^{(2)},\gamma)\right\rangle\mathbf{B}_2^{-1}\bar{\mathbf{d}}\right\}, \tag{11.6-17}$$

for material 2 in $x_2<0$. The solution in material 1 is identical to that in material 2 if π is replaced by $-\pi$ (Suo, 1990). It can be shown that the solution meets the boundary and interface conditions (11.5-1)–(11.5-3) if $\mathbf{t}_\Gamma$ there is replaced by $\mathbf{t}_\gamma$ and if the vector $\mathbf{d}$ defined in $(11.6\text{-}5)_1$ satisfies (Ting, 1992c)

$$\mathbf{d} + \bar{\mathbf{d}} = \sqrt{1-\beta^2}\,\mathbf{t}_\gamma. \tag{11.6-18}$$

To determine $\mathbf{d}$, (11.6-18) is multiplied by $\breve{\mathbf{S}}$ which yields

$$-i\beta(\mathbf{d} - \bar{\mathbf{d}}) = \sqrt{1-\beta^2}\,\breve{\mathbf{S}}\mathbf{t}_\gamma$$

so that

$$\mathbf{d} = \frac{\sqrt{1-\beta^2}}{2\beta}(\beta\mathbf{t}_\gamma + i\breve{\mathbf{S}}\mathbf{t}_\gamma). \tag{11.6-19}$$

Real form expressions for the displacement $\mathbf{u}$, the surface traction $\mathbf{t}_2$, and the hoop stress vector $\mathbf{t}_1$ are deduced from (11.6-16) and (11.6-17) by employing (11.1-2), (11.3-7), and (11.4-7). For $x_2=0$, $\pm x_1>a$,

$$\mathbf{u}_1 = 2(\cosh\gamma\pi)\mathrm{Re}\left\{\left(x_1 \mp e^{i\gamma X}\sqrt{x_1^2-a^2}\right)\left(\mathbf{S}_1\mathbf{L}_1^{-1} + i\beta\mathbf{L}_1^{-1}\right)\mathbf{d}\right\},$$

$$\mathbf{u}_2 = 2(\cosh\gamma\pi)\mathrm{Re}\left\{\left(x_1 \mp e^{i\gamma X}\sqrt{x_1^2-a^2}\right)\left(\mathbf{S}_2\mathbf{L}_2^{-1} - i\beta\mathbf{L}_2^{-1}\right)\mathbf{d}\right\},$$

$$\mathbf{t}_2 = 2(\cosh\gamma\pi)\mathrm{Re}\left\{[\pm\eta(x_1)e^{i\gamma X} - 1]\mathbf{d}\right\}, \tag{11.6-20}$$

$$\mathbf{t}_1^{(1)} = -2(\cosh\gamma\pi)\mathrm{Re}\left\{[\pm\eta(x_1)e^{i\gamma X} - 1][\mathbf{G}_1^{(1)} + i\beta\mathbf{G}_3^{(1)}]\mathbf{d}\right\},$$

$$\mathbf{t}_1^{(2)} = -2(\cosh\gamma\pi)\mathrm{Re}\left\{[\pm\eta(x_1)e^{i\gamma X} - 1][\mathbf{G}_1^{(2)} - i\beta\mathbf{G}_3^{(2)}]\mathbf{d}\right\},$$

where

$$\cosh\gamma\pi = \frac{1}{\sqrt{1-\beta^2}}, \quad \eta(x_1) = \frac{x_1 + 2i\gamma a}{\sqrt{x_1^2 - a^2}}.$$

For $x_2 = \pm 0$, $|x_1| < a$,

$$\mathbf{u}_1 = 2(\cosh\gamma\pi)x_1\,\mathrm{Re}\left\{\left(\mathbf{S}_1\mathbf{L}_1^{-1} + i\beta\mathbf{L}_1^{-1}\right)\mathbf{d}\right\} + 2\sqrt{a^2 - x_1^2}\,\mathbf{L}_1^{-1}\,\mathrm{Re}\left\{e^{i\gamma X}\mathbf{d}\right\},$$

$$\mathbf{u}_2 = 2(\cosh\gamma\pi)x_1\,\mathrm{Re}\left\{\left(\mathbf{S}_2\mathbf{L}_2^{-1} - i\beta\mathbf{L}_2^{-1}\right)\mathbf{d}\right\} - 2\sqrt{a^2 - x_1^2}\,\mathbf{L}_2^{-1}\,\mathrm{Re}\left\{e^{i\gamma X}\mathbf{d}\right\},$$

$$\mathbf{t}_2 = -\mathbf{t}_\Gamma, \tag{11.6-21}$$

$$\mathbf{t}_1^{(1)} = 2(\cosh\gamma\pi)\mathrm{Re}\left\{[\mathbf{G}_1^{(1)} + i\beta\mathbf{G}_3^{(1)}]\mathbf{d}\right\} - \frac{2}{\sqrt{a^2 - x_1^2}}\mathbf{G}_3^{(1)}\,\mathrm{Re}\left\{(x_1 + 2i\gamma a)e^{i\gamma X}\mathbf{d}\right\},$$

$$\mathbf{t}_1^{(2)} = 2(\cosh\gamma\pi)\mathrm{Re}\left\{[\mathbf{G}_1^{(2)} - i\beta\mathbf{G}_3^{(2)}]\mathbf{d}\right\} + \frac{2}{\sqrt{a^2 - x_1^2}}\mathbf{G}_3^{(2)}\,\mathrm{Re}\left\{(x_1 + 2i\gamma a)e^{i\gamma X}\mathbf{d}\right\}.$$

To distinguish quantities related to materials 1 and 2, the subscripts 1 and 2 are used for $\mathbf{u}$, $\mathbf{S}$, $\mathbf{L}$ while the superscripts (1) and (2) are employed for $\mathbf{t}_1$, $\mathbf{t}_2$, $\mathbf{G}_1$, and $\mathbf{G}_3$. It is easily verified that $\mathbf{u}_1 = \mathbf{u}_2$ in (11.6-20). $\mathbf{t}_2$ in (11.6-20) is $\mathbf{t}_2^{(1)}$ or $\mathbf{t}_2^{(2)}$. The solutions reduce to those given in (11.4-8) and (11.4-9) when β (and hence γ) vanishes.

The traction $\mathbf{t}_2$ along the interface $x_1 > a$ given by (11.6-20) depends on $\mathbf{d}$. Consequently $\mathbf{t}_2$ lies on the right eigenplane of $\breve{\mathbf{S}}$ as does the applied traction $\mathbf{t}_\gamma$ on the crack surface. As x_1 varies from ∞ to a or from $-\infty$ to $-a$, $\mathbf{t}_2$ rotates on the right eigenplane with increasing frequency and amplitude. The crack opening displacement $\Delta\mathbf{u}$ is obtained by subtracting $\mathbf{u}_2$ from $\mathbf{u}_1$ in (11.6-21). The first terms cancel each other due to $(11.6\text{-}5)_1$ and

$$\Delta\mathbf{u} = 2\sqrt{a^2 - x_1^2}\,\mathrm{Re}(e^{i\gamma X}\mathbf{D}\mathbf{d}). \tag{11.6-22}$$

Therefore the vector $\Delta\mathbf{u}$ is on $\mathbf{D}\mathbf{d}$, the left eigenplane of $\breve{\mathbf{S}}$. As x_1 varies from 0 to a or from 0 to $-a$, $\Delta\mathbf{u}$ rotates on the left eigenplane with increasing frequency and diminishing amplitude (Ting, 1992c). The displacement is oscillatory.

11.7 Multiple Interfacial Cracks with Nonuniform Tractions

Multiple interfacial cracks in general anisotropic elastic bimaterials have been studied by Suo (1990), Hwu (1991), and Ting (1992c). The derivation of the latter is given below.

The decomposition principle for a single interfacial crack with constant traction $\mathbf{t}_\Gamma$ remains valid for multiple interfacial cracks with variable traction $\mathbf{t}_\Gamma(x_1)$ with one minor modification. The component $\mathbf{t}_\gamma(x_1)$ of $\mathbf{t}_\Gamma(x_1)$ defined in (11.6-11) may not remain in the same direction for a different x_1. It is therefore necessary to select a fixed right eigenvector $\mathbf{d}=\mathbf{d}_1+i\mathbf{d}_2$ on the right eigenplane and decompose $\mathbf{t}_\Gamma(x_1)$ in the form

$$\mathbf{t}_\Gamma(x_1)=\mathbf{t}_o(x_1)+\mathbf{t}_\gamma(x_1)=t_o(x_1)\mathbf{d}_o+t_\gamma(x_1)\mathbf{d}+\bar{t}_\gamma(x_1)\bar{\mathbf{d}}. \tag{11.7-1}$$

In the above $\mathbf{t}_o(x_1)$ and $\mathbf{t}_\gamma(x_1)$ are real vectors, $t_o(x_1)$ is a real scalar, and $t_\gamma(x_1)$ and $\bar{t}_\gamma(x_1)$ are a pair of complex conjugate scalars. The vectors $\mathbf{d}$, $\bar{\mathbf{d}}$, and $\mathbf{d}_o$ are the eigenvectors of $\breve{\mathbf{S}}$ defined in (11.6-5). It is not necessary to solve the eigenrelation (11.6-5) to determine $\mathbf{d}$, $\bar{\mathbf{d}}$, and $\mathbf{d}_o$. They can be determined explicitly in terms of any real vectors by following the derivation of the eigenvectors for $\mathbf{S}$ presented in (6.6-55). Making use of the orthogonality relations (11.6-10), $(11.7\text{-}1)_2$ yields

$$t_\gamma(x_1)=\frac{\bar{\mathbf{d}}^T\mathbf{D}\mathbf{t}_\Gamma(x_1)}{\bar{\mathbf{d}}^T\mathbf{D}\mathbf{d}}=\frac{(\mathbf{d}_1-i\mathbf{d}_2)^T\mathbf{D}\mathbf{t}_\Gamma(x_1)}{\mathbf{d}_1^T\mathbf{D}\mathbf{d}_1+\mathbf{d}_2^T\mathbf{D}\mathbf{d}_2} \tag{11.7-2}$$

and $(11.7\text{-}1)_{1,2}$ furnish $\mathbf{t}_o(x_1)$. The decomposition of $\mathbf{t}_\Gamma(x_1)$ is now complete.

The solution for multiple interfacial cracks in the bimaterial contains two parts. The first part is

$$\begin{aligned}\mathbf{u}_1&=\mathrm{Re}\left\{\mathbf{A}_1\left\langle f_o(z_*^{(1)})\right\rangle\mathbf{B}_1^{-1}\right\}\mathbf{d}_o,\\ \boldsymbol{\phi}_1&=\mathrm{Re}\left\{\mathbf{B}_1\left\langle f_o(z_*^{(1)})\right\rangle\mathbf{B}_1^{-1}\right\}\mathbf{d}_o,\end{aligned} \tag{11.7-3}$$

for material 1 in $x_1>0$ and

$$\begin{aligned}\mathbf{u}_2&=\mathrm{Re}\left\{\mathbf{A}_2\left\langle f_o(z_*^{(2)})\right\rangle\mathbf{B}_2^{-1}\right\}\mathbf{d}_o,\\ \boldsymbol{\phi}_2&=\mathrm{Re}\left\{\mathbf{B}_2\left\langle f_o(z_*^{(2)})\right\rangle\mathbf{B}_2^{-1}\right\}\mathbf{d}_o,\end{aligned} \tag{11.7-4}$$

for material 2 in $x_2<0$. This is identical to (11.5-4) and (11.5-5) with $\mathbf{t}_\Gamma$ replaced by $\mathbf{d}_o$ and $f_o(z)$ is an unknown function to be determined. The second part is given by (11.6-16) and (11.6-17) in which $f(z,\gamma)$ is an unknown function to be determined. Let Γ denote the cracks which are located at

$$x_2 = 0, \quad a_k < x_1 < b_k, \quad k = 1, 2, \ldots, n,$$

$$a_1 < b_1 < a_2 < \cdots < a_n < b_n.$$

The functions $f_o(z)$ and $f(z,\gamma)$ are continuous everywhere except at Γ and vanish at infinity. Satisfaction of the conditions (11.5-1), (11.5-3), and the prescribed tractions at Γ leads to

$$f_o(z) = \bar{f}_o(z) \tag{11.7-5}$$

and

$$\begin{aligned} &g_o^+(x_1) + g_o^-(x_1) = -2t_o(x_1), \\ &e^{\gamma\pi} g^+(x_1,\gamma) + e^{-\gamma\pi} g^-(x_1,\gamma) = -2t_\gamma(x_1), \\ &e^{-\gamma\pi} \bar{g}^+(x_1,\gamma) + e^{\gamma\pi} \bar{g}^-(x_1,\gamma) = -2\bar{t}_\gamma(x_1), \end{aligned} \tag{11.7-6}$$

where

$$g_o(z) = \frac{d}{dz} f_o(z), \quad g(z,\gamma) = \frac{\partial}{\partial z} f(z,\gamma).$$

Equations (11.7-6) are the *Hilbert* problem for which the solutions are (Muskhelishvili, 1953)

$$g_o(z) = -\frac{\chi(z,0)}{\pi i} \int_\Gamma \frac{t_o(\lambda) d\lambda}{\chi^+(\lambda,0)(\lambda - z)} + P_o(z)\chi(z,0),$$

$$g(z,\gamma) = -\frac{\chi(z,\gamma)}{\pi i} \int_\Gamma \frac{e^{-\gamma\pi} t_\gamma(\lambda) d\lambda}{\chi^+(\lambda,\gamma)(\lambda - z)} + P_o(z,\gamma)\chi(z,\gamma),$$

$$\bar{g}(z,\gamma) = -\frac{\bar{\chi}(z,\gamma)}{\pi i} \int_\Gamma \frac{e^{\gamma\pi} \bar{t}_\gamma(\lambda) d\lambda}{\bar{\chi}^+(\lambda,\gamma)(\lambda - z)} + \bar{P}_o(z,\gamma)\bar{\chi}(z,\gamma).$$

In the above

$$\chi(z,\gamma) = \prod_{k=1}^{n} (z - a_k)^{-\frac{1}{2} - i\gamma} (z - b_k)^{-\frac{1}{2} + i\gamma}$$

and $P_o(z)$, $P(z,\gamma)$ are polynomials in z of order less than n. They are determined such that the crack opening at all crack tips vanishes. It can be shown that

$$ie^{\gamma\pi}\chi^+(\lambda,\gamma) \quad \text{and} \quad ie^{-\gamma\pi}\bar{\chi}^+(\lambda,\gamma)$$

are complex conjugates of each other, confirming that the functions

$g(z,\gamma)$ and $\bar{g}(z,\gamma)$ are indeed complex conjugates of each other. Likewise the solution for $g_o(z)$ satisfies the condition (11.7-5).

The traction along the interface is, by substituting (11.7-3), (11.7-4), (11.6-16), and (11.6-17) into $(11.3\text{-}7)_1$,

$$\mathbf{t}_2(x_1) = g_o(x_1)\mathbf{d}_o + 2(\cosh\gamma\pi)\mathrm{Re}\{g(x_1,\gamma)\mathbf{d}\}.$$

The traction from the first term is in the direction of the right null vector of $\breve{\mathbf{S}}$ while that from the second term lies on the right eigenplane of $\breve{\mathbf{S}}$. By writing

$$f_o^+(x_1) = [f_o^+(x_1) - f_o^-(x_1)] + f_o^-(x_1)$$

and similar expressions for $f^+(x_1,\gamma)$ and $\bar{f}^+(x_1,\gamma)$, the crack opening displacement $\Delta\mathbf{u}$ can be shown to be

$$\Delta\mathbf{u} = -\tfrac{1}{2}(\mathbf{W}+i\mathbf{D})\Big\{[f_o^+(x_1) - f_o^-(x_1)]\mathbf{d}_o + e^{\gamma\pi}[f^+(x_1,\gamma)$$

$$-f^-(x_1,\gamma)]\mathbf{d} + e^{-\gamma\pi}[\bar{f}^+(x_1,\gamma) - \bar{f}^-(x_1,\gamma)]\bar{\mathbf{d}}\Big\}. \qquad (11.7\text{-}7)$$

Observing that the complex conjugates of

$$f_o^+(x_1),\quad f^+(x_1,\gamma),\quad f^-(x_1,\gamma)$$

are, respectively,

$$f_o^-(x_1),\quad \bar{f}^-(x_1,\gamma),\quad \bar{f}^+(x_1,\gamma)$$

and setting

$$\mathbf{W}+i\mathbf{D} = \mathbf{D}(\breve{\mathbf{S}}+i\mathbf{I}),$$

equation (11.7-7) can be further simplified by using (11.6-5) as

$$\Delta\mathbf{u} = \mathrm{Im}\{f_o^+(x_1)\}\mathbf{D}\mathbf{d}_o + \sqrt{1-\beta^2}\,\mathrm{Im}\Big\{[f^+(x_1,\gamma) - f^-(x_1,\gamma)]\mathbf{D}\mathbf{d}\Big\}.$$

It is clear that the first term is in the direction of the left null vector of $\breve{\mathbf{S}}$ while the second term lie on the left eigenplane of $\breve{\mathbf{S}}$.

11.8 The Comninou Crack

The oscillatory solutions presented in the last two sections are mathematically correct. However they are physically invalid. The oscillatory displacement of the crack surfaces implies that the upper and lower crack surfaces interpenetrate each other. One may argue that the slit line is a mathematical idealization of a crack. A real crack in the material has a gap between the upper and lower surfaces. Since the amplitude of the oscillatory displacement diminishes as one approaches a crack tip, the upper and lower surfaces may not overlap if there is a gap in the crack. Nevertheless alternate solutions to remedy the problem have been considered. Achenbach et al. (1979) introduced an adhesion zones at the tip of crack. Knowles and Sternberg (1983) considered a neo-Hookian nonlinear elastic material and discovered that the displacements at the crack surface do not overlap. One could also introduce a plastic zone at the crack tip to alleviate the overlapping of the crack surfaces (Dugdale, 1960; Barenblatt, 1962). All these deviate from the assumption of linear elasticity.

The oscillatory displacement near an interfacial crack tip was due to the complex eigenvalue for the interfacial crack. Although the complex eigenvalue for an interfacial crack in isotropic bimaterials was obtained by Williams in 1959, the contradiction that the materials overlap was first pointed out by Malyshev and Salganik (1965) and England (1965). The zones over which the materials overlap are small in comparison with the crack length for tensile loading, but Willis (1971) found that the overlap can cover a large zone for shear loading. The dilemma was resolved by Comninou (1977a,b, 1978). She suggested that the crack does not fully open under an external load. The crack surfaces are in contact at the two ends near the crack tips. The normal stress inside the contact zones is compressive while the displacement at the remaining crack surfaces do not overlap. The problem was formulated in the form of two integral equations which are reduced to a single integral equation and solved numerically. It was shown by Shield (1982) that the solution to the Comninou crack is unique. The problem was also investigated by Atkinson (1982a,b) using a Mellin transform, but was not solved exactly. The contact zone may have friction (Comninou and Dundurs, 1979a). A general discussion of the Comninou crack and several interesting results that are not widely noticed are given by Comninou and Dundurs (1980). The governing integral equation was solved exactly by Gautesen and Dundurs (1987a,b; 1988) and by Dundurs and Gautesen (1988).

The Comninou crack in anisotropic bimaterials has been studied by Wang and Choi (1983), Wu and Hwang (1990) and Ni and Nemat-Nasser (1991, 1992).

It is not surprising that a crack can partially close subjected to an external load. A crack that runs into a compression part of stress field provides an example of partial closure (Bowie and Freese, 1976). Elber (1970) and Budiansky and Hutchinson (1978) showed that the tip of a fatigue crack closes over a part of the load cycle. Comninou and Schmueser (1979) demonstrated that the interfacial crack cannot be fully open for any combination of tension-compression and shear. Comninou and Dundurs (1979b) proved that, depending on the mismatch in the material constants, the tip of an oblique crack that terminates at an interface at an angle may remain closed for all loads. The results of Comninou and Schmueser, and of Comninou and Dundurs are for isotropic materials. Their results may not apply to anisotropic materials.

The determination of the extent of the separation and the contact zones in the Comninou crack is an integral part of the solution. A direct formulation of the problem is achieved by considering a distribution of dislocations on the interfacial crack. Let a single dislocation with Burgers vector **b** be located at $x_1=0$ on the interface. The stress function ϕ_2 for material 2 in $x_2<0$ is given by (8.8-3)$_2$ with $\theta_o=0$ in which **g** and **h** are given by (8.8-7b) with $\mathbf{f}=\mathbf{0}$. Hence

$$\phi_2 = \frac{1}{\pi}(\ln r)\tilde{\mathbf{L}}\mathbf{b} + \left\{\mathbf{L}_2(\theta)\tilde{\mathbf{S}} + \mathbf{S}_2^T(\theta)\tilde{\mathbf{L}}\right\}\mathbf{b} \tag{11.8-1}$$

where $\mathbf{S}_2(\theta)$ and $\mathbf{L}_2(\theta)$ are the tensors $\mathbf{S}(\theta)$ and $\mathbf{L}(\theta)$ associated with material 2. $\tilde{\mathbf{L}}$ and $\tilde{\mathbf{S}}$ are defined in (8.8-6). $\theta=0$ along the positive x_1-axis and $\theta=-\pi$ along the negative x_1-axis. Therefore ϕ_2 along the x_1-axis is

$$\phi_2 = \begin{cases} \frac{1}{\pi}(\ln r)\tilde{\mathbf{L}}\mathbf{b} & \text{for } x_1 > 0, \\ \frac{1}{\pi}(\ln r)\tilde{\mathbf{L}}\mathbf{b} - (\mathbf{L}\tilde{\mathbf{S}} + \mathbf{S}^T\tilde{\mathbf{L}})\mathbf{b} & \text{for } x_1 < 0. \end{cases}$$

From (8.8-34a),

$$\mathbf{L}_2\tilde{\mathbf{S}} + \mathbf{S}_1^T\tilde{\mathbf{L}} = \tilde{\mathbf{W}}$$

so that we may write ϕ_2 along the x_1-axis as

$$\phi_2 = \frac{1}{\pi}(\ln|x_1|)\tilde{\mathbf{L}}\mathbf{b} - H(-x_1)\tilde{\mathbf{W}}\mathbf{b}$$

where $H(\lambda)$ is the Heaviside step function. It vanishes for $\lambda<0$ and assumes the value of unity at $\lambda>0$. The traction vector **t** on the surface $x_2=0$ is

$$\mathbf{t} = \frac{\partial}{\partial x_1}\phi_2 = \frac{1}{\pi x_1}\tilde{\mathbf{L}}\mathbf{b} + \delta(x_1)\tilde{\mathbf{W}}\mathbf{b} \tag{11.8-2}$$

in which $\delta(\lambda)$ is the delta function defined in (11.2-6). It should be noted from (8.8-24) and (8.8-26) that

$$(\mathbf{M}_1^{-1} + \overline{\mathbf{M}}_2^{-1})^{-1} = \tilde{\mathbf{L}} + i\tilde{\mathbf{W}}, \tag{11.8-3a}$$

where

$$\begin{aligned} \tilde{\mathbf{W}} &= \mathbf{D}^{-1}\mathbf{W}\tilde{\mathbf{L}} = \tilde{\mathbf{L}}\mathbf{W}\mathbf{D}^{-1}, \\ \tilde{\mathbf{L}} &= (\mathbf{D} - \mathbf{W}^T\mathbf{D}^{-1}\mathbf{W})^{-1}. \end{aligned} \tag{11.8-3b}$$

$\tilde{\mathbf{L}} + i\tilde{\mathbf{W}}$ is a positive definite Hermitian. $\tilde{\mathbf{L}}$ is symmetric and positive definite while $\tilde{\mathbf{W}}$ is skew-symmetric. For a distribution of dislocations with density $\mathbf{b}(x_1)$ for $-a<x_1<a$ we have

$$\mathbf{t} = -\frac{1}{\pi}\int_{-a}^{a} \frac{\tilde{\mathbf{L}}\mathbf{b}(\xi)}{\xi - x_1} d\xi + \tilde{\mathbf{W}}\mathbf{b}(x_1).$$

Let $(\mathbf{t}^\infty)_i = \sigma_{i2}^\infty$ be the stress applied at infinity. Assuming that it can be translated into surface tractions $\mathbf{t}^\infty$ and $-\mathbf{t}^\infty$ applied at the lower and upper crack surfaces, respectively, the net traction $\mathbf{t}$ is

$$\mathbf{t} = \mathbf{t}^\infty - \frac{1}{\pi}\int_{-a}^{a} \frac{\tilde{\mathbf{L}}\mathbf{b}(\xi)}{\xi - x} d\xi + \tilde{\mathbf{W}}\mathbf{b}(x). \tag{11.8-4}$$

We have omitted the subscript 1 of x_1. In the case of a fully opened crack this leads to the following integral equation

$$\mathbf{t}^\infty = \frac{1}{\pi}\int_{-a}^{a} \frac{\tilde{\mathbf{L}}\mathbf{b}(\xi)}{\xi - x} d\xi - \tilde{\mathbf{W}}\mathbf{b}(x) \tag{11.8-5}$$

for $\mathbf{b}(x)$. The crack opening displacement $\Delta\mathbf{u}$ is

$$\Delta\mathbf{u}(x) = \int_{-a}^{x} \mathbf{b}(\xi) d\xi$$

so that (11.8-5) is subjected to the condition

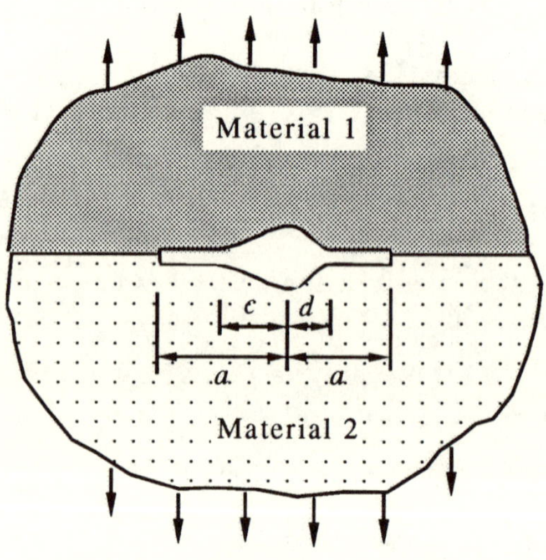

Fig. 11.6 Geometry of the Comninou crack.

$$\int_{-a}^{a} \mathbf{b}(\xi)d\xi = \mathbf{0}. \tag{11.8-6}$$

If the crack is partially open as shown in Fig. 11.6,

$$b_2(x) = 0 \quad \text{for} \quad -a < x < -c \quad \text{and} \quad d < x < a. \tag{11.8-7}$$

By letting

$$\tilde{\mathbf{W}} = \begin{bmatrix} 0 & \tilde{w}_3 & -\tilde{w}_2 \\ -\tilde{w}_3 & 0 & \tilde{w}_1 \\ \tilde{w}_2 & -\tilde{w}_1 & 0 \end{bmatrix}$$

(11.8-5) is written as

$$t_1^\infty = \frac{1}{\pi}\int_{-a}^{a} \frac{\{\tilde{\mathbf{L}}\mathbf{b}(\xi)\}_1}{\xi - x} d\xi - \tilde{w}_3 b_2(x) + \tilde{w}_2 b_3(x), \tag{11.8-8a}$$

$$t_3^\infty = \frac{1}{\pi}\int_{-a}^{a} \frac{\{\tilde{\mathbf{L}}\mathbf{b}(\xi)\}_3}{\xi - x} d\xi - \tilde{w}_2 b_1(x) + \tilde{w}_1 b_2(x), \tag{11.8-8b}$$

for $-a<x<a$ and

$$t_2^\infty = \frac{1}{\pi}\int_{-a}^{a} \frac{\{\tilde{\mathbf{L}}\mathbf{b}(\xi)\}_2}{\xi - x} d\xi + \tilde{w}_3 b_1(x) - \tilde{w}_1 b_3(x) \tag{11.8-8c}$$

for $-c<x<d$. Equations (11.8-8) are subjected to the conditions (11.8-6) and (11.8-7). The solution is physically acceptable if the normal traction obtained from (11.8-4) in the contact zones is compressive and the crack opening in the separation zone does not overlap. These means that

$$t_2(x) \le 0 \quad \text{for} \quad -a < x < -c \quad \text{and} \quad d < x < a, \tag{11.8-9}$$

$$\int_{-a}^{x} b_2(\xi)d\xi \ge 0 \quad \text{for} \quad -c < x < d. \tag{11.8-10}$$

In the special case of monoclinic materials with the symmetry plane at x_3=0 the structure of **S** and **L** has been given in (6.7-23). One can then show that

$$\tilde{w}_1 = \tilde{w}_2 = 0, \quad \tilde{L}_{13} = \tilde{L}_{23} = 0$$

and that (11.8-8) reduce to

$$t_1^\infty = \frac{1}{\pi}\int_{-a}^{a} \frac{\tilde{L}_{11}b_1(\xi) + \tilde{L}_{12}b_2(\xi)}{\xi - x} d\xi - \tilde{w}_3 b_2(x), \tag{11.8-11a}$$

$$t_3^\infty = \frac{1}{\pi}\int_{-a}^{a} \frac{\tilde{L}_{33}b_3(\xi)}{\xi - x}d\xi \tag{11.8-11b}$$

for $-a<x<a$ and

$$t_2^\infty = \frac{1}{\pi}\int_{-a}^{a} \frac{\tilde{L}_{12}b_1(\xi)+\tilde{L}_{22}b_2(\xi)}{\xi - x}d\xi + \tilde{w}_3 b_1(x) \tag{11.8-11c}$$

for $-c<x<d$. Equations (11.8-11b) is a Cauchy integral equation for the single unknown $b_3(x)$ which can be solved analytically. By eliminating $b_1(\xi)$ between (11.8-11a) and (11.8-11c) we have

$$\tilde{L}_{11}t_2^\infty - \tilde{L}_{12}t_1^\infty = \frac{1}{\pi}\int_{-c}^{d} \frac{(\tilde{L}_{11}\tilde{L}_{22} - \tilde{L}_{12}^2)b_2(\xi)}{\xi - x}d\xi$$
$$+\tilde{w}_3\{\tilde{L}_{11}b_1(x) + \tilde{L}_{12}b_2(x)\} \tag{11.8-12}$$

for $-c<x<d$. Equations (11.8-12) and (11.8-11a) are two integral equations for $\{\tilde{L}_{11}b_1(x)+\tilde{L}_{12}b_2(x)\}$ and $b_2(x)$. They are analogous to (2.1) and (2.2) in Gautesen and Dundurs (1988) for isotropic materials.

Another special case in which the integral equations are analogous to those for isotropic materials is when the material is monoclinic with the symmetry plane at $x_2=0$. With the structure of **S** and **L** presented in (6.7-32) we have

$$\tilde{w}_2 = 0, \quad \tilde{L}_{12} = \tilde{L}_{23} = 0.$$

Introducing the notations

$$\mathbf{t}^\text{o} = \begin{bmatrix} t_1^\infty \\ t_3^\infty \end{bmatrix}, \quad \mathbf{b}^\text{o} = \begin{bmatrix} b_1 \\ b_3 \end{bmatrix}, \quad \tilde{\mathbf{w}}^\text{o} = \begin{bmatrix} \tilde{w}_3 \\ -\tilde{w}_1 \end{bmatrix}, \quad \mathbf{X} = \begin{bmatrix} \tilde{L}_{11} & \tilde{L}_{13} \\ \tilde{L}_{13} & \tilde{L}_{33} \end{bmatrix},$$

(11.8-8a,b) can be written as

$$\mathbf{X}^{-1}\mathbf{t}^\text{o} = \frac{1}{\pi}\int_{-a}^{a} \frac{\mathbf{b}^\text{o}(\xi)}{\xi - x}d\xi - \mathbf{X}^{-1}\tilde{\mathbf{w}}^\text{o} b_2(x) \tag{11.8-13}$$

for $-a<x<a$. Equation (11.8-8c) is reduced to

$$t_2^\infty = \frac{1}{\pi}\int_{-c}^{d} \frac{\tilde{L}_{22}b_2(\xi)}{\xi - x}d\xi + \tilde{\mathbf{w}}^\text{o}\cdot\mathbf{b}^\text{o}(x) \tag{11.8-14}$$

for $-c<x<d$. Equations (11.8-13) is a Cauchy integral equation that can be solved for $\mathbf{b}^\text{o}(x)$ in terms of $b_2(x)$. Substitution of $\mathbf{b}^\text{o}(x)$ so obtained in (11.8-14) yields a single integral equation for $b_2(x)$ (Wu and Hwang, 1990). When multiplied by $(\tilde{\mathbf{w}}^\text{o})^T$, (11.8-13) leads to

$$(\tilde{\mathbf{w}}^{\circ})^T \mathbf{X}^{-1}\mathbf{t}^{\circ} = \frac{1}{\pi}\int_{-a}^{a} \frac{\tilde{\mathbf{w}}^{\circ}\cdot\mathbf{b}^{\circ}(\xi)}{\xi - x} d\xi - (\tilde{\mathbf{w}}^{\circ})^T \mathbf{X}^{-1}\tilde{\mathbf{w}}^{\circ} b_2(x) \qquad (11.8\text{-}15)$$

for $-a<x<a$. Equations (11.8-14) and (11.8-15) are two integral equations for $\tilde{\mathbf{w}}^{\circ}\cdot\mathbf{b}^{\circ}(x)$ and $b_2(x)$. Again, they are analogous to (2.1) and (2.2) of Gautesen and Dundurs (1988) for isotropic materials.

Chapter 12

STEADY STATE MOTION AND SURFACE WAVES

The Stroh formalism for two-dimensional elastostatics can be extended to elastodynamics when the problem is a *steady state motion*. Most of the identities in Chapters 6 and 7 remain applicable. The Barnett–Lothe tensors **S**, **H**, **L** now depend on the speed v of the steady state motion. However $\mathbf{S}(v)$, $\mathbf{H}(v)$, $\mathbf{L}(v)$ are no longer tensors because they do not obey the laws of tensor transformation when $v \neq 0$. Depending on the problems the speed v may not be prescribed arbitrarily. This is particularly the case for surface waves in a half-space where v is the surface wave speed. The problem of the existence and uniqueness of a surface wave speed in anisotropic materials is the crux of surface wave theory. It is a subject that has been extensively studied since the pioneer work of Stroh (1962). Excellent expositions on surface waves for anisotropic elastic materials have been given by Farnell (1970), Chadwick and Smith (1977), Barnett and Lothe (1985), and more recently, by Chadwick (1989d).

12.1 Reduction to Elastostatics Equations

In a fixed rectangular coordinate system x_i (i = 1,2,3), let u_i and σ_{ij} be, respectively, the displacement and stress in an anisotropic elastic material. The stress-strain laws and the equations of motion are

$$\sigma_{ij} = C_{ijks} u_{k,s}, \tag{12.1-1}$$

$$\sigma_{ij,j} = \rho \ddot{u}_i, \tag{12.1-2}$$

in which ρ is the mass density, the comma and dot denote differentiation with x_i and time t, respectively, and repeated indices imply summation. The C_{ijks} are the elastic stiffnesses that are assumed to possess the full symmetry (2.1-2). They are positive definite so that the strain energy is positive.

For a steady state motion in the x_1-direction with a constant speed $v>0$,

$$u_i = u_i(x_1 - vt,\ x_2,\ x_3)$$

from which we obtain

$$\dot{u}_i = -v u_{i,1}, \quad \ddot{u}_i = -v \dot{u}_{i,1}. \tag{12.1-3}$$

Equations (12.1-1) and (12.1-2) can then be written as

$$(\sigma_{ij} + \rho v \dot{u}_i \delta_{j1}) = (C_{ijks} - \rho v^2 \delta_{ik} \delta_{j1} \delta_{s1}) u_{k,s},$$

$$(\sigma_{ij} + \rho v \dot{u}_i \delta_{j1}),_j = 0,$$

where δ_{ij} is the Kronecker delta. If we set

$$\sigma^*_{ij} = \sigma_{ij} + \rho v \dot{u}_i \delta_{j1}, \qquad (12.1\text{-}4)$$

$$C^*_{ijks} = C_{ijks} - \rho v^2 \delta_{ik} \delta_{j1} \delta_{s1}, \qquad (12.1\text{-}5)$$

we have

$$\sigma^*_{ij} = C^*_{ijks} u_{k,s}, \qquad (12.1\text{-}6)$$

$$\sigma^*_{ij,j} = 0. \qquad (12.1\text{-}7)$$

The two equations are identical to the stress-strain laws and equations of equilibrium for elastostatics. However there are differences. First, the modified stiffnesses C^*_{ijks} are no longer fully symmetric. They depend on v and may not be positive definite. Secondly, the modified stress σ^*_{ij} is not the real stress. It depends on linear momentum $\rho \dot{u}_i$ in which u_i is a dependent variable. This causes no problems if u_i can be solved from

$$C^*_{ijks} u_{k,sj} = 0, \qquad (12.1\text{-}8)$$

resulting from insertion of (12.1-6) into (12.1-7). σ^*_{ij} is then determined from (12.1-6) and the stress σ_{ij} from (12.1-4). The results (12.1-6) and (12.1-7) are general in that they are not limited to two-dimensional deformations.

12.2 The Stroh Formalism

For two-dimensional deformations for which u_i $(i=1,2,3)$ are independent of x_3, the Stroh formalism presented in Chapter 5 applies here if x_1, σ_{ij}, and C_{ijks} are replaced, respectively, by $x_1 - vt$, σ^*_{ij}, and C^*_{ijks}. The displacement u_i is obtained from (5.1-4) as

$$u_i = a_i f(z), \quad \text{or} \quad \mathbf{u} = \boldsymbol{a} f(z), \qquad (12.2\text{-}1a)$$

$$z = x_1 - vt + p x_2, \qquad (12.2\text{-}1b)$$

where f is an arbitrary function of z, and p and a_i are determined by (5.1-8) which is written as

$$\left\{ \mathbf{Q}(v) + p\left(\mathbf{R}(v) + \mathbf{R}^T(v)\right) + p^2 \mathbf{T}(v) \right\} \boldsymbol{a} = \mathbf{0}. \qquad (12.2\text{-}2)$$

The dependence of p and $\boldsymbol{a}$ on v is understood. Notice that

$$Q_{ik}(v) = C^*_{i1k1} = C_{i1k1} - \rho v^2 \delta_{ik},$$

$$R_{ik}(v) = C^*_{i1k2} = C_{i1k2}, \quad T_{ik}(v) = C^*_{i2k2} = C_{i2k2}.$$

Thus **R** and **T** are unchanged while

$$\mathbf{Q}(v) = \mathbf{Q} - \rho v^2 \mathbf{I} \tag{12.2-3}$$

where **I** is the identity tensor. $\mathbf{Q}(v)$ is symmetric and we have

$$\left\{\mathbf{Q} - \rho v^2 \mathbf{I} + p(\mathbf{R} + \mathbf{R}^T) + p^2 \mathbf{T}\right\} \boldsymbol{a} = \mathbf{0}. \tag{12.2-4}$$

From (5.3-3) and (5.3-2) the stress function $\boldsymbol{\phi}$ is

$$\varphi_i = b_i f(z) \quad \text{or} \quad \boldsymbol{\phi} = \boldsymbol{b}\, f(z), \tag{12.2-5}$$

where

$$\boldsymbol{b} = \{\mathbf{R}^T + p\mathbf{T}\}\boldsymbol{a} = -\frac{1}{p}\{\mathbf{Q}(v) + p\mathbf{R}\}\boldsymbol{a}. \tag{12.2-6}$$

The vector $\boldsymbol{b}$ also depends on v. Equation (5.3-4) is replaced by

$$\sigma_{i1} + \rho v \dot{u}_i = -\varphi_{i,2}, \quad \sigma_{i2} = \varphi_{i,1}. \tag{12.2-7}$$

Since $\sigma_{21} = \sigma_{12}$ we have

$$\varphi_{1,1} + \varphi_{2,2} + \rho v \dot{u}_2 = 0 \tag{12.2-8a}$$

or, by (12.2-1a) and (12.2-5) (Barnett and Lothe, 1985),

$$b_1 + p b_2 - \rho v^2 a_2 = 0. \tag{12.2-8b}$$

The eigenrelation (12.2-4) and (12.2-6) provide six eigenvalues p_α and six associated eigenvectors $\boldsymbol{a}_\alpha$, $\boldsymbol{b}_\alpha$ (α=1,2,...6). The general solutions for the displacement **u** and the stress function $\boldsymbol{\phi}$ obtained by superposing six solutions associated with the six eigenvalues are

$$\mathbf{u} = \sum_{\alpha=1}^{6} \boldsymbol{a}_\alpha f_\alpha(z_\alpha), \quad \boldsymbol{\phi} = \sum_{\alpha=1}^{6} \boldsymbol{b}_\alpha f_\alpha(z_\alpha), \tag{12.2-9}$$

$$z_\alpha = x_1 - vt + p_\alpha x_2.$$

If the components of the stress vectors $\mathbf{t}_1$ and $\mathbf{t}_2$ are defined by

$$(\mathbf{t}_1)_i = \sigma_{i1}, \quad (\mathbf{t}_2)_i = \sigma_{i2}, \tag{12.2-10}$$

(12.2-7) can be written as

$$\mathbf{t}_1 = -\phi_{,2} - \rho v \dot{\mathbf{u}}, \quad \mathbf{t}_2 = \phi_{,1}. \tag{12.2-11}$$

The difference in ϕ between any two points on a curved boundary Γ no longer represents the resultant traction force acting between these two points unless Γ is a straight line parallel to the x_1-axis. Let ν_i be a unit vector normal to Γ. The surface traction $\mathbf{t}$ on Γ is

$$t_i = \sigma_{i1}\nu_1 + \sigma_{i2}\nu_2.$$

Making use of (5.7-1) and (12.2-7) we have

$$\mathbf{t} = \frac{d\phi}{ds} - \rho v \nu_1 \dot{\mathbf{u}} \tag{12.2-12}$$

where s is the arclength of Γ measured in the direction such that, when one faces the direction of increasing s, the material is on the right side. Thus

$$\int_{s_1}^{s_2} \mathbf{t}\, ds = \Delta\phi - \int_{s_1}^{s_2} \rho v \dot{\mathbf{u}} \nu_1 ds.$$

The difference $\Delta\phi$ in ϕ between any two points on Γ is not equal to the resultant traction exerted between these two points unless $\nu_1 = 0$.

If Γ is a closed curve the equation of motion for the mass inside the closed curve is

$$\int_\Gamma \mathbf{t}\, ds + \mathbf{f} = \int_A \rho \ddot{\mathbf{u}}\, dA$$

where $\mathbf{f}$ is a concentrated force inside the closed curve and A denotes the area integral. Invoking (12.1-3) to replace $\ddot{\mathbf{u}}$ by $-v\dot{\mathbf{u}}_{,1}$ and converting the area integral to a line integral, the right side is

$$-\int_\Gamma \rho v \dot{\mathbf{u}} \nu_1 ds.$$

Together with (12.2-12) we obtain

$$\int_\Gamma \frac{d\phi}{ds} ds = \Delta\phi = -\mathbf{f}. \tag{12.2-13}$$

The result is identical to the case of elastostatics in that, when we traverse Γ counterclockwise a complete circle, ϕ increases by the amount $\mathbf{f}$.

Following the derivation in Section 5.5 the two equations in (12.2-6) can be cast in the following standard eigenrelation

$$\mathbf{N}(v)\xi = p\,\xi, \tag{12.2-14}$$

$$\mathbf{N}(v)=\begin{bmatrix}\mathbf{N}_1 & \mathbf{N}_2\\ \mathbf{N}_3(v) & \mathbf{N}_1^T\end{bmatrix},\qquad \xi=\begin{bmatrix}\boldsymbol{a}\\ \boldsymbol{b}\end{bmatrix}, \tag{12.2-15}$$

$$\mathbf{N}_1=-\mathbf{T}^{-1}\mathbf{R}^T,\quad \mathbf{N}_2=\mathbf{T}^{-1},\quad \mathbf{N}_3(v)=\mathbf{R}\mathbf{T}^{-1}\mathbf{R}^T-\mathbf{Q}(v). \tag{12.2-16}$$

Thus $\mathbf{N}_1$ and $\mathbf{N}_2$ remain the same as in elastostatics while

$$\mathbf{N}_3(v)=\mathbf{N}_3+\rho v^2\mathbf{I}.$$

It should be noted that when the matrix $\mathbf{N}(0)$ is nonsemisimple, $\mathbf{N}(v)$ can be simple or semisimple for $v\neq0$. This is the case for isotropic materials for which $\mathbf{N}(v)$ is semisimple when $v\neq0$. The eigenvalues are $p_1^2=-1+(v^2/c_1^2)$ and $p_2^2=p_3^2=-1+(v^2/c_2^2)$ where c_1, c_2 are, respectively, the dilatational and shear wave speeds. Conversely, when $\mathbf{N}(0)$ is simple $\mathbf{N}(v)$ can be semisimple or nonsemisimple for a certain value of v (Section 12.8). The ξ in (12.2-14) is a right eigenvector of $\mathbf{N}(v)$. The left eigenvector η is

$$\eta=\begin{bmatrix}\boldsymbol{b}\\ \boldsymbol{a}\end{bmatrix}.$$

The orthogonality relation (5.4-10) remains valid, i.e.,

$$\eta_\alpha\cdot\xi_\beta=0,\qquad \text{if } p_\alpha\neq p_\beta.$$

If $\mathbf{N}(v)$ is simple or semisimple we normalize ξ_α such that

$$\eta_\alpha\cdot\xi_\beta=\delta\alpha\beta \tag{12.2-17a}$$

or

$$\boldsymbol{b}_\alpha^T\boldsymbol{a}_\beta+\boldsymbol{a}_\alpha^T\boldsymbol{b}_\beta=\delta_{\alpha\beta}. \tag{12.2-17b}$$

The eigenvalues p_α may be real or complex for $v\neq0$. The complex eigenvalues appear in pairs of complex conjugates. Let

$$\operatorname{Im}p_\alpha>0,\quad p_{\alpha+3}=\bar{p}_\alpha,\quad \boldsymbol{a}_{\alpha+3}=\bar{\boldsymbol{a}}_\alpha,\quad \boldsymbol{b}_{\alpha+3}=\bar{\boldsymbol{b}}_\alpha, \tag{12.2-18}$$

where α=1,2, or 3 and the overbar denotes the complex conjugate. If there is only one pair of complex conjugates p_α, (12.2-18) apply to α=1 only. If there are two or three pairs of complex conjugates p_α, α=1 and 2, or α = 1, 2, and 3, respectively. Let

$$\begin{gathered}\mathbf{A}=[\boldsymbol{a}_1,\boldsymbol{a}_2,\boldsymbol{a}_3],\quad \mathbf{B}=[\boldsymbol{b}_1,\boldsymbol{b}_2,\boldsymbol{b}_3],\\ \hat{\mathbf{A}}=[\boldsymbol{a}_4,\boldsymbol{a}_5,\boldsymbol{a}_6],\quad \hat{\mathbf{B}}=[\boldsymbol{b}_4,\boldsymbol{b}_5,\boldsymbol{b}_6].\end{gathered} \tag{12.2-19}$$

The 3×3 matrices $\hat{\mathbf{A}}$ and $\hat{\mathbf{B}}$ are not complex conjugates of $\mathbf{A}$ and $\mathbf{B}$

unless p_α are all complex. The *orthogonality relation* (12.2-17) is

$$\begin{bmatrix} \mathbf{B}^T & \mathbf{A}^T \\ \hat{\mathbf{B}}^T & \hat{\mathbf{A}}^T \end{bmatrix} \begin{bmatrix} \mathbf{A} & \hat{\mathbf{A}} \\ \mathbf{B} & \hat{\mathbf{B}} \end{bmatrix} = \begin{bmatrix} \mathbf{I} & \mathbf{0} \\ \mathbf{0} & \mathbf{I} \end{bmatrix} \tag{12.2-20}$$

or

$$\begin{aligned} \mathbf{B}^T\mathbf{A} + \mathbf{A}^T\mathbf{B} = \mathbf{I} = \hat{\mathbf{B}}^T\hat{\mathbf{A}} + \hat{\mathbf{A}}^T\hat{\mathbf{B}}, \\ \mathbf{B}^T\hat{\mathbf{A}} + \mathbf{A}^T\hat{\mathbf{B}} = \mathbf{0} = \hat{\mathbf{B}}^T\mathbf{A} + \hat{\mathbf{A}}^T\mathbf{B}. \end{aligned} \tag{12.2-21}$$

The two 6×6 matrices on the left of (12.2-20) are the inverses of each other. Their products commute so that

$$\begin{bmatrix} \mathbf{A} & \hat{\mathbf{A}} \\ \mathbf{B} & \hat{\mathbf{B}} \end{bmatrix} \begin{bmatrix} \mathbf{B}^T & \mathbf{A}^T \\ \hat{\mathbf{B}}^T & \hat{\mathbf{A}}^T \end{bmatrix} = \begin{bmatrix} \mathbf{I} & \mathbf{0} \\ \mathbf{0} & \mathbf{I} \end{bmatrix} \tag{12.2-22}$$

or

$$\begin{aligned} \mathbf{A}\mathbf{B}^T + \hat{\mathbf{A}}\hat{\mathbf{B}}^T = \mathbf{I} = \mathbf{B}\mathbf{A}^T + \hat{\mathbf{B}}\hat{\mathbf{A}}^T, \\ \mathbf{A}\mathbf{A}^T + \hat{\mathbf{A}}\hat{\mathbf{A}}^T = \mathbf{0} = \mathbf{B}\mathbf{B}^T + \hat{\mathbf{B}}\hat{\mathbf{B}}^T. \end{aligned} \tag{12.2-23}$$

These are the *closure relations*. Keeping in mind that **A**, **B**, $\hat{\mathbf{A}}$, and $\hat{\mathbf{B}}$ depend on v, we extend the definition of the Barnett–Lothe tensors **S**, **H**, and **L** to

$$\begin{aligned} \mathbf{S}(v) &= i(2\mathbf{A}\mathbf{B}^T - \mathbf{I}) = -i(2\hat{\mathbf{A}}\hat{\mathbf{B}}^T - \mathbf{I}), \\ \mathbf{H}(v) &= 2i\mathbf{A}\mathbf{A}^T = -2i\hat{\mathbf{A}}\hat{\mathbf{A}}^T, \\ \mathbf{L}(v) &= -2i\mathbf{B}\mathbf{B}^T = 2i\hat{\mathbf{B}}\hat{\mathbf{B}}^T. \end{aligned} \tag{12.2-24}$$

They are not necessarily real. Due to the second equalities in (12.2-24) we have the alternate expressions (see also Section 6.5)

$$\begin{aligned} \mathbf{S}(v) &= i(\mathbf{A}\mathbf{B}^T - \hat{\mathbf{A}}\hat{\mathbf{B}}^T), \\ \mathbf{H}(v) &= i(\mathbf{A}\mathbf{A}^T - \hat{\mathbf{A}}\hat{\mathbf{A}}^T), \\ \mathbf{L}(v) &= -i(\mathbf{B}\mathbf{B}^T - \hat{\mathbf{B}}\hat{\mathbf{B}}^T), \end{aligned} \tag{12.2-25}$$

which can be rewritten as

$$\begin{bmatrix} \mathbf{S}(v) & \mathbf{H}(v) \\ -\mathbf{L}(v) & \mathbf{S}^T(v) \end{bmatrix} = \begin{bmatrix} \mathbf{A} & \hat{\mathbf{A}} \\ \mathbf{B} & \hat{\mathbf{B}} \end{bmatrix} \begin{bmatrix} i\mathbf{I} & \mathbf{0} \\ \mathbf{0} & -i\mathbf{I} \end{bmatrix} \begin{bmatrix} \mathbf{B}^T & \mathbf{A}^T \\ \hat{\mathbf{B}}^T & \hat{\mathbf{A}}^T \end{bmatrix}.$$

It follows from (12.2-20) and (12.2-22) that

$$\begin{bmatrix} \mathbf{S}(v) & \mathbf{H}(v) \\ -\mathbf{L}(v) & \mathbf{S}^T(v) \end{bmatrix} \begin{bmatrix} \mathbf{S}(v) & \mathbf{H}(v) \\ -\mathbf{L}(v) & \mathbf{S}^T(v) \end{bmatrix} = -\begin{bmatrix} \mathbf{I} & \mathbf{0} \\ \mathbf{0} & \mathbf{I} \end{bmatrix}, \tag{12.2-26}$$

or

$$\mathbf{H}(v)\mathbf{L}(v) - \mathbf{S}(v)\mathbf{S}(v) = \mathbf{I} = \mathbf{L}(v)\mathbf{H}(v) - \mathbf{S}^T(v)\mathbf{S}^T(v), \tag{12.2-27}$$

$$\mathbf{S}(v)\mathbf{H}(v) + \mathbf{H}(v)\mathbf{S}^T(v) = \mathbf{0} = \mathbf{L}(v)\mathbf{S}(v) + \mathbf{S}^T(v)\mathbf{L}(v). \tag{12.2-28}$$

Therefore $\mathbf{L}(v)\mathbf{S}(v)$ and $\mathbf{S}(v)\mathbf{H}(v)$ are skew-symmetric, although they may be complex. If $\mathbf{H}^{-1}(v)$ and $\mathbf{L}^{-1}(v)$ exist, it can be shown that $\mathbf{S}(v)\mathbf{L}^{-1}(v)$ and $\mathbf{H}^{-1}(v)\mathbf{S}(v)$ are also skew-symmetric. We may extend the definition of the impedance tensor $\mathbf{M}$ and its inverse $\mathbf{M}^{-1}$ to (Section 5.6)

$$\begin{aligned} \mathbf{M}(v) &= -i\mathbf{B}\mathbf{A}^{-1} = \mathbf{H}^{-1}(v) + i\mathbf{H}^{-1}(v)\mathbf{S}(v), \\ \mathbf{M}^{-1}(v) &= i\mathbf{A}\mathbf{B}^{-1} = \mathbf{L}^{-1}(v) - i\mathbf{S}(v)\mathbf{L}^{-1}(v). \end{aligned} \tag{12.2-29}$$

They are not necessarily Hermitian.

When the eigenvalues p_α are all complex, $\mathbf{S}(v)$, $\mathbf{H}(v)$, $\mathbf{L}(v)$ are real and $\mathbf{M}(v)$ and $\mathbf{M}^{-1}(v)$ are Hermitian. All quantities reduce to the corresponding quantities in elastostatics when $v=0$. Hence $\mathbf{H}(0)$ and $\mathbf{L}(0)$ are positive definite, and $\mathbf{M}(0)$ and $\mathbf{M}^{-1}(0)$ are positive definite Hermitian. It can be shown that $\mathbf{S}(v)$, $\mathbf{H}(v)$, $\mathbf{L}(v)$, and $\mathbf{M}(v)$ for $v \neq 0$ do not obey the laws of tensor transformation when the coordinate system is rotated an angle about the x_3-axis. Hence they are not tensors when $v \neq 0$.

12.3 Identities for $v < \hat{v}$

Since $v=0$ corresponds to elastostatics, the six p_α in (12.2-4) are all complex when $v=0$. As v increases from zero one or more pairs of the p_α become real. If $\hat{v}$ is the *limiting wave speed* at which one pair of the p_α first becomes real, the eigenvalues are all complex for v in $[0, \hat{v})$. They appear in three pairs of complex conjugates. A steady motion is *subsonic* when $v < \hat{v}$. It is *supersonic* when $v > \hat{v}$ and *transonic* when $v = \hat{v}$. The *identities* presented in Chapters 6 and 7 for elastostatics remain valid for $v < \hat{v}$ if the dependence on v is properly introduced. In fact many of the identities remain valid for $v > \hat{v}$, but we will restrict the speed v to the subsonic range $[0, \hat{v})$. We begin with $\mathbf{Q}(\theta, v)$, $\mathbf{R}(\theta, v)$, and $\mathbf{T}(\theta, v)$ whose components are, by (7.2-12),

$$Q_{ik}(\theta, v) = C^*_{ijks} n_j n_s, \quad R_{ik}(\theta, v) = C^*_{ijks} n_j m_s, \quad T_{ik}(\theta, v) = C^*_{ijks} m_j m_s,$$

$$\mathbf{n}(\theta) = \begin{bmatrix} \cos\theta \\ \sin\theta \\ 0 \end{bmatrix}, \quad \mathbf{m}(\theta) = \begin{bmatrix} -\sin\theta \\ \cos\theta \\ 0 \end{bmatrix}. \tag{12.3-1}$$

Insertion of C^*_{ijks} from (12.1-5) leads to

$$\begin{aligned} \mathbf{Q}(\theta, v) &= \mathbf{Q}(\theta) - \rho v^2 \cos^2\theta \mathbf{I}, \\ \mathbf{R}(\theta, v) &= \mathbf{R}(\theta) + \rho v^2 \sin\theta\cos\theta \mathbf{I}, \\ \mathbf{T}(\theta, v) &= \mathbf{T}(\theta) - \rho v^2 \sin^2\theta \mathbf{I}, \end{aligned} \tag{12.3-2}$$

in which

$$Q_{ik}(\theta) = C_{ijks} n_j n_s, \quad R_{ik}(\theta) = C_{ijks} n_j m_s, \quad T_{ik}(\theta) = C_{ijks} m_j m_s. \tag{12.3-3}$$

$\mathbf{Q}(\theta)$ is the *acoustic tensor* which is positive definite for all θ. Carrying out the summation on the indices j and s for $Q_{ik}(\theta)$ in $(12.3\text{-}3)_1$, $\mathbf{Q}(\theta, v)$ in $(12.3\text{-}2)_1$ can be written as

$$\mathbf{Q}(\theta, v) = (\cos\theta)^2 \left\{ \mathbf{Q} - \rho v^2 \mathbf{I} + \tan\theta(\mathbf{R} + \mathbf{R}^T) + (\tan\theta)^2 \mathbf{T} \right\}. \tag{12.3-4}$$

Comparison with (12.2-4) suggests that the zeros of the determinant of $\mathbf{Q}(\theta, v)$ are related to the eigenvalue p by

$$p = \tan\theta. \tag{12.3-5}$$

When p is complex, so is θ. For v in $[0, \hat{v})$ the p_α are all complex so that $\mathbf{Q}(\theta, v)$ is nonsingular for v in $[0, \hat{v})$ for any real θ. In fact $\mathbf{Q}(\theta, v)$ is positive definite. Observing that

$$\mathbf{T}(\theta, v) = \mathbf{Q}(\theta + \tfrac{\pi}{2}, v), \tag{12.3-6}$$

$\mathbf{T}(\theta, v)$ is also positive definite for v in $[0, \hat{v})$ for any real θ.

The eigenrelation (7.3-14) for elastostatics is replaced by

$$\mathbf{N}(\theta, v)\boldsymbol{\xi} = p(\theta)\boldsymbol{\xi}, \tag{12.3-7}$$

$$\mathbf{N}(\theta, v) = \begin{bmatrix} \mathbf{N}_1(\theta, v) & \mathbf{N}_2(\theta, v) \\ \mathbf{N}_3(\theta, v) & \mathbf{N}_1^T(\theta, v) \end{bmatrix}, \quad \boldsymbol{\xi} = \begin{bmatrix} \mathbf{a} \\ \mathbf{b} \end{bmatrix}, \tag{12.3-8}$$

$$\begin{aligned} \mathbf{N}_1(\theta, v) = -\mathbf{T}^{-1}(\theta, v)\mathbf{R}^T(\theta, v), \quad \mathbf{N}_2(\theta, v) = \mathbf{T}^{-1}(\theta, v), \\ \mathbf{N}_3(\theta, v) = \mathbf{R}(\theta, v)\mathbf{T}^{-1}(\theta, v)\mathbf{R}^T(\theta, v) - \mathbf{Q}(\theta, v). \end{aligned} \tag{12.3-9}$$

In the above, dependence of $p(\theta)$ on v is understood and

$$p(\theta) = \frac{p\cos\theta - \sin\theta}{p\sin\theta + \cos\theta}. \tag{12.3-10}$$

Equations (7.6-4) and (7.6-6) are replaced by

$$\mathbf{S}(v)=\frac{1}{\pi}\int_0^{\pi}\mathbf{N}_1(\theta,v)d\theta,$$

$$\mathbf{H}(v)=\frac{1}{\pi}\int_0^{\pi}\mathbf{N}_2(\theta,v)d\theta, \tag{12.3-11}$$

$$\mathbf{L}(v)=-\frac{1}{\pi}\int_0^{\pi}\mathbf{N}_3(\theta,v)d\theta,$$

and

$$\mathbf{S}(\theta,v)=\frac{1}{\pi}\int_0^{\theta}\mathbf{N}_1(\lambda,v)d\lambda,$$

$$\mathbf{H}(\theta,v)=\frac{1}{\pi}\int_0^{\theta}\mathbf{N}_2(\lambda,v)d\lambda, \tag{12.3-12}$$

$$\mathbf{L}(\theta,v)=-\frac{1}{\pi}\int_0^{\theta}\mathbf{N}_3(\lambda,v)d\lambda.$$

Clearly, $\mathbf{H}(v)$ and $\mathbf{H}(\theta,v)$ are both symmetric and positive definite. Equations (12.3-11) are the integral formalism of $\mathbf{S}(v)$, $\mathbf{H}(v)$, and $\mathbf{L}(v)$ (Barnett and Lothe, 1973). Unlike the algebraic representation (12.2-24), (12.3-11) do not require computation of the eigenvalues p and the eigenvectors $\boldsymbol{a}$ and $\boldsymbol{b}$ Therefore (12.3-11) apply to degenerate materials for which $\mathbf{N}(\theta,v)$ is nonsemisimple.

Integration of (12.3-7) with θ yields (Section 7.6)

$$\begin{bmatrix}\mathbf{S}(v) & \mathbf{H}(v)\\ -\mathbf{L}(v) & \mathbf{S}^T(v)\end{bmatrix}\begin{bmatrix}\boldsymbol{a}_\alpha\\ \boldsymbol{b}_\alpha\end{bmatrix}=i\begin{bmatrix}\boldsymbol{a}_\alpha\\ \boldsymbol{b}_\alpha\end{bmatrix} \tag{12.3-13}$$

for α=1,2,3, or

$$\begin{aligned}\mathbf{S}(v)\boldsymbol{a}_\alpha+\mathbf{H}(v)\boldsymbol{b}_\alpha&=i\boldsymbol{a}_\alpha,\\ -\mathbf{L}(v)\boldsymbol{a}_\alpha+\mathbf{S}^T(v)\boldsymbol{b}_\alpha&=i\boldsymbol{b}_\alpha.\end{aligned} \tag{12.3-14}$$

They can be combined into one equation as

$$\begin{aligned}\mathbf{S}(v)\mathbf{A}+\mathbf{H}(v)\mathbf{B}&=i\mathbf{A},\\ -\mathbf{L}(v)\mathbf{A}+\mathbf{S}^T(v)\mathbf{B}&=i\mathbf{B}.\end{aligned} \tag{12.3-15}$$

Noting that $\mathbf{H}(v)$ is nonsingular, we write $(12.2\text{-}28)_1$ as

$$\mathbf{S}(v)=-\mathbf{H}(v)\mathbf{S}^T(v)\mathbf{H}^{-1}(v).$$

On taking the trace and determinant we have

$$\mathrm{tr}\,\mathbf{S}(\upsilon)=0 \quad \text{and} \quad |\mathbf{S}(\upsilon)|=0. \tag{12.3-16}$$

Therefore $\mathbf{S}(\upsilon)$ is singular and traceless. The eigenvalues of $\mathbf{S}(\upsilon)$ are 0 and $\pm is(\upsilon)$ where (Section 6.6)

$$s(\upsilon)=\left\{-\tfrac{1}{2}\mathrm{tr}[\mathbf{S}^2(\upsilon)]\right\}^{1/2}>0. \tag{12.3-17}$$

12.4 Properties of A, B, and L(υ)

We will first show that the vectors $\boldsymbol{a}_\alpha$ (α=1,2,3) are linearly independent when p_α are complex and distinct. Equation (12.2-6) for p=p_1, p_2, p_3 can be written in one equation as

$$\mathbf{B}=\mathbf{R}^T\mathbf{A}+\mathbf{TAP}=-\{\mathbf{Q}(\upsilon)\mathbf{AP}^{-1}+\mathbf{RA}\} \tag{12.4-1}$$

in which the 3×3 matrices $\mathbf{A}$ and $\mathbf{B}$ are defined in (12.2-19) and $\mathbf{P}$ is the diagonal matrix

$$\mathbf{P}=\mathrm{diag}[p_1,\ p_2,\ p_3].$$

Likewise (12.2-4) for p=p_1, p_2, p_3 can be written as

$$(\mathbf{Q}-\rho\upsilon^2\mathbf{I})\mathbf{A}+(\mathbf{R}+\mathbf{R}^T)\mathbf{AP}+\mathbf{TAP}^2=\mathbf{0}. \tag{12.4-2}$$

Suppose that $\boldsymbol{a}_\alpha$ are linearly dependent. This means that there exists a nonzero vector $\mathbf{q}$ such that

$$\mathbf{Aq}=\mathbf{0}. \tag{12.4-3}$$

If we postmultiply $(12.4\text{-}1)_1$ by $\mathbf{q}$, premultiply by $\overline{\mathbf{q}}^T\overline{\mathbf{PA}}^T$, and invoke the relation

$$\overline{\mathbf{A}}^T\mathbf{B}=-\overline{\mathbf{B}}^T\mathbf{A} \tag{12.4-4}$$

deduced from $(12.2\text{-}21)_4$ by setting $\hat{\mathbf{A}}=\overline{\mathbf{A}}$ and $\hat{\mathbf{B}}=\overline{\mathbf{B}}$, we have

$$(\overline{\mathbf{AP}\mathbf{q}})^T\mathbf{T}(\mathbf{APq})=\mathbf{0}. \tag{12.4-5}$$

The fact that $\mathbf{T}$ is positive definite implies that

$$\mathbf{APq}=\mathbf{0}. \tag{12.4-6}$$

Postmultiplying (12.4-2) by $\mathbf{q}$ and using (12.4-3) and (12.4-6) yields

$$\mathbf{AP}^2\mathbf{q}=\mathbf{0}, \tag{12.4-7}$$

again observing that $\mathbf{T}$ is positive definite. For a fixed α (12.4-3), (12.4-6), and (12.4-7) can be written as

$$\begin{bmatrix} 1 & 1 & 1 \\ p_1 & p_2 & p_3 \\ p_1^2 & p_2^2 & p_3^2 \end{bmatrix} \begin{bmatrix} A_{\alpha 1} q_1 \\ A_{\alpha 2} q_2 \\ A_{\alpha 3} q_3 \end{bmatrix} = \mathbf{0}.$$

The determinant of the 3×3 matrix is

$$(p_1 - p_2)(p_2 - p_3)(p_3 - p_1)$$

which is nonzero when p_α are distinct. Thus

$$A_{\alpha 1} q_1 = A_{\alpha 2} q_2 = A_{\alpha 3} q_3 = 0$$

for α=1,2,3. They are equivalent to

$$q_1 \boldsymbol{a}_1 = q_2 \boldsymbol{a}_2 = q_3 \boldsymbol{a}_3 = \mathbf{0}.$$

But the vectors $\boldsymbol{a}_\alpha$ are nonzero vectors, so we must have $\mathbf{q}=\mathbf{0}$. This leads to a contradiction, proving that $\boldsymbol{a}_\alpha$ are linearly independent.

The line of proof presented above is not identical to the one given by Stroh (1958) but it follows his idea. We present below a simpler proof that does not require that p_α be distinct.

The eigenrelation (12.2-14) is

$$\begin{aligned} \mathbf{N}_1 \boldsymbol{a} + \mathbf{N}_2 \boldsymbol{b} &= p\, \mathbf{a}, \\ (\mathbf{N}_3 + \rho v^2 \mathbf{I}) \boldsymbol{a} + \mathbf{N}_1^T \boldsymbol{b} &= p\, \boldsymbol{b}. \end{aligned} \tag{12.4-8}$$

For $p=p_1, p_2, p_3$ they can be cast in one equation as

$$\begin{aligned} \mathbf{N}_1 \mathbf{A} + \mathbf{N}_2 \mathbf{B} &= \mathbf{AP}, \\ (\mathbf{N}_3 + \rho v^2 \mathbf{I}) \mathbf{A} + \mathbf{N}_1^T \mathbf{B} &= \mathbf{BP}. \end{aligned} \tag{12.4-9}$$

Suppose that $\boldsymbol{a}_\alpha$ are linearly dependent so that (12.4-3) holds. Postmultiplying $(12.4\text{-}9)_1$ by $\mathbf{q}$, premultiplying by $\overline{\mathbf{q}}^T \overline{\mathbf{B}}^T$, and invoking (12.4-4) leads to

$$(\overline{\mathbf{B}} \overline{\mathbf{q}})^T \mathbf{N}_2 (\mathbf{B}\mathbf{q}) = \mathbf{0}.$$

The positive definiteness of $\mathbf{N}_2$ tells us that

$$\mathbf{B}\mathbf{q} = \mathbf{0}.$$

This and (12.4-3) imply that

$$q_1 \boldsymbol{\xi}_1 + q_2 \boldsymbol{\xi}_2 + q_3 \boldsymbol{\xi}_3 = \mathbf{0}. \tag{12.4-10}$$

This is not possible if the 6×6 matrix $\mathbf{N}(v)$ is simple because $\boldsymbol{\xi}_\alpha$ are then linearly independent. If $\mathbf{N}(v)$ is semisimple, say $p_1=p_2\neq p_3$, there exist two independent eigenvectors $\boldsymbol{\xi}_1$ and $\boldsymbol{\xi}_2$ that satisfy the orthogonality condition (12.2-17). Equation (12.4-10) again is violated. Consequently, $\boldsymbol{a}_\alpha$ are linearly independent when the matrix $\mathbf{N}(v)$ is simple or semisimple.

If $\mathbf{N}(v)$ is nonsemisimple, there exists only one independent eigenvector ξ_1 for $p_1=p_2$ and a generalized eigenvector ξ_2' takes the place of ξ_2. Let $\boldsymbol{a}_1$, $\boldsymbol{a}_2'$, $\boldsymbol{a}_3$ and $\boldsymbol{b}_1$, $\boldsymbol{b}_2'$, $\boldsymbol{b}_3$ form the columns of $\mathbf{A}$ and $\mathbf{B}$, respectively. The proof remains valid if the matrix $\mathbf{P}$ in (12.4-9) is replaced by (Section 5.9)

$$\begin{bmatrix} p_1 & 1 & 0 \\ 0 & p_1 & 0 \\ 0 & 0 & p_3 \end{bmatrix}.$$

Again the linear dependence of $\boldsymbol{a}_1$, $\boldsymbol{a}_2'$, $\boldsymbol{a}_3$ leads to the linear dependence of ξ_1, ξ_2', ξ_3, which is a contradiction. Therefore $\boldsymbol{a}_1$, $\boldsymbol{a}_2'$, $\boldsymbol{a}_3$ are linearly independent when $\mathbf{N}(v)$ is nonsemisimple.

Instead of $(12.4\text{-}9)_1$ we could postmultiply $(12.3\text{-}15)_1$ by $\mathbf{q}$. Due to the positive definiteness of $\mathbf{H}(v)$ the assumption that $\mathbf{Aq}=\mathbf{0}$ leads to $\mathbf{Bq}=\mathbf{0}$, and hence to (12.4-10) which is a contradiction.

Can $\boldsymbol{b}_\alpha$ (α=1,2,3) be linearly dependent? Assume that they are linearly dependent so that there exists a nonzero vector $\mathbf{q}$ such that

$$\mathbf{Bq}=\mathbf{0}. \tag{12.4-11}$$

Postmultiplying $(12.4\text{-}9)_2$ by $\mathbf{q}$, premultiplying by $\overline{\mathbf{q}}^T\overline{\mathbf{A}}^T$, and by virtue of (12.4-4) we have

$$(\overline{\mathbf{A}}\overline{\mathbf{q}})^T(\mathbf{N}_3+\rho v^2\mathbf{I})(\mathbf{Aq})=\mathbf{0}. \tag{12.4-12}$$

The inverse of the eigenrelation (12.2-14) is

$$\mathbf{N}^{-1}(v)\xi=p^{-1}\,\xi$$

or (Section 6.2)

$$\begin{aligned} \mathbf{N}_1^{(-1)}(v)\boldsymbol{a} + \mathbf{N}_2^{(-1)}(v)\boldsymbol{b} &= p^{-1}\boldsymbol{a}, \\ \mathbf{N}_3^{(-1)}(v)\boldsymbol{a} + \mathbf{N}_1^{(-1)^T}(v)\boldsymbol{b} &= p^{-1}\boldsymbol{b}, \end{aligned} \tag{12.4-13}$$

where

$$\begin{aligned} \mathbf{N}_1^{(-1)}(v)=-(\mathbf{Q}-\rho v^2\mathbf{I})^{-1}\mathbf{R},\quad \mathbf{N}_2^{(-1)}(v)=-(\mathbf{Q}-\rho v^2\mathbf{I})^{-1}, \\ \mathbf{N}_3^{(-1)}(v)=\mathbf{T}-\mathbf{R}^T(\mathbf{Q}-\rho v^2\mathbf{I})^{-1}\mathbf{R}. \end{aligned} \tag{12.4-14}$$

Equation $(12.4\text{-}13)_2$ for p=p_1, p_2, p_3 can be combined into one equation as

$$\mathbf{N}_3^{(-1)}(v)\mathbf{A}+\mathbf{N}_1^{(-1)^T}(v)\mathbf{B}=\mathbf{BP}^{-1}.$$

Following the same procedure as in deriving (12.4-12) one obtains

$$(\overline{\mathbf{A}}\overline{\mathbf{q}})^T \mathbf{N}_3^{(-1)}(v)(\mathbf{A}\mathbf{q}) = \mathbf{0}. \tag{12.4-15}$$

Consider now v=0. The matrices $-\mathbf{N}_3$ and $\mathbf{N}_3^{(-1)}(0)$ are both positive semidefinite. From the structure of $-\mathbf{N}_3$ and $\mathbf{N}_3^{(-1)}(0)$ presented in (6.1-15) and (6.2-12) it is not difficult to conclude from (12.4-12) and (12.4-15) that

$$\mathbf{A}\mathbf{q} = \mathbf{0}.$$

But we have proved that $\mathbf{A}$ is nonsingular. Hence $\mathbf{q}=\mathbf{0}$ which leads to a contradiction. This was essentially the argument used by Stroh (1958) who showed that, for elastostatics, the vectors $\boldsymbol{b}_\alpha$ are linearly independent.

When $v \neq 0$, the vectors $\boldsymbol{b}_\alpha$ can be linearly dependent and (12.4-11) holds. This is the case for surface waves in a half-space $x_2>0$. Equation (12.4-11) then represents the condition that the surface traction at $x_2=0$ vanishes. We will return to (12.4-12) and (12.4-15) in Section 12.9.

We next show that $-\mathbf{L}'(v)$ where the prime denotes differentiation with v is positive definite. From (12.3-2) one obtains

$$\begin{aligned}
\mathbf{Q}'(\theta, v) &= -2\rho v \cos^2\theta\,\mathbf{I},\\
\mathbf{R}'(\theta, v) &= 2\rho v \sin\theta\cos\theta\,\mathbf{I},\\
\mathbf{T}'(\theta, v) &= -2\rho v \sin^2\theta\,\mathbf{I}.
\end{aligned} \tag{12.4-16}$$

Differentiating the identity

$$\mathbf{T}^{-1}(\theta, v)\mathbf{T}(\theta, v) = \mathbf{I}$$

with v and employing $(12.4\text{-}16)_3$ yields

$$\{\mathbf{T}^{-1}(\theta, v)\}' = 2\rho v \sin^2\theta\,\mathbf{T}^{-2}(\theta, v). \tag{12.4-17}$$

Differentiating $\mathbf{N}_3(\theta,v)$ in $(12.3\text{-}9)_3$ and making use of (12.4-16) and (12.4-17) provides the result

$$\mathbf{N}_3'(\theta, v) = 2\rho v\,\mathbf{X}^T\mathbf{X}, \quad \mathbf{X} = \cos\theta\,\mathbf{I} - \sin\theta\,\mathbf{N}_1(\theta, v). \tag{12.4-18}$$

For any nonzero vector $\mathbf{y}$,

$$\mathbf{y}^T\mathbf{N}_3'(\theta, v)\mathbf{y} = 2\rho v(\mathbf{X}\mathbf{y})^T(\mathbf{X}\mathbf{y}) \geq 0$$

in which the equality holds when $\mathbf{X}$ is singular and $\mathbf{y}$ is a null vector of $\mathbf{X}$. Hence $\mathbf{N}_3'(\theta, v)$ is positive semidefinite. To show that $-\mathbf{L}'(v)$ is positive definite we obtain from $(12.3\text{-}11)_3$

$$-\mathbf{y}^T\mathbf{L}'(v)\mathbf{y} = \frac{1}{\pi}\int_0^{\pi} \mathbf{y}^T\mathbf{N}_3'(\theta, v)\mathbf{y}d\theta.$$

The matrix $\mathbf{X}$ is not singular for all θ, an example being when θ=0. Hence the integrand is not zero for all θ, implying that the right side is positive and nonzero. Hence $-\mathbf{L}'(v)$ is positive definite.

The direct proof of positive definiteness of $-\mathbf{L}'(v)$ given above is due to Chadwick and Jarvis (1979c). An indirect proof was provided earlier by Barnett et al. (1973) and Chadwick and Smith (1977).

Let λ and $\mathbf{e}$ be, respectively, an eigenvalue and eigenvector of $\mathbf{L}(v)$, i.e.,

$$\mathbf{L}(v)\mathbf{e} = \lambda\,\mathbf{e}.$$

Differentiation with v yields

$$\mathbf{L}'(v)\mathbf{e} + \mathbf{L}(v)\mathbf{e}' = \lambda'\mathbf{e} + \lambda\mathbf{e}'$$

from which we have

$$\mathbf{e}^T\mathbf{L}'(v)\mathbf{e} = \lambda'\mathbf{e}^T\mathbf{e}.$$

The left side is negative, implying that $\lambda'<0$. Since $\mathbf{L}(0)$ is positive definite the eigenvalues of $\mathbf{L}(v)$ are positive and nonzero at v=0. They decrease monotonically as v increases from zero. At a certain value of v, say v_R, one of the eigenvalues may vanish and $\mathbf{L}(v)$ becomes singular at $v=v_R$.

Following the same procedure it can be shown that the derivative of $(\mathbf{Q}-\rho v^2\mathbf{I})^{-1}$ with v is positive definite. The eigenvalues of $(\mathbf{Q}-\rho v^2\mathbf{I})^{-1}$, which are positive and nonzero at v=0, increase monotonically with v. Likewise it can be shown that the derivative of $\mathbf{N}_3^{(-1)}(v)$ in (12.4-14) with v is negative definite. Consequently the eigenvalues of $\mathbf{N}_3^{(-1)}(v)$ decrease monotonically with v.

12.5 Steady Motion of a Line Singularity

A line dislocation moving at a constant speed $v<\hat{v}$ along the x_1-axis in an infinite anisotropic elastic medium has been studied by Bullough and Bilby (1954). They presented explicit solutions for monoclinic materials with the symmetry plane at x_3=0 for which the edge and screw dislocations can be considered separately. For a general anisotropic solid with a line force and/or a line dislocation moving along the x_1-axis the solution is given by (8.1-16) and (8.1-18) in which z is replaced by (12.2-1b). The $\mathbf{f}$ and $\mathbf{b}$ in (8.1-18) are the concentrated force and the Burgers vector, respectively. The real form solution is given by (8.1-21) and (8.1-20b) in which $\mathbf{S}$, $\mathbf{H}$, $\mathbf{L}$, $\mathbf{S}(\theta)$, $\mathbf{H}(\theta)$, and $\mathbf{L}(\theta)$ are replaced by $\mathbf{S}(v)$, $\mathbf{H}(v)$, $\mathbf{L}(v)$,

$\mathbf{S}(\theta, v)$, $\mathbf{H}(\theta, v)$, and $\mathbf{L}(\theta, v)$, respectively. The r and θ are related to x_1 and x_2 by

$$x_1 - vt = r\cos\theta, \quad x_2 = r\sin\theta.$$

For a moving line singularity in a half-space the solution is given by (8.6-2). The $\mathbf{q}_\beta$ in (8.6-2) is presented in (8.6-11) if the boundary x_2=0 of the half-space is traction-free. Since (8.6-11) contains the matrix $\mathbf{B}^{-1}$, the solution is valid if $\mathbf{B}$ is nonsingular. From $(12.2\text{-}24)_5$ this is equivalent to the requirement that $\mathbf{L}(v)$ be nonsingular. We will see in Section 12.9 that $\mathbf{L}(v)$ is singular when v is a surface wave speed for the half-space. Thus the solution (8.6-2) for a moving line singularity in the half-space with a free boundary is not valid when the speed v of the steady motion is equal to a surface wave speed of the half-space (Malen and Lothe, 1971).

If the boundary x_2=0 of the half-space is fixed, $\mathbf{q}_\beta$ in (8.6-2) is given by (8.6-12) which contains the matrix $\mathbf{A}^{-1}$. We have shown in Section 12.4 that $\mathbf{A}$ is nonsingular for any subsonic speed v.

For a moving line singularity in a bimaterial the solution is given by (8.8-14), (8.8-15), and (8.8-21) if the singularity is moving along the interface of the two materials. If the singularity is moving parallel to the interface, the solution is given by (8.9-3), (8.9-4), and (8.9-8). In both cases the solution is valid if

$$\mathbf{M}_1(v) + \overline{\mathbf{M}}_2(v)$$

is not singular where the subscripts 1 and 2 refer to materials 1 and 2 in the bimaterial. We will show in Section 12.9 that the value of v for which $\mathbf{M}_1(v) + \overline{\mathbf{M}}_2(v)$ is singular is a wave speed for the propagation of an interfacial wave in the bimaterial.

Extension of the Stroh formalism to steady state motion is not limited to steady motion of a line singularity. It can be applied to a propagating crack in a homogeneous medium or an extending interfacial crack in a bimaterial (K. C. Wu, 1989b, 1991a), harmonic waves in anisotropic layered composites (Braga and Herrmann, 1988; Ting and Chadwick, 1988, 1990; Norris, 1991), surface waves, and interfacial waves. We will study surface waves and interfacial waves in Section 12.9. In the next sections we present various subjects that are intimately related to surface waves and interfacial waves.

12.6 Homogeneous Plane Waves or Body Waves

The displacement given by

$$u_i = a_i f(z), \quad \text{or} \quad \mathbf{u} = \boldsymbol{a} f(z), \tag{12.6-1a}$$

$$z=\mathbf{n}\cdot\mathbf{x}-ct, \tag{12.6-1b}$$

where $c>0$ and $\mathbf{n}$ is a real unit vector represents a steady wave moving in the direction of $\mathbf{n}$ at a constant speed c. It is a *body wave* or a *homogeneous plane wave* because the displacement and the stress are homogeneous on the plane $\mathbf{n}\cdot\mathbf{x}$=constant. The vector $\boldsymbol{a}$ is the *polarization* vector. We consider here a fully three-dimensional motion so that all three components of the vectors $\mathbf{n}$ and $\mathbf{x}$ can be nonzero. If $\mathbf{n}$ is restricted to the $x_3=0$ plane, the dependence on x_3 disappears and the deformation is two-dimensional, although the u_3 component of the displacement may still be nonzero due to the anisotropy of the material. The connection between the solutions (12.6-1) and (12.2-1) will be given in Section 12.7. The stress obtained from (12.1-1) is

$$\sigma_{ij}=C_{ijks}a_k n_s f'(z) \tag{12.6-2}$$

where the prime denotes differentiation with z. The equation of motion (12.1-2) reduces to

$$(C_{ijks}n_j n_s)a_k=(\rho c^2)a_i \tag{12.6-3}$$

or,

$$\mathbf{Q}(\mathbf{n})\boldsymbol{a}=(\rho c^2)\boldsymbol{a} \tag{12.6-4}$$

where

$$Q_{ik}(\mathbf{n})=C_{ijks}n_j n_s. \tag{12.6-5}$$

$\mathbf{Q}(\mathbf{n})$ is the acoustic tensor. It reduces to $\mathbf{Q}(\theta)$ in (12.3-3) when $\mathbf{n}$ is on the $x_3=0$ plane defined in (12.3-1). Equation (12.6-4) tells us that ρc^2 and $\boldsymbol{a}$ are, respectively, an eigenvalue and eigenvector of $\mathbf{Q}(\mathbf{n})$. Since $\mathbf{Q}(\mathbf{n})$ is positive definite the three eigenvalues are positive and nonzero. Therefore there are three real wave speeds c for a given $\mathbf{n}$. The associated polarization vector $\boldsymbol{a}$ are also real and mutually orthogonal.

In the (x_1,x_2,x_3) space the surface V generated by the vectors

$$v_i=cn_i \tag{12.6-6}$$

for all choices of $\mathbf{n}$ is called the *velocity surface*. Its inverse surface S generated by the vectors

$$s_i=c^{-1}n_i \tag{12.6-7}$$

for all choices of $\mathbf{n}$ is the *slowness surface*. Since there are three wave speeds for c we have a three-sheeted velocity surface and a three-sheeted slowness surface. The sheets may cross each other. From (12.6-3) ρc^2 is a root of the determinant

$$\left|C_{ijkq}n_j n_q - \rho c^2 \delta_{ik}\right| = 0 \tag{12.6-8}$$

or

$$S = \left|C_{ijkq}s_j s_q - \rho \delta_{ik}\right| = 0. \tag{12.6-9}$$

This is an algebraic equation of degree six in s_1, s_2, s_3. Any straight line can intersect the slowness surface in at most six points. Consequently, the inner most sheet of the slowness surface must be convex. The slowness surface is centro-symmetric because (12.6-9) is invariant with the replacement of s_i by $-s_i$.

Another surface W, the *wave surface*, is the wave front due to a point source. A point source generates infinitesimal plane waves in all directions. A plane wave with a unit normal vector $\mathbf{n}$ moves to, after a unit time, the location

$$\mathbf{n}\cdot\mathbf{x} = c.$$

The envelope of the above plane wave fronts for all choices of $\mathbf{n}$ is obtained by differentiating the equation with n_j, i.e.,

$$x_j = \frac{\partial c}{\partial n_j}.$$

Therefore the wave surface W is generated by the vectors

$$r_j = \frac{\partial c}{\partial n_j} \tag{12.6-10}$$

for all choices of $\mathbf{n}$.

Without loss in generality we normalize the vector $\boldsymbol{a}$ to a unit vector. Differentiation with n_p of the equation

$$C_{ijks}n_j n_s a_i a_k = \rho c^2 \tag{12.6-11}$$

obtained from (12.6-3) yields

$$2C_{ipks}n_s a_i a_k + 2C_{ijks}n_j n_s a_k \frac{\partial a_i}{\partial n_p} = 2\rho c \frac{\partial c}{\partial n_p}$$

in which the symmetry relation $C_{ijks} = C_{ksij}$ has been employed. The second term on the left is, by (12.6-3),

$$2\rho c^2 a_i \frac{\partial a_i}{\partial n_p} = \rho c^2 \frac{\partial}{\partial n_p}(\boldsymbol{a}\cdot\boldsymbol{a})$$

which vanishes because $\boldsymbol{a}$ is a unit vector. Hence r_j of (12.6-10) can be written as

$$r_j = \frac{1}{\rho c} C_{ijks} a_i a_k n_s. \tag{12.6-12}$$

It follows immediately from (12.6-7), (12.6-12), and (12.6-11) that

$$r_j s_j = 1. \tag{12.6-13}$$

The wave surface and the slowness surface are the *polar reciprocals* of each other (Duff, 1960). An interesting property of the wave surface due to a point source is that there may exist a region called a *lacuna* within the wave surface in which there is no disturbance (Burridge, 1967).

We will show that the vector r_j is normal to the slowness surface. We write (12.6-3) as

$$(C_{ijkq} s_j s_q - \rho \delta_{ik}) a_k = 0. \tag{12.6-14}$$

The variation of this equation with the vector s_j is

$$\{C_{ijkq}(ds_j)s_q + C_{ijkq}s_j(ds_q)\}a_k + (C_{ijkq}s_j s_q - \rho\delta_{ik})(da_k) = 0.$$

Premultiplied by a_i and with the use of (12.6-14) and the symmetry relation $C_{ijks} = C_{ksij}$ this equation yields

$$C_{ijkq} a_i a_k s_q (ds_j) = 0.$$

In view of (12.6-12) this is equivalent to

$$r_j(ds_j) = 0. \tag{12.6-15}$$

On the other hand from the equation for the slowness surface (12.6-

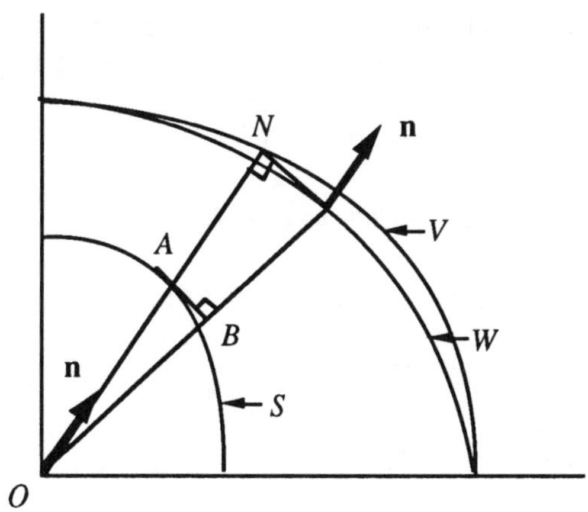

Fig. 12.1. The velocity surface V, slowness surface S, and wave surface W.

9) we obtain

$$dS = \frac{\partial S}{\partial s_j}(ds_j) = 0$$

in which $\partial S/\partial s_j$ is normal to the slowness surface. Comparison with (12.6-15) verifies that r_j is proportional to $\partial S/\partial s_j$ and hence is normal to the slowness surface.

Figure 12.1 shows the cross section of the velocity surface V, the slowness surface S, and the wave surface W at a plane containing the origin. The lines ON and OA represent the vectors v_i and s_i, respectively. The line NM denotes the plane wave front with a unit normal vector **n**. Since the wave surface W is the envelope of plane wave fronts, NM is tangential to the wave surface at M. The normal to the wave surface at M is therefore **n**. The line OM is the *ray velocity* r_i. The normal to the slowness surface at A is in the direction of r_i. Hence the tangent to the slowness surface at A is perpendicular to the line OM.

In the theory of characteristics for hyperbolic partial differential equations the vector r_i is in the direction of the *ray*, or the *bicharacteristics* (Courant and Hilbert, 1962), and is the *ray velocity*, or the *bicharacteristic velocity*. The velocity surface is a cross section of the *normal cone* while the wave surface is a cross section of the *Monge cone*.

From (12.6-1) the kinetic energy T of a body wave is

$$T = \tfrac{1}{2}\rho\dot{u}_i\dot{u}_i = \tfrac{1}{2}\rho c^2 a_i a_i (f')^2 .$$

The strain energy ε is

$$\varepsilon = \tfrac{1}{2}C_{ijks}u_{i,j}u_{k,s} = \tfrac{1}{2}C_{ijks}a_i n_j a_k n_s (f')^2 = \tfrac{1}{2}\rho c^2 a_i a_i (f')^2 ,$$

use being made of (12.6-3) in the last equality. Thus the total energy is equally divided between the strain energy and the kinetic energy for a body wave.

The total energy U in a volume is

$$U = \frac{1}{2}\int_V (\rho\dot{u}_i\dot{u}_i + C_{ijks}u_{i,j}u_{k,s})dV$$

in which V here denotes the volume integral. Hence

$$\frac{dU}{dt} = \int_V (\rho\ddot{u}_i\dot{u}_i + C_{ijks}u_{k,s}\dot{u}_{i,j})dV = \int_V (\sigma_{ij}\dot{u}_i)_{,j}dV$$

appeal being made to (12.1-1), (12.1-2), and the symmetry relation $C_{ijks} = C_{ksij}$. If there is no energy source inside the volume,

$$-\sigma_{ij}\dot{u}_i = -\sigma_{ji}\dot{u}_i$$

represents the *energy flux* flowing out of the volume. By (12.6-1), (12.6-2), and (12.6-12),

$$-\sigma_{ij}\dot{u}_i = c\,C_{ijks}a_k n_s a_i\{f'(z)\}^2 = \rho c^2 r_j\{f'(z)\}^2$$

so that the energy flux is co-directional with the ray velocity.

It is seen that for a body wave with the unit normal vector **n** the energy is propagated in the direction of the ray velocity that is in general different from **n**. For a curved wave front propagating into a uniform region, the discontinuity across the wave front and the energy are also propagated in the direction of the ray velocity, not in the direction normal to the wave front. For isotropic materials the slowness surface consists of three concentric spheres, two of which coalesce into the same sphere. The ray direction and the normal to the slowness surface coincide. For more information on homogeneous plane waves or body waves the article by Lothe (1991) should be consulted.

12.7 Limiting Wave Speed $\hat{\upsilon}$ and Exceptional Waves

When the unit normal vector **n** in (12.6-1b) is on the x_3=0 plane defined in (12.3-1),

$$z = \mathbf{n}\cdot\mathbf{x} - ct = \cos\theta\{x_1 - \upsilon t + \tan\theta\, x_2\}, \qquad (12.7\text{-}1)$$

$$c = \upsilon\cos\theta. \qquad (12.7\text{-}2)$$

The plane wave solution (12.6-1) is then equivalent to (12.2-1) if we accept (12.7-2) and

$$p = \tan\theta. \qquad (12.7\text{-}3)$$

Indeed, with (12.7-2) and (12.7-3), (12.6-3) reduces to (12.2-4). Equations (12.7-2) and (12.7-3) suggest a graphical solution for a real p of (12.2-4) with the use of the slowness surface. Figure 12.2 shows the intersection with the x_3=0 plane of a three-sheeted slowness surface. For a given υ, draw a vertical line at perpendicular distance υ^{-1} from the origin. Let P_3 and P_6 be the points of intersection with the slowness surface and θ_3 and θ_6 be, respectively, the angles OP_3 and OP_6 make with the x_1-axis. We then have

$$p_3 = \tan\theta_3, \quad p_6 = \tan\theta_6.$$

In this case there are only two real roots for p. The remaining four roots are two pairs of complex conjugates.

It is clear that, as υ increases from zero, the p_α are complex until $\upsilon = \hat{\upsilon}$ at which the vertical line AA' first touches the slowness surface. $\hat{v}$ is the *limiting wave speed*, and the plane wave with the speed $\hat{v}$ is the *limiting wave*. The steady wave motion (12.2-1) is *subsonic* when $0 \le v < \hat{v}$. It is *supersonic* if $v > \hat{v}$. At $v = \hat{v}$ the motion is *transonic*. It represents the first transonic state. The vertical lines *BB′* and *CC′* represent the *subsequent transonic states*.

A homogeneous plane wave or a body wave is said to be *exceptional* when there exists a plane with unit normal **d** on which the surface traction vanishes (Al'shits and Lothe, 1979a,b,c). This means

$$\sigma_{ij} d_j = 0$$

or, by (12.6-2),

$$C_{ijks} a_k n_s d_j = 0.$$

From (12.6-3), (12.6-12), and the symmetry of C_{ijks},

$$a_i d_i = \frac{1}{\rho c^2} C_{ijks} n_j n_s a_k d_i = 0, \quad r_j d_j = \frac{1}{\rho c} C_{ijks} a_i a_k n_s d_j = 0. \qquad (12.7\text{-}4)$$

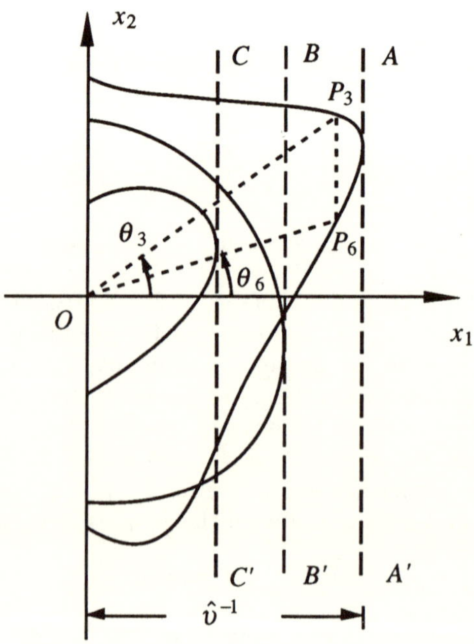

Fig. 12.2. The vertical line *AA′* tangential to the slowness surface is the first transonic state. The vertical lines *BB′* and *CC′* represent subsequent transonic states.

Therefore d_i is perpendicular to both a_i and r_i. In other words, a_i and r_i are parallel to the traction-free surface. Since r_i is normal to the slowness surface, d_i must be tangential to the slowness surface. For plane waves in a half-space $x_2>0$ with the surface $x_2=0$ being traction-free, it is clear that the plane wave represented by P_3P_6 cannot be exceptional whereas the transonic states represented by AA', BB', CC' can be exceptional. An exceptional wave is a generalization of the classical SH-wave in isotropic materials. A body wave that is not exceptional is said to be *normal*.

Figure 12.3 shows six possible types of first transonic states depending on whether the vertical line AA' in Fig. 12.2 is tangential to the slowness surface at one, two, or three points and whether the points of tangency coalesce (Chadwick and Smith, 1977). The first transonic state shown in Fig. 12.2 is of Type 1. At a transonic state the eigenvalue p_α corresponding to the point of tangency becomes real and the associated body wave is a limiting wave. It can be exceptional or normal. Barnett and Lothe (1985) have shown that an exceptional transonic state is necessarily of Type 1. Thus the first transonic state for Types 2 through 6 is normal. The first transonic state for Type 1 can be either normal or exceptional.

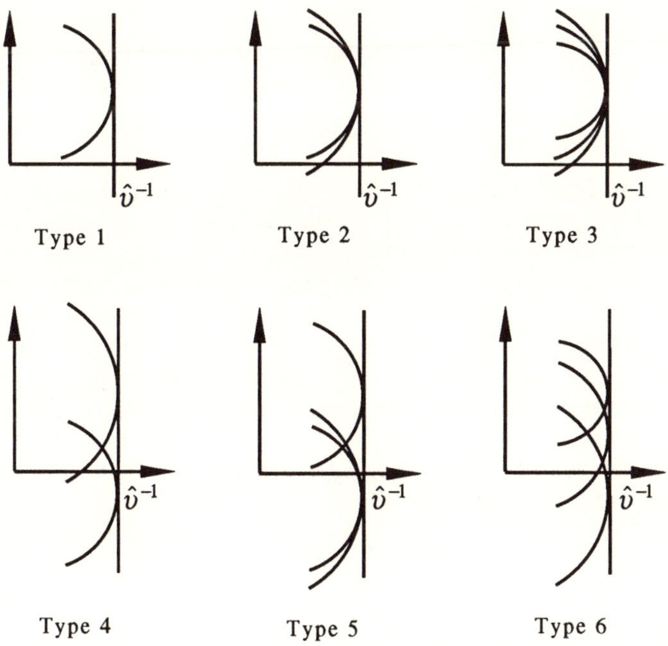

Fig. 12.3. The six possible types of first transonic state.

12.8 One-Component Surface Waves

We have shown in Section 12.6 that the steady wave motion given by (12.2-1) is a body wave when p is real. When p is complex, it is an *inhomogeneous plane wave*. The displacement and the stress are no longer constants on a plane. Consider the inhomogeneous plane wave

$$\mathbf{u} = \boldsymbol{a}\, e^{ikz}, \qquad \phi = \boldsymbol{b}\, e^{ikz}, \tag{12.8-1a}$$

$$z = x_1 - \upsilon t + p x_2, \tag{12.8-1b}$$

where $k>0$ is the real wave number, $\upsilon>0$ is the wave speed, and p, $\boldsymbol{a}$, and $\boldsymbol{b}$ are complex. They are determined by (12.2-4) and (12.2-6). When (12.8-1) is a solution, its complex conjugate is also a solution. By a linear combination of the complex conjugate solutions one can obtain a real form solution for $\mathbf{u}$ and ϕ. This is equivalent to taking the real or imaginary part of (12.8-1) as the solution. Let the real and imaginary parts of p and $\boldsymbol{a}$ be given by

$$p = p' + ip'', \qquad \boldsymbol{a} = \boldsymbol{a}' + i\boldsymbol{a}''.$$

Equation $(12.8\text{-}1\text{a})_1$ can then be written as

$$\begin{aligned} \mathbf{u} &= e^{-kp''x_2}\{\mathbf{a}'\cos\Theta + \mathbf{a}''\sin\Theta\} \\ &+ ie^{-kp''x_2}\{\mathbf{a}'\cos(\Theta+\tfrac{\pi}{2}) + \mathbf{a}''\sin(\Theta+\tfrac{\pi}{2})\} \end{aligned} \tag{12.8-2}$$

where

$$\Theta = k(\upsilon t - x_1 - p'x_2). \tag{12.8-3}$$

The real and imaginary parts of $\mathbf{u}$ differ by a phase angle of $\pi/2$. At a fixed position (x_1,x_2) the motion is sinusoidal. The locus of the displacement is an ellipse with $\boldsymbol{a}'$ and $\boldsymbol{a}''$ being a pair of conjugate radii of the ellipse. As t increases the displacement vector on the ellipse moves from $\boldsymbol{a}'$ to $\boldsymbol{a}''$ (Section 1.8). We therefore take (12.8-1a) as the solution without specifying the real or the imaginary part of (12.8-1a) because they differ only in the phase angle.

Consider now the one-component inhomogeneous plane wave (12.8-1) propagating in the half-space $x_2 \geq 0$. The boundary conditions are the vanishing of displacement at $x_2=\infty$ and the surface traction at $x_2=0$. This means

$$\mathbf{u} = \mathbf{0} \quad \text{at} \quad x_2=\infty, \tag{12.8-4}$$

$$\phi = \mathbf{0} \quad \text{at} \quad x_2=0. \tag{12.8-5}$$

It is known that one-component surface waves cannot propagate in an isotropic elastic solid. The possibility of one-component surface waves propagating in an anisotropic elastic material was first pointed out by Barnett et al. (1991) who established the conditions for the existence of these waves. It was demonstrated by Barnett and Chadwick (1991) through a numerical example that one-component surface waves can exist in an anisotropic elastic medium. Further studies on one-component surface waves were made by Chadwick (1992), Norris (1992), Ting (1992f), and Wang and Gundersen (1992).

The eigenrelation (12.4-8) can be written as

$$\begin{aligned}(\mathbf{N}_1 - p\mathbf{I})\boldsymbol{a} + \mathbf{T}^{-1}\boldsymbol{b} &= \mathbf{0},\\ (\mathbf{N}_3 + \rho v^2\mathbf{I})\boldsymbol{a} + (\mathbf{N}_1 - p\mathbf{I})^T\boldsymbol{b} &= \mathbf{0}.\end{aligned} \tag{12.8-6}$$

Equation (12.8-1) satisfies (12.8-4) and (12.8-5) if we set

$$\operatorname{Im} p > 0, \quad \boldsymbol{b} = \mathbf{0}. \tag{12.8-7}$$

Equations (12.8-6) then reduce to

$$(\mathbf{N}_1 - p\mathbf{I})\boldsymbol{a} = \mathbf{0}, \tag{12.8-8}$$

$$(\mathbf{N}_3 + \rho v^2\mathbf{I})\boldsymbol{a} = \mathbf{0}. \tag{12.8-9}$$

Thus p and $-\rho v^2$ are, respectively, eigenvalues of $\mathbf{N}_1$ and $\mathbf{N}_3$. It was shown in Section 6.1 that the matrices $-\mathbf{N}_1$ and $-\mathbf{N}_3$ have the structures

$$-\mathbf{N}_1 = \begin{bmatrix} r_6 & 1 & s_6 \\ r_2 & 0 & s_2 \\ r_4 & 0 & s_4 \end{bmatrix}, \qquad -\mathbf{N}_3 = \begin{bmatrix} \kappa & 0 & \zeta \\ 0 & 0 & 0 \\ \zeta & 0 & \mu \end{bmatrix} \tag{12.8-10}$$

where κ, μ, ζ, r_i and s_i $(i=6,2,4)$ are possibly nonzero elements. Equations (12.8-8) and (12.8-9) can therefore be written as

$$\begin{bmatrix} r_6+p & 1 & s_6 \\ r_2 & p & s_2 \\ r_4 & 0 & s_4+p \end{bmatrix} \begin{bmatrix} a_1 \\ a_2 \\ a_3 \end{bmatrix} = \mathbf{0}, \tag{12.8-11}$$

$$\begin{bmatrix} \kappa-\rho v^2 & 0 & \zeta \\ 0 & -\rho v^2 & 0 \\ \zeta & 0 & \mu-\rho v^2 \end{bmatrix} \begin{bmatrix} a_1 \\ a_2 \\ a_3 \end{bmatrix} = \mathbf{0}. \tag{12.8-12}$$

Equation (12.8-12) consists of three equations of which the second equation yields

$$a_2 = 0. \tag{12.8-13}$$

Therefore the displacement is polarized on the $x_2=0$ plane. With $a_2=0$, the first and third equations of (12.8-11) are

$$\begin{bmatrix} r_6+p & s_6 \\ r_4 & s_4+p \end{bmatrix}\begin{bmatrix} a_1 \\ a_3 \end{bmatrix}=\mathbf{0} \tag{12.8-14}$$

from which we obtain

$$(r_6+p)(s_4+p)-r_4 s_6=0. \tag{12.8-15}$$

The eigenvalue p with Imp>0 is

$$p=\varepsilon+i\delta,$$

$$\varepsilon=-\tfrac{1}{2}(r_6+s_4),\quad \delta=\{-r_4 s_6-\tfrac{1}{4}(r_6-s_4)^2\}^{1/2}.$$

The requirement that Imp>0 means that

$$-r_4 s_6-\tfrac{1}{4}(r_6-s_4)^2>0. \tag{12.8-16}$$

The inequality tells us that r_4 and s_6 cannot be zero, and must have opposite signs. The vector $\boldsymbol{a}$ in (12.8-11) can be chosen as

$$\boldsymbol{a}=\begin{bmatrix} a_1 \\ a_2 \\ a_3 \end{bmatrix}=\begin{bmatrix} -s_6 \\ 0 \\ r_6+p \end{bmatrix} \tag{12.8-17}$$

That s_6 and Imp cannot vanish establishes the result that a_1 and a_3 are nonzero and a_3 is complex. By (12.8-12) and the second equation of (12.8-11) we conclude that

$$r_2=s_2=\zeta=0,\quad \kappa=\mu=\rho v^2. \tag{12.8-18}$$

Equations (12.8-16) and (12.8-18) are the conditions for the existence of a one-component surface wave. With the expressions of κ, μ, ζ, r_i and s_i (i=6,2,4) given in (6.1-18) and (6.1-16), they can be written explicitly as

$$\begin{gathered} s'_{12}=s'_{15}=s'_{25}=0,\quad s'_{11}=s'_{55}=\rho v^2, \\ -s'_{14}s'_{56}-\tfrac{1}{4}(s'_{45}-s'_{16})^2>0, \end{gathered} \tag{12.8-19}$$

in which $s'_{\alpha\beta}$ are the reduced elastic compliances. r_4, r_6, s_4, and s_6 are simplified to

$$r_\alpha=-s'_{1\alpha}/s'_{11},\quad s_\alpha=-s'_{5\alpha}/s'_{11}\quad (\alpha=4,6).$$

An important consequence of the inequality in (12.8-19) is that s'_{14} and s'_{56} cannot vanish and must have opposite signs. Thus one-component surface waves cannot propagate in monoclinic materials with the symmetry plane at $x_2=0$ or $x_3=0$ because s'_{14} and s'_{56} vanish for these materials (Barnett et al., 1991; Chadwick, 1992).

The stress function ϕ vanishes because $\boldsymbol{b}=\mathbf{0}$. This does not mean that the stresses are zero. From (12.2-11) and (12.8-1),

$$\mathbf{t}_1 = -\rho v \dot{\mathbf{u}} = ik\rho v^2 \boldsymbol{a}\, e^{ikz},$$

$$\mathbf{t}_2 = \mathbf{0}.$$

The nonzero stresses are σ_{11} and σ_{31}, in addition to σ_{33}. Consequently, one-component surface waves have the unique feature that the deformation is both plane strain and plane stress polarized on the (x_1,x_3)-plane.

The vanishing of the vector $\boldsymbol{b}$ means that the eigenvector $\xi^T=(\boldsymbol{a},\boldsymbol{b})$ cannot be normalized, i.e.,

$$2\boldsymbol{a}\cdot\boldsymbol{b}= 1$$

obtained from (12.2-17b) is violated. This implies that the 6×6 matrix $\mathbf{N}(v)$ must be nonsemisimple. It could also be semisimple (Wang and Gundersen, 1992). However the eigenvector $\xi^T=(\boldsymbol{a},\mathbf{0})$ for a semisimple $\mathbf{N}(v)$ is a linear combination of a pair of orthonormalized eigenvectors (Section 12.11). Therefore it is more appropriate to classify the surface wave associated with a semisimple $\mathbf{N}(v)$ as a two-component surface wave. In any case, the pair of complex conjugates p obtained from (12.8-15) are double roots. This accounts for four p. The remaining two p can be determined as follows. From (12.8-18) and (12.8-10),

$$(\mathbf{N}_1 - p\mathbf{I}) = -\begin{bmatrix} r_6+p & 1 & s_6 \\ 0 & p & 0 \\ r_4 & 0 & s_4+p \end{bmatrix}, \tag{12.8-20}$$

$$\mathbf{N}_3 + \rho v^2 \mathbf{I} = \rho v^2 \mathbf{I}_2, \quad \mathbf{I}_2 = \begin{bmatrix} 0 & 0 & 0 \\ 0 & 1 & 0 \\ 0 & 0 & 0 \end{bmatrix}.$$

It is easily shown that

$$(\mathbf{N}_1 - p\mathbf{I})^T \mathbf{I}_2 = -p\mathbf{I}_2 = \mathbf{I}_2(\mathbf{N}_1 - p\mathbf{I}), \quad \mathbf{I}_2\mathbf{T}^{-1}\mathbf{I}_2 = (\mathbf{T}^{-1})_{22}\mathbf{I}_2. \tag{12.8-21}$$

Elimination of $\boldsymbol{b}$ between the two equations in (12.8-6) leads to

$$\{(\mathbf{N}_1 - p\mathbf{I})^T \mathbf{T}(\mathbf{N}_1 - p\mathbf{I}) - \rho v^2 \mathbf{I}_2\}\boldsymbol{a} = \mathbf{0}$$

or, by virtue of (12.8-21),

$$(\mathbf{N}_1 - p\mathbf{I})^T\left(\mathbf{T} - \frac{\rho v^2}{p^2}\mathbf{I}_2\right)(\mathbf{N}_1 - p\mathbf{I})\boldsymbol{a} = \mathbf{0}. \qquad (12.8\text{-}22)$$

The vanishing of the determinant of $(\mathbf{N}_1\text{-}p\mathbf{I})$ is given in (12.8-15) and is identical to the vanishing of the determinant of $(\mathbf{N}_1\text{-}p\mathbf{I})^T$. This confirms that the pair of complex conjugates p of (12.8-15) are double roots. Since

$$\mathbf{T} - \frac{\rho v^2}{p^2}\mathbf{I}_2 = \left(\mathbf{I} - \frac{\rho v^2}{p^2}\mathbf{I}_2\mathbf{T}^{-1}\right)\mathbf{T}$$

in which

$$\mathbf{I} - \frac{\rho v^2}{p^2}\mathbf{I}_2\mathbf{T}^{-1} = \begin{bmatrix} 1 & 0 & 0 \\ -\frac{\rho v^2}{p^2}(\mathbf{T}^{-1})_{21} & 1 - \frac{\rho v^2}{p^2}(\mathbf{T}^{-1})_{22} & -\frac{\rho v^2}{p^2}(\mathbf{T}^{-1})_{23} \\ 0 & 0 & 1 \end{bmatrix},$$

the remaining two p are

$$p = \pm\{\rho v^2(\mathbf{T}^{-1})_{22}\}^{1/2}. \qquad (12.8\text{-}23)$$

They are real.

The two waves associated with the real p are body waves that do not leave the surface x_2=0 traction-free. However, (12.8-22) with p given by (12.8-23) is satisfied by

$$\boldsymbol{a} = -(\mathbf{N}_1 - p\mathbf{I})^{-1}\mathbf{T}^{-1}\mathbf{i}_2$$

where $\mathbf{i}_2^T$=(0,1,0) and we obtain from $(12.8\text{-}6)_1$

$$\boldsymbol{b} = \mathbf{i}_2.$$

Hence the traction vectors at x_2=0 due to the two body waves are co-directional with the x_2-axis. One of the body waves represents the incident wave while the other is a reflected wave. By a linear combination of the incident and reflected waves the surface x_2=0 can be left traction-free. The pair thus forms a *simple reflection*, i.e., one incident and one reflected wave alone are able to leave the boundary traction-free (Al'shits and Lothe, 1981).

In deriving (12.8-22) the inequality in (12.8-16) was not employed. If the inequality is reversed the roots of p in (12.8-15) become real and remain double roots. We then have a subsequent transonic state of Type 4. Since this body wave leaves the surface x_2=0 traction-free, it is an exceptional limiting wave. If the inequality is replaced by an equality, the p of (12.8-15) is a root of multiplicity four. We then have an exceptional subsequent transonic state of Type 2 or Type E1 that has zero curvature of the slowness surface at the point of contact with the vertical line (Barnett et al., 1990).

In closing this section we derive the condition for $\mathbf{N}(v)$ to be semisimple. When $\mathbf{N}(v)$ is semisimple there exists a vector $\boldsymbol{a}$ different from the one given in (12.8-17). Let $\mathbf{y}$ be the left eigenvector of $(\mathbf{N}_1 - p\mathbf{I})$, i.e.,

$$(\mathbf{N}_1 - p\mathbf{I})^T \mathbf{y} = \mathbf{0}.$$

From (12.8-20) we may take

$$\mathbf{y}^T = (-r_4,\ r_4/p,\ r_6+p) \tag{12.8-24}$$

with p given by (12.8-15). Equation (12.8-22) provides a new $\boldsymbol{a}$ if

$$(\mathbf{N}_1 - p\mathbf{I})\boldsymbol{a} = \left(\mathbf{T} - \frac{\rho v^2}{p^2}\mathbf{I}_2\right)^{-1} \mathbf{y} \tag{12.8-25}$$

which has a solution for $\boldsymbol{a}$ if

$$\mathbf{y}^T \left(\mathbf{T} - \frac{\rho v^2}{p^2}\mathbf{I}_2\right)^{-1} \mathbf{y} = 0. \tag{12.8-26}$$

This is the condition for $\mathbf{N}(v)$ to be semisimple. With the aid of $(12.8\text{-}21)_2$ it can be shown that

$$\left(\mathbf{T} - \frac{\rho v^2}{p^2}\mathbf{I}_2\right)\left(\mathbf{T}^{-1} - \gamma \mathbf{T}^{-1}\mathbf{I}_2\mathbf{T}^{-1}\right) = \mathbf{I},$$

$$\gamma = \left((\mathbf{T}^{-1})_{22} - \frac{p^2}{\rho v^2}\right)^{-1}.$$

Hence

$$\left(\mathbf{T} - \frac{\rho v^2}{p^2}\mathbf{I}_2\right)^{-1} = \mathbf{T}^{-1} - \gamma \mathbf{T}^{-1}\mathbf{I}_2\mathbf{T}^{-1}, \tag{12.8-27}$$

and (12.8-26) is written more explicitly as

$$\mathbf{y}^T(\mathbf{T}^{-1} - \gamma \mathbf{T}^{-1}\mathbf{I}_2\mathbf{T}^{-1})\mathbf{y} = \mathbf{0}. \tag{12.8-28}$$

The eigenvector $\boldsymbol{b}$ is obtained from $(12.8\text{-}6)_1$. By (12.8-25) and (12.8-27),

$$\boldsymbol{b} = -(\mathbf{I} - \gamma \mathbf{I}_2\mathbf{T}^{-1})\mathbf{y}.$$

If the vector $\mathbf{k}$ denotes the second column of $\mathbf{T}^{-1}$,

$$\mathbf{T}^{-1}\mathbf{I}_2\mathbf{T}^{-1} = \mathbf{k}\mathbf{k}^T, \quad \mathbf{I}_2\mathbf{T}^{-1}\mathbf{y} = (\mathbf{k}^T\mathbf{y})\mathbf{i}_2.$$

Explicit expressions of the components of $\mathbf{T}^{-1}$ (which is $\mathbf{N}_2$) in terms of $s'_{\alpha\beta}$ are given in (6.1-17).

12.9 Surface Waves and Interfacial Waves

Surface waves in an isotropic elastic half-space are known as Rayleigh (1885) waves. The one-component surface wave presented in the previous section consists of one inhomogeneous plane wave. It can propagate only in certain special anisotropic elastic materials. For general anisotropic elastic media the surface wave in general consists of two or three inhomogeneous plane waves. Let v_R denote the surface wave speed. While two-component surface waves can occur for $v_R \leq \hat{v}$ or $v_R > \hat{v}$, three-component surface waves can occur only for a subsonic speed $v_R < \hat{v}$. Isotropic materials are an example for which the surface wave is subsonic, and consists of two inhomogeneous plane waves. We discuss separately surface waves with $v_R < \hat{v}$, $v_R > \hat{v}$, and interfacial waves below.

(i) Subsonic Surface Waves with $v_R < \hat{v}$

The following presentation is largely based on the work of Barnett and Lothe (1974a, 1978, 1985, 1992), Lothe and Barnett (1976a), Chadwick and Smith (1977), and Chadwick (1989d, 1990). Let the surface wave in a half-space $x_2 \geq 0$ be given by

$$\mathbf{u} = \sum_{\alpha=1}^{3} q_\alpha \boldsymbol{a}_\alpha e^{ikz_\alpha}, \quad \boldsymbol{\phi} = \sum_{\alpha=1}^{3} q_\alpha \boldsymbol{l}_\alpha e^{ikz_\alpha}, \tag{12.9-1a}$$

$$z_\alpha = x_1 - vt + p_\alpha x_2, \tag{12.9-1b}$$

in which $k>0$ is the real wave number, $0<v<\hat{v}$ is the steady wave speed, and q_α $(\alpha=1,2,3)$ are complex constants to be determined. If one of the q_α vanishes we have a two-component surface wave. The condition that displacement at $x_2=\infty$ vanishes means that

$$\text{Im}\, p_\alpha > 0 \quad (\alpha=1,2,3).$$

It remains to satisfy the boundary condition at $x_2=0$. At $x_2=0$ (12.9-1) gives

$$\mathbf{u}(x_1,0) = \boldsymbol{a}_R e^{ik(x_1 - vt)}, \quad \boldsymbol{\phi}(x_1,0) = \boldsymbol{b}_R e^{ik(x_1 - vt)}, \tag{12.9-2}$$

$$\boldsymbol{a}_R = \mathbf{A}\mathbf{q}, \quad \boldsymbol{b}_R = \mathbf{B}\mathbf{q}, \tag{12.9-3}$$

where the matrices **A** and **B** are defined in (12.2-19). Thus $\boldsymbol{a}_R$ is the polarization vector of the surface wave at the boundary $x_2=0$. It is a linear combination of $\boldsymbol{a}_\alpha$ ($\alpha=1,2,3$). We have shown in Section 12.4 that $\boldsymbol{a}_\alpha$ are linearly independent for $v<\hat{v}$. Hence a subsonic surface wave cannot propagate in a half-space with a rigidly clamped boundary.

Let $x_2=0$ be a traction-free boundary. This means that $\phi(x_1,0)=\mathbf{0}$ or

$$\mathbf{Bq}=\mathbf{0}. \tag{12.9-4}$$

Consequently, a subsonic surface wave exists if

$$|\mathbf{B}|=0 \quad \text{at} \quad v=v_R<\hat{v}. \tag{12.9-5}$$

By (12.2-24),

$$|\mathbf{L}(v)|=8i|\mathbf{B}|^2 \tag{12.9-6}$$

so that

$$|\mathbf{L}(v_R)|=0. \tag{12.9-7}$$

Conversely, when (12.9-7) holds $\mathbf{L}(v_R)$ is singular and, by (12.9-6), so is **B** at $v=v_R$. This implies that there exists a nonzero vector **q** such that (12.9-4) holds. Equation (12.9-5) or (12.9-7) is the *secular equation* for the surface wave speed v_R. Equation (12.9-6) suggests that two of the eigenvalue of $\mathbf{L}(v)$ vanish together at $v=v_R$.

Since **B** is complex, the secular equation (12.9-5) is equivalent to two real equations for v_R. This led Synge (1956) to speculate that the existence of a surface wave in an anisotropic elastic solid is rather restrictive. It was Stroh (1962) who pointed out that (12.9-5) is equivalent to (12.9-7). As a matter of fact the two real equations associated with (12.9-5) are identical because, by (12.9-6),

$$|\mathbf{B}|=\pm\frac{1-i}{4}|\mathbf{L}(v)|^{1/2}$$

so that the real and imaginary parts of $|\mathbf{B}|$ are equal and opposite in signs.

We next show that if a subsonic surface wave exists it is unique. Postmultiplying (12.3-15) by **q** and using $(12.9\text{-}3)_1$ and (12.9-4) yields

$$\mathbf{S}(v_R)\boldsymbol{a}_R=i\boldsymbol{a}_R, \qquad \mathbf{L}(v_R)\boldsymbol{a}_R=\mathbf{0}. \tag{12.9-8}$$

Equation $(12.9\text{-}8)_2$ tells us that the rank of $\mathbf{L}(v_R)$ is at most two. If the rank is two, there is only one null vector $\boldsymbol{a}_R$ which can be taken to be real. This contradicts $(12.9\text{-}8)_1$ because $\boldsymbol{a}_R$ is nonzero and $\mathbf{S}(v_R)$ is real. If the rank is zero $\mathbf{L}(v_R)=\mathbf{0}$ or, by (12.2-27), $\mathbf{S}^2(v_R)=-\mathbf{I}$. This contradicts $(12.3\text{-}16)_2$ that $\mathbf{S}(v_R)$ is singular. We therefore have the result that

$$\text{rank}\ \mathbf{L}(v_R)=1, \tag{12.9-9}$$

assuring us that $\boldsymbol{a}_R$ of $(12.9\text{-}8)_2$ is genuinely complex. The real and imaginary parts of $\boldsymbol{a}_R$ both satisfy $(12.9\text{-}8)_2$, and are independent of each other. The displacement at the boundary $x_2=0$ describes an elliptical path. Equation (12.9-9) confirms the statement made earlier that two of the eigenvalues of $\mathbf{L}(v)$ vanish together at $v=v_R$. We have shown in Section 12.4 that the eigenvalues of $\mathbf{L}(v)$ are positive and nonzero at $v=0$, and decrease monotonically for $0\le v<\hat{v}$. If there were two subsonic surface waves, one of the three eigenvalues of $\mathbf{L}(v)$ would have to vanish twice for $0\le v<\hat{v}$. This contradicts the monotonicity of the eigenvalues of $\mathbf{L}(v)$. Hence a subsonic surface wave is unique.

When the first transonic state is normal, Chadwick and Smith (1977) have shown that one of the eigenvalues of $\mathbf{L}(v)$ becomes $-\infty$ at $v=\hat{v}$. The three eigenvalues of $\mathbf{L}(v)$ for $0\le v<\hat{v}$ are sketched in Fig. 12.4(a). It tells us that the sum of the eigenvalues at $v=\hat{v}$ is negative, and hence

$$\text{tr}\ \mathbf{L}(\hat{v}) < 0. \tag{12.9-10}$$

When the first transonic state is exceptional, and hence is of Type 1, it can be shown (Chadwick and Smith, 1977) that the eigenvalues are bounded and one of them vanishes at $v=\hat{v}$. They are shown in Fig. 12.4(b) and (c). For Fig. 12.4(b), (12.9-10) applies while for Fig. 12.4(c),

$$\text{tr}\ \mathbf{L}(\hat{v}) > 0. \tag{12.9-11}$$

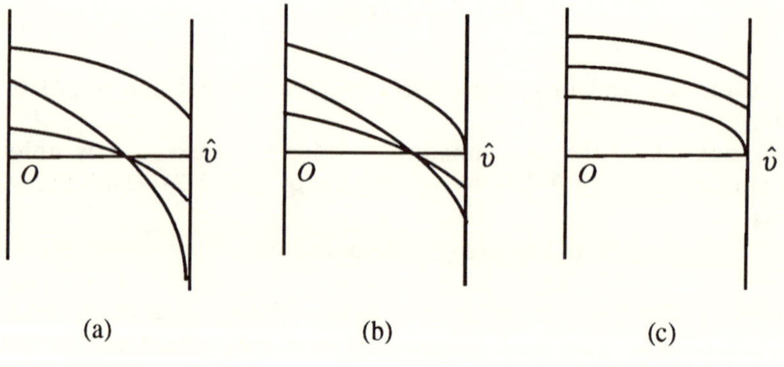

(a) (b) (c)

Fig. 12.4. The eigenvalues of $\mathbf{L}(v)$ for $0\le v<\hat{v}$ are shown in (a) when the first transonic state is normal, and in (b), (c) when it is exceptional. An exceptional transonic state is necessarily of Type 1.

Consequently we may state the existence theorem as follows: A unique subsonic surface wave exists if (12.9-10) holds. No such wave exists if (12.9-11) holds.

A subsonic surface wave always exists if the transonic state is normal. If the transonic state is exceptional, and hence of Type 1, the existence of a subsonic surface wave depends on whether (12.9-10) or (12.9-11) holds.

The above existence theorem is based on the assumption that only one pair each of complex conjugates p_α become real at the first transonic state so that the curvature of the slowness surface at the point of tangency with the slowness surface is nonzero. Barnett and Lothe (1989) and Barnett et al. (1990) have shown that a subsonic surface wave always exists for a zero curvature transonic state.

There are alternatives to the secular equation given in (12.9-5) and (12.9-7). Equation $(12.9\text{-}8)_1$ is equivalent to

$$\{\mathbf{I} + i\mathbf{S}(v_R)\}\boldsymbol{a}_R = \mathbf{0}. \tag{12.9-12}$$

The eigenvalues of $\mathbf{S}(v)$ are 0 and $\pm is(v)$ where $s(v)>0$ is given in (12.3-17). The eigenvalues of $\mathbf{I}\pm i\mathbf{S}(v)$ are therefore 1 and $1\pm s(v)$, and the determinant of $\mathbf{I}\pm i\mathbf{S}(v)$ is $1-s^2(v)$. Hence

$$|\mathbf{I}\pm i\mathbf{S}(v_R)| = F(v_R) = 1+\tfrac{1}{2}\mathrm{tr}\{\mathbf{S}^2(v_R)\} = 0. \tag{12.9-13}$$

Equation $(12.9\text{-}13)_3$ is favored by Chadwick and Smith (1982). By $(12.2\text{-}27)_1$ and (12.9-13) we obtain

$$\begin{gathered} F(v) = -\tfrac{1}{2}+\tfrac{1}{2}\mathrm{tr}\{\mathbf{H}(v)\mathbf{L}(v)\}, \\ |\mathbf{L}(v)| = F^2(v)|\mathbf{H}(v)|^{-1}. \end{gathered} \tag{12.9-14}$$

Equation $(12.9\text{-}14)_2$ also suggests that the zero eigenvalue of $\mathbf{L}(v)$ at $v=v_R$ is a double root. Postmultiplying $(12.3\text{-}15)_1$ by $\mathbf{q}$ and using (12.9-3) yields

$$\boldsymbol{b}_R = i\mathbf{M}(v)\boldsymbol{a}_R \tag{12.9-15}$$

where

$$\mathbf{M}(v) = \mathbf{H}^{-1}(v)\{\mathbf{I}+i\mathbf{S}(v)\}. \tag{12.9-16}$$

Equation (12.9-15) shows that $\mathbf{M}(v)$ effectively transforms the surface displacement $\boldsymbol{a}_R$ into the surface traction $\boldsymbol{b}_R$. Hence $\mathbf{M}(v)$ is called the *surface impedance matrix* or simply the *impedance matrix* (Lothe and Barnett, 1976a; see also Ingebrigtsen and Tonning, 1969). $\mathbf{M}(v)$ is a tensor when $v=0$ (Section 7.2). The vanishing of the surface traction $\boldsymbol{b}_R$ at $v=v_R$ means that

$$\mathbf{M}(v_R)\boldsymbol{a}_R = \mathbf{0}. \tag{12.9-17}$$

This leads to another secular equation

$$|\mathbf{H}(v_R)|^{-1}F(v_R)=|\mathbf{M}(v_R)|=0 \tag{12.9-18}$$

favored by Barnett and Lothe (1985, 1989). The first equality follows from (12.9-16) and $(12.9\text{-}13)_1$. Equation (12.9-17) could have been obtained by multiplying (12.9-12) by the nonsingular matrix $i\mathbf{H}^{-1}(v_R)$. Using a very different approach Taylor (1978) has derived a secular equation similar to $(12.9\text{-}18)_2$.

Simple relations exist between the polarization vector $\boldsymbol{a}_R$ defined in $(12.9\text{-}3)_1$ and the surface wave speed v_R. Equation (12.4-12) can be written as

$$\rho v_R^2\,\bar{\boldsymbol{a}}_R^T\,\boldsymbol{a}_R=-\bar{\boldsymbol{a}}_R^T\mathbf{N}_3\,\boldsymbol{a}_R \tag{12.9-19}$$

while (12.4-15) is, with the use of $(12.4\text{-}14)_3$,

$$\bar{\boldsymbol{a}}_R^T\left\{\mathbf{R}^T(\mathbf{Q}-\rho v_R^2\mathbf{I})^{-1}\mathbf{R}\right\}\boldsymbol{a}_R=\bar{\boldsymbol{a}}_R^T\mathbf{T}\,\boldsymbol{a}_R. \tag{12.9-20}$$

Equation (12.9-19) tells us that ρv_R^2 is bounded by the largest and smallest eigenvalues of $-\mathbf{N}_3$ (Section 1.6). An explicit expression of $-\mathbf{N}_3$ has been given in (6.1-16). For surface waves polarized on the x_3=0 plane (12.9-19) is simplified with the use of (6.1-16) to

$$\rho v_R^2(\bar{a}_{R1}a_{R1}+\bar{a}_{R2}a_{R2})=\kappa\bar{a}_{R1}a_{R1}, \tag{12.9-21a}$$

$$\kappa=s'_{55}\,/\,(s'_{11}s'_{55}-s'_{15}s'_{15}). \tag{12.9-21b}$$

The real and imaginary parts of $\boldsymbol{a}_R$ are a pair of conjugate radii of the elliptic path described by the displacement at x_2=0. If the principal axes of the ellipse coincide with the x_1-, x_2-axes we may set

$$\boldsymbol{a}_R^T=[a,\ -ib,\ 0] \tag{12.9-22}$$

where a, b are the principal radii of the ellipse. We then have the remarkably simple result

$$(b/a)^2=(\kappa/\rho v_R^2)-1. \tag{12.9-23}$$

Application of (12.9-22) in (12.9-20) does not give an equally simple expression for $(b/a)^2$ for general anisotropic materials. For orthotropic materials with the symmetry planes coinciding with the coordinate planes, (12.9-20) does give a simple expression and, by eliminating $(b/a)^2$ from (12.9-23), leads to the *secular equation* for v_R (Section 12.10). It is remarkable that, at least for orthotropic materials, the secular equation can be determined from (12.9-19) and (12.9-20) without computing $\mathbf{B}$, $\mathbf{L}(v)$, or $\mathbf{S}(v)$. In principle,

(12.9-19) and (12.9-20) alone are sufficient to provide the secular equation for any anisotropic materials whenever $\boldsymbol{a}_R$ is given by (12.9-22), (12.9-22) multiplied by a complex factor, or when $\boldsymbol{a}_R$ can be reduced to a scalar multiple of a vector that depends on one composite parameter (b/a).

Equation (12.9-19) or (12.9-20) does not give information on the direction of motion on the elliptical path described by the displacement at the boundary $x_2=0$. The vector $\boldsymbol{a}_R$ in $(12.9\text{-}8)_1$ is the eigenvector of $\mathbf{S}(v_R)$ associated with the eigenvalue i. Let

$$\boldsymbol{a}_R=\mathbf{e}_1-i\mathbf{e}_2 \tag{12.9-24}$$

where $\mathbf{e}_1$, $\mathbf{e}_2$ are real vectors. They depend on v_R. By equating the real and imaginary parts of $(12.9\text{-}8)_1$ one has

$$\mathbf{S}(v_R)\mathbf{e}_1=\mathbf{e}_2, \quad \mathbf{S}(v_R)\mathbf{e}_2=-\mathbf{e}_1. \tag{12.9-25}$$

Hence

$$\boldsymbol{a}_R=\{\mathbf{I}-i\mathbf{S}(v_R)\}\mathbf{e}_1=-i\{\mathbf{I}-i\mathbf{S}(v_R)\}\mathbf{e}_2.$$

The eigenvector $\boldsymbol{a}_R$ given by (12.9-24) is unique up to a complex multiplier, say $e^{i\psi}$, where ψ is real. The real and imaginary parts of $e^{i\psi}\boldsymbol{a}_R$ lie on the eigenplane spanned by $\mathbf{e}_1$ and $\mathbf{e}_2$, and form a pair of conjugate radii of an ellipse. Any vector on the eigenplane can be taken as $\mathbf{e}_1$ (Section 1.8). $\mathbf{e}_2$ is then determined from $(12.9\text{-}25)_1$. We have shown in Section 12.8 that the direction of motion is, as time increases, from the real part of $\boldsymbol{a}_R$ to the imaginary part of $\boldsymbol{a}_R$. Hence the motion is from $\mathbf{e}_1$ to $-\mathbf{e}_2$. If $\mathbf{e}_1$, $\mathbf{e}_2$, $\mathbf{e}_3$ form a right-handed triad, the motion on the ellipse is clockwise on the $(\mathbf{e}_1, \mathbf{e}_2)$ plane.

As an illustration, consider a monoclinic material with the symmetry plane at $x_3=0$. The structure of $\mathbf{S}(v)$ is identical to the one for elastostatics given in (6.4-12), i.e.,

$$\mathbf{S}(v)=S_{11}(v)\mathbf{i}_1\mathbf{i}_1^T+S_{12}(v)\mathbf{i}_1\mathbf{i}_2^T+S_{21}(v)\mathbf{i}_2\mathbf{i}_1^T+S_{22}(v)\mathbf{i}_2\mathbf{i}_2^T$$

in which $S_{22}(v)=-S_{11}(v)$ by (12.3-16) and $\mathbf{i}_1$, $\mathbf{i}_2$ are unit vectors along the x_1-, x_2-axes, respectively. $\mathbf{e}_1$ and $\mathbf{e}_2$ are on the (x_1,x_2) plane. Without loss in generality let

$$\mathbf{e}_1=a\mathbf{i}_1, \quad \mathbf{e}_2=d\mathbf{i}_1+b\mathbf{i}_2 \tag{12.9-26a}$$

where $a>0$, and d and b are constants. They are determined by $(12.9\text{-}25)_1$ which yields

$$d=aS_{11}(v_R), \quad b=aS_{21}(v_R). \tag{12.9-26b}$$

It can be shown that $S_{21}(v_R)>0$. Hence $\mathbf{e}_2\cdot\mathbf{i}_2>0$, and the motion is retrograde (clockwise) on the (x_1,x_2) plane (Chadwick, 1990). If $d=0$, a and b are the principal radii of the ellipse.

We saw in Section 12.8 that the displacement $\mathbf{u}$ on the boundary $x_2=0$ is polarized on the plane $x_2=0$ for one-component surface waves. Barnett (1992) has shown that the displacement $\mathbf{u}$ on $x_2=0$ cannot be polarized on the plane $x_2=0$ for subsonic surface waves.

The theory is quite complete for subsonic surface waves in general anisotropic materials. Surface waves in cubic materials were investigated by Chadwick and Smith (1982), Barnett and Lothe (1987), and Chadwick and Wilson (1992b) who also considered orthotropic materials. Extensive study was made for transversely isotropic materials by Chadwick (1988, 1989a,b,c,d), and for monoclinic materials by Chadwick (1990) and Chadwick and Wilson (1992a,b). In these studies the planes of material symmetry need not coincide with the coordinate planes. Mention should be made of work on allied problems of surface waves in a prestressed elastic body (Chadwick and Jarvis, 1979a,b) and in an inextensible transversely isotropic material (Captain and Chadwick, 1986; Chadwick, 1989c). Chadwick (1993, 1994) and Chadwick and Wilson (1994) have studied extensively plane waves and surface waves in transversely isotropic incompressible materials. Surface waves in piezoelectric materials were considered by Taylor and Crampin (1978) and by Lothe and Barnett (1976b, 1977, 1979).

(ii) Surface Waves with $v_R > \hat{v}$

Surface waves with $v_R > \hat{v}$ do exist (Farnell, 1970). In this case the vertical line in Fig. 12.2 intersects the slowness surface. If the vertical line intersects the slowness surface in two points such as P_3 and P_6, two of the eigenvalues p_α are real. They represent body waves which should be excluded from the surface wave solution because the displacement due to a body wave does not vanish at $x_2=\infty$. The remaining four p_α consist of two pairs of complex conjugates. Only p_α with a positive imaginary part is admissible so that the surface wave in this case consists of two inhomogeneous plane waves. One would assume that a one-component surface wave would have four real roots and one pair of complex conjugates for p_α. That this is not the case was shown in Section 12.8. The p_α for one-component surface waves have two real roots and two identical pairs of complex conjugates. Consequently, regardless of one- or two-component surface waves, there are only two real p_α when $v_R > \hat{v}$.

Surface waves with $v_R > \hat{v}$ are known as *supersonic surface waves*. When $v_R > \hat{v}$, the characteristic wave speeds relative to which v_R is supersonic are from the body waves that are associated with the real roots of p_α. The characteristic wave speed c associated with a real p_α is, by (12.7-3) and (12.7-2), $c=v_R\cos\theta<v_R$ if v_R is not a limiting wave speed. Therefore v_R is *supersonic* relative to these body wave speeds. But these body waves are excluded in the surface wave solution. Consequently v_R is not supersonic with respect to any characteristic wave speed that is present in the surface wave solution. Precisely for this reason shock waves and head waves that are common in supersonic flow in fluid dynamics do not appear in surface waves when $v_R>\hat{v}$. The following example illustrates the ambiguity that may arise in defining surface waves with $v_R>\hat{v}$ as supersonic. Consider surface waves in an orthotropic material whose planes of symmetry coincide with the coordinate planes. The surface wave is a two-component wave polarized on the $x_3=0$ plane. The solution depends only on p_1 and p_2, not on p_3 which is (Section 12.10)

$$p_3 = i\sqrt{(C_{55}-\rho v_R^2)/C_{44}}\,.$$

The surface wave solution is completely independent of the elastic stiffness C_{55}. By changing C_{55}, p_3 can be made complex or real, i.e., $v_R<\hat{v}$ or $v_R>\hat{v}$. In other words, the surface wave can be subsonic or supersonic without changing the surface wave solution. The same situation occurs for monoclinic materials with the symmetry plane at $x_3=0$. From (5.2-10),

$$p_3 = \frac{1}{C_{44}}\left\{-C_{45} + i\sqrt{C_{44}(C_{55}-\rho v_R^2)-C_{45}^2}\right\}.$$

By changing the elastic constants C_{44}, C_{45}, and C_{55} which do not appear in the surface wave solution, the surface wave can be subsonic or supersonic. This shows the inappropriateness of employing the term supersonic for surface waves with $v_R>\hat{v}$. We hasten to add that truly supersonic steady waves with shock waves and head waves do exist in an anisotropic elastic medium when one considers, for example, a moving line singularity.

Unlike surface wave theory for $v_R<\hat{v}$, there is no fully developed existence theory for surface waves with $v_R>\hat{v}$. However, an existence theory for the special case when the wave is polarized on a plane of material symmetry has been developed by Chadwick (1990). Gundersen (1992) has studied the classification of the subsequent transonic states. Al'shits and Lothe (1981) have shown that, if there is a surface wave with $v_R>\hat{v}$, the conditions for simple reflection are automatically satisfied. One can therefore focus the search of surface waves with $v_R>\hat{v}$ in the space of simple reflection (Gundersen et al., 1991; Wang and Lothe, 1992). It appears that the space of simple reflection is of fundamental importance in understanding surface waves with $v_R>\hat{v}$.

(iii) **Interfacial Waves**

The impedance matrix approach for surface waves extends naturally to interfacial waves in two bonded anisotropic elastic half-spaces. Interfacial waves in isotropic bimaterials are known as Stoneley (1924) waves. Let the half-space $x_2>0$ be occupied by material 1 and the other half-space $x_2<0$ be occupied by material 2. The basic solution (12.9-1) applies to material 1 in $x_2>0$. For material 2 in $x_2<0$ we should take the complex conjugate of (12.9-1) so that the displacement vanishes at $x_2=-\infty$. The two materials are rigidly bonded at the interface $x_2=0$. This means that the displacement $\boldsymbol{a}_R$ and the surface traction $\boldsymbol{b}_R$ at $x_2=0$ are continuous. Using the superscripts (1) and (2) to label materials 1 and 2, we have

$$\boldsymbol{a}_R^{(1)} = \bar{\boldsymbol{a}}_R^{(2)}, \quad \boldsymbol{b}_R^{(1)} = \bar{\boldsymbol{b}}_R^{(2)}.$$

The two conditions are satisfied if, by (12.9-15),

$$\left\{\mathbf{M}_1(v) + \overline{\mathbf{M}}_2(v)\right\} \boldsymbol{a}_R^{(1)} = \mathbf{0} \tag{12.9-27}$$

where the subscripts 1 and 2 refer to materials 1 and 2, respectively. An interfacial wave exists when there is a v_S for which (Barnett et al., 1985; see also Chadwick and Currie, 1974),

$$\left|\mathbf{M}_1(v_S) + \overline{\mathbf{M}}_2(v_S)\right| = \mathbf{0}. \tag{12.9-28}$$

The interfacial wave is subsonic if v_S is less than $\hat{v}^{(1)}$ and $\hat{v}^{(2)}$. Barnett et al. (1985) prove that a subsonic interfacial wave is unique if it exists. Its speed v_S is not less that the smaller of $v_R^{(1)}$ and $v_R^{(2)}$. If neither half-space by itself transmits a subsonic surface wave, no subsonic interfacial wave exists.

Although $\mathbf{M}_1(v) + \overline{\mathbf{M}}_2(v)$ is complex, it is Hermitian so that the three eigenvalues are real (Section 1.7). The left side of (12.9-28) is identical to the products of the three eigenvalues, and hence it is real. An explicit real equation for (12.9-28) can be obtained as follows. With the use of (12.2-29), (12.9-27) is written as

$$\{\mathbf{I} + i\hat{\mathbf{S}}^T(v)\}\, \boldsymbol{a}_R^{(1)} = \mathbf{0} \tag{12.9-29}$$

where

$$\hat{\mathbf{S}}(v) = \hat{\mathbf{W}}(v)\hat{\mathbf{D}}^{-1}(v),$$

$$\hat{\mathbf{D}}(v) = \mathbf{H}_1^{-1}(v) + \mathbf{H}_2^{-1}(v), \quad \hat{\mathbf{W}}(v) = \mathbf{S}_1^T(v)\mathbf{H}_1^{-1}(v) - \mathbf{S}_2^T(v)\mathbf{H}_2^{-1}(v).$$

$\hat{\mathbf{S}}(v)$, $\hat{\mathbf{D}}(v)$, $\hat{\mathbf{W}}(v)$ reduce to $\hat{\mathbf{S}}$, $\hat{\mathbf{D}}$, $\hat{\mathbf{W}}$ defined in Section 8.8 when $v=0$. They satisfy the identity (8.8-43). Equation (12.9-29) is identical to (12.9-12). $\hat{\mathbf{S}}(v)$ is the product of a skew-symmetric matrix $\hat{\mathbf{W}}(v)$ and a symmetric positive definite matrix $\hat{\mathbf{D}}^{-1}(v)$. It has the same properties as $\mathbf{S}(v)$ and $\mathbf{S}(0)$ have (Section 6.6). Following the derivation of $(12.9\text{-}13)_3$ a real expression alternate to (12.9-28) is

$$1+\tfrac{1}{2}\mathrm{tr}\{\hat{\mathbf{S}}^2(v_S)\}=0. \tag{12.9-30}$$

Extension of interfacial waves to piezoelectric materials was investigated by Abbudi and Barnett (1990). The related problem of slip waves in two half-spaces that are in sliding contact was studied by Barnett et al. (1988).

12.10 Secular Equation and S(v), H(v), L(v) for Orthotropic Materials

There are several forms of the secular equation for a subsonic surface wave speed v_R:

$$|\mathbf{B}|=\mathbf{0}, \tag{12.10-1}$$

$$|\mathbf{L}(v_R)|=0, \tag{12.10-2}$$

$$F(v_R)=1+\tfrac{1}{2}\mathrm{tr}\{\mathbf{S}^2(v_R)\}=0, \tag{12.10-3}$$

$$|\mathbf{M}(v_R)|=|\mathbf{H}(v_R)|^{-1}\cdot|\mathbf{I}+i\mathbf{S}(v_R)|=0. \tag{12.10-4}$$

They are related to each other as can be seen from (12.9-6), $(12.9\text{-}13)_1$, $(12.9\text{-}14)_2$, and (12.9-18) (see also Chadwick and Ting, 1987). With the exception of (12.10-1) which requires the complex matrix $\mathbf{B}$, the rest of the equations are in terms of $\mathbf{S}(v)$, $\mathbf{H}(v)$, $\mathbf{L}(v)$ which are real. Unfortunately explicit expressions of $\mathbf{S}(v)$, $\mathbf{H}(v)$, $\mathbf{L}(v)$ are not available for general anisotropic elastic materials. They are available for orthotropic materials. Using the integral formalism (12.3-11), Chadwick and Wilson (1992a,b) formulate the expressions of $\mathbf{S}(v)$, $\mathbf{H}(v)$, $\mathbf{L}(v)$ for monoclinic materials with the symmetry plane at x_3=0 in terms of integrals that can be evaluated numerically. Explicit expressions are then deduced for the special case of orthotropic and cubic materials. Earlier, Dongye and Ting (1989) employed the algebraic relations (12.2-24) and obtained explicit expressions of $\mathbf{S}(v)$, $\mathbf{H}(v)$, $\mathbf{L}(v)$ for orthotropic materials. The final expressions of $\mathbf{S}(v)$, $\mathbf{H}(v)$, $\mathbf{L}(v)$ were communicated to them by Chadwick.

If the secular equation is all one needs for orthotropic materials with the symmetry planes coinciding with the coordinate planes, we have pointed out in Section 12.9 that there is no need to employ any of the equations in (12.10-1)–(12.10-4). We will show in this section that (12.9-19) and (12.9-20) alone are sufficient to provide the secular equation. However, explicit expressions of $\mathbf{S}(v)$, $\mathbf{H}(v)$, $\mathbf{L}(v)$ are useful not only for the secular equation, but also for other steady wave problems such as a moving line singularity.

We present briefly below the derivation of $\mathbf{S}(v)$, $\mathbf{H}(v)$, $\mathbf{L}(v)$ for orthotropic materials whose planes of symmetry coincide with the coordinate planes. We employ the algebraic approach but use $(12.2\text{-}29)_4$ instead, i.e.,

$$-\mathbf{AB}^{-1} = \mathbf{S}(v)\mathbf{L}^{-1}(v) + i\mathbf{L}^{-1}(v). \tag{12.10-5}$$

Thus $\mathbf{S}(v)\mathbf{L}^{-1}(v)$ and $\mathbf{L}^{-1}(v)$ are determined from the real and imaginary parts of $-\mathbf{AB}^{-1}$. $\mathbf{L}(v)$ is then obtained from $\mathbf{L}^{-1}(v)$, $\mathbf{S}(v)$ from

$$\mathbf{S}(v) = \{\mathbf{S}(v)\mathbf{L}^{-1}(v)\}\mathbf{L}(v), \tag{12.10-6}$$

and $\mathbf{H}(v)$ is, by (12.2-27),

$$\mathbf{H}(v) = \mathbf{L}^{-1}(v) - \mathbf{S}(v)\{\mathbf{S}(v)\mathbf{L}^{-1}(v)\}. \tag{12.10-7}$$

For orthotropic materials with the symmetry planes coinciding with the coordinate planes (12.2-4) can be written as

$$\begin{bmatrix} C_{11}^*+p^2C_{66} & p(C_{12}+C_{66}) & 0 \\ p(C_{12}+C_{66}) & C_{66}^*+p^2C_{22} & 0 \\ 0 & 0 & C_{55}^*+p^2C_{44} \end{bmatrix} \boldsymbol{a} = \mathbf{0} \tag{12.10-8}$$

where

$$C_{ij}^* = C_{ij} - \rho v^2 \delta_{ij}.$$

If we let

$$p_3 = i\sqrt{C_{55}^* / C_{44}}, \tag{12.10-9}$$

the roots p_1 and p_2 are determined from

$$(C_{11}^*+p^2C_{66})(C_{66}^*+p^2C_{22}) - p^2(C_{12}+C_{66})^2 = 0. \tag{12.10-10}$$

This is a quadratic equation in p^2. When the two roots are real they must be negative so that $p_1=i\beta_1$, $p_2=i\beta_2$ where β_1, β_2 are positive and nonzero. If the two roots for p^2 are complex conjugates, taking the square roots and keeping the ones with a positive imaginary part we have $p_1=\beta_1+i\beta_2$, $p_2=-\beta_1+i\beta_2$. In either case p_1p_2 is real and negative while p_1+p_2 is purely imaginary. In computing $-\mathbf{AB}^{-1}$, p_1 and p_2 appear in the form of p_1p_2 and p_1+p_2. From the coefficients of the quadratic equation (12.10-10) one obtains

$$p_1^2 + p_2^2 = -\left\{(C_{11}^* C_{22} + C_{66}^* C_{66}) - (C_{12} + C_{66})^2\right\} / (C_{22} C_{66}),$$
$$p_1 p_2 = -\sqrt{(C_{11}^* C_{66}^*) / (C_{22} C_{66})}. \tag{12.10-11}$$

Making use of the relation

$$(p_1 \pm p_2)^2 = (p_1^2 + p_2^2) \pm 2(p_1 p_2)$$

we have

$$p_1 + p_2 = i\sqrt{Y/(C_{22} C_{66})},$$
$$(p_1 - p_2)^2 = -Z/(C_{22} C_{66}), \tag{12.10-12}$$

where

$$Y = \left(\sqrt{C_{11}^* C_{22}} + \sqrt{C_{66}^* C_{66}}\right)^2 - (C_{12} + C_{66})^2,$$
$$Z = \left(\sqrt{C_{11}^* C_{22}} - \sqrt{C_{66}^* C_{66}}\right)^2 - (C_{12} + C_{66})^2. \tag{12.10-13}$$

The conditions for p_α to be complex are

$$C_{55}^* > 0, \quad C_{11}^* C_{66}^* > 0, \quad \left(\sqrt{C_{11}^* C_{22}} + \sqrt{C_{66}^* C_{66}}\right)^2 > (C_{12} + C_{66})^2. \tag{12.10-14}$$

The first inequality is for p_3 to be complex. It can be ignored for surface waves because the surface wave solution in this case does not depend on p_3.

The vector $\boldsymbol{a}$ in (12.10-8) can be chosen as

$$\boldsymbol{a}_3 = \frac{1}{p_3 C_{44}} \begin{bmatrix} 0 \\ 0 \\ 1 \end{bmatrix} \quad \text{for } p = p_3,$$

$$\boldsymbol{a}_\alpha = \begin{bmatrix} -(C_{66}^* + p_\alpha^2 C_{22}) \\ p_\alpha (C_{12} + C_{66}) \\ 0 \end{bmatrix} \quad \text{for } \alpha = 1,2.$$

They form the columns of **A**. The vectors $\boldsymbol{b}_\alpha$ that form the columns of **B** are computed from $(12.2\text{-}6)_1$. It can be shown that

$$\mathbf{B} = \begin{bmatrix} \xi_1 & \xi_2 & 0 \\ \eta_1 & \eta_2 & 0 \\ 0 & 0 & 1 \end{bmatrix}$$

in which

$$\xi_\alpha = p_\alpha C_{66}(C_{12} + \rho v^2 C_{66} - p_\alpha^2 C_{22}),$$

$$\eta_\alpha = -C_{12}C_{66}^* + p_\alpha^2 C_{22} C_{66}.$$

Hence

$$\mathbf{B}^{-1} = \frac{1}{|\mathbf{B}|}\begin{bmatrix} \eta_2 & -\xi_2 & 0 \\ -\eta_1 & \xi_1 & 0 \\ 0 & 0 & |\mathbf{B}| \end{bmatrix}.$$

It is redundant to normalize the vectors $\boldsymbol{a}_\alpha\ \boldsymbol{b}_\alpha$ in computing $-\mathbf{AB}^{-1}$ (Section 6.4). With the use of (12.10-11) and (12.10-12) it can be shown that

$$|\mathbf{B}| = -(p_1 - p_2)(C_{12} + C_{66})\sqrt{C_{66}^*/C_{66}}\ X, \tag{12.10-15}$$

$$X = (C_{11}^* C_{22} - C_{12}^2)\sqrt{C_{66}^* C_{66}} - \rho v^2 C_{66}\sqrt{C_{11}^* C_{22}}. \tag{12.10-16}$$

Notice that $|\mathbf{B}|$ is the determinant of the first 2×2 principal submatrix of $\mathbf{B}$. Notice also that if the vectors $\boldsymbol{a}_\alpha\ \boldsymbol{b}_\alpha$ were normalized, $|\mathbf{B}|$ of (12.10-15) would be multiplied by a complex factor. Following the procedure outlined in (12.10-5)–(12.10-7) the nonzero components of $\mathbf{S}(v)$, $\mathbf{H}(v)$, $\mathbf{L}(v)$ are

$$L_{33}(v) = H_{33}^{-1}(v) = \sqrt{C_{55}^* C_{44}},$$

$$\sqrt{C_{66}^* C_{22}}\, L_{11}(v) = \sqrt{C_{11}^* C_{66}}\, L_{22}(v) = XY^{-1/2},$$

$$-\sqrt{C_{11}^* C_{66}}\, S_{12}(v) = \sqrt{C_{66}^* C_{22}}\, S_{21}(v) = \left(\sqrt{C_{11}^* C_{22}}\, C_{66} - C_{12}\sqrt{C_{66}^* C_{66}}\right) Y^{-1/2},$$

$$\sqrt{C_{11}^* C_{66}}\, H_{11}(v) = \sqrt{C_{66}^* C_{22}}\, H_{22}(v) = \left(\sqrt{C_{11}^* C_{22}} + \sqrt{C_{66}^* C_{66}}\right) Y^{-1/2}.$$

If needed, explicit expressions of p_1 and p_2 are obtained from (12.10-12) and (12.10-13) as

$$\left.\begin{aligned} p_1 &= i(\sqrt{Y} + \sqrt{Z})\big/\sqrt{4C_{22}C_{66}} \\ p_2 &= i(\sqrt{Y} - \sqrt{Z})\big/\sqrt{4C_{22}C_{66}} \end{aligned}\right\} \text{ when } Z \ge 0,$$

$$\left.\begin{aligned} p_1 &= (\sqrt{-Z} + i\sqrt{Y})\big/\sqrt{4C_{22}C_{66}} \\ p_2 &= (-\sqrt{-Z} + i\sqrt{Y})\big/\sqrt{4C_{22}C_{66}} \end{aligned}\right\} \text{ when } Z \le 0.$$

$p_1 = p_2$ if $Z=0$, i.e., if

$$\left(\sqrt{C_{11}^* C_{22}} - \sqrt{C_{66}^* C_{66}}\right)^2 = (C_{12} + C_{66})^2.$$

From (12.10-15), (12.10-16), and (12.10-14), the secular equation $|\mathbf{B}|=0$ reduces to (Stoneley, 1963)

$$X = (C_{11}^* C_{22} - C_{12}^2)\sqrt{C_{66}^* C_{66}} - \rho v^2 C_{66}\sqrt{C_{11}^* C_{22}} = 0. \qquad (12.10\text{-}17)$$

The secular equations (12.10-2) through (12.10-4) also reduce to $(12.10\text{-}17)_2$ when nonzero factors are ignored. Let

$$x = \frac{\rho v^2}{C_{66}}, \quad \alpha = \frac{C_{22}}{C_{11}}, \quad \gamma = \frac{C_{66}}{C_{11}}, \quad \beta = \frac{C_{11}C_{22} - C_{12}^2}{C_{22}C_{66}}.$$

They are all positive and nonzero quantities. The secular equation (12.10-17) takes the dimensionless form

$$f(x) = \sqrt{\alpha}(\beta - x)\sqrt{1-x} - x\sqrt{1-\gamma x} = 0. \qquad (12.10\text{-}18)$$

Since

$$0 < \beta = \frac{C_{11}C_{22} - C_{12}^2}{\gamma C_{11} C_{22}} < \frac{1}{\gamma},$$

the second inequality in (12.10-14) is satisfied if

$$0 < x < \hat{x}, \quad \hat{x} = \min(1, \ \beta). \qquad (12.10\text{-}19)$$

From (12.10-18), $f(0)>0$ and $f(\hat{x})<0$ so that there exists a surface wave speed. It must satisfy the third inequality in (12.10-14) because the existence of a surface wave for the material under consideration has been proved by Chadwick (1976b).

The κ in (12.9-21b) reduces to $1/s'_{11}$ because $s'_{15}=0$ for orthotropic materials. By (2.4-11),

$$\begin{bmatrix} s'_{11} & s'_{12} \\ s'_{12} & s'_{22} \end{bmatrix}\begin{bmatrix} C_{11} & C_{12} \\ C_{12} & C_{22} \end{bmatrix} = \mathbf{I}$$

for orthotropic materials from which we have

$$s'_{11} = C_{22}/(C_{11}C_{22} - C_{12}^2) = (C_{66}\beta)^{-1}.$$

Hence (12.9-23) reduces to

$$(\beta/x) - 1 = (b/a)^2 = \{S_{21}(v_R)\}^2, \qquad (12.10\text{-}20)$$

giving the ratio b/a of the principal radii of the elliptical path described by the displacement at the boundary $x_2=0$. The last equality follows from (12.9-26) with $d=0$. It is a special case of equations (2.50) and (2.51) of Chadwick and Wilson (1992a). It can be shown that application of (12.9-20) to orthotropic materials with the $\boldsymbol{a}_R$ given by (12.9-22) yields

$$(b/a)^2 = \frac{x(1-\gamma x)}{\alpha(1-x)(\beta - x)}. \tag{12.10-21}$$

In view of the secular equation (12.10-18), (12.10-21) is equivalent to $(12.10\text{-}20)_1$. This confirms the statement made earlier that one could obtain the secular equation (12.10-18) by applying (12.9-19) and (12.9-20) to orthotropic materials and eliminating $(b/a)^2$ between the two equations without computing the complex matrix **B**.

For isotropic materials $\alpha=1$, $1<\beta=4(1-\gamma)<4$, and (12.10-18) is

$$[4(1-\gamma)-x]\sqrt{1-x} - x\sqrt{1-\gamma x} = 0. \tag{12.10-22}$$

To put it in a form familiar in the literature, we employ the relations

$$x[4(1-\gamma)-x] = 4(1-\gamma x)-(2-x)^2,$$

$$x^2 = (2-x)^2 - 4(1-x).$$

Equation (12.10-22), after multiplication by x, can be factorized into two parts. The first part is

$$\sqrt{1-x} + \sqrt{1-\gamma x}$$

which is nonzero, and hence can be ignored. The second part is

$$(2-x)^2 - 4\sqrt{1-x}\sqrt{1-\gamma x} = 0. \tag{12.10-23}$$

This is the Rayleigh (1885) equation for isotropic materials.

To determine analytically the root of (12.10-18), (12.10-22), or (12.10-23) we move the second term to the right side and square both sides of the equation. This operation leads to a cubic equation in x that provides us at least one real root. The operation on (12.10-23) leads to a quartic equation, but the constant term vanishes. By the uniqueness of subsonic surface waves, there is only one root that satisfies the inequalities $(12.10\text{-}14)_{2,3}$.

If the secular equation is all one needs for an anisotropic elastic material, it is perhaps simpler to use $|\mathbf{B}|=0$ when explicit expressions of $\mathbf{S}(v)$, $\mathbf{H}(v)$, $\mathbf{L}(v)$ are not available. For monoclinic materials with the symmetry plane at $x_3=0$, the surface wave is polarized on the $x_3=0$ plane. One needs only the first 2×2 principal submatrix of **B** in computing $|\mathbf{B}|$.

12.11 Semisimple and Extraordinary Semisimple Matrix $\mathbf{N}(\upsilon)$

When the 6×6 matrix $\mathbf{N}(\upsilon)$ is semisimple with $p_1=p_2$, there exist two independent eigenvectors ξ_1, ξ_2 associated with the repeated eigenvalue p. To be consistent with the case when $\mathbf{N}(\upsilon)$ is simple, the left eigenvectors η_1 and η_2 are given by

$$\eta_\alpha = \hat{\mathbf{I}}\xi_\alpha \quad (\alpha=1,2), \quad \hat{\mathbf{I}} = \begin{bmatrix} \mathbf{0} & \mathbf{I} \\ \mathbf{I} & \mathbf{0} \end{bmatrix}.$$

The orthonormalized relations (12.2-17a) for $\alpha,\beta=1,2$ are

$$\xi_1^T\hat{\mathbf{I}}\xi_1 = 1 = \xi_2^T\hat{\mathbf{I}}\xi_2, \quad \xi_1^T\hat{\mathbf{I}}\xi_2 = 0 = \xi_2^T\hat{\mathbf{I}}\xi_1. \tag{12.11-1}$$

Equation $(12.11\text{-}1)_4$ is redundant since it is the transpose of $(12.11\text{-}1)_3$. Hence only the first three equations need to be considered. The eigenvectors ξ_1, ξ_2 are not unique. If ξ_1, ξ_2 satisfy (12.11-1) so do

$$\hat{\xi}_1 = \cos\theta\,\xi_1 + \sin\theta\,\xi_2, \quad \hat{\xi}_2 = -\sin\theta\,\xi_1 + \cos\theta\,\xi_2, \tag{12.11-2}$$

where θ is an arbitrary parameter.

We will show that, when two independent eigenvectors ξ_1, ξ_2 are chosen arbitrarily, one can always deduce two orthonormalized eigenvectors that satisfy (12.11-1). There are three possibilities which are discussed separately below.

(i) If $\xi_1^T\hat{\mathbf{I}}\xi_2=0$, $\xi_1^T\hat{\mathbf{I}}\xi_1$ and $\xi_2^T\hat{\mathbf{I}}\xi_2$ cannot vanish. Suppose that $\xi_1^T\hat{\mathbf{I}}\xi_1=0$. This means that the left eigenvector $\eta_1=\hat{\mathbf{I}}\xi_1$ is orthogonal to ξ_1 and ξ_2. But η_1 is also orthogonal to ξ_α $(\alpha=3,4,5,6)$. Since ξ_α $(\alpha=1,2,\dots,6)$ form an independent set, no vectors can be orthogonal to all ξ_α. This leads to a contradiction, proving that $\xi_1^T\hat{\mathbf{I}}\xi_1$ is nonzero. Likewise $\xi_2^T\hat{\mathbf{I}}\xi_2$ is nonzero when $\xi_1^T\hat{\mathbf{I}}\xi_2=0$. One can then normalize ξ_1 and ξ_2 such that $(12.11\text{-}1)_{1,2}$ hold.

(ii) Let $\xi_1^T\hat{\mathbf{I}}\xi_2\neq 0$ and at least one of $\xi_1^T\hat{\mathbf{I}}\xi_1$ and $\xi_2^T\hat{\mathbf{I}}\xi_2$ be nonzero. Without loss in generality let $\xi_2^T\hat{\mathbf{I}}\xi_2$ be nonzero and ξ_2 be normalized to satisfy $(12.11\text{-}1)_2$. If we replace ξ_1 by $\hat{\xi}_1$ given by

$$\hat{\xi}_1 = \xi_1 - k\xi_2, \quad k = \xi_1^T\hat{\mathbf{I}}\xi_2,$$

$(12.11\text{-}1)_3$ is satisfied. We have reduced the problem to Case (i).

(iii) Let $(\xi_1^\circ)^T\hat{\mathbf{I}}\xi_1^\circ = (\xi_2^\circ)^T\hat{\mathbf{I}}\xi_2^\circ = 0$. Following the same argument in Case (i) it can be shown that $(\xi_1^\circ)^T\hat{\mathbf{I}}\xi_2^\circ$ and $(\xi_1^\circ)^T\hat{\mathbf{I}}\xi_1^\circ$ cannot vanish together if ξ_1°, ξ_2° are independent of each other. The orthonormalized eigenvectors are given by

$$\xi_1 = \tfrac{1}{2}(k_1^{-1}\xi_1^\circ - ik_2^{-1}\xi_2^\circ), \quad \xi_2 = \tfrac{1}{2}(k_2^{-1}\xi_2^\circ - ik_1^{-1}\xi_1^\circ), \tag{12.11-3}$$

where k_1, k_2 are arbitrary nonzero constants subjected to the condition

$$2ik_1k_2 = (\xi_1^o)^T \hat{\mathbf{I}} \xi_2^o \neq 0. \tag{12.11-4}$$

The eigenvector ξ^o that satisfies the identity

$$(\xi^o)^T \hat{\mathbf{I}} \xi^o = 0 \tag{12.11-5}$$

is an *unnormalizable* eigenvector. It cannot be normalized according to $(12.11\text{-}1)_{1,2}$. We will show that there are two unique unnormalizable eigenvectors. Let ξ_1 and ξ_2 be orthonormalized eigenvectors that satisfy (12.11-1). It is easily shown that

$$\xi_1^o = k_1(\xi_1 + i\xi_2), \quad \xi_2^o = k_2(\xi_2 + i\xi_1), \tag{12.11-6}$$

are unnormalizable eigenvectors in which k_1, k_2 are arbitrary nonzero complex multipliers. Replacing i by $-i$ in ξ_1^o does not provide a new ξ^o because

$$\xi_1 - i\xi_2 = -i(\xi_2 + i\xi_1),$$

which is essentially ξ_2^o. Likewise, replacing i by $-i$ in $\hat{\xi}_2^o$ does not give a new ξ^o. Equations (12.11-6) are the inverses of (12.11-3). If $\hat{\xi}_1$, $\hat{\xi}_2$ are another set of orthonormalized eigenvectors, the associated $\hat{\xi}_1^o$, $\hat{\xi}_2^o$ are

$$\hat{\xi}_1^o = \hat{k}_1(\hat{\xi}_1 + i\hat{\xi}_2) = \hat{k}_1 e^{-i\theta}(\xi_1 + i\xi_2),$$

$$\hat{\xi}_2^o = \hat{k}_2(\hat{\xi}_2 + i\hat{\xi}_1) = \hat{k}_2 e^{i\theta}(\xi_2 + i\xi_1),$$

in which the second equalities follow from (12.11-2). Thus ξ_1^o, ξ_2^o are unique except a complex multiplier. If ξ_1^o is an unnormalizable eigenvector while ξ_2 is a normalized eigenvector, ξ_2^o can be easily obtained from

$$\xi_2^o = \xi_1^o - k\xi_2, \quad k = 2\xi_2^T \mathbf{J} \xi_1^o.$$

Let

$$\xi_1^o = \begin{bmatrix} \boldsymbol{a}_1 \\ \boldsymbol{b}_1 \end{bmatrix}, \quad \xi_2^o = \begin{bmatrix} \boldsymbol{a}_2 \\ \boldsymbol{b}_2 \end{bmatrix}.$$

This means, by (12.11-5),

$$\boldsymbol{a}_1 \cdot \boldsymbol{b}_1 = 0 = \boldsymbol{a}_2 \cdot \boldsymbol{b}_2.$$

In particular if $\boldsymbol{b}_1 = \mathbf{0}$, $\boldsymbol{b}_2 \neq \mathbf{0}$. Otherwise $(12.11\text{-}4)_2$ is violated. Hence if ξ_1^o provides a solution for a free surface wave, ξ_2^o does not. It should be noted from (12.11-6) that ξ_1^o is a linear combination of two orthonormalized eigenvectors ξ_1 and ξ_2. The surface wave associated with ξ_1^o is a two-component wave when referred to a set of orthonormalized eigenvectors.

The matrix $\mathbf{N}(v)$ is *extraordinary semisimple* when it has three repeated eigenvalues p and three associated eigenvectors $\xi=(\boldsymbol{a},\boldsymbol{b})$. The vector $\boldsymbol{a}$ obtained from (12.2-4) would have three independent solutions for the same p (see also Section 6.10). This implies that

$$\mathbf{Q}-\rho v^2\mathbf{I}+p(\mathbf{R}+\mathbf{R}^T)+p^2\mathbf{T}=\mathbf{0}. \tag{12.11-7}$$

Let α and β be the real and imaginary parts of p, and let

$$p=\alpha+i\beta, \quad \lambda^2=\alpha^2+\beta^2. \tag{12.11-8}$$

Equating the real and imaginary parts of (12.11-7) leads to

$$\mathbf{R}+\mathbf{R}^T=-2\alpha\mathbf{T}, \quad \mathbf{Q}-\lambda^2\mathbf{T}=\rho v^2\mathbf{I}\,. \tag{12.11-9}$$

Written explicitly, we have

$$\begin{gathered}
C_{16}=-\alpha C_{66}, \quad C_{26}=-\alpha C_{22}, \quad C_{45}=-\alpha C_{44}, \\
C_{16}=\lambda^2 C_{26}, \quad C_{15}=\lambda^2 C_{46}, \quad C_{56}=\lambda^2 C_{24}, \\
C_{12}+C_{66}=-2\alpha C_{26}, \quad C_{14}+C_{56}=-2\alpha C_{46}, \quad C_{25}+C_{46}=-2\alpha C_{24}, \\
C_{11}-\lambda^2 C_{66}=C_{55}-\lambda^2 C_{44}=C_{66}-\lambda^2 C_{22}=\rho v^2.
\end{gathered} \tag{12.11-10}$$

Elimination of C_{16} and C_{26} among the three equations in $(12.11\text{-}10)_{1,2,4}$ yields

$$\alpha(C_{66}-\lambda^2 C_{22})=0.$$

Case I. Let $C_{66}=\lambda^2 C_{22}$. Then α is arbitrary. The last equation in (12.11-10) tells us that

$$\rho v^2=0.$$

The 5×5 matrix $\mathbf{C}^o$ without the elements C_{3s} $(s=1,2,\ldots,6)$ obtained with the use of (12.11-10) can be shown to be singular. Hence $\mathbf{N}$ cannot be extraordinary semisimple for the static case

Case II. Let $C_{66}\neq\lambda^2 C_{22}$. Then $\alpha=0$ and $p=i\beta$. The last equation in (12.11-10) tells us that

$$\rho v^2\neq 0.$$

The 5×5 matrix $\mathbf{C}^o$ obtained with the use of (12.11-10) can be positive definite. Hence $\mathbf{N}(v)$ can be extraordinary semisimple for a steady state motion.

Chapter 13

DEGENERATE AND NEAR DEGENERATE MATERIALS

The Stroh formalism presented in Sections 5.3 and 5.5 assumes that the 6×6 fundamental elasticity matrix **N** is simple, i.e., the three pairs of eigenvalues p_α are distinct. The eigenvectors ξ_α (α=1,2,3) are independent of each other, and the general solution (5.3-10) consists of three independent solutions. The formalism remains valid when **N** is semisimple. In this case there is a repeated eigenvalue, say $p_2=p_1$, but there exist two independent eigenvectors ξ_2 and ξ_1 associated with the repeated eigenvalue. The general solution (5.3-10) continues to consist of three independent solutions. Moreover one can always choose ξ_2 and ξ_1 such that the orthogonality relations (5.5-11) and the subsequent relations (5.5-13)–(5.5-17) hold. When **N** is nonsemisimple with $p_2=p_1$, there exists only one independent eigenvector associated with the repeated eigenvalue. The general solution (5.3-10) now contains only two independent solutions. The orthogonality relations (5.5-11) do not hold for α,β=1,2 and 4,5, and the relations (5.5-13)–(5.5-17) are not valid. Anisotropic elastic materials with a nonsemisimple **N** are called *degenerate materials*. They are degenerate in the mathematical sense, not necessarily in the physical sense. Isotropic materials are a special group of degenerate materials that happen to be degenerate also in the physical sense. There are degenerate anisotropic materials that have no material symmetry planes (Ting, 1994). It should be mentioned that the breakdown of the formalism for degenerate materials is not limited to the Stroh formalism. Other formalisms have the same problem.

13.1 General Solution for Degenerate and Near Degenerate Materials

We have seen in Chapters 8 through 12 that in many applications the arbitrary constant **q** that appears in the general solution (5.3-10) can be determined analytically using the relations (5.5-13)–(5.5-17). These solutions are consequently not valid for degenerate materials. Alternate to the algebraic representation of **S**, **H**, **L** in (5.5-17), it is shown in Section 7.6 that one can use an integral representation to determine **S**, **H**, **L** without computing the eigenvalues p_α and the eigenvectors ξ_α. If the final solution is expressed in terms of **S**, **H**, and **L** the solution is valid for degenerate materials. However not all solutions can be expressed entirely in terms of **S**, **H**, and **L**. It is therefore necessary to modify the formalism for degenerate materials.

The difficulty with the general solution (5.3-10) occurs not only when the material is degenerate. If p_2 is nearly equal to p_1 and ξ_2 is also nearly equal to ξ_1, the magnitudes of the orthonormalized eigenvectors ξ_2 and ξ_1 are very large, and become infinity as p_2 approaches p_1 (Ting and Hwu, 1988). This causes problems in numerical calculations. Therefore we consider in this chapter materials that are degenerate or almost degenerate.

In this section we study how to restore three independent solutions for degenerate materials. The degenerate case $p_2=p_1$ and $\xi_2=\xi_1$ will be studied. We will derive a general solution that is valid also for almost degenerate materials. The *extraordinary degenerate* case $p_3=p_2=p_1$ and $\xi_3=\xi_2=\xi_1$ will not be considered here. It has been an open question if there exists a material for which the matrix **N** is extraordinary degenerate. For surface waves in an anisotropic half-space Barnett (1992) has considered the possibility that the matrix $\mathbf{N}(v)$ where v is the surface wave speed is extraordinary degenerate. However no examples have been found for which $\mathbf{N}(v)$ is extraordinary degenerate. It has been proved recently by Ting (1996b) that an extraordinary degenerate **N** exists for $v=0$. Thus while isotropic material is the most degenerate material in the physical sense, it is not the most degenerate material in the mathematical sense. Ting also pointed out that an extraordinary degenerate $\mathbf{N}(v)$ exists for surface waves.

When p_α are distinct, (13.1-12) consists of three independent solutions. A linear combination of two independent solutions can be substituted for an independent solution. Therefore let

$$\mathbf{u} = \mathrm{Re}\left\{q_1 f(z_1)\boldsymbol{a}_1 + \frac{q_2'}{p_2-p_1}[f(z_2)\boldsymbol{a}_2 - f(z_1)\boldsymbol{a}_1] + q_3 f(z_3)\boldsymbol{a}_3\right\}$$

in which q_2' is an arbitrary constant. A similar equation can be written for ϕ. Since

$$f(z_2)\boldsymbol{a}_2 - f(z_1)\boldsymbol{a}_1 = f(z_2)(\boldsymbol{a}_2 - \boldsymbol{a}_1) + [f(z_2) - f(z_1)]\boldsymbol{a}_1$$

we have

$$\mathbf{u} = \mathrm{Re}\left\{q_1 f(z_1)\boldsymbol{a}_1 + q_2'[f(z_2)\boldsymbol{a}_2' + x_2 f'(z_1)\boldsymbol{a}_1] + q_3 f(z_3)\boldsymbol{a}_3\right\}$$

or, with the same derivation for ϕ,

$$\mathbf{u} = \mathrm{Re}\{\mathbf{A}'\mathbf{F}(z)\mathbf{q}'\}, \quad \phi = \mathrm{Re}\{\mathbf{B}'\mathbf{F}(z)\mathbf{q}'\}. \tag{13.1-1}$$

In the above,

$$\mathbf{A}' = [\boldsymbol{a}_1, \boldsymbol{a}_2', \boldsymbol{a}_3], \quad \mathbf{B}' = [\boldsymbol{b}_1, \boldsymbol{b}_2', \boldsymbol{b}_3], \tag{13.1-2}$$

$$\mathbf{F}(z) = \begin{bmatrix} f(z_1) & x_2 f'(z_1) & 0 \\ 0 & f(z_2) & 0 \\ 0 & 0 & f(z_3) \end{bmatrix}, \quad \mathbf{q}' = \begin{bmatrix} q_1 \\ q_2' \\ q_3 \end{bmatrix}, \tag{13.1-3}$$

$$\boldsymbol{a}_2' = \frac{\boldsymbol{a}_2 - \boldsymbol{a}_1}{p_2 - p_1}, \quad \boldsymbol{b}_2' = \frac{\boldsymbol{b}_2 - \boldsymbol{b}_1}{p_2 - p_1}, \tag{13.1-4}$$

$$f'(z_1) = \frac{f(z_1) - f(z_2)}{z_1 - z_2}. \tag{13.1-5}$$

The x_2 in $\mathbf{F}(z)$ may be replaced by

$$x_2 = \frac{z_1 - \bar{z}_1}{p_1 - \bar{p}_1} = \frac{z_1 - \bar{z}_1}{2i \operatorname{Im} p_1}. \tag{13.1-6}$$

The general solution (13.1-1) applies regardless of whether $p_2=p_1$ or not. In fact $\delta=p_2-p_1$ need not be small, i.e., the material needs not be almost degenerate. Thus the general solution (13.1-1) apply to simple and nonsemisimple $\mathbf{N}$. It does not apply to semisimple $\mathbf{N}$.

The vectors $\boldsymbol{a}_2'$ and $\boldsymbol{b}_2'$ are determined by writing (5.1-8), (5.3-2) for $p=p_2$, p_1 and subtracting the resulting equations. We have

$$\begin{gathered} \left\{\mathbf{Q} + p_2(\mathbf{R} + \mathbf{R}^T) + p_2^2\mathbf{T}\right\} \boldsymbol{a}_2' = -\left\{(\mathbf{R} + \mathbf{R}^T) + (p_1 + p_2)\mathbf{T}\right\}\boldsymbol{a}_1 \\ \boldsymbol{b}_2' = (\mathbf{R}^T + p_2\mathbf{T})\,\boldsymbol{a}_2' + \mathbf{T}\boldsymbol{a}_1 = -\frac{1}{p_2}(\mathbf{Q} + p_1\mathbf{R})\,\boldsymbol{a}_2' + \frac{1}{p_1 p_2}\mathbf{Q}\boldsymbol{a}_1. \end{gathered} \tag{13.1-7}$$

Alternately, the same operation on (5.5-3) leads to

$$\mathbf{N}\boldsymbol{\xi}_2' = p_2\boldsymbol{\xi}_2' + \boldsymbol{\xi}_1, \tag{13.1-8}$$

$$\boldsymbol{\xi}_2' = \frac{\boldsymbol{\xi}_2 - \boldsymbol{\xi}_1}{p_2 - p_1}. \tag{13.1-9}$$

In the limit $p_2=p_1$, $\boldsymbol{\xi}_2'$ is the generalized eigenvector of $\mathbf{N}$ associated with the repeated eigenvalue $p_2=p_1$ (Section 1.4). We will refer $\boldsymbol{\xi}_2'$ as the *generalized eigenvector* regardless of whether $p_2=p_1$ or not.

An important property inherited in the Stroh formalism is the orthogonality (5.5-13) and closure (5.5-15) relations connecting $\mathbf{A}$ and $\mathbf{B}$. We will study in the next section the orthogonality and closure relations that connect the matrices $\mathbf{A}'$ and $\mathbf{B}'$ appearing in the general solution (13.1-1) for degenerate and almost degenerate materials.

13.2 Orthogonality and Closure Relations

When $p_2=p_1$ the orthogonality relations (5.3-10) are not valid for α, $\beta=1,2$ and 4,5. It suffices to study the modification of (5.3-10) for $\alpha,\beta=1,2$. The presentation below follows essentially that of Ting and Hwu (1988) (see also Sections 1.4 and 5.9).

The right eigenvectors ξ_1 and ξ_2' are, by (5.5-3) and (13.1-8),

$$\begin{aligned} \mathbf{N}\xi_1 &= p_1\xi_1, \\ \mathbf{N}\xi_2' &= p_2\xi_2' + \xi_1. \end{aligned} \tag{13.2-1}$$

The left eigenvectors η_1' and η_2 are

$$\begin{aligned} \mathbf{N}^T\eta_1' &= p_1\eta_1' + \eta_2, \\ \mathbf{N}^T\eta_2 &= p_2\eta_2, \end{aligned} \tag{13.2-2}$$

where

$$\eta_1' = \frac{\eta_1 - \eta_2}{p_1 - p_2}. \tag{13.2-3}$$

Equation $(13.2\text{-}2)_1$ is obtained by writing (5.5-6) for $p=p_1,p_2$ and subtracting the resulting equations. From (5.5-9), (13.2-3), and using the relation

$$\xi_2 = \xi_1 + \delta\xi_2' \tag{13.2-4}$$

where $\delta=p_2-p_1$ obtained from (13.1-9) we have

$$\eta_1' = \hat{\mathbf{I}}\xi_2', \quad \eta_2 = \hat{\mathbf{I}}(\xi_1 + \delta\xi_2'). \tag{13.2-5}$$

One could also obtain (13.2-5) by premultiplying (13.2-1) by $\hat{\mathbf{I}}$, using (5.5-8), (5.5-9) and, after some manipulation, comparing the results with (13.2-2).

The new vectors satisfy the following identities

$$\eta_2^T\xi_2' - \eta_1'^T\xi_1 = \delta\,\eta_1'^T\xi_2', \tag{13.2-6}$$

$$\eta_2^T\xi_1 = 0. \tag{13.2-7}$$

Equation (13.2-6) and (13.2-7) are obtained when we premultiply $(13.2\text{-}1)_{2,1}$ by $\eta_1'^T$ and use $(13.2\text{-}2)_1$. To form an orthonormal system we must have in addition to (13.2-7)

$$\eta_1'^T\xi_1 = 1, \quad \eta_2^T\xi_2' = 1, \quad \eta_1'^T\xi_2' = 0. \tag{13.2-8}$$

In view of (13.2-6) we do not need all three equations in (13.2-8).

Let $\hat{\xi}_1$, $\hat{\xi}_2'$, $\hat{\eta}_1'$, and $\hat{\eta}_2$ be solutions to (13.2-1) and (13.2-2).

They satisfy (13.2-6) and (13.2-7). It is clear that

$$\xi_1 = k_1\hat{\xi}_1, \quad \xi_2' = k_1\hat{\xi}_2' + k_2'\hat{\xi}_2,$$

are also solutions where k_1, k_2' are arbitrary constants. In the above $\hat{\xi}_2$ is a solution of (5.5-3) for $p=p_2$ which can be eliminated by using (13.2-4). Hence

$$\xi_1 = k_1\hat{\xi}_1, \quad \xi_2' = k_2\hat{\xi}_2' + k_2'\hat{\xi}_1, \tag{13.2-9}$$

$$k_2 = k_1 + \delta k_2'. \tag{13.2-10}$$

By (13.2-9) and (13.2-5),

$$\eta_1' = k_1\hat{\eta}_1' + k_2'\hat{\eta}_2, \quad \eta_2 = k_2\hat{\eta}_2. \tag{13.2-11}$$

Our objective is to find k_1, k_2 and k_2' such that (13.2-8) are satisfied. Imposition of $(13.2\text{-}8)_{1,2}$ and use of (13.2-7) leads to

$$k_1^{-2} = \hat{\eta}_1'^T\hat{\xi}_1, \quad k_2^{-2} = \hat{\eta}_2^T\hat{\xi}_2', \tag{13.2-12}$$

while $(13.2\text{-}8)_3$ yields with the aid of (13.2-12)

$$k_2' = \frac{-k_1^2k_2^2}{k_1 + k_2}\hat{\eta}_1'^T\hat{\xi}_2'. \tag{13.2-13}$$

Insertion of the right sides of (13.2-12) and (13.2-13) into (13.2-6) recovers (13.2-10). It should be noted that k_1, k_2 obtained from (13.2-12) are not unique. If k_1 is a solution, so is $-k_1$. When $k_1^2 = k_2^2$, one can always choose the sign such that the denominator of (13.2-13) is nonzero. This completes the orthonormalization of ξ_1, ξ_2', η_1', and η_2. It should be emphasized that the ξ_1 and η_2 so determined are different from ξ_1 and η_2 determined by applying (5.5-11) for $\alpha,\beta=1,2$.

With ξ_2, η_1 replaced by ξ_2', η_1', respectively, the orthogonality relations (5.5-11) hold for all α,β. Let

$$\mathbf{Y} = \begin{bmatrix} 0 & 1 & 0 \\ 1 & \delta & 0 \\ 0 & 0 & 1 \end{bmatrix}, \qquad \mathbf{Y} = \mathbf{Y}^T. \tag{13.2-14}$$

In view of (13.2-5) the orthogonality relations can be written as

$$\begin{bmatrix} \mathbf{YB}'^T & \mathbf{YA}'^T \\ \overline{\mathbf{YB}}'^T & \overline{\mathbf{YA}}'^T \end{bmatrix} \begin{bmatrix} \mathbf{A}' & \overline{\mathbf{A}}' \\ \mathbf{B}' & \overline{\mathbf{B}}' \end{bmatrix} = \begin{bmatrix} \mathbf{I} & \mathbf{0} \\ \mathbf{0} & \mathbf{I} \end{bmatrix}. \tag{13.2-15}$$

The two 6×6 matrices on the left are the inverses of each other, and hence their products commute, i.e.,

$$\begin{bmatrix} \mathbf{A}' & \overline{\mathbf{A}}' \\ \mathbf{B}' & \overline{\mathbf{B}}' \end{bmatrix} \begin{bmatrix} \mathbf{YB}'^T & \mathbf{YA}'^T \\ \overline{\mathbf{YB}}'^T & \overline{\mathbf{YA}}'^T \end{bmatrix} = \begin{bmatrix} \mathbf{I} & \mathbf{0} \\ \mathbf{0} & \mathbf{I} \end{bmatrix} \tag{13.2-16}$$

or

$$\begin{aligned} \mathbf{A}'\mathbf{YB}'^T + \overline{\mathbf{A}}'\overline{\mathbf{YB}}'^T = \mathbf{I} = \mathbf{B}'\mathbf{YA}'^T + \overline{\mathbf{B}}'\overline{\mathbf{YA}}'^T, \\ \mathbf{A}'\mathbf{YA}'^T + \overline{\mathbf{A}}'\overline{\mathbf{YA}}'^T = \mathbf{0} = \mathbf{B}'\mathbf{YB}'^T + \overline{\mathbf{B}}'\overline{\mathbf{YB}}'^T. \end{aligned} \tag{13.2-17}$$

These are the *closure* relations. Using the same argument in reaching (5.5-17) one has

$$\mathbf{S} = i(2\mathbf{A}'\mathbf{YB}'^T - \mathbf{I}), \quad \mathbf{H} = 2i\mathbf{A}'\mathbf{YA}'^T, \quad \mathbf{L} = -2i\mathbf{B}'\mathbf{YB}'^T. \tag{13.2-18}$$

Ting and Hwu (1988) have shown that

$$\mathbf{AA}^T = \mathbf{A}'\mathbf{YA}'^T, \quad \mathbf{BB}^T = \mathbf{B}'\mathbf{YB}'^T, \quad \mathbf{AB}^T = \mathbf{A}'\mathbf{YB}'^T, \tag{13.2-19}$$

so that **S**, **H**, and **L** in (13.2-18) are indeed the Barnett–Lothe tensors.

Insertion of (13.2-19) into the right sides of (5.6-5) yields

$$\mathbf{BA}^{-1} = \mathbf{B}'\mathbf{A}'^{-1}, \quad \mathbf{AB}^{-1} = \mathbf{A}'\mathbf{B}'^{-1},$$

and (5.6-6) can be written as

$$\mathbf{M} = -i\mathbf{B}'\mathbf{A}'^{-1} = \mathbf{H}^{-1} + i\mathbf{H}^{-1}\mathbf{S},$$

$$\mathbf{M}^{-1} = i\mathbf{A}'\mathbf{B}'^{-1} = \mathbf{L}^{-1} - i\mathbf{SL}^{-1}.$$

We note that when $\delta = p_2 - p_1 = 0$, **Y** of (13.2-14) reduces to

$$\Gamma = \begin{bmatrix} 0 & 1 & 0 \\ 1 & 0 & 0 \\ 0 & 0 & 1 \end{bmatrix}. \tag{13.2-20}$$

Equations (13.2-15)–(13.2-18) reduce to (5.9-15)–(5.9-18) for degenerate materials.

For isotropic materials $p_1 = p_2 = p_3 = i$, $\xi_1 = \xi_2 \neq \xi_3$. The procedures outlined above in finding the orthonormalized eigenvectors ξ_1 and ξ_2', and the ξ_3 from (5.5-3) yields (Ting and Hwu, 1988),

$$\mathbf{A}' = \psi \begin{bmatrix} 1 & -i\gamma & 0 \\ i & -\gamma & 0 \\ 0 & 0 & \varepsilon \end{bmatrix}, \quad \mathbf{B}' = \mu\psi \begin{bmatrix} 2i & 1 & 0 \\ -2 & -i & 0 \\ 0 & 0 & i\varepsilon \end{bmatrix}, \tag{13.2-21}$$

$$\psi = [8\mu(1-\nu)]^{-1/2}, \quad \gamma = \tfrac{1}{2}(3-4\nu), \quad \varepsilon = (1-i)\sqrt{2(1-\nu)}.$$

In the above μ and ν are, respectively, the shear modulus and Poisson's ratio. The inverse of $\mathbf{B}'$ is

$$\mathbf{B}'^{-1} = \frac{1}{4\mu\psi}\begin{bmatrix} -i & -1 & 0 \\ 2 & 2i & 0 \\ 0 & 0 & -4i/\varepsilon \end{bmatrix}. \tag{13.2-22}$$

Substitution of (13.2-21) into (13.2-18) with the $\mathbf{Y}$ replaced by Γ in (13.2-20) recovers the Barnett–Lothe tensors **S**, **H**, and **L** for isotropic materials given in (6.4-13).

13.3 A Crack in a Degenerate Material

As an illustration consider a crack of length $2a$ centrally located at $x_2=0$, $|x_1|<a$ in a homogeneous anisotropic degenerate material. A uniform stress σ_{ij}^∞ is applied at infinity. The crack surfaces are traction-free. The solution consists of a uniform stress solution due to the uniform loading σ_{ij}^∞ and a disturbed solution due to the presence of the crack. For the disturbed solution a uniform traction $\mathbf{t}_\Gamma$ is applied at the upper crack surface and $-\mathbf{t}_\Gamma$ is applied at the lower crack surface. The stress vanishes at infinity. Thus the boundary conditions for the stress function ϕ are

$$\phi = \mathbf{0} \quad \text{for} \quad |\mathbf{x}| = \infty, \tag{13.3-1}$$

$$\phi = -x_1\,\mathbf{t}_\Gamma \quad \text{for} \quad x_2=\pm 0,\ |x_1|<a. \tag{13.3-2}$$

From the general solution (13.1-1), and following the derivation of the solution (11.4-3) for non-degenerate materials, it is easily shown that the solution is

$$\mathbf{u} = \mathrm{Re}\{\mathbf{A}'\mathbf{F}(z)\mathbf{B}'^{-1}\}\mathbf{t}_\Gamma, \quad \phi = \mathrm{Re}\{\mathbf{B}'\mathbf{F}(z)\mathbf{B}'^{-1}\}\mathbf{t}_\Gamma, \tag{13.3-3}$$

where

$$\mathbf{F}(z) = \begin{bmatrix} f(z_1) & x_2 f'(z_1) & 0 \\ 0 & f(z_2) & 0 \\ 0 & 0 & f(z_3) \end{bmatrix}, \tag{13.3-4}$$

$$f(z) = \sqrt{z^2 - a^2} - z, \quad f'(z) = \frac{z}{\sqrt{z^2 - a^2}} - 1.$$

The solution (13.3-3) applies to both degenerate and non-degenerate materials.

For isotropic materials which are degenerate, $p_1=p_2=p_3=i$ and substitution of (13.2-21) and (13.2-22) into (13.3-3) leads to

$$\mathbf{u} = \frac{1}{2\mu}\mathrm{Re}\{f(z)\mathbf{E} - ix_2 f'(z)\mathbf{V}\}\mathbf{t}_\Gamma,$$

$$\boldsymbol{\phi} = \mathrm{Re}\{f(z)\mathbf{I} + x_2 f'(z)\mathbf{V}\}\mathbf{t}_\Gamma, \tag{13.3-5}$$

where $z=x_1+ix_2$ and

$$\mathbf{E} = \begin{bmatrix} -2i(1-\nu) & 1-2\nu & 0 \\ -(1-2\nu) & -2i(1-\nu) & 0 \\ 0 & 0 & -2i \end{bmatrix}, \quad \mathbf{V} = \begin{bmatrix} i & -1 & 0 \\ -1 & -i & 0 \\ 0 & 0 & 0 \end{bmatrix}. \tag{13.3-6}$$

13.4 Conversion of Solutions from Non-degenerate Materials

The generalized eigenvectors introduced in Section 13.1 restore three independent solutions for degenerate materials. They are also useful, for instance, in identifying the three one-component Green's functions that form the Green's function for the infinite space (Section 8.2). However, the generalized eigenvectors are not simple to compute. Orthonormalization of the generalized eigenvectors also takes some effort. In this section we show how one can modify the solutions for non-degenerate materials so that they can be applied for degenerate or almost degenerate materials. With this process we circumvent the need for computing the generalized eigenvectors. The following presentation is an extension of the derivation given by Ting (1992e) who studied the Green's function for degenerate anisotropic half-space.

We write the solution (5.3-10) for non-degenerate materials as

$$\mathbf{u} = \mathrm{Re}\{(\mathbf{A}\mathbf{B}^{-1})\mathbf{X}(z)(\mathbf{B}\mathbf{q})\}, \quad \boldsymbol{\phi} = \mathrm{Re}\{\mathbf{X}(z)(\mathbf{B}\mathbf{q})\}, \tag{13.4-1}$$

where

$$\mathbf{X}(z) = \mathbf{B}\langle f(z_*)\rangle\mathbf{B}^{-1}. \tag{13.4-2}$$

The quantity $\mathbf{Bq}$ in most applications presented in Chapters 8 through 11 is given in terms of

$$\mathbf{A}\mathbf{A}^T,\ \mathbf{B}\mathbf{A}^T,\ \mathbf{B}\mathbf{B}^T,\ \mathbf{B}\mathbf{A}^{-1}, \text{ and } \mathbf{A}\mathbf{B}^{-1}.$$

They can be expressed in terms of the Barnett–Lothe tensors $\mathbf{S}$, $\mathbf{H}$, and $\mathbf{L}$ through (5.5-19) and (5.6-6). $\mathbf{S}$, $\mathbf{H}$, and $\mathbf{L}$ can be computed using the integral representation (Section 7.6) which is valid regardless of whether the material is degenerate or not. It suffices therefore to study the 3×3 matrix $\mathbf{X}(z)$ for degenerate and almost degenerate materials. For problems involving bimaterials, $\mathbf{Bq}$ may also contain the terms

$$\mathbf{B}\mathbf{I}_\alpha\mathbf{B}^{-1},\quad \mathbf{B}\mathbf{I}_\alpha\mathbf{A}^{-1},\quad \mathbf{A}\mathbf{I}_\alpha\mathbf{B}^{-1},\quad \mathbf{A}\mathbf{I}_\alpha\mathbf{A}^{-1},$$

where

$$\mathbf{I}_1 = \text{diag}[1,\ 0,\ 0],\quad \mathbf{I}_2 = \text{diag}[0,\ 1,\ 0],\quad \mathbf{I}_3 = \text{diag}[0,\ 0,\ 1].$$

A trivial identity is

$$\mathbf{I}_1 + \mathbf{I}_2 + \mathbf{I}_3 = \mathbf{I}. \tag{13.4-3}$$

By writing

$$\mathbf{B}\mathbf{I}_\alpha\mathbf{A}^{-1} = (\mathbf{B}\mathbf{I}_\alpha\mathbf{B}^{-1})(\mathbf{B}\mathbf{A}^{-1}),$$

$$\mathbf{A}\mathbf{I}_\alpha\mathbf{B}^{-1} = (\mathbf{A}\mathbf{B}^{-1})(\mathbf{B}\mathbf{I}_\alpha\mathbf{B}^{-1}),$$

$$\mathbf{A}\mathbf{I}_\alpha\mathbf{A}^{-1} = (\mathbf{A}\mathbf{B}^{-1})(\mathbf{B}\mathbf{I}_\alpha\mathbf{B}^{-1})(\mathbf{B}\mathbf{A}^{-1}),$$

it suffices to investigate

$$\mathbf{U}_\alpha = \mathbf{B}\mathbf{I}_\alpha\mathbf{B}^{-1},\quad (\alpha{=}1,2,3). \tag{13.4-4}$$

Consider the matrices $\mathbf{U}_\alpha$. From (13.4-3),

$$\mathbf{U}_1 + \mathbf{U}_2 + \mathbf{U}_3 = \mathbf{I}. \tag{13.4-5}$$

Explicit expressions of $\mathbf{B}$ and $\mathbf{B}^{-1}$ were given in (6.3-2) and (6.3-4). In computing $\mathbf{U}_\alpha$ it is more convenient to write $\mathbf{B}$ and $\mathbf{B}^{-1}$ as

$$\mathbf{B} = -[k_1\mathbf{g}_1,\ k_2\mathbf{g}_2,\ k_3\mathbf{g}_3],\quad (\mathbf{B}^{-1})^T = \frac{1}{\Delta}\left[-k_1^{-1}\mathbf{h}_2,\ k_2^{-1}\mathbf{h}_1,\ k_3^{-1}\mathbf{h}_3\right]$$

where k_α are the normalization factors,

$$\mathbf{g}_\alpha = \begin{bmatrix} p_\alpha \\ -1 \\ \lambda_\alpha \end{bmatrix},\quad \mathbf{h}_\alpha = \begin{bmatrix} 1-\lambda_\alpha\lambda_3 \\ p_\alpha-\lambda_\alpha\lambda_3 p_3 \\ (p_\alpha-p_3)\lambda_3 \end{bmatrix}, \tag{13.4-6}$$

for α=1,2, and

$$\mathbf{g}_3 = \begin{bmatrix} p_3\lambda_3 \\ -\lambda_3 \\ 1 \end{bmatrix},\quad \mathbf{h}_3 = \begin{bmatrix} \lambda_1-\lambda_2 \\ \lambda_1 p_2-\lambda_2 p_1 \\ -(p_1-p_2) \end{bmatrix}.$$

In the above

$$\Delta = (p_1-p_2) + \lambda_3[(\lambda_1 p_2 - \lambda_2 p_1) - p_3(\lambda_1-\lambda_2)],$$

and λ_α (α=1,2,3) are defined in (4.6-9). Therefore

$$\mathbf{U}_1 = \frac{1}{(p_1-p_2)\gamma}\mathbf{g}_1\mathbf{h}_2^T, \quad \mathbf{U}_2 = \frac{-1}{(p_1-p_2)\gamma}\mathbf{g}_2\mathbf{h}_1^T, \tag{13.4-7}$$

$$\mathbf{U}_3 = \frac{-1}{(p_1-p_2)\gamma}\mathbf{g}_3\mathbf{h}_3^T = \frac{1}{\gamma}\begin{bmatrix} p_3\lambda_3 \\ -\lambda_3 \\ 1 \end{bmatrix}\left[-\lambda',\ \tilde{\lambda}-\tilde{p}\lambda',\ 1\right] \tag{13.4-8}$$

in which

$$\tilde{p} = \tfrac{1}{2}(p_1+p_2), \quad \tilde{\lambda} = \tfrac{1}{2}(\lambda_1+\lambda_2),$$

$$\lambda' = \frac{\lambda_1-\lambda_2}{p_1-p_2}, \quad \gamma = \frac{\Delta}{p_1-p_2} = 1+\lambda_3[\lambda'(\tilde{p}-p_3)-\tilde{\lambda}].$$

The normalization factors k_α drop out. As $p_2 \to p_1$, λ' is bounded, so is $\mathbf{U}_3$. $\mathbf{U}_1$ and $\mathbf{U}_2$ are not bounded but $\mathbf{U}_1+\mathbf{U}_2$ is according to (13.4-5). Let

$$\mathbf{V} = \tfrac{1}{2}(p_1-p_2)(\mathbf{U}_1-\mathbf{U}_2). \tag{13.4-9}$$

Substituting from (13.4-7) and using the relation

$$\mathbf{g}_1\mathbf{h}_2^T + \mathbf{g}_2\mathbf{h}_1^T = \tfrac{1}{2}(\mathbf{g}_1+\mathbf{g}_2)(\mathbf{h}_1^T+\mathbf{h}_2^T) - \tfrac{1}{2}(\mathbf{g}_1-\mathbf{g}_2)(\mathbf{h}_1^T-\mathbf{h}_2^T)$$

one has

$$\mathbf{V} = \frac{1}{\gamma}\begin{bmatrix} \tilde{p} \\ -1 \\ \tilde{\lambda} \end{bmatrix}\left[1-\tilde{\lambda}\lambda_3,\ \tilde{p}-\tilde{\lambda}\lambda_3 p_3,\ (\tilde{p}-p_3)\lambda_3\right]$$

$$-\frac{\delta^2}{4\gamma}\begin{bmatrix} 1 \\ 0 \\ \lambda' \end{bmatrix}\left[-\lambda'\lambda_3,\ 1-\lambda'\lambda_3 p_3,\ \lambda_3\right] \tag{13.4-10}$$

where $\delta = p_2 - p_1$. $\mathbf{V}$ is bounded when $p_2 = p_1$.

We now turn to the matrix $\mathbf{X}(z)$ of (13.4-2) which is written as

$$\mathbf{X}(z) = f(z_1)\mathbf{U}_1 + f(z_2)\mathbf{U}_2 + f(z_3)\mathbf{U}_3.$$

Since

$$f(z_1)\mathbf{U}_1 + f(z_2)\mathbf{U}_2 = \tfrac{1}{2}[f(z_1)+f(z_2)][\mathbf{U}_1+\mathbf{U}_2]$$
$$+\tfrac{1}{2}[f(z_1)-f(z_2)][\mathbf{U}_1-\mathbf{U}_2]$$

we have, by (13.1-5), (13.4-5), and (13.4-9),

$$\mathbf{X}(z) = \tfrac{1}{2}[f(z_1) + f(z_2)](\mathbf{I} - \mathbf{U}_3) + x_2 f'(z_1)\mathbf{V} + f(z_3)\mathbf{U}_3. \qquad (13.4\text{-}11)$$

The function $\mathbf{X}(z)$ is in terms of $\mathbf{U}_3$ and $\mathbf{V}$ which are bounded when $p_2=p_1$.

The expressions for $\mathbf{U}_3$ and $\mathbf{V}$ are simplified considerably for monoclinic materials with the symmetry plane at $x_3=0$. For these materials $\lambda_1=\lambda_2=\lambda_3=\tilde{\lambda}=\lambda'=0$, $\gamma=1$ so that

$$\mathbf{U}_3 = \mathbf{I}_3, \quad \mathbf{V} = \begin{bmatrix} \tilde{p} & p_1 p_2 & 0 \\ -1 & -\tilde{p} & 0 \\ 0 & 0 & 0 \end{bmatrix}. \qquad (13.4\text{-}12)$$

In the limit $p_2=p_1$ (13.4-11) reduces to

$$\mathbf{X}(z) = \begin{bmatrix} f(z_1) & 0 & 0 \\ 0 & f(z_1) & 0 \\ 0 & 0 & f(z_3) \end{bmatrix} + x_2 f'(z_1) \begin{bmatrix} p_1 & p_1^2 & 0 \\ -1 & -p_1 & 0 \\ 0 & 0 & 0 \end{bmatrix}. \qquad (13.4\text{-}13)$$

As an illustration consider a finite crack in an infinite isotropic medium subjected to a uniform traction $\mathbf{t}_\Gamma$ on the crack surfaces studied in Section 13.3. From $(5.6\text{-}6)_4$ and (13.4-1) with $\mathbf{Bq}=\mathbf{t}_\Gamma$,

$$\mathbf{u} = \mathrm{Re}\left\{-(\mathbf{SL}^{-1} + i\mathbf{L}^{-1})\mathbf{X}(z)\right\}\mathbf{t}_\Gamma, \quad \boldsymbol{\phi} = \mathrm{Re}\{\mathbf{X}(z)\}\mathbf{t}_\Gamma. \qquad (13.4\text{-}14)$$

For isotropic materials $p_1=p_2=p_3=i$ and $\mathbf{V}$ is given by (13.3-6). Explicit expressions of $\mathbf{S}$ and $\mathbf{L}$ for isotropic materials are presented in (6.4-13) from which we obtain

$$\mathbf{L}^{-1} = \frac{1}{\mu}\begin{bmatrix} 1-\nu & 0 & 0 \\ 0 & 1-\nu & 0 \\ 0 & 0 & 1 \end{bmatrix}, \quad \mathbf{SL}^{-1} = \frac{1-2\nu}{2\mu}\begin{bmatrix} 0 & -1 & 0 \\ 1 & 0 & 0 \\ 0 & 0 & 0 \end{bmatrix}.$$

By setting $p_1=p_3=i$ in (13.4-13), (13.4-14) recovers (13.3-5). There is no need to compute the generalized eigenvectors in this derivation.

Chapter 14

GENERALIZATION OF THE STROH FORMALISM

It was shown in Chapter 12 that the Stroh formalism can be extended to elastodynamics when the problem is a steady state motion. In this chapter we will see several areas where the Stroh formalism can be generalized. First, the formalism is extended to a group of general boundary conditions that include as special cases the boundary conditions for a traction-free, a rigid, and a slippery surface. Explicit solutions are obtained for the Green's function and the image force in a half-space with a slippery boundary surface subjected to a line dislocation in the half-space. The Stroh formalism is then extended to thermo-anisotropic elasticity where we point out that applications to the interfacial crack may lead to a new and higher order stress singularity. Generalization of the formalism to piezoelectric materials results in an octet formalism for which there are four pairs of complex eigenvalues. Extensions of the Stroh formalism to three-dimensional deformations of anisotropic body require a special attention and are investigated in the next chapter.

14.1 The Stroh Formalism for General Boundary Conditions

The Stroh formalism is in terms of the displacement $\mathbf{u}$ and the stress function $\boldsymbol{\phi}$, both 3-vectors. It is therefore most suitable for boundary conditions that are in terms of the displacement or the surface traction. For a slippery surface the normal component of the displacement and the two tangential components of the surface traction vanish. The boundary conditions are not in terms of $\mathbf{u}$ or $\boldsymbol{\phi}$, but a mixture of one component from $\mathbf{u}$ and two components from $\boldsymbol{\phi}$. This destroys the elegant sextic formalism of Stroh. We present here a generalized formalism that is applicable to a class of general boundary conditions. They include as special cases the boundary conditions for a traction-free, a rigid, and a slippery surface. The material presented below is taken from Ting and Wang (1992).

Let $x_2=0$ be the boundary surface on which some components of u_i and σ_{2i} vanish. There are eight possible choices:

$$\sigma_{21} = 0, \quad \sigma_{22} = 0, \quad \sigma_{23} = 0, \tag{14.1-1a}$$

$$\sigma_{21} = 0, \quad \sigma_{22} = 0, \quad u_3 = 0, \tag{14.1-1b}$$

$$\sigma_{21}=0, \quad u_2=0, \quad \sigma_{23}=0, \tag{14.1-1c}$$

$$\sigma_{21}=0, \quad u_2=0, \quad u_3=0, \tag{14.1-1d}$$

$$u_1=0, \quad \sigma_{22}=0, \quad \sigma_{23}=0, \tag{14.1-1e}$$

$$u_1=0, \quad \sigma_{22}=0, \quad u_3=0, \tag{14.1-1f}$$

$$u_1=0, \quad u_2=0, \quad \sigma_{23}=0, \tag{14.1-1g}$$

$$u_1=0, \quad u_2=0, \quad u_3=0. \tag{14.1-1h}$$

Equations (14.1-1a,h,c) are, respectively, the conditions for a traction-free, a rigid, and a slippery surface. Since $\sigma_{2i}=\varphi_{i,1}$, $\sigma_{2i}=0$ at $x_2=0$ can be replaced by φ_i=constant. In most cases it can be replaced by $\varphi_i=0$ without loss in generality. In terms of the 3-vectors $\mathbf{u}$ and $\boldsymbol{\phi}$ (14.1-1) can be represented by one equation as

$$\mathbf{I}_u\mathbf{u}+\mathbf{I}_\phi\boldsymbol{\phi}=\mathbf{0} \tag{14.1-2}$$

where $\mathbf{I}_u$, $\mathbf{I}_\phi$ are diagonal matrices whose elements are either one or zero, and satisfy the condition

$$\mathbf{I}_u+\mathbf{I}_\phi=\mathbf{I}, \tag{14.1-3a}$$

$\mathbf{I}$ being the unit matrix. It is clear that $\mathbf{I}_u$ and $\mathbf{I}_\phi$ also satisfy the relation

$$\mathbf{I}_u\,\mathbf{I}_\phi=\mathbf{0}. \tag{14.1-3b}$$

The special cases (14.1-1a,h) correspond to $(\mathbf{I}_u,\mathbf{I}_\phi)=(\mathbf{0},\ \mathbf{I})$ and $(\mathbf{I},\ \mathbf{0})$, respectively. As to (14.1-1c) we have

$$\mathbf{I}_u=\text{diag}[0,\ 1,\ 0], \quad \mathbf{I}_\phi=\text{diag}[1,\ 0,\ 1]$$

Equation (14.1-2) is a linear combination of $\mathbf{u}$ and $\boldsymbol{\phi}$ in which $\mathbf{I}_u$ and $\mathbf{I}_\phi$ can be interchanged. We consider the following more general linear combinations

$$\mathbf{X}\mathbf{u}+\mathbf{Y}\boldsymbol{\phi}=\mathbf{0} \quad \text{or} \quad \mathbf{Y}\mathbf{u}+\mathbf{X}\boldsymbol{\phi}=\mathbf{0},$$

where $\mathbf{X}$ and $\mathbf{Y}$ are 3×3 constant real matrices subjected to the conditions

$$\begin{bmatrix}\mathbf{X} & \mathbf{Y}\\ \mathbf{Y} & \mathbf{X}\end{bmatrix}\begin{bmatrix}\mathbf{X}^T & \mathbf{Y}^T\\ \mathbf{Y}^T & \mathbf{X}^T\end{bmatrix}=\mathbf{I}=\begin{bmatrix}\mathbf{X}^T & \mathbf{Y}^T\\ \mathbf{Y}^T & \mathbf{X}^T\end{bmatrix}\begin{bmatrix}\mathbf{X} & \mathbf{Y}\\ \mathbf{Y} & \mathbf{X}\end{bmatrix} \tag{14.1-4}$$

or

$$\mathbf{X}\mathbf{X}^T + \mathbf{Y}\mathbf{Y}^T = \mathbf{I} = \mathbf{X}^T\mathbf{X} + \mathbf{Y}^T\mathbf{Y},$$
$$\mathbf{X}\mathbf{Y}^T + \mathbf{Y}\mathbf{X}^T = \mathbf{0} = \mathbf{X}^T\mathbf{Y} + \mathbf{Y}^T\mathbf{X}. \tag{14.1-5}$$

They reduce to (14.1-3a,b) when $(\mathbf{X},\mathbf{Y})=(\mathbf{I}_u, \mathbf{I}_\phi)$ or $(\mathbf{I}_\phi, \mathbf{I}_u)$. Equations $(14.1\text{-}5)_{3,4}$ tell us that $\mathbf{X}\mathbf{Y}^T$, $\mathbf{Y}\mathbf{X}^T$, $\mathbf{X}^T\mathbf{Y}$, and $\mathbf{Y}^T\mathbf{X}$ are skew-symmetric. They all vanish if any one of them does. Suppose that $\mathbf{X}\mathbf{Y}^T=\mathbf{0}$. Then $\mathbf{Y}\mathbf{X}^T=\mathbf{0}$ by $(14.1\text{-}5)_3$, and

$$\mathbf{X}^T\mathbf{Y} = \mathbf{X}^T\mathbf{Y}(\mathbf{X}^T\mathbf{X} + \mathbf{Y}^T\mathbf{Y}) = \mathbf{X}^T(\mathbf{Y}\mathbf{X}^T)\mathbf{X} - \mathbf{Y}^T(\mathbf{X}\mathbf{Y}^T)\mathbf{Y} = \mathbf{0}.$$

$\mathbf{Y}^T\mathbf{X}=\mathbf{0}$ follows from $(14.1\text{-}5)_4$.

The general solutions for $\mathbf{X}$ and $\mathbf{Y}$ can be obtained from the following identities deduced from (14.1-5),

$$(\mathbf{X} \pm \mathbf{Y})(\mathbf{X} \pm \mathbf{Y})^T = \mathbf{I} = (\mathbf{X} \pm \mathbf{Y})^T(\mathbf{X} \pm \mathbf{Y}).$$

If Ω_1, Ω_2 are any orthogonal matrices,

$$\mathbf{X} = \tfrac{1}{2}(\Omega_1 + \Omega_2), \quad \mathbf{Y} = \tfrac{1}{2}(\Omega_1 - \Omega_2).$$

By letting $\Omega_o = \Omega_1^T\Omega_2$ which is also an orthogonal matrix,

$$\mathbf{X} = \tfrac{1}{2}\Omega_1(\mathbf{I} + \Omega_o), \quad \mathbf{Y} = \tfrac{1}{2}\Omega_1(\mathbf{I} - \Omega_o). \tag{14.1-6}$$

The special case $(\mathbf{X},\mathbf{Y})=(\mathbf{I}_u, \mathbf{I}_\phi)$ corresponds to $\Omega_1=\mathbf{I}$, $\Omega_o=\mathbf{I}_u-\mathbf{I}_\phi$.

Now consider the following generalized solution

$$\hat{\mathbf{u}} = \mathbf{X}\mathbf{u} + \mathbf{Y}\boldsymbol{\phi}, \quad \hat{\boldsymbol{\phi}} = \mathbf{Y}\mathbf{u} + \mathbf{X}\boldsymbol{\phi}. \tag{14.1-7}$$

When $\mathbf{X}$ and $\mathbf{Y}$ are interchanged $\hat{\mathbf{u}}$ and $\hat{\boldsymbol{\phi}}$ are interchanged. The special cases are

$$(\hat{\mathbf{u}}, \hat{\boldsymbol{\phi}}) = \begin{cases} (\mathbf{u}, \boldsymbol{\phi}) & \text{when } (\mathbf{X}, \mathbf{Y}) = (\mathbf{I}, \mathbf{0}), \\ (\boldsymbol{\phi}, \mathbf{u}) & \text{when } (\mathbf{X}, \mathbf{Y}) = (\mathbf{0}, \mathbf{I}). \end{cases}$$

Consequently $\hat{\mathbf{u}}$ (and $\hat{\boldsymbol{\phi}}$) can assume the role of $\mathbf{u}$ or $\boldsymbol{\phi}$. An interesting relation arising from (14.1-7) and (14.1-5) is

$$\hat{\mathbf{u}}^T\hat{\boldsymbol{\phi}} = \mathbf{u}^T\boldsymbol{\phi}.$$

This relation is obtained by inserting (14.1-7) into the above equation and observing that $\mathbf{X}^T\mathbf{Y}$ and $\mathbf{Y}^T\mathbf{X}$ are skew-symmetric, and that $\boldsymbol{\phi}^T(\mathbf{Y}^T\mathbf{Y})\mathbf{u}$ is identical to its transpose $\mathbf{u}^T(\mathbf{Y}^T\mathbf{Y})\boldsymbol{\phi}$. From the general solution of the Stroh formalism

$$\mathbf{u} = \mathrm{Re}\{\mathbf{A}\langle f(z_*)\rangle\mathbf{q}\}, \quad \boldsymbol{\phi} = \mathrm{Re}\{\mathbf{B}\langle f(z_*)\rangle\mathbf{q}\}, \tag{14.1-8}$$

we have

$$\hat{\mathbf{u}} = \mathrm{Re}\left\{\hat{\mathbf{A}}\langle f(z_*)\rangle\mathbf{q}\right\}, \quad \hat{\boldsymbol{\phi}} = \mathrm{Re}\left\{\hat{\mathbf{B}}\langle f(z_*)\rangle\mathbf{q}\right\}, \tag{14.1-9}$$

where

$$\hat{\mathbf{A}} = \mathbf{XA} + \mathbf{YB}, \quad \hat{\mathbf{B}} = \mathbf{YA} + \mathbf{XB}. \tag{14.1-10}$$

In particular,

$$(\hat{\mathbf{A}}, \hat{\mathbf{B}}) = \begin{cases} (\mathbf{A}, \mathbf{B}) & \text{when } (\mathbf{X}, \mathbf{Y}) = (\mathbf{I}, \mathbf{0}), \\ (\mathbf{B}, \mathbf{A}) & \text{when } (\mathbf{X}, \mathbf{Y}) = (\mathbf{0}, \mathbf{I}). \end{cases} \tag{14.1-11}$$

Hence $\hat{\mathbf{A}}$ (and $\hat{\mathbf{B}}$) can assume the role of **A** or **B**.

Writing (14.1-10) and its complex conjugate as

$$\begin{bmatrix} \hat{\mathbf{A}} & \overline{\hat{\mathbf{A}}} \\ \hat{\mathbf{B}} & \overline{\hat{\mathbf{B}}} \end{bmatrix} = \begin{bmatrix} \mathbf{X} & \mathbf{Y} \\ \mathbf{Y} & \mathbf{X} \end{bmatrix} \begin{bmatrix} \mathbf{A} & \overline{\mathbf{A}} \\ \mathbf{B} & \overline{\mathbf{B}} \end{bmatrix}, \quad \begin{bmatrix} \hat{\mathbf{B}}^T & \hat{\mathbf{A}}^T \\ \overline{\hat{\mathbf{B}}}^T & \overline{\hat{\mathbf{A}}}^T \end{bmatrix} = \begin{bmatrix} \mathbf{B}^T & \mathbf{A}^T \\ \overline{\mathbf{B}}^T & \overline{\mathbf{A}}^T \end{bmatrix} \begin{bmatrix} \mathbf{X}^T & \mathbf{Y}^T \\ \mathbf{Y}^T & \mathbf{X}^T \end{bmatrix} \tag{14.1-12}$$

and employing (14.1-4) and the orthogonality relations (5.5-13), we obtain the generalized *orthogonality relations*

$$\begin{bmatrix} \hat{\mathbf{B}}^T & \hat{\mathbf{A}}^T \\ \overline{\hat{\mathbf{B}}}^T & \overline{\hat{\mathbf{A}}}^T \end{bmatrix} \begin{bmatrix} \hat{\mathbf{A}} & \overline{\hat{\mathbf{A}}} \\ \hat{\mathbf{B}} & \overline{\hat{\mathbf{B}}} \end{bmatrix} = \mathbf{I} \tag{14.1-13}$$

identical to (5.5-13) connecting **A** and **B**. Thus the *closure relations* (5.5-15) apply to $\hat{\mathbf{A}}$ and $\hat{\mathbf{B}}$, and the three matrices defined by

$$\hat{\mathbf{S}} = i(2\hat{\mathbf{A}}\hat{\mathbf{B}}^T - \mathbf{I}), \quad \hat{\mathbf{H}} = 2i\hat{\mathbf{A}}\hat{\mathbf{A}}^T, \quad \hat{\mathbf{L}} = -2i\hat{\mathbf{B}}\hat{\mathbf{B}}^T, \tag{14.1-14}$$

are real. By (14.1-11),

$$(\hat{\mathbf{S}}, \hat{\mathbf{H}}, \hat{\mathbf{L}}) = \begin{cases} (\mathbf{S}, \mathbf{H}, \mathbf{L}) & \text{when } (\mathbf{X}, \mathbf{Y}) = (\mathbf{I}, \mathbf{0}), \\ (\mathbf{S}^T, -\mathbf{L}, -\mathbf{H}) & \text{when } (\mathbf{X}, \mathbf{Y}) = (\mathbf{0}, \mathbf{I}). \end{cases} \tag{14.1-15}$$

Unlike **H** and **L** which are positive definite, it is clear that $\hat{\mathbf{H}}$ and $\hat{\mathbf{L}}$ may not be positive definite or negative definite for general **X**, **Y**.

Since $\hat{\mathbf{S}}$, $\hat{\mathbf{H}}$, and $\hat{\mathbf{L}}$ are real, one-halves of the right sides of (14.1-14) can be replaced by their complex conjugates. The result is written as

$$\begin{bmatrix} \hat{\mathbf{S}} & \hat{\mathbf{H}} \\ -\hat{\mathbf{L}} & \hat{\mathbf{S}}^T \end{bmatrix} = \begin{bmatrix} \hat{\mathbf{A}} & \overline{\hat{\mathbf{A}}} \\ \hat{\mathbf{B}} & \overline{\hat{\mathbf{B}}} \end{bmatrix} \begin{bmatrix} i\mathbf{I} & \mathbf{0} \\ \mathbf{0} & -i\mathbf{I} \end{bmatrix} \begin{bmatrix} \hat{\mathbf{B}}^T & \hat{\mathbf{A}}^T \\ \overline{\hat{\mathbf{B}}}^T & \overline{\hat{\mathbf{A}}}^T \end{bmatrix}. \tag{14.1-16}$$

This is identical (6.5-3) for **S**, **H**, and **L**. Substitution from (14.1-12) yields

$$\begin{bmatrix} \hat{\mathbf{S}} & \hat{\mathbf{H}} \\ -\hat{\mathbf{L}} & \hat{\mathbf{S}}^T \end{bmatrix} = \begin{bmatrix} \mathbf{X} & \mathbf{Y} \\ \mathbf{Y} & \mathbf{X} \end{bmatrix} \begin{bmatrix} \mathbf{S} & \mathbf{H} \\ -\mathbf{L} & \mathbf{S}^T \end{bmatrix} \begin{bmatrix} \mathbf{X}^T & \mathbf{Y}^T \\ \mathbf{Y}^T & \mathbf{X}^T \end{bmatrix} \tag{14.1-17}$$

and, by (14.1-4) and (6.5-4),

$$\begin{bmatrix} \hat{\mathbf{S}} & \hat{\mathbf{H}} \\ -\hat{\mathbf{L}} & \hat{\mathbf{S}}^T \end{bmatrix} \begin{bmatrix} \hat{\mathbf{S}} & \hat{\mathbf{H}} \\ -\hat{\mathbf{L}} & \hat{\mathbf{S}}^T \end{bmatrix} = -\mathbf{I}. \tag{14.1-18}$$

This is identical to (6.5-4) for **S**, **H**, and **L**. Hence $\hat{\mathbf{S}}\hat{\mathbf{H}}$, $\hat{\mathbf{L}}\hat{\mathbf{S}}$, $\hat{\mathbf{H}}^{-1}\hat{\mathbf{S}}$, and $\hat{\mathbf{S}}\hat{\mathbf{L}}^{-1}$ are skew-symmetric and

$$\hat{\mathbf{H}}\hat{\mathbf{L}} - \hat{\mathbf{S}}\hat{\mathbf{S}} = \mathbf{I}. \tag{14.1-19}$$

Equation (14.1-17) provides explicit expressions of $\hat{\mathbf{S}}$, $\hat{\mathbf{H}}$, $\hat{\mathbf{L}}$ in terms of **S**, **H**, **L**. They are

$$\begin{aligned} \hat{\mathbf{S}} &= \mathbf{XSX}^T + \mathbf{YS}^T\mathbf{Y}^T + \mathbf{XHY}^T - \mathbf{YLX}^T, \\ \hat{\mathbf{H}} &= \mathbf{XHX}^T - \mathbf{YLY}^T + \mathbf{XSY}^T + \mathbf{YS}^T\mathbf{X}^T, \\ \hat{\mathbf{L}} &= \mathbf{XLX}^T - \mathbf{YHY}^T - \mathbf{XS}^T\mathbf{Y}^T - \mathbf{YSX}^T. \end{aligned} \tag{14.1-20}$$

The special cases (14.1-15) can be deduced from (14.1-20).

The matrices $\hat{\mathbf{H}}$ and $\hat{\mathbf{L}}$ are symmetric. As we stated earlier they may not be positive definite or negative definite. We will show that they are nonsingular if

$$\mathbf{XY}^T = \mathbf{0}. \tag{14.1-21}$$

It can be shown from (14.1-6) that this is equivalent to the condition that Ω_o is symmetric. Equation (14.1-21) is automatically satisfied by $(\mathbf{X},\mathbf{Y})=(\mathbf{I}_u,\mathbf{I}_\phi)$ according to (14.1-3b). To show that $\hat{\mathbf{H}}$ and $\hat{\mathbf{L}}$ are nonsingular, it suffices to show that $\hat{\mathbf{A}}$ and $\hat{\mathbf{B}}$ are nonsingular. We first show that $\hat{\mathbf{A}}=\mathbf{XA}+\mathbf{YB}$ is nonsingular. Consider

$$\mathbf{K} = 2i(\mathbf{XA} + \mathbf{YB})(\mathbf{XA} - \mathbf{YB})^T. \tag{14.1-22}$$

Carrying out the matrix products and using (14.1-21) we have

$$\mathbf{K} = \mathbf{XHX}^T + \mathbf{YLY}^T - \mathbf{XSY}^T + \mathbf{YS}^T\mathbf{X}^T$$

which is real. If **K** were singular with **w** being a right null vector,

$$\mathbf{wKw} = \boldsymbol{\xi}^T\mathbf{H}\boldsymbol{\xi} + \boldsymbol{\eta}\mathbf{L}\boldsymbol{\eta} = \mathbf{0}, \tag{14.1-23}$$

$$\boldsymbol{\xi} = \mathbf{X}^T\mathbf{w}, \quad \boldsymbol{\eta} = \mathbf{Y}^T\mathbf{w}.$$

Equation (14.1-23) holds only if $\boldsymbol{\xi}$ and $\boldsymbol{\eta}$ vanish identically. But

$$\xi^T\xi + \eta^T\eta = \mathbf{w}^T(\mathbf{X}\mathbf{X}^T + \mathbf{Y}\mathbf{Y}^T)\mathbf{w} = \mathbf{w}^T\mathbf{w} \neq 0,$$

appeal being made to $(14.1\text{-}5)_1$. This leads to a contradiction, proving that $\mathbf{K}$, and hence $\hat{\mathbf{A}}$, is nonsingular. By interchanging $\mathbf{X}$ and $\mathbf{Y}$ in (14.1-22) it can be shown that $\hat{\mathbf{B}}$ is nonsingular.

Analogous to (5.6-6) we define the generalized impedance matrix and its inverse by

$$\begin{aligned} \hat{\mathbf{M}} &= -i\hat{\mathbf{B}}\hat{\mathbf{A}}^{-1} = \hat{\mathbf{H}}^{-1} + i\hat{\mathbf{H}}^{-1}\hat{\mathbf{S}}, \\ \hat{\mathbf{M}}^{-1} &= i\hat{\mathbf{A}}\hat{\mathbf{B}}^{-1} = \hat{\mathbf{L}}^{-1} - i\hat{\mathbf{S}}\hat{\mathbf{L}}^{-1}. \end{aligned} \tag{14.1-24}$$

They are Hermitian but not necessarily positive definite or negative definite.

It is useful to know the inverses of (14.1-7) and (14.1-10). They are obtained with the aid of (14.1-4) as

$$\mathbf{u} = \mathbf{X}^T\hat{\mathbf{u}} + \mathbf{Y}^T\hat{\boldsymbol{\phi}}, \quad \boldsymbol{\phi} = \mathbf{Y}^T\hat{\mathbf{u}} + \mathbf{X}^T\hat{\boldsymbol{\phi}}, \tag{14.1-25}$$

$$\mathbf{A} = \mathbf{X}^T\hat{\mathbf{A}} + \mathbf{Y}^T\hat{\mathbf{B}}, \quad \mathbf{B} = \mathbf{Y}^T\hat{\mathbf{A}} + \mathbf{X}^T\hat{\mathbf{B}}. \tag{14.1-26}$$

Likewise, the inverses of (14.1-20) obtained from (14.1-17) are

$$\begin{aligned} \mathbf{S} &= \mathbf{X}^T\hat{\mathbf{S}}\mathbf{X} + \mathbf{Y}^T\hat{\mathbf{S}}^T\mathbf{Y} + \mathbf{X}^T\hat{\mathbf{H}}\mathbf{Y} - \mathbf{Y}^T\hat{\mathbf{L}}\mathbf{X}, \\ \mathbf{H} &= \mathbf{X}^T\hat{\mathbf{H}}\mathbf{X} - \mathbf{Y}^T\hat{\mathbf{L}}\mathbf{Y} + \mathbf{X}^T\hat{\mathbf{S}}\mathbf{Y} + \mathbf{Y}^T\hat{\mathbf{S}}^T\mathbf{X}, \\ \mathbf{L} &= \mathbf{X}^T\hat{\mathbf{L}}\mathbf{X} - \mathbf{Y}^T\hat{\mathbf{H}}\mathbf{Y} - \mathbf{X}^T\hat{\mathbf{S}}^T\mathbf{Y} - \mathbf{Y}^T\hat{\mathbf{S}}\mathbf{X}. \end{aligned} \tag{14.1-27}$$

14.2 Green's Function for Half-Space for a General Boundary Condition

As an application of the generalized formalism presented in the previous section, consider an anisotropic elastic half-space $x_2>0$ that is subjected to a line force $\mathbf{f}$ and a line dislocation with Burgers vector $\mathbf{b}$ applied at

$$(x_1, x_2) = (0, d), \quad d > 0.$$

The solution is given in (8.6-2) which is

$$\begin{aligned} \mathbf{u} &= \frac{1}{\pi}\text{Im}\{\mathbf{A}\langle\ln(z_* - p_* d)\rangle\mathbf{q}^\infty\} + \frac{1}{\pi}\text{Im}\sum_{\beta=1}^{3}\{\mathbf{A}\langle\ln(z_* - \bar{p}_\beta d)\rangle\mathbf{q}_\beta\}, \\ \boldsymbol{\phi} &= \frac{1}{\pi}\text{Im}\{\mathbf{B}\langle\ln(z_* - p_* d)\rangle\mathbf{q}^\infty\} + \frac{1}{\pi}\text{Im}\sum_{\beta=1}^{3}\{\mathbf{B}\langle\ln(z_* - \bar{p}_\beta d)\rangle\mathbf{q}_\beta\}, \end{aligned} \tag{14.2-1}$$

where $\mathbf{q}_\beta$ (β=1,2,3) are complex vectors to be determined and

$$\mathbf{q}^\infty = \mathbf{A}^T\mathbf{f} + \mathbf{B}^T\mathbf{b}. \tag{14.2-2}$$

Let the boundary conditions at x_2=0 be given by

$$\hat{\boldsymbol{\phi}} = \mathbf{Y}\mathbf{u} + \mathbf{X}\boldsymbol{\phi} = \mathbf{0} \text{ at } x_2 = 0. \tag{14.2-3}$$

Insertion of (14.2-1) into (14.1-7) and use of (14.1-10) leads to

$$\begin{aligned}\hat{\mathbf{u}} &= \frac{1}{\pi}\mathrm{Im}\left\{\hat{\mathbf{A}}\langle \ln(z_* - p_* d)\rangle \mathbf{q}^\infty\right\} + \frac{1}{\pi}\mathrm{Im}\sum_{\beta=1}^{3}\left\{\hat{\mathbf{A}}\langle \ln(z_* - \bar{p}_\beta d)\rangle \mathbf{q}_\beta\right\},\\ \hat{\boldsymbol{\phi}} &= \frac{1}{\pi}\mathrm{Im}\left\{\hat{\mathbf{B}}\langle \ln(z_* - p_* d)\rangle \mathbf{q}^\infty\right\} + \frac{1}{\pi}\mathrm{Im}\sum_{\beta=1}^{3}\left\{\hat{\mathbf{B}}\langle \ln(z_* - \bar{p}_\beta d)\rangle \mathbf{q}_\beta\right\}.\end{aligned} \tag{14.2-4}$$

Following the derivation of (8.6-11) it is easily shown that condition (14.2-3) is satisfied if

$$\mathbf{q}_\beta = \hat{\mathbf{B}}^{-1}\overline{\hat{\mathbf{B}}}\mathbf{I}_\beta \overline{\mathbf{q}}^\infty \tag{14.2-5}$$

where

$$\mathbf{I}_1 = \mathrm{diag}[1,\ 0,\ 0],\quad \mathbf{I}_2 = \mathrm{diag}[0,\ 1,\ 0],\quad \mathbf{I}_3 = \mathrm{diag}[0,\ 0,\ 1].$$

A trivial identity is

$$\mathbf{I}_1+\mathbf{I}_2+\mathbf{I}_3=\mathbf{I}. \tag{14.2-6}$$

If the boundary at x_2=0 is traction-free we let $(\mathbf{X},\mathbf{Y})=(\mathbf{I},\mathbf{0})$ so that $\hat{\boldsymbol{\phi}}=\boldsymbol{\phi}=\mathbf{0}$ by (14.2-3). From (14.1-11) $\hat{\mathbf{B}}=\mathbf{B}$ and (14.2-5) recovers (8.6-11). If the boundary is a rigid surface we let $(\mathbf{X},\mathbf{Y})=(\mathbf{0},\mathbf{I})$ so that $\hat{\boldsymbol{\phi}}=\mathbf{u}=\mathbf{0}$ by (14.2-3). Equation (14.1-11) gives $\hat{\mathbf{B}}=\mathbf{A}$ and (14.2-5) reduces to (8.6-12). For a slippery surface,

$$\mathbf{X}=\mathrm{diag}[1,\ 0,\ 1],\quad \mathbf{Y}=\mathrm{diag}[0,\ 1,\ 0], \tag{14.2-7}$$

and the $\hat{\mathbf{B}}$ appearing in (14.2-5) is

$$\hat{\mathbf{B}} = \begin{bmatrix} A_{11} & A_{12} & A_{13} \\ B_{21} & B_{22} & B_{23} \\ A_{31} & A_{32} & A_{33} \end{bmatrix}.$$

Real form expressions can be obtained for $\hat{\mathbf{u}}$ at x_2=0. From (14.2-4) and following the derivation of (8.6-32) it can be shown that (Ting and Wang, 1992)

$$\hat{\mathbf{u}}(x_1,0) = 2\left\{\hat{\mathbf{u}}^\infty(x_1,0) + \hat{\mathbf{S}}\hat{\mathbf{L}}^{-1}\hat{\boldsymbol{\phi}}^\infty(x_1,0)\right\}$$

where $\hat{\mathbf{u}}^\infty$, $\hat{\phi}^\infty$ are the first terms on the right of (14.2-4). They represent the Green's function for the infinite space subjected to a line force $\mathbf{f}$ and a line dislocation $\mathbf{b}$ at $(x_1,x_2)=(0,d)$. With the use of (14.1-7) $\hat{\mathbf{u}}^\infty$, $\hat{\phi}^\infty$ are expressed in terms of $\mathbf{u}^\infty$, ϕ^∞ whose real form solutions are given in Section 8.1.

Real form expressions can also be obtained from $(14.2\text{-}1)_2$ for the image force F that applies at the line dislocation. When the half-space is subjected to the line dislocation $\mathbf{b}$ only, (14.2-2) with $\mathbf{f}=\mathbf{0}$ and (14.2-5) gives

$$\mathbf{q}_\beta = \hat{\mathbf{B}}^{-1}\overline{\hat{\mathbf{B}}}\mathbf{I}_\beta\overline{\mathbf{B}}^T\mathbf{b}. \tag{14.2-8}$$

The image force F that exerted on the dislocation is presented in $(8.6\text{-}16)_2$. Substitution of (14.2-8) yields

$$F = \frac{-1}{2\pi d}\mathrm{Re}\sum_{\beta=1}^{3}\sum_{\alpha=1}^{3}\mathbf{b}^T\left\{\frac{p_\alpha}{p_\alpha-\bar{p}_\beta}(\mathbf{B}\mathbf{I}_\alpha\hat{\mathbf{B}}^{-1})\hat{\mathbf{L}}(\overline{\mathbf{B}}\mathbf{I}_\beta\overline{\hat{\mathbf{B}}}^{-1})^T\right\}\mathbf{b} \tag{14.2-9}$$

where use has been made of the relations $\hat{\mathbf{L}}=-2i\hat{\mathbf{B}}\hat{\mathbf{B}}^T=2i\overline{\hat{\mathbf{B}}}\overline{\hat{\mathbf{B}}}^T$. We now perform the following three operations on the right of (14.2-9) which leaves F unchanged. First the α and β are interchanged. Next the right-hand side is replaced by its complex conjugate. Finally we apply the matrix transpose on the right. The result is

$$F = \frac{-1}{2\pi d}\mathrm{Re}\sum_{\beta=1}^{3}\sum_{\alpha=1}^{3}\mathbf{b}^T\left\{\frac{-\bar{p}_\beta}{p_\alpha-\bar{p}_\beta}(\mathbf{B}\mathbf{I}_\alpha\hat{\mathbf{B}}^{-1})\hat{\mathbf{L}}(\overline{\mathbf{B}}\mathbf{I}_\beta\overline{\hat{\mathbf{B}}}^{-1})^T\right\}\mathbf{b}.$$

Addition of the above equation to (14.2-9) and use of (14.2-6) yields

$$F = \frac{-1}{4\pi d}\mathbf{b}^T(\mathbf{B}\hat{\mathbf{B}}^{-1})\hat{\mathbf{L}}(\overline{\mathbf{B}}\overline{\hat{\mathbf{B}}}^{-1})^T\mathbf{b}. \tag{14.2-10}$$

The right-hand side is identical to its complex conjugate, assuring us that it is indeed real. From $(14.1\text{-}26)_2$ and $(14.1\text{-}24)_4$,

$$\mathbf{B}\hat{\mathbf{B}}^{-1} = \mathbf{X}^T - \mathbf{Y}^T(\hat{\mathbf{S}}\hat{\mathbf{L}}^{-1} + i\hat{\mathbf{L}}^{-1}).$$

Equation (14.2-10) can be simplified to, after invoking (14.1-19) and $(14.1\text{-}27)_3$,

$$F = \frac{-1}{4\pi d}\mathbf{b}^T(\mathbf{L} + 2\mathbf{Y}^T\hat{\mathbf{L}}^{-1}\mathbf{Y})\mathbf{b}. \tag{14.2-11}$$

The dislocation is attracted to the slippery surface or repelled by it if $F<0$ or $F>0$.

When $x_2=0$ is a free surface $\mathbf{Y}=\mathbf{0}$ and (14.2-11) recovers (8.6-

19). The image force is attracted to the free surface. If $x_2=0$ is a rigid surface $\mathbf{Y}=\mathbf{I}$, $\hat{\mathbf{L}}=-\mathbf{H}$, and (14.2-11) reduces to (8.6-24). It was shown in (8.6-25) that $F>0$ so that the dislocation is repelled by the rigid surface. In general it is clear from (14.2-11) that if the Burgers vector **b** is such that $\mathbf{Yb}=\mathbf{0}$ the image force is identical to that for a free surface.

For a slippery surface **Y** is given by $(14.2\text{-}7)_2$, and (14.2-11) reduces to

$$F=\frac{-1}{4\pi d}\left\{\mathbf{b}^T\mathbf{L}\mathbf{b}+2(\hat{\mathbf{L}}^{-1})_{22}b_2^2\right\}. \qquad (14.2\text{-}12)$$

When $b_2=0$ the image force is identical to that for a free surface. From (14.2-7) and $(14.1\text{-}20)_3$,

$$\hat{\mathbf{L}}=\begin{bmatrix} L_{11} & -S_{21} & L_{13} \\ -S_{21} & -H_{22} & -S_{23} \\ L_{13} & -S_{23} & L_{33} \end{bmatrix} \qquad (14.2\text{-}13)$$

and hence

$$(\hat{\mathbf{L}}^{-1})_{22}=(L_{11}L_{33}-L_{13}^2)\big/\left|\hat{\mathbf{L}}\right|,$$

$$\left|\hat{\mathbf{L}}\right|=2L_{13}S_{21}S_{23}-(L_{11}S_{23}^2+L_{33}S_{21}^2)-H_{22}(L_{11}L_{33}-L_{13}^2).$$

Since **H** and **L** are positive definite, $\left|\hat{\mathbf{L}}\right|<0$ in the special case when $S_{23}=0$ or $L_{13}=0$. We conclude that $\left|\hat{\mathbf{L}}\right|<0$ in general. Indeed, if $\left|\hat{\mathbf{L}}\right|>0$ for a certain anisotropic material, there exists a material for which $\left|\hat{\mathbf{L}}\right|=0$, contradicting the results obtained in Section 14.1 that $\hat{\mathbf{L}}$ is nonsingular. Therefore $(\hat{\mathbf{L}}^{-1})_{22}<0$, and the second part of the image force in (14.2-12) is positive. Thus F can be positive or negative for a slippery surface.

When the material is isotropic, use of the explicit expressions of **S**, **H**, and **L** given in (6.4-13) leads to

$$\mathbf{L}=\mu\,\mathrm{diag}[\kappa,\ \kappa,\ 1],\quad (\hat{\mathbf{L}}^{-1})_{22}=-\mu\kappa,\quad \kappa=(1-\nu)^{-1}$$

where μ and ν are, respectively, the shear modulus and Poisson's ratio. Equation (14.2-12) is simplified to

$$F=\frac{-\mu}{4\pi d}\left\{\kappa(b_1^2-b_2^2)+b_3^2\right\}.$$

$F<0$ for b_1 and b_3 while $F>0$ for b_2, a well known result in isotropic elasticity (Dundurs, 1969).

With the generalized formalism introduced in the previous section (14.2-11) covers at least the eight different boundary conditions listed in (14.1-1). This represents a great saving in deriving the solutions if one is interested in the solutions to all eight different boundary conditions. The generalized formalism has been applied to the problem of stress decay in an anisotropic elastic strip that occupies the region $x_1 \geq 0$, $-1 \leq x_2 \leq 1$. One can have eight different boundary conditions each at $x_1=0$ and $x_2=\pm 1$, resulting in a total of 8^3 different sets of problems (Wang et al., 1993). With the use of the generalized formalism, one derivation of the solution encompasses all 8^3 different boundary conditions with little additional effort.

14.3 Thermo-Anisotropic Elasticity

The influence of heat flow on the stress concentration in an anisotropic medium is an important problem, especially for composite materials. We present in this section the extension of Stroh's sextic formalism to thermo-anisotropic elasticity due to Clements (1973) and Hwu (1990b). Clements' approach was a generalization of Sharma (1958) and Akroz and Tauchart (1972). New conservation laws related to thermo-anisotropic elasticity have been recently formulated by Yeh et al. (1993a,b).

The basic equations for thermo-anisotropic elasticity have been well documented (see, for example, Carslaw and Jaeger, 1959; Boley and Weiner, 1960; Nowacki, 1962; and Carlson, 1972). Let u_i, σ_{ij}, h_i, and T be, respectively, the displacement, stress, heat flux, and temperature in an anisotropic elastic medium. The equations for the stress-strain laws, the equilibrium, the heat conduction, and the balance of energy are

$$\sigma_{ij} = C_{ijks} u_{k,s} - \beta_{ij} T, \tag{14.3-1}$$

$$\sigma_{ij,j} = 0, \tag{14.3-2}$$

$$h_i = -\kappa_{ij} T_{,j}, \tag{14.3-3}$$

$$h_{i,i} = 0, \tag{14.3-4}$$

where β_{ij}, κ_{ij} are the thermal moduli and the heat conduction coefficients, respectively, and C_{ijks} are the elastic stiffnesses which are assumed to be fully symmetric and positive definite. Repeated indices imply summation and the comma denotes differentiation. For two-dimensional deformations for which T and u_i depend on x_1 and x_2 only, a general solution for T and u_i can be written as

$$T = g'(z_\tau), \quad \mathbf{u} = \boldsymbol{a}\, f(z) + \mathbf{c} g(z_\tau), \tag{14.3-5}$$

$$z = x_1 + p x_2, \qquad z_\tau = x_1 + \tau x_2,$$

where $\mathbf{u}^T=[u_1,u_2,u_3]$, f and g are arbitrary functions of their arguments, and the prime denotes differentiation with the argument. The p, τ, $\boldsymbol{a}$, and $\mathbf{c}$ are constants which, upon substituting (14.3-5) into (14.3-1) through (14.3-4), satisfy the equations

$$\kappa_{22}\tau^2+(\kappa_{12}+\kappa_{21})\tau+\kappa_{11}=0, \tag{14.3-6}$$

$$\left\{\mathbf{Q}+p(\mathbf{R}+\mathbf{R}^T)+p^2\mathbf{T}\right\}\boldsymbol{a}=\mathbf{0}, \tag{14.3-7}$$

$$\left\{\mathbf{Q}+\tau(\mathbf{R}+\mathbf{R}^T)+\tau^2\mathbf{T}\right\}\mathbf{c}=\boldsymbol{\beta}_1+\tau\,\boldsymbol{\beta}_2. \tag{14.3-8}$$

The superscript T denotes the transpose, and the components of the 3-vectors $\boldsymbol{\beta}_1$ and $\boldsymbol{\beta}_2$ and the 3×3 matrices $\mathbf{Q}$, $\mathbf{R}$, $\mathbf{T}$ are

$$(\boldsymbol{\beta}_1)_i=\beta_{i1},\quad (\boldsymbol{\beta}_2)_i=\beta_{i2},$$

$$Q_{ik}=C_{i1k1},\quad R_{ik}=C_{i1k2},\quad T_{ik}=C_{i2k2}.$$

The matrices $\mathbf{Q}$ and $\mathbf{T}$ are symmetric and positive definite. The heat flux $\mathbf{h}$ and the stresses σ_{ij} obtained from (14.3-1) and (14.3-3) are

$$\mathbf{h}=-(\boldsymbol{\kappa}_1+\tau\boldsymbol{\kappa}_2)g''(z_\tau), \tag{14.3-9}$$

$$\sigma_{i1}=-\varphi_{i,2},\quad \sigma_{i2}=\varphi_{i,1}. \tag{14.3-10}$$

The components of the 3-vectors $\boldsymbol{\kappa}_1$ and $\boldsymbol{\kappa}_2$ are

$$(\boldsymbol{\kappa}_1)_i=\kappa_{i1},\quad (\boldsymbol{\kappa}_2)_i=\kappa_{i2},$$

and φ_i, a 3-vector, is the stress function given by

$$\boldsymbol{\phi}=\boldsymbol{b}f(z)+\mathbf{d}g(z_\tau), \tag{14.3-11}$$

$$\boldsymbol{b}=(\mathbf{R}^T+p\mathbf{T})\boldsymbol{a}=-\frac{1}{p}(\mathbf{Q}+p\mathbf{R})\boldsymbol{a}, \tag{14.3-12}$$

$$\mathbf{d}=(\mathbf{R}^T+\tau\mathbf{T})\mathbf{c}-\boldsymbol{\beta}_2=-\frac{1}{\tau}[(\mathbf{Q}+\tau\mathbf{R})\mathbf{c}-\boldsymbol{\beta}_1]. \tag{14.3-13}$$

The second equalities for $\boldsymbol{b}$ and $\mathbf{d}$ follow from (14.3-7) and (14.3-8).

Equation (14.3-7) for the eigenvalue p and the eigenvector $\boldsymbol{a}$ is identical to the case when there is no heat flow. Therefore there are three pairs of complex conjugates for p (Section 5.1). If p_α $(\alpha=1,2,\ldots,6)$ are the eigenvalues, let

$$\operatorname{Im} p_\alpha>0,\quad p_{\alpha+3}=\bar{p},\quad \boldsymbol{a}_{\alpha+3}=\bar{\boldsymbol{a}}_\alpha,\quad \boldsymbol{b}_{\alpha+3}=\bar{\boldsymbol{b}}_\alpha,\quad (\alpha=1,2,3).$$

The fact that heat flows from higher temperature to lower temperature means that

$$h_i T_{,i} < 0.$$

It follows from (14.3-3) that

$$\kappa_{ij} y_i y_j > 0$$

for any real nonzero vector y_i. Thus κ_{ij} is positive definite. Equation (14.3-6) written as

$$\kappa_{ij}(\delta_{i1} + \tau\delta_{i2})(\delta_{j1} + \tau\delta_{j2}) = 0$$

where δ_{ij} is the Kronecker delta tells us that τ cannot be real. If τ is the root of (14.3-6) with positive imaginary part,

$$\tau = -\frac{\kappa_{12} + \kappa_{21}}{2\kappa_{22}} + i\sqrt{\frac{\kappa_{11}}{\kappa_{22}} - \left(\frac{\kappa_{12} + \kappa_{21}}{2\kappa_{22}}\right)^2}. \tag{14.3-14}$$

When the Onsager's (1931) reciprocal relations are invoked κ_{ij} is symmetric. Many experiments have been conducted, and the results invariably supported the validity of the Onsager's reciprocal relations (Carslaw and Jaeger, 1959; Truesdell, 1965; Fung, 1965). Equation (14.3-14) for $\kappa_{12} = \kappa_{21}$ reduces to

$$\tau = -\frac{\kappa_{12}}{\kappa_{22}} + i\sqrt{\frac{\kappa_{11}}{\kappa_{22}} - \left(\frac{\kappa_{12}}{\kappa_{22}}\right)^2}.$$

We assume that either τ is not equal to one of the p's or, if $\tau = p$, $\mathbf{c}$ of (14.3-8) has a solution. Modification required when $\tau = p$ and $\mathbf{c}$ of (14.3-8) has no solutions have been considered by Wu (1984). The general solutions for T, $\mathbf{u}$, and $\boldsymbol{\phi}$ are obtained from (14.3-5) and (14.3-11) by superposing solutions associated with six p's and two τ's. We have

$$T = g_1'(z_\tau) + g_2'(\bar{z}_\tau), \tag{14.3-15}$$

$$\mathbf{u} = \sum_{\alpha=1}^{3} [\mathbf{a}_\alpha f_\alpha(z_\alpha) + \bar{\mathbf{a}}_\alpha f_{\alpha+3}(\bar{z}_\alpha)] + \mathbf{c}\, g_1(z_\tau) + \bar{\mathbf{c}}\, g_2(\bar{z}_\tau), \tag{14.3-16}$$

$$\boldsymbol{\phi} = \sum_{\alpha=1}^{3} [\mathbf{b}_\alpha f_\alpha(z_\alpha) + \bar{\mathbf{b}}_\alpha f_{\alpha+3}(\bar{z}_\alpha)] + \mathbf{d}\, g_1(z_\tau) + \bar{\mathbf{d}}\, g_2(\bar{z}_\tau). \tag{14.3-17}$$

The heat flow is, from (14.3-9),

$$\mathbf{h} = -(\boldsymbol{\kappa}_1 + \tau\boldsymbol{\kappa}_2)\, g_1''(z_\tau) - (\boldsymbol{\kappa}_1 + \bar{\tau}\boldsymbol{\kappa}_2)\, g_2''(\bar{z}_\tau). \tag{14.3-18}$$

The two equations in (14.3-12) can be cast in the standard eigen-relation (Section 5.5)

$$\mathbf{N}\boldsymbol{\xi}=p\boldsymbol{\xi}, \tag{14.3-19}$$

$$\mathbf{N}=\begin{bmatrix}\mathbf{N}_1 & \mathbf{N}_2\\ \mathbf{N}_3 & \mathbf{N}_1^T\end{bmatrix}, \qquad \boldsymbol{\xi}=\begin{bmatrix}\boldsymbol{a}\\ \boldsymbol{b}\end{bmatrix},$$

$$\mathbf{N}_1=-\mathbf{T}^{-1}\mathbf{R}^T, \quad \mathbf{N}_2=\mathbf{T}^{-1}, \quad \mathbf{N}_3=\mathbf{R}\mathbf{T}^{-1}\mathbf{R}^T-\mathbf{Q}.$$

Likewise the two equations in (14.3-13) can be written as

$$\mathbf{N}\begin{bmatrix}\mathbf{c}\\ \mathbf{d}\end{bmatrix}=\tau\begin{bmatrix}\mathbf{c}\\ \mathbf{d}\end{bmatrix}-\begin{bmatrix}\mathbf{0} & \mathbf{N}_2\\ \mathbf{I} & \mathbf{N}_1^T\end{bmatrix}\begin{bmatrix}\boldsymbol{\beta}_1\\ \boldsymbol{\beta}_2\end{bmatrix}. \tag{14.3-20}$$

When the coordinate system is rotated about the x_3-axis an angle θ, (14.3-20) takes the form (Hwu, 1990b)

$$\begin{bmatrix}\mathbf{N}_1(\theta) & \mathbf{N}_2(\theta)\\ \mathbf{N}_3(\theta) & \mathbf{N}_1^T(\theta)\end{bmatrix}\begin{bmatrix}\mathbf{c}\\ \mathbf{d}\end{bmatrix}=\tau(\theta)\begin{bmatrix}\mathbf{c}\\ \mathbf{d}\end{bmatrix}-\lambda^{-1}(\theta)\begin{bmatrix}\mathbf{0} & \mathbf{N}_2(\theta)\\ \mathbf{I} & \mathbf{N}_1^T(\theta)\end{bmatrix}\begin{bmatrix}\boldsymbol{\beta}_1(\theta)\\ \boldsymbol{\beta}_2(\theta)\end{bmatrix}, \tag{14.3-21}$$

where

$$\mathbf{N}_1(\theta)=-\mathbf{T}^{-1}(\theta)\mathbf{R}^T(\theta), \quad \mathbf{N}_2(\theta)=\mathbf{T}^{-1}(\theta),$$

$$\mathbf{N}_3(\theta)=\mathbf{R}(\theta)\mathbf{T}^{-1}(\theta)\mathbf{R}^T(\theta)-\mathbf{Q}(\theta),$$

$$\tau(\theta)=\frac{\tau\cos\theta-\sin\theta}{\tau\sin\theta+\cos\theta}, \quad \lambda(\theta)=\tau\sin\theta+\cos\theta,$$

$$[\boldsymbol{\beta}_1(\theta)]_i=\beta_{ij}n_j, \quad [\boldsymbol{\beta}_2(\theta)]_i=\beta_{ij}m_j.$$

In the above,

$$n_i=[\cos\theta,\ \sin\theta,\ 0], \quad m_i=[-\sin\theta,\ \cos\theta,\ 0],$$

and the components of $\mathbf{Q}(\theta)$, $\mathbf{R}(\theta)$, $\mathbf{T}(\theta)$ are

$$Q_{ik}(\theta)=C_{ijks}n_jn_s, \quad R_{ik}(\theta)=C_{ijks}n_jm_s, \quad T_{ik}(\theta)=C_{ijks}m_jm_s.$$

The derivation of (14.3-21) is similar to that of (7.3-14). Integration of (14.3-21) with the use of (7.6-4) yields

$$\begin{bmatrix}\mathbf{S} & \mathbf{H}\\ -\mathbf{L} & \mathbf{S}^T\end{bmatrix}\begin{bmatrix}\mathbf{c}\\ \mathbf{d}\end{bmatrix}=i\begin{bmatrix}\mathbf{c}\\ \mathbf{d}\end{bmatrix}-\begin{bmatrix}\boldsymbol{\gamma}_1\\ \boldsymbol{\gamma}_2\end{bmatrix} \tag{14.3-22}$$

where

$$\gamma_1 = \frac{1}{\pi}\int_0^\pi \lambda^{-1}(\theta)\mathbf{N}_2(\theta)\boldsymbol{\beta}_2(\theta)d\theta$$

$$\gamma_2 = \frac{1}{\pi}\int_0^\pi \lambda^{-1}(\theta)[\boldsymbol{\beta}_1(\theta) + \mathbf{N}_1^T(\theta)\boldsymbol{\beta}_2(\theta)d\theta \tag{14.3-23}$$

The constant vectors γ_1 and γ_2 play key roles in thermo-anisotropic elasticity (Ting and Hwu, 1991; Ting and Yan, 1992). Equations (14.3-23) provide an integral representation for determining γ_1 and γ_2. An algebraic representation is presented below. We write (14.3-20) as

$$(\mathbf{N} - \tau\mathbf{I})\begin{bmatrix}\mathbf{c}\\ \mathbf{d}\end{bmatrix} = -\begin{bmatrix}\mathbf{0} & \mathbf{N}_2\\ \mathbf{I} & \mathbf{N}_1^T\end{bmatrix}\begin{bmatrix}\boldsymbol{\beta}_1\\ \boldsymbol{\beta}_2\end{bmatrix}. \tag{14.3-24}$$

Multiplication of (14.3-22) by $(\mathbf{N}-\tau\mathbf{I})$ and making use of (4.3-31) and (14.3-24) leads to

$$(\mathbf{N} - \tau\mathbf{I})\begin{bmatrix}\gamma_1\\ \gamma_2\end{bmatrix} = \begin{bmatrix}\mathbf{y}_1\\ \mathbf{y}_2\end{bmatrix} \tag{14.3-25}$$

where

$$\begin{bmatrix}\mathbf{y}_1\\ \mathbf{y}_2\end{bmatrix} = \left\{\begin{bmatrix}\mathbf{S} & \mathbf{H}\\ -\mathbf{L} & \mathbf{S}^T\end{bmatrix} - i\mathbf{I}\right\}\begin{bmatrix}\mathbf{0} & \mathbf{N}_2\\ \mathbf{I} & \mathbf{N}_1^T\end{bmatrix}\begin{bmatrix}\boldsymbol{\beta}_1\\ \boldsymbol{\beta}_2\end{bmatrix}.$$

Since we have assumed that $\tau \neq p$, (14.3-25) can be solved for γ_1 and γ_2 (Yan and Ting, 1993).

The stress singularities at the tip of an interfacial crack between two dissimilar elastic materials can be determined by assuming that $f(z)$ and $g(z_\tau)$ in (14.3-15)–(14.3-17) are proportional to $z^{\delta+1}$, $z_\tau^{\delta+1}$, respectively. The analysis is similar to the case studied in Section 9.6 when the heat flow is ignored. One obtains four stress singularities δ, three of which are given by (9.6-16), namely,

$$\delta = -\frac{1}{2} \quad \text{and} \quad -\frac{1}{2} \mp i\gamma. \tag{14.3-26}$$

The fourth δ, denoted by δ_κ, is (Ting and Hwu, 1991)

$$\cos\delta_\kappa = \frac{\hat{\kappa}_I' - \hat{\kappa}_I}{\hat{\kappa}_{II}' + \hat{\kappa}_{II}} \tag{14.3-27}$$

in which the prime denotes the second material and

$$\hat{\kappa}_I = \frac{\kappa_I}{\kappa_I^2 + \kappa_{II}^2}, \quad \hat{\kappa}_{II} = \frac{\kappa_{II}}{\kappa_I^2 + \kappa_{II}^2},$$

$$\kappa_I = \tfrac{1}{2}(\kappa_{21} - \kappa_{12}), \quad \kappa_{II} = \sqrt{\kappa_{11}\kappa_{22} - \tfrac{1}{4}(\kappa_{12} + \kappa_{21})^2}\,.$$

It is seen that $\delta_\kappa = -\frac{1}{2}$ if $\hat{\kappa}_I = \hat{\kappa}'_I$ or, in particular, when $\hat{\kappa}_I = 0 = \hat{\kappa}'_I$. The latter is the case when the Onsager's reciprocal relations are invoked. When $\delta_\kappa = -\frac{1}{2}$, the stress singularities $\delta = -\frac{1}{2}$ is a root of multiplicity two if $\gamma \neq 0$, and of multiplicity four if $\gamma = 0$. The stress singularities associated with the repeated root of $\delta = -\frac{1}{2}$ may have the form $r^{-\frac{1}{2}}(\ln r)$, which is a new and higher order singularity (Yan and Ting, 1993).

Investigation of the stress distribution in isotropic and anisotropic materials with the presence of a crack subjected to heat flow has been given by Olesiak and Sneddon (1960), Sih (1962), Florence and Goodier (1963), Atkinson and Clements (1977), Tsai (1983, 1984), Sturla and Barber (1988), and Hwu (1990b, 1992a). Studies of stresses disturbed by an insulated hole were given by Florence and Goodier (1960), Chen (1967a), and Hwu (1990b).

14.4 Piezoelectric Materials

Piezoelectric materials deform when subjected to an electric field and polarize when stressed (Fatuzzo and Merz, 1967; Jaffe et al., 1971; Mikhailov and Parton, 1990). The elastic stress σ_{ij}, the elastic displacement u_i, the electric displacement D_i and the electric field E_i satisfy the constitutive equations

$$\begin{aligned} \sigma_{ij} &= C_{ijks} u_{k,s} - e_{sij} E_s, \\ D_i &= e_{iks} u_{k,s} + \omega_{is} E_s, \end{aligned} \tag{14.4-1}$$

in which repeated indices imply summation, a comma stands for differentiation, C_{ijks} are the elastic stiffnesses, and e_{iks} and ω_{ij} are the piezoelectric stress constants and permittivity constants, respectively. The material constants have the following symmetries

$$\begin{aligned} C_{ijks} &= C_{jiks} = C_{ksij}, \\ e_{sij} &= e_{sji}, \quad \omega_{ij} = \omega_{ji}. \end{aligned} \tag{14.4-2}$$

Moreover C_{ijks} and ω_{ij} are positive definite in the sense that

$$C_{ijks} u_{i,j} u_{j,s} > 0, \quad \omega_{ij} E_i E_j > 0, \tag{14.4-3}$$

for arbitrary real nonzero $u_{i,j}$ and E_i. Letting

$$E_i = -u_{4,i} \tag{14.4-4}$$

where u_4 is the electric potential (14.4-1) is

$$\begin{aligned} \sigma_{ij} &= C_{ijks}u_{k,s} + e_{sij}u_{4,s}, \\ D_i &= e_{iks}u_{k,s} - \omega_{is}u_{4,s}. \end{aligned} \tag{14.4-5}$$

In the absence of body force and free charges the balance laws require

$$\sigma_{ij,j} = 0, \quad D_{i,i} = 0. \tag{14.4-6}$$

For two-dimensional deformations for which u_i (i=1,2,3) and u_4 depend on x_1 and x_2 only, a general solution to (14.4-6) is

$$u_J = a_J f(z) \quad \text{or} \quad \mathbf{u}=\boldsymbol{a}f(z), \tag{14.4-7}$$

$$z = x_1 + px_2,$$

in which J=1,2,3,4, f is an arbitrary function of z, and p and a_J are determined as follows. Insertion of (14.4-7) into (14.4-5) yields

$$\begin{aligned} \sigma_{ij} &= (C_{ijks}a_k + e_{sij}a_4)(\delta_{s1} + p\delta_{s2})f'(z), \\ D_i &= (e_{iks}a_k - \omega_{is}a_4)(\delta_{s1} + p\delta_{s2})f'(z), \end{aligned} \tag{14.4-8}$$

where the prime denotes differentiation with z and δ_{ij} is the Kronecker delta. Substitution in (14.4-6) leads to

$$\begin{aligned} (C_{ijks}a_k + e_{sij}a_4)(\delta_{j1} + p\delta_{j2})(\delta_{s1} + p\delta_{s2}) = 0, \\ (e_{iks}a_k - \omega_{is}a_4)(\delta_{i1} + p\delta_{i2})(\delta_{s1} + p\delta_{s2}) = 0. \end{aligned} \tag{14.4-9}$$

In matrix notation we have

$$\begin{aligned} &\left\{\mathbf{Q}^E+p[\mathbf{R}^E+(\mathbf{R}^E)^T]+p^2\mathbf{T}^E\right\}\boldsymbol{a}^E+\left\{\mathbf{e}_{11}+p(\mathbf{e}_{21}+\mathbf{e}_{12})+p^2\mathbf{e}_{22}\right\}a_4 = \mathbf{0}, \\ &\left\{\mathbf{e}_{11}^T+p(\mathbf{e}_{21}^T+\mathbf{e}_{12}^T)+p^2\mathbf{e}_{22}^T\right\}\boldsymbol{a}^E-\left\{\omega_{11}+p(\omega_{12}+\omega_{21})+p^2\omega_{22}\right\}a_4 = 0, \end{aligned} \tag{14.4-10}$$

in which the elements of 3×3 matrices $\mathbf{Q}^E$, $\mathbf{R}^E$, $\mathbf{T}^E$, and 3×1 matrices $\mathbf{e}_{ij}$ and $\boldsymbol{a}^E$ are

$$Q_{ik}^E = C_{i1k1}, \quad R_{ik}^E = C_{i1k2}, \quad T_{ik}^E = C_{i2k2},$$

$$(\mathbf{e}_{ij})_s = e_{ijs}, \quad \boldsymbol{a}^E = \begin{bmatrix} a_1 \\ a_2 \\ a_3 \end{bmatrix}.$$

The 3×3 matrices $\mathbf{Q}^E$ and $\mathbf{T}^E$ are symmetric and positive definite. The two equations in (14.4-10) can be combined as

$$\left\{\mathbf{Q}+p(\mathbf{R}+\mathbf{R}^T)+p^2\mathbf{T}\right\}\mathbf{a}=\mathbf{0} \qquad (14.4\text{-}11)$$

where

$$\mathbf{Q}=\begin{bmatrix}\mathbf{Q}^E & \mathbf{e}_{11}\\ \mathbf{e}_{11}^T & -\omega_{11}\end{bmatrix},\quad \mathbf{R}=\begin{bmatrix}\mathbf{R}^E & \mathbf{e}_{21}\\ \mathbf{e}_{12}^T & -\omega_{12}\end{bmatrix},\quad \mathbf{T}=\begin{bmatrix}\mathbf{T}^E & \mathbf{e}_{22}\\ \mathbf{e}_{22}^T & -\omega_{22}\end{bmatrix}. \qquad (14.4\text{-}12)$$

Introducing the generalized stress function φ_J (J=1,2,3,4) by

$$\varphi_J=b_Jf(z) \quad \text{or} \quad \boldsymbol{\phi}=\boldsymbol{b}\,f(z), \qquad (14.4\text{-}13)$$

$$\boldsymbol{b}=(\mathbf{R}^T+p\mathbf{T})\boldsymbol{a}=-\frac{1}{p}(\mathbf{Q}+p\mathbf{R})\boldsymbol{a}, \qquad (14.4\text{-}14)$$

in which the second equality for $\boldsymbol{b}$ follows from (14.4-11), the stress σ_{ij} and the electric displacement D_i shown in (14.4-8) can be written as

$$\begin{aligned}\sigma_{i1}&=-\varphi_{i,2}, \quad \sigma_{i2}=\varphi_{i,1},\\ D_1&=-\varphi_{4,2}, \quad D_2=\varphi_{4,1}.\end{aligned} \qquad (14.4\text{-}15)$$

It is therefore sufficient to consider the generalized stress function $\boldsymbol{\phi}$ because the stress σ_{ij} and the electric displacement D_i can be obtained from $\boldsymbol{\phi}$ by differentiation.

We show that the eigenvalue p of (14.4-11) cannot be real. If p were real, multiplication of $(14.4\text{-}9)_{1,2}$ by a_i and a_4, respectively, and subtraction of the results leads to (Suo et al., 1992)

$$\begin{aligned}&C_{ijks}[a_i(\delta_{j1}+p\delta_{j2})][a_k(\delta_{s1}+p\delta_{s2})]\\ &\quad+\omega_{is}[a_4(\delta_{i1}+p\delta_{i2})][a_4(\delta_{s1}+p\delta_{s2})]=0.\end{aligned}$$

This violates conditions (14.4-3). Hence p cannot be real.

There are eight eigenvalues for p which consist of four pairs of complex conjugates. If p_α (α=1,2,...,8) are the eigenvalues we let

$$\operatorname{Im} p_\alpha>0,\quad p_{\alpha+4}=\bar{p}_\alpha,\quad \boldsymbol{a}_{\alpha+4}=\bar{\boldsymbol{a}}_\alpha,\quad \boldsymbol{b}_{\alpha+4}=\bar{\boldsymbol{b}}_\alpha \quad (\alpha=1,2,3,4).$$

The general solutions obtained by superposing eight solutions of the form (14.4-7) and (14.4-13) are

$$\mathbf{u}=\sum_{\alpha=1}^{4}\left\{\boldsymbol{a}_\alpha f_\alpha(z_\alpha)+\bar{\boldsymbol{a}}_\alpha f_{\alpha+4}(\bar{z}_\alpha)\right\}, \qquad (14.4\text{-}16)_1$$

$$\phi=\sum_{\alpha=1}^{4}\left\{l_\alpha f_\alpha(z_\alpha)+\bar{l}_\alpha f_{\alpha+4}(\bar{z}_\alpha)\right\}, \tag{14.4-16$_2$}$$

where f_α (α=1,2,...,8) are arbitrary functions of their arguments and

$$z_\alpha = x_1 + p_\alpha x_2.$$

In most applications f_α have the same function form. If we let

$$f_\alpha(z_\alpha)=f(z_\alpha)q_\alpha, \quad f_{\alpha+3}(\bar{z}_\alpha)=\bar{f}(\bar{z}_\alpha)\bar{q}_\alpha, \quad (\alpha=1,2,3)$$

where q_α are arbitrary complex constants, (14.4-16) can be written as

$$\mathbf{u}=2\,\mathrm{Re}\left\{\mathbf{A}\left\langle f(z_*)\right\rangle\mathbf{q}\right\}, \quad \boldsymbol{\phi}=2\,\mathrm{Re}\left\{\mathbf{B}\left\langle f(z_*)\right\rangle\mathbf{q}\right\}. \tag{14.4-17}$$

In the above Re stands for the real part, **A** and **B** are 4×4 complex matrices defined by

$$\mathbf{A}=\left[\boldsymbol{a}_1,\ \boldsymbol{a}_2,\ \boldsymbol{a}_3,\ \boldsymbol{a}_4\right], \quad \mathbf{B}=\left[\boldsymbol{b}_1,\ \boldsymbol{b}_2,\ \boldsymbol{b}_3,\ \boldsymbol{b}_4\right], \tag{14.4-18}$$

and $\left\langle f(z_*)\right\rangle$ is the diagonal matrix

$$\left\langle f(z_*)\right\rangle=\mathrm{diag}\left[f(z_1),\ f(z_2),\ f(z_3),\ f(z_4)\right].$$

When **q** is replaced by $-i\mathbf{q}$, (14.4-17) leads to the alternate form

$$\mathbf{u}=2\,\mathrm{Im}\left\{\mathbf{A}\left\langle f(z_*)\right\rangle\mathbf{q}\right\}, \quad \boldsymbol{\phi}=2\,\mathrm{Im}\left\{\mathbf{B}\left\langle f(z_*)\right\rangle\mathbf{q}\right\}. \tag{14.4-19}$$

The derivations from (14.4-7) to (14.4-19) parallel Stroh's sextic formalism presented in Section 5.2 and 5.3 except that we are employing the eight dimensional space. We can also extend the orthogonality and closure relations of Section 5.5 to the present *octet formalism*. To this end we have to be sure that the inverse of the 4×4 matrix **T** shown in (14.4-12) exists.

The 4×4 matrices **T** and **Q** in (14.4-12) are symmetric but no longer positive definite. The matrix **T** has the property that its first 3×3 submatrix $\mathbf{T}^E$ is positive definite while its last element $T_{44}=-\omega_{22}$ is negative and nonzero. A matrix of this type will be called *Type* **T** *matrix*. It is clear that the matrix **Q** is also of Type **T**.

A matrix of Type **T** is not singular. It **T** were singular so is the matrix

$$\hat{\mathbf{T}}=\begin{bmatrix}\mathbf{I} & \mathbf{0}\\ \mathbf{0} & -1\end{bmatrix}\mathbf{T}=\begin{bmatrix}\mathbf{T}^E & \mathbf{e}_{22}\\ -\mathbf{e}_{22}^T & \omega_{22}\end{bmatrix}.$$

By virtue of the positive definiteness of $\mathbf{T}^E$ and $\omega_{22}>0$, it is easily shown that

$$\mathbf{w}^T\hat{\mathbf{T}}\mathbf{w} > 0$$

for any nonzero 4×1 real vector $\mathbf{w}$. If $\hat{\mathbf{T}}$ were singular with $\mathbf{w}$ being a right null vector,

$$\hat{\mathbf{T}}\mathbf{w}=\mathbf{0} \quad \text{or} \quad \mathbf{w}^T\hat{\mathbf{T}}\mathbf{w} = 0.$$

This leads to a contradiction, proving that $\hat{\mathbf{T}}$, and hence $\mathbf{T}$, is nonsingular. Likewise it can be shown that $\mathbf{Q}$ is nonsingular.

An alternate proof that $\mathbf{T}$ is nonsingular is to find the determinant of $\mathbf{T}$. It can be shown from $(14.4\text{-}12)_3$ that

$$|\mathbf{T}| = -\omega_{22}|\mathbf{T}^E| - \mathbf{e}_{22}^T\widehat{\mathbf{T}}^E\mathbf{e}_{22}$$

where $(\widehat{\mathbf{T}}^E)^T$ is the adjoint of $\mathbf{T}^E$ that satisfies the relation (Section 1.1)

$$\mathbf{T}^E(\widehat{\mathbf{T}}^E)^T = |\mathbf{T}^E|\mathbf{I}.$$

Hence, noticing that $\widehat{\mathbf{T}}^E$ is symmetric,

$$|\mathbf{T}| = -|\mathbf{T}^E|\{\omega_{22} + \mathbf{e}_{22}^T(\mathbf{T}^E)^{-1}\mathbf{e}_{22}\} < 0. \qquad (14.4\text{-}20)$$

Therefore $\mathbf{T}$ is nonsingular.

More information on the Type $\mathbf{T}$ matrix is gained by considering the eigenvalues of $\mathbf{T}$. If λ is an eigenvalue of $\mathbf{T}$, following the derivation of (14.4-20) we have

$$F(\lambda) = |\mathbf{T} - \lambda\mathbf{I}| = -|\mathbf{T}^E - \lambda\mathbf{I}|\{(\omega_{22}+\lambda) + \mathbf{e}_{22}^T(\mathbf{T}^E-\lambda\mathbf{I})^{-1}\mathbf{e}_{22}\} = 0. \quad (14.4\text{-}21)$$

Let

$$\lambda_1^E \geq \lambda_2^E \geq \lambda_3^E > 0$$

be the eigenvalues of $\mathbf{T}^E$ and $\mathbf{q}_1$, $\mathbf{q}_2$, $\mathbf{q}_3$ be the orthonormalized eigenvectors. This means that

$$|\mathbf{T}^E - \lambda\mathbf{I}| = (\lambda_1^E - \lambda)(\lambda_2^E - \lambda)(\lambda_3^E - \lambda),$$

$$\mathbf{T}^E - \lambda\mathbf{I} = (\lambda_1^E - \lambda)\mathbf{q}_1\mathbf{q}_1^T + (\lambda_2^E - \lambda)\mathbf{q}_2\mathbf{q}_2^T + (\lambda_3^E - \lambda)\mathbf{q}_3\mathbf{q}_3^T,$$

$$(\mathbf{T}^E-\lambda\mathbf{I})^{-1} = (\lambda_1^E-\lambda)^{-1}\mathbf{q}_1\mathbf{q}_1^T + (\lambda_2^E-\lambda)^{-1}\mathbf{q}_2\mathbf{q}_2^T + (\lambda_3^E-\lambda)^{-1}\mathbf{q}_3\mathbf{q}_3^T.$$

Hence

$$F(\lambda) = -(\lambda_1^E-\lambda)(\lambda_2^E-\lambda)(\lambda_3^E-\lambda)(\omega_{22}+\lambda) - (\lambda_2^E-\lambda)(\lambda_3^E-\lambda)(\mathbf{e}_{22}\cdot\mathbf{q}_1)^2$$
$$-(\lambda_3^E-\lambda)(\lambda_1^E-\lambda)(\mathbf{e}_{22}\cdot\mathbf{q}_2)^2 - (\lambda_1^E-\lambda)(\lambda_2^E-\lambda)(\mathbf{e}_{22}\cdot\mathbf{q}_3)^2 = 0.$$

It is clear that

$$F(\lambda)\begin{cases} \geq 0 \text{ for } \lambda = -\infty,\ \lambda_2^E,\ \infty, \\ \leq 0 \text{ for } \lambda = -\omega_{22},\ \lambda_1^E,\ \lambda_3^E. \end{cases}$$

If λ_i (i=1,2,3,4) are the eigenvalues of **T** we conclude that

$$\lambda_4 \leq -\omega_{22} < 0 < \lambda_3^E \leq \lambda_3 \leq \lambda_2^E \leq \lambda_2 \leq \lambda_1^E \leq \lambda_1. \tag{14.4-22}$$

Thus three of the four eigenvalues are positive and nonzero while the fourth eigenvalue is negative and nonzero. The same statement applies to the 4×4 matrix **Q**.

With **T** being nonsingular, let

$$\mathbf{T}^{-1} = \begin{bmatrix} \mathbf{X} & \mathbf{y} \\ \mathbf{y}^T & -w \end{bmatrix}. \tag{14.4-23}$$

It is not difficult to show that

$$\begin{gathered} w = [\omega_{22} + \mathbf{e}_{22}^T (\mathbf{T}^E)^{-1} \mathbf{e}_{22}]^{-1}, \\ \mathbf{y} = w(\mathbf{T}^E)^{-1} \mathbf{e}_{22}, \\ \mathbf{X} = \left\{\mathbf{T}^E + \tfrac{1}{\omega_{22}} \mathbf{e}_{22} \mathbf{e}_{22}^T\right\}^{-1}. \end{gathered} \tag{14.4-24}$$

Clearly, w>0 and **X** is symmetric and positive definite. Therefore if a matrix is of Type **T**, so is its inverse.

The two equations in (14.4-14) can now be recast in the standard eigenrelation as (Section 5.4)

$$\mathbf{N}\boldsymbol{\xi} = p\boldsymbol{\xi}, \tag{14.4-25}$$

$$\mathbf{N} = \begin{bmatrix} \mathbf{N}_1 & \mathbf{N}_2 \\ \mathbf{N}_3 & \mathbf{N}_1^T \end{bmatrix}, \quad \boldsymbol{\xi} = \begin{bmatrix} \boldsymbol{a} \\ \boldsymbol{l} \end{bmatrix},$$

$$\mathbf{N}_1 = -\mathbf{T}^{-1}\mathbf{R}^T, \quad \mathbf{N}_2 = \mathbf{T}^{-1}, \quad \mathbf{N}_3 = \mathbf{R}\mathbf{T}^{-1}\mathbf{R}^T - \mathbf{Q}.$$

Following the derivation in Section 5.5 the left eigenvector $\boldsymbol{\eta}$ of **N** is

$$\boldsymbol{\eta} = \begin{bmatrix} \boldsymbol{b} \\ \boldsymbol{a} \end{bmatrix}$$

and the *orthogonality* and *closure relations* shown in (5.4-10)–(5.4-16) hold. As in (5.4-17) the 4×4 matrices defined by

$$\mathbf{S} = i(2\mathbf{A}\mathbf{B}^T - \mathbf{I}), \quad \mathbf{H} = 2i\mathbf{A}\mathbf{A}^T, \quad \mathbf{L} = -2i\mathbf{B}\mathbf{B}^T \tag{14.4-26}$$

are real. They are generalized Barnett–Lothe tensors for piezoelectric materials.

From the symmetry property (14.4-2) one can show that

$$Q_{I2} = R_{I1}, \quad R_{2K} = T_{1K}, \quad (I,K = 1,2,3,4). \tag{14.4-27}$$

Hence (Section 6.1)

$$(\mathbf{N}_1)_{I2} = -\delta_{I1}, \quad (\mathbf{N}_3)_{I2} = 0. \tag{14.4-28}$$

Consequently $\mathbf{N}_1$ and $\mathbf{N}_3$ have the structures

$$\mathbf{N}_1 = \begin{bmatrix} * & -1 & * & * \\ * & 0 & * & * \\ * & 0 & * & * \\ * & 0 & * & * \end{bmatrix}, \quad \mathbf{N}_3 = \begin{bmatrix} * & 0 & * & * \\ 0 & 0 & 0 & 0 \\ * & 0 & * & * \\ * & 0 & * & * \end{bmatrix}$$

in which the * denotes a possibly nonzero element.

The integral formalism for the 4×4 matrices **S**, **H**, **L** presented in (14.4-26) can be obtained by following the derivation presented in Sections 7.3 and 7.6 for non-piezoelectric materials. The solutions (14.4-7) and (14.4-13) say that $\mathbf{u}$, ϕ are proportional to $\boldsymbol{a}$, $\boldsymbol{b}$, respectively, with the proportional factor $f(z)$. z is a linear combination of x_1 and x_2 in which we took the coefficient of x_1 to be unity without loss in generality. We will replace z by the following more general linear combination of x_1, x_2,

$$z = \mathbf{n}\cdot\mathbf{x} + p(\theta)\mathbf{m}\cdot\mathbf{x} \tag{14.4-29}$$

where θ is an arbitrary constant, $p(\theta)$ depends on θ, and

$$\mathbf{n} = \begin{bmatrix} \cos\theta \\ \sin\theta \\ 0 \end{bmatrix}, \quad \mathbf{m} = \begin{bmatrix} -\sin\theta \\ \cos\theta \\ 0 \end{bmatrix}.$$

It should be noted that with z replaced by (14.4-29), $\boldsymbol{a}$ in (14.4-7) and $\boldsymbol{b}$ in (14.4-13) remain independent of θ. Equations (14.4-8) and (14.4-9) are valid if δ_{s1} and δ_{s2} are replaced by n_s and m_s, respectively. Equations (14.4-11) and (14.4-12) become

$$\left\{\mathbf{Q}(\theta) + p(\theta)\left(\mathbf{R}(\theta) + \mathbf{R}^T(\theta)\right) + p^2(\theta)\mathbf{T}(\theta)\right\}\boldsymbol{a} = \mathbf{0},$$

$$\mathbf{Q}(\theta) = \begin{bmatrix} \mathbf{Q}^E(\theta) & \mathbf{e}_{11}(\theta) \\ \mathbf{e}_{11}^T(\theta) & -\omega_{11}(\theta) \end{bmatrix}, \quad \mathbf{R}(\theta) = \begin{bmatrix} \mathbf{R}^E(\theta) & \mathbf{e}_{21}(\theta) \\ \mathbf{e}_{12}^T(\theta) & -\omega_{12}(\theta) \end{bmatrix},$$

$$\mathbf{T}(\theta)=\begin{bmatrix}\mathbf{T}^{E}(\theta) & \mathbf{e}_{22}(\theta)\\ \mathbf{e}_{22}^{T}(\theta) & -\omega_{22}(\theta)\end{bmatrix}.$$

In the above,

$$Q_{ik}^{E}(\theta)=C_{ijks}n_j n_s,\quad R_{ik}^{E}(\theta)=C_{ijks}n_j m_s,\quad T_{ik}^{E}(\theta)=C_{ijks}m_j m_s,$$

$$[\mathbf{e}_{11}(\theta)]_s=e_{ijs}n_i n_j,\quad [\mathbf{e}_{12}(\theta)]_s=e_{ijs}n_i m_j,$$

$$[\mathbf{e}_{21}(\theta)]_s=e_{ijs}m_i n_j,\quad [\mathbf{e}_{22}(\theta)]_s=e_{ijs}m_i m_j,$$

$$\omega_{11}(\theta)=\omega_{ij}n_i n_j,\quad \omega_{12}(\theta)=\omega_{ij}n_i m_j,\quad \omega_{22}(\theta)=\omega_{ij}m_i m_j.$$

Equation (14.4-25) is replaced by

$$\mathbf{N}(\theta)\xi=p(\theta)\xi, \tag{14.4-30}$$

$$\mathbf{N}(\theta)=\begin{bmatrix}\mathbf{N}_1(\theta) & \mathbf{N}_2(\theta)\\ \mathbf{N}_3(\theta) & \mathbf{N}_1^{T}(\theta)\end{bmatrix},\quad \xi=\begin{bmatrix}a\\ l\end{bmatrix},$$

$$\mathbf{N}_1(\theta)=-\mathbf{T}^{-1}(\theta)\mathbf{R}^{T}(\theta),\quad \mathbf{N}_2(\theta)=\mathbf{T}^{-1}(\theta),$$

$$\mathbf{N}_3(\theta)=\mathbf{R}(\theta)\mathbf{T}^{-1}(\theta)\mathbf{R}^{T}(\theta)-\mathbf{Q}(\theta).$$

From the diagonalization of $\mathbf{N}(\theta)$ and upon integration with θ from 0 to π shown in Section 7.6, we obtain the integral representation of **S**, **H**, **L** as (Lothe and Barnett, 1976b)

$$\mathbf{S}=\frac{1}{\pi}\int_0^{\pi}\mathbf{N}_1(\theta)d\theta,\quad \mathbf{H}=\frac{1}{\pi}\int_0^{\pi}\mathbf{N}_2(\theta)d\theta,\quad \mathbf{L}=-\frac{1}{\pi}\int_0^{\pi}\mathbf{N}_3(\theta)d\theta. \tag{14.4-31}$$

It is clear that $\mathbf{T}^{E}(\theta)$ is symmetric and positive definite, and that $\omega_{22}(\theta)>0$. Hence $\mathbf{T}(\theta)$ is of Type **T**, so is its inverse which is $\mathbf{N}_2(\theta)$. Following (14.4-23) let

$$\mathbf{H}=\begin{bmatrix}\mathbf{H}_{11} & \mathbf{H}_{14}\\ \mathbf{H}_{14}^{T} & -H_{44}\end{bmatrix},\quad \mathbf{L}=\begin{bmatrix}\mathbf{L}_{11} & \mathbf{L}_{14}\\ \mathbf{L}_{14}^{T} & -L_{44}\end{bmatrix}$$

where $\mathbf{H}_{11}$ and $\mathbf{L}_{11}$ are 3×3 matrices while $\mathbf{H}_{14}$ and $\mathbf{L}_{14}$ are 4×1 column matrices. From $(14.4\text{-}31)_2$ it is clear that $\mathbf{H}_{11}$ is symmetric and positive definite, and $H_{44}>0$. The matrix **H** is of Type **T**. The four eigenvalues of **H** and the three real and positive eigenvalues of $\mathbf{H}_{11}$ are related by the inequalities similar to the ones shown in (14.4-22). By considering a line defect in the infinite piezoelectric material and the fact that the prelogarithmic factor of the work done in the material is positive definite, Lothe and Barnett (1976b) showed that $\mathbf{L}_{11}$ is positive definite and $L_{44}>0$. Thus the matrix **L** is also of Type **T**. This is particularly useful for surface waves in a piezoelectric half-space for which **L** depends on the surface wave speed v, and v is determined by the vanishing of the determinant of $\mathbf{L}(v)$.

We have shown that Stroh's sextic formalism can be extended to an octet formalism for piezoelectric materials. Applications to a line force and a line charge in a piezoelectric material were studied by Barnett and Lothe (1975b). The problem of a line force, line dislocation, and a line charge in an infinite composite space or an angularly inhomogeneous space was investigated by Chung and Ting (1995a,b). Other applications that have been considered are surface waves (Lothe and Barnett, 1977, 1979), Stoneley waves (Abbudi and Barnett, 1990), the stress singularities at an interfacial crack tip (Kuo and Barnett, 1991), and fracture mechanics for piezoelectric ceramics (Suo et al., 1992). Al'shits et al. (1992) have extended the Stroh formalism to a ten-dimensional formalism that includes piezoelectric and piezomagnetic properties of materials.

Chapter 15

THREE-DIMENSIONAL DEFORMATIONS

There appears to be very little study, if any, on the extension of Stroh's formalism to three-dimensional deformations of anisotropic elastic materials. In most three-dimensional problems the analyses employ approaches that are remotely related to Stroh's two-dimensional formalism. This is not unexpected, since this has been the situation between two-dimensional and three-dimensional isotropic elasticity. However it needs not be the case for three-dimensional anisotropic elasticity. Much can be gained if a connection to the Stroh formalism can be established. Barnett and Lothe (1975a) appeared to be the only ones who made a connection between a three-dimensional solution and Stroh's two-dimensional formalism. Earlier, several investigators obtained the Green's function for the infinite anisotropic medium in term of a line integral on an oblique plane in the three-dimensional space. That line integral, as we will see here, is one of Barnett–Lothe tensors on an oblique plane. We propose in this chapter extensions and applications of Stroh's two-dimensional formalism to certain three-dimensional deformations of anisotropic elastic solids.

15.1 Stroh's Eigenrelation for an Oblique Plane

The first step is to generalize Stroh's eigenrelation for the (x_1,x_2) plane to an oblique plane in the three-dimensional space. It was shown in Sections 5.1 and 5.3 that the eigenvalue p and the Stroh eigenvectors $\boldsymbol{a}$, $\boldsymbol{b}$ for the (x_1,x_2)-plane satisfy the relation

$$\left\{\mathbf{Q}+p(\mathbf{R}+\mathbf{R}^T)+p^2\mathbf{T}\right\}\boldsymbol{a}=\mathbf{0}, \tag{15.1-1}$$

$$\boldsymbol{b}=(\mathbf{R}^T+p\mathbf{T})\boldsymbol{a}=-\frac{1}{p}(\mathbf{Q}+p\mathbf{R})\boldsymbol{a}, \tag{15.1-2}$$

where

$$Q_{ik}=C_{i1k1}, \quad R_{ik}=C_{i1k2}, \quad T_{ik}=C_{i2k2}. \tag{15.1-3}$$

Consider now an oblique plane with a normal vector $\mathbf{k}$ in the three-dimensional space. Let $\mathbf{n}$ and $\mathbf{m}$ be any two orthogonal unit vectors on this plane. The vectors $[\mathbf{n}, \mathbf{m}, \mathbf{k}]$ form a right-handed triad. $\mathbf{n}$ and $\mathbf{m}$ may be regarded as the new x_1- and x_2- axes. Accordingly we define the new $\mathbf{Q}$, $\mathbf{R}$, and $\mathbf{T}$ by

$$Q_{ik} = C_{ijks}n_j n_s, \quad R_{ik} = C_{ijks}n_j m_s, \quad T_{ik} = C_{ijks}m_j m_s. \tag{15.1-4}$$

They reduce to (15.1-3) when **n** and **m** coincide with the x_1- and x_2-axes, respectively. The new **Q** and **T** remain symmetric and positive definite, and p remains complex if the strain energy is positive definite. If p_α, (α = 1,2,...,6) are the eigenvalues we let

$$\operatorname{Im} p_\alpha > 0, \quad p_{\alpha+3} = \bar{p}, \quad \mathbf{a}_{\alpha+3} = \bar{\mathbf{a}}_\alpha, \quad \boldsymbol{b}_{\alpha+3} = \bar{\boldsymbol{b}}_\alpha \quad (\alpha = 1,2,3),$$

where Im stands for the imaginary part and the overbar denotes the complex conjugate. As in Section 5.5 the two equations in (15.1-2) can be rewritten in a standard eigenrelation

$$\begin{bmatrix} \mathbf{N}_1 & \mathbf{N}_2 \\ \mathbf{N}_3 & \mathbf{N}_1^T \end{bmatrix} \begin{bmatrix} \boldsymbol{a} \\ \boldsymbol{b} \end{bmatrix} = p \begin{bmatrix} \boldsymbol{a} \\ \boldsymbol{b} \end{bmatrix}, \tag{15.1-5}$$

$$\mathbf{N}_1 = -\mathbf{T}^{-1}\mathbf{R}^T, \quad \mathbf{N}_2 = \mathbf{T}^{-1}, \quad \mathbf{N}_3 = \mathbf{R}\mathbf{T}^{-1}\mathbf{R}^T - \mathbf{Q}.$$

$\mathbf{N}_1$ is symmetric and positive definite while $\mathbf{N}_3$ is symmetric and positive semidefinite. It can be shown that (Section 7.2)

$$\mathbf{Qm} = \mathbf{Rn}, \quad \mathbf{R}^T\mathbf{m} = \mathbf{Tn},$$

and

$$\mathbf{N}_1\mathbf{m} = -\mathbf{n}, \quad \mathbf{N}_3\mathbf{m} = \mathbf{0}. \tag{15.1-6}$$

Hence $\mathbf{N}_3$ is singular. The 6×1 matrix $(\boldsymbol{a},\boldsymbol{b})$ in (15.1-5) is a right eigenvector. The left eigenvector is $(\boldsymbol{b},\boldsymbol{a})$. The orthogonality and closure relations between the right and left eigenvectors delineated in Section 5.5 hold. As a consequence the generalized Barnett–Lothe tensors **S**, **H**, **L** defined by

$$\mathbf{S}[\mathbf{k}] = i(2\mathbf{AB}^T - \mathbf{I}), \quad \mathbf{H}[\mathbf{k}] = 2i\mathbf{AA}^T, \quad \mathbf{L}[\mathbf{k}] = -2i\mathbf{BB}^T, \tag{15.1-7}$$

where

$$\mathbf{A} = [\boldsymbol{a}_1, \boldsymbol{a}_2, \boldsymbol{a}_3], \quad \mathbf{B} = [\boldsymbol{b}_1, \boldsymbol{b}_2, \boldsymbol{b}_3],$$

are real. The appearance of **k** is merely to identify the oblique plane on which **S**, **H**, **L** are computed. Hence **k** need not be a unit vector. **H** and **L** are symmetric and positive definite while **LS**, $\mathbf{S}\mathbf{L}^{-1}$, **SH**, and $\mathbf{H}^{-1}\mathbf{S}$ are skew-symmetric. The impedance tensor **M** and its inverse $\mathbf{M}^{-1}$ defined by

$$\begin{aligned} \mathbf{M} &= -i\mathbf{BA}^{-1} = \mathbf{H}^{-1} + i\mathbf{H}^{-1}\mathbf{S}, \\ \mathbf{M}^{-1} &= i\mathbf{AB}^{-1} = \mathbf{L}^{-1} - i\mathbf{SL}^{-1} \end{aligned} \tag{15.1-8}$$

are positive definite Hermitian.

We will show that $\boldsymbol{a}$ and $\boldsymbol{b}$, and hence **S**, **H**, **L**, are independent of the choice of **n** and **m** on the oblique plane. Let $\mathbf{n}^*$ and $\mathbf{m}^*$ be a new set of mutually orthogonal unit vectors on the oblique plane defined by

$$\mathbf{n}^* = \mathbf{n}\cos\psi + \mathbf{m}\sin\psi, \quad \mathbf{m}^* = -\mathbf{n}\sin\psi + \mathbf{m}\cos\psi, \tag{15.1-9}$$

where ψ is the angle between **n** and $\mathbf{n}^*$. The inverse of (15.1-9) is

$$\mathbf{n} = \mathbf{n}^*\cos\psi - \mathbf{m}^*\sin\psi, \quad \mathbf{m} = \mathbf{n}^*\sin\psi + \mathbf{m}^*\cos\psi. \tag{15.1-10}$$

Inserting (15.1-10) into (15.1-4), (15.1-1) and (15.1-2) yield

$$\left\{\mathbf{Q}(\psi) + p(\psi)[\mathbf{R}(\psi) + \mathbf{R}^T(\psi)] + p^2(\psi)\mathbf{T}(\psi)\right\}\boldsymbol{a} = \mathbf{0}, \tag{15.1-11}$$

$$\boldsymbol{b} = [\mathbf{R}^T(\psi) + p(\psi)\mathbf{T}(\psi)]\boldsymbol{a} = -\frac{1}{p(\psi)}[\mathbf{Q}(\psi) + p(\psi)\mathbf{R}(\psi)]\boldsymbol{a}, \tag{15.1-12}$$

where

$$Q_{ik}(\psi) = C_{ijks}n_j^*n_s^*, \quad R_{ik}(\psi) = C_{ijks}n_j^*m_s^*, \quad T_{ik}(\psi) = C_{ijks}m_j^*m_s^*, \tag{15.1-13}$$

$$p(\psi) = \frac{p\cos\psi - \sin\psi}{p\sin\psi + \cos\psi} = \frac{d}{d\psi}\ln(p\sin\psi + \cos\psi). \tag{15.1-14}$$

The second equality in (15.1-12) follows from (15.1-11). The first equality is reached with the aid of $(15.1\text{-}12)_2$ and $(15.1\text{-}14)_1$.

It is seen that $\boldsymbol{a}$ and $\boldsymbol{b}$ in (15.1-11) and (15.1-12) remain the same with the new $\mathbf{n}^*$ and $\mathbf{m}^*$. Thus the Stroh eigenvectors $\boldsymbol{a}$ and $\boldsymbol{b}$ are invariant with the choice of **n** and **m** on the oblique plane. This point should not be confused with the results obtained in Section 7.2 where the components of $\boldsymbol{a}$ and $\boldsymbol{b}$ change when they are referred to a rotated coordinate system. No rotation of the coordinate system is involved here. The components of the vectors $\boldsymbol{a}$ and $\boldsymbol{b}$ are referred to the (x_1, x_2, x_3) coordinate system regardless of the choice of **n** and **m** on the oblique plane.

As a consequence of the invariance of the Stroh eigenvectors $\boldsymbol{a}$ and $\boldsymbol{b}$, the matrices **A**, **B**, and the tensors **S**, **H**, **L** are also invariant with the choice of **n** and **m** on the oblique plane. The two equations in (15.1-12) can be cast in the standard eigenrelation (15.1-5) where $\mathbf{N}_i$ $(i=1,2,3)$ now depend on ψ. Following the derivation in Section 7.6 it is easily shown that an alternate integral representation for **S**, **H**, **L** in (15.1-7) is (Barnett and Lothe, 1973)

$$\mathbf{S}[\mathbf{k}] = \frac{1}{\pi}\int_0^{\pi} \mathbf{N}_1(\psi)d\psi,$$

$$\mathbf{H}[\mathbf{k}] = \frac{1}{\pi}\int_0^{\pi} \mathbf{N}_2(\psi)d\psi, \qquad (15.1\text{-}15)$$

$$\mathbf{L}[\mathbf{k}] = -\frac{1}{\pi}\int_0^{\pi} \mathbf{N}_3(\psi)d\psi,$$

where

$$\mathbf{N}_1(\psi) = -\mathbf{T}^{-1}(\psi)\mathbf{R}^T(\psi), \quad \mathbf{N}_2(\psi) = \mathbf{T}^{-1}(\psi),$$

$$\mathbf{N}_3(\psi) = \mathbf{R}(\psi)\mathbf{T}^{-1}(\psi)\mathbf{R}^T(\psi) - \mathbf{Q}(\psi).$$

It should be pointed out that $\mathbf{Q}(\psi)$, $\mathbf{R}(\psi)$, $\mathbf{T}(\psi)$ are periodic in ψ with periodicity π, and that $\mathbf{T}^{-1}(\psi) = \mathbf{Q}^{-1}(\psi + \frac{\pi}{2})$. Hence $(15.1\text{-}15)_2$ is equivalent to

$$\mathbf{H}[\mathbf{k}] = \frac{1}{\pi}\int_0^{\pi} \mathbf{Q}^{-1}(\psi)d\psi. \qquad (15.1\text{-}16)$$

The components of $\mathbf{H}$ has a physical meaning when $\mathbf{k}$ is the position vector $\mathbf{x}$. We will show in Section 15.3 that $H_{ik}(\mathbf{x})/(4\pi r)$ where $r = |\mathbf{x}|$ is the Green's function for the infinite space for the displacement u_i at $\mathbf{x}$ when a unit force in the direction of x_k is applied at $\mathbf{x} = \mathbf{0}$.

15.2 The Tensors S[x], H[x], L[x]

It has been shown in the previous section that the tensors $\mathbf{S}[\mathbf{k}]$, $\mathbf{H}[\mathbf{k}]$, $\mathbf{L}[\mathbf{k}]$ given by (15.1-7) or (15.1-15) do not depend on the choice of the unit vectors $\mathbf{n}$ and $\mathbf{m}$ on the oblique plane. They depend on the location of the oblique plane which is identified by its normal vector $\mathbf{k}$. If the oblique plane is normal to the position vector $\mathbf{x}$, we write $\mathbf{S}[\mathbf{x}]$, $\mathbf{H}[\mathbf{x}]$, $\mathbf{L}[\mathbf{x}]$. If the normal to the oblique plane lies on the (x_1, x_2) plane making an angle θ with the x_1-axis, we write $\mathbf{S}[\theta]$, $\mathbf{H}[\theta]$, $\mathbf{L}[\theta]$. The square brackets $[\theta]$ are used instead of (θ) to avoid confusion with $\mathbf{S}(\theta)$, $\mathbf{H}(\theta)$, $\mathbf{L}(\theta)$ employed in Section 7.6.

When the position vector $\mathbf{x}$ is normal to the oblique plane, the unit vectors $\mathbf{n}$ and $\mathbf{m}$ are chosen such that $(\mathbf{n}, \mathbf{m}, \mathbf{x})$ from a right-handed triad. If $\mathbf{x}$ is replaced by $-\mathbf{x}$, we may replace $\mathbf{m}$ by $-\mathbf{m}$ (or $\mathbf{n}$ by $-\mathbf{n}$) to maintain a right-handed triad. It is easily shown that p, b, and $\mathbf{N}_1$ change their signs. Hence

$$\mathbf{S}(\mathbf{x}) = -\mathbf{S}(-\mathbf{x}), \quad \mathbf{H}(\mathbf{x}) = \mathbf{H}(-\mathbf{x}), \quad \mathbf{L}(\mathbf{x}) = \mathbf{L}(-\mathbf{x}). \qquad (15.2\text{-}1)$$

In the spherical coordinate system (r,θ,φ) we may choose

$$\mathbf{x}=r\begin{bmatrix}\sin\varphi\cos\theta\\ \sin\varphi\sin\theta\\ \cos\varphi\end{bmatrix},\quad \mathbf{n}=\begin{bmatrix}\cos\varphi\cos\theta\\ \cos\varphi\sin\theta\\ -\sin\varphi\end{bmatrix},\quad \mathbf{m}=\begin{bmatrix}-\sin\theta\\ \cos\theta\\ 0\end{bmatrix},\tag{15.2-2}$$

where $r=|\mathbf{x}|$. To compute $\mathbf{S}[\mathbf{x}]$, $\mathbf{H}[\mathbf{x}]$, $\mathbf{L}[\mathbf{x}]$ from (15.1-7) we find the columns of $\mathbf{A}$ and $\mathbf{B}$ from (15.1-1) and (15.1-2) in which $\mathbf{Q}$, $\mathbf{R}$, and $\mathbf{T}$ are determined from (15.1-4). The columns of $\mathbf{A}$ and $\mathbf{B}$ must be normalized according to (5.5-12). To employ the integral formalism (15.1-15), we compute $\mathbf{n}^*$ and $\mathbf{m}^*$ from (15.1-9), and $\mathbf{Q}(\psi)$, $\mathbf{R}(\psi)$, $\mathbf{T}(\psi)$ from (15.1-13).

For isotropic materials,

$$C_{ijks}=\lambda\delta_{ij}\delta_{ks}+\mu(\delta_{is}\delta_{jk}+\delta_{ik}\delta_{js})$$

where λ and μ are the Lamé constants. Hence

$$Q_{ik}=C_{ijks}n_jn_s=\lambda\,n_in_k+\mu\,(n_kn_i+\delta_{ik})$$

or, with a similar derivation for R_{ik} and T_{ik},

$$\begin{aligned}\mathbf{Q}&=(\lambda+\mu)\mathbf{n}\mathbf{n}^T+\mu\mathbf{I},\\ \mathbf{R}&=\lambda\,\mathbf{n}\mathbf{m}^T+\mu\,\mathbf{m}\mathbf{n}^T,\\ \mathbf{T}&=(\lambda+\mu)\mathbf{m}\mathbf{m}^T+\mu\mathbf{I}.\end{aligned}\tag{15.2-3}$$

The eigenrelation (15.1-1) reduces to

$$\left\{\mu(1+p^2)\mathbf{I}+(\lambda+\mu)(\mathbf{n}+p\mathbf{m})(\mathbf{n}+p\mathbf{m})^T\right\}\boldsymbol{a}=\mathbf{0}.\tag{15.2-4}$$

Using the identity (1.1-21), the vanishing of the determinant of the coefficient matrix of $\boldsymbol{a}$ in (15.2-4) is

$$\mu^3(1+p^2)^3+\mu^2(1+p^2)^2(\lambda+\mu)(\mathbf{n}+p\mathbf{m})^T(\mathbf{n}+p\mathbf{m})=0.$$

Since $(\mathbf{n}+p\mathbf{m})^T(\mathbf{n}+p\mathbf{m})=1+p^2$ we obtain

$$\mu^2(\lambda+2\mu)(1+p^2)^3=0$$

and hence $p=\pm i$ are roots of multiplicity three. Setting $p=i$, (15.2-4) becomes

$$(\mathbf{n}+i\mathbf{m})(\mathbf{n}+i\mathbf{m})^T\boldsymbol{a}=\mathbf{0}.$$

The eigenvectors associated with $p=i$ are

$$a_1 = \mathbf{x}, \quad a_2 = \mathbf{n} + i\mathbf{m},$$

where $\mathbf{x}$ is normal to $\mathbf{n}$ and $\mathbf{m}$. There are only two independent eigenvectors associated with the triple eigenvalues $p=i$, proving that isotropic materials remain degenerate materials when referred to an oblique plane. Therefore (15.1-7) is not valid for computing $\mathbf{S}[\mathbf{x}]$, $\mathbf{H}[\mathbf{x}]$, $\mathbf{L}[\mathbf{x}]$ for isotropic materials.

We will employ (15.1-15) for isotropic materials. The $\mathbf{Q}(\psi)$, $\mathbf{R}(\psi)$, $\mathbf{T}(\psi)$ are obtained from (15.2-3) by replacing $\mathbf{n}$, $\mathbf{m}$ by $\mathbf{n}^*$, $\mathbf{m}^*$, respectively, i.e.,

$$\mathbf{Q}(\psi) = (\lambda + \mu)\mathbf{n}^*\mathbf{n}^{*T} + \mu\mathbf{I},$$

$$\mathbf{R}(\psi) = \lambda\,\mathbf{n}^*\mathbf{m}^{*T} + \mu\,\mathbf{m}^*\mathbf{n}^{*T},$$

$$\mathbf{T}(\psi) = (\lambda + \mu)\mathbf{m}^*\mathbf{m}^{*T} + \mu\mathbf{I}.$$

The identity matrix $\mathbf{I}$ has the representations

$$\begin{aligned}\mathbf{I} &= \mathbf{n}\mathbf{n}^T + \mathbf{m}\mathbf{m}^T + \mathbf{k}\mathbf{k}^T \\ &= \mathbf{n}^*\mathbf{n}^{*T} + \mathbf{m}^*\mathbf{m}^{*T} + \mathbf{k}\mathbf{k}^T\end{aligned} \qquad (15.2\text{-}5)$$

where

$$\mathbf{k} = \mathbf{x}/r. \qquad (15.2\text{-}6)$$

Equations (15.2-5) are diagonalization of the unit matrix $\mathbf{I}$ (Section 1.3). By writing $\mathbf{Q}(\psi)$ as

$$\mathbf{Q}(\psi) = (\lambda + 2\mu)\mathbf{n}^*\mathbf{n}^{*T} + \mu(\mathbf{m}^*\mathbf{m}^{*T} + \mathbf{k}\mathbf{k}^T),$$

and a similar expression for $\mathbf{T}(\psi)$, the inverses of $\mathbf{Q}(\psi)$ and $\mathbf{T}(\psi)$ are

$$\mathbf{Q}^{-1}(\psi) = (\lambda + 2\mu)^{-1}\mathbf{n}^*\mathbf{n}^{*T} + \mu^{-1}(\mathbf{m}^*\mathbf{m}^{*T} + \mathbf{k}\mathbf{k}^T),$$

$$\mathbf{T}^{-1}(\psi) = (\lambda + 2\mu)^{-1}\mathbf{m}^*\mathbf{m}^{*T} + \mu^{-1}(\mathbf{n}^*\mathbf{n}^{*T} + \mathbf{k}\mathbf{k}^T).$$

The matrices $\mathbf{N}_1(\psi)$, $\mathbf{N}_2(\psi)$, $\mathbf{N}_3(\psi)$ are

$$\mathbf{N}_1(\psi) = -\left\{\mathbf{n}^*\mathbf{m}^{*T} + \frac{\nu}{1-\nu}\mathbf{m}^*\mathbf{n}^{*T}\right\},$$

$$\mathbf{N}_2(\psi) = -\frac{1}{\mu}\left\{\mathbf{I} - \frac{1}{2(1-\nu)}\mathbf{m}^*\mathbf{m}^{*T}\right\}, \qquad (15.2\text{-}7)$$

$$\mathbf{N}_3(\psi) = -\mu\left\{\frac{2}{1-\nu}\mathbf{n}^*\mathbf{n}^{*T} + \mathbf{k}\mathbf{k}^T\right\},$$

where ν is the Poisson ratio, and use has been made of the relation

$$\lambda = \frac{2\mu\nu}{1-2\nu}.$$

It can be verified with the aid of (15.1-9) that

$$\frac{1}{\pi}\int_0^\pi \mathbf{n}^*\mathbf{n}^{*T} d\psi = \tfrac{1}{2}(\mathbf{n}\mathbf{n}^T + \mathbf{m}\mathbf{m}^T) = \tfrac{1}{2}(\mathbf{I} - \mathbf{k}\mathbf{k}^T),$$

$$\frac{1}{\pi}\int_0^\pi \mathbf{m}^*\mathbf{m}^{*T} d\psi = \tfrac{1}{2}(\mathbf{n}\mathbf{n}^T + \mathbf{m}\mathbf{m}^T) = \tfrac{1}{2}(\mathbf{I} - \mathbf{k}\mathbf{k}^T),$$

$$\frac{1}{\pi}\int_0^\pi \mathbf{n}^*\mathbf{m}^{*T} d\psi = \tfrac{1}{2}(\mathbf{n}\mathbf{m}^T - \mathbf{m}\mathbf{n}^T).$$

Therefore the integrals in (15.1-15) for isotropic materials are easily obtained as

$$\mathbf{S}[\mathbf{x}] = \frac{1-2\nu}{2(1-\nu)}(\mathbf{m}\mathbf{n}^T - \mathbf{n}\mathbf{m}^T),$$

$$\mathbf{H}[\mathbf{x}] = \frac{1}{4\mu(1-\nu)}\left\{(3-4\nu)\mathbf{I} + \frac{\mathbf{x}\mathbf{x}^T}{r^2}\right\}, \qquad (15.2\text{-}8)$$

$$\mathbf{L}[\mathbf{x}] = \frac{\mu}{(1-\nu)}\left\{\mathbf{I} - \nu\frac{\mathbf{x}\mathbf{x}^T}{r^2}\right\}.$$

15.3 Green's Function for Infinite Space

We present in this and the following sections examples of three-dimensional deformations of anisotropic elastic solids for which the solutions can be expressed in terms of **S**[**x**], **H**[**x**], **L**[**x**] on oblique planes. The equations of equilibrium are

$$C_{ijks}u_{k,sj} = 0 \qquad (15.3\text{-}1)$$

where the differentiation applies to all three variables x_1, x_2, x_3.

Consider the Green's function for the infinite anisotropic medium subjected to a concentrated force **f** at the origin **x**=**0**. The right-hand side of (15.3-1) is replaced by

$$-\delta(x_1)\delta(x_2)\delta(x_3)f_i$$

where $\delta(x)$ is the Dirac's delta function. On taking Fourier transforms of (15.3-1) with respect to x_1, x_2, x_3 we have

$$C_{ijks}y_jy_s\tilde{u}_k = f_i \tag{15.3-2}$$

where y_i (i=1,2,3) are the transform parameters and

$$\tilde{u}_k = \iiint u_k(x_1,x_2,x_3)e^{i\mathbf{y}\cdot\mathbf{x}}dx_1dx_2dx_3$$

is the Fourier transform of u_k. The vector **y** in the integrand is

$$\mathbf{y}^T = [y_1, y_2, y_3].$$

Solving for $\tilde{u}_k$ from (15.3-2) and performing the inverse transforms leads to

$$\mathbf{u} = \frac{1}{(2\pi)^3}\iiint \mathbf{Q}^{-1}(\mathbf{y})\mathbf{f}\ e^{-i\mathbf{y}\cdot\mathbf{x}}\ dy_1dy_2dy_3 \tag{15.3-3}$$

where

$$Q_{ik}(\mathbf{y}) = C_{ijks}y_jy_s. \tag{15.3-4}$$

Considerable efforts have been made in finding the integrals in (15.3-3) (Willis, 1965; Indenbom and Orlov, 1968; Mura and Kinoshita, 1971). Explicit expressions for the Green's function for transversely isotropic materials are obtained by Kröner (1953), Willis (1965), and Pan and Chou (1976).

It was shown by Fredholm (1900), Lifshitz and Rozenzweig (1947), and Synge (1957) that (15.3-3) is reducible to

$$\mathbf{u} = \frac{1}{8\pi^2 r}\int_s \mathbf{Q}^{-1}(\mathbf{y})\mathbf{f}\,ds \tag{15.3-5}$$

where $r=|\mathbf{x}|$ and the integral is taken around the unit circle $|\mathbf{y}|=1$ on the plane normal to **x**. By (15.1-16),

$$\frac{1}{2\pi}\int_s \mathbf{Q}^{-1}(\mathbf{y})ds = \mathbf{H}[\mathbf{x}]$$

is the tensor **H** on the oblique plane whose normal is **x**. Consequently (15.3-5) can be written in the remarkably compact form

$$\mathbf{u}(\mathbf{x}) = \frac{1}{4\pi r}\mathbf{H}[\mathbf{x}]\mathbf{f}. \tag{15.3-6}$$

With **H**[**x**] given by $(15.2\text{-}8)_2$ for isotropic materials (15.3-6) recovers the Green's function for an infinite isotropic medium first obtained by Kelvin (1848, 1882).

Interesting properties can be read off from (15.3-6). When $\mathbf{x}=[0,0,x_3]$, $\mathbf{H}[\mathbf{x}]$ is computed on the (x_1,x_2) plane and is independent of C_{ij33}. Hence the displacement $\mathbf{u}$ on the x_3-axis is independent of C_{ij33}. Likewise, the displacement on the x_1- and x_2-axes are independent of C_{ij11} and C_{ij22}, respectively. By virtue of the positive definiteness of $\mathbf{H}[\mathbf{x}]$, (15.3-6) tells us that

$$\mathbf{u}(\mathbf{x})\cdot\mathbf{f}>0 \tag{15.3-7}$$

for any point $\mathbf{x}$ in the infinite medium. Therefore the angle between $\mathbf{u}$ and $\mathbf{f}$ is less than $\pi/2$. $\mathbf{u}$ and $\mathbf{f}$ are co-directional when $\mathbf{f}$ is an eigenvector of $\mathbf{H}[\mathbf{x}]$. Since there are three mutually orthogonal eigenvectors for $\mathbf{H}[\mathbf{x}]$, there are at least three possible directions of $\mathbf{f}$ for which the displacement $\mathbf{u}$ at $\mathbf{x}$ is co-directional with $\mathbf{f}$. For isotropic materials the possible directions of $\mathbf{f}$ for which $\mathbf{u}$ and $\mathbf{f}$ are co-directional are the direction $\mathbf{x}$, and any direction orthogonal to $\mathbf{x}$.

15.4 Surface Green's Function for Half-Space

Let the half-space $x_3\geq 0$ be subjected to a concentrated force $\mathbf{f}$ at the origin $\mathbf{x}=\mathbf{0}$. The surface $x_3=0$ is otherwise traction-free. In this case we apply Fourier transforms to (15.3-1) with x_1 and x_2 only. The result is (Willis, 1966)

$$C_{i\alpha k\beta}y_\alpha y_\beta \tilde{u}_k + i(C_{i\alpha k3}+C_{i3k\alpha})y_\alpha \tilde{u}_{k,3} - C_{i3k3}\tilde{u}_{k,33}=0 \tag{15.4-1}$$

where α, β=1,2. If we let

$$\mathbf{y}=\eta\mathbf{n},\qquad \mathbf{n}=\begin{bmatrix}n_1\\n_2\\0\end{bmatrix},\qquad \mathbf{m}=\begin{bmatrix}0\\0\\1\end{bmatrix}, \tag{15.4-2}$$

a general solution is

$$\tilde{\mathbf{u}}(y_1,y_2,x_3)=\boldsymbol{a}\,e^{-ip\eta x_3} \tag{15.4-3}$$

in which p and $\boldsymbol{a}$ satisfy the eigenrelation

$$\left\{\mathbf{Q}+p(\mathbf{R}+\mathbf{R}^T)+p^2\mathbf{T}\right\}\boldsymbol{a}=\mathbf{0}, \tag{15.4-4}$$

$$Q_{ik}=C_{ijks}n_j n_s,\quad R_{ik}=C_{ijks}n_j m_s,\quad T_{ik}=C_{ijks}m_j m_s.$$

Equation (15.4-4) is the Stroh eigenrelation for the oblique plane spanned by $\mathbf{n}$ and $\mathbf{m}$ defined in (15.4-2). The normal to the oblique plane lies on the (x_1,x_2)-plane. Let

$$t_i = \sigma_{i3} = C_{i3ks}u_{k,s}. \tag{15.4-5}$$

Application of Fourier transforms with x_1 and x_2 and use of (15.4-3) yields

$$\begin{aligned}\tilde{t}_i &= -iy_\alpha C_{i3k\alpha}\tilde{u}_k + C_{i3k3}\tilde{u}_{k,3}\\ &= -i\eta(C_{ijks}m_j n_s + pC_{ijks}m_j m_s)a_k e^{-ip\eta x_3}.\end{aligned}$$

Hence

$$\tilde{\boldsymbol{t}} = -i\eta \boldsymbol{b} e^{-ip\eta x_3} \tag{15.4-6}$$

where

$$\boldsymbol{b} = (\mathbf{R}^T + p\mathbf{T})\boldsymbol{a} = -\frac{1}{p}(\mathbf{Q} + p\mathbf{R})\boldsymbol{a}.$$

The second equality follows from (15.4-4).

To satisfy the condition that $\mathbf{u}=\mathbf{0}$ as $x_3 \to \infty$, we take the p_α with negative imaginary part. Therefore the general solutions obtained by superposing three solutions of (15.4-3) and (15.4-6) associated with $\bar{p}_1$, $\bar{p}_2$, $\bar{p}_3$ can be written as

$$\begin{aligned}\tilde{\mathbf{u}}(y_1, y_2, x_3) &= i\eta^{-1}\overline{\mathbf{A}}\left\langle e^{-i\bar{p}_*\eta x_3}\right\rangle\overline{\mathbf{q}},\\ \tilde{\mathbf{t}}(y_1, y_2, x_3) &= \overline{\mathbf{B}}\left\langle e^{-i\bar{p}_*\eta x_3}\right\rangle\overline{\mathbf{q}},\end{aligned} \tag{15.4-7}$$

where $\overline{\mathbf{q}}$ is a complex vector to be determined and

$$\left\langle e^{-i\bar{p}_*\eta x_3}\right\rangle = \mathrm{diag}[e^{-i\bar{p}_1\eta x_3},\ e^{-i\bar{p}_2\eta x_3},\ e^{-i\bar{p}_3\eta x_3}].$$

In particular when $x_3=0$,

$$\tilde{\mathbf{u}}(y_1, y_2, 0) = i\eta^{-1}\overline{\mathbf{A}}\overline{\mathbf{q}}, \quad \tilde{\mathbf{t}}(y_1, y_2, 0) = \overline{\mathbf{B}}\overline{\mathbf{q}}. \tag{15.4-8}$$

A concentrated force $\mathbf{f}$ applied at $\mathbf{x}=\mathbf{0}$ implies that

$$\mathbf{t}(x_1, x_2, 0) = -\delta(x_1)\delta(x_2)\mathbf{f},$$

or

$$\tilde{\mathbf{t}}(y_1, y_2, 0) = -\mathbf{f}.$$

By $(15.4\text{-}8)_2$ we determine the unknown vector

$$\overline{\mathbf{q}} = -\overline{\mathbf{B}}^{-1}\mathbf{f}.$$

Therefore the surface Green's function is, from (15.4-7),

$$\tilde{\mathbf{u}}(y_1, y_2, x_3) = -i\eta^{-1}\overline{\mathbf{A}}\left\langle e^{-i\bar{p}_*\eta x_3}\right\rangle \overline{\mathbf{B}}^{-1}\mathbf{f},$$
$$\tilde{\mathbf{t}}(y_1, y_2, x_3) = -\overline{\mathbf{B}}\left\langle e^{-i\bar{p}_*\eta x_3}\right\rangle \overline{\mathbf{B}}^{-1}\mathbf{f}. \tag{15.4-9}$$

The displacement at $x_3=0$ can be obtained in a real form. At $x_3=0$,

$$\tilde{\mathbf{u}}(y_1, y_2, 0) = -i\eta^{-1}\overline{\mathbf{A}}\overline{\mathbf{B}}^{-1}\mathbf{f}$$

and

$$\mathbf{u}(x_1, x_2, 0) = \frac{-i}{(2\pi)^2}\iint \frac{1}{\eta}\overline{\mathbf{A}}\overline{\mathbf{B}}^{-1}\mathbf{f}\, e^{-i(y_1x_1+y_2x_2)}dy_1dy_2. \tag{15.4-10}$$

From (15.1-8), $-i\overline{\mathbf{A}}\overline{\mathbf{B}}^{-1} = \mathbf{L}^{-1} + i\mathbf{S}\mathbf{L}^{-1}$. By letting

$$x_1 = r\cos\theta_o, \quad x_2 = r\sin\theta_o,$$

and setting $dy_1dy_2 = \eta d\theta d\eta$ Barnett and Lothe (1975a) have shown that

$$\mathbf{u}(x_1, x_2, 0) = \frac{1}{2\pi r}\left\{\mathbf{L}^{-1}[\theta_o] - \frac{1}{\pi}\int_0^{\pi} \frac{\mathbf{S}[\theta]\mathbf{L}^{-1}[\theta]}{\sin(\theta-\theta_o)}d\theta\right\}\mathbf{f}. \tag{15.4-11}$$

Here $\mathbf{S}[\theta]$ and $\mathbf{L}[\theta]$ are computed on the plane whose normal vector lies on the (x_1, x_2) plane making an angle θ with the x_1-axis. This is one of very few three-dimensional problems that have made connection with Stroh's two-dimensional formalism.

The surface Green's function for isotropic materials was obtained by Boussinesq (1885) for the force normal to the free surface.

15.5 Interfacial Green's Function

Let the half-spaces $x_3>0$ and $x_3<0$ be occupied by materials 1 and 2, respectively. They are bonded together at the interface $x_3=0$. A concentrated force $\mathbf{f}$ is applied at the origin $\mathbf{x}=\mathbf{0}$. The continuity in displacements and surface tractions at $x_3=0$ demands that

$$\mathbf{u}_1 = \mathbf{u}_2, \quad \mathbf{t}_1 - \mathbf{t}_2 = -\delta(x_1)\delta(x_2)\mathbf{f}, \quad \text{at } x_3=0, \tag{15.5-1}$$

where the subscripts 1 and 2 for $\mathbf{u}$ and $\mathbf{t}$ denote materials 1 and 2, respectively. Making use of (15.4-7) let the solution be

$$\tilde{\mathbf{u}}_1(y_1, y_2, x_3) = i\eta^{-1}\overline{\mathbf{A}}_1\left\langle e^{-i\bar{p}_*^{(1)}\eta x_3}\right\rangle \overline{\mathbf{A}}_1^{-1}\mathbf{q},$$
$$\tilde{\mathbf{t}}_1(y_1, y_2, x_3) = \overline{\mathbf{B}}_1\left\langle e^{-i\bar{p}_*^{(1)}\eta x_3}\right\rangle \overline{\mathbf{A}}_1^{-1}\mathbf{q}, \tag{15.5-2}$$

for material 1 in $x_3>0$ and

$$\tilde{\mathbf{u}}_2(y_1,y_2,x_3) = i\eta^{-1}\mathbf{A}_2\left\langle e^{-ip_*^{(2)}\eta x_3}\right\rangle \mathbf{A}_2^{-1}\mathbf{q},$$

$$\tilde{\mathbf{t}}_2(y_1,y_2,x_3) = \mathbf{B}_2\left\langle e^{-ip_*^{(2)}\eta x_3}\right\rangle \mathbf{A}_2^{-1}\mathbf{q}, \tag{15.5-3}$$

for material 2 in $x_3<0$. $\mathbf{q}$ is an unknown vector to be determined and the superscripts (1) and (2) for p_α denote materials 1 and 2, respectively. We employed $p_\alpha^{(2)}$, not $\bar{p}_\alpha^{(2)}$, in (15.5-3) so that $\tilde{\mathbf{u}}_2$ vanishes at $x_3=-\infty$. Equations $(15.5\text{-}2)_1$ and $(15.5\text{-}3)_1$ automatically satisfy the continuity condition $(15.5\text{-}1)_1$. To satisfy $(15.5\text{-}1)_2$, substitution from $(15.5\text{-}2)_2$ and $(15.5\text{-}3)_2$ yields

$$(\overline{\mathbf{B}}_1\overline{\mathbf{A}}_1^{-1} - \mathbf{B}_2\mathbf{A}_2^{-1})\mathbf{q} = -\mathbf{f}$$

or, by (15.1-8),

$$\mathbf{q} = -i(\overline{\mathbf{M}}_1 + \mathbf{M}_2)^{-1}\mathbf{f} = -i(\hat{\mathbf{D}} + i\hat{\mathbf{W}})^{-1}\mathbf{f} = -i(\breve{\mathbf{H}} - i\breve{\mathbf{W}})\mathbf{f}. \tag{15.5-4}$$

In the above,

$$\hat{\mathbf{D}} = \mathbf{H}_1^{-1} + \mathbf{H}_2^{-1}, \quad \hat{\mathbf{W}} = \mathbf{S}_1^T\mathbf{H}_1^{-1} - \mathbf{S}_2^T\mathbf{H}_2^{-1},$$

$$\breve{\mathbf{H}} = (\hat{\mathbf{D}} - \hat{\mathbf{W}}^T\hat{\mathbf{D}}^{-1}\hat{\mathbf{W}})^{-1}, \quad \breve{\mathbf{W}} = \hat{\mathbf{D}}^{-1}\hat{\mathbf{W}}\breve{\mathbf{H}} = \breve{\mathbf{H}}\hat{\mathbf{W}}\hat{\mathbf{D}}^{-1}.$$

The $\breve{\mathbf{H}}$ here is identical to the $\tilde{\mathbf{H}}$ in (8.8-29).

The displacement at $x_3=0$ can be obtained in a real form as follows. From $(15.5\text{-}3)_1$ and $(15.5\text{-}4)_3$,

$$\mathbf{u}(x_1,x_2,0) = \frac{1}{(2\pi)^2}\iint \frac{1}{\eta}(\breve{\mathbf{H}} - i\breve{\mathbf{W}})\mathbf{f}\, e^{-i(y_1x_1+y_2x_2)}dy_1dy_2.$$

Following the derivation from (15.4-10) to (15.4-11) we have

$$\mathbf{u}(x_1,x_2,0) = \frac{1}{2\pi r}\left\{\breve{\mathbf{H}}[\theta_o] + \frac{1}{\pi}\int_0^\pi \frac{\breve{\mathbf{W}}[\theta]}{\sin(\theta-\theta_o)}d\theta\right\}\mathbf{f}. \tag{15.5-5}$$

The surface traction at $x_3=0$ are, from $(15.5\text{-}2)_2$, $(15.5\text{-}3)_2$, and $(15.5\text{-}4)_2$,

$$\tilde{\mathbf{t}}_1 = -\tfrac{1}{2}\mathbf{f} + \tilde{\mathbf{t}}, \quad \tilde{\mathbf{t}}_2 = \tfrac{1}{2}\mathbf{f} + \tilde{\mathbf{t}},$$

$$\tilde{\mathbf{t}} = -\tfrac{1}{2}(\overline{\mathbf{M}}_1 - \mathbf{M}_2)(\overline{\mathbf{M}}_1 + \mathbf{M}_2)^{-1}\mathbf{f}.$$

Hence,

$$\mathbf{t}_1(x_1,x_2,0) = -\tfrac{1}{2}\delta(x_1)\delta(x_2)\mathbf{f} + \mathbf{t},$$
$$\mathbf{t}_2(x_1,x_2,0) = \tfrac{1}{2}\delta(x_1)\delta(x_2)\mathbf{f} + \mathbf{t}, \tag{15.5-6}$$

$$\mathbf{t} = \frac{-1}{8\pi^2}\iint (\overline{\mathbf{M}}_1 - \mathbf{M}_2)(\overline{\mathbf{M}}_1 + \mathbf{M}_2)^{-1}\mathbf{f}\, e^{-i(y_1x_1+y_2x_2)}dy_1dy_2. \tag{15.5-7}$$

When the two materials are identical $\breve{\mathbf{W}}=\mathbf{0}$, $\breve{\mathbf{H}}=\frac{1}{2}\mathbf{H}=i\mathbf{A}\mathbf{A}^T=-i\overline{\mathbf{A}}\overline{\mathbf{A}}^T$ and (15.5-5) reduces to

$$\mathbf{u}(x_1,x_2,0) = \frac{1}{4\pi r}\mathbf{H}[\theta_o]\mathbf{f}.$$

This agrees with the result obtained in (15.3-6) when the latter is evaluated at $x_3=0$. Equations (15.5-2) and (15.5-3) are simplified to, with the use of $(15.5\text{-}4)_2$,

$$\tilde{\mathbf{u}}(y_1,y_2,x_3) = -i\eta^{-1}\overline{\mathbf{A}}\left\langle e^{-i\bar{p}_*\eta x_3}\right\rangle\overline{\mathbf{q}}^{\infty},$$
$$\tilde{\mathbf{t}}(y_1,y_2,x_3) = -\overline{\mathbf{B}}\left\langle e^{-i\bar{p}_*\eta x_3}\right\rangle\overline{\mathbf{q}}^{\infty}, \tag{15.5-8}$$

for $x_3>0$ and

$$\tilde{\mathbf{u}}(y_1,y_2,x_3) = i\eta^{-1}\mathbf{A}\left\langle e^{-ip_*\eta x_3}\right\rangle\mathbf{q}^{\infty},$$
$$\tilde{\mathbf{t}}(y_1,y_2,x_3) = \mathbf{B}\left\langle e^{-ip_*\eta x_3}\right\rangle\mathbf{q}^{\infty}, \tag{15.5-9}$$

for $x_3<0$ where

$$\mathbf{q}^{\infty} = \mathbf{A}^T\mathbf{f}. \tag{15.5-10}$$

Equations (15.5-8) and (15.5-9) are an alternate expression for the Green's function for the infinite space. As we will see in the next two sections they are more convenient in treating Green's functions for half-spaces and bimaterials.

15.6 Green's Function for Half-Space

In an anisotropic elastic half-space $x_3>0$ let a concentrated force $\mathbf{f}$ be applied at

$$\mathbf{x} = (0,\ 0,\ d), \quad d>0. \tag{15.6-1}$$

We will employ (15.5-8) and (15.5-9) to construct the Green's function. Let

$$\tilde{\mathbf{u}}(y_1,y_2,x_3) = -i\eta^{-1}\overline{\mathbf{A}}\left\langle e^{-i\bar{p}_*\eta(x_3-d)}\right\rangle\overline{\mathbf{q}}^{\infty} - i\eta^{-1}\overline{\mathbf{A}}\left\langle e^{-i\bar{p}_*\eta x_3}\right\rangle\overline{\mathbf{q}}, \quad (15.6\text{-}2)_1$$

$$\tilde{\mathbf{t}}(y_1,y_2,x_3) = -\overline{\mathbf{B}}\left\langle e^{-i\bar{p}_*\eta(x_3-d)}\right\rangle\overline{\mathbf{q}}^\infty - \overline{\mathbf{B}}\left\langle e^{-i\bar{p}_*\eta x_3}\right\rangle\overline{\mathbf{q}}, \qquad (15.6\text{-}2)_2$$

for $x_3>d$ and

$$\begin{aligned}\tilde{\mathbf{u}}(y_1,y_2,x_3) &= i\eta^{-1}\mathbf{A}\left\langle e^{-ip_*\eta(x_3-d)}\right\rangle\mathbf{q}^\infty - i\eta^{-1}\overline{\mathbf{A}}\left\langle e^{-i\bar{p}_*\eta x_3}\right\rangle\overline{\mathbf{q}},\\ \tilde{\mathbf{t}}(y_1,y_2,x_3) &= \mathbf{B}\left\langle e^{-ip_*\eta(x_3-d)}\right\rangle\mathbf{q}^\infty - \overline{\mathbf{B}}\left\langle e^{-i\bar{p}_*\eta x_3}\right\rangle\overline{\mathbf{q}},\end{aligned} \qquad (15.6\text{-}3)$$

for $0 \le x_3 < d$ where $\mathbf{q}^\infty$ is defined in (15.5-10) and $\overline{\mathbf{q}}$ is to be determined. The first terms in (15.6-2) and (15.6-3) are the Green's function for the infinite space due to the concentrated force $\mathbf{f}$ applied at the location (15.6-1). The second terms are added to satisfy the boundary conditions at $x_3=0$.

From (15.6-3) $\tilde{\mathbf{u}}$ and $\tilde{\mathbf{t}}$ at $x_3=0$ are

$$\begin{aligned}\tilde{\mathbf{u}}(y_1,y_2,0) &= i\eta^{-1}\left\{\mathbf{A}\left\langle e^{ip_*\eta d}\right\rangle\mathbf{q}^\infty - \overline{\mathbf{A}}\overline{\mathbf{q}}\right\},\\ \tilde{\mathbf{t}}(y_1,y_2,0) &= \mathbf{B}\left\langle e^{ip_*\eta d}\right\rangle\mathbf{q}^\infty - \overline{\mathbf{B}}\overline{\mathbf{q}}.\end{aligned} \qquad (15.6\text{-}4)$$

When the surface $x_3=0$ is traction-free $(15.6\text{-}4)_2$ gives

$$\overline{\mathbf{q}} = \overline{\mathbf{B}}^{-1}\mathbf{B}\left\langle e^{ip_*\eta d}\right\rangle\mathbf{q}^\infty. \qquad (15.6\text{-}5)$$

Equation $(15.6\text{-}4)_1$ then reduces to

$$\tilde{\mathbf{u}}(y_1,y_2,0) = i\eta^{-1}\left(\mathbf{A}\mathbf{B}^{-1} - \overline{\mathbf{A}}\overline{\mathbf{B}}^{-1}\right)\mathbf{B}\left\langle e^{ip_*\eta d}\right\rangle\mathbf{q}^\infty,$$

or, by (15.1-8),

$$\tilde{\mathbf{u}}(y_1,y_2,0) = 2\eta^{-1}\mathbf{L}^{-1}\tilde{\mathbf{t}}^\infty(y_1,y_2,0) \qquad (15.6\text{-}6)$$

where $\tilde{\mathbf{t}}^\infty(y_1,y_2,x_3)$ represents the first term on the right of $(15.6\text{-}3)_2$. The Fourier inverse of (15.6-6) provides the displacements at the free surface in terms of the tractions at $x_3=0$ of the Green's function for the infinite space with the force $\mathbf{f}$ applied at $\mathbf{x}=(0,0,d)$.

If the surface $x_3=0$ is a rigid surface, $(15.6\text{-}4)_1$ gives

$$\overline{\mathbf{q}} = \overline{\mathbf{A}}^{-1}\mathbf{A}\left\langle e^{ip_*\eta d}\right\rangle\mathbf{q}^\infty, \qquad (15.6\text{-}7)$$

and $(15.6\text{-}4)_2$ can be written as

$$\tilde{\mathbf{t}}(y_1,y_2,0) = 2\eta\mathbf{H}^{-1}\tilde{\mathbf{u}}^\infty(y_1,y_2,0). \qquad (15.6\text{-}8)$$

$\tilde{\mathbf{u}}^\infty(y_1,y_2,x_3)$ denotes the first term on the right of $(15.6\text{-}3)_1$. The

Fourier inverse of (15.6-8) furnishes tractions at the rigid surface $x_3=0$ in terms of the displacements at $x_3=0$ of the Green's function for the infinite space with the force $\mathbf{f}$ applied at $\mathbf{x}=(0,0,d)$.

The Green's function for an isotropic half-space was obtained by Mindlin (1936, 1953) for a free surface and by Rongved (1955a) for a fixed boundary. Lee (1979) and Pan and Chou (1979a) give explicit expressions of the Green's function for a transversely isotropic material whose axis of symmetry is normal to the free-surface.

15.7 Green's Function for Bimaterials

Let the half-spaces $x_3>0$ and $x_3<0$ be occupied by materials 1 and 2, respectively. They are bonded together at the interface $x_3=0$. A concentrated force $\mathbf{f}$ is applied at

$$\mathbf{x}=(0,\ 0,\ d),\quad d>0. \tag{15.7-1}$$

The continuity in displacements and surface tractions at $x_3=0$ imply that

$$\mathbf{u}_1=\mathbf{u}_2,\quad \mathbf{t}_1=\mathbf{t}_2,\quad \text{at } x_3=0. \tag{15.7-2}$$

For the material 1 in $x_3>0$ we employ the solution given in (15.6-2) and (15.6-3) in which p, $\mathbf{q}$, $\mathbf{A}$, $\mathbf{B}$ are replaced by $p^{(1)}$, $\mathbf{q}_1$, $\mathbf{A}_1$, $\mathbf{B}_1$, respectively. For material 2 in $x_3<0$ let the solution be

$$\begin{aligned}\tilde{\mathbf{u}}_2(y_1,y_2,x_3)&=i\eta^{-1}\mathbf{A}_2\left\langle e^{-ip_*^{(2)}\eta x_3}\right\rangle\mathbf{q}_2,\\ \tilde{\mathbf{t}}_2(y_1,y_2,x_3)&=\mathbf{B}_2\left\langle e^{-ip_*^{(2)}\eta x_3}\right\rangle\mathbf{q}_2,\end{aligned} \tag{15.7-3}$$

where $\mathbf{q}_2$ is an unknown vector to be determined. Substitution of (15.6-3) and (15.7-3) into (15.7-2) yields

$$\begin{aligned}\mathbf{A}_1\left\langle e^{ip_*^{(1)}\eta d}\right\rangle\mathbf{q}^\infty-\overline{\mathbf{A}}_1\overline{\mathbf{q}}_1&=\mathbf{A}_2\mathbf{q}_2,\\ \mathbf{B}_1\left\langle e^{ip_*^{(1)}\eta d}\right\rangle\mathbf{q}^\infty-\overline{\mathbf{B}}_1\overline{\mathbf{q}}_1&=\mathbf{B}_2\mathbf{q}_2.\end{aligned} \tag{15.7-4}$$

Equations (15.7-4) can be solved for $\overline{\mathbf{A}}_1\overline{\mathbf{q}}_1$, $\mathbf{A}_2\mathbf{q}_2$, or $\overline{\mathbf{B}}_1\overline{\mathbf{q}}_1$, $\mathbf{B}_2\mathbf{q}_2$. To solve for $\overline{\mathbf{A}}_1\overline{\mathbf{q}}_1$, $\mathbf{A}_2\mathbf{q}_2$, $(15.7\text{-}4)_2$ is replaced by

$$(\mathbf{B}_1\mathbf{A}_1^{-1})\mathbf{A}_1\left\langle e^{ip_*^{(1)}\eta d}\right\rangle\mathbf{q}^\infty-(\overline{\mathbf{B}}_1\overline{\mathbf{A}}_1^{-1})\overline{\mathbf{A}}_1\overline{\mathbf{q}}_1=(\mathbf{B}_2\mathbf{A}_2^{-1})\mathbf{A}_2\mathbf{q}_2.$$

By virtue of (15.1-8) it can be shown that

$$\overline{\mathbf{A}}_1\overline{\mathbf{q}}_1 = -(\overline{\mathbf{M}}_1 + \mathbf{M}_2)^{-1}(\mathbf{M}_1 - \mathbf{M}_2)\,\mathbf{A}_1\left\langle e^{ip_*^{(1)}\eta d}\right\rangle\mathbf{q}^\infty,$$
$$\mathbf{A}_2\mathbf{q}_2 = (\overline{\mathbf{M}}_1 + \mathbf{M}_2)^{-1}(\mathbf{M}_1 + \overline{\mathbf{M}}_1)\,\mathbf{A}_1\left\langle e^{ip_*^{(1)}\eta d}\right\rangle\mathbf{q}^\infty. \tag{15.7-5}$$

To solve for $\overline{\mathbf{B}}_1\overline{\mathbf{q}}_1$, $\mathbf{B}_2\mathbf{q}_2$, $(15.7\text{-}4)_1$ is replaced by

$$(\mathbf{A}_1\mathbf{B}_1^{-1})\mathbf{B}_1\left\langle e^{ip_*^{(1)}\eta d}\right\rangle\mathbf{q}^\infty - (\overline{\mathbf{A}}_1\overline{\mathbf{B}}_1^{-1})\overline{\mathbf{B}}_1\overline{\mathbf{q}}_1 = (\mathbf{A}_2\mathbf{B}_2^{-1})\mathbf{B}_2\mathbf{q}_2,$$

and one obtains

$$\overline{\mathbf{B}}_1\overline{\mathbf{q}}_1 = -(\overline{\mathbf{M}}_1^{-1} + \mathbf{M}_2^{-1})^{-1}(\mathbf{M}_1^{-1} - \mathbf{M}_2^{-1})\,\mathbf{B}_1\left\langle e^{ip_*^{(1)}\eta d}\right\rangle\mathbf{q}^\infty,$$
$$\mathbf{B}_2\mathbf{q}_2 = (\overline{\mathbf{M}}_1^{-1} + \mathbf{M}_2^{-1})^{-1}(\mathbf{M}_1^{-1} + \overline{\mathbf{M}}_1^{-1})\,\mathbf{B}_1\left\langle e^{ip_*^{(1)}\eta d}\right\rangle\mathbf{q}^\infty. \tag{15.7-6}$$

As in (15.6-6) and (15.6-8) $\tilde{\mathbf{u}}$ and $\tilde{\mathbf{t}}$ at $x_3=0$ can be expressed in terms of $\tilde{\mathbf{u}}_1^\infty$ and $\tilde{\mathbf{t}}_1^\infty$ at $x_3=0$. From $(15.7\text{-}5)_2$ and $(15.7\text{-}6)_2$ we have

$$\tilde{\mathbf{u}}(y_1, y_2, 0) = (\overline{\mathbf{M}}_1 + \mathbf{M}_2)^{-1}(\mathbf{M}_1 + \overline{\mathbf{M}}_1)\,\tilde{\mathbf{u}}_1^\infty(y_1, y_2, 0),$$
$$\tilde{\mathbf{t}}(y_1, y_2, ,0) = (\overline{\mathbf{M}}_1^{-1} + \mathbf{M}_2^{-1})^{-1}(\mathbf{M}_1^{-1} + \overline{\mathbf{M}}_1^{-1})\,\tilde{\mathbf{t}}_1^\infty(y_1, y_2, 0). \tag{15.7-7}$$

The Green's function for isotropic bimaterials was obtained by Rongved (1955b) and Aderogba (1977). The allied problem of isotropic bimaterials with a slippery interface was considered by Dundurs and Hetenyi (1965). Pan and Chou (1979b) have obtained the Green's function for transversely anisotropic bimaterials whose axes of symmetry are normal to the interface.

15.8 Remarks

The results presented in this chapter are rather rudimentary in that, with a few exceptions, the solutions are in terms of the Fourier transform. Fourier inverse transform has to be carried out to recover physical quantities. Nevertheless we have shown the potential for extending the Stroh formalism to three-dimensional anisotropic elasticity problems. Recently, Wang and Achenbach (1993) have employed the Radon transform to obtain elastodynamics Green's functions for three-dimensional anisotropic elastic materials due to a point source. The *Radon transform* (see John 1955; Courant and Hilbert, 1962; Helgason, 1980; and Deans, 1983) is an integration of $f(\mathbf{x})$ over all planes defined by $\mathbf{n}\cdot\mathbf{x}=s$ where $\mathbf{n}$ is a unit vector. The inverse Radon transform is an integration over a unit sphere.

Not many analytical solutions for three-dimensional elasticity problems are available for isotropic materials, much less for anisotropic elastic materials. The reader is referred to the book by Mura (1987) that has extensive references on three-dimensional problems, and to a recent book by Barber (1992) on *Elasticity*. For more recent work on three-dimensional anisotropic elasticity problems, we mention the work by Yu and Sanday (1994) and Yu et al. (1994, 1995), and the references listed therein.

REFERENCES

Abbudi, M., and Barnett, D. M. (1990) "On the existence of interfacial (Stoneley) waves in bonded piezoelectric half-spaces," *Proc. Roy. Soc. London* A**429**, 587–611.

Achenbach, J. D., Keer, L. M., Khetan, R. P., and Chen, H. S. (1979) "Loss of adhesion at the tip of an interface crack," *J. Elasticity* **9**, 397–424.

Aderogba, K. (1977) "On eigenstresses in dissimilar media," *Phil. Mag.* **35**, 281–292.

Akroz, A. Y., and Tauchert, T. R. (1972) "Thermal stresses in an orthotropic elastic semispace," *J. Appl. Mech.* **39**, 87–90.

Al'shits, V. I., and Lothe, J. (1979a) "Elastic waves in triclininc crystals. I. general theory and the degeneracy problem," *Sov. Phys. Crystallogr.* **24**, 387–392.

Al'shits, V. I., and Lothe, J. (1979b) "Elastic waves in triclininc crystals. II. topology of polarization fields and some general theorems," *Sov. Phys. Crystallogr.* **24**, 393–398.

Al'shits, V. I., and Lothe, J. (1979c) "Elastic waves in triclininc crystals. III. the problem of existence of exceptional surface waves and some of their general properties," *Sov. Phys. Crystallogr.* **24**, 644–648.

Al'shits, V. I., and Lothe, J. (1981) "Comments on the relation between surface wave theory and the theory of reflection," *Wave Motion* **3**, 297–310.

Al'shits, V. I., Darinskii, A. N., and Lothe, J. (1992) "On the existence of interface waves in half-infinite anisotropic elastic media with piezoelectric and piezomagnetic properties," *Wave Motion* **16**, 265–283.

Antman, S. S. (1995) *Nonlinear Problems of Elasticity*, Springer-Verlag, New York.

Asaro, R. J., Hirth, J. P., Barnett, D. M., and Lothe, J. (1973) "A further synthesis of sextic and integral theories for dislocations and line forces in anisotropic media," *Phys. Status Solidi* B**60**, 261–271.

Atkinson, C. (1973) "Some ribbon-like inclusion problems," *Int. J. Eng. Sci.* **11**, 243–265.

Atkinson, C. (1982a) "The interface crack with a contact zone (an analytical treatment)," *Int. J. Fracture* **18**, 161–177.

Atkinson, C. (1982b) "The interface crack with a contact zone (the crack of finite length)," *Int. J. Fracture* **19**, 131–138.

Atkinson, C., and Clements, D. L. (1977) "On some crack problems in anisotropic thermoelasticity," *Int. J. Solids Structures* **13**, 855–864.

Auld, B. A. (1973) *Acoustic Fields and Waves in Solids*, Vol. I, Wiley, New York, 73–75.

Bacon, D. J., Barnett, D. M., and Scattergood, R. O. (1978) "The anisotropic continuum theory of lattice defects," *Prog. Mater. Sci.* **23**, 51–262.

Ballarini, R. (1987) "An integral equation approach for rigid line inhomogeneity problems," *Int. J. Fracture* **33**, R23–R26.

Ballarini, R. (1995) "A certain mixed boundary value problem for a bimaterial interface," *Int. J. Solid Structures* **32**, 279-289.

Barber, J. R. (1992) *Elasticity*, Kluwer. Dordrecht, Netherlands.

Barenblatt, G. I. (1962) "The mathematical theory of equilibrium cracks in brittle fracture," *Adv. Appl. Mech.* **7**, 55–129.

Barnett, D. M. (1969) "Edge dislocation stress concentrations in two-phase media," *Scripta Metall.* **3**, 543–546.

Barnett, D. M. (1992) "On the non-existence of subsonic boundary-polarized two-component free surface waves," in *The Jens Lothe Symposium Volume, Physica Scripta* T**44**, 98–103.

Barnett, D. M., and Asaro, R. J. (1972) "The fracture mechanics of slit-like cracks in anisotropic elastic media," *J. Mech. Phys. Solids* **20**, 353–366.

Barnett, D. M., and Chadwick, P. (1991) "The existence of one-component surface waves and exceptional subsequent transonic states of types 2, 4, and E1 in anisotropic elastic media," in *Modern Theory of Anisotropic Elasticity and Applications*, J. J. Wu, T. C. T. Ting and D. M. Barnett, eds., SIAM Proceedings Series, SIAM, Philadelphia, 199–214.

Barnett, D. M., and Lothe, J. (1973) "Synthesis of the sextic and the integral formalism for dislocations, Greens function and surface waves in anisotropic elastic solids," *Phys. Norv.* **7**, 13–19.

Barnett, D. M., and Lothe, J. (1974a) "Consideration of the existence of surface wave (Rayleigh wave) solutions in anisotropic elastic crystals," *J. Phys. F.* **4**, 671–686.

Barnett, D. M., and Lothe, J. (1974b) "An image force theorem for dislocations in bicrystals," *J. Phys. F.* **4**, 1618–1635.

Barnett, D. M., and Lothe, J. (1975a) "Line force loadings on anisotropic half-spaces and wedges," *Phys. Norv.* **8**, 13–22.

Barnett, D. M., and Lothe, J. (1975b) "Dislocations and line charges in anisotropic piezoelectric insulators," *Phys. Status Solidi* B**67**, 105–111.

Barnett, D. M., and Lothe, J. (1978) "The existence of Rayleigh (surface) wave solutions in anisotropic elastic half-spaces," in *Modern Problems in Elastic Wave Propagation*, J. Miklowitz and J. D. Achenbach, eds., Wiley, New York, 445–457.

Barnett, D. M., and Lothe, J. (1985) "Free surface (Rayleigh) waves in anisotropic elastic half-spaces: The surface impedance methods," *Proc. Roy. Soc. London* A**402**, 135–152.

Barnett, D. M., and Lothe, J. (1987) "On the existence of Type 6 transonic states in linear elastic media of cubic symmetry," *Proc. Roy. Soc. London* A**411**, 251–263.

Barnett, D. M., and Lothe, J. (1989) "Surface wave existence theory for the case of zero curvature transonic states," in *Elastic Wave Propagation*, M. F. McCarthy and M. A. Hayes, eds., North–Holland, Amsterdam, 33–38.

Barnett, D. M., and Lothe, J. (1992) "Conditions admitting the existence of Type 3 transonic states in stable anisotropic linear elastic media," in *The P. Chadwick Symposium Volume, J. Mech. Phys. Solids* **40**, 1433–1448.

Barnett, D. M., and Lothe, J. (1995) "Mutual attraction of a dislocation to a bi-metallic interface and a theorem on proportional anisotropic bimaterials," *Int. J. Solids Structures* **32**, 291-301.

Barnett, D. M., and Swanger, L. A. (1971) "The elastic energy of a straight dislocation in an infinite anisotropic elastic medium," *Phys. Status Solidi* B**48**, 419–428.

Barnett, D. M., and Tetelman, A. S. (1966) "The stress distribution produced by screw dislocation pile-up at rigid circular cylindrical inclusions," *J. Mech. Phys. Solids* **14**, 329–348.

Barnett, D. M., and Tetelman, A. S. (1967) "The stresses produced by a screw dislocation pile-up at a circular inclusion of finite rigidity," *Can. J. Phys.* **45**, 841–863.

Barnett, D. M., Chadwick, P., and Lothe, J. (1991) "The behaviour of elastic surface waves polarized in a plane of material symmetry. I. Addendum," *Proc. Roy. Soc. London* A**433**, 699–710.

Barnett, D. M., Gavazza, S. D., and Lothe, J. (1988) "Slip waves along the interface between two anisotropic elastic half-spaces in sliding contact," *Proc. Roy. Soc. London* A**415**, 389–419.

Barnett, D. M., Lothe, J., and Gundersen, S. A. (1990) "Zero curvature transonic states and free surface waves in anisotropic elastic media," *Wave Motion*, **12**, 341–360.

Barnett, D. M., Lothe, J., Musgrave, M. J. P., and Gavazza, S. D. (1985) "Interfacial (Stoneley) waves in bonded anisotropic elastic half-spaces," *Proc. Roy. Soc. London* A**402**, 153–166.

Barnett, D. M., Lothe, J., Nishioka, K., and Asaro, R. J. (1973) "Elastic surface waves in anisotropic crystals: A simplified method for calculating Rayleigh wave velocities using dislocation theory," *J. Phys. F.*, **3**, 1083–1096.

Bary, N. K. (1964) *A Treatise on Trigonometric Series*, Macmillan, New York, 51–52.

Bassani, J. L., and Qu, Jianmin (1989) "Finite cracks on bimaterial and bicrystal interfaces," *J. Mech. Phys. Solids* **37**, 435–453.

Bazant, Z. P. (1974) "Three-dimensional harmonic functions near termination or intersection of gradient singularity lines: a general numerical method," *Int. J. Eng. Sci.* **12**, 221–243.

Bazant, Z. P., and Estenssoro, L. F. (1977) "General numerical method for three dimensional singularities in cracks or notched elastic solids," *Fracture 1977, Proc 4th Int. Conf. Fracture*, D. M. R. Taplin, ed., Univ. Waterloo, Ontario, Canada, **3**, 371–385.

Bazant, Z. P., and Estenssoro, L. F. (1979) "Surface singularity and crack propagation," *Int. J. Solids Structures* **15**, 405–426.

Bazant, Z. P., and Keer, L. M. (1974) "Singularities of elastic stresses and of harmonic functions at conical notches and inclusion," *Int. J. Solids Structure* **10**, 957–964.

Belov, A. Y., and Kirchner, H. O. K. (1995) "Critical angles in bending of rotationally inhomogeneous elastic wedges," *J. Appl. Mech.* **62**, 429-440.

Belov, A. Y., Chamrov, V. A., Indenbom, V. L., and Lothe, J. (1983) "Elastic fields of dislocations piercing the interface of an anisotropic bicrystal," *Phys. Status Solidi* **B119**, 565–578.

Benthem, J. P. (1977) "State of stress at the vertex of a quarter-infinite crack in a half-space," *Int. J. Solids Structure* **13**, 479–492.

Benthem, J. P. (1980) "The quarter-infinite crack in a half-space: alternative and additional solutions," *Int. J. Solids Structure* **16**, 119–130.

Bhargava, R. D., and Radhakrishna, H. C. (1964) "Elliptic inclusion in orthotropic medium," *J. Phys. Soc. Japan* **19**, 396–405.

Bilby, B. A. (1960) "Continuous distributions of dislocations," in *Progress in Solid Mechanics*, **1**, I. N. Sneddon and R. Hill, eds., North–Holland, 329–398.

Bilby, B. A., and Gardner, L. R. T. (1958) "Continuous distributions of dislocations V. Twisting under conditions of single glide," *Proc. Roy. Soc. London* A**247**, 92–108.

Bilby, B. A., and Smith, E. (1956) "Continuous distributions of dislocations III," *Proc. Roy. Soc. London* A**236**, 481–505.

Bilby, B. A., Bullough, R., Gardner, L. R. T., and Smith, E. (1958) "Continuous distributions of dislocations IV. Single glide and plane strain," *Proc. Roy. Soc. London* A**244**, 538–557.

Bilby, B. A., Bullough, R., and Smith, E. (1955) "Continuous distributions of dislocations: A new application of the methods of non-Riemannian geometry," *Proc. Roy. Soc. London* A**231**, 263–273.

Bogy, D. B. (1970) "On the problem of edge-bonded elastic quarter plane loaded at the boundary," *Int. J. Solids Structures* **6**, 1287–1313.

Bogy, D. B. (1971a) "Two edge-bonded elastic wedges of different materials and wedge angles under surface tractions," *J. Appl. Mech.* **38**, 377–386.

Bogy, D. B. (1971b) "On the plane elastostatics problem of a loaded crack terminating at a material interface," *J. Appl. Mech.*

38, 911–918.

Bogy, D. B. (1972) "The plane solution for anisotropic elastic wedges under normal and shear loading," *J. Appl. Mech.* **39**, 1103–1109.

Bogy, D. B. (1974) "Plane solutions for traction problems on orthotropic unsymmetrical wedges and symmetrically twinned wedges," *J. Appl. Mech.* **41**, 203–208.

Bogy, D. B., and Wang, K. C. (1971) "Stress singularities at interface corners in bonded dissimilar isotropic elastic materials," *Int. J. Solids Structures* **7**, 993–1005.

Boley, B. A., and Weiner, J. H. (1960) *Theory of Thermal Stresses*, John Wiley, New York.

Bond, W. L. (1943) *Bell System Tech. J.* **22**, 1–72.

Boulanger, Ph., and Hayes, M. (1991) "Bivectors and inhomogeneous plane waves in anisotropic elastic bodies," in *Modern Theory of Anisotropic Elasticity and Applications*, J. J. Wu, T. C. T. Ting and D. M. Barnett, eds., SIAM Proceedings Series, SIAM, Philadelphia, 280–289.

Boussinesq, J. (1885) *Applications des Potentiels a l'etude de l'equilibre et du mouvement des solides elastiques.* Gauthier–Villars, Paris.

Bowie, O. E., and Freese, C. E. (1976) "On the overlapping problem in a crack analysis," *Eng. Frac. Mech.* 373–379.

Braekhus, J., and Lothe, J. (1971) "Dislocations at and near planar interfaces," *Phys. Status Solidi* B**43**, 651–657.

Braga, A. M. B., and Herrmann, G. (1988) "Plane waves in anisotropic layered composites," in *Symp. on Wave Propagation in Structural Composites*, A. K. Mal and T. C. T. Ting, eds., ASME, AMD-Vol. **90**, 81–98.

Budiansky, B., and Hutchinson, J. W. (1978) "Analysis of closure in fatigue crack growth," *J. Appl. Mech.* **45**, 267–276.

Bullough, R., and Bilby, B. A. (1954) "Uniformly moving dislocations in anisotropic media," *Proc. Phys. Soc.* **B**67, 615–624.

Burridge, R. (1967) "Lacunas in two-dimensional wave propagation," *Proc. Camb. Phil. Soc.* **63**, 819–825.

Captain, V. S., and Chadwick, P. (1986) "Surface waves in a constrained elastic body," *Q. J. Mech. Appl. Math.* **39**, 309–325.

Carlson, D. E. (1972) *Linear thermoelasticity*, Handbuch der Physik, VIa/2, Springer, Berlin.

Carothers, S. D. (1912) "Plane strain in a wedge," *Proc. Roy. Soc.* Edingburgh **23**, 292–306.

Carslaw, H. S., and Jaeger, J. C. (1959) *Conduction of Heat in Solids*, Clarendon Press, Oxford.

Chadwick, P. (1976a) *Continuum Mechanics. Concise Theory and Problems.* Allen & Unwin, London.

Chadwick, P. (1976b) "The existence of pure surface modes in elastic materials with orthorhombic symmetry," *J. Sound Vib.*, **47**, 39–52.

Chadwick, P. (1988) "Aspects of the behaviour of surface waves in transversely isotropic elastic media," in *Recent Developments in Surface Acoustic Waves*, D. F. Parker and G. A. Maugin, eds., Springer, Berlin, 80–90.

Chadwick, P. (1989a) "Wave propagation in transversely isotropic elastic media. I Homogeneous plane waves," *Proc. Roy. Soc. London* A**422**, 23–66.

Chadwick, P. (1989b) "Wave propagation in transversely isotropic elastic media. II Surface waves," *Proc. Roy. Soc. London* A**422**, 67–101.

Chadwick, P. (1989c) "Wave propagation in transversely isotropic elastic media. III The special case $a_5 = 0$ and the inextensible limit," *Proc. Roy. Soc. London* A**422**, 103–121.

Chadwick, P. (1989d) "Recent developments in the theory of elastic surface and interfacial waves," in *Elastic Wave Propagation*, M. F. McCarthy and M. A. Hayes, eds., North–Holland, Amsterdam, 3–16.

Chadwick, P. (1990) "The behaviour of elastic surface waves polarized in a plane of material symmetry I. General analysis," *Proc. Roy. Soc. London* A**430**, 213–240.

Chadwick, P. (1992) "Some remarks on the existence of one-component surface waves in elastic materials with symmetry," in *The Jens Lothe Symposium Volume, Physica Scripta* T**44**, 94–97.

Chadwick, P. (1993) "Wave propagation in incompressible transversely isotropic elastic media I. Homogeneous plane waves," *Proc Roy. Irish Acad.* **93**A, 231-253.

Chadwick, P. (1994) "Wave propagation in incompressible transversely isotropic elastic media II. Inhomogeneous plane waves," *Proc Roy. Irish Acad.* **94**A, 85-104.

Chadwick, P., and Currie, P. K. (1974) "Stoneley waves at an interface between elastic crystals," *Q. J. Mech. Appl. Math.* **27**, 497–503.

Chadwick P., and Jarvis, D. A. (1979a) "Interfacial waves in a pre-strained neo-Hookean body I. Biaxial states of strain," *Q. J. Mech. Appl. Math.* **32**, 387–399.

Chadwick P., and Jarvis, D. A. (1979b) "Interfacial waves in a pre-strained neo-Hookean body II. Triaxial states of strain," *Q. J. Mech. Appl. Math.* **32**, 387–399.

Chadwick P., and Jarvis, D. A. (1979c) "Surface waves in a pre-stressed elastic body," *Proc. Roy. Soc. London* A**366**, 507–536.

Chadwick, P., and Smith, G. D. (1977) "Foundations of the theory of surface waves in anisotropic elastic materials," *Adv. Appl. Mech.* **17**, 303–376.

Chadwick, P., and Smith, G. D. (1982) "Surface waves in cubic elastic materials," in *Mechanics of Solids, The Rodney Hill 60th Anniversary Volume*, H. G. Hopkins and M. J. Sewell, eds. Pergamon, Oxford, 47–100.

Chadwick, P., and Ting, T. C. T. (1987) "On the structure and invariance of the Barnett–Lothe tensors," *Q. Appl. Math.* **45**, 419–427.

Chadwick, P., and Wilson, N. J. (1992a) "The behaviour of elastic surface waves polarized in a plane of material symmetry II. Monoclinic media," *Proc. Roy. Soc. London* A**438**, 207–223.

Chadwick, P., and Wilson, N. J. (1992b) "The behaviour of elastic surface waves polarized in a plane of material symmetry III. Orthorhombic and cubic media," *Proc. Roy. Soc. London* A**438**, 225–247.

Chadwick, P., and Wilson, N. J. (1994) "Wave propagation in incompressible transversely isotropic elastic media III. Surface waves," *Proc Roy. Irish Acad.* **94**A, 255-282.

Chen, W. T. (1967a) "Plane thermal stress at an insulated hole under uniform heat flow in an orthotropic medium," *J. Appl. Mech.* **34**, 133–136.

Chen, W. T. (1967b) "On an elliptic elastic inclusion in an anisotropic medium," *Q. J. Mech. Appl. Math.* **20**, 307–313.

Cherkaev, A. V., Lurie, K. A., and Milton, G. (1992) "Invariant properties of the stress in plane elasticity and equivalence classes of composites," *Proc. Roy. Soc. London* A**438**, 519-529.

Choi, I., and Horgan, C. O. (1977) "Saint-Venant's principle and end effects in anisotropic elasticity," *J. Appl. Mech.* **44**, 424–430.

Choi, I., and Horgan, C. O. (1978) "Saint-Venant end effects for plane deformation of sandwich strips," *Int. J. Solids Structures* **14**, 187–195.

Chou, Y. T. (1966) "On dislocation-boundary interaction in an anisotropic aggregate," *Phys. Status Solidi* B**15**, 123–127.

Christensen, R. M. (1979) *Mechanics of Composite Materials*, Wiley, New York.

Chung, M. Y., and Ting, T. C. T. (1995a) "Line force, charge, and dislocation in anisotropic piezoelectric composite wedges and spaces," *J. Appl. Mech.* **62**, 423-428.

Chung, M. Y., and Ting, T. C. T. (1995b) "Line force, charge, and dislocation in angularly inhomogeneous anisotropic piezoelectric wedges and spaces," *Phil. Mag.* A**71**, 1335-1343.

Churchill, R. V.(1948) *Introduction to Complex Variables and Applications*, McGraw-Hill, New York.

Clements, D. L. (1971) "A crack between dissimilar anisotropic media," *Int. J. Eng. Sci.* **9**, 257–265.

Clements, D. L. (1973) "Thermal stress in an anisotropic elastic half-space," *SIAM J. Appl. Math.* **24**, 332–337.

Comninou, M. (1977a) "A property of interface dislocations," *Phil. Mag.* **36**, 1281–1283.
Comninou, M. (1977b) "The interface crack," *J. Appl. Mech.* **44**, 631–636.
Comninou, M. (1978) "The interface crack in a shear field," *J. Appl. Mech.* **45**, 287–290.
Comninou, M., and Dundurs, J. (1979a) "On the frictional contact in crack analysis," *Eng. Frac. Mech.* **12**, 117–123.
Comninou, M., and Dundurs, J. (1979b) "A closed crack tip terminating at an interface," *J. Appl. Mech.* **46**, 97–100.
Comninou, M., and Dundurs, J. (1980) "On the behavior of interface cracks," *Res. Mechanica* **1**, 249–264.
Comninou, M., and Schmueser, D. (1979) "The interface crack in a combined tension-compression and shear field," *J. Appl. Mech.* **46**, 345–358.
Cook, T. S., and Erdogan, F. (1972) "Stresses in bonded materials with a crack perpendicular to the interface," *Int. J. Eng. Sci.* **10**, 677–697.
Cottrell, A. H. (1949) *Dislocations and Plastic Flow in Crystals*, Clarendon Press.
Courant, R., and Hilbert, D. (1962) *Methods of mathematical Physics*, Vol. II, Interscience, New York.
Cowin, S. C. (1989) "Properties of the anisotropic elasticity tensor," *Q, J. Mech. Appl. Math.* **42**, 249–266.
Cowin, S. C., and Mehrabadi, M. M. (1987) "On the identification of material symmetry for anisotropic elastic materials," *Q. J. Mech. Appl. Math.* **40**, 451–476.
Cowin, S. C., Mehrabadi, M. M., and Sadegh, A. M. (1992) "Kelvin's formulation of the anisotropic Hook's law," in *Modern Theory of Anisotropic Elasticity and Applications*, J. J. Wu, T. C. T. Ting and D. M. Barnett, eds., SIAM Proceedings Series, SIAM, Philadelphia, 340–356.
Crafter, E. C., Heise, R. M., Horgan, C. O., and Simmonds, J. G. (1993) "The eigenvalues for a self-equilibrated, semi-infinite, anisotropic elastic strip," *J. Appl. Mech.* **60**, 276-281.
Deans, S. R. (1983) *The Radon Transform and Some of its Applications*, Wiley-Interscience.
Delale, F., and Erdogan, F. (1979) "Bonded orthotropic strips with cracks," *Int. J. Fractures* **15**, 343–364.
Delale, F., Kishore, N. N., and Wang, A. S. D. (1984) "Stress analysis of a composite plate with circular hole under axisymmetric bending," *J. Composite Materials* **18**, 420–433.
Dempsey, J. P. (1981) "The wedge subjected to tractions: a paradox resolved," *J. Elasticity* **11**, 1–10.
Dempsey J. P., and Sinclair, G. B. (1979) "On the stress singularities in the plane elasticity of the composite wedge," *J. Elasticity*

9, 373–391.

Dempsey J. P., and Sinclair, G. B. (1981) "On the singular behavior at the vertex of a bi-material wedge," *J. Elasticity* **11**, 317–327.

Dongye, Changsong, and Ting, T. C. T. (1989) "Explicit expressions of Barnett–Lothe tensors and their associated tensors for orthotropic materials," *Q. Appl. Math.* **47**, 723–734.

Duff, G. F. D. (1960) "The Cauchy problem for elastic waves in an anisotropic medium," *Phil. Trans. Roy. Soc. London* A**252**, 249–273

Dugdale, D. S. (1960) "Yielding of steel sheets containing slits," *J. Mech. Phys. Solids* **8**, 100–108.

Dundurs, J. (1967a) "On the interaction of a screw dislocation with inhomogeneities," in *Recent Advances in Engineering Science*, Vol. 2, 223–233.

Dundurs, J. (1967b) "Dependence of stress on Poisson's ratio in plane elasticity," *Int. J. Solids Structures* **3**, 1013–1021.

Dundurs, J. (1969) "Elastic interaction of dislocations with inhomogeneities," in *Mathematical Theory of Dislocations*, T. Mura, ed., ASME, New York, 70–115.

Dundurs, J. (1970) "Some properties of elastic stresses in a composite," in *Recent Advances in Engineering*, ed. by A. C. Eringen, Vol. 5, Gordon and Breach, NY, 203–216.

Dundurs, J., and Gautesen, A. K. (1988) "An opportunistic analysis of the interface crack," *Int. J. Fracture* **36**, 151–159.

Dundurs, J., and Hetényi, M. (1965) "Transmission of force between two semi-infinite solids," *J. Appl. Mech.* **32**, 671–674.

Dundurs, J., and Markenscoff, X. (1993) "Invariance of stresses under a change in elastic compliances," *Proc. Roy. Soc. London* A**443**, 289-300.

Dundurs, J., and Mura, T. (1964) "Interaction between an edge dislocation and a circular inclusion," *J. Mech. Phys. Solids* **12**, 177–189.

Dundurs, J., and Sendeckyj, G. P. (1965a) "Behavior of an edge dislocation near a bimetallic interface," *J. Appl. Phys.* **36**, 3353–3354.

Dundurs, J., and Sendeckyj, G. P. (1965b) "Edge dislocation inside a circular inclusion," *J. Mech. Phys. Solids* **13**, 141–147.

Dundurs, J., and Ting, T. C. T. (1994) Private communications.

Dvorak, G. J., and Benveniste, Y. (1992) "On transformation strains and uniform fields in multiphase elastic media," *Proc. Roy. Soc. London* A**437**, 291–310.

Elber, W. (1970) "Fatigue crack closure under cyclic tension," *Eng. Frac. Mech.* **2**, 37–45.

Elliott, H. A. (1948) "Three-dimensional stress distributions in hexagonal aeolotropic crystals," *Proc. Camb. Phil. Soc.* **44**,

522–533.

Elliott, H. A. (1949) "Axial symmetric stress distributions in aeolotropic hexagonal crystals. The problem of the plane and related problems," *Proc. Camb. Phil. Soc.* **45**, 621–630.

England, A. H. (1965) "A crack between dissimilar media," *J. Appl. Mech.* **32**, 400–402.

England, A. H. (1971) "On stress singularities in linear elasticity," *Int. J. Solids Structures* **9**, 571–585.

Erdogan, F. (1963) "Stress singularities in a nonhomogeneous elastic plane with cracks," *J. Appl. Mech.* **30**, 232–236.

Erdogan, F. (1965) "Stress distribution in a nonhomogeneous elastic materials with cracks," *J. Appl. Mech.* **32**, 403–410.

Erdogan, F. (1983) "Stress intensity factor," *J. Appl. Mech.* **50**, 992–1002.

Erdogan, F., and Gupta, G. D. (1972) "Stresses near a flat inclusion in bonded dissimilar materials," *Int. J. Solids Structures* **8**, 533–547.

Eshelby, J. D. (1956) "The continuum theory of lattice defects," *Solid State Phys.* **3**, 79-144.

Eshelby, J. D. (1957) "The determination of the elastic field of an ellipsoidal inclusion, and related problems," *Proc. Roy. Soc. London* A**241**, 376–396.

Eshelby, J. D. (1959) "The elastic field outside an ellipsoidal inclusion," *Proc. Roy. Soc. London* A**252**, 561–569.

Eshelby, J. D. (1961) "Elastic inclusions and inhomogeneities," in *Progress in Solid Mechanics* Vol. 2, I. N. Sneddon and R. Hill, eds., North-Holland, Amsterdam, 89–140.

Eshelby, J. D., Read, W. T., and Shockley, W. (1953) "Anisotropic elasticity with applications to dislocation theory," *Acta Metall.* **1**, 251–259.

Fadle, J. (1941) "Die selbstspannungs-eigenwertfunktionen der quadratischen scheibe," *Ingenieur-Arch.* **11**, 125–148.

Farnell, G. W. (1970) "Properties of elastic surface waves," *Phys. Acoust.* **6**, 109–166.

Fatuzzo, E., and Merz, W. J. (1967) *Ferroelectricity*. North Holland, New York.

Florence, A. L., and Goodier, J. N. (1960) "Thermal stresses due to disturbance of uniform heat flow by an insulated ovaloid hole," *J. Appl. Mech.* **27**, 635–639.

Florence, A. L., and Goodier, J. N. (1963) "The linear thermoelastic problem of uniform heat flow disturbed by a penny-shaped insulated crack," *Int. J. Eng. Sci.* **1**, 533–540.

Fredholm, I. (1900) "Sur les equations de l'equilibre d'un corps solide elastique," *Act.* Math. **23**, 1–42.

Fung, Y. C. (1965) *Foundations of solid Mechanics*, Prentice-Hall, Englewood Cliffs, NJ. 365.

Gao, Huajian (1992) "Stress analysis of holes in anisotropic elastic solids: conformal mapping and boundary perturbation," *Q. J. Mech. Appl. Math.* **45**, 149–168.

Gao, Huajian., Abbudi, M., and Barnett, D. M. (1992) "On interfacial crack-tip field in anisotropic solids," *J. Mech. Phys. Solids* **40**, 393–416.

Gautesen, A. K., and Dundurs, J. (1987a) "On the solution to a Cauchy principal value integral equation which arises in fracture mechanics," *SIAM J. Appl. Math.* **47**, 109–116.

Gautesen, A. K., and Dundurs, J. (1987b) "The interface crack in a tension field," *J. Appl. Mech.* **54**, 93–98.

Gautesen, A. K., and Dundurs, J. (1988) "The interface crack under combined loading," *J. Appl. Mech.* **55**, 580–586.

Gibbs, J. W. (1961) *Elements of Vector Analysis*, Dover, New York.

Gotoh, M. (1967) "Some problems of bonded anisotropic plates with cracks along the bond," *Int. J. Frac. Mech.* **3**, 253–260.

Green, A. E., and Zerna, W. (1960) *Theoretical Elasticity*, Oxford Univ. Press, London.

Griffith, A. A. (1920) "The phenomena of rupture and flow in solids," *Phil. Trans. Roy. Soc.* **A221**, 163–198.

Gundersen, S. A. (1992) "Classification of the subsequent transonic states in anisotropic elasticity theory," in *The Jens Lothe Symposium Volume, Physica Scripta* T**44**, 80–84.

Gundersen, S. A., and Lothe, J. (1987) "A new method for numerical calculations in anisotropic elasticity problems," *Phys. Status Solidi* B**143**, 73–85.

Gundersen, S. A., Barnett, D. M., and Lothe, J. (1987) "Rayleigh wave existence theory. A supplementary remark," *Wave Motion* **9**, 319–321.

Gundersen, S. A., Wang, L., and Lothe, J. (1991) "Secluded supersonic elastic surface waves," in *Modern Theory of Anisotropic Elasticity and Applications*, J. J. Wu, T. C. T. Ting and D. M. Barnett, eds., SIAM Proceedings Series, SIAM, Philadelphia, 186–198.

Gurtin, M. E. (1972) *The Linear Theory of Elasticity*, Handbuch der Physik, VIa/2, Springer, Berlin.

Hansen, M. T. (1995) "Elastic fields resulting from concentrated loading on a three-dimensional incompressible wedge," *J. Appl. Mech.* **62**, in press.

Hardiman, N. J. (1954) "Elliptic elastic inclusion in an infinite elastic plate," *Q. J. Mech. Appl. Math.* **7**, 226–230.

Hayes, M. A. (1984) "Inhomogeneous plane waves," *Arch. Rat. Mech. Anal.* **85**, 41–79.

Hayes, M. A., and Norris, A. N. (1992) "Static implications of the existence of a plane of symmetry in an anisotropic elastic solid,"

Q. J. Mech. Appl. Math. **45**, 141-147.

Head, A. K. (1953) "Interaction of dislocations and boundaries," *Phil. Mag.* **44**, 92–94.

Head, A. K. (1965) "The dislocation image force in cubic polycrystals," *Phys. Status Solidi* **10**, 481–484.

Head, A. K. (1979a) "The Galois unsolvability of the sextic equation of anisotropic elasticity," *J. Elasticity*, **9**, 9–20.

Head, A. K. (1979b) "The monodromic Galois groups of the sextic equation of anisotropic elasticity," *J. Elasticity*, **9**, 321–324.

Hearmon, R. F. S. (1961) *An Introduction to Applied Anisotropic elasticity*, Oxford University Press, New York, 13.

Helgason, S. (1980) *The Radon Transform*, Birkhauser, Boston Basel Stuttgart.

Hildebrand, F. B. (1954) *Methods of Applied Mathematics*, Prentice-Hall, New York.

Hill, R. (1983) "Interfacial operators in the mechanics of composite media," *J. Mech. Phys. Solids* **31**, 347–357.

Hirth, J. P., and Lothe, J. (1982) *Theory of dislocations*, Wiley, New York.

Hoenig, A. (1982) "Near-tip behavior of a crack in a plane anisotropic elastic body," *Eng. Frac. Mech.* **16**, 393–403.

Hohn, Franz E. (1965) *Elementary Matrix Algebra*, Macmillan, New York.

Horgan, C. O. (1972a) "On Saint-Venant's principle in plane anisotropic elasticity," *J. Elasticity* **2**, 169–180.

Horgan, C. O. (1972b) "Some remarks on Saint-Venant's principle for transversely isotropic composite," *J. Elasticity* **2**, 335–339.

Horgan, C. O., and Miller, K. L. (1994) "Antiplane shear deformations for homogeneous and inhomogeneous anisotropic linearly elastic solids," *J. Appl. Mech.* **61**, 23–29.

Horgan. C. O., and Simmonds, J. G. (1991) "Asymptotic analysis of an end-loaded, transversely isotropic, elastic semi-infinite strip weak in shear," *Int. J. Solids Structures* **27**, 1895–1914.

Hwu, Chyanbin (1990a) "Anisotropic plates with various openings under uniform loading or pure bending," *J. Appl. Mech.* **57**, 700–706.

Hwu, Chyanbin (1990b) "Thermal stresses in an anisotropic plate disturbed by an insulated elliptic hole or crack," *J. Appl. Mech.* **57**, 916–922.

Hwu, Chyanbin (1991) "Collinear cracks in anisotropic bodies," *Int. J. Fracture* **54**, 239–256.

Hwu, Chyanbin (1992a) "Thermoelastic interface crack problems in dissimilar anisotropic media," *Int. J. Solids Structures*, **29**, 2077–2090.

Hwu, Chyanbin (1992b) "Polygonal holes in anisotropic media," *Int. J. Solids Structures* **29**, 2369–2384.

Hwu, Chyanbin (1993) "Explicit solutions for the collinear interface crack problems," *Int. J. Solids Structures* **30**, 301–312.

Hwu, Chyanbin, and Ting, T. C. T. (1989) "Two-dimensional problems of the anisotropic elastic solid with an elliptic inclusion," *Q. J. Mech. Appl. Math.* **42**, 553–572.

Hwu, Chyanbin, and Ting, T. C. T. (1990) "Solutions for the anisotropic elastic wedges at critical wedge angles," *J. Elasticity* **24**, 1–20.

Hwu, Chyanbin, and Yen, Wen J. (1991) "Green's functions of two-dimensional anisotropic plates containing an elliptic hole," *Int. J. Solids Structures* **27**, 1705–1719.

Hwu, Chyanbin, and Yen, Wen J. (1992) "Plane problems for anisotropic bodies with an elliptic hole subjected to arbitrary loadings," *The Chinese Journal of Mechanics* **8**, 123–129.

Hwu, Chyanbin, and Yen, Wen J. (1993) "On anisotropic elastic inclusions in plane elastostatics," *J. Appl. Mech.* **60**, 626–632.

Indenbom, V. L., and Lothe, J. (1992) *Elastic Strain Fields and Dislocation Mobility*, Modern Problems in Condensed Matter Sciences, Vol. 31, North-Holland, Amsterdam.

Indenbom, V. L., and Orlov, S. S. (1968) "Construction of Green's function in terms of Green's function of lower dimension," *J. Appl. Mech.* **32**, 414–420.

Ingebrigtsen, K. A., and Tonning, A. (1969) "Elastic surface waves in crystal," *Phys. Rev.* **184**, 942–951.

Inglis, C. E. (1922) "Some special cases of two-dimensional stress and strain," *Trans. Inst. Naval Architects*. London, **64**, 253.

Irwin, G. R. (1957) "Analysis of stresses and strains near the end of crack traversing a plate," *J. Appl. Mech.* **24**, 361–364.

Jaffe, B., Cook, W. R., and Jaffe, H. (1971) *Piezoelectric Ceramics*. Academic Press, London.

Jaswon, M. A., and Bhargava, R. D. (1961) "Two-dimensional elastic inclusion problems," *Proc. Camb. Phil. Soc.* **57**, 669–680.

Jin, Yijian, and Ting, T. C. T. (1988) "Modified eigenfunctions for transversely isotropic composites with applications to stress analysis of a broken fiber," *J. Composite Materials* **22**, 224–244.

John, F. (1955) *Plane Waves and Spherical Means Applied to Partial Differential Equations*, Interscience, New York.

Johnson. M. W., and Little, R. W. (1965) "The semi-infinite strip," *Q. Appl. Math.* **1**, 335–344.

Jones, R. M. (1975) *Mechanics of Composite Materials*, McGraw-Hill, New York.

Juretschke, H. J. (1974) *Crystal Physica*, Benjamin, Menlo Park, CA, p. 31.

Kanninen, M. F., and Popelar, C. H. (1985) *Advance Fracture*

Mechanics, Oxford University Press, New York.

Kassir, M. K., and Sih, G. C. (1975) *Mechanics of Fracture 2—Three Dimensional Crack Problems*, Noordhoff Int. Pub. Leyden.

Kawai, T., Fujitani, Y., and Kumagai, K. (1977) "Analysis of singularity at the root of the surface crack problem," *Proc. Int. Conf. Fracture Mechanics and Technology*, G. C. Sih and C. L. Chow, eds., **11**, 1157–1163.

Kelvin, Lord (1848) *Cambridge and Dublin Math. J.*

Kelvin, Lord (1882) "Note on the integration of the equations of equilibrium of an elastic solid," *Mathematical and Physical Papers*, **1**, Cambridge Univ. Press, 97–98.

Kirchner, H. O. K. (1987) "Line defects along the axis of rotationally inhomogeneous media," *Phil. Mag.* A**55**, 537–542.

Kirchner, H. O. K. (1989) "Elastically anisotropic angularly inhomogeneous media I. a new formalism," *Phil. Mag.* B**60**, 423–432.

Kirchner, H. O. K., and Lothe, J. (1986) "On the redundancy of the $\overline{N}$ matrix of anisotropic elasticity," *Phil. Mag.* A**53**, L7–L10.

Kirchner, H. O. K., and Lothe, J. (1987) "Displacements and tractions along interfaces," *Phil. Mag.* A**56**, 583–594.

Klingbeil, W. W., and Shield, R. T. (1966) "On a class of solutions in plane finite elasticity," *Zeit. Ang. Math. Phys.* **17**, 489–511.

Knein, M. (1926) "Zur theorie des druckversuchs," *Zeit. Ang. Math. Mech.* **6**, 414–416.

Knowles, J. K., and Sternberg, E. (1983) "Large deformations near a tip of an interface-crack between two neo-hookean sheets," *J. Elasticity* **13**, 257–293.

Kolodner, I. I. (1966) "Existence of longitudinal waves in anisotropic media," *J. Acoust. Soc. Am.* **40**, 730–731.

Kröner, E. (1953) "Das Fundamentalintegral der anisotropen elastischen Differentialgleichungen," *Zeit. Phys.* **136**, 402–410.

Kuo, Chen-Ming, and Barnett, D. M. (1991) "Stress singularities of interfacial cracks in bonded piezoelectric half-spaces," in *Modern Theory of Anisotropic Elasticity and Applications*, J. J. Wu, T. C. T. Ting and D. M. Barnett, eds., SIAM Proceedings Series, SIAM, Philadelphia, 33–50.

Kuo, M. C., and Bogy, D. B. (1974) "Plane solutions for the displacement and traction-displacement problems for anisotropic elastic wedges," *J. Appl. Mech.* **41**, 197–203.

Lee, N. G. (1979) "Elastic Green's function for an infinite half-space of a hexagonal continuum with its basal plane as surface," *Int. J. Eng. Sci.* **17**, 681–689.

Lekhnitskii, S. G. (1950) *Theory of Elasticity of an Anisotropic Body*, Gostekhizdat, Moscow (in Russian). *Theory of Elasticity of an Anisotropic Elastic Body*, Holden-Day, San Francisco (in

English, 1963), and Mir Pub. Moscow (in English, 1981).

Lekhnitskii, S. G. (1957) *Anisotropic Plates*, 2nd ed., Gostekhizdat, Moscow (in Russian); transl. by S. W. Tsai and T. Cheron, Gordon and Breach, New York (1968, 1984, 1987).

Levy, M. (1899) "Sur la legitimite de la regle dite du trapeze dans l'etude de la resistance des barrages en maconnerie," *Compt. Rend.* **126**, 1235–1240.

Li, Qianqian, and Ting, T. C. T. (1989) "Line inclusions in anisotropic elastic solids," *J. Appl. Mech.* **56**, 556–563.

Lifshitz, I. M., and Rozenzweig, L. N. (1947) "On the construction of the Green's tensor for the basic equation of the theory of elasticity of an anisotropic infinite medium," *Zh. Eksp. Teor. Fiz.* **17**, 783–791.

Lin, K. Y., and Mar, J. W. (1976) "Finite element analysis of stress intensity factors for cracks in bi-material interface," *Int. J. Frac. Mech.* **12**, 521–531.

Lo, K. K. (1978) "Analysis of branched cracks," *J. Appl. Mech.* **45**, 797–802.

Lothe, J. (1991) "Body waves in anisotropic elastic media," in *modern theory of anisotropic elasticity and applications*, J. J. Wu, T. C. T. Ting and D. M. Barnett, eds., SIAM Proceedings Series, SIAM, Philadelphia, 173–185.

Lothe, J., and Barnett, D. M. (1976a) "On the existence of surface-wave solutions for anisotropic half-spaces with free surface," *J. Appl. Phys.* **47**, 428–433.

Lothe, J., and Barnett, D. M. (1976b) "Integral formalism for surface waves in piezoelectric crystals. Existence considerations," *J. Appl. Phys.* **47**, 1799–1807.

Lothe, J., and Barnett, D. M. (1977) "Further development of the theory of surface waves in piezoelectric crystals," *Physica Norv.* **8**, 239–254.

Lothe, J., and Barnett, D. M. (1979) "On the existence of surface wave solutions in piezoelectric crystals: An example of nonexistence," *Wave Motion*, **1**, 107–112.

Love, A. E. H. (1927) *A Treatise on the mathematical Theory of Elasticity*, University Press, Cambridge, 157–160.

Malen, K. (1971) "A unified six-dimensional treatment of elastic Green's functions and dislocations," *Phys. Status Solidi* B**44**, 661–672.

Malen, K., and Lothe, J. (1970) "Explicit expressions for dislocation derivatives," *Phys. Status Solidi* B**39**, 287–296.

Malen, K., and Lothe, J. (1971) "Comment on dislocation moving with the Rayleigh wave velocity," *Phys. Status Solidi* B**43**, K139–K142.

Malyshev, B. M., and Salganik, R. L. (1965) "The strength of adhesive joints using the theory of fracture," *Int. J. Frac. Mech.*

1, 114–128.

Mehrabadi, M. M., and Cowin, S. C. (1990) "Eigentensors of linear anisotropic elastic materials," *Q. J. Mech. Appl. Math.* **43**, 15–41.

Michell, J. H. (1899) "On the direct determination of stress in an elastic solid, with application to the theory of plates," *Proc. Lond. Math. Soc.* **31**, 100.

Mikhailov, G. K., and Parton, V. Z. (1990) *Electromagneto-elasticity.* Hemisphere, New York.

Mindlin, R. D. (1936) "Note on the Galerkin and Papkovich stress functions," *Bull. Amer. Math. Soc.* **42**, 373–376.

Mindlin, R. D. (1953) "Force at a point in the interior of a semi-infinite solid," *Proc. Midwest. Conf. Solid Mech.* 56–59.

Moffatt, H. K., and Duffy, B. R. (1980) "Local similarity solutions and their limitations," *J. Fluid Mech.* **96**, 299–313.

Mura, T. (1975) "A note on the strain field of a dislocation line in anisotropic media," *Phys. Status Solidi* B**70**, K1-K6.

Mura, T. (1987) *Micromechanics of Defects in Solids*, 2nd ed. Martinus Nijhoff. Dordrecht.

Mura, T. (1988) "Inclusion problems," *Appl. Mech. Rev.* **41**, 15–20.

Mura, T., and Kinoshita, N. (1971) "Green's functions for anisotropic elasticity," *Phys. Status Solidi* B**47**, 607–618.

Musgrave, M. J. P. (1970) *Crystal Acoustics*, Holden-Day, Cambridge.

Muskhelishvili, N. I. (1953) *Some Basic Problems of the Mathematical Theory of Elasticity*, transl. by J. R. M. Radok, Noordhoff, Groningen.

Nakahara, S., and Willis, J. R. (1973) "Some remarks on interfacial dislocations," *J. Phys. F.* **3**, L249–254.

Neuber, H. (1963) "Losung des Carothers problems mittels prinzipien der kraffubertragung (keil mit moment an der spitze)," *Zeit. Ang. Math. Mech.* **43**, 211–228.

Ni, L., and Nemat-Nasser, S. (1991) "Interface cracks in anisotropic dissimilar materials: an analytic solution," *J. Mech. Phys. Solids* **39**, 113–144.

Ni, L., and Nemat-Nasser, S. (1992) " Interface cracks in anisotropic dissimilar materials: general case," *Q. Appl. Math.* **50**, 305-322.

Nishioka, K., and Lothe, J. (1972a) "Isotropic limiting behaviour of the six-dimensional formalism of anisotropic dislocation theory and anisotropic Green's function theory. I. Sum rules and their applications," *Phys. Status Solidi* B**51**, 645–656.

Nishioka, K., and Lothe, J. (1972b) "Isotropic limiting behaviour of the six-dimensional formalism of anisotropic dislocation theory and anisotropic Green's function theory. II. Perturbation theory on the isotropic N-matrix," *Phys. Status Solidi* B**52**, 45–54.

Norris, A. N. (1989) "On the acoustic determination of the elastic moduli of anisotropic solids and acoustic conditions for the existence of symmetry planes," *Q. J. Mech. Appl. Math.* **42**, 413–426.

Norris, A. N. (1991) "Low frequency wave propagation in periodically layered anisotropic elastic solids," in *Modern Theory of Anisotropic Elasticity and Applications*, J. J. Wu, T. C. T. Ting and D. M. Barnett, eds., SIAM Proceedings Series, SIAM, Philadelphia, 255–262.

Norris, A. N. (1992) "One-component surface waves in materials with symmetry," in *The P. Chadwick Symposium Volume, J. Mech. Phys. Solids* **40**, 1569–1582.

Nowacki, W. (1962) *Thermoelasticity*, Addison–Wesley, Reading, MA.

Okubo, H. (1952) "The stress distribution in an aeolotropic circular disc compressed diametrically," *J. Math. Phys.*, **31**, 75-83.

Olesiak, Z., and Sneddon, I. N. (1960) "The distribution of thermal stress in an infinite elastic solid containing a penny-shaped crack," *Arch. Rat. Mech. Anal.* **4**, 238–254.

Olver, P. J. (1988) "Canonical elastic moduli," *J. Elasticity*, **19**, 189-212.

Olver, P. J. (1992) "Canonical anisotropic elastic moduli," in *Modern Theory of Anisotropic Elasticity and Applications*, J. J. Wu, T. C. T. Ting and D. M. Barnett, eds., SIAM Proceedings Series, SIAM, Philadelphia, 325–339.

Onsager, L. (1931) "Reciprocal relations in irreversible processes. I." *Phys. Rev.* **37**, 4, 405–426. *Phys. Rev.* **38**, 12, 2265–2279.

Pan, Y. C., and Chou, T. W. (1976) "Point force solution for an infinite transversely isotropic solid," *J. Appl. Mech.* **43**, 608–612.

Pan, Y. C., and Chou, T. W. (1979a) "Green's function solutions for semi-infinite transversely isotropic materials," *Int. J. Eng. Sci.* **17**, 545–551.

Pan, Y. C., and Chou, T. W. (1979b) "Green's functions for two-phase transversely isotropic materials," *J. Appl. Mech.* **46**, 551–556.

Papkovich, P. F. (1941) "The questions of the theory of bending of thin elastic plates," *Prik. Mat. Mek.* **5**, 359–374.

Peach, M. D., and Koehler, J. S. (1950) "The force exerted on dislocations and the stress field produced by them," *Phys. Rev.* **80**, 436–439.

Pease, Marshall C. III (1965) *Methods of Matrix Algebra*, Academic Press, New York.

Qu, Jianmin (1991) "Green's functions in anisotropic bimaterials," in *Modern Theory of Anisotropic Elasticity and Applications*, J. J. Wu, T. C. T. Ting and D. M. Barnett, eds., SIAM Proceedings

Series, SIAM, Philadelphia, 62–73.

Qu, Jianmin, and Bassani, J. L. (1989) "Cracks on bimaterial and bicrystal interfaces," *J. Mech. Phys. Solids* **37**, 417–433.

Qu, Jianmin, and Bassani, J. L. (1993) "Interfacial fracture mechanics for anisotropic bimaterials," *J. Appl. Mech.* **60**, 422-431.

Qu, Jianmin, and Li, Qianqian (1991) "Interfacial dislocation and its application to interface crack in anisotropic bimaterials," *J. Elasticity* **26**, 167–195.

Rayleigh, Lord (John William Strutt) (1885) "On waves propagated along the plane surface of an elastic solid," *Proc. Lond. Math. Soc.* **17**, 4–11, or *Scientific Papers*, Vol. 2, Cambridge University Press, London, (1900), 441–447.

Reissner, E. (1944) "Note on the theorem of the symmetry of the stress tensor," *J. Math. Phys.* **23**, 192–194.

Rice, J. R. (1968) "A path independent integral and the approximate analysis of strain concentration by notches and cracks," *J. Appl. Mech.* **35**, 379–386.

Rice, J. R. (1985) "Conserved integrals and energetic forces," in *Fundamentals of Deformation and Fracture: Eshelby Memorial Symposium*, ed. by B. A. Bilby, K. J. Miller, and J. R. Willis, Cambridge Univ. Press, Cambridge, 33–56.

Rice, J. R., and Sih, G. C. (1965) "Plane problems of cracks in dissimilar media," *J. Appl. Mech.* **32**, 418–423.

Rongved, R. (1955a) "Point force in the interior of a semi-infinite solid with fixed boundary," *J. Appl. Mech.* **22**, 545–546.

Rongved, R. (1955b) "Force interior to one of two joined semi-infinite solids," *Proc. 2nd Midwest. Conf. Appl. Mech.* 1–13.

Saint-venant, Barre de (1855) *Mémoires de l'académie des sciences des savants étrangers*, **14**, 233–560.

Santare, M. H., and Keer, L. M. (1986) "Interaction between an edge dislocation and a rigid elliptical inclusion," *J. Appl. Mech.* **53**, 382–385.

Savin, G. N. (1961) *Stress Concentration Around Holes*, Pergamon Press, London.

Scott, N. H. (1993a) "A theorem on isotropic null vectors and its application to thermoelasticity," *Proc. Roy. Soc. London* A**440**, 431-442.

Scott, N. H. (1993b) "A new canonical form for complex symmetric matrices," *Proc. Roy. Soc. London* A**441**, 625-640.

Selvadurai, A. S. P. (1980) "The displacement of a flexible inhomogeneity embedded in a transversely isotropic elastic medium," *Fiber Science and Technology* **14**, 251–259.

Sendeckyj, D. P. (1970) "Elastic inclusion problems in plane elastostatics," *Int. J. Solids Structures* **6**, 1535–1543.

Sharma, B. (1958) "Thermal stresses in transversely isotropic semi-infinite elastic solids," *J. Appl. Mech.* **25**, 86–88.

Shield, R. T. (1982) "Uniqueness for elastic crack and punch problems," *J. Appl. Mech.* **49**, 516–518.

Sih, G. C. (1962) "On the singular character of thermal stresses near a crack tip," *J. Appl. Mech.* **29**, 587–589.

Sih, G. C. (1971) "A review of the three dimensional stress problem for a cracked plate," *Int. J. Frac. Mech.* **9**, 39–61.

Sih, G. C., and Chen E. P. (1981) *Cracks in Composite Materials*, Martinus Nijhoff, The Hague, 87–97.

Sih, G. C., and Liebowitz, H. (1968) "Mathematical theories of brittle fracture," in *Fracture*, ed. by H. Liebowitz, Vol. II, Academic Press, New York, 115–125.

Sih, G. C., Paris, P. C., and Irwin, G. R. (1965) "On cracks in rectilinearly anisotropic bodies," *Int. J. Frac. Mech.* **1**, 189–302.

Sneddon, I. N. (1951) *Fourier Transform*, McGraw-Hill, New York.

Sokolnikoff, I. S. (1956) *Mathematical Theory of Elasticity*, McGraw-Hill, New York.

Somaratna, Nihal, and Ting, T. C. T. (1986a) "Three dimensional stress singularities at conical notches and inclusions in transversely isotropic materials," *J. Appl. Mech.* **53**, 89–96.

Somaratna, Nihal, and Ting, T. C. T. (1986b) "Three dimensional stress singularities in anisotropic materials and composites," *Int. J. Eng. Sci.* **24**, 1115–1134.

Spencer, A. J. M. (1980) *Continuum Mechanics*, Longmans, London.

Stagni, L. (1982) "On the elastic field perturbation by inhomogeneities in plane elasticity," *ZAMP*. **33**, 315–325.

Stagni, L., and Lizzio, R. (1983) "Shape effects in the interaction between an edge dislocation and an elliptic inhomogeneity," *Appl. Phys.* A**30**, 217–221.

Sternberg, E., and Koiter, W. T. (1958) "The wedge under a concentrated couple: a paradox in the two-dimensional theory of elasticity," *J. Appl. Mech.* **25**, 575–581.

Stolarski, H. K., and Chiang, M. Y. M. (1989) "On the significance of the logarithmic term in the free edge stress singularity of composite laminates," *Int. J. Solids Structures* **25**, 75–93.

Stone, J. M. (1963) *Radiation and Optics*, McGraw-Hill, New York.

Stoneley, R. (1924) "Elastic waves at the surface of separation of two solids," *Proc. Roy. Soc. London* A**106**, 416–428.

Stoneley, R. (1963) "The propagation of surface waves in an elastic medium with orthorhombic symmetry," *Geophys. J. Astr. Soc.*, **8**, 176–186.

Stroh, A. N. (1958) "Dislocations and cracks in anisotropic elasticity," *Phil. Mag.* **3**, 625–646.

Stroh, A. N. (1962) "Steady state problems in anisotropic elastic ity," *J. Math. Phys.* **41**, 77–103.

Sturla, F. A., and Barber, J. R. (1988) "Thermal stresses due to a plane crack in general anisotropic material," *J. Appl. Mech.* **55**, 372–376.

Suo, Z (1990) "Singularities, interfaces and cracks in dissimilar anisotropic media," *Proc. Roy. Soc. London* A**427**, 331–358.

Suo, Z., Kuo, Chen-Ming, Barnett, D. M., and Willis, J. R. (1992) "Fracture mechanics for piezoelectric ceramics," *J. Mech. Phys. Solids* **40**, 739–765.

Synge, J. L. (1956) "Elastic waves in anisotropic media," *J. Math. Phys.* **35**, 323–334.

Synge, J. L. (1957) *The Hypercircle in Mathematical Physics*, Cambridge Univ. Press, Cambridge, 411–413.

Synge, J. L. (1966) *The Petrov Classification of Gravitational Fields*, Dublin Institute for Advanced Studies, Dublin.

Tanuma, K. (1996) "Surface impedance tensors of transversely isotropic elastic materials," *Q. J. Mech. Appl. Math.* **49**, No.1.

Taylor, D. B. (1978) "On the existence of elastic surface waves," *Q. J. Mech. Appl. Math.* **31**, 335–347.

Taylor, D. B., and Crampin, S. (1978) "Surface waves in anisotropic media: Propagation in a homogeneous piezoelectric halfspace," *Proc. Roy. Soc. London* A**364**, 161–179.

Tewary, V. K., and Kritz, R. D. (1991) "Stress distribution in a bimaterial composite containing a free surface," in *Modern Theory of Anisotropic Elasticity and Applications*, J. J. Wu, T. C. T. Ting and D. M. Barnett, eds., SIAM Proceedings Series, SIAM, Philadelphia, 74–86.

Tewary, V. K., Wagoner, R. H., and Hirth, J. P. (1989a) "Elastic Green's function for a composite solids with a planar interface," *J. Mater. Res.* **4**, 113–123.

Tewary, V. K., Wagoner, R. H., and Hirth, J. P. (1989b) "Elastic Green's function for a composite solids with a planar crack in the interface," *J. Mater. Res.* **4**, 124–136.

Thompson. T. R., and Little, R. W. (1970) "End effects in a truncated semi-infinite cone," *Q. J. Mech. Appl. Math.* **23**, 185–195.

Thorpe, M. F., and Jasiuk, I. (1992) "New results in the theory of elasticity for two-dimensional composites," *Proc. Roy. Soc. London* A**438**, 531-534.

Timoshenko, S. P., and Goodier, J. N. (1970) *Theory of Elasticity*, McGraw-Hill, New York.

Ting, T. C. T. (1970) "Elastic-plastic boundaries in the propagation of plane and cylindrical waves of combined stress," *Q. Appl. Math.* **27**, 441-449.

Ting, T. C. T. (1982) "Effects of change of reference coordinates on the stress analyses of anisotropic elastic materials," *Int. J. Solids Structures* **18**, 139–152.

Ting, T. C. T. (1984) "The wedge subjected to tractions: A paradox re-examined," *J. Elasticity* **14**, 235–247.

Ting, T. C. T. (1985a) "A paradox on the elastic wedge subjected to a concentrated couple and on the Jeffery–Hammel viscous flow problem," *Zeit. Ang. Math. Mech.* **65**, 188–190.

Ting, T. C. T. (1985b) "Elastic wedge subjected to anti-plane shear tractions–a paradox explained," *Q. J. Mech. Appl. Math.* **38**, 245–255.

Ting, T. C. T. (1985c) "Asymptotic solution near the apex of an elastic wedge with curved boundaries," *Q. Appl. Math.* **42**, 467–476.

Ting, T. C. T. (1986) "Explicit solution and invariance of the singularities at an interface crack in anisotropic composites," *Int. J. Solids Structures* **22**, 965–983.

Ting, T. C. T. (1987) "Invariants of anisotropic elastic constants," *Q. J. Mech. Appl. Math.* **40**, 431–448.

Ting, T. C. T. (1988a) "Some identities and the structure of $\mathbf{N}_i$ in the Stroh formalism of anisotropic elasticity," *Q. Appl. Math.* **46**, 109–120.

Ting, T. C. T. (1988b) "Line forces and dislocations in anisotropic elastic composite wedges and spaces," *Phys. Status Solidi* B **146**, 81–90.

Ting, T. C. T. (1988c) "The critical angle of the anisotropic elastic wedge subject to uniform tractions," *J. Elasticity* **20**, 113–130.

Ting, T. C. T. (1988d) "The anisotropic elastic wedge under a concentrated couple," *Q. J. Mech. Appl. Math.* **41**, 563–578.

Ting, T. C. T. (1989) "Line forces and dislocations in angularly inhomogeneous anisotropic elastic wedges and spaces," *Q. Appl. Math.* **47**, 123–128.

Ting, T. C. T. (1990a) "The eigenvectors of the **S** matrix and their relations with line dislocations and forces in anisotropic elastic solids," in *Micromechanics and Inhomogeneity*, The Toshio Mura Anniversary Volume, Springer-Verlag, New York, 449–467.

Ting, T. C. T. (1990b) "Interface cracks in anisotropic bimaterials," *J. Mech. Phys. Solids* **38**, 505–513.

Ting, T. C. T. (1991) "The Stroh formalism and certain invariances in two-dimensional anisotropic elasticity," in *Modern Theory of Anisotropic Elasticity and Applications*, J. J. Wu, T. C. T. Ting and D. M. Barnett, eds., SIAM Proceedings Series, SIAM, Philadelphia, 3–32.

Ting, T. C. T. (1992a) "Barnett–Lothe tensors and their associated tensors for monoclinic materials with the symmetry plane at $x_3 = 0$," *J. Elasticity* **27**, 143–165.

Ting, T. C. T. (1992b) "Image singularities of Green's functions for anisotropic elastic half-spaces and bimaterials," *Q. J. Mech.*

Appl. Math. **45**, 119–139.

Ting, T. C. T. (1992c) "Interface cracks in anisotropic elastic bimaterials – a decomposition principle," in *George Herrmann 70th Anniversary Issue, Int. J. Solids Structures* **29**, 1989–2003.

Ting, T. C. T. (1992d) "Synergism of mechanics, mathematics and anisotropic elastic materials," *Proc. 12th Army Symposium on solid Mechanics*, S. C. Chou, ed., U.S. Army Materials Technology Laboratory, Watertown, MA, 17–35.

Ting, T. C. T. (1992e) "Anatomy of Green's functions for line forces and dislocations in anisotropic media and degenerate materials," in *The Jens Lothe Symposium Volume, Physica Scripta* **T44**, 137–144.

Ting, T. C. T. (1992f) "The motion of one-component surface waves," in *The P. Chadwick Symposium Volume, J. Mech. Phys. Solids* **40**, 1637–1650.

Ting, T. C. T. (1993) "On the orthogonal, Hermitian and positive definite properties of the matrices $i\mathbf{B}^{-1}\overline{\mathbf{B}}$ and $-i\mathbf{A}^{-1}\overline{\mathbf{A}}$ in anisotropic elasticity," *J. Elasticity* **30**, 277–284.

Ting, T. C. T. (1994) "On anisotropic elastic materials that possess three identical Stroh eigenvalues as do isotropic materials," *Q. Appl. Math.* **52**, 363-375.

Ting, T. C. T. (1995a) "Generalized Dundurs constants for anisotropic bimaterials," *Int. J. Solids Structures* **32**, 483-500.

Ting, T. C. T. (1995b) "Green's functions for an anisotropic elliptic inclusion under antiplane deformations," in *Anisotropy, Inhomogeneity, & Nonlinearity in Solid Mechanics*, D. F. Parker and A. H. England, eds., Kluwer Acad. Pub., 321-326.

Ting, T. C. T. (1995c) "Antiplane deformations of anisotropic elastic materials," in *Recent Advances in Elasticity, Viscoelasticity, and Inelasticity*, K. R. Rajagopal, ed., World Sci. Pub. (Festschrift Vol. for Prof. T. C. Woo on his 70th birthday) 150-179.

Ting, T. C. T. (1996a) "Green's functions for an anisotropic elliptic inclusion under generalized plane strain deformations," *Q. J. Mech. Appl. Math.* **49**, No.1.

Ting, T. C. T. (1996b) "Existence of an extraordinary degenerate matrix **N** for anisotropic elastic materials," *Q. J. Mech. Appl. Math.*, in press.

Ting, T. C. T., and Barnett, D. M. (1993) "Image force on line dislocations in anisotropic elastic half-spaces with a fixed boundary," *Int. J. Solids Structures* **30**, 313–323.

Ting, T. C. T., and Chadwick, P. (1988) "Harmonic waves in periodically layered anisotropic elastic composites," in *Symp. on Wave Propagation in Structural Composites*, A. K. Mal and T. C. T. Ting, eds., ASME, AMD-Vol. **90**, 69–79.

Ting, T. C. T., and Chadwick, P. (1990) "Harmonic waves in a periodically bilaminated anisotropic elastic composite," in *Elastic*

Waves and Ultrasonic Nondestructive Evaluation, S. K. Datta, J. D. Achenbach, and Y. S. Rajapakse, eds., Elsevier, The Netherlands, 261–266.

Ting, T. C. T., and Chou, S. C. (1981) "Edge singularities in anisotropic composites," *Int. J. Solids Structures* **17**, 1057–1068.

Ting, T. C. T., and Chou, S. C. (1985) "Logarithmic singularity of an elastic composite wedge subjected to out-of-the-plane extensional strain," *J. Theoretical and Applied Fracture Mechanics* **4**, 223–231.

Ting, T. C. T., and Hoang, P. H. (1984) "Singularities at the tip of a crack normal to the interface of an anisotropic layered composite," *Int. J. Solids Structures* **20**, 439–454.

Ting, T. C. T., and Hwu, Chyanbin (1988) "Sextic formalism in anisotropic elasticity for almost non-semisimple matrix **N**," *Int. J. Solids Structures* **24**, 65–76.

Ting, T. C. T., and Hwu, Chyanbin (1991) "Effects of heat flow on stress singularities at the interface crack," in *Mechanics of composites at Elevated and Cryogenic Temperatures*, S. N. Singhal, W. F. Jones and C. T. Herakovitch, eds., ASME, AMD-Vol. **118**, 231–238.

Ting, T. C. T., and Wang, M. Z. (1992) "Generalized Stroh formalism for anisotropic elasticity for general boundary conditions," *Acta Mechanica Sinica* **8**, 193–207.

Ting, T. C. T., and Yan, Gongpu (1991) "The anisotropic elastic solids with an elliptic hole or rigid inclusion," *Int. J. Solids Structures* **27**, 1879–1894.

Ting, T. C. T., and Yan, Gongpu (1992) "The $r^{-1/2}(\ln r)$ singularities at interface cracks due to heat flow," in *The Bruno A. Boley Symposium Volume, J. Thermal Stresses* **15**, 85–99.

Ting, T. C. T., Jin, Yijian, and Chou, S. C. (1985) "Eigenfunctions at a singular point in transversely isotropic materials under axisymmetric deformations," *J. Appl. Mech.* **52**, 565–570.

Tong, P., Pian, T. H. H., and Lasry, S. J. (1973) "A hybrid-element approach to crack problems in plane elasticity," *Int. J. Numer. Meth. Eng.* **7**, 297–308.

Truesdell, C. (1965) *Rational Thermodynamics*, McGraw-Hill, Chapter 7.

Truesdell, C., and Noll, W. (1965) *The Non-Linear Field Theories of Mechanics*, Handbuch der Physik, Vol. III/3, Springer-Verlag, New York.

Tsai, S. W., and Hahn, H. T. (1980) *Introduction to Composite Materials*, Technomic Pub., Westport, CT, 75–76, 92–93.

Tsai, S. W., and Pagano, N. J. (1968) *Composite Materials*, Technomic Pub., Westport, CT, 233–253.

Tsai, Y. M. (1983) "Transversely isotropic thermoelastic problem of uniform heat flow disturbed by a penny-shaped crack," *J.*

Thermal Stresses **6**, 379–389.

Tsai, Y. M. (1984) "Orthotropic thermoelastic problem of uniform heat flow disturbed by a central crack," *J. Composite Materials* 122–131.

Tucker, M. O. (1969) "Plane boundaries and straight dislocations in elastically anisotropic materials," *Phil. Mag.* **19**, 1141–1159.

Vitek, V. and Hirth, J. P. (1977) "Line singularity fields in the presence of a crack of elliptic cross section," *Scripta Metall.* **11**, 339–340.

Voigt, W. (1910) *Lehrbuch der Kristallphysik*, Leipzig, 560.

Wang, C.-Y., and Achenbach, J. D. (1993) "A new method to obtain 3-D Green's functions for anisotropic solids," *Wave Motion* **18**, 273-289.

Wang, L., and Gundersen, S. A. (1992) "Existence of one component surface waves in anisotropic elastic media," *Physica Scripta* T**47**, 394–404.

Wang, L., and Lothe, J. (1992) "Simple reflection in anisotropic elastic media and its relation to exceptional waves and supersonic surface waves," *Wave Motion*, **16**, 89–112.

Wang, M. Z., Ting, T. C. T., and Yan, Gongpu (1993) "The anisotropic elastic semi-infinite strip," *Q. Appl. Math.* **51**, 283–297.

Wang, S. S. (1984) "Edge delamination in angle-ply composite laminates," *AIAA Journal* **22**, 256–264.

Wang, S. S., and Choi, I. (1982) "Boundary-layer effects in composite laminates: Part I—free-edge stress singularities," *J. Appl. Mech.* **49**, 541–548.

Wang, S. S., and Choi, I. (1983) "The interface crack between dissimilar anisotropic composite materials," *J. Appl. Mech.* **50**, 169–178.

Wang, Y. M., and Tarn, J. Q. (1993) "Green's functions for generalized plane problems of anisotropic bodies with a hole or rigid inclusion," *J. Appl. Mech.* **60**, 583–588.

Wang, Z. Y., Zhang, H. T., and Chou, Y. T. (1985) "Characteristic of the elastic field of rigid line inhomogeneity," *J. Appl. Mech.* **52**, 818–822.

Wang, Z. Y., Zhang, H. T., and Chou, Y. T. (1986) "Stress singularity at the tip of a rigid line inhomogeneity under antiplane shear loading," *J. Appl. Mech.* **53**, 459–462.

Warren, W. E. (1983) "The edge dislocation inside an elliptical inclusion," *Mech. Materials* **2**, 319–330.

Wei, Lixin, and Ting, T. C. T. (1994) "Explicit expressions of Barnett–Lothe tensors for anisotropic materials," *J. Elasticity* **36**, 67-83.

Williams, M. L. (1952) "Stress singularities resulting from various boundary conditions in angular corners of plates in extension," *J. Appl. Mech.* **19**, 526–528.

Williams, M. L. (1959) "The stresses around a fault or crack in dissimilar media," *Bull. Seis. Soc. America* **49**, 199–204.

Willis, J. R. (1964) "Anisotropic elastic inclusion problems," *Q. J. Mech. Appl. Math.* **17**, 157–174.

Willis, J. R. (1965) "The elastic interaction energy of dislocation loops in anisotropic media," *Q. J. Mech. Appl. Math.* **18**, 419–433.

Willis, J. R. (1966) "Hertzian contact of anisotropic bodies," *J. Mech. Phys. Solids* **14**, 163–176.

Willis, J. R. (1970) "Stress field produced by dislocations in anisotropic media," *Phil Mag.* **21**, 931–949.

Willis, J. R. (1971) "Fracture mechanics of interfacial cracks," *J. Mech. Phys. Solids* **19**, 353–368.

Woo, T. C., and Shield, R. T. (1962) "Fundamental solutions for small deformations superposed on finite biaxial extension of an elastic body," *Arch. Rat. Mech. Anal.* **9**, 196–224.

Wu, C. H. (1982a) "Unconventional internal cracks, Part I: Symmetric variations of a straight crack," *J. Appl. Mech.* **49**, 62–68.

Wu, C. H. (1982b) "Unconventional internal cracks, Part II: Methods of generating simple cracks," *J. Appl. Mech.* **49**, 383–388.

Wu, C. H. (1984) "Plane anisotropic thermoelasticity," *J. Appl. Mech.* **51**, 724–726.

Wu, J. J., Ting, T. C. T., and Barnett, D. M. (1991) *Modern Theory of Anisotropic Elasticity and Applications.* SIAM Proceedings Series, SIAM, Philadelphia.

Wu, K. C. (1989a) "Representations of stress intensity factors by path-independent integrals," *J. Appl. Mech.* **56**, 780–785.

Wu, K. C. (1989b) "On the crack-tip fields of a dynamically propagating crack in an anisotropic elastic solid," *Int. J. Fracture* **41**, 253–266.

Wu, K. C. (1990) "Stress intensity factor and energy release rate for interfacial cracks between dissimilar anisotropic materials," *J. Appl. Mech.* **57**, 882–886.

Wu, K. C. (1991a) "Explicit crack tip fields of an extending interface crack in anisotropic bimaterial," *Int. J. Solids Structures* **27**, 455–466.

Wu, K. C. (1991b) "Explicit solutions for interface cracks in anisotropic bimaterials," in *Modern Theory of Anisotropic Elasticity and Applications*, J. J. Wu, T. C. T. Ting and D. M. Barnett, eds., SIAM Proceedings Series, SIAM, Philadelphia, 51–61.

Wu, K. C. (1991c) "Anti-plane shear interface cracks in anisotropic bimaterials," *J. Appl. Mech.* **58**, 399-403.

Wu, K. C. (1992) "On the contact zone model for interface cracks," *J. Appl. Mech.* **59**, 92–94.

Wu, K. C., and Hwang, Shyh-Jye (1990) "Correspondence relations for the interface crack in monoclinic composites under mixed loading," *J. Appl. Mech.* **57**, 894–900.

Yan, Gongpu, and Ting, T. C. T. (1993) "The $r^{-1/2}(\ln r)$ singularities at interface cracks in monoclinic and isotropic bimaterials due to heat flow," *J. Appl. Mech.* **60**, 432–437.

Yang, H. C., and Chou, Y. T. (1976) "Generalized plane problems of elastic inclusions in anisotropic solids," *J. Appl. Mech.* **43**, 424–430.

Yang, H. C., and Chou, Y. T. (1977) "Anisotropic strain problems of an elliptic inclusion in an anisotropic medium," *J. Appl. Mech.* **44**, 437–444.

Yang, H. C., and Chou, Y. T. (1978) "On rigid elliptic inclusions in orthotropic solids," *J. Appl. Mech.* **45**, 683–685.

Yang, H. C., and Chou, Y. T. (1982) "The <111> elliptic inclusion in an anisotropic solid of cubic symmetry," *J. Appl. Mech.* **49**, 353–360.

Yeh, C. S., Shu, Y. C., and Wu, K. C. (1993a) "Conservation laws in anisotropic elasticity I. Basic frame work," *Proc. Roy. Soc. London* A**443**, 139-151.

Yeh, C. S., Shu, Y. C., and Wu, K. C. (1993b) "Conservation laws in anisotropic elasticity II. Extension and application to thermoelasticity," *Proc. Roy. Soc. London* A**443**, 153-161.

Yen, Wen J., and Hwu, Chyanbin (1994) "Interactions between dislocations and anisotropic elastic elliptical inclusions," *J. Appl. Mech.* **61**, 548-554.

Yen, Wen J., Hwu, Chyanbin, and Liang, Y. K. (1995) "Dislocation inside, outside, or on the interface of an anisotropic elliptical inclusion," *J. Appl. Mech.* **62**, 306-311.

Yu, H. Y., and Sanday, S. C. (1994) "Dislocations and disclinations in transversely isotropic elastic solids," *Phil. Mag.* A**70**, 725-738.

Yu, H. Y., Sanday, S. C., and Chang, C. I. (1994) "Elastic inclusions and inhomogeneities in transversely isotropic solids," *Proc. Roy. Soc. London* A**444**, 239-252.

Yu, H. Y., Sanday, S. C., and Chang, C. I. (1995) "Elastic fields due to defects in transversely isotropic bimaterials," *Proc. Roy. Soc. London*, A**449**, 1-30.

Zak, A. K., and Williams, M. L. (1963) "Crack point stress singularities at a bi-materials interface," *J. Appl. Mech.*, **30**, 142–143.

Zwiers, R. I., and Ting, T. C. T. (1983) "Singularity of contact-edge stress in laminated composites under uniform extension," *J. Composite Materials* **17**, 49–63.

Zwiers, R. I., Ting, T. C. T., and Spilker, R. L. (1982) "On the logarithmic singularity of free-edge stress in laminated composites under uniform extension," *J. Appl. Mech.* **49**, 561–569.

AUTHOR INDEX

SUBJECT INDEX